Proceedings in Life Sciences

Biosynthesis, Metabolism and Mode of Action of Invertebrate Hormones

Edited by
J. Hoffmann and M. Porchet

With 154 Figures

Springer-Verlag
Berlin Heidelberg New York Tokyo 1984

Professor Dr. JULES HOFFMANN
Laboratoire de Biologie Générale
de l'Université L. Pasteur et
Unité de Recherche Associeé au CNRS
«Endocrinologie et Immunologie des Insectes»
12, Rue de l'Universite
67000 Strasbourg, France

Professor MAURICE PORCHET
Université des Sciences et Techniques
de Lille
59655 Villeneuve D'Asq, France

ISBN-13: 978-3-642-69924-5 e-ISBN-13: 978-3-642-69922-1
DOI: 10.1007/978-3-642-69922-1

Library of Congress Cataloging in Publication Data. Main entry under title: Biosynthesis, metabolism, and mode of action of invertebrate hormones. (Proceedings in life sciences). 1. Invertebrates – Physiology. 2. Hormones. I. Hoffmann, J. (Jules), 1941– . II. Porchet, M. (Maurice), 1943– . III. Series. QL364.B56. 1984. 592'.01927. 84–13959.

Softcover reprint of the hardcover 1st edition 1984

2131/3130-543210

Preface

The neuroendocrine control of reproduction and development of invertebrates has a long tradition as an important area of research in France. The reader of this volume is certainly familiar with the significant contributions to this field made by such outstanding scientific personalities as Jean-Jacques Bounhiol, Jean Panouse, Bernard Possompes, Pierre and Line Joly, Hélène Charniaux-Cotton, Maurice Durchon, Manfred Gabe, Guy Echalier, Marie Raabe, and others.

It is therefore not surprising that the Centre National de la Recherche Scientifique (CNRS) sponsored, in 1975 and 1983, two major international meetings devoted to this subject. The organizers of the 1975 meeting, which was held in Lille, decided to concentrate on the biosynthesis, metabolism, and mode of action of the invertebrate hormones. To some extent, Professors Durchon and Joly wanted to convey the message that they felt that the period of classical invertebrate endocrinology had come to an end and that traditional approaches were being replaced by biochemical analyses. How right they were is illustrated by the present volume. Today biochemistry, molecular biology, and analytical chemistry are tools of the invertebrate endocrinologist, who now starts his morning work by homogenizing his worm, mollusc, insect ... and by extracting DNA, steroids, or peptides.

The aim of the second meeting, which brought some 200 scientists together in Strasbourg during one sunny August week in 1983, was both to summarize the state of the art and to outline the trends of future research in the field of invertebrate hormones.

The progress made between the two meetings is indeed considerable. The structures of half a dozen peptide hormones have been elucidated and many others are close to identification, whereas in 1975 the amino acid sequence of only one invertebrate peptide hormone, the Red Pigment Concentrating Hormone of prawns, was known. The list of juvenile hormones has grown from 3 to 5, and that of Zooecdysteroids has increased at least five-fold and now exceeds 30. Radioimmunoassays, in their infancy at Lille, have since allowed the recording of minute fluctuations of ecdysteroid titers and, more recently, of juvenile hormone titers; high pressure liquid chromatography has allowed the accurate separation of closely related hormone precursors or metabolites, which have then been identified by impressive batteries of spectroscopic techniques. Although the Lille meeting listed amongst its objectives the study of the mode of action of invertebrate hormones, few precise results were available, whereas 8 years later a long series of lectures gave significant information on ecdysteroid and juvenile hormone receptors and on the mode of action of these hormones at the genome level.

The Editors have asked the various authors to survey progress in their respective fields and to highlight the perspectives and prospects. It is hoped that the reader will find this book useful and that it will stimulate young scientists to enter the rapidly expanding field of research on invertebrate hormones.

JULES HOFFMANN

Contents

Part I Peptide Hormones

Introduction
J. Hoffmann 2

I.1 Hormonal Control of Regeneration in Hydra
H.C. Schaller, S. Hoffmeister, and H. Bodenmüller 5

I.2 Peptidic Hormones in Annelids
M. Durchon 10

I.3 Recent Progress in Endocrinology of Molluscs
J. Joosse (With 1 Figure) 19

I.4 The Structure of the Commands for a Neuropeptide-Mediated Behavior, Egg-Laying, in an Opisthobranch Mollusc
F. Strumwasser 36

I.5 Studies on Release Activities of the Neurosecretory Caudo-Dorsal Cells of *Lymnaea stagnalis*
W.P.M. Geraerts, A. ter Maat, and Th.M. Hogenes (With 4 Figures) 44

I.6 Roles of Calcium and cAMP in Biosynthesis and Release of the Ovulation Hormone of the Freshwater Snail *Lymnaea stagnalis*
E.W. Roubos, P. Buma, and A.M.H. van de Ven (With 2 Figures) 51

I.7 The Egg-Laying Behavior of the Pond Snail: Electrophysiological Aspects
A. ter Maat and R.F. Jansen (With 4 Figures) 57

I.8 Prothoracicotropic Hormone of *Bombyx mori*
H. Ishizaki and A. Suzuki (With 11 Figures) 63

I.9 Neurohormonal Regulation of Molting and Metamorphosis in the Tobacco Hornworm, *Manduca sexta*
W.E. Bollenbacher, N.A. Granger, W.A. Smith, and L.I. Gilbert 78

I.10 Isolation of an Allatotropic Factor in *Locusta migratoria* and Its Effect on Corpus Allatum Activity in Vitro
H.J. Ferenz (With 2 Figures) 92

I.11 Neurosecretion in Insects: Multisynthesis, Mode of Action, and Multiregulation
A. Girardie, J. Girardie, J. Proux, C. Rémy, and J. Vieillemaringe (With 1 Figure) 97

I.12 Neuropeptides of Vertebrate Type in the Blowfly *Calliphora*
A. Thorpe and H. Duve (With 3 Figures) 106

I.13 Peptides from the Corpora cardiaca and CNS of Locusts
W. Mordue and P.J. Morgan (With 2 Figures) 118

I.14 Adipokinetic Hormones in Locusts
G.J. Goldsworthy and C.H. Wheeler (With 3 Figures) 126

I.15 Ecdysteroids Regulate the Release and Action of Eclosion Hormone in the Moth *Manduca sexta*
J.W. Truman (With 3 Figures) 136

I.16 Studies on Crustacean Eyestalk Neuropeptides by Use of High Performance Liquid Chromatography
R. Keller and G. Kegel (With 7 Figures) 145

I.17 Present State of Knowledge on the Chemical Nature of the Androgenic Hormone in Higher Crustaceans
P. Juchault, J.J. Legrand, and J. Maissiat 155

Part II Ecdysteroids

Introduction
P. Karlson and J. Hoffmann 162

II.1 Ecdysteroids in Ovaries and Embryos of *Locusta migratoria*
M. Lagueux, J.A. Hoffmann, F. Goltzené, C. Kappler, G. Tsoupras, C. Hetru, and B. Luu (With 5 Figures) 168

II.2 Biosynthesis of Ovarian Ecdysteroid Phosphates and Their Metabolic Fate During Embryogenesis in *Schistocerca gregaria*
H.H. Rees and R.E. Isaac (With 2 Figures) 181

II.3 Ecdysone Metabolism
R. Lafont and J. Koolman (With 7 Figures) 196

II.4 Ecdysone 20-Monooxygenases
G.F. Weirich, J.A. Svoboda, and M.J. Thompson
(With 2 Figures) 227

II.5 Ecdysteroids and Integuments in Locusts
P. Porcheron, J.-P. Caruelle, J.-C. Baehr, and P. Cassier
(With 8 Figures) 234

II.6 Relationships Between Hormones and Epidermal
Cell Cycles During the Metamorphosis of *Tenebrio molitor*
J. Delachambre, M.T. Besson, A. Quennedey, and
J.P. Delbecque (With 9 Figures) 245

II.7 On the Mode of Action of Ecdysteroids in Crustaceans
K.-D. Spindler, L. Dinan, and M. Londershausen
(With 4 Figures) 255

II.8 Hormonal Control of Sequential Gene Expression in
Insect Epidermis
L.M. Riddiford (With 2 Figures) 265

II.9 Ecdysteroid Regulation of the Major Transcripts of
Drosophila melanogaster Larval Salivary Glands
and Fat Bodies
G. Richards and J.-A. Lepesant (With 3 Figures) 273

II.10 Ecdysteroid-Regulated Puffs and Genes in *Drosophila*
O. Pongs (With 1 Figure) 285

II.11 Pleiotropic Specific Responses Induced by Ecdysterone
in the Cultured Kc Cells of *Drosophila melanogaster*
J.-L. Couderc, M.-L. Sobrier, G. Giraud, D. Micard,
B. Dastugue, M. Ropp, J.-L. Becker, C. Maisonhaute,
F. Peronnet, A.-M. Courgeon, and M. Best-Belpomme
(With 2 Figures) 293

II.12 The Action of Ecdysone in Imaginal Discs and Kc Cells of
Drosophila melanogaster
L. Cherbas, J.W. Fristrom, and J.D. O'Connor
(With 2 Figures) 305

II.13 Ecdysteroids in Helminths
J. Koolman, J. Walter, and H. Zahner 323

II.14 Ecdysteroids in Schistosomes and Host-Parasite Relationship
P. Nirdé, G. Torpier, A. Capron, M. Delaage, and M.L. de Reggi (With 4 Figures) 331

II.15 The Occurrence and Possible Physiological Significance of Ecdysteroids During Nematode and Cestode Development
H.H. Rees and A.H.W. Mendis 338

II.16 Ecdysteroids in Annelids
M. Porchet, N. Gaillet, F. Sauber, M. Charlet, and J.A. Hoffmann (With 1 Figure) 346

Part III Juvenile Hormones

Introduction
L.I. Gilbert 350

III.1 Radioimmunoassays of Juvenile Hormones. State of the Methods and Recent Data on Validation
C. Strambi, A. Strambi, M. de Reggi, and M. Delaage (With 5 Figures) 355

III.2 The Qualitative and Quantitative Determination of Juvenile Hormones by Mass Spectrometry
B. Mauchamp, M. Zander, and R. Wolff (With 4 Figures) . . 363

III.3 Juvenile Hormones 0, I, and II Exist Only in Lepidoptera
D.A. Schooley, F.C. Baker, L.W. Tsai, C.A. Miller, and G.C. Jamieson (With 2 Figures) 373

III.4 Biochemical Aspects of Juvenile Hormone Biosynthesis in Hemimetabola
G.E. Pratt (With 3 Figures) 384

III.5 Regulation of Juvenile Hormone Titer in the Adult Female Cockroach *Diploptera punctata*
S.S. Tobe, B.A. Stay, F.C. Baker, and D.A. Schooley (With 4 Figures) 397

III.6 Environmentally Controlled Changes in Juvenile Hormone Titers: The Colorado Potato Beetle as a Model System
C.A.D. de Kort and M.A. Khan (With 8 Figures) 407

III.7 The Role of Juvenile Hormone Metabolism in the Metamorphosis of Selected Lepidoptera
B.D. Hammock, Y.A.I. Abdel-Aal, T. Hanzlik, D. Jones, G. Jones, R.M. Roe, M. Rudnicka, T.C. Sparks, and K.D. Wing 416

III.8 Hemolymph Transport of the Juvenile Hormones
W.G. Goodman, J.T. Trost, C.T. Reiter, D.W. Bean, and R.O. Carlson (With 3 Figures) 426

III.9 Photoaffinity Labelling of Juvenile Hormone-Binding Proteins in *Leucophaea maderae*
J.K. Koeppe, G.E. Kovalick, and G.D. Prestwich (With 3 Figures) 438

III.10 Regulation of Vitellogenesis in Insects: The Pleiotropic Role of Juvenile Hormones
F. Engelmann (With 6 Figures) 444

III.11 On Titers, Origin, and Functions of Juvenile Hormone III, Methylfarnesoate, and Ecdysteroids in Embryonic Development of the Ovoviviparous Cockroach *Nauphoeta cinerea*
B. Lanzrein, H. Imboden, C. Bürgin, E. Brüning, and H. Gfeller (With 5 Figures) 454

III.12 Defining the Anti-Metamorphic Action of Juvenile Hormone
J.H. Willis and D.L. Cox (With 4 Figures) 466

III.13 Vitellogenin Synthesis in Locust Fat Body: Juvenile Hormone-Stimulated Gene Expression
G.R. Wyatt, T.S. Dhadialla, and P.E. Roberts (With 4 Figures) 475

Part IV Hormonal Mechanisms in Echinoderms

Introduction
J.A. Hoffmann 486

IV.1 Early Steps in Transduction of the Hormonal Signal Controlling Starfish Oocyte Maturation
P. Guerrier, L. Meijer, M. Moreau, and J.A. Render (With 3 Figures) 487

IV.2 Production of the Maturation-Promoting Factor (MPF) Following Hormonal Stimulation of Starfish Oocytes
M. Dorée, G. Peaucellier, A. Picard, and F. Le Bouffant (With 5 Figures) 497

IV.3 Hormonal Mechanism of Starfish Spawning
H. Kanatani 509

Genera Index 511

Subject Index 513

Contributors

You will find the addresses at the beginning of the respective contributions

Abdel-Aal, Y.A.I. 416
Baehr, J.-C. 234
Baker, F.C. 373, 397
Bean, D.W. 426
Becker, J.-L. 293
Besson, M.T. 245
Best-Belpomme, M. 293
Bodenmüller, H. 5
Bollenbacher, W.E. 78
Brüning, E. 454
Bürgin, C. 454
Buma, P. 51
Capron, A. 331
Carlson, R.O. 426
Caruelle, J.-P. 234
Cassier, P. 234
Charlet, M. 346
Cherbas, L. 305
Couderc, J.-L. 293
Courgeon, A.-M. 293
Cox, D.L. 466
Dastugue, B. 293
de Kort, C.A.D. 407
Delaage, M. 331, 355
Delachambre, J. 245
Delbecque, J.-P. 245
de Reggi, M.L. 331, 355
Dhadialla, T.S. 475
Dinan, L. 255
Dorée, M. 497
Durchon, M. 10
Duve, H. 106
Engelmann, F. 444
Ferenz, H.J. 92
Fristrom, J.W. 305
Gaillet, N. 346
Geraerts, W.P.M. 44
Gfeller, H. 454
Gilbert, L.I. 78, 350
Girardie, A. 97
Girardie, J. 97
Giraud, G. 293
Goldsworthy, G.J. 126
Goltzené, F. 168
Goodman, W.G. 426
Granger, N.A. 78
Guerrier, P. 487
Hammock, B.D. 416
Hanzlik, T. 416
Hetru, C. 168
Hoffmann, J.A. 2, 162, 168, 346, 486
Hoffmeister, S. 5
Hogenes, Th.M. 44
Imboden, H. 454
Isaac, R.E. 181
Ishizaki, H. 63
Jamieson, G.C. 373
Jansen, R.F. 57
Jones, D. 416
Jones, G. 416
Joosse, J. 19
Juchault, P. 155
Kanatani, H. 509
Kappler, C. 168
Karlson, P. 162
Kegel, G. 145
Keller, R. 145
Khan, M.A. 407
Koeppe, J.K. 438
Koolman, J. 196, 323
Kovalick, G.E. 438
Lafont, R. 196

Lagueux, M. 168
Lanzrein, B. 454
Le Bouffant, F. 497
Legrand, J.-J. 155
Lepesant, J.-A. 273
Londershausen, M. 255
Luu, B. 168
Maisonhaute, C. 293
Maissiat, J. 155
Mauchamp, B. 363
Meijer, L. 487
Mendis, A.H.W. 338
Micard, D. 293
Miller, C.A. 373
Mordue, W. 118
Moreau, M. 487
Morgan, P.J. 118
Nirdé, P. 331
O'Connor, J.D. 305
Peaucellier, G. 497
Peronnet, F. 293
Picard, A. 497
Pongs, O. 285
Porcheron, P. 234
Porchet, M. 346
Pratt, G.E. 384
Prestwich, G.D. 438
Proux, J. 97
Quennedey, A. 245
Rees, H.H. 181, 338
Reiter, C.T. 426
Rémy, C. 97
Render, J.A. 487
Richards, G. 273
Riddiford, L.M. 265
Roberts, P.E. 475
Roe, R.M. 416
Ropp, M. 293
Roubos, E.W. 51
Rudnicka, M. 416
Sauber, F. 346
Schaller, H.C. 5
Schooley, D.A. 373, 397
Smith, W.A. 78
Sobrier, M.-L. 293
Sparks, T.C. 416
Spindler, K.-D. 255
Stay, B.A. 397
Strambi, A. 355
Strambi, C. 355
Strumwasser, F. 36
Suzuki, A. 63
Svoboda, J.A. 227
ter Maat, A. 44, 57
Thompson, M.J. 227
Thorpe, A. 106
Tobe, S.S. 397
Torpier, G. 331
Trost, J.T. 426
Truman, J.W. 136
Tsai, L.W. 373
Tsoupras, G. 168
van de Ven, A.M.H. 51
Vieillemaringe, J. 97
Walter, J. 323
Weirich, G.F. 227
Wheeler, C.H. 126
Willis, J.H. 466
Wing, K.D. 416
Wolff, R. 363
Wyatt, G.R. 475
Zahner, H. 323
Zander, M. 363

Part I Peptide Hormones

Introduction

J. HOFFMANN

The first Part of this Volume is devoted to hormones of peptidic nature and summarizes 18 papers given at the Symposium. The first article by H.C. Schaller and associates deals with the hormonal control of regeneration in an Acoelomate, *Hydra*. This model system for the study of the control of growth and differentiation in a primitive metazoan was first presented by H.C. Schaller at the meeting in Lille. The progress made in this area since that symposium is illustrated by the recent elucidation of the structure of the "head activator" which is a peptide consisting of 11 amino acids. The "head activator" of *Hydra* is also found in human and rat hypothalami and rat intestines and certainly has a wide distribution in the animal kingdom. The authors discuss possible functions of this molecule in mammals and suggest that this is the first example of a neuropeptide which has conserved its structure and role in the control of growth and differentiation during evolution.

The paper by M. Durchon summarizes our information on the neuroendocrine control of reproduction, development, homeostasis, and regeneration in annelid worms. The contribution highlights the recent advances in the study of the control of nereid reproduction and gives preliminary results concerning the structure of nereidin, a peptide hormone inhibiting gametogenesis and epitokal metamorphosis. Until very recently, it was a common belief that the major events in the development and reproduction of annelids were under the exclusive control of neurohormones; the situation appears now to be more complex, and ecdysteroids are probably also involved in this control, as explained in the third section of this volume (see M. Porchet et al.).

The neuroendocrinology of molluscs was the subject of five contributions which were introduced to the Symposium by the general lecture of J. Joosse. The major part of the papers given on molluscs was devoted to the spectacular progress made in the last years in the study of the neuroendocrine control of egg-laying in the marine gastropod *Aplysia californica* ("bag cell system") and the freshwater snail *Lymnaea stagnalis* ("caudo-dorsal cell system"). F. Strumwasser's paper deals with the first system, whereas the three articles by the Amsterdam group (Geraerts et al., Roubos et al., ter Maat et al.) concern the *Lymnaea* model. These contributions on molluscs raised a series of interesting points which are also essential in several contributions on insects in this Part, and which were main topics during the round table discussions, namely (1) the presence and physiological significance of peptides which cross-react with antibodies raised against mammalian peptide hormones; 29 such "vertebrate-like" factors have been recorded so far in molluscs and almost as many in insects; (2) the observation that one neuron can release multiple secretory products. This is exemplified in molluscs by the

Biosynthesis, Metabolism and Mode of Action of Invertebrate Hormones
(ed. by J. Hoffmann and M. Porchet)

caudo-dorsal cells of *Lymnaea* and also by the bag-cells of *Aplysia*; in the latter case, one common precursor molecule yields 4 peptides (Egg Laying Hormone, 36 amino acids; α-Bag Cell Peptide, 9 amino acids; β-Bag Cell Peptide, Acidic Peptide, 27 amino acids) which are secreted together; of these it is assumed that one, the Egg-Laying Hormone, probably serves both as a neurohormone and as a neurotransmitter, whereas two others (α and β-Bag Cell Proteins) probably act as neurotransmitters. These various problems are discussed in this volume in the contributions of J. Joosse and also in those of A. Girardie and associates and A. Thorpe and H. Duve.

The remaining articles of this Part are concerned with peptide hormones in arthropods. Three papers are devoted to neurohormones which control the activity of ecdysone or juvenile hormone-synthesizing glands. The contribution by H. Ishizaki and A. Suzuki gives the history and the present status of the efforts of the Japanese scientists to isolate and elucidate the structure of the *Bombyx* prothoracicotropic hormone. The neurohormonal regulation of both the prothoracic glands and the corpora allata in another lepidopteran, *Manduca sexta*, is the subject of the article by the Chapel Hill groups. This paper includes essential aspects of the development of in vitro assays for qualitative and quantitative studies on prothoracicotropic, allatotropic, and allatoinhibiting (allatohibins) neurohormones. H.J. Ferenz, in a short communication, reports on a pronase-sensitive, heat-stable allatotropic factor present in the brain and corpora cardiaca of the orthopteran *Locusta migratoria* which significantly stimulates juvenile hormone synthesis in corpora allata under in vitro conditions.

The potential of high performance liquid chromatography for analytical separation and characterization of neuropeptides is elegantly illustrated by the article of R. Keller and G. Kegel on Crustacean eyestalk neuropeptides. Only two of the apparently numerous eyestalk neuropeptides have so far had their structure elucidated: a red pigment-concentrating hormone (RPCH) and a melanophore-dispersing hormone (MDH), but it is expected that this list will rapidly increase in the near future. A major problem results from the fact that a molecular heterogeneity may exist for a given eyestalk hormone from various species, as evidenced for crustacean hyperglycemic hormone (CHH). HPLC analyses, as shown in Keller's report, can clearly separate CHH from different crustacean species such as *Carcinus, Orconectes,* and *Eriocheir*, and during evolution the changes in the structure of CHH are such in crustaceans that CHH from the isopod *Porcellio* is not recognized by target tissues from the brachyueran *Carcinus* and vice versa.

P. Juchault and colleagues summarize our information on the androgenic gland hormone(s) of higher crustaceans, and provide evidence in favor of the protein character of an androgenic factor extracted from these glands in isopods.

The next paper, by A. and J. Girardie and their associates, highlights recent developments in the study of neurosecretion in insects. As in molluscs and vertebrates, neurosecretory cells of insects can be both aminergic and peptidergic as illustrated in the median neurosecretory cells of *Locusta* in this report. The major part of the contribution is devoted to an arginine-vasopressine-like substance associated in the neurosecretory cells of the suboesophageal ganglion with a neurophysine-like protein. This substance has a diuretic effect in *Locusta* and its production and secretion is shown by the Bordeaux group to be controlled by stimulating (subocellar) and inhibiting (median) neurosecretory cells. This is probably the first demonstration of a multineuroendocrine regulation of a neurosecretory center in insects; it is evocative of the multineuroendocrine control of an endocrine gland such as the corpora allata, which is discussed in the article of E. Bollenbacher and associates.

A. Thorpe and H. Duve examine vertebrate type neuropeptides in the blowfly *Calliphora* and more particularly focus on the gastrin-cholecystokinin and endorphin-enkephalin families. These authors have included in their article a section of *Concluding Comments* which is a welcome summary of the prevailing opinions at the Symposium's round table discussions on the presence and functional significance of vertebrate type peptides in invertebrates.

The next two papers are devoted to diuretic and adipokinetic peptide hormones of *Locusta*. W. Mordue and P.J. Morgan provide an account of the recent isolation techniques and bioassays for these hormones which are of critical importance for the control of metabolism, namely flight metabolism, and homeostasis. The contribution also briefly reviews current information on diuretic, hyperglycemic, and hyperlipemic peptides from several insect species. In their contribution, G.J. Goldsworthy and C.H. Wheeler review the known actions of adipokinetic hormones in adult locusts; they analyze the roles of these hormones in the integrated utilization of metabolic fuels during flight and describe the lipoprotein-protein interactions during adipokinetic hormone-stimulated lipid transport in *Locusta*.

In the last paper of the session on peptide hormones, J.W. Truman analyzes the intimate relationship between ecdysteroids and the peptidic eclosion hormone in coordinating events at the end of the molt of *Manduca sexta*.

I.1 Hormonal Control of Regeneration in Hydra

H.C. SCHALLER, S. HOFFMEISTER, and H. BODENMÜLLER[1]

1. Introduction into the Hormonal Control of Regeneration in Hydra

Within the animal kingdom cells specialized for nervous function first appear among coelenterates. We have chosen to work with hydra, because its nervous system and its general organization appeared simple. The animal contains only two cell layers, ectoderm and entoderm, consisting of epithelio-muscular cells among which luxury cells, such as nerve cells and others, are interspersed. The nervous system is organized as a continuous net with a higher density in the head and in the foot region.

The nervous system in hydra serves two functions: it is necessary for the coordination of fast processes like body movements, and it controls slow processes like cellular growth and differentiation. So far not much is known how and with which transmitter substances the fast interneuronal communication system in hydra works. Accumulating evidence suggests that conventional transmitters, such as acetylcholine or catecholamines, are not present in nerve cells of hydra. It seems more likely that peptides may play that role since more and more of them are found coexisting or separately in nerve cells of hydra [1, 2]. We know more about the second function of the nervous system and about the substances which control cellular growth and differentiation and, thus, pattern formation in hydra. To understand pattern formation we want to know why a certain structure is induced at a definite position and time during development, and why it subsequently inhibits the formation of an identical such structure nearby. Hydra has two centers of organization, head and foot, from both of which gradients of induction and inhibition extend towards the opposite ends. This has been deduced mostly from transplantation, but also from other biological experiments. One basic experiment is the following: if a hydra is cut into two pieces of equal length, each piece will regenerate the missing structure such that two perfect hydra result. The cut surface is identical in both cases. However, from the same piece of tissue in the upper half a foot is regenerated, in the lower half a head. Transplantation experiments have shown that the induction of the respective structures, head or foot, is a relatively fast process, that within a few hours after cutting (3–6 h) the regenerating gastric tissue has changed its program with the result that it behaves more like a foot or a head tissue. Induction is a very local process and restricted to the tissue close to the original cut. At the cellular level this means that only cells close to the original cut surface are changed in their growth and

1 Max-Planck-Institut für Medizinische Forschung, Abteilung Biophysik, 6900 Heidelberg, FRG

Biosynthesis, Metabolism and Mode of Action of Invertebrate Hormones
(ed. by J. Hoffmann and M. Porchet)

differentiation characteristics, i.e., become head-specific of foot-specific cells, respectively. Inhibition, on the other hand, is long range: an existing head usually inhibits that another head is formed along the whole body column. Only if a certain size is reached, if the distance to the existing head is large enough, a secondary head induction is allowed. The potency to induce or inhibit a head is highest in or close to the head region decreasing in a graded fashion to the foot. Similarly, the potency of foot induction or foot inhibition decreases from foot to head.

2. Substances Controlling Regeneration

We could show that four substances are responsible for induction and inhibition in hydra: an activator and an inhibitor of head formation and an activator and an inhibitor of foot formation [3]. The gradients in biological properties are caused by a graded distribution along the body axis of the animal of these four morphogenetic substances that induce or inhibit head or foot formation. The effect of substances which influence head or foot formation in hydra can most easily be measured as acceleration or inhibition of head or foot regeneration [3]. Such a response is dose-dependent, and allows the quantitation of the respective factor per animal or in a fraction during the purification procedure. By means of such assays we have purified the four substances, the head activator to homogeneity (10^9-fold enrichment), the other three at least 10^5-fold [3]. The two activators are destroyed by proteases and acid hydrolysis (6 N HCl, 110°, 24 h), the two inhibitors survive such a treatment. The two activators under neutral conditions absorb to anion exchangers, i.e., have an overall negative charge, the two inhibitors have an overall positive change. The head activator has a molecular weight of 1,124, the foot activator between 500 and 1,000. The two inhibitors have molecular weights probably below 500. The inhibitors are very hydrophilic molecules, the activators are more hydrophobic. Thus, the head activator at high concentrations (10^{-3} *M*) is insoluble in water, soluble in organic solvents, and it has a very high surface adhesivity. The parallelism in chemical properties of the inhibitors with the activators is striking and is probably of biological relevance. In particular, the difference in hydrophobicity or surface adhesivity resulting in completely different diffusion rates provides a plausible explanation for the biological finding that the action of the inhibitors is long-range, whereas that of the activators is short-range.

From the degree of purity and the molecular weight we can calculate that all four substances act at very low concentrations. According to the present state of purity the two foot factors and the head inhibitor act at concentrations below 10^{-8} *M*, the head activator at 10^{-13} *M*. This also means that a hydra needs very little of these substances and explains our difficulties in obtaining in pure form workable quantities from hydra for a chemical analysis. For example, we were able to isolate from 3×10^6 hydra (3 kg), accumulated over several years, 0.5 μg of the pure head activator. Fortunately, we discovered that other coelenterates, in particular sea anemones, which are 10^4-fold larger and abundant in the ocean, contain the same set of substances as hydra in similar concentrations. Meanwhile, we have processed 200 kg of the sea anemone, *Anthopleura elegantissima*, and used it so far for the chemical analysis of the head activator [4].

3. Isolation and Sequence Analysis of the Head Activator

The sources from which the head activator was isolated and sequenced are listed in Table 1.

Table 1. Head activator yield from different animal sources

Animal source	Starting material	Yield of head activator
Coelenterates:		
Hydra attenuate	3 kg	500 pmol (0.5 μg)
Anthopleura elegantissima	200 kg	20 nmol (20 μg)
Mammals:		
Rat intestine	3,000 pieces	20 nmol (20 μg)
Human hypothalamus	2 pieces	2 nmol (2 μg)

To obtain workable quantities of the peptide we had to start with relatively large amounts of tissue. In general, we could enrich the head activator 10^8- to 10^9-fold with a 10%–20% yield. Purification steps included organic solvent extraction, molecular sieve and ion exchange chromatography, and HPLC methods [4, 5]. The amounts of head activator available for a sequence analysis were relatively minute and required use and improvement of micromethods. To determine the amino acid composition of the head activator the purified peptide was hydrolyzed, dansylated and the dansylated amino acids were separated for a qualitative analysis on microthin-layer plates and for quantitation on HPLC. We found that the head activator contained the following amino acids: Gly (2), Glu (1), Ile (1), Leu (1), Lys (1), Phe (1), Pro (2), Ser (1), Val (1). The sequence analysis was complicated by the fact that the head activator, like many other biologically active peptides, had no free amino terminus, that its amino end was blocked by pyroglutamic acid. Therefore, most of the sequence was obtained by enzymatic degradation methods and in part by a chemical degradation of the biologically inactive Glu-peptide which contained glutamic acid instead of pyroglutamic acid as amino terminal amino acids. The head activator was found to have the sequence pGlu–Pro–Pro–Gly–Gly–Ser–Lys–Val–Ile–Phe [4]. No difference in sequence was found for the head activator from the sea anemone, *Anthopleura elegantissima*, from *Hydra attenuata*, from rat intestine, and from bovine and human hypothalamus [4, 5]. This sequence of the head activator was confirmed by synthesis [6]. We could show that there is no difference between the purified native and the synthetic peptides in chemical, enzymatic, and biological properties. We and others also produced a series of peptides with minor modifications. So far none of those was as biologically active as the native molecule. Together with the evolutionary conservation of the head activator from hydra to mammals we interprete this to mean that the native molecule at least in what regards its action on hydra receptors has very strict conformational requirements.

The chemistry of the head activator predicts or explains the following outstanding properties: (1) It is very lipophilic leading to a relatively low solubility in aqueous solvents. (2) In common with many other biologically active peptides it has a very high surface adhesivity resulting in adsorption in vitro to glass, plastic, etc. and in vivo probably to membranes and to hydrophobic parts of proteins rendering it relatively immobile or preventing diffusion to distant areas. (3) The head activator occurs in two distinguishable conformations, one with an apparent molecular weight of about 1,500 and one of about 700. Preliminary evidence supports the idea that the low molecular weight conformation is a monomeric peptide, whereas the larger one is dimeric. The freshly isolated, relatively impure, and dilute native peptide has the monomeric conformation. The synthetic molecule and the purified head activator in physiological media converts to the more stable dimeric conformation. The two conformations are interconvertible. Since only the monomer is biologically active we think that this conversion may be of physiological relevance.

4. Effects of the Head Activator on Morphogenesis, Cellular Growth, and Differentiation

The head activator and its counterpart, the head inhibitor, are present in hydra in an inactive structure-bound form. The head activator is stored in neurosecretory granules [7]. This has the advantage that the release can be regulated. The release of the head factors is controlled by the head inhibitor which acts as a release-inhibiting factor for its own release and for that of the head activator as well [8]. The head activator has no effect on its own release nor on that of the head inhibitor. Thus, release of both factors is triggered by the absence of the head inhibitor. The head inhibitor, as a small very hydrophilic molecule, is able to diffuse once released into the intercellular space over long distances. The head activator due to its hydrophobic character will adhere to membranes of neighboring cells or to the extracellular matrix and will therefore be restricted locally. The distribution of sources for both substances within hydra is relatively similar, both having their maximal concentration in the head area (hypostome). Thus, although both substances may be released from the same location, at the release site there will be a local predominance of head activator over head inhibitor and at a greater distance predominance of head inhibitor over head activator. At the cellular level head activator and head inhibitor act antagonistically: the head activator stimulates cells to divide and is necessary for head-specific differentiation, the head inhibitor blocks such processes. Thus, an interstitial stem cell will become a head-specific nerve cell if the concentration of the head activator is higher than that of the head inhibitor. The stem cell will remain stem cell or center another differentiation pathway (e.g., to nematocytes) if head inhibitor prevails [8-11]. This means that close to a common release site stem cells differentiate to nerve cells, whereas further away they enter another differentiation pathway. The new nerve cells will then produce head activator and/or head inhibitor, thus, reinforcing or amplifying the head-specific influence. Such a feedback mechanism may ensure that wherever head activator predominates over head inhibitor, head-like structures will be induced and maintained. Release control and events at the cellular level have recently been incorporated into a computer simulation of head regeneration in hydra [12].

5. Action of the Head Activator in Mammals

The head activator was also isolated and sequenced from human and bovine hypothalami and from rat intestines. Immunological methods (HPLC–RIA or ELISA) [13, 14] indicate that immune-like head activator also occurs in other animals (drosophila, artemia, frog, rabbit, guinea pig) and in other locations. Thus, we found that it is especially abundant in rat retina and frog skin.

The sequence of the head activator differs from that of other biologically active peptides. The only structural homology we observed is with bradykinins, in the sequence X–Pro–Pro–Gly–X–Ser–. Bradykinins are known as smooth muscle stimulants. The head activator at a concentration of 10^{-5} *M* did stimulate uterus contraction in rats (Ganten, unpubl.). This may imply a very weak structural and functional homology. From the localization of the head activator in the brain and intestinal tract, from its appearance early in mammalian development, and in analogy to its action in hydra, the following speculation on its action may be allowed. Like other neuropeptides it may act as a transmitter-like messenger in the brain. The high concentration in the upper part of the intestine may hint at a regulatory function in the digestive system [15]. In embryonal life and maybe also in the adult the head activator may act on some neuronal stem cell either as growth factor or as differentiation signal. Its presence in tumors or tumor cell lines derived from early stages of neuronal differentiation hints at such a possible trophic or mitotic role in nerve cell development. This may be the first example that a neuropeptide conserved its structure, and maybe also its growth or differentiation controlling functions during evolution from coelenterates to humans.

Acknowledgment. H.C. Schaller is a recipient of a Heisenberg fellowship. We are supported by the Deutsche Forschungsgemeinschaft (Scha 253/8).

References

1. Grimmelikhuijzen CJP, Dierickx K, Boer GJ (1982) Neuroscience 7:3191–3199
2. Grimmelikhuijzen CJP (1983) Neuroscience 9:837–845
3. Schaller HC, Schmidt T, Grimmelikhuijzen CJP (1979) Wilhelm Roux's Arch 186:139–149
4. Schaller HC, Bodenmüller H (1981) PNAS 78:7000–7004
5. Bodenmüller H, Schaller HC (1981) Nature (London) 293:579–580
6. Birr C, Zachmann B, Bodenmüller H, Schaller HC (1981) FEBS Lett 131:317–321
7. Schaller HC, Gierer A (1973) J Embryol Exp Morphol 29:39–52
8. Kemmner W, Schaller HC (1984) Differentiation 26:91–96
9. Schaller HC, Bodenmüller H, Kemmner W (1982) Verh Dtsch Zool Ges:81–90
10. Schaller HC (1976) Cell Differ 5:1–11
11. Schaller HC (1976) Cell Differ 5:13–20
12. Kemmner W (1984) Differentiation 26:83–90
13. Bodenmüller H, Zachmann B (1983) FEBS Lett 159:237–240
14. Schaller HC, Bodenmüller H, Zachmann B, Schilling E (1984) Eur J Biochem 183:365–371
15. Feuerle GE, Bodenmüller H, Báca J (1983) Neurosci Lett 38:287–289

I.2 Peptidic Hormones in Annelids

M. DURCHON[1]

Since the first colloquium organized by C.N.R.S. on *Biosynthesis, Metabolism, and Mode of Action of Invertebrate Hormones,* noteworthy progress has been registered in annelids. Results concerning oligochaetes, hirudinea, and polychaetes will be successively considered.

1. Oligochaetes

It has been experimentally demonstrated by several authors [1,2,14] that in Lumbricids the cerebral ganglia are hormonally involved in the differentiation of the somatic sexual characteristics at the time of reproduction. The presence of the cerebral ganglia is required for the maintenance of the clitellum turgescence.

Moreover, Marcel [30] has shown that the cerebral ganglia are the site of production of a factor inhibiting regeneration. Investigations have now been focused on the isolation of the substances responsible for these physiological actions [31-33]. The first result obtained [31] has established that the cephalic regeneration inhibitor is a peptide with a molecular weight of ca. 2,000, including 18 amino acids residues.

The isolation of the clitellogenic factor and investigations related to its biological activity led to the conclusion that this factor is identical to the cephalic regeneration inhibitor.

Moreover [20-23], organotypic cultures and in vivo implantations experiments have shown that in *Eisenia foetida* the cerebral ganglia are implicated in the synthesis of an androgenic factor by the testes. Experiments were designed by Lattaud and Marcel (unpubl. data) in organotypic cultures to investigate the effect of cerebral ganglia extracts. Different types of cultures were realized:

- Control associations: testes + ovaries
- Control associations: cerebral + ganglia + testes + ovaries
- Testes associated with ovaries in the presence of fraction A originating from a brain homogenate.

Culture time is 7-17 days.

1 Laboratoire Associé au C.N.R.S. no. 148, Endocrinologie comparée des Invertébrés, Université de Lille I, 59655 Villeneuve d'Ascq Cedex, France

Biosynthesis, Metabolism and Mode of Action of Invertebrate Hormones
(ed. by J. Hoffmann and M. Porchet)

The results obtained indicate that ovaries transform into ovotestes if they are cultured either in the presence of testes and cerebral ganglia or with the brain homogenates fraction. Moreover, biochemical analyses with HPLC, now in progress, allow the assumption that the clitellogenic factor and the factor inhibiting cephalic regeneration are one and the same factor. Thus, in *E. foetida,* one and the same peptidic neurohormone would control testicular differentiation, somatic sexual characteristic development, and cephalic morphogenesis. These results will be explicated with the determination of the amino acid sequence of this peptide and investigations related to the mechanisms of action at the level of target cells.

2. Hirudinea

The first data related to the endocrinology of Hirudinea were brought by Hagadorn [13] and concerned the microscopic anatomy, the cerebral nervous centers ultrastructure, and the secretory cycle in correlation with the annual cycle in *Theromyzon rude* Baird. Moreover, this author demonstrated experimentally in the same species that in the absence of "brain", spermatogenesis does not reach its term and that the ovary does not undergo activation. From these results, it was concluded that a gonadotropic hormone of cerebral origin was involved in the sexual cycle of Hirudinea. Malecha has supported these conclusions with in vitro data obtained from organ cultures. More recently, Van Damme [44–46] demonstrated that the buccal nervous ganglia exert in the leech *Erpobdella octoculata* L. an inhibiting effect on genital maturation.

The buccal ganglia is composed of two neuronal follicles, each including a neurosecretory cell. Their bilateral destruction provokes a precocious genital development whenever the peripharyngeal complex is preserved. This operation triggers spermatogenesis and at the level of the ovary, a unique fit of oocytes developing directly in the follicles.

Thus, the single neurosecretory cell, situated in each buccal ganglion, would exert, at least in this species of Hirudinea, an inhibition on the genital maturity in the young animals.

Another aspect of the endocrinology of Hirudinea has just been discovered in our laboratory by Malecha [24, 25]; it concerns a neuroendocrine control of osmoregulation in the Rhynchobdella Hirudinea *Theromyzon tessulatum* O.F.M.

It is well known that Hirudinea have the faculty to regulate their osmoregulation [5, 41]. Rosca et al. [42] put forward the hypothesis that in *Hirudo medicinalis*, osmoregulation was regulated by the whole nerve cord. Kulkari and Nagabhushanam [19] conclude after experiments on *Poecilobdella viriolis* that in this species water loss is under the dependence of a cerebral neurohormone.

Malecha took upon the study of this problem in *Theromyzon tessulatum*, Hirudinea which only needs three meals in the course of its life, with the condition that these are quantitatively sufficient. After the last one, growth of *T. tessulatum* is characterized by a considerable increase of mass due to an important retention of water. Removal of the brain after the last meal has the following consequences:

– a loss of mass, imputed to an excessive diuresis:
– a considerable slowing down of the digestion of the digestive content;
– an increase in longevity compared to the controls.

An implantation of brains to previously decerebrated leeches restores their growth. Moreover, parabiosis experiments between a stage 3 leech, decerebrated (after the third meal), and a stage 1 leech (after the first meal) indicate that the effects of decerebration are partially or totally suppressed in these associations.

All these experiments and elective ablations give evidence of the existence in *T. tessulatum* of a factor regulating hydric balance (FRBH), secreted by the brain at the level of the posterior paramedial follicles (follicles 4). Isolation of this cerebral neurohormone is in progress.

3. Polychaetes

The start of the neuroendocrinology of polychaete annelids dates back to 1948, with experimental studies on nereids. Since this time, cellular and biochemical mechanisms have been studied and isolation of the neurohormone is in progress. Moreover, this neuroendocrinology has been extended to other fields: meiotic reinitiation in the lugworm and neuroendocrine determinism of sex reversal in the Eunicid *Ophryotrocha puerilis*.

3.1 Neuroendocrinology of Nereids

3.1.1 Purification of the Cerebral Neurohormone of Nereis

The results were obtained in team [9, 10] and with the collaboration of Luu Bang, Maître de Recherche at the Neurochemistry Laboratory of Strasbourg.

After methanolic extraction, followed by a chromatography on gel of Sephadex G25 and G25 superfine, an active fraction called "fraction 5" has been obtained. It has been purified by high performance chromatography (HPLC).

The bioassays of the different fractions are done in organ culture. The in vitro test used is the inhibition of the meiotic divisions of spermatogonia present in the parapodia of male *Nereis diversicolor*.

Concerning the isolation by means of HPLC, the best separation of the hormonal fractions was obtained with μ-bondapak C_{18} (Waters) in radial pressure (module RCM 100) in eluting with 40% methanol-Absorbance at 206 nm.

The final fraction obtained (peak A) has been analyzed. It possesses the same hormonal activity as that of the control culture. An amino acid analysis followed by a sequential analysis according to the technique of Edman modified by Chang has been coupled to mass spectrometry.

These investigations carried out by Luu Bang and Jiu-Yu-Chu (Institute of Neurochemistry, University Louis Pasteur, Strasbourg) gave the following results:

- Amino acid analysis:
  Pro (4), Gly (3), Asp (0.4), Thr (0.4), Ser (0.36), Glu (0.54), Ala (0.23);
- Fraction A, biologically active, contains several peptides (probably 4).
  One of them has as a sequence:
  $$\underset{1}{\text{Pro}} - \underset{2}{\text{Pro}} - \underset{3}{\text{Gly}} - \underset{4}{\text{Pro}} - \underset{5}{\text{Pro}} - \underset{6}{\text{Gly}}$$
  N – Pro – Pro – Gly – Pro – Pro – Gly – C .

From a biological point of view, no proof allows affirmation that this hexapeptide supports the hormonal activity. The organic synthesis has been done but this product does not block the meiotic division of spermatogonia.

Nevertheless, the result obtained in *Nereis* may be paralleled with the one registered by C. Schaller in *Hydra* (this volume) where for the cephalic activatory peptide, the link $\underset{2}{\text{Pro}} - \underset{3}{\text{Pro}} - \underset{4}{\text{Gly}} - \underset{5}{\text{Gly}}$ is also found.

The "genealogy" of neurohormones in the animal kingdom begins to be written.

3.1.2 Regulation of the Cerebral Neurosecretion

We know that the cerebral neuroendocrine activity, high in young *Nereis*, gradually decreases to zero at the time of reproduction. It has also been established that photoperiod and a "feedback" substance originating from the oocytes are involved in the regulation of the cerebral neurosecretory activity. More recently, Porchet, Dhainaut, and Porchet-Hennere [40] gave evidence of the existence, in the coelomic fluid, of two peptides (B_1 and B_2). These peptides, absent during the phase of sexual indifferentiation, appear at the time of the oocyte submaturity stage.

When these peptides B_1 and B_2 are injected in the coelom of very young females, they stimulate an important biosynthesis of the glycoconjugates of the oocyte cortical alveoli; a precocious oocyte maturation being then registered. Correlatively, the cerebral neuroendocrine activity becomes null. This effect is alike to the one obtained after an intracoelomic injection of the feedback substance. From this result, it is feasible that peptides B_1 and B_2 play an activatory role in the biosynthesis of the feedback substance and, indirectly, in the regulation of the cerebral neuroendocrine activity.

3.1.3 Target Cellular Mechanisms of the Neurohormone

Previous investigations (Porchet et al. [37–40]) having shown the importance of glycoproteins and glycolipids in the oocyte differentiation controlled by the neurohormone, attention has been paid to the glycosyltransferases, enzymes catalyzing the biosynthesis of all the known glycoconjugates (glycoproteins, glycolipids, glycosaminoglycans).

It has to be reminded that glucose is the precursor monosaccharide in the coelomic fluid of *Nereis*. After incorporation in the oocytes and in the coelomocytes, glucose undertakes numerous transformations leading to the cellular production of nine other monosaccharides (galactose, mannose, xylose, fucose, rhamnose, glucosamine, galactosamine, glucuronic acid). All these monosaccharides are conjugated.

In the young oocyte, when the hormonal rate is high, the conjugated monosaccharides, synthesized in the yolk are essentially hexoses (galactose, mannose) and hexosamines.

The oocyte more advanced in age (diameter 140–250 μm) elaborates cortical alveoli, whereas the neurohormone rate progressively decreases to zero when the oocytes are submature (180–200 μm). The glycoconjugates characteristic of the alveolar mucous substances are fucose, glucuronic acid, and immediately prior to the eggs maturation sialic acid.

The investigations on glycosyltransferases have been carried out in our laboratory by Bulet in collaboration with LA 217 (Prof. A. Verbert team: R. Cacan and B. Hoflack).

To measure the enzymatic activities, a separation of oocytes and coelomocytes was requested. A technique was designed based on a sedimentation of the coelomic content at 1 g through 5% Ficoll [16]. Moreover, it was necessary to check in vitro the concentration in precursors (nucleotides, sugars), in acceptors (endogenous glycoconjugates) the temperature of the incubation medium, the time of incubation, the ionic force, the concentration in detergent (Triton X 100), and the presence of bivalent ions (Mn^{++}, Mg^{++}).

From research in progress, two results will be put forward. The first concerns the optimal temperatures; indeed, it has to be checked if for each enzyme the optimum stayed constant the whole year long or if it changed according to the seasons. It was found that the temperature of functioning of a given type of glycosyltransferase varies; consequently, the variation of the optimal temperatures of functioning are regulatory systems of the oocyte differentiation. The second result deals with the degradation of the precursor. The transformation of GDP-mannose in GDP-fucose was studied [7]. The products of degradation of GDP-mannose are of special interest in the case of *Perinereis cultrifera* oocytes. Successively, mannose-1-phosphate and then mannose appear. Thereafter, unknown products are detected; they result from a very rapid conversion in GDP-fucose. This finding presents a great biological interest; indeed, the aged oocyte needs an important quantity of fucose, whereas the young oocyte requests much mannose. The kinetic of the transformation of GDP-mannose shows that up to the size of 90 μm, the oocyte does not perform such a transformation, but that beyond this stage the major part of GDP-mannose turns into GDP-fucose; the enzymatic system becomes active when the oocyte is going to stop its yolk syntheses.

Thus, such an enzymatic system is a biochemical marker of the oocyte maturation. It has now to be researched if the neurohormone exerts an influence and if in the affirmative at what stage of the reaction chain.

3.1.4 Action of the Neurohormone at the Level of Nucleic Acids

It is known that in every male of *Nereis*, removal of the prostomium, source of the neurohormone, induces a precocious spermatogenesis; moreover, in organ culture, in isolated parapodia, on an anhormonal medium, spermatogonia give rise in 6 days at 20 °C to spermatozoa.

Cultures on a medium labelled with tritiated thymidine gave evidence that spermatogonia are then experimenting a strong DNA synthesis [11]. Nevertheless, it has to be pointed out that these DNA syntheses, resulting from the experimental lifting of the inhibition are followed by a meiosis in cells whose normal destiny would be to undergo many gonial mitoses. Bertout [3] investigated if the cerebral hormone exerts an influence on RNA syntheses during spermatogenesis. In organotypic culture and with tritiated thymidine and uridine labellings, working at 15 °C, he found that the phase of DNA synthesis (S phase) occurs during the seventh day of culture. The syntheses of RNA begin at the end of the G1 phase (fifth day), continue during the S phase and decrease during the G2 phase, the latter being short in time. The preponderance of the nucleolar labelling seems to indicate that the synthesized RNAs are mostly of a ribosomial nature. In the presence of hormone, no modifications in the rate of RNA synthesis are registered. Alone, the absence of neurohormone provokes a considerable increase of RNA syntheses.

Investigations have also been carried out on the oocytes. In order to give evidence of an influence of the neurohormone on RNAs, *P. cultrifera* females were either cut in two, the anterior part being considered as a control, or deprived of prostomium; in this case, control animals were reared in the same conditions. The experiments, of a 10 to 16 days length, have included a quantitative analysis of RNAs and proteins, an autoradiographic study and a biochemical analysis.

The results obtained [8] show that the neurohormone stimulates the synthesis of "stable" RNA; in young females, where the neurohormonal rate is high, removal of the prostomium induces, compared to the controls, a decrease of RNAs rate, whereas the one of proteins remains stable. Qualitative modifications are also registered, especially a storage of the RNA synthesized in the nucleus. In summary, after experimental removal of the hormonal source, an alteration in the maturation of rRNA is registered. It is quite possible that in *Nereidae*, the hormonal control of the storage of oocyte rRNA is not direct at the level of the transcription, but would be the result of a regulation of the synthesis of the mRNA necessary to the production of 45S RNA. These biochemical data were complemented by an ultrastructural and autoradiographic study of the nucleus. The results obtained show that more specially in the young oocytes, the nucleoplasmic granulations are highly radioactive after a short period of incorporation of 3H uridine; it is, therefore, probable that a synthesis of extranucleolar RNA takes place. For technical reasons, it has proved impossible to determine the exact nature of these RNAs. Moreover, it has to be pointed out that in previtellogenic oocytes, the synthesis of extranucleolar RNA is more significant than that of nucleolar RNA origin. This preponderance is more pronounced in vitellogenic oocytes (60-100 μm in diameter) and disappears during the phase of synthesis of the cortical alveoli.

The results obtained with autoradiographic and biochemical studies show that RNAs of a nucleolar origin (18 and 28S rRNA) are synthesized during the different oocyte stages, but at different rates. The maximum seems to occur during vitellogenesis.

3.2 Neuroendocrine Determinism of Meiotic Reinitiation in the Lugworm

Howie [15-17] studied the neuroendocrine influence of the cerebral ganglia on gametes maturation in *Arenicola marina* L. More recently, Meijer [26-29] studied this problem and demonstrated the hormonal control of meiotic reinitiation in the oocytes of this polychaete.

In the lugworm, the oocytes are arrested in the first prophase of meiosis until spawning occurs. Free in the coelom, their diameter is 180 μm and they have a lenticular shape, with a convex (vegetal) and concave (animal) face. The germinal vesicle is masked by the cytoplasmic inclusions.

An experimental study of the oocyte maturation has been carried out by Meijer. When the oocytes are incubated in seawater containing an extract of cerebral ganglia (0.05 prostomium/ml), the germinal vesicle breakdown starts between the 30th and 35th min. The meiotic spindle (20 μm length) is oriented perpendicular to the plasma membrane. Chromosomes appear around the 40th min and they arrange in a metaphase plate.

The control oocytes, placed in pure seawater, do not show any maturation. It is possible to obtain a fertilization of the oocytes which underwent an experimental maturation and to induce an activation with ionophore A 23187. It is highly probable that the neurohormone acts on the oocyte in increasing the concentration of intracellular free calcium.

The chemical nature of the neurohormone responsible for the meiotic reinitiation is not yet known. Howie [17] experimentally demonstrated that the producing neurosecretory cells were located in the posterior part of the brain. Moreover, this author observed [18] in an ultrastructural study four types of neurosecretory cells, two of which having an aminopeptidergic secretion product.

3.3 Neuroendocrine Determinism of Sex Reversal in *Ophryotrocha puerilis*

Ophryotrocha puerilis is a small eunicid whose sexual behavior has been intensively investigated. When reared, a male phase is registered when the animals have ten segments, it is followed by a female phase (16 metameres) which persists if the annelids are cultured in isolation and are well fed. If two females are put together, one of them expels its oocytes in a delay of 3–5 days and is masculinized in one week.

Evidence of the neuroendocrine determinism of this sex reversal was given by Pfannenstiel [34–36].

Removal of the prostomium done in young females induces sex reversal. Young males, decapitated, do not reach the female phase.

When decapitation is followed by an implantation in the coelom of a prostomium taken on a female, the female sex is maintained in females and established in males. From these results, it is concluded that the prostomium is involved in the passage to the female phase and in its maintenance. Neurosecretory cells have been described in the cerebral ganglia of *O. puerilis*. A study of their cycle shows that the neurosecretory products are released at the time of the reversal in females. Therefore, these cells would produce an ootrophic neurohormone.

On the other hand, the influences between female specimens which intervene in sexual differentiation would be explained by the existence of a pheromone.

Indeed, it has been experimentally demonstrated that contacts between females were necessary to sex reversal and that the prostomium would be both the site of production and of reception of a pheromone. The rate of this pheromone has to change during oogenesis. When two females are put together, one of them must exert a greater influence than the other; the latter would then experience an inhibition of the secretion of the ootrophic neurohormone and would present a masculinization. Conversely, when a male would not be submitted to a sufficient pheromone rate, neurosecretion being inhibited, a female phase would establish.

4. Conclusion

From the results obtained in annelids, two essential facts are to be retained:

1. In these yet primitive invertebrates, peptidic neurohormones may be of a stimulatory or inhibitory type; but these two modes of action have to be considered according to the target cell; such is the case, particularly, of polychaete nereids.
   In Hirudinea, it seems that the two neuroendocrine systems have simultaneously differentiated.
2. The second fact is that investigations are now at the molecular level.

References

1. Avel M (1929) Bull Biol Fr Belg 63:149–318
2. Berjon JJ (1965) C R Acad Sci 260:6212–6214
3. Bertout M (1976) C R Acad Sci Ser D 282:1613–1616
4. Bertout M, Caner F, Dhainaut A, Krembel J (1981) Int J Invertebr Reprod 3:121–132
5. Boroffka I (1968) Osmo- und Volumenregulation bei *Hirudo medicinalis*. Z Vergl Physiol 57: 348–375
6. Bulet Ph, Hoflack B, Verbert A, Porchet M (1983) 3rd Int Symp Invertebr Reprod, Tübingen
7. Bulet Ph, Hoflack B, Verbert A, Porchet M (1983) Experientia 39:436–437
8. Caner F, Bertout M, Krembel J, Dhainaut A (1981) Comp Biochem Physiol B 708:493–498
9. Cardon C, Durchon M, Porchet M (1981) Reprod Nutr Dev 21:383–390
10. Cardon C, Durchon M, Porchet M (1981) 12 th Colloq Soc Fr Neuroendocrinol Exp
11. Dhainaut A (1964) C R Acad Sci 259:461–464
12. Durchon M (1948) C R Acad Sci 227:157–158
13. Hagadorn IR (1962) Gen Comp Endocrinol 2:516–540
14. Herlant-Meewis H (1958/1959) Ann Soc R Zool Belg 89:281–336
15. Howie DID (1962) Gen Comp Endocrinol 2:264
16. Howie DID (1963) Gen Comp Endocrinol 3:660–668
17. Howie DID (1966) Gen Comp Endocrinol 6:347–361
18. Howie DID (1977) Gen Comp Endocrinol 31:350–363
19. Kulkari GK, Nagabhushanam R (1978) Hydrobiologica 59:197–201
20. Lattaud C (1974) C R Acad Sci 279:935–938
21. Lattaud C (1975) In: Reinboth R (ed) Intersexuality in the animal kingdom. Springer, Berlin Heidelberg New York, pp 64–71
22. Lattaud C (1980) Int J Invertebr Reprod 2:23–36
23. Lattaud C (1982) Int J Invertebr Reprod 4:223–237
24. Malecha J (1979) C R Acad Sci Ser D 288:693–696
25. Malecha J (1983) Gen Comp Endocrinol 49:344–351
26. Meijer L (1979) Dev Growth Differ 21:303–314
27. Meijer L (1979) Dev Growth Differ 21:315–329
28. Meijer L (1980) Dev Growth Differ 22:33–38
29. Meijer L, Durchon M (1977) C R Acad Sci 285:377–380
30. Marcel R (1967) C R Acad Sci 265:693–694

31. Marcel R, Cardon C (1979) Comp Biochem Physiol 63B:233–237
32. Marcel R, Cardon C (1982) J Physiol 78:574–578
33. Marcel R, Cardon C (in press)
34. Pfannenstiel MD (1971) Naturwissenschaften 58:367
35. Pfannenstiel MD (1974) In: Intersexuality in the animal kingdom. Springer, Berlin Heidelberg New York, pp 48–56
36. Pfannenstiel MD (1975) Pubbl Stn Zool Napoli Suppl 39:542–549
37. Porchet M (1976) Ann Biol 15:330–377
38. Porchet M, Spik G (1978) Comp Biochem Physiol 59b:175–181
39. Porchet M, Cardon C (1976) Gen Comp Endocrinol 30:378–390
40. Porchet M, Dhainaut A, Porchet-Hennere E (1979) Wilhelm Roux's Arch 186:129–137
41. Reynoldson TB, Davies RW (1980) Comp Biochem Physiol A 66:711–714
42. Rosca DI, Wittenberger C, Rusdea D (1958) Cercet Biol Ser Zool 9:113–136
43. Reynoldson TB, Davies RW (1976) Can Zool 54:1908–1911
44. Van Damme N (1976) C R Acad Sci Ser D 283:967–969
45. Van Damme N (1977) Arch Biol 88:31–52
46. Van Damme N (1978) Arch Biol 89:451–482

I.3 Recent Progress in Endocrinology of Molluscs

J. JOOSSE[1]

1. Introduction

Among the invertebrate groups the study of the endocrinology of molluscs has had a much slower start than annelid, crustacean, and insect endocrinology. The number of functions of hormones/neurohormones of molluscs known at the moment is still rather small. In contrast, the basic mechanisms of some types of neurosecretory cells (NSC) are known in greater detail than in any other invertebrate group. This is especially due to the fact that in some gastropods these neurons, like their conventional neurons, are giant cells. For this reason, they are highly suitable for experimental endocrinological, neurophysiological, and molecular biological studies. Great progress has recently been made in this field and the results are of importance for neuroendocrinology in general. This is particularly true for research on the bag cell system of the marine gastropod *Aplysia californica* [1] and the caudo-dorsal cell system of the freshwater snail *Lymnaea stagnalis* [2-4] which are both involved in control of egg-laying. Therefore, in this volume the various aspects of these model systems for neuroendocrinology are presented separately in more detail (see Geraerts et al., Roubos et al., Strumwasser, Ter Maat, and Jansen, this volume).

In this paper attention is focused on a number of other aspects of molluscan endocrinology, where rapid progress is made at the moment. For a more complete picture of the present status of molluscan endocrinology [3, 5].

2. Structural Aspects of Molluscan Neurosecretory Systems

The size of the perikarya of the molluscan neurons shows a great variability. In gastropods their diameter ranges from less than 10-100 μm: for comparison, the largest (motor)neurons of vertebrates are 100 μm in diameter. The size of the large neurons increases with increasing body size. This increase is discontinuous by doubling of the volume, paralleled by increasing polyploidy. In the freshwater pulmonate snail *Lymnaea stagnalis* the largest neurons appeared to have 4,096 times the haploid amount of DNA [6].

1 Biological Laboratory, Vrije Universiteit, P.O. Box 7161, 1007 MC Amsterdam, The Netherlands

Biosynthesis, Metabolism and Mode of Action of Invertebrate Hormones
(ed. by J. Hoffmann and M. Porchet)

The neurosecretory neurons share the above mentioned characteristics with the conventional neurons. In some cases cells of the same type are located in groups. Their axons form bundles on their way to the neurohemal areas. The structure of the release sites of molluscan neurohormones is exceptional. In contrast to crustaceans, insects and vertebrates, molluscs have no special neurohemal organs for the release of the neurohormones in the hemolymph. In some cases they make use of the wall of blood vessels. But most frequently release sites are located at the periphery of nerves, connectives, commissures and even ganglia.

The axons seem to branch frequently. The number of neurohemal endings may be as high as 800 per neuron [7]. As a result the release areas are of enormous size compared to those of non-molluscan groups. It has been suggested that the absence of a concentration of the release sites might be related to the fact that molluscs lack a blood-brain barrier [8]. Since peptides do not easily pass blood-brain barriers, release of peptide hormones in crustaceans, insects and vertebrates is concentrated in special organs outside the barrier. Further research is needed to prove that in molluscs a blood-brain barrier is lacking.

3. Diversity of Identified Neurosecretory Cell Types

The application of the classical Gomori methods, i.e., the chrome-hematoxylin and paraldehyde fuchsin staining techniques, based on the reactions with protein-bound cysteine, appeared to be rather specific for NSC. Gomori-positive and Gomori-negative cell types could be distinguished in this way. Further differentiation was based on the position of the cells. Gabe [9] has summarized the descriptive literature of this starting period of molluscan neuroendocrinology.

The number of identified NSC types increased considerably after the introduction of the alcian blue-alcian yellow technique [10]. This technique is based on the reactivity of the secretory product in the neurons with these stains at different pH. In *L. stagnalis* 11 NSC types could be identified (Table 1). However, it appeared that the application of this technique needs great skill and seems not to be applicable in some molluscan species.

From knowledge gained in other animal groups it became clear that many NSC produce neurohormones of a peptide nature. This is most probably also true for the molluscan neurohormones. Gradually, this characteristic has been used also in the reverse way: all "peptidergic" neurons were considered to be NSC, i.e., cells that release a neurohormone.

Two important recent developments have changed the classical picture of neuroendocrinology in general, and also that of molluscs. First, techniques have been developed to raise antibodies against purified neurohormones and other biologically active peptides. These antibodies are now available, mostly for peptides from vertebrate origin. The antibodies are used to test immunoreactivity in neurons. When a positive reaction is shown in the cytoplasm or the axons of neurons, such cells are considered to be peptidergic. In this way, *immunocytochemistry* has become a new means for the identification of different types of neurosecretory (peptidergic) cells [11–17]. Since the

Table 1. Diversity of peptidergic neurons in *Lymnaea stagnalis*

	Classical staining techniques[a]		Immunocytochemistry[b]
+	Bright green cells		Thyrotropin releasing hormone (TRH)
+	Dark green cells	–	ACTH
+	Light green cells	–	FMRFamide
+	Light yellow cells		αMSH
	Yellow brown cells	–	Arginine vasopressin (AVP)
	Yellow cells	–	Arginine vasotocin (AVT)
	Yellow green cells		Calcitonin
+	Caudo-dorsal cells		Gastrin
+	B-cells	–	Gastrointestinal peptide (GIP)
+	Canopy cells	+	Glucagon
+	Droplet cells	+	Insulin
		+	Met-enkephalin
		+	Oxytocin
		+	Pancreatic polypeptide (PP)
			Secretin
		–	Somatostatin
			Substance P
			Vasoactiveintestinal peptide (VIP)

a Cell types identified with classical histological staining techniques (chrome-hematoxylin phloxin, alcian blue/alcian yellow

b Cell types identified immunocytochemically by using antibodies raised against the peptides indicated

\+ Cell types with identified neurohemal release areas

– Cell type in which the peptide might act as a neurotransmitter

exact nature of the secretory product remains to be identified, these neurons are indicated by designations such as insulin-like cells. The presently available immunocytochemical data indicate that the diversity of the peptidergic system is much greater than presumed before. In *L. stagnalis* the number of identified peptidergic cell types increased from 11 to 29, which at the moment is the maximum for any invertebrate [16] (Table 1).

The second development concerns the observations on *multiple secretory products* released by one type of neuron. Firm biochemical evidence has been presented by Strumwasser (this volume) for release of various peptides by the bag cells of *A. californica* and by Geraerts [18] (see also this volume) for the caudo-dorsal cells of *L. stagnalis*. These peptides are released simultaneously. Also immunocytochemical evidence has been given for the presence of more than one peptide in neurons of *L. stagnalis* (H.H. Boer and L.P.C. Schot, pers. commun.). Finally, there are clear indications for the presence of a biogenic amine and a peptide in one and the same neuron [19]. It is not yet clear whether these latter products are also released simultaneously.

4. Peptides as Neurotransmitters

Neurophysiological studies on NSC of various gastropods have demonstrated that these cells share the great variability in electrical characteristics known from conventional neurons [1-3]. There is good evidence that action potentials induce release of peptides from NSC [20, 21]. From studies on mammals it has been suggested that peptides might also have a role in neurotransmission [22]. This role has been attributed to the molluscan tetrapeptide FMRFamide in *L. stagnalis* [23]. Most recently circumstantial evidence for the role of FMRFamide as a "motoneuron transmitter substance" was given by Cottrell et al. [24] for neurons innervating tentacular muscles in *Helix*.

These results raise the question whether peptidergic neurons can be distinguished in cells in which the peptide functions as a transmitter substance and others in which the peptide has the role of a neurohormone. Morphological studies on the patterns of axonal branchings of the peptidergic cells partly answer this question. The horseradish peroxidase injection technique is successfully applied on molluscan neurons for that purpose [25]. Electron microscopical studies are also useful as they can demonstrate the presence of characteristic elementary granules of identified peptidergic cells in axonal endings as observed in neurohemal areas or in synapse-like contacts with other neurons or target cells. Based on the latter criteria Schot et al. [16] suggested that a number of the "vertebrate-like" peptidergic neurons of *L. stagnalis* will use peptides as a neurotransmitter (Table 1).

Whether individual neurons are able to release a peptide as a transmitter substance as well as a neurohormone is still a matter of debate. In this respect it is of interest to note that some authors use the term transmitter action for the effect of hemolymph transported peptides on centrally located neurons involved in control of behavior [26].

5. Molecular Structure of Peptides

5.1 FMRFamide Family

FMRFamide is one of the numerous cardioactive peptides known to occur in molluscs that has been purified, identified, and synthesized [27]. It is a tetrapeptide isolated from the central nervous system of the bivalve *Macrocallista nimbosa*. Structure-activity relationship studies strongly suggest that the C-terminal with the free amide group is the operational portion of the FMRFamide molecule [28]. Immunocytochemical studies in the snails *L. stagnalis* [12] and *Helix aspersa* [29] provide clear evidence for a neuronal localization of a FMRFamide-like peptide in these species. Moreover, FMRFamide immunoreactivity has been demonstrated in a large variety of nonmolluscan animals, e.g., the coelenterate *Hydra*, the CNS of crustaceans and insects, and the CNS, pituitary, gastrointestinal tract, and pancreas of various vertebrates [3]. This indicates that there is a family of FMRFamide and related peptides with a wide distribution in the animal kingdom [30].

The molecular structure within the FMRFamide family might show considerable variation. The molecule extracted from the CNS of the snail *Helix aspersa* is already different from that of the bivalve *M. nimbosa* [31] (Table 2).

There is a wide variety of actions known from FMRFamide and related peptides. These include not only heart muscles, but also other muscles and nerve cells (Table 2). Therefore, it is suggested that these peptides will play a role in neurotransmission. However, the physiological function is not known.

Table 2. Molecular structures of FMRFamide and an FMRFamide-like peptide, and the actions of these peptides

Species	Peptide	Action
Macrocallista nimbosa [27]	*FMRFamide* Phe–Met–Arg–Phe–NH_2	Cardio-excitation Cardio-inhibition Muscle contraction Muscle relaxation Nerve cell hyper-polarization Nerve cell depolar-ization
Helix aspersa [31]	*pQNFIRFa* pGlu–Asx–Phe–Ile–Arg–Phe–NH_2	Cardio-excitation Muscle contraction

Apart from the "small" FMRFamide-like cardioactive peptide, the presence of "large" cardioactive peptides has been demonstrated in *H. aspersa* [32–34] and *L. stagnalis* [35]. Apart from their molecular weight (about 7,000) and peptide nature, nothing is known about their molecular structure.

5.2 Opioid Peptides

There is a rapidly increasing interest in research on the presence of opioid and opioid-like peptides in molluscs. This work is greatly facilitated by the fact that for these peptides all important techniques are available for their identification: various antibodies, bioassays, receptor binding studies, HPLC techniques and amino acid sequencing methods.

The molluscan opioids of which the molecular structure is now known, are listed in Table 3. In the CNS of *Mytilus* both met- and leu-enkephalin occur [36], which are well-known opioids of vertebrates. Moreover, the brain of *Mytilus* appears to contain a heptapeptide, similar to met-enkephalin, with Arg–Phe added to the C-terminal [37].

Table 3. The occurrence of opioid peptides in molluscs

Mytilus	– CNS [36]	Met-enkephalin Tyr–Gly–Gly–Phe–Met Leu-enkephalin Tyr–Gly–Gly–Phe–Leu
Mytilus	– CNS [37]	Heptapeptide Tyr–Gly–Gly–Phe–Met–Arg–Phe
Octopus	– vena cava neuropil [38]	Heptapeptideamide Tyr–Gly–Gly–Phe–Met–Arg–Phe–NH_2 FMRFamide (Phe–Met–Arg–Phe–NH_2) Met-enkephalin (Tyr–Gly–Gly–Phe–Met)

From *Octopus* vena cava neuropil a heptapeptideamide has been isolated. This molecule is very interesting, as it shares part of the molecule with met-enkephalin as well as with FMRFamide (Table 3). Therefore, this molecule contains the ancestral structure of both these peptides. Greenberg and Price [30], although not being aware of the discovery of this molecule in molluscs at the time, suggest that both met-enkephalin and FMRFamide diverged from heptapeptideamide, which acted at a single receptor having a pair of binding sites, each complementary to one end of the peptide. Kiehling et al. [38], who isolated this peptide, demonstrated that indeed heptapeptideamide has both opioid and cardioexcitatory potency. They have also found three other peptides in which probably the C-terminal portion –Arg–Phe-amide is contained. However, these peptides are obviously structurally different from the three mammalian opioid systems, since the octopus material was inactive in RIAs specific for met-enkephalin, β-endorphin, leu-enkephalin, and dynorphin 1-8.

There is no doubt that these studies on opioid peptides in molluscs will have a rapid follow-up in the next years. The most fascinating aspect of this work is that it demonstrates that the molecular structure of certain opioid peptides has been conserved for millions of years during animal history.

5.3 The Egg-Laying Hormones of Gastropods

Neurohormones inducing egg-laying and egg-laying behavior have been found in all subclasses of the gastropods: in prosobranchs by Ram [39] in 1977, in opisthobranchs by Kupfermann [40] in 1967 and in pulmonates by Geraerts and Bohlken [41] in 1976. As mentioned already, great progress has been made in our knowledge of the two latter systems, the bag cell system of various species of *Aplysia*, particularly *A. californica* [1] and the caudo-dorsal cell system of *L. stagnalis* [3, 4].

The molecular structure of the egg-laying hormone (ELH) of *A. californica* was determined by Chiu et al. [42] in 1979 (Table 4). ELH consists of 36 amino acid residues. The pI is about 9.3, so that the molecule is highly basic [43]. ELH shows hardly any resemblance to other vertebrate or invertebrate peptides, the chemical structure of which is known. It appeared that at 20 °C, 2.5 nm of pure ELH suffice to induce egg-laying in *A. californica*.

Evidence has been presented by Arch et al. [44] that another organ of *Aplysia*, the atrial gland, may also play a role in the regulation of egg-laying. The atrial gland is a densely infolded gland in the most rostral part of the large hermaphrodite duct. After injection of atrial gland extract in *Aplysia* egg-laying is induced and the response latency for egg-laying is identical to that after ELH injection. It was, therefore, of great interest to know the molecular structure of the active principle in this extract.

Three peptides have now been isolated from atrial gland extract and sequenced. Heller et al. [45] found two bioactive peptides (A and B), which consist of 34 amino acids and which are identical except for four amino acid positions (4, 7, 8, and 9) (Table 4). Thus, although the number of amino acids in these atrial gland peptides nearly equals that of ELH, their composition is quite different. The atrial gland peptides A and B cause egg-laying in an indirect way, because in vitro they induce an afterdischarge, and, thus, release of ELH, in bag cells. This suggests that in vivo these peptides

Table 4. Amino acid sequence of the bag cell [42] and atrial gland [45, 46] peptides of *Aplysia californica*

	1	2	3	4	5	6	7	8	9	10	11	12	13	14
Bag cell ELH	NH_2-Ile	-Ser	-Ile	-Asn	-Gln	-Asp	-Leu	-Lys	-Ala	-Ile	-Thr	-Asp	-Met	-Leu
Atrial gland ERH	NH_2-Ile	-Ser/Val	-Ile	-Val	-Ser	-Leu	-Pne	-Lys	-Ala	-Ile	-Thr	-Asp	-Met	-Leu
Peptide A	NH_2-Ala	-Val	-Lys	-Leu	-Ser	-Ser	-Asp	-Gly	-Asn	-Tyr	-Pro	-Phe	-Asp	-Leu
Peptide B	NH_2-Ala	-Val	-Lys	-Ser	-Ser	-Ser	-Tyr	-Glu	-Lys	-Tyr	-Pro	-Phe	-Asp	-Leu

	15	16	17	18	19	20	21	22	23	24	25	26	27	28
Bag cell ELH	-Leu	-Thr	-Glu	-Gln	-Ile	-Arg	-Glu	-Arg	-Glu	-Arg	-Tyr	-Leu	-Ala	-Asp-
Atrial gland ERH	-Leu	-Thr	-Glu	-Gln	-Ile	-Tyr	-Ala	-Asn	-Tyr	-Phe	-Ser	-Thr	-Pro	-Arg-
Peptide A	-Ser	-Lys	-Glu	-Asp	-Gly	-Ala	-Gln	-Pro	-Tyr	-Phe	-Met	-Thr	-Pro	-Arg-
Peptide B	-Ser	-Lys	-Glu	-Asp	-Gly	-Ala	-Gln	-Pro	-Tyr	-Phe	-Met	-Thr	-Pro	-Arg-

	29	30	31	32	33	34	35	36
Bag cell ELH	-Leu	-Arg	-Gln	-Arg	-Leu	-Leu	-Gly	-Lys-COOH
Atrial gland ERH	-Leu	-Arg	-Phe	-Tyr	-Pro	-Ile-COOH		
Peptide A	-Leu	-Arg	-Phe	-Tyr	-Pro	-Ile-COOH		
Peptide B	-Leu	-Arg	-Phe	-Tyr	-Pro	-Ile-COOH		

will activate the ELH system. Most probably this activation is indirect via centers in the central nervous system outside the parietovisceral ganglion [45].

The third bioactive peptide isolated from atrial gland extracts by Schlesinger et al. [46] in 1981 is called egg releasing hormone (ERH). This peptide has a very interesting structure. It consists of 34 amino acids, like peptides A and B, and shares the greater part of amino acids 1-19 with ELH and 23-34 with peptides A and B. ERH can induce egg-laying in the absence of ELH.

At the moment it is not clear in what way the atrial gland peptides are involved in egg-laying. There is as yet no proof that these peptides are released in the blood. Therefore, their hormonal nature needs further confirmation.

The gene coding for ELH and the other peptides released by the bag cells of *Aplysia* is now known. For further details, see Strumwasser (this volume).

In the freshwater pulmonate snail *L. stagnalis* the caudo-dorsal cells produce a neurohormone (CDCH) that induces egg-laying [41]. This hormone is a peptide with a molecular weight of about 4,700 and a p_I of about 9.3 [47]. These properties are strikingly similar to those of the egg-laying hormone of *A. californica*. However, most probably the molecules will be structurally different, since the homologous hormones within the freshwater pulmonates appeared to be specific at the genus level [48]. For a recent review of the CDC system and CDCH, see [4].

5.4 Insulin-Like Factors/Hormones in Molluscs

Evidence for the occurrence of an insulin-like substance (ILS) in molluscs dates back as far as 1923 [49]. Collip extracted an ILS from the tissues of the marine bivalve *Mya arenaria* and injected it into a rabbit. This caused the typical convulsions of a hypoglycemic condition, which could be relieved by an injection of dextrose.

The presence of cells producing ILS in molluscs is now convincingly demonstrated. These cells are identified with the classical staining method with paraldehyde fuchsin as well as with immunocytochemical methods by using antibodies against mammalian insulin. ILS cells have been identified in the gut epithelium in the region of the hepatopancreas of the marine bivalve *Mytilus edulis* [50, 51] and the freshwater bivalves *Unio pictorum* and *Anodonta cygnea* [52], in the intestine of the prosobranch *Buccinum undatum* [53, 54], in the digestive gland and pylorus of the terrestrial pulmonate *Achatina fulica* [55], and in the CNS of *L. stagnalis*. In the latter species neurons reacting with antiglucagon have also been observed [16].

With regard to the physiological effects of ILS in molluscs, experiments have been performed by Plisetskaya and associates in the freshwater bivalves *U. pictorum* and *A. cygnea*. A glucose load caused a loss of secretion product from the ILS cells in the small intestine. At the same time, the insulin titer in the hemolymph, determined with a radioimmunoassay test for mammalian insulin, increased. Injections of mammalian insulin in the adductor muscles were followed by a decrease in the blood glucose level and an increase in the activity of the enzyme glycogen synthetase in these muscles [3, 56, 57]. Similar reactions were obtained in the marine bivalves *M. galloprovincialis* and *Chlamys glaber ponticus* [58].

Various attempts have been undertaken to characterize the ILS of molluscs. ILS extracted from the intestine of *Unio* and *Anodonta* appeared to be more potent in these species than mammalian insulin. In the marine bivalve *M. galloprovincialis* the extract appeared to stimulate the activity of glycogen synthetase, whereas mammalian insulin had little effect [58]. Moreover, application of an antiserum against molluscan insulin revealed a 10 × higher insulin titer in the radioimmunoassay test compared with that for mammalian insulin [59]. These results clearly show that the insulin molecule of bivalves differs from that of mammals.

In gastropods, Marques and Falkmer [60] have studied the effects of insulin. They saw a more rapid disappearance of a glucose load from the hemolymph of the terrestrial pulmonate *Strophocheilus oblongus* when the animals had been injected with a high dose of mammalian insulin. Normally, the glucose concentration in the blood of this snail is too low for the effect of insulin to be detected. In the slugs *Arion ater* and *Ariolimax columbianus* the hemolymph glucose concentration is high, up to 250 μg/ml. An injection of mammalian insulin in these species was followed by a significant decrease of the glucose levels in the blood [57].

In the freshwater pulmonate snail *L. stagnalis* the hemolymph glucose concentration shows considerable variation in laboratory bred as well as field specimens [61, 62]. Injection of mammalian insulin did not change the hemolymph glucose concentration even after a glucose load. On the other hand, an ILS fraction could be extracted from the intestine of this snail, which is identified by its positive reaction in the radioimmunoassay for mammalian insulin. This ILS, however, has no hypoglycemic effect in the same species. In contrast, the *Lymnaea* ILS stimulates glucose uptake in isolated rat fat cells, which indicates that it can bind successfully to insulin receptors of mammals (M.A. Hemminga, pers. commun.). Fractionation of the *Lymnaea* ILS with HPLC procedures showed that this molecule behaves very similar to mammalian insulin, but the molecules are not identical. The *Lymnaea* insulin molecule is slightly more hydrophobic (R.H.M. Ebberink and M.A. Hemminga, pers. commun.).

The above mentioned data indicate that further research on ILS of molluscs is very promising for the demonstration of the (small) molecular differences of these substances compared to the various vertebrate insulins.

5.5 Vasopressin/Vasotocin-Like Substances in Molluscs

With immunocytochemical techniques the presence of vasopressin- and vasotocin-like material in neurons of *L. stagnalis* has been demonstrated [16]. The possible role of these peptides in hydromineral regulation is suggested by the observation of immunoreactive material in the neurohemal areas of cardioactive peptides in *O. vulgaris* [63] and *L. stagnalis* (L.P.C. Schot and J. Wijdenes, unpublished results). However, synthetic vasopressin and vasotocin appeared to have no effects on the in vitro heart preparations of *L. stagnalis* (De With and Geraerts, unpublished results) and of *Octopus* [64].

Recently, Sawyer et al. [65] demonstrated the presence of an arginine-vasotocin-like substance in the pleuropedal ganglia of the terrestrial pulmonate slugs *A. columbianus* and *L. maximus*. This substance resembles, but clearly differs from, vertebrate arginine-vasotocin. Both isolated peptides enhance the water permeability of the body wall of slugs in vitro.

6. Morphogenetic Brain Factors Involved in Sex Reversal

The marine snail *Crepidula fornicata* is a representative of the Prosobranchia. These animals have the peculiar habit of living in colonies or chains of up to 15 individuals. The snails at the top of the chain are younger small males, whereas those at the bottom are large females. The intermediate animals are in the hermaphroditic phase of sex reversal (for review see [3]).

During the sex reversal phase the male accessory sex organs, consisting of a sperm duct, seminal vesicles, an external sperm groove, and an elaborate penis, are replaced by an oviduct, a gonopericardial canal, a receptaculum seminis, a uterus and a vagina. Simultaneously the testis changes into an ovarium.

In the laboratories of Streiff and Lubet at Caen in a long-term research program the mechanisms involved in control of sex reversal of *C. fornicata* have been clarified. P. Le Gall [66] has shown that placing a free-living small male specimen in a chain, always at the top, induces rapid body growth and a gradual transition to the female sex in the specimen in the former top position. He observed that contacts between the members of the chain occur mainly by the tentacles and the pallial border, and he demonstrated that a masculinizing factor is released by the pallial border, which is registered by the tentacles of the higher specimens. A feminizing factor is released by the tentacles of the higher specimens and registered by the pallial border of the lower specimens. In the chain, these effects are cumulative. Finally, a growth-accelerating factor affects the growth of the lower specimens, but this effect is not cumulative. It is a trigger mechanism. The chemical character of the external messengers is not yet known. They are not carried by the water. Isolated specimens are also able to show sex reversal, but this change is then delayed. Thus, the social contact in a chain is not a condition for the start of the change of sex.

The masculinizing and feminizing factors induce the release of endocrine factors by the central nervous system involved in sex reversal. S. Le Gall [67-69] has shown that different factors are involved in the differentiation and lysis of the penis. A neurohormone produced by the pedal ganglia stimulates the differentiation of the penis in a morphogenetic territory located near the right tentacle. This hormone is released into the hemolymph and seems to accumulate in specific hemal lacunae near the right tentacle and the morphogenetic penial territory. This peripheral accumulation seems well established, but is problematic since this phenomenon has not been observed in other endocrine studies. The release of the pedal ganglion neurohormone can be stimulated by a factor released by the cerebropleural ganglia. With this factor inactive pedal ganglia of snails which are, after sex reversal, in the female phase can be reactivated in organ culture conditions.

The differentiation or lysis of the penis during the transition to the female phase is controlled by another neurohormone produced by neuroendocrine centers located in the mediodorsal area of the pleural ganglia. These cells release their products from the cerebropleural connectives. The lysis inducing action of this dedifferentiation factor is clearly demonstrated in organ cultures. After extirpation of the mediodorsal area in vivo, the regression of the penis in animals during sex reversal stops. This indicates that the necrotic action is under continuous control of this factor.

This work deserves great attention. The demonstration of specific morphogenetic factors of neuroendocrine origin is unique. Since in vitro bioassays for these factors are available, it is hoped that biochemical studies on these hormones will soon start. The *Crepidula* sex reversal can be considered as a model system for morphogenetic studies.

7. Photoperiodic Effects in Pulmonate Gastropods

In both a freshwater and a terrestrial pulmonate, detailed studies have been performed on the effects of the photoperiod on reproductive activity (for reviews see [70, 71]).

In the pond snail *L. stagnalis* a Long-Day (LD) photoperiod induces an earlier start of egg-laying in animals bred in mass culture conditions. Moreover, in adult snails the number of egg masses produced is considerably higher in LD snails compared to Medium-Day (MD) and Short-Day (SD) snails. Thus, LD conditions favor female maturation and ovipository activity, but the LD photoperiod is not conditional for female reproductive activity [72]. This reaction to the photoperiod fits in the life cycle of this snail. Young snails hatching in June and July are rapidly brought to maturity by the LD photoperiod and will show egg-laying before autumn, thus giving rise to a second generation. In spring the LD conditions will activate female reproductive activity again, after the stop during autumn and winter. Thereafter, the snails will die, since *Lymnaea* is an annual species.

In contrast, in the terrestrial slug, *Limax maximus*, the reproductive cycle is seasonally synchronized by a photoperiodic mechanism that responds to lengthening days in the spring. Experiments with artificial photoperiods have shown that slugs raised on SD conditions will remain immature indefinitely, while a transition to LD conditions results after about 4 weeks in the initiation of full, male and female, reproductive maturation [73, 74]. Therefore, in *Limax* the LD photoperiod is conditional for sex maturation.

Both *Lymnaea* and *Limax* are now used to study the interaction of the effects of the photoperiod with other environmental factors and to clarify the pathways of the light/dark information to the endocrine centers involved in control of reproduction.

In *L. stagnalis* it appeared that the photoperiodic effects show an interaction with those of feeding and temperature. At SD conditions starving snails and snails kept at temperatures below 10 °C rapidly stop ovipository activity. However, at an LD photoperiod the snails continue to lay egg masses during these adverse conditions. The number of egg masses and their size, however, are much smaller compared to those of fed snails kept at 20 °C.

Female reproductive activity of *L. stagnalis* is controlled by two gonadotropic hormones. The Dorsal Body Hormone (DBH) which is produced by the endocrine Dorsal Bodies (DB) stimulates vitellogenesis in the oocytes and cellular differentiation, growth and synthetic activity of the female accessory sex organs. The Caudo-Dorsal Cell Hormone (CDCH) is produced by the neuroendocrine Caudo-Dorsal Cells (CDC). This hormone induces ovulation, egg mass formation and egg-laying behavior [3, 75]. In starving snails at MD conditions the CDC are hyperpolarized [76]. Apparently the stop in oviposition of these snails is caused by a block in the transition from the resting to the active state of the CDC [25]. Preliminary studies of the CDC of starving LD snails revealed that their membrane potential had not changed [77]. Therefore, it is plausible that LD conditions prevent the depolarization of the CDC in snails which are starving or kept at low temperature. However, the small number of eggs produced under the latter conditions indicates that the LD photoperiod can not prevent the decrease in the release of DBH.

As regards the pathway of the photoperiodic stimuli to these (neuro-)endocrine centers, little is known at the moment. The paired eyes are not necessary. The Lateral Lobes (LL) are known to contain center(s) that stimulate the activity of the DB and CDC [78]. Therefore, the LL were expected to be part of the pathway. However, in the absence of the LL the snails are able to adapt female reproductive activity after transfer from an MD to an LD photoperiod [79].

In *Limax maximus* the photoperiodic stimuli induce the release of two hormones from the cerebral ganglia. The hormones may, in fact, be identical, but their activity is assayed differently. When cerebral ganglia of slugs exposed for 4 weeks to LD conditions are transferred to SD snails, the latter animals start sexual maturation. Therefore, the LD cerebral ganglia are expected to release a Maturation Hormone (MH). Similarly, these cerebral ganglia are able to stimulate the incorporation of tritiated thymidine in male gonadal cells. This Male Gonadotropic Factor (MGF) may be identical to MH, but this is not clear at the moment [74].

The source of the MGF has been most probably identified. This is a paired group of neurons in the cerebral ganglia near the cerebral commissure. This area is called area Z. It contains 50 neurons on either side of the commissure. Axonal branches of these cells have been traced to the pedal ganglia and neurohemal structures were observed in a pedal nerve [74].

Like in *L. stagnalis*, in *L. maximus* the paired eyes are not essential for photoperiodic light perception. However, the isolated nervous system exhibits a response to direct low level illumination, and the action spectrum of this neural response falls off rapidly above 600 nm. It is, therefore, possible that extraocular photoperiod receptors reside within the CNS.

7.1 Steroid Hormones in Gastropods

Numerous authors have demonstrated that the gonad of molluscs is the source of hormone(s). As in the vertebrates, the function of these hormones may be control of gamete production, synchronization of gonadal and accessory sex gland activities, feedback control of the gonadotropic center(s), and control of reproductive behavior. These suggestions included that the gonadal hormones would have a steroid nature. Many papers report about the capacity of gonadal tissue or cells for steroid synthesis or the presence of steroids (for literature, see [3]). This concerns gastropods, as well as bivalves and cephalopods.

Identification of specific hormone-producing cell types in the gonad of molluscs appears to be very difficult. Nonsexual elements, such as the Sertoli cells and the follicle cells, might be involved in hormone production. A distinct correlation between biochemical and histological data could be shown by De Jong-Brink et al. [80]. They found simultaneously diurnal activity changes of steroid-synthesizing enzymes and of the extension of the smooth endoplasmic reticulum in Sertoli cells carrying late male phase cells and in Sertoli cells after spermiation in the ovotestis of *L. stagnalis*. However, castration of the related planorbid snail *Bulinus truncatus* did not prevent the continuation of the production of egg masses (without egg cells) [81]. Therefore, it has been suggested that possible gonadal steroids of freshwater pulmonates would have an intragonadal function [80].

More recently a series of detailed experiments has been published by Takeda [82–84] in which the role of steroids in control of reproductive activity of the terrestrial pulmonate snail *Euhadra peliomphala* is studied. In this snail castration leads to atrophy of the accessory sex organs and of the head-wart, which is an organ located between the optic tentacles known to produce a substance that stimulates copulatory behavior. In radioimmunoassays the presence of steroids, such as testosterone, estradiol, and progesterone in the hermaphroditic gonad could be demonstrated [83]. Injection of testosterone into castrated snails restored the atrophy of the male accessory sex organs, whereas injection of estradiol had a positive effect on the female accessory sex organs. In organ culture conditions, the development of the head-wart and the male accessory sex organs were stimulated by progesterone, whereas female parts developed under the influence of estradiol. These effects were inhibited by anti-steroid substances. Finally, the brain as well as the optic gland contain centers that stimulate the steroid release by the gonad and, therefore, are considered to be gonadotropic centers [84].

These results strongly suggest that in this snail steroids are involved in control of reproductive activity in a way surprisingly similar to that in vertebrates.

7.2 Steroids and Parasitic Infection in Gastropods

Many snail species are used as an intermediate host by trematodes. These digenean parasites show asexual multiplication in a mollusc, and sexual maturation and reproductive activity in a vertebrate species, which is the definitive host. These parasites are now studied in relation to the presence and metabolism of steroids in two ways.

First, it has been demonstrated that the human parasite *Schistosoma mansoni* produces ecdysteroids which can be found in the blood serum and urine of infected humans and experimental animals. The presence of ecdysteroids is therefore indicative for the infection. There are indications that treatment of experimentally infected rodents with antiecdysone antibodies leads to a significant reduction of the worm population in these hosts [85].

Second, a new and highly sensitive technique for steroid detection has been employed in normal and parasitically infected *L. stagnalis*. This freshwater snail is the intermediate host of a number of trematode parasites. We are interested in the effects of the parasites on the snail host. For experimental studies in the laboratory we use the parasite *Trichobilharzia ocellata*, which has *L. stagnalis* as an intermediate and the duck *Anas platyrhynchos* as the definitive host. Infection of juvenile pond snails with miracidia of *T. ocellata* causes important physiological changes in the snail: body growth rate increases, whereas growth of the gonad and of the male and female accessory sex organs is highly retarded [86–88]. Measurements on the change in the proportional wet weights of the accessory sex organs of juvenile snails during the development of the parasite indicate that it is the maturation of these organs which is inhibited in the infected snails (Fig. 1).

T. ocellata does not cause any mechanical damage to the tissues of the host. The parasites develop in the blood lacunae between the lobes of the digestive gland. Thus, obviously the host is affected via the hemolymph. Concerning the mechanism of this parasitic effect it is hypothesized that the parasite affects the snail by intervening in the functioning of the hormones controlling reproductive activity, whereas the effect on body growth is secondary to this. In previous studies the effects on the DBH and CDCH functioning have been investigated. These led to the conclusion that DBH, although most probably released in higher quantities in infected snails, is less effective in its actions on the targets [88-90].

For the determination of the effects of the parasite on possible steroids in the hemolymph, an advanced technique for the identification of steroids has been employed. Identification occurred after preparing the O-pentafluorobenzyl-oxime-heptafluorobutyrylester-derivatives of the hemolymph extracts, followed by negative ion chemical ionization capillary gas chromatography-mass spectrometry. The results are presented in Table 5. In the hemolymph of adult control snails 14 steroids have been identified. A number of androgens and progestagens are found, but estrogens are lacking. It is obvious that the relative concentrations of the majority of these steroids has changed in parasitically infected pond snails. Particularly the androgens occur in higher concentrations (J. Joosse, H. Diederick, and R. van Elk, unpubl. results). Apparently the parasite affects steroid metabolism. These preliminary results will stimulate us to try to elucidate the functions of these steroids in freshwater snails.

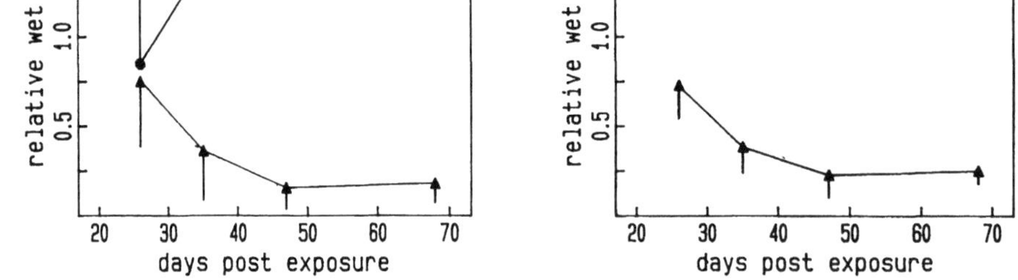

Fig. 1. The effect of infection of *Lymnaea stagnalis* with the trematode parasite *Trichobilharzia ocellata* on the maturation of the accessory reproductive system of the host.
Maturation (increase in relative wet weight) of the female albumen gland of the control snails (*dots*) occurred between days 25 and 35 of the experiment, whereas male maturation had taken place before day 25. In the infected snails maturation of these organs (*triangles*) does not occur. In contrast the glands even show a regression

Table 5. Steroids identified in the hemolymph of adult control and parasitically infected (70 days after infection) specimens of *L. stagnalis*[a]

Steroids	Control (C)	Infected (I)	I/C
Pregnenolone	++	+	0.42
17 α-Hydroxypregnenolone	+	+	0.97
Dehydroepiandrosterone	+	++++	4.5
Androstenedione	+	++++	5.4
Testosterone	+	++++	5.8
5 α-Dihydrotestosterone	+	++++	4.2
11 β-Hydroxytestosterone	+	++	1.6
11 Ketotestosterone	+	++++	5.0
Progesterone	+	++	2.0
17 α-Hydroxyprogesterone	+	++	1.8
17 α-Hydroxy-20β-dihydroprogesterone	+	+++	3.6
Androsterone	+	+	
Androstanediol	+	+	
Androstenediol	+	+	

[a] The relative concentrations are indicated (+)

References

1. Strumwasser F, Kaczmarek LK, Chiu AY, Heller E, Jennings KR, Viele DP (1980) In: Bloom FE (ed) Peptides: Integrators of cell and tissue function. Raven Press, New York, pp 197–218
2. Joosse J, de Vlieger TA, Roubos EW (1982) In: Buijs et al. (eds) Chemical transmission in the brain. Progress in brain research, vol 55. Elsevier, Amsterdam New York, pp 379–404
3. Joosse J, Geraerts WPM (1983) In: Wilbur KM (ed) The mollusca, vol IV. Physiology, part 1. Academic Press, London New York, pp 317–406
4. Roubos EW (in press) Int Rev Cytol 89:295–346
5. Joosse J (1979) In: Barrington EJW (ed) Hormones and evolution. Academic Press, London New York, pp 119–157
6. Boer HH, Jong-Brink M de, Cornelisse CJ (1977) Neth J Zool 27:245–252
7. Wendelaar Bonga SE (1971) Z Zellforsch 113:490–517
8. Maddrell SHP, Nordmann JJ (1979) Neurosecretion. Blackie and Son, Glasgow London
9. Gabe M (1966) Neurosecretion. Pergamon Press, Oxford, GB
10. Wendelaar Bonga SE (1970) Z Zellforsch 108:190–224
11. Boer HH, Schot LPC, Roubos EW, ter Maat A, Lodder JC, Reichelt D (1979) Cell Tissue Res 202:231–240
12. Boer HH, Schot LPC, Veenstra JA, Reichelt D (1980) Cell Tissue Res 213:21–27
13. Grimm-Jørgensen Y (1978) Gen Comp Endocrinol 35:387–389
14. Grimm-Jørgensen Y (1979) J Exp Zool 208:169–175
15. Grimm-Jørgensen Y, Jackson IMD (1975) Nature (London) 254:620
16. Schot LPC, Boer HH, Swaab DF, van Noorden S (1981) Cell Tissue Res 216:273–291
17. van Noorden S, Fritsch HAR, Grillo TAI, Polak JM, Pearse AGE (1980) Gen Comp Endocrinol 40:375–376
18. Geraerts WPM, Tensen CP, Hogenes ThM (1983) Neurosci Lett 41:151–155
19. Osborne NN (1982) Biology of sertonergic transmission. Wiley, London New York
20. Kits KS (1981) Adv Physiol Sci 23:35–54
21. Geraerts WPM, Asselt A van, Maat A ter, Hogenes ThM (this vol)
22. Buijs RM, Swaab DF (1980) Cell Tissue Res 204:335–367

23. Schot LPC, Boer HH, Wijdenes J (1983) In: Lever J, Boer HH (eds) Molluscan neuro-endocrinology. North-Holland Publ, Amsterdam Oxford New York, pp 203–208
24. Cottrell GA, Schot LPC, Dockray GJ (1983) Nature London 304:638–640
25. de Vlieger TA, Kits KS, Maat A ter, Lodder JC (1980) J Exp Biol 84:259–271
26. Rothman BS, Mayeri E, Brown RO, Yuan PM, Shively JE (1983) Proc Natl Acad Sci USA 80
27. Price DA, Greenberg MJ (1978) Prep Biochem 7:261–281
28. Price DA, Greenberg MJ (1980) Gen Pharmacol 11:237–241
29. Marchand C-R, Wijdenes J, Schot LPC (1982) CR Acad Sci 294:34–44
30. Greenberg MJ, Price DA (1983) Annu Rev Physiol 45:271–288
31. Price DA (1982) Comp Biochem Physiol 72C:325–328
32. Lloyd PE (1978a) J Comp Physiol 128:269–276
33. Lloyd PE (1978b) J Comp Physiol 128:277–283
34. Lloyd PE (1980) J Comp Physiol 138:265–270
35. Geraerts WPM, de With ND, Tan BTG, van Hartingsveldt W, Hogenes ThM (in press) Comp Biochem Biophys
36. Leung M, Stefano GB (1984) Gen Comp Endocrinol (in press)
37. Stefano GB, Leung N (1984) Gen Comp Endocrinol (in press)
38. Kiehling C, Martin R, Geis R, Bickel U, Voigt KH (1984) Gen Comp Endocrinol (in press)
39. Ram JL (1977) Biol Bull 152:221–232
40. Kupfermann I (1967) Nature London 216:814–815
41. Geraerts WPM, Bohlken S (1976) Gen Comp Endocrinol 28:350–357
42. Chiu AY, Hunkapiller MW, Heller E, Stuart DK, Hood LE, Strumwasser F (1979) Proc Natl Acad Sci USA 76:6656–6660
43. Arch S (1976) Am Zool 16:167–175
44. Arch S, Smock T, Gurvis R, McCarthy C (1978) J Comp Physiol 128:67–70
45. Heller E, Kaczmarek K, Hunkapiller MW, Hood LE, Strumwasser F (1980) Proc Natl Acad Sci USA 77:2328–2332
46. Schlesinger DH, Babirak SB, Blankenship JE (1981) In: Schlesinger DH (ed) Neurohypophysial peptide hormones and other biologically active peptides. Elsevier/North-Holland, New York Amsterdam Oxford, pp 137–150
47. Geraerts WPM, Cheeseman P, Ebberink RHM, Nuyt K, Hogenes ThM (1983) Gen Comp Endocrinol 51:471–476
48. Dogterom GE, van Loenhout H (1983) Gen Comp Endocrinol 52:121–125
49. Collip JP (1923) J Biol Chem 55:39
50. Fritsch HAR, Sprang R (1977) Cell Tissue Res 177:407–413
51. Fritsch HAR, van Noorden S, Pearse AGE (1976) Cell Tissue Res 165:365–369
52. Plisetskaya E, Kazakov VK, Soltitskaya L, Leibson LG (1978b) Gen Comp Endocrinol 35: 133–145
53. Boquist L, Falkmer S, Mehrotra BK (1971) Gen Comp Endocrinol 17:236–239
54. Davidson IK, Falkmer S, Mehrotra BK, Wilson S (1971) Gen Comp Endocrinol 17:388–401
55. Gomih YK, Grillo TAI (1976) In: Grillo TAI, Leibson L, Epple S (eds) The evolution of pancreatic islets. Pergamon Press, Oxford, GB, pp 153–162
56. Plisetskaya E, Soltitskaya L, Rusacov YI (1978a) In: Gaillard PJ, Boer HH (eds) Comparative endocrinology. Elsevier/North Holland Biomed Press, Amsterdam, pp 449–453
57. Plisetskaya E, Joosse J (in press) In: Lofts B (ed) Advances in comparative endocrinology. Honk Kong Univ Press, Hong Kong
58. Plisetskaya E, Soltitskaya LP (1979) Zh Evol Biokhim Fiziol 15:288–294
59. Rusacov YI, Kazakov VK (1979) Zh Evol Biokhim Fiziol 15:617–619
60. Marques M, Falkmer S (1976) Gen Comp Endocrinol 29:522–530
61. Scheerboom JEM, van Elk R (1978) Proc K Ned Akad Wet C81:365–376
62. Scheerboom JEM, Hemminga MA, Doderer A (1978) Proc K Ned Akad Wet C81:335–346
63. Martin R, Froesch D, Voigt KH (1980) Gen Comp Endocrinol 42:235–243
64. Kiehling C, Froesch D, Martin R, Voigt KH (1981) Abstr. 11 Conf Eur Soc Comp Endocrinol, Jerusalem
65. Sawyer WH, Pang PKT, Deyrup-Olsen I, Martin AW (in press) In: Lofts B (ed) Advances in comparative endocrinology. Hong Kong Univ Press, Hong Kong

66. Le Gall P (1980) Etude expérimentale de l'association en chaine et de son influence sur la croissance et la sexualité chez la crépidule *Crepidula fornicata* (Mollusque mésogastéropode). Thesis, Univ Caen, pp 1–251
67. Le Gall S (1974) Déterminisme de la morphogénèse et du cycle du tractus génitale mâle externe chez *Crepidula fornicata* Phil (Mollusque hermaphrodite protandre). Thesis, Univ Caen, pp 1 to 267
68. Le Gall S (1978) CR Acad Sci 287:1305–1307
69. Le Gall S (1981) Gen Comp Endocrinol 43:51–62
70. Joosse J (1984) In: Photoperiodic regulation of insect and molluscan hormones. CIBA Symp, 104:204–220
71. Sokolove PG: In: Photoperiodic regulation of insect and molluscan hormones. CIBA Symp 104:189–203
72. Bohlken S, Joosse J (1982) Int J Invertebr Reprod 4:213–222
73. Sokolove PG, McCrone EJ (1978) J Comp Physiol 125:317–325
74. Sokolove PG, Minnen J van (1983) In: Lever J, Boer HH (eds) Molluscan neuro-endocrinology. North-Holland Publ, Amsterdam Oxford New York, pp 101–105
75. Geraerts WPM, Joosse J (1983) In: Wilbur K, Saleuddin ASM (eds) Biology of molluscs. Academic Press, London New York, pp 141–207
76. ter Maat A, Lodder JC, Veenstra J, Goldschmeding JT (1982) Brain Res 239:535–542
77. de Vlieger TA, Lodder JC, Joosse J, van Elk R (1983) In: Lever J, Boer HH (eds) Molluscan neuro-endocrinology. North-Holland Publ, Amsterdam Oxford New York, pp 182–183
78. Roubos EW, Geraerts WPM, Boerrigter GH, van Kampen GPJ (1980) Gen Comp Endocrinol 40:446–454
79. Bohlken S, van der Fluit M, Joosse J (1984) Proc K Ned Akad Wet C87:25–31
80. de Jong-Brink M, Schot LPC, Schoenmaker HJN, Bergamin-Sassen MJM (1981) Gen Comp Endocrinol 45:30–38
81. Boer HH, Mohamed AM, van Minnen J, de Jong-Brink M (1976) Neth J Zool 26:94–105
82. Takeda N (1979) Comp Biochem Physiol A62:273–278
83. Takeda N (1980) Embryol Exp Morphol 60:57–69
84. Takeda N (1983) In: Lever J, Boer HH (eds) Molluscan neuro-endocrinology. North-Holland Publ, Amsterdam Oxford New York, pp 106–111
85. Nirde P (1983) FEBS Lett 151:223–227
86. Sluiters JF (1981) Z Parasitenkd 64:303–319
87. Sluiters JF, Brussaard-Wüst CCM, Meuleman EA (1980) Z Parasitenkd 63:13–26
88. Joosse J, van Elk R (1983) In: Lever J, Boer HH (eds) Molluscan neuroendocrinology. North-Holland Publ, Amsterdam Oxford New York, pp 118–120
89. Sluiters JF, Roubos EW, Joosse J (1984) Z Parasitenkd 70:67–72
90. Sluiters JF, Dogterom GE (in press) Z Parasitenkd

I.4 The Structure of the Commands for a Neuropeptide-Mediated Behavior, Egg-Laying, in an Opisthobranch Mollusc

F. STRUMWASSER[1]

1. Introduction

This paper is concerned with the structure of the commands and the neural program for the innate reproductive act of egg-laying in the opisthobranch mollusc *Aplysia*. Our lab has purified three peptides that initiate egg-laying and the primary structure of these peptides has been determined [1, 2]. Recently, the genes for these three peptides have been cloned by Scheller et al. [3]. The nucleotide sequences confirm the primary structure data obtained on the three peptides and show that there are relationships among the three genes.

The final command for egg-laying is the neurohormone, egg-laying hormone (ELH). ELH is a basic peptide (pI 9.1) with molecular weight, determined from its primary structure, of 4,385 [1]. It is synthesized by the approx. 1,000 bag cell neurons of the abdominal ganglion [4]. The bag cell neurons are organized into two symmetrical clusters which occur at the junction of the abdominal ganglion and the two pleurovisceral connectives which connect this ganglion with the pleural ganglia in the head of the animal. Antibodies made in rabbits to ELH conjugated to thyroglobulin, after removal of the anti-thyroglobulin fraction, selectively stain each of the bag cells within the two clusters, using the peroxidase anti-peroxidase second antibody technique [5].

The neural program for releasing ELH is a long-lasting synchronous pacemaker-like discharge of all the bag cells. This 30 min discharge is easily evoked in vitro by brief electrical stimulation of either of the pleurovisceral connectives [6] and has been shown by in vivo recordings to always precede "spontaneous" egg-laying in *Aplysia braziliana* [7]. Our studies have shown that this 30 min program of pacemaker-discharge is mediated by a cyclic AMP-dependent protein phosphorylation mechanism [8, 9, 10].

The precommand for egg-laying is only partly worked out. Exactly what triggers the bag cells to discharge and release ELH in the intact *Aplysia* is not presently known. Isolated *Aplysia* are known to lay eggs [11], so that there is some mechanism, which we can term spontaneous egg-laying, that does not appear to depend on any presently known external signal. Presumably mating, which occurs in these hermaphrodites, plays a role in reproductive function, but it is given a minor role as a causative factor in egg-laying in *Aplysia braziliana* [12]. However, it is known that there are several peptides, extractable from a special region of the reproductive tract, the atrial gland, that can

1 Department of Physiology, Boston University, School of Medicine, 80 East Concord Street, Boston, MA 02118, USA, and Marine Biological Laboratory, Woods Hole, MA 02543, USA

Biosynthesis, Metabolism and Mode of Action of Invertebrate Hormones
(ed. by J. Hoffmann and M. Porchet)

produce egg-laying when injected into intact animals. Two of these peptides, termed peptides A and B, which bear no resemblance to ELH [2], mediate egg-laying by acting through the abdominal ganglion since these peptides are inactive in the absence of the ganglion [13]. Furthermore, peptides A and B are known to initiate bag cell discharge in vitro at moderately low concentrations [2]. A third peptide, extracted from the atrial gland, called "egg-releasing hormone", resembles both ELH and peptides A and B, and can induce egg-laying in the absence of the abdominal ganglion [14]. When an *Aplysia* genomic library was repeatedly screened with a radioactive cDNA, coding for ELH-like sequences, a clone for this putative gene could not be found although the same probe turned up clones for the three other peptide genes [3].

2. The Command for Egg-Laying

2.1 Purification and Structure of ELH

The command for egg-laying is the release of ELH from the bag cells. ELH can be purified to homogeneity by a two-step procedure, involving cation exchange chromatography on SP C25 (Sephadex) followed by gel filtration on a column of Bio-Rad P6 material [1]. The absorbance profile (at 215 nm) of fractions from the P6 column shows a major peak at Kav 0.2, the only region from which egg-laying could be induced. This two-step procedure provides a 100-fold enrichment of ELH and on amino-acid sequence analysis shows a single isoleucine NH2-terminus. When 2.5 nmol of this purified material is injected into *Aplysia californica*, it consistently induces egg-laying [1].

ELH is a basic peptide (pI 9.1) containing 36 amino acid residues [1]. The primary structure of ELH is:

H–Ile–Ser–Ile–Asn–Gln–Asp–Leu–Lys–Ala–Ile–Thr–Asp–Met–Leu–Leu–
Thr–Glu–Gln–Ile–Arg–Glu–Arg–Gln–Arg–Tyr–Leu–Ala–Asp–Leu–Arg–
Gln–Arg–Leu–Leu–Glu–Lys–OH

Since carboxypeptidases A and B were unable to cleave ELH, we suggested that the carboxy-terminal lysine was probably blocked by amidation, but other possibilities were not excluded at the time.

2.2 Synthetic ELH

Recently, in collaborative experiments with Stephen B.H. Kent and Suzanna J. Horvath (of Caltech), various forms of ELH have been synthesized. Our preliminary tests of ELH-lysine-amide show that it induces egg-laying as well as the normal behaviors associated with egg-laying (inhibition of locomotor activity, head weaving movements to wind the egg string and mouth movements to attach the egg string to the substrate). Furthermore, the latencies for egg-laying in the two test animals were normal (33 and 36 min) as well as the weight of the egg masses (1.1 and 1.8 g). The egg strings were examined under a dissecting microscope and they both contained eggs indicating that the synthetic material did indeed release eggs normally from the gonad into the small hermaphroditic duct. Packaging of the eggs, within the egg string, was also normal in both cases. Thus, we can conclude that ELH-lysine-amide is sufficient to cause normal egg-laying and associated behaviors in intact *Aplysia*.

2.3 Gene for ELH

The induction of egg-laying with synthetic as well as native ELH is of great interest in view of the structure of the gene for ELH. Scheller and colleagues [3] have sequenced a clone (ELH–1R) that codes for one complete copy of ELH as well as the 27 amino acid residue called acidic peptide (pI 4.8). This gene, if fully transcribed and translated would code for a 41,800 mol. wt. precursor. Furthermore, from the presence of seven dibasic residues, excluding those related to the signal sequence, ELH and the acidic peptide, at least seven peptide fragments could be generated by action of endopeptidases. Rothman and colleagues [15] have provided evidence that one of these peptide fragments, an octapeptide, alpha bag cell peptide, can be purified from bag cells and has inhibitory actions on certain autoactive identifiable neurons in the upper left quadrant of the abdominal ganglion.

2.4 Role of the Abdominal Ganglion in Egg-Laying

This latter result raises the question as to whether the abdominal ganglion itself plays any role in the overt behaviors of egg-laying, as well as the generation and extrusion of the egg string, except as a source of ELH. Some years ago we showed that surgical removal of the entire abdominal ganglion, including the two bag cell clusters, did not prevent the induction of egg-laying and associated behaviors by crude extracts of the abdominal ganglion containing ELH [16]. While these results should be repeated with synthetic ELH, they suggest that the abdominal ganglion does not play an essential role in the various aspects of egg-laying including the associated behaviors, referred to above, other than as a source of ELH. In conclusion, ELH by itself is a sufficient command to cause egg-laying and associated behaviors in *Aplysia*.

3. The Neural Program for Releasing ELH

3.1 Bag Cell Discharge Precedes Egg-Laying

It is known from in vivo recordings from the pleurovisceral connectives in *Aplysia braziliana* that bag cells undergo a long lasting electrical discharge (average of 21 min) prior to spontaneous bouts of egg-laying [7]. It is also known that electrical stimulation of the pleurovisceral connectives either in vivo [7] or in vitro with the isolated abdominal ganglion [6, 8] causes an afterdischarge lasting about 30 min.

3.2 ELH, Acidic and Other Peptides Are Released During Bag Cell Afterdischarge

By allowing isolated abdominal ganglia to incorporate radioactive amino acids into newly synthesized proteins, it has been possible to subsequently study the release of radiolabeled peptides from a single isolated bag cell cluster during an electrically initiated afterdischarge. Using this approach, it has been demonstrated that ELH, the acidic peptide, and at least two other smaller peptides are released during an afterdischarge [17]. ELH was identified by its basic pI and comigration with purified radiolabeled marker ELH on isoelectric focusing (IEF) gels. The acidic peptide was identified by its pI of 4.8 on IEF gels. The two smaller peptides were not identified, but recent results by Rothman and colleagues [15] would suggest that alpha bag cell peptide, an octapeptide, was one of the remaining two peptides released.

3.3 Afterdischarge is a Cyclic AMP-Mediated Event

We have shown that the long lasting afterdischarge in bag cells, after brief electrical stimulation of the pleurovisceral connectives, is a cyclic-AMP (cAMP) dependent mechanism. cAMP rises within 1 min after initiation of an afterdischarge and peaks by 2 min at approx. 2.5 times the basal level before afterdischarge [8]. This rise in cAMP appears to play a causal role in afterdischarge since membrane permeant, phosphodiesterase-resistant, cAMP analogs such as 8-benzylthio or N6-*n*-butyl-8-(benzylthio)-cAMP initiate discharge of bag cells with similar time-varying spike frequency characteristics as does electrically induced afterdischarge [8, 10]. Forskolin, a diterpene activator of adenylate cyclase [18], when combined with theophylline, also initiates the characteristic bag cell discharge [10] and phosphodiesterase inhibitors, such as isobutylmethyl xanthine, caffeine, and papaverine, significantly prolong electrically initiated afterdischarges [8].

3.4 Bag Cell Afterdischarge Is Associated with Phosphorylation of Specific Proteins

Since bag cell afterdischarge is cAMP dependent, we examined protein phosphorylation at 2 and 20 min into afterdischarge. Two different procedures were used (pre- and postlabeling). In the first procedure, abdominal ganglia were incubated in ^{32}P-orthophosphate for periods around 24 h. Subsequently, an after-discharge was initiated in one hemi-ganglion, while the other hemi-ganglion served as a control. The bag cell clusters were removed and homogenized at one of the two experimental time points into afterdischarge. Proteins were separated on SDS-polyacrylamide slab gels and autoradiograms of the dried gels were obtained as described in the full publication [9]. With this procedure two phosphoproteins in particular showed consistent changes during afterdischarge. Both proteins showed enhanced phosphorylation: BC–1 (apparent mol.wt. of 33,000) was increased by 82% ± 14% at 2 min and 69% ± 43% at 20 min relative to controls. BC–2 (apparent mol.wt. of 21,000) was increased at 20 min (92% ± 23%), but not at 2 min (−19% ± 28%).

In the second procedure, bag cell clusters were homogenized at the same two time points into afterdischarge, but phosphorylation was assayed in the cell-free preparation by using gamma-labeled ^{32}P–ATP and exogenous cAMP-dependent catalytic subunit of protein kinase. In this procedure the degree of phosphorylation observed in autoradiograms bears an inverse relationship to the endogenous phosphorylation since only the remaining available sites for phosphorylation can be occupied by radioactive P. We found that there was a 73% ± 9% reduction in BC–2 at 20 min and no significant change at 2 min.

These results suggest that one or more proteins, such as BC–1 that shows an enhanced phosphorylation early in afterdischarge, could be playing a role in the enhanced membrane excitability of afterdischarge. Likewise, proteins such as BC–2, that show enhanced phosphorylation late in afterdischarge may play some role in the termination of or recovery processes from afterdischarge. One approach to testing these hypotheses would be to raise antibodies, preferably monoclonal, to these proteins and use them as functional reagents to interfere with the normal processes of afterdischarge. There are currently examples of the use of monoclonal antibodies that interfere with various membrane enzymes, receptors, and channels such as Na–K–ATPase [19], transferrin receptors, and the growth of mammalian cells [20] and sodium channels and nerve impulse conduction [21].

3.5 The Program for Afterdischarge Is Stored in Individual Bag Cells

In order to further analyze the membrane mechanisms altered by cAMP-dependent protein phosphorylation, we developed techniques for dissociating bag cells and maintaining them in primary culture [22, 23]. Such isolated bag cells primarily produce calcium spikes, which are insensitive to tetrodotoxin and sensitive to cobalt ions [13]. The amplitude and duration of these calcium spikes increase with repetitive transmembrane stimulation.

When cAMP analogs (see Sect. 3.3) are added to such primary cultures the bag cells undergo spontaneous discharge [24]. Interestingly, when spikes are added for a few seconds by transmembrane stimulation, there is an increase in the background rate of spike discharge for many seconds after the applied current [24]. We tested for the afterdischarge property in a more stringent manner. After adding the cAMP analog, an isolated bag cell was immediately hyperpolarized in order to prevent the spontaneous phase of afterdischarge. When the hyperpolarization was momentarily released, about 15 min later, in order to allow spike discharge for a few seconds, an afterdischarge lasting for 12–20 min occurred. These results demonstrate that the program of afterdischarge is intrinsic to each bag cell and furthermore that cAMP priming followed by calcium action potentials releases the program. Recently we have demonstrated by measuring absorbance changes due to the calcium indicator dye, arsenazo III, that action potentials in bag cells do indeed increase calcium intracellularly [25].

3.6 Membrane Changes with Enhanced Protein Phosphorylation

Using isolated bag cells in primary culture, we have injected the catalytic subunit of cAMP-dependent protein kinase (PKC), purified from beef heart, in order to further test the hypothesis that protein phosphorylation is responsible for the enhanced membrane excitability associated with afterdischarge [26]. Intracellular PKC injections enhanced the rise time, amplitude, and duration of the calcium action potentials in bag cells. Associated with these changes was an increase in membrane resistance.

Voltage clamp studies of isolated bag cells show that net outward current is decreased during the action of external cAMP analogs [27] which is consistent with the enhancement of calcium action potentials seen with both PKC injections as well as with external cAMP analogs. In addition it has been recently observed that a transient outward current, identified as the "A" current of Connor and Stevens [28], is diminished during cAMP action (Strumwasser, unpubl.). Since there are at least three different potassium channels in molluskan neurons, one approach to further studies of the action of protein phosphorylation on the membrane of bag cells is the recording of single ion channels by the gigaseal patch recording technique [29]. We have successfully recorded single channel activity in isolated bag cells and are currently classifying the nature of these channels (Strumwasser and McIntyre, unpubl.). Siegelbaum and colleagues [30] have recorded single channel activity in sensory neurons of the intact abdominal ganglion and have found serotonin-sensitive outward current channels that are turned off by external cAMP analogs.

4. The Precommand for Egg-Laying

4.1 Peptides A and B of the Atrial Gland

As indicated in the introduction there is incomplete information on the factors that initiate bag cell discharge in vivo. Arch and colleagues [31] first showed that extracts of the atrial gland, a part of the large hermaphroditic duct near the gonopore, could induce egg-laying. We purified two factors from the atrial gland of *Aplysia californica*, peptides A and B, which induced egg-laying. These peptides each contain 34 amino acid residues and differ from one another in only four residues [2]. Peptide A has a calculated mol.wt. of 3,924 and a pI of 8.0; peptide B has a calculated mol.wt. of 4,032 and a pI of 9.1. The amino acid sequences of peptides A and B (differences in B residues shown in parentheses) are:

H–Ala–Val–Lys–Leu(Ser)–Ser–Ser–Asp(Tyr)–Gly(Glu)–Asn(Lys)–Tyr–Pro–
Phe–Asp–Leu–Ser–Lys–Glu–Asp–Gly–Ala–Gln–Pro–Tyr–Phe–Met–Thr–
Pro–Arg–Leu–Arg–Phe–Tyr–Pro–Ile–OH .

Peptides A and B induce egg-laying in *Aplysia* only when the abdominal ganglion containing the bag cells is left intact [13]. Peptides A and B also induce bag cell discharge in the in vitro abdominal ganglion preparation at concentrations as low as 0.1 μmol [2].

4.2 The Nature of the Atrial Gland

These findings raise the issue as to the nature of the atrial gland. In recent studies, Beard and colleagues [32] conclude that the atrial gland is an exocrine gland. They find that the gland is a highly infolded stratified epithelium consisting of two cell types. The "capping" cell has cilia that border on the lumen of the hermaphroditic duct, while the "goblet-like" exocrine cell has 1–2 μm electron dense granules. It is known that neighboring regions of the large hermaphroditic duct do not contain factors that induce egg-laying. The structure of cells in this latter region are rather similar, but the large electron dense granules have crista-like infoldings and the granule is bounded by a double limiting membrane, whereas the atrial gland granule has a single limiting membrane and no crista [32].

There is hence an enigma. The atrial gland contains at least three peptides that can induce egg-laying (peptides A, B, and egg-releasing "hormone"). All three peptides activate the bag cells and in addition ERH can induce egg-laying in the absence of bag cells. However, the contents of the secretory cells in the atrial gland presumably enter the lumen of the large hermaphroditic duct in the intact animal close to the gonopore and the salty ocean just external to this orifice. It is possible then that the three peptides of the atrial gland never get into the circulation and serve some other function, presumably reproductive, yet to be discovered. On the other hand, it is possible that these reproductive tract peptides are similar to or identical with the putative transmitters that turn on the bag cells. Thus, antibodies to these three peptides might lead us to the group of neurons that trigger the bag cells to discharge. It is noteworthy that antibodies to ELH besides staining the bag cells in the abdominal ganglion [5] also stain discrete and symmetrical groups of neurons in the pleural and cerebral ganglia [33].

Acknowledgment. Original research supported by a NIH grant (NS 15183).

References

1. Chiu AY, Hunkapiller MW, Heller E, Stuart DK, Hood LE, Strumwasser F (1979) Proc Natl Acad Sci USA 76:6656–6660
2. Heller E, Kaczmarek LK, Hunkapiller MW, Hood LE, Strumwasser F (1980) Proc Natl Acad Sci USA 77:2328–2332
3. Scheller RH, Jackson JF, McAllister LB, Rothman BS, Mayeri E, Axel R (1983) Cell 32:7–22
4. Arch S (1972) J Gen Physiol 60:102–119
5. Chiu AY, Strumwasser F (1981) J Neurosci 1:812–826
6. Kupfermann I, Kandel ER (1970) J Nerophysiol 33:865–876
7. Dudek FE, Cobbs JS, Pinsker HM (1979) J Neurophysiol 42:804–817
8. Kaczmarek LK, Jennings KR, Strumwasser F (1978) Proc Natl Acad Sci USA 75:5200–5204
9. Jennings KR, Kaczmarek LK, Hewick RM, Dreyer WJ, Strumwasser F (1982) J Neurosci 2: 158–168
10. Strumwasser F, Kaczmarek LK, Jennings KR (1982) Fed Proc 41:2933–2939
11. MacGinitie GE (1934) Biol Bull 67:300–303
12. Blankenship JE, Rock MK, Robbins LC, Livingston CA, Lehman HK (1983) Fed Proc 42: 96–100
13. Strumwasser F, Kaczmarek LK, Jennings K, Chiu AY (1981) In: Farner DS, Lederis K (eds) Neurosecretion: Molecules, cells, systems. Plenum Press, New York, pp 249–268
14. Schlesinger DH, Babirak SP, Blankenship JE (1981) In: Schlesinger DH (ed) Symposium on neurohypophyseal peptide hormones and other biologically active peptides. Elsevier/North Holland Publ, Amsterdam New York, pp 137–150
15. Rothman BS, Mayeri E, Brown RO, Yuan PM, Shively JE (1983) Proc Natl Acad Sci USA 80:
16. Strumwasser F, Schlechte FR, Bower S (1972) Fed Proc 31:405
17. Stuart DK, Chiu AY, Strumwasser F (1980) J Neurophysiol 43:488–498

18. Seamon KB, Padgett W, Daly JW (1981) Proc Natl Acad Sci USA 78:3363–3367
19. Schenk DB, Leffert HL (1983) Proc Natl Acad Sci USA 80:5281–5285
20. Trowbridge IS, Lopez F (1982) Proc Natl Acad Sci USA 79:1175–1179
21. Meiri H, Zeitoun I, Grunhagen HH, Lev-Ram V, Cohen Y, Eshhar Z, Schlessinger J (1983) Soc Neurosci Abstr 9:20
22. Strumwasser F, Kaczmarek LK, Viele D (1978) Soc Neurosci Abstr 4:207
23. Kaczmarek LK, Finbow M, Revel JP, Strumwasser F (1979) J Neurobiol 10:535–550
24. Kaczmarek LK, Strumwasser F (1981) J Neurosci 1:626–634
25. Woolum JC, Strumwasser F (1983) Biophys J 41:59a
26. Kaczmarek LK, Jennings KR, Strumwasser F, Nairn AC, Walter U, Wilson FD, Greengard P (1980) Proc Natl Acad Sci USA 77:7487–7491
27. Kaczmarek LK, Strumwasser F (1981) Soc Neurosci Abstr 7:932
28. Connor JA, Stevens CF (1971) J Physiol 213:21–30
29. Hamill OP, Marty A, Neher E, Sakmann B, Sigworth FJ (1981) Pfluegers Arch 391:85–100
30. Siegelbaum SA, Camardo JS, Kandel ER (1982) Nature (London) 299:413–417
31. Arch S, Smock T, Gurvis R, McCarthy C (1978) J Comp Physiol 128:67–70
32. Beard M, Millecchia L, Masuoka C, Arch S (1982) Tissue Cell 14:297–308
33. Chiu AY, Strumwasser F (1984) Brain Res 294:83–93

I.5 Studies on Release Activities of the Neurosecretory Caudo-Dorsal Cells of *Lymnaea stagnalis*

W.P.M. GERAERTS, A. ter MAAT, and Th.M. HOGENES[1]

1. Introduction

Egg-laying of gastropod molluscs involves a complex series of overt and covert behavioral events that result in the oviposition of the egg mass [1]. These processes are initiated and coordinated by one type of neurosecretory cell. In the freshwater pulmonate *Lymnaea stagnalis* the cells form paired groups of together ca. 100 caudo-dorsal cells (CDC), located in both cerebral ganglia. They release to ovulation hormone (CDCH) [2], a 4,700 mol.wt. peptide [3], from the periphery of the intercerebral commissure (COM) [4]. Electrophysiological [5] and ultrastructural [6] studies provide indirect evidence that CDCH is released from neurosecretory granules of CDC axon terminals during CDC discharges (synchronous series of action potentials in all CDC lasting ca. 50 min). The presence of the hormone in the blood and the kinetics of its release are, however, as yet not investigated. This chapter provides direct quantitative evidence that CDCH is contained by neurosecretory granules of the CDC and, furthermore, is released into the blood during CDC discharges. We also have attempted to analyse the nervous pathways involved in its release. Finally, we present data on release of various peptides, besides CDCH, by electrically active CDC.

2. CDC Neurosecretory Granules Contain the Ovulation Hormone

The pheriphery of the COM consists mainly of CDC axon terminals that contain numerous neurosecretory granules (Fig. 1), the contents of which are thought to be released during CDC discharges. (Only few axons of neurosecretory cells other than the CDC are present in the COM.) In an attempt to demonstrate the association of CDCH activity with CDC neurosecretory granules, COMs were dissected and subjected to density gradient centrifugation as outlined in Table 1. The in vivo assay of Dogterom et al. [7] was used to quantify CDCH bioactivity in each fraction. Bioactivity is expressed as ovulation inducing units, one OIU being the threshold dose for ovulation in adult recipient snails. Various dilutions were injected into recipients (n = 8 per dose), and after 30 min the presence of egg cells in the spermoviduct was checked. The criterion used in determining the threshold was a response of 1-3 animals out of 8 injected animals.

1 Biological Laboratory, Vrije Universiteit, P.O. Box 7161, 1007 MC Amsterdam, The Netherlands

Biosynthesis, Metabolism and Mode of Action of Invertebrate Hormones
(ed. by J. Hoffmann and M. Porchet)

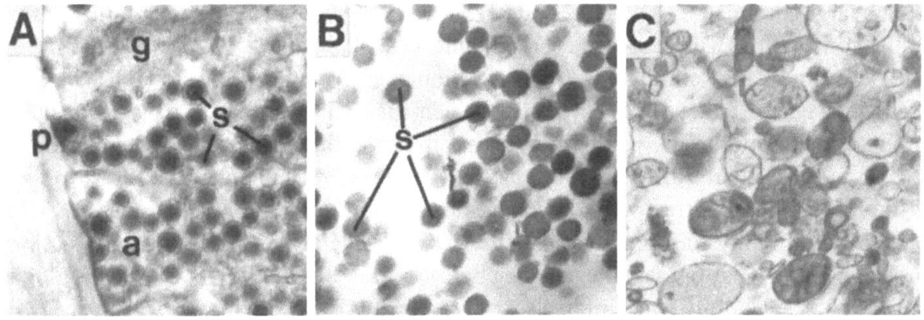

Fig. 1 A–C. Electron micrographs of sections of intact COM tissue and of subcellular fractions obtained by density gradient centrifugation. **A** CDC axon terminals; *a* in the periphery of the COM; *g* glial cell; *p* perineurium; *s* neurosecretory granules. x 20,000. **B** Fraction 6 (1.5 *M* sucrose, see Table 1) with numerous CDC neurosecretory granules (*s*). x 20,000. Many of the isolated granules do not possess a clear electron-lucent halo (cf. **A**). This may be the result of swelling of the osmiophilic contents during isolation [8]. **C** Inhomogeneous fraction 3 (1.0 *M* sucrose, see Table 1) of COM with many different vesicular structures. x 20,000

Table 1. Distribution of CDCH-activity (expressed as OIU[a]) after density-gradient centrifugation of COM-homogenate

Source	Original CDCH-activity (%)		Spec. CDCH activity (OIU/μg protein)
0.3 *M* sucrose homogenate	100.0		183.5
3,000 g (5min, 4 °C) sediment	9.6		93.8
Density-gradient centrifugation of 3,000 g supernatant (100,000 g, 3 h, 4 °C)		96.5	
1.0 *M* sucrose, 1.2 *M* sucrose, 1.4 *M* sucrose	25.4		267.0
1.6 *M* sucrose	53.7		588.9
1.8 *M* sucrose	7.8		287.3

[a] OIU = ovulation inducing unit

The 1.6 *M* sucrose fraction appeared to contain most of the hormone activity (more than 50% of the total activity present on the entire gradient, see Table 1). Electron microscopy revealed that this fraction consisted of a homogeneous preparation of CDC neurosecretory granules [8] (Fig. 1). In conclusion, these results suggest that the ovulation hormone is contained within CDC neurosecretory granules.

3. Release of CDCH During Egg-Laying

There is indirect evidence for the in vivo release of CDCH by electrically active CDC: (1) CDC are active around the time of ovulation only [5], (2) in vitro, active CDC release large amounts of CDCH [5], and (3) active CDC show massive exocytotic release of the contents of CDC neurosecretory granules [6].

The water change stimulus was used to induce egg-laying. It consists in leaving the snails in stagnant water for 5 days, after which they are transferred to clean jars with fresh aerated water. This procedure results in egg-laying within 3 h after the water change in more than 90% of the animals [9]. We have attempted to associate changes in the CDCH blood titres with (1) changes in the amount of CDCH in the CDC/COM system, (2) the physiological cycle of the CDC, (3) the stages of egg mass production process, and (4) the phases of egg-laying behavior (see Fig. 2). CDCH activity in blood collected after various time intervals after the water change stimulus was quantified (see Sect. 2). Similarly, changes in hormonal content of the CDC/COM system were studied by determining hormone activities in homogenates of the cerebral ganglia and COM. Monitoring of spiking activity of the CDC [5], observations on egg mass formation [10], and egg-laying behavior [11] were performed as described before.

Figure 2 shows that the onset of CDC discharges precedes the appearance of CDCH in the blood and, furthermore, that ovulations do not begin until the hormone has appeared in the blood. This sequence of events indicates a causal relationship between CDC discharges, CDCH release, and induction of ovulation. Figure 2 also shows that CDCH release takes place almost exclusively during the active state of the CDC and is accompanied by a decrease in the CDCH content of the CDC/COM system. The blood level of CDCH peaks at about the end of the active state and then decreases slowly to zero at 4 h. The hormonal content of the CDC/COM system is slowly replenished (in ca. 24 h). Ovulations occur when CDCH titres are low. Most likely CDCH has a direct effect on the ovotestis. The "resting" behavior starts immediately after application of the water change and overlaps largely with ovulation. It is not dependent on the release of CDCH into the blood [12]. Formation of eggs and the egg mass, as well as "turning" behavior occur at high CDCH titres. The effects of the hormone on these processes very likely involve the nervous network of the female reproductive organs [13]. CDCH titres again are low during "oviposition" and "inspection".

Raised oxygen content of the water, clean water, as well as a clean substrate are the most potent factors of the water change stimulus in evoking oviposition [9]. In nature, sensing these aspects of water quality might very well be relevant for the animal in finding suitable places to deposit the egg mass and, thus, secure optimal development of offspring. Indeed, *L. stagnalis* preferentially oviposits on leaves of plants in the upper region of ditches, a microhabitat with favorable conditions for the development of embryos and hatchlings. This control mechanism must be considered a typical neuroendocrine reflex mechanism.

4. Nervous Pathways Involved in the Release of CDCH

Lesion studies [14] have shown that sectioning of all nerves that originate from the parietal and visceral ganglia causes a complete cessation of egg-laying. This indicates that information is mediated via these nerves, as well as via both "sides" of the CNS (see Fig. 3), that causes a CDC discharge under normal laboratory conditions. We per-

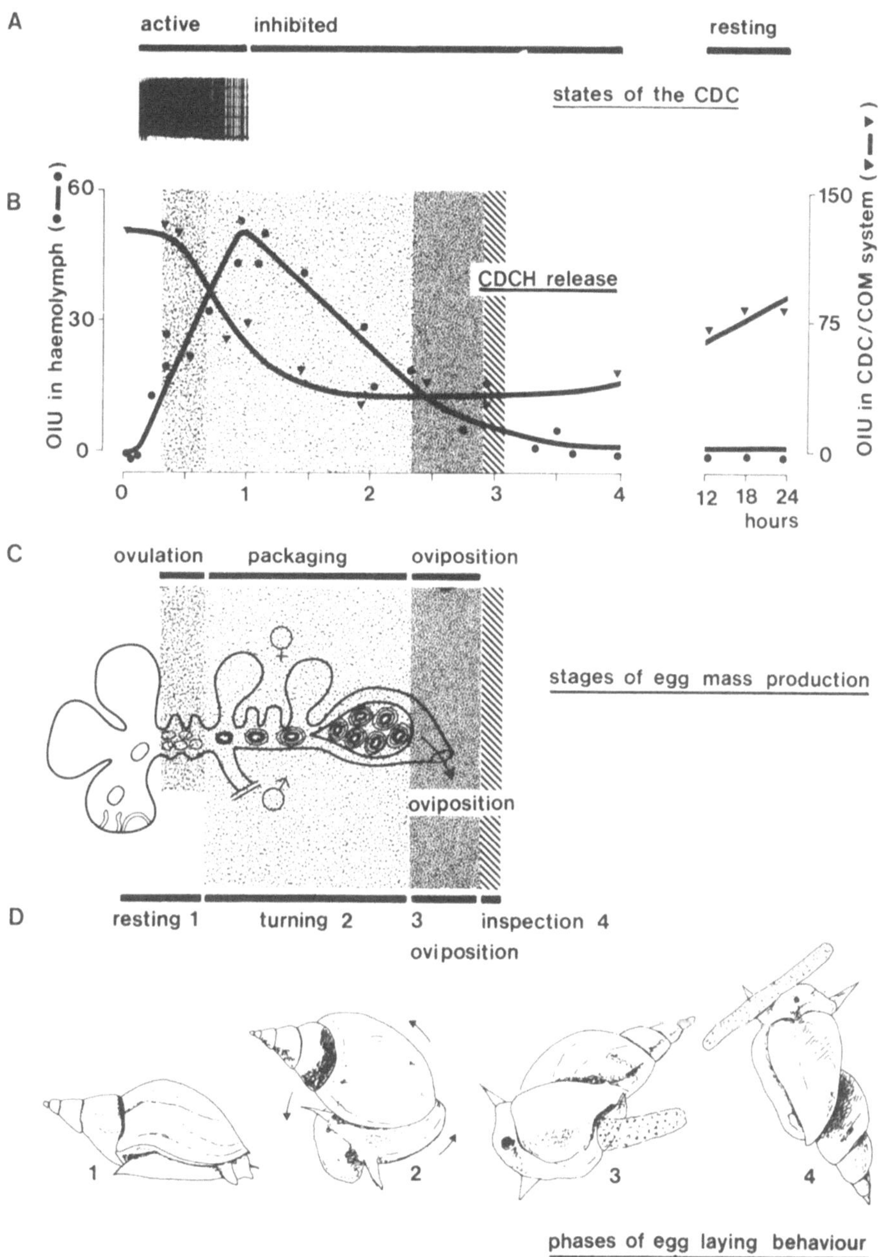

Fig. 2A–D. CDCH release during egg-laying (mean size of the snails: 27 mm). **A** The physiological cycle of the CDC (active, inhibited, and resting state). **B** CDCH titres in the blood and CDCH contents of the CDC/COM system. **C** Stages of egg mass production (simultaneous ovulation of numerous oocytes, egg formation, egg mass formation, and oviposition). **D** Phases of egg-laying behavior (resting, turning, oviposition, and inspection)

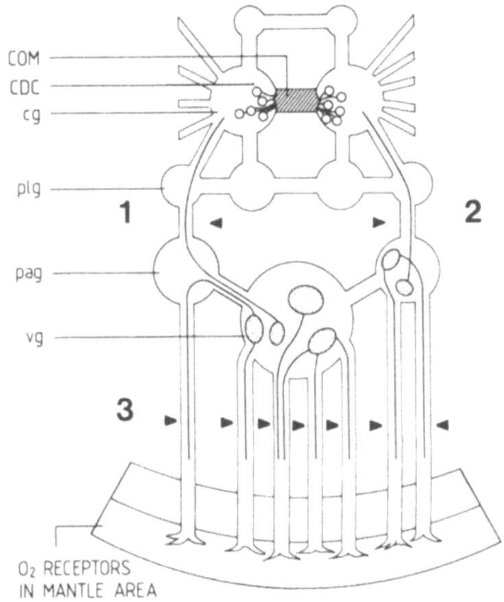

Fig. 3. The nervous pathways involved in CDCH release. *1–3* Position of lesions. *cg* Cerebral ganglia; *plg* pleural ganglia; *pag* parietal ganglia; *vg* visceral ganglia; *CDC* caudo-dorsal cells; *COM* commissure. Indicated are the positions of oxygen-sensitive neurones together with their axons in the visceral and right parietal ganglia

formed lesion experiments to assess whether this information pertains to the state of the ovotestis and/or accessory sex organs, or to water quality. In addition to keeping track of spontaneous ovipository activity under constant conditions, we applied the water change stimulus and injected the animals with CDCH. Sectioning of both the pleuro-parietal connectives at 1 and 2 (see Fig. 3) or of all nerves at 3 produces an immediate stop of spontaneous ovipository activity. Such lesioned snails are unable to oviposit or to release CDCH in response to the water change stimulus. However, upon injection of CDCH ten animals out of ten operated animals laid eggs. This means that the female reproductive system is in working order, but that somehow information concerning the state of either the female reproductive apparatus or the environment does not reach the CDC. Behavioral and electrophysiological experiments [15, 16] have demonstrated the occurrence of oxygen receptors in the mantle area, and neurons in the CNS have been shown to be sensitive to changes in oxygen that occur in the mantle region (see Fig. 3). Because oxygen is one of the component factors of the water change stimulus, it might be that, at least part of, our results can be explained by the interruption of the pathway from the mantle to the cerebral ganglia.

5. Multiple Release of Peptides by Electrically Active CDC

The complexity of the process of egg-laying in *L. stagnalis* justifies the question of whether CDC release other peptides, besides CDCH, with a role in egg-laying. This possibility was investigated in in vitro experiments. Pairs of cerebral ganglia (CG) connected by the COM (CG/COM) were dissected (all nerves of the CG were cut at the origin) and prelabelled for 20 h in Ringer (pH 7.8 at 20 °C) with radioactive amino

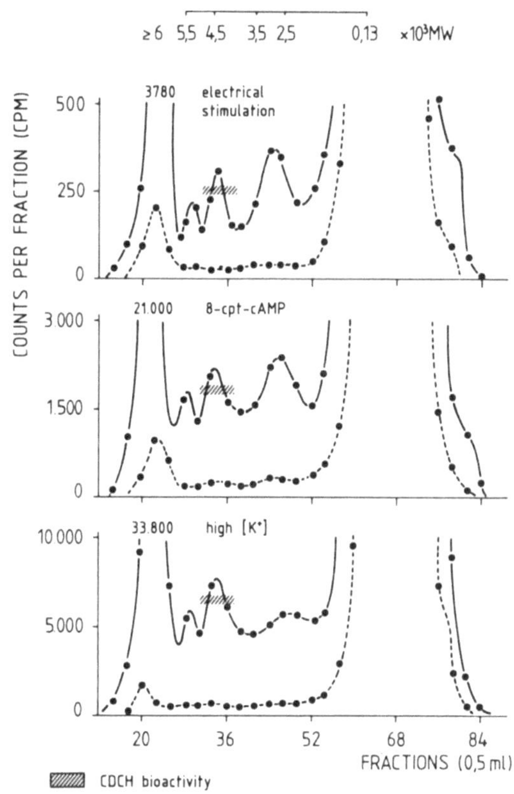

Fig. 4. Chromatography of consecutive 1 h perfusates from CG/COM (20 preparations per experiment) on Bio–Gel P–6 columns (dimensions 58 × 0.9 cm; elution buffer: 50 m*M* ammonium acetate of pH 7.0; flow: 3.0 ml/h; fraction size: 0.5 ml; calibrated with blue dextran, glucagon, insulin A, and [^{14}C] Leu). The radioactivity profiles of the control solution (from the control hour) and the stimulation solution (second hour, where CDC discharges were induced) of each experiment are shown. Radioactivity of every third fraction is indicated. End concentrations of 8-cpt-cAMP and high [K^+] were 2 and 20 mmol, respectively

acids. They then were rinsed and CDC discharges were induced by brief electrical stimulation, or by 8-cpt-cAMP or high [K^+]. Control and stimulation solutions were chromatographed, and radioactivity and bioactivity in fractions were determined as described [17].

The three types of stimulus gave identical results; four factors (presumably peptides) of different molecular weight are released during a CDC discharge, compared with the preceding control period (Fig. 4). The labelled material, present in peak 3 of the stimulation solution shows a manyfold increase over the control period. The fact that this peak also contains released CDCH suggests that labelled peptide material in this peak is identical to CDCH. Besides CDCH, other peptides are released during a CDC discharge: peaks 1, 2, and 4 corresponding to molecular weights of ⩾ 6,000, ca. 5,500, and ca. 2,500, respectively. The role of these peptides in egg-laying is presently being studied.

References

1. Joosse J, Geraerts WPM (1983) In: Saleuddin ASM, Wilbur KM (eds) The mollusca, vol IV. Physiology, part I. Academic Press, London New York, pp 317–406
2. Geraerts WPM, Bohlken S (1976) Gen Comp Endocrinol 28:350–357
3. Geraerts WPM, Ebberink RHM, Cheeseman P, Nuyt C, Hogenes ThM (1983) Gen Comp Endocrinol 51:471–476
4. Roubos EW, Schmidt ED, Moorer-van Delft CM (1981) Cell Tissue Res 215:63–73
5. Kits KS (1980) J Neurobiol 11:397–410
6. Roubos EW, Boer HH, Schot LPC (1981) In: Farner DS, Lederis K (eds) Neurosecretion: Molecules, cells, systems. Plenum Publishing, New York London, pp 119–127
7. Dogterom GE, Bohlken S, Geraerts WPM (1983) Gen Comp Endocrinol 50:476–482
8. Geraerts WPM, Buma P, Hogenes ThM (1984) Gen Comp Endocrinol 53:212–217
9. Kits KS, Maat A ter (1983) In: Lever J, Boer HH (eds) Molluscan neuro-endocrinology. North-Holland Publ, Amsterdam Oxford New York, pp 60–68
10. Dogterom GE, Bohlken S, Joosse J (1983) Gen Comp Endocrinol 49:255–260
11. Goldschmeding JT, Wilbrink M, Maat A ter (1983) In: Lever J, Boer HH (eds) Molluscan neuro-endocrinology. North-Holland Publ, Amsterdam Oxford New York, pp 251–255
12. Maat A ter, Jansen R (1984) This volume
13. Jong-Brink M de, Goldschmeding JT (1983) In: Lever J, Boer HH (eds) Molluscan neuro-endocrinology. North-Holland Publ, Amsterdam Oxford New York, pp 126–131
14. Unpublished observations
15. Janse C (1982) J Comp Physiol 142:51–59
16. Janse C (1982) J Comp Physiol 145:311–319
17. Geraerts WPM, Tensen CP, Hogenes ThM (1983) Neurosci Lett 41:151–155

I.6 Roles of Calcium and cAMP in Biosynthesis and Release of the Ovulation Hormone of the Freshwater Snail *Lymnaea stagnalis*

E.W. ROUBOS, P. BUMA, and A.M.H. van de VEN[1]

1. Introduction

Because of their large size and easy accessibility for various morphological, physiological, and pharmacological experimental approaches, neurons in the central nervous system of molluscs are very suitable objects for studies of fundamental problems in neurobiology. The bag cells of the sea hare *Aplysia californica* [1, 2] and the caudo-dorsal cells (CDC) of the freshwater snail *Lymnaea stagnalis* [3, 4] belong to the most extensively studied neuronal systems of molluscs. Investigations of these cell types have provided considerable insight into the significance of peptides in neural and neuroendocrine communication, not only as to their diverse functions, but also with respect to the mechanisms that control peptide biosynthesis and release. In this paper we will consider recent neurobiological studies of the CDC with particular reference to functional-morphological aspects.

The CDC occur in two clusters, one in the left (ca. 30 cells), the other in the right cerebral ganglion. Their somata are large, measuring upon 90 μm in diameter [5]. Their axons terminate in the periphery of the intercerebral commissure. From here the CDC release an ovulation stimulating neurohormone (CDCH) into the hemolymph [6]. CDCH is located in secretory elementary granules with a mean diameter of 150 nm [7, 8]. The hormone is a peptide with a molecular weight of ca. 4,700, an isoelectric point of 9.3, and is rich of arginine, leucine, aspartate, and glutamate [9, 10]. In addition to the induction of ovulation, the CDC are involved in the control of some other aspects of female reproductive activity, such as stimulation of synthetic activity of the albumen gland [11], an accessory sex gland that secretes a perivitellin fluid that serves as nutrient for the developing embryo. Furthermore, the CDC trigger *Lymnaea*'s stereotyped egg-laying behavior [12], probably by having combined neurohormonal and local effects on peripheral targets and central motorneurons, respectively [13, 14]. Possibly, the different functions of the CDC are not exerted by CDCH alone since the cells release at least three other peptides [15].

1 Department of Histology and Electron Microscopy, Subfaculty of Biology, Free University, De Boelelaan 1087, 1007 MC Amsterdam, The Netherlands

Biosynthesis, Metabolism and Mode of Action of Invertebrate Hormones
(ed. by J. Hoffmann and M. Porchet)

2. Control of Exocytosis of Secretory Material

The CDC exhibit three states of electrical activity [16, 17]. Usually they are electrically silent ("resting state", lasting ca. 24 h). Prolonged intracellular stimulation with depolarizing pulses initiates a period of continuous spiking activity called the discharge ("active state"; ca. 45 min). In vivo this state can be induced by placing snails that have stayed for some days in dirty, oxygen-poor water into fresh, oxygenated water (freshwater stimulation). After the active state the CDC enter the inhibited state, during which they are silent again and no discharge can be evoked. This state lasts 4–6 h. There appears to be a clear relation between the electrical states and the egg-laying cycle. At the moment of ovulation the CDC are in the active state; ca. 2 h thereafter, when oviposition takes place, the cells are inhibited. About 3–5 h later transition to the resting state occurs. The CDC remain in this state until the next discharge starts. This suggests that the CDC induce ovulation by releasing CDCH during the active state. This idea was substantiated by in vitro and in vivo studies in which CDCH-release from discharging CDC was demonstrated in physiological saline [18] and hemolymph [15], respectively.

There is convincing ultrastructural evidence that CDCH-release occurs by exocytosis of the contents of CDC elementary granules. These exocytozing contents can be demonstrated very clearly and selectively by fixing tissues according to the TAGO-method (tannic acid-glutaraldehyde-osmium fixation) [19]. During fixation tannic acid forms metal complexes with extracellular secretory materials and as a result the exocytozed granule contents appear highly electron dense. In vitro studies of isolated cerebral ganglia with quantitative electron microscopic methods have shown that the number of exocytotic phenomena in a cross section halfway along the cerebral commissure is low during the resting state (ca. 60), but rapidly increases during the active state: 5 min after the onset of the discharge it is ca. 9 times higher as in resting CDC, after 15 min of discharge even 36 times [18, 20]. In vivo studies [21, 22] substantiate the idea that CDCH is released by exocytosis during the active state. Activation of the CDC by the physiological freshwater stimulus results into a depletion of elementary granules from the axon terminals and a 140-fold increase of the number of exocytotic phenomena compared to unstimulated (resting) CDC (Fig. 1). Interestingly, in the subsequent inhibited state (3 h after stimulation) the number of exocytotic phenomena is clearly lower than during the discharge, but still considerably higher (x40) than during rest. This indicates that the CDC release a fair amount of secretory material during electrical silence, i.e., when the cells are hyperpolarized again. The nature nor the function(s) of the substance(s) released during inhibition are known. Possibly, not CDCH but other peptides are concerned, controlling the activity of accessory sex glands.

Obviously, calcium ions are a prerequisite for the occurrence of high exocytotic activity in the CDC. This appears from a number of studies. For instance, depolarization of the axolemma by incubation in high potassium containing Ringer induces mass exocytosis [23] and CDCH-release [15]. However, hardly any increase in exocytosis activity is found when potassium stimulation is carried out in Ringer without calcium ions [23] or in Ringer containing 5.10^{-5} M $CoCl_2$, a blocker of calcium channels [24]. Thus, during the active state extracellular calcium probably enters the CDC axon terminals and induces exocytosis. As was noted above, exocytosis also occurs during the

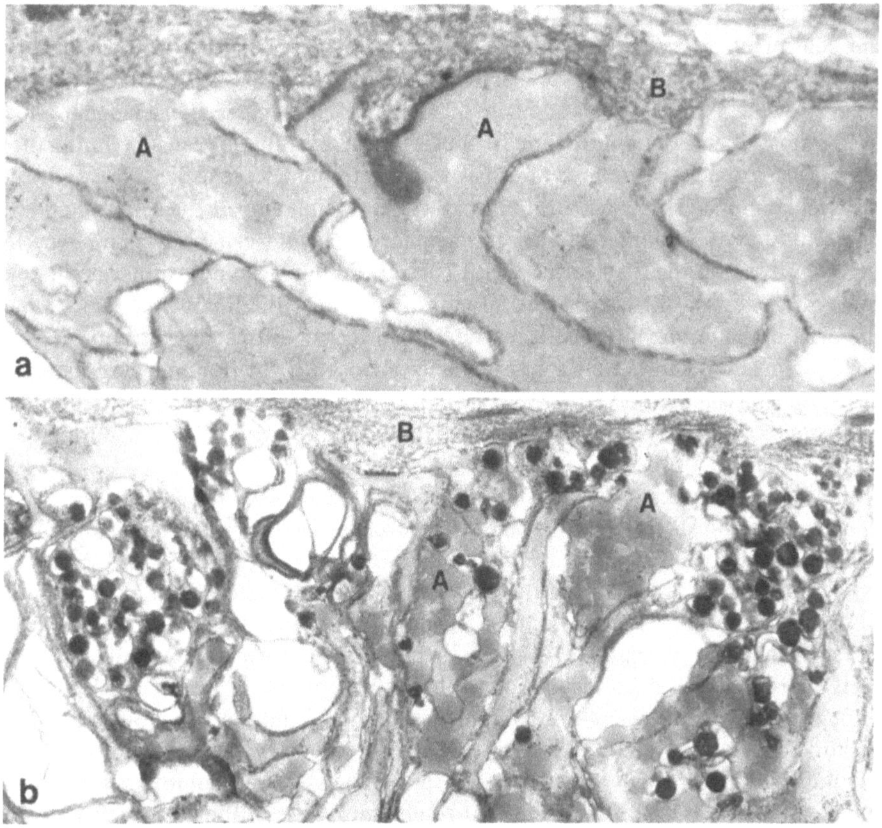

Fig. 1 a,b. Relation between electrical states and exocytosis in CDC axon terminals (*A*); *B* basal lamina. TAGO-fixation. **a** Resting state: no exocytosis. x35,000. **b** Active state: numerous exocytoses (highly electron dense granule contents). x20,000

inhibited and, though at a very low level, during the resting state. Since in these states, compared to the active state, the CDC are hyperpolarized, it is likely that influx of extracellular calcium will hardly or not take place. Therefore, the possibility can be raised that intracellular calcium, entered during the active state, is involved in the induction of exocytosis during inhibition and rest. The intracellular dynamics of calcium have been studied to some detail in the CDC, using the ultrastructural K-pyroantimonate technique [21, 22]. It has been shown that during the active state calcium is taken up by two types of cell organelle, viz. vesicular structures and mitochondria. As to vesicular structures, various types have been distinguished: small clear vesicles, vacuoles, cup-shaped vesicles, double-membraned vesicles, and extracellular clear vesicle-like structures. The vesicular structures are preferentially located near the axolemma and are thought to be involved in the process of recapture and intra-axonal sequestration of parts of the axolemma after high exocytotic activity. All of these structures contain, in variable amounts, calcium deposits. Various possibilities exist as to the origin of this calcium (e.g., micropinocytotic uptake of extracellular calcium, intracellular calcium uptake). It remains to be seen whether this vesicle-bound calcium plays a role in the control of exocytosis. Involvement in exocytosis seems likely for the calcium found in mitochondria. During the active state mitochondria are located near the axolemma of the axon terminals and contain an increasing number of calcium deposits [21, 22]. Most likely, these mitochondria buffer the large influx of calcium during the discharge and protect the more central parts of the terminals against an excessive rise of the axoplasmic calcium concentration that would severely impair the cell's functioning. Recent studies demonstrate that mitochondrial calcium can be released experimentally by treating the CDC in vitro with 10^{-5} *M* carbonylcyanid-p-trifluormethoxyphenylhydrazone (FCCP) [24]. After 30 min of treatment the number of exocytotic phenomena in the terminals is about ten times as high as in untreated cells. Bioassay of the bathing medium shows that a considerable part of the ovulation inducing material in the neurohemal area has been released. This strongly favors the idea that calcium stored in mitochondria can play a role in the stimulation of hormone release, e.g., during the inhibited state.

Various studies indicate that cAMP is involved in the process of CDCH-release. The key enzyme for cAMP production, adenylate cyclase, has been demonstrated ultracytochemically to occur at the axolemma of the CDC axon terminals [21, 25]. Enzyme activity is particularly high during the active state. Indeed, radioimmunoassay studies suggest that the cAMP contents of the CDC neurohemal area are about twice as high in the active as in the resting state [26]. Apparently, a rise of axoplasmic cAMP contents triggers the discharge, since incubation with 1 mmol of the cAMP-analogue 8-parachlorophenylthio-cAMP (8-cpt-cAMP) leads within half a minute to a discharge, high exocytotic activity, and mass release of CDCH. Moreover, the same effects are found when the CDC are treated with the potent phosphodiesterase inhibitor 3-isobutyl-1-methylxanthine (IBMX), a procedure that results into a rise of the intracellular cAMP concentration [26]. Presumably, such a rise stimulates in the CDC terminals phosphorylation of proteins that control the permeability of the axolemma for particular ions, which may result into the start of a discharge. Obviously, cAMP may also be involved in various other aspects of the release process, such as exocytotic fusion of the elementary granule membrane with the axolemma, membrane recapture after exocytosis, and intracellular sequestration of vesicular and mitochondrial calcium.

3. Control of Biosynthesis of Secretory Material

Detailed ultrastructural morphometric studies [27, 28] have shown that not only release activity, but also biosynthetic activity of the CDC has a cyclic character: high release of CDC secretory material is followed by an increased activity of the Golgi apparatus. This results into an increased rate of formation of elementary granules which accumulate in the somata and are subsequently transported to the neurohemal area to refill the axon terminals. Recently we studied the possibility that depolarization of the CDC and an intracellular rise of the cAMP concentration are determining factors in the activation of the process of synthesis of secretory material. This appears to be the case (Fig. 2). Thus, when the CDC are depolarized by incubation in high potassium

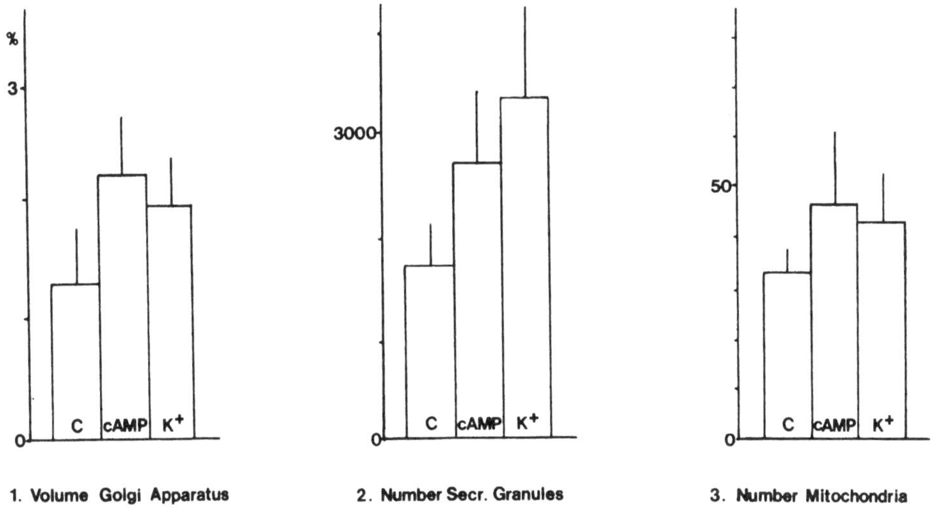

Fig. 2. Quantitative electron microscopic data (means derived from five snails, and standard errors of the means) of CDC somata after 3 h of incubation in control Ringer (*C*), 1 mmol 8-cpt-cAMP Ringer (*cAMP*), and high potassium (10 mmol KCl) Ringer (K^+). Parameter *1* expressed as % of 1,000 μm^3 cytoplasm, parameters *2* and *3* as numbers per 1,000 μm^2 area of cytoplasm

Ringer (10 mmol KCl) not only a discharge starts, but after 3 h there is a considerable increase of the volume of the Golgi apparatus (+50%) and of the number of stored elementary granules (+95%). Moreover, the volume of the mitochondria in the somata is increased by 29%, suggesting an increased cellular respiratory activity. Essentially the same effects are found after treatment with 1 mmol 8-cpt-cAMP.

4. Conclusions

The above presented data strongly indicate that both biosynthesis and release of CDC secretory material (in particular of CDCH, but possibly also of other peptides) are related to the different states of electrical activity. Little is known about the mechanisms that control the transitions between these states, but an important role of the adenylate cyclase-cAMP system seems obvious. CDC activity is known to be controlled by various, external and internal, factors such as water quality, photoperiod, tactile stimulation, food supply, parasitical infection, synaptic inputs from other brain regions, and neuropeptides [4, 14]. We are inclined to think that many of the external factors control synthetic and/or release activity of the CDC by acting via (peptidergic) messengers on adenylate cyclase-coupled receptors in the CDC axolemma.

Acknowledgments. Part of this study was made possible by a grant from the Foundation for Fundamental Biological Research (BION), which is subsidized by the Netherlands Organization for the Advancement of Pure Research (Z.W.O.).

References

1. Kandel ER (1979) Behavioral biology of *Aplysia*. Freeman, San Francisco
2. Strumwasser F, Kaczmarek LK, Heller E, Jennings KR, Viele DP (1980) In: Bloom FE (ed) Peptides: Integrators of cell and tissue function. Raven Press, New York, pp 197–218
3. Joosse J, de Vlieger TA, Roubos EW (1982) Prog Brain Res 55:379–404
4. Roubos EW (1984) Int Rev Cytol 89:295–346
5. Joosse J (1964) Arch Néerl Zool 16:1–103
6. Geraerts WPM, Bohlken S (1976) Gen Comp Endocrinol 28:350–357
7. Wendelaar Bonga SE (1970) Z Zellforsch 108:190–224
8. Geraerts WPM, Buma P, Hogenes TM (1984) Gen Comp Endocrinol 53:212–217
9. Ebberink RHM, Loenhout H van, Geraerts WPM, Hogenes TM, Hoogland H (1983) In: Lever J, Boer HH (eds) Molluscan neuro-endocrinology. North-Holland Publ, Amsterdam Oxford New York, pp 56–58
10. Ebberink RHM (pers. commun.)
11. Wijdenes J, Elk R van, Joosse J (1983) Gen Comp Endocrinol 51:263–271
12. Goldschmeding JT, Wilbrink M, Maat A ter (1983) In: Lever J, Boer HH (eds) Molluscan neuro-endocrinology. North-Holland Publ, Amsterdam Oxford New York, pp 251–256
13. Jansen RF, Bos NPA (1983) In: Lever J, Boer HH (eds) Molluscan neuro-endocrinology. North-Holland Publ, Amsterdam Oxford New York, pp 78–81
14. Maat A ter, Jansen RF (this vol.)
15. Geraerts WPM, Maat A ter, Hogenes TM (this vol.)
16. Vlieger TA de, Kits KS, Maat A ter, Lodder JC (1980) J Exp Biol 84:259–271
17. Kits KS (1980) J Neurobiol 11:397–410
18. Kits KS (1981) Adv Physiol Sci 23:35–54
19. Roubos EW, Wal-Divendal RM van der (1980) Cell Tissue Res 207:267–275
20. Roubos EW, Boer HH, Schot LPC (1981) In: Farner DS, Lederis K (eds) Neurosecretion: Molecules, cells, systems. Plenum Publ, New York London, pp 119–127
21. Roubos EW, Buma P (1982) Prog Brain Res 55:185–194
22. Buma P, Roubos EW (1983) Cell Tissue Res 233:143–159
23. Roubos EW, Schmidt ED, Moorer-van Delft CM (1981) Cell Tissue Res 215:63–73
24. Buma P, Grootenhuis A, Roubos EW (1984) Gen Comp Endocrinol 53:488
25. Roubos EW, Keijzer AN de, Buma P (1981) Cell Tissue Res 220:665–668
26. Buma P, Roubos EW, Pieters FAL (1983) In: Lever J, Boer HH (eds) Molluscan neuro-endocrinology. North Holland Publ, Amsterdam Oxford New York, pp 74–77
27. Wendelaar Bonga SE (1971) Z Zellforsch 113:490–517
28. Roubos EW (1975) Cell Tissue Res 160:291–314

I.7 The Egg-Laying Behavior of the Pond Snail: Electrophysiological Aspects

A. ter MAAT and R.F. JANSEN[1]

The ovulation hormone of the pond snail *Lymnaea stagnalis* is produced by the caudo-dorsal cells (CDCs) of the cerebral ganglia [1]. The CDCs secrete their products (see also Geraerts et al., this volume) during a 30 min period of repetitive spiking activity termed the afterdischarge [2]. This discharge can be elicited by a sudden change of environmental conditions [3] and triggers a stereotyped pattern of overt and covert behaviors that result in the deposition of an egg mass containing ±100 eggs on a suitable substrate [4].

The overt egg-laying behavior comprises three distinct phases. Directly after the water change, the animal is quiescent, the shell pulled over the head region, tentacles drooping. After about 30 min the animal becomes more active and turns the shell till it points forward over the head region.

The eating movements of the buccal mass increase in frequency, but their intensity has decreased, so as not to result in the consumption of the substrate which, in nature, consists of the leaves of water plants. About 2 h after the stimulus the deposition phase starts. The shell returns to its normal position, pointing backwards, but is now tilted such that the egg mass that is extruded from the right hand side of the snail will be pressed on to the substrate. Then the egg mass appears and is laid precisely in the area that has been cleaned by the eating movements during the second phase. After it has finished egg deposition, the animal usually turns around and crawls along the egg mass while touching it with the lip.

During the discharge of the CDCs, a neuron of the right cerebral ganglion [5] increases its firing rate (Fig. 1A). The effect is probably local, though it is not fast. The effect of the CDC on this neuron is not apparent until several (1–5) minutes after the initial bursts. In a preparation where a ring neuron was filled with horseradish peroxydase, serial EM-sections (Roubos and Jansen, unpubl.) revealed that the CDCs and the ring neuron approach each other closely in the intercerebral commissure.

In an earlier study [4] we found that the behavior after injection of crude extracts of intercerebral commissures – the neurohemal area of the CDC – did not contain the first (resting) phase, and lasted about 50 min less than the behavior after water change. This is in keeping with our finding that, although the later phases of the egg-laying behavior may be caused by the hormone that circulates in the blood, at least part of the first phase is mediated by the local effect of the CDCs on the ring neuron. Saline

1 Biological Laboratory, Vrije Universiteit, P.O. Box 7161, 1007 MC Amsterdam, The Netherlands

Biosynthesis, Metabolism and Mode of Action of Invertebrate Hormones
(ed. by J. Hoffmann and M. Porchet)

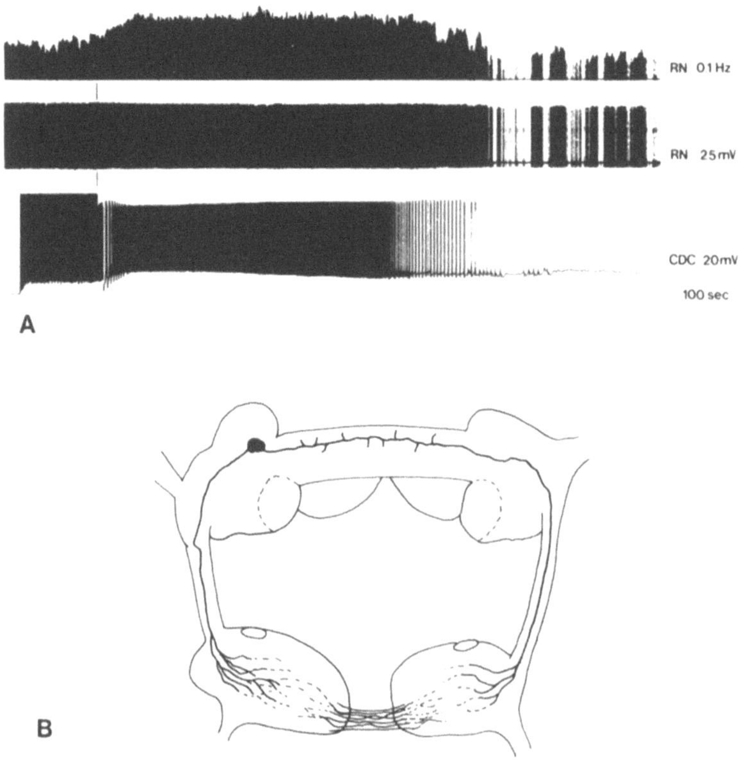

Fig. 1A,B. Some morphological and electrical characteristics of the ring neuron. **A** Intracellular stimulation of a single CDC evokes spontaneous electrical activity in the whole group, which persists after the stimulation is stopped. At the onset of the discharge the firing rate of the ring neuron increases to approx, twice the original value, and decreases again at the end of the discharge. **B** A drawing after a lucifer yellow fill of the ring neuron shows the two axons to the pedal ganglia (the other ganglia have been left out for clarity), the massive branching, and the axons crossing the pedal commissure

in which an afterdischarge has run its course, when added to another preparation, has an inhibitory effect on the ring neuron as opposed to the excitation described above. This means that the CDCs exert their influence on the ring neuron in two ways: directly within the CNS and indirectly by the hormone(s) secreted into the blood.

The morphology of the ring neuron was studied by injection of Lucifer yellow. The neuron is unipolar, having two massive axonal branches, one traveling through the right cerebro-pedal connective, and a second crossing the intercerebral commissure and then projecting towards the left pedal ganglion. The axons arborize extensively in the pedal ganglia, many fine branches crossing the pedal commissure, thus completing the ring (Fig. 1B).

The ring neuron contacts many cells in the pedal ganglia. Using the map of Slade and Winlow [6] neurons in the G-, F-, and E-clusters of these ganglia that could be indentified as motor neurons innervating the columellar muscle were found. This muscle is responsible for the stance that characterizes the first part of the egg-laying behavior. Figure 2 shows the firing pattern and identification of one such neuron in the G-cluster. Firing the ring neuron causes excitation (Fig. 2B). In addition to the ring neuron, the hormone(s) secreted by the CDCs also control the motor neuron: saline in which the CDCs have secreted their hormone has an inhibitory effect on the motor neurons tested so far (Fig. 2C).

On other motor neurons, firing the ring neuron produces inhibition. We are currently sorting out these effects, but it is clear that the control of egg-laying behaviors comprises both transmission within the CNS towards the motor neurons and hormonal actions on neurons in the CNS via the hemolymph (Fig. 2D).

Paradoxically, when the ring neuron is fired at relatively high rates, the CDCs are inhibited, and an ongoing discharge can be arrested (Fig. 3). To investigate whether the normal increase in firing frequency of the ring neuron during the afterdischarge causes the termination of the CDC activity, the lengths of discharges with free running ring neurons were compared with discharges during which the ring neurons were hyperpolarized to prevent spiking. It did not make any difference with regards to the lengths of the discharges whether the ring neuron fired or not. Therefore, the inhibition of the CDCs by the ring neuron must serve another purpose.

On the basis of the findings of Kovac and Davis [7] on behavioral choice, we hypothesized that the ring neuron might be involved in a behavior other than egg-laying. If this behavior were incompatible with egg-laying, the ring neuron would serve to prevent the generation of a CDC afterdischarge. In this hermaphroditic animal an obvious possibility is mating as *Lymnaea* never lays eggs during copulation (van Duivenboden, pers. comm.). We, therefore, stimulated the penis nerve with electrical pulses and found that this had no effect on the CDCs, but elicited a large EPSP in the ring neuron (Fig. 4). From all other nerves tested a biphasic response could be obtained in both the CDCs and the ring neuron (Fig. 4). This response in the CDCs has been described in detail and is also shared by the growth hormone producing light green cells [8-10]. Our result that the penis nerve is the only nerve that produces different effects in CDCs and ring neuron is in keeping with our conjecture that the ring neuron not only mediates a part of the egg-laying behavior, but is also involved in copulation behavior, one of its actions being the prevention of an afterdischarge.

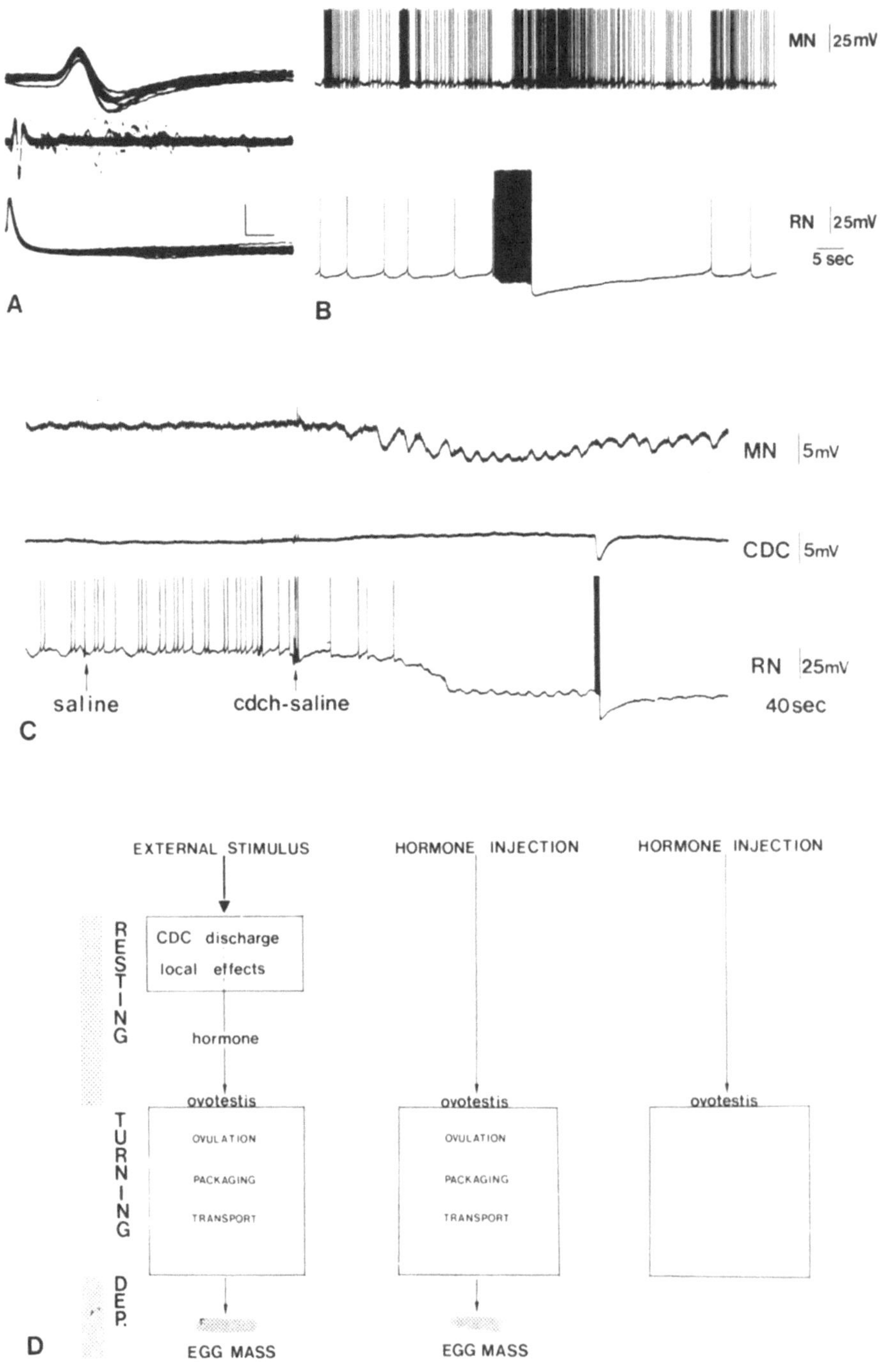
MN 25mV
RN 25mV
5 sec
A
B
MN 5mV
CDC 5mV
RN 25mV
40sec
saline
cdch-saline
C
EXTERNAL STIMULUS
HORMONE INJECTION
HORMONE INJECTION
RESTING
CDC discharge
local effects
hormone
ovotestis
ovotestis
ovotestis
TURNING
OVULATION
PACKAGING
TRANSPORT
OVULATION
PACKAGING
TRANSPORT
DEP.
EGG MASS
EGG MASS
D

Fig. 2 A–D

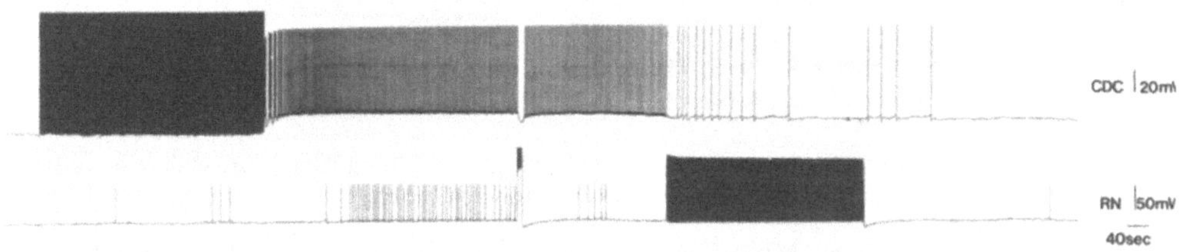

Fig. 3. Interaction of the ring neuron and the CDC. As in Fig. 1A, the ring neuron shows a (moderate) increase in firing rate as a response to the discharge. A short DC-stimulation of the ring neuron, however, interrupts the discharge and when sustained (with 10 Hz square wave pulses), the discharge can be stopped

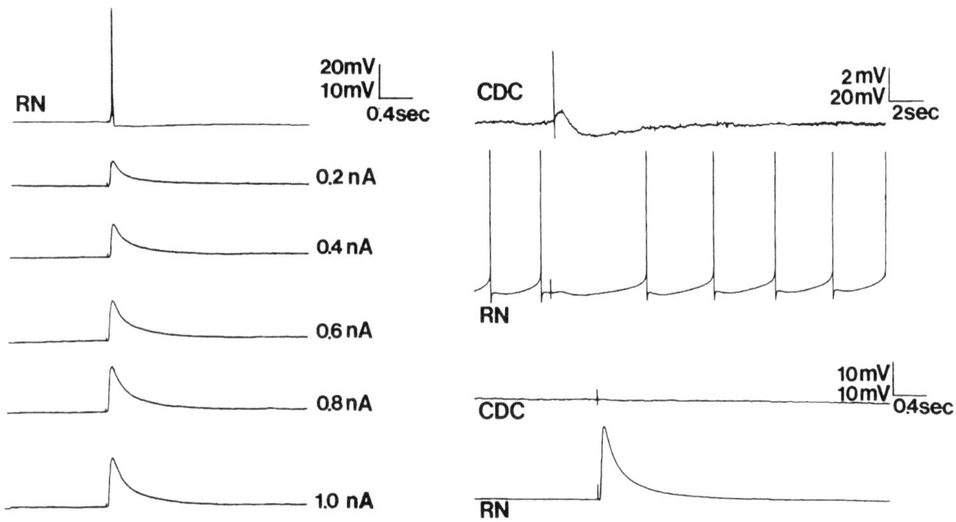

Fig. 4. Ring neuron excitation by penis nerve stimulation. Nerve stimulation generally causes the well-described biphasic response in both CDC and ring neuron (*upper right:* 2 traces). When the penis nerve is stimulated the CDC show no response (*lower right:* 2 traces) and the ring neuron is excited and spikes (*left*). When the ring neuron is hyperpolarized (*left and lower right*) the psp increases, indicating the response is membrane potential-dependent

◄

Fig. 2A–D. The responses of motorneurons to ring neuron firing and ovulation hormone. **A** Identification of a motoneuron. The soma spikes in the penetrated neuron are followed by 1 for 1, constant latency, elements in the nerve and muscle potentials in the columellar muscle. (Cal. *upper trace*: 50 mV, *middle trace*: 20 mV, *lower trace*: 25 mV; time base 20 ms). **B** The same motorneuron, showing prolonged extation on ring neuron stimulation with square wave current pulses at 5 Hz. **C** When ovulation hormone is added to a CNS in which the ring neuron and a motorneuron, identified as in **A**, are recorded, both respond with a slow hyperpolarization, as opposed to the response of the ring neuron to a discharge (see Fig. 1A). **D** The current model shows that a part of the behavior is caused by the hormone and the occurrence of packaging and transport. The resting phase, however, is caused by local effects in the CNS that may be a direct effect of the external stimulus and/or of the CDC discharge which, via the ring neuron, can modulate motorneurons

References

1. Geraerts WPM, Bohlken S (1976) The control of ovulation in the hermaphroditic freshwater snail *Lymnaea stagnalis* by the neurohormone of the caudo-dorsal cells. Gen Comp Endocrinol 28:350–357
2. Kits KS (1980) States of excitability in ovulation hormone producing neuro-endocrine cells of *Lymnaea stagnalis* (Gastropoda) and their relation to the egg laying cycle. J Neurobiology 11:397–410
3. ter Maat A, Lodder JC, Wilbrink M (in press) Induction of egg laying in the pond snail *Lymnaea stagnalis* by environmental stimulation of the release of ovulation hormone from the caudo dorsal cells. Int J Invertebr Reprod
4. Goldschmeding JT, Wilbrink W, Maat A ter (1983) The role of the ovulation hormone in the control of egg laying behavior in *Lymnaea stagnalis*. In: Lever J, Boer HH (eds) Molluscan neuro-endocrinology. North Holland Publ, Amsterdam Oxford New York
5. Jansen RF, Bos NPA (1984) An identified neuron modulating the ovulation hormone caudo-dorsal cells of the pond snail *Lymnaea stagnalis*. J Neurobiol 15(2):161–169
6. Slade CT, Winlow W (1981) The neuronal organisation of the paired pedal ganglia of *Lymnaea stagnalis* (L.). Comp Biochem Physiol 69A:789–803
7. Kovac MP, Davis WJ (1980) Neuronal mechanism underlying behavioral choice in Pleurobranchaea. J Neurophysiol 43:469–487
8. ter Maat A (1979) Neuronal input on the ovulation hormone producing neuroendocrine caudo-dorsal cells of the freshwater snail *Lymnaea stagnalis*. Proc K Ned Akad Wet 82C:333–342
9. ter Maat A, Lodder JC (1980) A biphasic cholinergic input on the ovulation hormone producing caudo-dorsal cells of the freshwater snail *Lymnaea stagnalis.* Comp Biochem Physiol 66C: 115–119
10. ter Maat A, Roubos EW, Lodder JC, Buma P (1983) Integration of biphasic synaptic input by electrotonically coupled neuroendocrine caudodorsal cells in the pond snail. J Neurophysiol 49:1392–1409

I.8 Prothoracicotropic Hormone of *Bombyx mori*

H. ISHIZAKI[1] and A. SUZUKI[2]

1. Introduction

In 1960 one of us (H. Ishizaki) began purifing the prothoracicotropic hormone (PTTH) from brains of the silkworm *Bombyx mori*, in collaboration with Prof. M. Ichikawa of Kyoto University, who died 7 years later. A few years before this time, that famous work by Butenandt and Karlson appeared which reported crystallization and partial characterization of ecdysone from *Bombyx* pupae which were imported from Japan. Ichikawa deeply regretted that this work had not been done by Japanese workers who had the advantage of abundant silkworm supply favored by the prosperity of sericultural industry in this country. He then became very enthusiastic about purifying the brain hormone, now called PTTH. He spoke often warmly of the desire that this time Japanese workers should achieve this work. It was 15 years after World War II had ended and Japan was a poor country on the way to recovery from great war damage. Everybody thought that it was a hopeless attempt. It was apparent that a project like PTTH purification needed well-equipped facilities and plentiful financial support. Ichikawa's enthusiasm surpassed the surrounding pessimism, however, and Ishizaki was engaged in this project from its beginning after just starting his graduate course at Kyoto.

Progress was satisfactory in the early phase of the research. The protein nature of PTTH was soon elucidated [1] and a considerable degree of purification was achieved [2]. As the purification work proceeded, however, it became clearer that PTTH purification was a really difficult task, far more difficult than had been initially anticipated. The most serious problem was a formidably minute content of PTTH in *Bombyx* brain. Several hundred thousands of brains were not satisfactory at all as starting material. After several purification steps using brains in the number of this magnitude, the purified sample did not contain any recognizable material, yet a clear-cut activity was detected. It was apparent that a huge amount of the starting material was needed if we aimed at obtaining the pure PTTH in an amount enough to permit structural analyses. Various other difficulties common to purification of peptide hormones, e.g., inactivation when purity was elevated, nonspecific adsorption to containers, and nonreproducible appearance of false heterogeneity in molecular forms presumably produced by stickiness of PTTH to other molecules, were also met. Thus, after a 10-year

1 Biological Institute, Faculty of Science, Nagoya University, Chikusa-ku, Nagoya 464, Japan
2 Department of Agricultural Chemistry, The University of Tokyo, Bunkyo-ku, Tokyo 113, Japan

Biosynthesis, Metabolism and Mode of Action of Invertebrate Hormones
(ed. by J. Hoffmann and M. Porchet)

struggle, Ishizaki reached the conclusion that any more efforts by a biologist were reckless and fruitless and decided to give up this project.

Just one week after this decision had been made, Prof. S. Tamura of Tokyo University, Department of Agricultural Chemistry, who soon retired and was succeeded by the present author Suzuki, offered to cooperate with Ishizaki on the PTTH purification. For a long time Ishizaki looked for a biochemist who would cooperate on this project, but nobody proposed to do so. Ironically enough, such a person appeared immediately after the resolution to discontinue this work had been made. Tamura spoke enthusiastically that he, as one of Japanese workers, wished to complete this work, in exactly the same way as Ichikawa while alive. Tamura's enthusiasm was enough to make Ishizaki break his resolution and once again started with refreshed energy.

A powerful joint research effort between the Nagoya group (formerly Kyoto) and the Tokyo group thus started in 1971. After thorough discussions we decided to divide work according to our respective specialities. All purifications were done at Tokyo and the samples to be bioassayed were lyophilized and sent to Nagoya. At Nagoya, all efforts were concentrated on rearing the saturniid moth *Samia cynthia ricini* to prepare brain-removed assay pupae. Aqueous extracts of *Bombyx* brains can activate brainless dormant pupae of *Bombyx* to cause them to develop to moths after injection [3-5]. The extracts similarly activate brainless pupae of *Samia* [6]. From these facts we judged that the biological activity of the *Bombyx* PTTH was species-nonspecific in activating pupae of these two species. Because of several technical advantages we used *Samia* brainless pupae as an assay animal for PTTH purification [7]. We prepared some ten thousands of brainless pupae every year so that the assay went on without delay no matter how many samples were sent from Tokyo. This labor division was very effective in enabling the performance of a tremendous number of experiments which are impracticable if one does both purification and bioassay alone.

Another important decision was the use of silkmoth heads rather than larval or pupal brains. From the experience we had had by that time it was apparent that the crucial factor for success of this work was how we can get enough starting material. We preferred collecting enough material by a simple operation of cutting heads which needs no technical skill over the advantage of "mechanical purification" associated with dissecting out brains. The *Bombyx* adult brain had been shown to contain a high titer of PTTH activity [8]. PTTH accumulates in the brain throughout development with no clear-cut fluctuation correlated to the stages of its physiological demands [8, 9]. More than 10 millions male *Bombyx* can be supplied free of charge by sericultural companies after having been used for breeding. The Tokyo group exploited a route by which such an amount of heads could be obtained. Moths were obtained from Gunma-Kenze Sanshu Co. Ltd. at Maebashi, a 4 h drive from Tokyo, and stored temporarily in a freezing room of an ice-cream company near Gunma-Kenze. During the off-season of sericulture we employed housewives of sericultural farmers to cut the heads which were then packed with dry ice to send to Tokyo.

Naturally the question arises whether the molecule with PTTH activity is identical with that contained at the younger stages of *Bombyx*. The function of PTTH is to activate the prothoracic glands which degenerate promptly after adult ecdysis, hence, PTTH seems to have no apparent function at the adult stage. We assumed, however, that the adult PTTH must be very close to the PTTH from immature stages in its structure if not identical, and decided to purify first the PTTH from adults, the only feasible way because of the ease with which abundant material is supplied. We shall continue work on the PTTH of the immature stages after obtaining more knowledge about the PTTH nature which will make the work feasible on a far smaller scale of extraction material.

2. Isolation and Characterization of a PTTH from *Bombyx* Adult Heads

Voluminous data have accumulated since this joint research project started and the results of partial purification [10, 11] and the PTTH nature as inferred from chemical and enzymatic inactivation made on partially purified preparations [11, 12] were published. Finally in 1982, one of the PTTH multimolecular forms was purified to homogeneity [13, 14]. The final scheme of purification of this PTTH is shown in Fig. 1. This purification was contributed mainly by H. Nagasawa, and partly by A. Isogai, T. Horii, and H. Hori.

The details of these purification procedures have been described [10–14] and only the following comments will be given here. The gradient chromatography on DEAE-Sepharose CL-6B, the third step from the last of the scheme shown in Fig. 1, resolved the PTTH activity into two peaks (Fig. 2). The central portion of the peak eluting earlier was subjected to a high performance liquid chromatography (HPLC) on Partisil-10 ODS-3, the last step of purification, to obtain a sharply defined peak of PTTH which was clearly separated from other inactive peaks and judged to be homogeneous (Fig. 3). Our unpublished results show that the remaining active fractions in the DEAE-Sepharose

648,000 Male adult *Bombyx* heads (4.86 kg)
↓
Acetone powder
↓
Washing with 80% EtOH
↓
Extraction with 2% NaCl
↓
Heat treatment
↓
Ppt. with 80% satd. $(NH_4)_2SO_4$
↓
Ppt. with 50-75% acetone
↓
Ppt. with 90% satd. picric acid
↓
Ppt. with 80% acetone ("crude PTTH")
↓
Sephadex G-50 (fine)
↓
DEAE-Sepharose CL-6B (stepwise elution)
↓
SP-Sephadex C-25 (stepwise elution)
↓
Sephadex G-50 (superfine)
↓
DEAE-Sepharose CL-6B (gradient elution)
↓
Sephadex G-50 (superfine)
↓
Reversed phase HPLC (gradient elution)
↓
PTTH (50 μg)

Fig. 1. Purification scheme of PTTH from *Bombyx* adult heads [14]

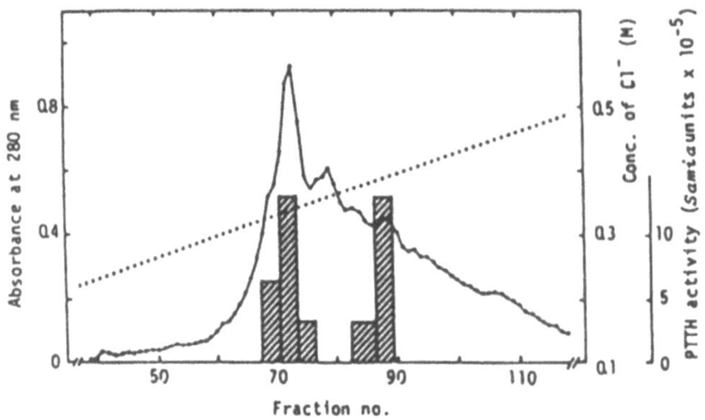

Fig. 2. Gradient chromatography (0.1–0.5 *M* NaCl) on DEAE-Sepharose CL-6B [14], the step third from the last of the purification scheme shown in Fig. 1. *Cross-hatched bars* indicate PTTH activity

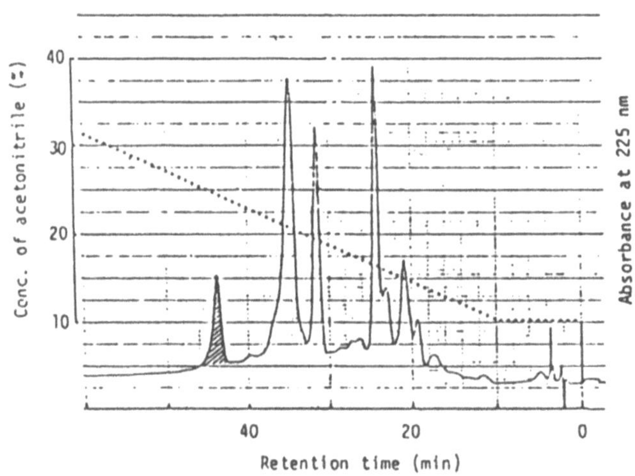

Fig. 3. Reverse-phase high-performance liquid chromatography (Partisil-10 ODS-3), the final step of PTTH purification [14]. *Cross-hatched peak* represents PTTH

gradient chromatography were resolved further into two components of PTTH by HPLC. Thus, at least three PTTHs have been resolved by these purification procedures, two of which still remain to be isolated.

Purification along this scheme yielded ca. 50 μg of the pure PTTH from a batch of ca. 6×10^5 *Bombyx* heads. The recovery of the PTTH activity was 8% and a 2×10^6-fold enrichment in the specific activity was attained.

This pure PTTH enabled adult development in a brainless *Samia* pupa by injection of 0.1 ng. Taking into account the hemolymph volume of a brainless pupa and a molecular weight of 4,400, this effective dose is calculated to give a concentration of 3×10^{-11} *M*. The PTTH content in the *Bombyx* brain as assessed by bioassay ranges from 1 to 30 *Samia* units depending on developmental stages [8], so that 0.1–3 ng of this PTTH are contained in a *Bombyx* brain.

This pure PTTH could also activate the prothoracic glands in vitro [14]. Thus, a three fold increase in ecdysone release by a prothoracic gland taken from newly ecdysed *Samia* pupae was observed when 5 pg (1×10^{-15} mol) of the pure PTTH were added to 100 μl of Grace's medium, compared to the control incubation (Fig. 4).

Amino acid composition obtained under three conditions of acid hydrolysis of the pure PTTH is shown in Table 1. The PTTH consists of 40–43 amino acid residues. The molecular weight calculated from this amino acid composition is 4,330–4,740. Only DNS-glycine and *O*-DNS-tyrosine were detected after dansyl analysis, indicating that PTTH is a single-chain peptide containing a tyrosine residue(s), whose N-terminus is glycine. Detection of a single N-terminus amino acid, together with a reasonable integral ratio of amino acid residues, strongly supports the homogeneity of this PTTH preparation.

The amino acid composition thus determined is generally consistent with our previous estimates made by chemical and enzymatic inactivation experiments [11, 12]. Inactivation of PTTH by trypsin and α-chymotrypsin is consistent with the presence of arginine, phenylalanine, and tyrosine. The presence of a disulfide bond as estimated by inactivation with mercaptoethanol or dithiothreitol is substantiated by the detection of cystines. Ineffectiveness of cyanogen bromide on PTTH activity can be attributed to the absence of methionine residue. That PTTH possesses more acidic amino acids than basic ones accounts for its acidic nature (pI 4.1–4.3). With respect to the following two points, on the other hand, disagreement was observed. Previously the N-terminus of PTTH was thought to be blocked because neither leucine aminopeptidase nor aminopeptidase M inactivated PTTH. The discrepancy of this conclusion with the presence of N-terminal glycine as demonstrated by dansyl analysis, however, may be reconciled by the facts that leucine aminopeptidase and aminopeptidase M cut N-terminal glycine extremely slowly [15] and relatively slowly [16], respectively. Alternatively it could be possible that the other PTTHs that remain to be isolated have the blocked N-terminus and the inactivation results were obscured since the PTTH preparations including all

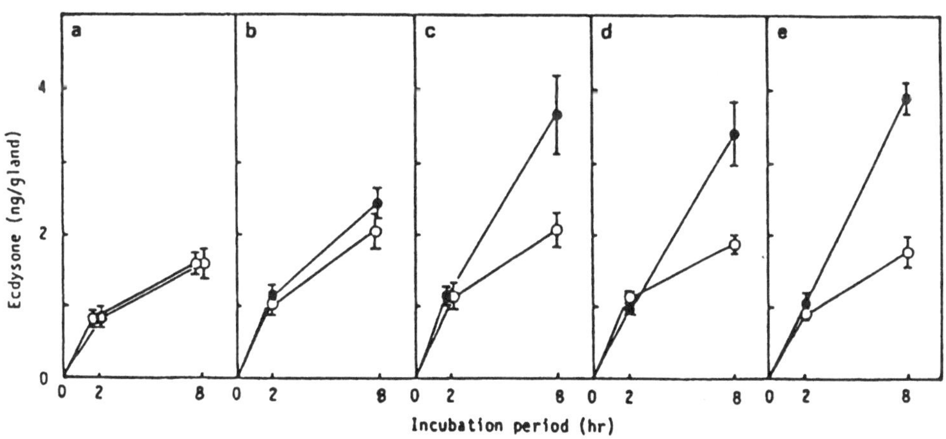

Fig. 4 a–e. Time course of ecdysone secretion into the medium by newly ecdysed *Samia* pupae prothoracic glands in vitro [14]. PTTH was added after a 2 h incubation: **a** control, **b** 0.5 pg, **c** 5 pg, **d** 50 pg, **e** 500 pg. –•– with PTTH; –○–, without PTTH. Each point is the mean ± SE (n = 4)

Table 1. Amino acid analyses of hydrolyzates of PTTH

Amino acid	Amino acid ratio[a]			Nearest integer
	A	B	C	
Asx	3.8	3.7	4.1	4
Thr	2.7	2.5	3.3	3
Ser	1.1	1.4	1.2	1
Glx	4.0	3.9	4.1	4
Pro	1.1	1.0	1.0	1
Gly	4.1	4.1	4.3	4
Ala	3.0	3.0	3.0	3
1/2 Cys_2	3.8	0.2	0.0	4
Val	3.6	4.6	3.5	4–5
Met	0.1	0.0	0.1	0
Ile	0.1	0.2	0.1	0
Leu	5.5	5.3	5.4	5–6
Tyr	0.7	0.2	1.6	1–2
Phe	1.8	1.8	1.7	2
His	0.8	0.8	0.9	1
Lys	0.1	0.1	0.2	0
Arg	2.8	2.5	3.2	3
Trp	0.0	0.0	0.0	0

Note. PTTH (6 μg) was hydrolyzed at 110° (A) for 20 h, (B) for 72 h in *5.7 N* HCl, and (C) for 20 h in *5.7 N* HCl containing 4% thioglycolic acid

[a] Moles of amino acid per 3 mol of Ala

PTTH forms were used for inactivation experiments. The other deviation from the presently shown amino acid composition is the presence of tryptophan which was inferred from inactivation of PTTH by N-bromosuccinimide(NBS) in 8 *M* urea. Now we consider that the tyrosine residue was responsible for this inactivation since NBS can react with it, although 250 times slower than with the tryptophan residue in 6 *M* urea [17].

Amino acid sequencing of this pure PTTH has not yet been completed because of its limited amount. The sequence thus far determined is H–Gly–Val–Val–Asp–.

We now add some comments on so far undescribed efforts which were made for establishing the final scheme of purification. Purification is essentially a trial and error work and what has been published is generally just the tip of an iceberg. From the very early phase of this research we have been seriously troubled with highly heterogeneous profiles of PTTH activity in almost all kinds of chromatographies, only a few of which appeared in publications [2, 10]. Figures 5–7 depict some such chromatographies to show how greatly and irregularly the distribution of activity varied depending on the materials applied and the conditions of chromatography. Figure 5 shows three chromatographies on DEAE-Sephadex A-25. A and B differed in the materials applied, but were chromatographed with about the same protocol, yielding completely different activity profiles. B and C were the same in the material, but differed in the eluting conditions, yielding again considerably different profiles. Figure 6 shows that the same mate-

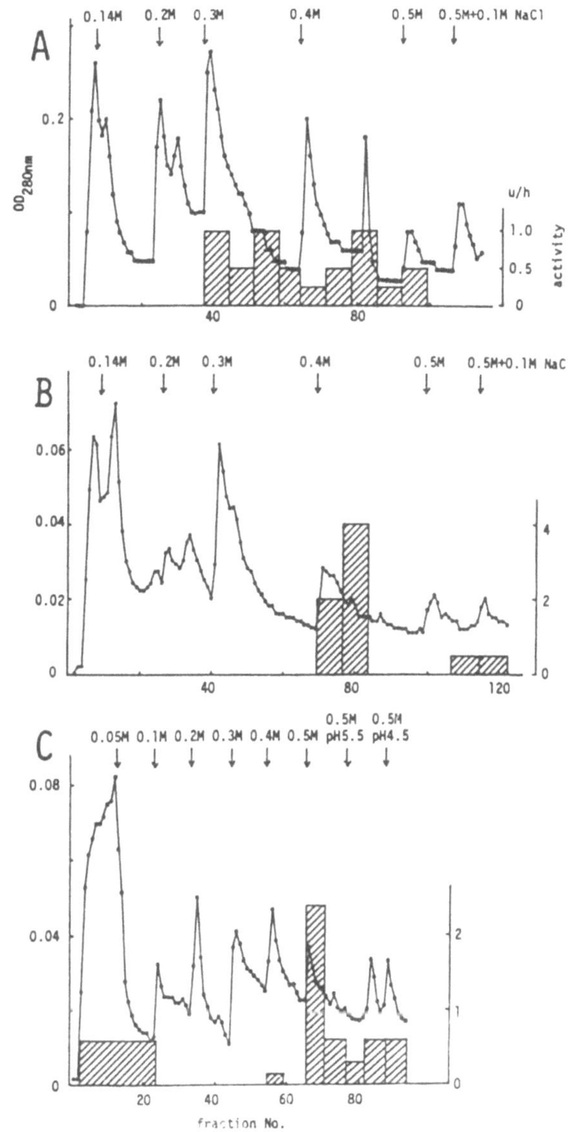

Fig. 5 A–C. DEAE-Sephadex A-25 chromatographies showing varying heterogeneous profiles of PTTH (unpublished). Materials applied are **A** "crude PTTH" [10] and **B**, **C** the active fractions after Sephadex G-50 gel filtration of "crude PTTH". Elution was made with **A**, **B** Tris–HCl buffer, pH 7.8, NaCl and **C** ammonium acetate, acetic acid/ammonium acetate buffer

rial produced considerably different activity profiles on SP-Sephadex C-25 chromatographies by changing only the concentration of the buffer used. Figure 7 shows that Sephadex G-50 gel filtration also produced varying activity profiles depending on the state of the materials applied. In B, the activity was detected in the void fractions, giving the appearance as if PTTH had a much larger molecular weight than it really has. We once estimated the erroneous values for a PTTH molecular weight ranging from 9,000 to 31,000[2], presumably owing to this kind of false chromatographic behavior of PTTH.

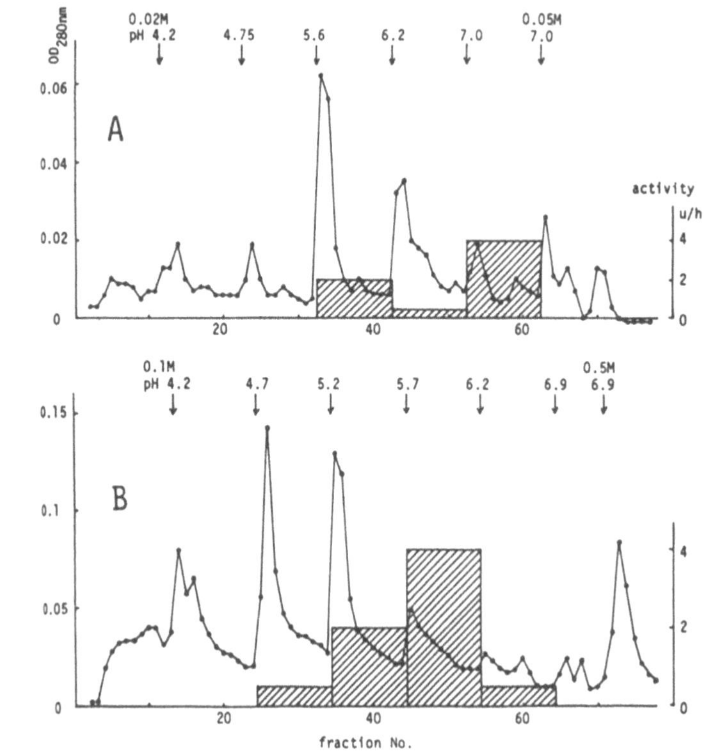

Fig. 6 A,B. SP-Sephadex C-25 chromatographies showing the variability of PTTH heterogeneous profile by a simple change in eluting buffer (unpublished). The materials applied were the same, but elution was made with **A** 0.02 *M*, **B** 0.1 *M* ammonium acetate

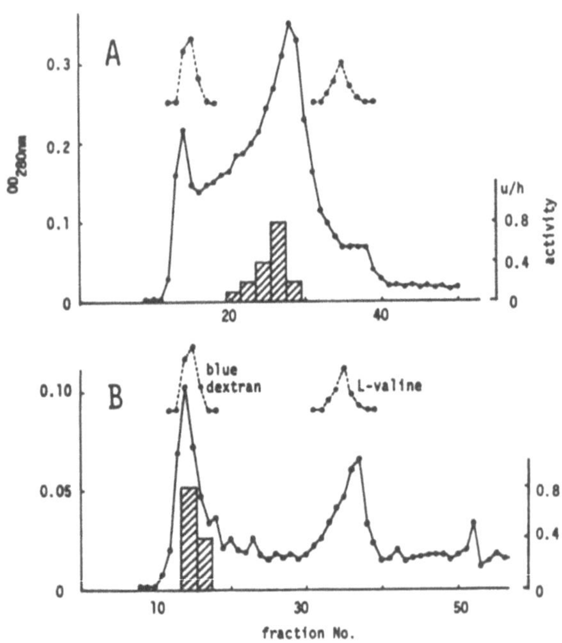

Fig. 7 A,B. Sephadex G-50 gel filtration of the two different PTTH preparations under the same eluting condition (unpublished). The material from which both samples were derived previously behaved as in **A**, but now PTTH in **B** was eluted in void fractions

As is apparent from these examples, it seemed unlikely that heterogeneity in PTTH activity observed at the early phase of purification represented the true multimolecular state of PTTH. Rather, it seemed probable that such heterogeneities were caused by nonspecific adhesion of PTTH(s) to other inactive substances. We judged, therefore, that it was not wise to work on each of the heterogeneous active fractions separately at the early stage of purification. The strategy we chose was to seek appropriate chromatographic conditions which gather the activity into a single fraction, with high yield, purification efficiency, and reproducibility, leaving the work to treat each of the heterogeneous fractions until the preparation became pure enough to manifest a true profile of PTTH heterogeneity. This was not an easy task, however, because of a strong tendency of PTTH to scatter over a wide area of chromatograms as emphasized above. Before establishing just one procedure of stepwise chromatography on DEAE-Sepharose CL-6B [11] (Fig. 8), for example, we had to try ca. 100 chromatographies and work for over 2 years. Similar efforts were made on establishing each of the other steps. How to combine these steps in due order was also critically examined. Actually we used ca. 10 million *Bombyx* heads during the course of these time-consuming preliminary experiments.

It is impressive that some invertebrate neurohormones have been isolated by extremely simple procedures, contrasting to our tedious purification scheme for PTTH. Table 2 depicts some data concerning purification of the invertebrate neurohormones so far isolated, including egg-laying hormone from *Aplysia* bag cells [18], Peptide A and B from *Aplysia* atrial gland [19], adipokinetic hormone 1 from locust corpora cardiaca [20], red pigment concentrating hormone (RPCH) from shrimp eyestalks [21], Hydra head activator from sea anemones [22], and PTTH [14]. It is surprising that the first four hormones were isolated by very simple procedures which are composed of only two to five steps. The glands producing these hormones contain exceedingly high amounts of the hormones, ranging from 0.5 to 150 μg per gland. This high content of these hormones in the glands must be the very reason for the success of purification by such simple procedures, though we should not neglect the great efforts hidden behind the publications which have been paid by these workers. RPCH, Hydra head activator, and PTTH, on the other hand, are contained in a very low amount in the extraction materials. Their purification procedures were composed of many steps; exceedingly

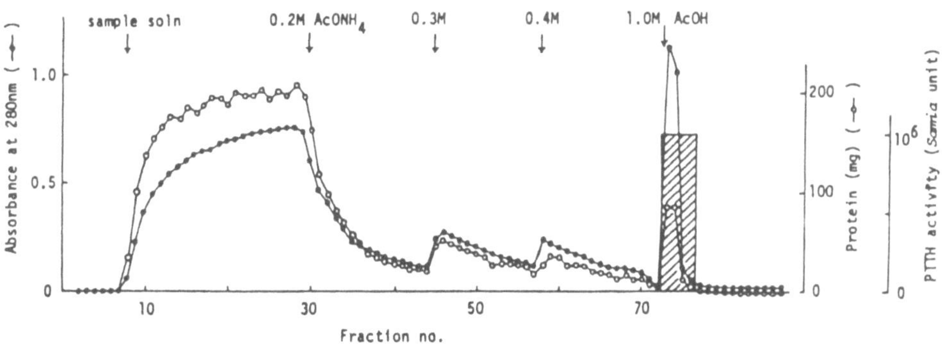

Fig. 8. DEAE-Sepharose CL-6B stepwise chromatography which has been finally fixed in PTTH purification scheme [11]

Table 2. Some data concerning purification of seven invertebrate neurohormones

Hormone	Purification steps to isolation	Purification fold	% of hormone extractable material	Hormone content/ gland or individual
Adipokinetic hormone 1 from locust corpora cardiaca [20]	2	–	–	0.5 μg/pr. of corpora cardiaca
Egg-laying hormone from *Aplysia* bag cells [18]	2	100	1	19–39 μg/pr. of bag cell clusters
Peptide A from *Aplysia* atrial glands [19]	5	–	10	> 150 μg/atrial gland
Peptide B from *Aplysia* atrial glands [19]	4	–	0.3	> 45 μg/atrial gland
Red pigment concentrating hormone from shrimp eyestalks [21]	7	3.9×10^5	0.00025	– (100 g eyestalks yielded 20 μg RPCH)
Hydra head activator from sea anemone [22]	9	2×10^9	–	0.1 pg/Hydra (200 kg sea anemone yielded 20 μg HA)
PTTH from *Bombyx* heads [14]	12	2×10^6	0.000058	0.1–3 ng/brain

high purification fold was attained; and the use of unusual adsorption of RPCH to Sephadex LH-20 and of head activator to Sephadex G-10 was a key to the successful isolation. The fact that more than half of the above hormones were isolated by simple procedures, largely favored by their exceptionally high content in the gland, may lead one to underestimate the difficulty which is associated with isolation of ordinary hormones. Many invertebrate neurohormones have been identified so far for their biological entity and partially purified, but remain to be isolated. A special trick or luck and patience must be necessary for the isolation of these hormones.

3. Presence of 4K-PTTH and 22K-PTTH in the Bombyx Brain: A Prelude to a New Area of the Functional and Structural Divergence of PTTH

From the results described above we once felt that the work of isolation and structural determination of the *Bombyx* PTTH was at the stage close to the final goal. Unexpected findings, however, were found. As already mentioned, we used *Samia* brainless pupae for assaying PTTH. Yamazaki and Kobayashi [3], on the other hand, performed purification of PTTH from *Bombyx* brains using *Bombyx* brainless pupae as an assay animal. We were then interested in comparing the sensitivity to PTTH of these *Samia* and

Bombyx assay systems. Peculiarly enough, the ratio of the PTTH units contained in a given material as determined by *Bombyx* assay to the PTTH units detected in the same material by *Samia* assay varied greatly from sample to sample [5]. As shown in Table 3, this ratio ranged from 0.73 as obtained with an extract from the brains of newly ecdysed pupae to 12.5 in the case of an extract from the brains of 2-day-old 5th-instar larvae. An even more enhanced difference was obtained when pupal brains were subjected to a successive extraction with distilled water and then with 2% NaCl and each extract was assayed. The ratios of *Bombyx* units to *Samia* units were 0.17 and 16.7 for the water extract and NaCl extract, respectively. The hypothesis then emerged that two kinds of PTTH might exist, one which specifically activates *Bombyx* pupae and the other specific to *Samia* pupae. If these two molecules are present in different proportions in various preparations, such a variation in the value of *Bombyx* units/*Samia* units would result. Meanwhile, a further striking and decisive finding was obtained when we tested the activity of the isolated PTTH on *Bombyx* brainless pupae. This pure PTTH which was active on *Samia* brainless pupae was completely inactive on *Bombyx* brainless pupae. We immediately began to reexamine the purification procedures to check if the two kinds of PTTH as postulated above had been separated from one another at a certain step.

Table 3. PTTH titers in the brain of *Bombyx mori* from various developmental stages as determined by *Bombyx* assay and *Samia* assay

Brain source	Extractant	PTTH titer / brain		*Bombyx* units / *Samia* units
		Bombyx unit	*Samia* unit	
Newly ecdysed 5th-instar larva	2% NaCl	14.2	2	7.1
2-day-old 5th-instar larva	2% NaCl	12.5	1	12.5
Newly ecdysed pupa	2% NaCl	7.3	10	0.73
Newly ecdysed adult	2% NaCl	8.6	10	0.86
Newly ecdysed pupa	Distilled water	1.7	10	0.17
	2% NaCl after water extraction	4.2	0.25	16.7

A clear result was soon obtained [23]. When a *Bombyx* brain extract was subjected to a fractional precipitation with acetone, the fraction precipitated at an acetone concentration of 40%–50% contained mainly the activity enabling adult development in *Bombyx* brainless pupae, whereas the precipitate at a 60%–80% concentration and the supernatant after addition of acetone up to 80% contained mainly the activity stimulating *Samia* pupae (Fig. 9A). Similar results were obtained with an extract from adult *Bombyx* heads (Fig. 9B). Gel filtration through Sephadex G-50 produced a complete separation of these two activities from one another (Fig. 10). The fractions eluted slightly after the void fractions exhibited the activity to cause only *Bombyx* development, whereas those eluted later were active only on *Samia*. The latter molecule, ob-

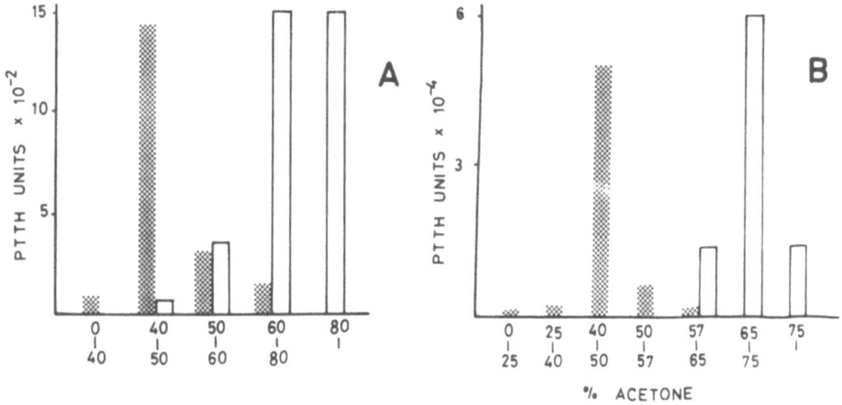

Fig. 9 A,B. Fractional precipitation with acetone of **A** pupal brain extract, **B** adult head extract of *Bombyx* [23]. *Shaded bars* and *white bars* show PTTH amounts as determined by *Bombyx* assay and *Samia* assay respectively

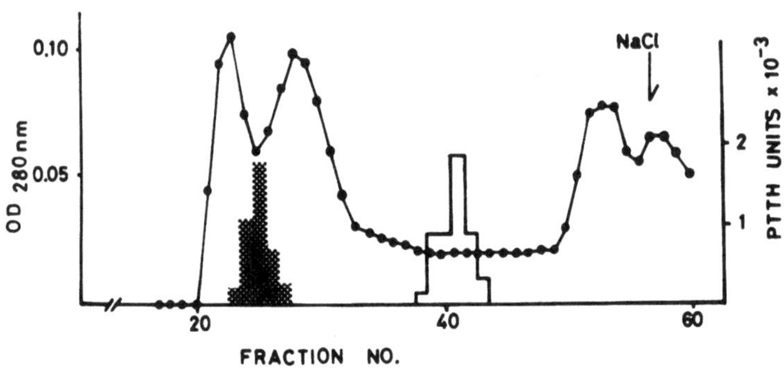

Fig. 10. Sephadex G-50 gel filtration of an extract of *Bombyx* pupal brains [23]. *Shaded bars* and *white bars* represent PTTH amounts as determined by *Bombyx* assay and *Samia* assay respectively

viously identical with the isolated PTTH as described in the previous section, has a molecular weight of ca. 4,400, hence we tentatively named this molecule(s) as 4K-PTTH (formerly called PTTH-S [23]). The other PTTH which specifically activates *Bombyx* pupae is tentatively named 22K-PTTH (formerly named PTTH-B [23]) because of its molecular weight of 22,000 as described later. The complete separation of these two PTTHs was also observed after DEAE-Sepharose CL-6B chromatography (Fig. 11). The 22K-PTTH passed through the column when applied in 0.1 *M* ammonium acetate, whereas 4K-PTTH was adsorbed on the column and eluted with 1 *M* acetic acid. Thus, these results clearly demonstrate that the chemically and biologically distinct two PTTHs are present in the *Bombyx* brain. Yamazaki and Kobayashi [3] purified PTTH from *Bombyx* brains using *Bombyx* brainless pupae for assay and reported that the PTTH they obtained was not bound to DEAE-cellulose and had a molecular weight of ca. 20,000. The discrepancies between the results obtained by Yamazaki and Kobayashi and those obtained by us, which remained unexplained for a long time, are now reconciled by the present finding which demonstrated the presence of two different PTTH

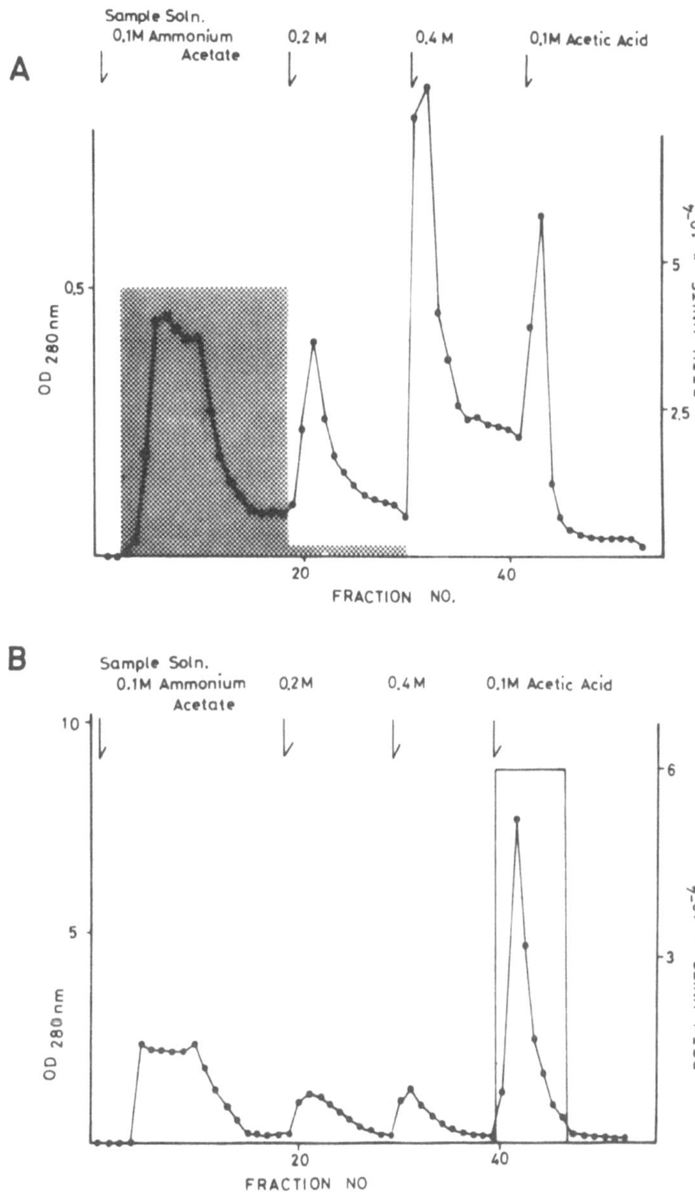

Fig. 11 A,B. DEAE-Sepharose CL-6B chromatography of 22K-PTTH(A) and 4K-PTTH(B) [23]. Respective PTTHs separated by acetone fractional precipitation of *Bombyx* adult head extract were used

molecules. Obviously, what we isolated is the 4K-PTTH, while that obtained by Yamazaki and Kobayashi is the 22K-PTTH.

We are now intensively purifying the 22K-PTTH. H. Kataoka has been mainly contributing to this purification. So far we have established a purification scheme consisting of heating, ammonium sulfate precipitation, acetone precipitation which separated

22K-PTTH from 4K-PTTH, DEAE-Sepharose CL-6B stepwise chromatography, CM-Sepharose CL-6B stepwise chromatography, Octyl-Sepharose CL-4B stepwise chromatography, Sephadex G-75 gel filtration, and CM-Sepharose CL-6B gradient chromatography. By these procedures, 160 μg of 22K-PTTH were obtained from a batch of 96,000 *Bombyx* heads with a yield of 10%. The purified sample possessed a specific activity of 1.8 ng/*Bombyx* unit and the purification fold was 9.5×10^4. We suspect that one more step of HPLC will suffice to isolate this PTTH.

4. Perspectives for Future Studies

It is strange that the *Bombyx* brain contains PTTH which is not concerned with its own development, in addition to that which activates its own prothoracic glands. Many questions immediately arise with regards to the significance of the presence of these two PTTHs. These include: (1) How are 4K-PTTH and 22K-PTTH chemically related? (2) Is 4K-PTTH secreted in *Bombyx*? If so, what is its function in *Bombyx*? (3) Is the PTTH secreted by the *Samia* brain identical with 4K-PTTH?

For the first question, purification and amino acid sequencing of these two PTTHs are the prerequisite and we hope these will soon be achieved. Current knowledge accumulated with vertebrate peptide hormones shows that the processing of a large precursor molecule to small hormone peptides and other biologically active peptides is a ubiquitous phenomenon. Though no evidence is available about PTTH on this line at the present time, it seems highly possible that the 4K-PTTH is a processing product of the 22K-PTTH.

As to the second question, our unpublished experiments demonstrated that the hemolymph of *Bombyx* pupae shortly after ecdysis contained 4K-PTTH and *Bombyx* pupal brains in vitro released the 4K-PTTH into culture medium. We have to ask then whether 4K-PTTH might function somehow in egg development. It is also worth examining if the 4K-PTTH activates the prothoracic glands at younger stages of *Bombyx*. Gilbert et al. [9] revealed the presence of 7K- and 22K-PTTH in *Manduca sexta* and proposed the "two forms–two functions hypothesis" to account for the differential appearance of these PTTHs in hemolymph during development. It is interesting to ask whether there would be any relation between their PTTH variation and our PTTH variation.

The answer to the third question must await purification and characterization of PTTH from *Samia* brains. This approach has been hampered by the fact that the *Samia* brain contains a far lower titer of PTTH activity compared to *Bombyx* brain (unpublished). We recall that an extract of the brains of *Hyalophora cecropia* failed to cause adult development in Cecropia assay pupae which could be activated by an extract of *Antherae pernyi* brains [24], indicating a significantly low content of PTTH in the Cecropia brain. A large difference among species in the amount of PTTH contained in the brain might suggest a possible variation in the site of the regulation mechanism which mediates the conversion of inactive precursor molecule to the active PTTH molecule, e.g., before or after the storage of neurosecretory granules in the PTTH-producing cell.

Finally, a comparative study on PTTH chemistry among various insect species may be fascinating. Our unpublished results show that diapausing pupae of the swallowtail *Papilio xuthus* L. could be caused to develop by injection of 4K-PTTH, but not of 22K-PTTH derived from *Bombyx* brains. Other preliminary experiments on diapausing pupae of various lepidopteran species suggest that each species uses either one of 4K- or 22K-PTTH in breaking diapause. Comparative studies on this line may offer an interesting field with respect to the evolution of PTTH-prothoracic gland interaction.

References

1. Ichikawa M, Ishizaki H (1963) Nature (London) 198:308–309
2. Ishizaki H, Ichikawa M (1967) Biol Bull 133:355–368
3. Yamazaki M, Kobayashi M (1969) J Insect Physiol 15:1981–1990
4. Nishiitsutsuji-Uwo J (1972) Botyu-Kagaku 37:93–102
5. Ishizaki H et al. (1983) Dev Growth Differ 25:595–602
6. Ichikawa M, Ishizaki H (1961) Nature (London) 191:933–934
7. Ishizaki H, Suzuki A (1980) In: Miller TA (ed) Neurohormonal techniques in insects. Springer, Berlin Heidelberg New York, p 244
8. Ishizaki H (1969) Dev Growth Differ 11:1–17
9. Gilbert LI et al. (1981) Am Zool 21:641–653
10. Suzuki A et al. (1975) Agric Biol Chem 39:2157–2162
11. Nagasawa H et al. (1979) Dev Growth Differ 21:29–38
12. Ishizaki H (1977) J Insect Physiol 23:1219–1222
13. Suzuki A et al. (1982) Agric Biol Chem 46:1107–1109
14. Nagasawa H et al. (1984) Gen Comp Endocrinol 53:143–152
15. Smith EL, Spackman DH (1955) J Biol Chem 212:271–299
16. Wachsmuth ED et al. (1966) Biochemistry 5:175–182
17. Funatsu M et al. (1964) J Am Chem Soc 86:1846–1848
18. Chiu AY et al. (1979) Proc Natl Acad Sci USA 76:6656–6660
19. Heller E et al. (1980) Proc Natl Acad Sci USA 77:2328–2332
20. Stone JV, Mordue W (1976) Nature (London) 263:207–211
21. Fernlund P, Josefsson L (1968) Biochem Biophys Acta 158:262–273
22. Schaller HC, Bodenmüller H (1981) Proc Natl Acad Sci USA 78:7000–7004
23. Ishizaki H et al. (1983) Dev Growth Differ 25:603–610
24. Williams CM (1967) In: Beament JWL, Treherne JE (ed) Insects and physiology. Oliver & Boyd, Edinburg, p 133

I.9 Neurohormonal Regulation of Molting and Metamorphosis in the Tobacco Hornworm, *Manduca sexta*

W.E. BOLLENBACHER[1], N.A. GRANGER[2], W.A. SMITH[1], and L.I. GILBERT[1]

1. Introduction

The endocrine regulation of insect postembryonic development is generally thought to involve three hormones: the prothoracicotropic hormone (PTTH), a cerebral neuropeptide; the steroid hormone 20-hydroxyecdysone; and juvenile hormone, an aliphatic sequiterpene [1]. PTTH is produced by one neurosecretory cell in each hemisphere of the insect brain [2], while 20-hydroxyecdysone is derived from the ecdysone produced by the prothoracic glands (PG) in response to PTTH [3, 4]. Juvenile hormone exists in four homologous forms (JH 0, I, II, and III) and is produced by corpora allata (CA) [5]. Although the integration of release of the hormones is critical for their roles in the regulation of insect molting and metamorphosis, the pivot of this regulatory mechanism is the periodic release of PTTH in response to environmental cues. Ecdysone is synthesized and released by the PG in response to PTTH and is then hydroxylated by peripheral tissues to 20-hydroxyecdysone, the hormone that elicits molting. JH acts in conjunction with 20-hydroxyecdysone, presumably at the level of the target tissues to determine the character of the molt [1].

In contrast to what is known of the neuroendocrine regulation of ecdysone synthesis by the PG, there is a lack of quantitative information available regarding the regulatory mechanisms involved in the control of CA activity [6]. Just recently however, evidence has been obtained that supports the concept of neuroendocrine regulation of the CA during larval-pupal development (Sect. 3). The complexity of this regulatory system may prove to be greater than that of the PTTH-PG axis, since both stimulatory neurohormones (allatotropins) and inhibitory neurohormones (allatohibins) have been implicated in the control of the CA. The demonstration that cerebral neurohormones control CA function would make the brain the control center for the regulation of molting and metamorphosis. In response to specific cues, either extrinsic or intrinsic, the brain would regulate the synthesis, transport, and release of the neurohormones acting on the PG and CA, and thus control the progression of postembryonic development. There is already a considerable body of information concerning the control of PTTH release by extrinsic and intrinsic cues [7], and it is likely that similar, if not identical, cues could control the release of neurohormones regulating the CA [6].

1 Department of Biology, The University of North Carolina at Chapel Hill, Chapel Hill, NC 27514, USA

2 Department of Anatomy, The University of North Carolina at Chapel Hill, Chapel Hill, NC 27514, USA

Biosynthesis, Metabolism and Mode of Action of Invertebrate Hormones
(ed. by J. Hoffmann and M. Porchet)

In addition to primary regulation by cerebral neurohormones, there is compelling evidence that other, so-called secondary effectors, such as temperature, photoperiod, and hormones also modulate the activity of the PG and CA [8]. Regulation of the glands is, therefore, the end result of the interactions of both primary and secondary effectors that act in a temporal, quantitative, and qualitative manner. Although the secondary regulators are clearly important, neurohormones remain the primary elements in control of glandular activity in insects, as they are in vertebrate and other comparable invertebrate systems. It is because of their predominant contribution to the regulation of insect postembryonic development that we are interested in elucidating their physiological function and chemical nature. This contribution summarizes some of our research on those neurohormones in the tobacco hornworm, *Manduca sexta*, which act as primary effectors of molting and metamorphosis, and this will cover: (1) purification of the prothoracicotropins (PTTH); (2) studies of the mechanism of PTTH action; (3) development of an experimental paradigm for probing putative neuroendocrine regulation of the CA; and (4) evidence arguing for the existence of a cerebral allatotropin and allatohibin.

2. The Prothoracicotropins

2.1 In Vitro Assay for PTTH

In the field of endocrinology, progress has often been dictated by available assays for detecting and quantifying a particular hormone. This has clearly been the case with insect neurohormones, and thus, for our investigations of the neuroendocrine regulation of insect postembryonic development, the aim was to devise assays that would facilitate the quantitative assessment of neurohormone activity. For reasons previously elaborated we chose an in vitro approach and developed an in vitro assay for PTTH [4, 9, 10]. Since this assay is specific, rapid, and yields reproducible and accurate results, it has proven invaluable in our pursuit of an understanding of the chemistry and endocrine physiology of PTTH. Larvae and pupae of the sphingid moth *Manduca sexta* were used to develop the PTTH assay because so much is known about the developmental endocrinology of this species compared to other insects and because certain structural features of the brain-PG axis facilitated the development of the in vitro assay. Basically, the assay consists of the incubation of one PG of a pair with PTTH (the experimental gland) and the contralateral gland without PTTH (the control gland). Activation of the PG by PTTH is then detected as an increased rate of ecdysone synthesis by the experimental PG. This paired gland approach is possible because rates of ecdysone synthesis by the members of a gland pair from the same animal are equivalent. Activation of a PG by PTTH is expressed as an activation ratio (A_r), which represents the quantity of ecdysone synthesized by the experimental gland divided by the amount synthesized by the control gland during the incubation period. This approach was implemented to minimize the inherent variation in the activities of the PG of different animals, since this variation dampened the detectable activation response of the experimental glands. Since originally developed, the PTTH assay has been modified [10] to enable quantification of ecdysone synthesis in medium directly by a "macro" ecdysone RIA having a standard range of 0.25–32 ng/ecdysone, in contrast to 0.025–2 ng/ecdysone with a "micro" RIA. The glands used in the assay are usually taken from day 0 pupae or day 3 last instar larvae, stage selection being dictated by the particular experimental design.

The activation of pupal PG by PTTH in vitro occurs in a characteristic dose-dependent manner over a very narrow dynamic range (two- to three-fold) of PTTH concentration. The maximum activation ratio (A_{max}) obtained in a dose response typically ranges from 3–8, while the concentration of neurohormone (ED_{50}) necessary to half maximally activate the gland (A_{50}) remains relatively constant for a given extract. Because of the constancy of the ED_{50}, it has been possible to use this value to quantify the amount of PTTH present in a test sample and between test samples [2, 4, 9–13].

Although the in vitro assay for PTTH has proven to have a clear advantage over standard in situ bioassays for studies of the endocrinology of this neurohormone, the assay is still biological in that it quantifies PTTH in terms of its stimulation of the synthesis of another hormone. Because of this, the assay has its own particular limitations [7]. However, the in vitro assay for PTTH is being used to resolve the chemistry of this neurohormone and to accomplish its purification, which should make possible the generation of antibodies. With antibodies to PTTH, a radioimmunoassay can be developed, and this should permit a comprehensive investigation of the endocrinology of PTTH. Thus, one of our main objectives at the present time is the purification of the PTTH of *Manduca*.

2.2 Purification of PTTH

The first information on the chemistry of *Manduca* PTTH was provided by Gibbs and Riddiford [14] who employed a larval in situ bioassay for PTTH and demonstrated that the neurohormone was a heat stable peptide. Subsequently, both Bollenbacher and Gilbert [13, 15] and Kingan [16] reported on their purification studies of *Manduca* PTTH. We used the in vitro *Manduca* PG assay to quantify PTTH activity and day 1 pupal brains as a source of the neurohormone, while Kingan utilized the *Manduca* larval bioassay and the heads of young fifth instar *Manduca* larvae as the source of PTTH. The results of these two studies indicated that PTTH was a large peptide and Bollenbacher and Gilbert [13] also raised the possibility that the *Manduca* PTTH existed in at least two molecular weight forms (big and small PTTH) [15]. Since these first reports we have extensively investigated the hypothesis that PTTH in *Manduca* exists in two forms, and substantial evidence has been obtained to support the existence of two peptide prothoracicotropins that may have specific and discrete neuroendocrine functions.

The procedures used in the purification of the PTTH from the day 1 pupal brain of *Manduca* were both less complex and less harsh than those used by other laboratories. A conservative approach to the extraction of PTTH was deemed especially critical for these studies since this would maximize the yield of PTTH activity and at the same time minimize the loss of possible large molecular forms of the neurohormone. In addition, the extraction method, as well as all other methods employed in the purification, had to be interfaced directly with the in vitro assay so that PTTH activity could be assessed rapidly. Homogenization of brains in Grace's culture medium: distilled water

(1:3, v/v) at 4 °C, followed by centrifugation of the extract and heat treatment of the resulting supernatant for 2 min at 100 °C, with a final centrifugation, proved to be an excellent extraction method. This procedure resulted in approximately a ten fold purification with nearly complete recovery of activity.

Gel filtration chromatography of this extract and assay of the resulting fractions with both larval and pupal PG revealed two peaks of PTTH activity. With day 1 pupal PG a major peak with an M_r of 29 kd was detected and less frequently, a second minor peak with an M_r of 6,000–7,000 was noted. Assays using larval PG revealed these same two peaks of PTTH activity, but the low molecular weight moiety was more clearly and consistently observed when assayed on larval glands. These initial molecular weight data were confirmed by the results of further size separation chromatography with different gel types and buffers at both low and high salt molarity, and by HPLC with a series of Waters I protein columns. The composite data suggested that on the basis of PTTH size and capacity to activate PG from different developmental stages, the pupal brain of *Manduca* possessed two factors with prothoracicotropic activity. For clarity, these two factors are rèferred to as neurohormones, big and small PTTH, but this terminology must be qualified, since neither form has been shown unequivocally to be a neurohormone.

To study these two moieties further, a means of quantifying the amount of each was required because: (1) the PTTH activity measured in the pupal brain represented the combined activity of both moieties; (2) big and small PTTH are present in different amounts in the brain; (3) they are purified with different efficiencies; and (4) the response of the PG to the two hormones is stage specific. Therefore, a standard PTTH unit was defined and then redefined for big and small PTTH, once they were separated. A standard PTTH unit is the amount of neurohormone present in a day 1 pupal brain of *Manduca* [10]. The unit is determined with an in vitro dose response of PG activation from which the amount of day 1 pupal brain extract (ED_{50}) needed to half maximally activate the PG (A_{50}) is obtained. For a day 1 pupal brain, the ED_{50} equals 0.06 brain equivalents and, therefore, 16.7 (1/0.06) in vitro PG assays can be conducted at the A_{50}. This amount of neurohormone is one PTTH unit. The definitions of big and small units retain as their basis the ED_{50} value derived from a dose response of PG activation in vitro obtained with a whole brain extract. Thus, one big PTTH unit is the amount of hormone supporting 16.7 in vitro assays at the A_{50} when either day 0 pupal or day 3 larval PG are used. One small PTTH unit is the amount of hormone supporting 16.7 in vitro assays at the A_{50} with day 3 larval PG. To quantify PTTH units in an unknown sample, a dose response of PG activation is generated with the sample to obtain the number of A_{50} assays that can be conducted with that sample. This number is then divided by 16.7 to yield the number of PTTH units in the sample.

Once a means of quantifying the two molecular weight forms of PTTH was established, their characterization and purification could proceed. To determine the chemical nature of the activities it was reasoned that if the big and small PTTHs were indeed neurohormones, they should be peptides; proof of this was provided by demonstrating their sensitivity to hydrolysis by protease and trypsin. Quantification of big and small PTTH units present before and after proteolytic hydrolysis revealed a nearly complete loss of activity after hydrolysis, indicating that the PTTHs were proteins.

During the molecular weight analyses it had been observed that while PG were activated by both forms in the dose-dependent manner characteristic of neurohormones, the amount of the small form necessary to activate the PG depended on the developmental stage of the glands. Thus, the ED_{50}s obtained for big PTTH with day 3 larval and day 0 pupal PG were comparable (0.075 U), but the ED_{50} for small PTTH obtained with pupal glands (3.8 U) was approx. 20 times that found with larval glands (0.15 U). This dramatic difference in gland sensitivity led to the hypothesis that small PTTH and big PTTH may act on the PG via different receptors and/or different second messengers (see below).

An additional characteristic of the response of larval PG to small PTTH was a decreased activation (less than half the A_{max}) at high concentrations of the hormone. A similar inhibitory effect was not observed in the activation of larval glands by high concentrations of big PTTH. The apparent stage-specific sensitivities of the PG to the PTTHs in vitro may have considerable significance in vivo, since these results suggest that there are changes in the sensitivity of the PG to the PTTHs during postembryonic development, and by implication that there are stage-specific physiological roles for the small and big PTTHs.

Further chemical characterizations of big and small PTTH revealed differences in their isoelectric points as well. Big PTTH had a pI of 5.4 and small PTTH had a pI of 5.2. Minor activity at a second pI was found for each tactor, at 5.07 for big PTTH and at 3.7 for small PTTH. The activities eluted from the isoelectric focusing (IEF) gels were shown to be big and small PTTH by their kinetics of activation of larval and pupal PG, which were essentially identical to those obtained with the postgel filtration fractions, and which again demonstrated the unique properties of small PTTH. The difference in their observed isoelectric points provided further support for the concept that two PTTHs exist in the *Manduca* brain.

While the results to this point strongly suggested that the moieties being monitored with the in vitro assay were prothoracicotropins, proof that they were actually neurohormones required a demonstration of their function in situ. This was accomplished by testing their biological activities in the *Manduca* larval bioassay for PTTH [14]. In this assay, injection of one unit of big PTTH elicited molting in 67% of the test animals, essentially equivalent to the 70% response obtained with a standard PTTH unit (1 brain equivalent). By contrast, the injection of 1-5 units of small PTTH elicited molting in 10%-20% of the test animals, a number only marginally supporting the conclusion that this moiety functioned in situ as a PTTH. However, the *Manduca* PTTH bioassay assesses only a molting response, and since the sensitivity of the PG to small PTTH appears to change during postembryonic development, it may not be possible to assess the true biological activity of the small PTTH with this bioassay. Further tests are required using a different type of bioassay, e.g., a commitment response, before the physiological function of small PTTH can be explained.

Although the studies conducted thus far on big and small PTTH indicate that they are neurohormones, final proof requires that both forms be isolated from the hemolymph of larvae at times during postembryonic development when PTTH release is known to occur (a head critical period). These studies have been initiated, and preliminary results indicate that both hormones are released into the hemolymph. Big PTTH is found in the hemolymph during the head critical period (HCP) of the fourth larval instar, while small PTTH may be released during the first HCP of the fifth larval instar (time of pupal commitment). The criteria used to identify the hemolymph neurohormones were identical to those used to analyze and characterize the brain PTTHs. Due to the observations that relatively small amounts of these neurohormones are present in the hemolymph and that the in vitro assay has certain sensitivity limiations, the release of small PTTH into the hemolymph during the fourth instar and big PTTH during the fifth instar remains a possibility. Definitive studies await the development of more sophisticated methods (RIAs) for detecting these neurohormones. Nevertheless, the investigation of the hemolymph PTTHs argue for the existence of two prothoracicotropins, and on the basis of present information, poses the intriguing possibility that they have different roles in regulating the postembryonic development of *Manduca*, big PTTH initiating events leading to a larval-larval molt, and small PTTH initiating events resulting in pupal commitment.

The generation of antibodies for the eventual development of new methods for monitoring and identifying PTTH requires the purification of these hormones. Big PTTH has been purified by a three step procedure that includes gel filtration, DEAE ion exchange chromatography, and isoelectric focusing. The only criterion of purity established thus far is the demonstration of a single band with an M_r of 28,500 by SDS gel electrophoresis and silver staining. Currently, this material is being used to generate monoclonal and polyclonal antibodies. These antibodies can then be used to develop an RIA, as well as affinity columns, with which further purification and immunocytochemical analyses can be conducted. Purification of the small PTTH has been initiated. Because of its greater stability in comparison to big PTTH, its purification principally involves HPLC. Once pure small PTTH has been obtained, antibodies will also be generated to this form. We will then be in a position to probe the synthesis and processing of these neurohormones, and perhaps determine whether they are derived from a common precursor produced in one NSC, or if they are the products of different genes and different NSC.

2.3 Mechanism of Action of PTTH

By analogy with the action of vertebrate steroidogenic hormones, such as adrenocorticotropin and luteinizing hormone [17], the mechanism by which PTTH activates the PG should involve a cyclic nucleotide second messenger. Thus, our working hypothesis for investigating the mechanism of PTTH action has been: (1) that the neurohormone(s) binds to a PG membrane receptor resulting in the activation of adenylate cyclase and a subsequent increase in the intracellular concentration of (cAMP); and (2) that cAMP subsequently binds to the regulatory subunit of a cAMP-dependent protein kinase, the activation of which leads to protein phosphorylation, increased enzyme activity, and enhanced steroidogenesis. If the PTTH system is indeed similar to comparable vertebrate systems, then mobilization of intracellular steroid substrate (e.g., cholesterol), the putative rate-limiting step in ecdysone synthesis, would result from PG activation by PTTH [18].

In initial studies on the mechanism by which PTTH activates the PG of *Manduca*, an in vitro approach was taken in which two inhibitors of phosphodiesterase activity, aminophylline, and the more potent methylisobutylxanthine (MIX), were shown to stimulate ecdysone synthesis [19, 20]. The cAMP analog dibutyryl cAMP (10 m*M*) appeared to enhance steroidogenesis by the PG, but its effect was not significantly different from controls. Cyclic AMP alone (10 m*M*) had no effect on the activity of the glands, a result consistent with the fact that cell membranes are relatively impermeable to cAMP. Thus, while the analog experiments were equivocal, it appeared from the inhibitor studies that a cyclic nucleotide(s) was involved in PTTH action.

In our earlier study of *Manduca* PG [20], an attempt was made to correlate developmental changes in steroidogenesis with alterations in the intracellular level of cAMP. Significant and dramatic developmental changes were observed in both PG activity and cAMP levels, but an expected close temporal correlation was not found. Instead, increases in cAMP levels preceded PG activity by as much as 1 day. One explanation for this lack of correlation was that the animals used were of mixed gates and, therefore, not developmentally synchronous. However, when these studies were repeated recently using synchronously developing (Gate II) larvae, similar results were obtained [21, 22]. A dramatic increase in cAMP levels in the PG was observed between days 2 and 3 of the fifth instar with a maximum 30-fold increase occurring on day 4. Ecdysone synthesis by the PG did not increase until day 4. It is puzzling that this rise in cAMP levels preceded the release of PTTH and onset of steroid secretion by 12–24 h, since in most analogous vertebrate systems (ovaries, adrenals), steroidogenesis is initiated within minutes of an increase in cAMP.

Between days 6 and 8 of the fifth instar, a second period of ecdysone secretion by the PG occurs, but no rise in cAMP was found at this time. This observation led to the hypothesis of a cAMP-dependent and cAMP-independent mechanism of PTTH activation [20]. We have recently readdressed this possibility, focusing on the mechanism of PTTH action on day 3 last instar larval and day 0 pupal PGs [22]. Confirming the earlier report, the phosphodiesterase inhibitor MIX activated both larval and pupal PG two- to six-fold in vitro at doses from 10^{-5} to 10^{-3} *M*. In contrast to the earlier study, however, dibutyryl cAMP (5 m*M*) significantly activated larval and pupal PG by three- to four-fold, and the 8-bromo analog of cAMP (10 m*M*) elicited a two- to four-fold stimulation of ecdysone synthesis by glands from both developmental stages.

Since big and small PTTH preparations were available (Sect. 2.2) we decided to test their individual abilities to elicit increases in the cAMP levels. In a preliminary in vitro study, day 3 fifth instar PG responded to mixed PTTHs in a crude brain extract by producing large amounts of cAMP (10- to 30-fold increase over controls). Day 0 pupal PG exhibited little or no accumulation of intracellular cAMP in response to the brain extract, despite the acquisition of a similar two-fold increase in steroid secretion. These results could support the hypothesis that ecdysone synthesis by the PG is stimulated by both cAMP-dependent (day 3 larval PG) and cAMP-independent (day 0 pupal PG) mechanisms, and that the hormones present in the crude extract (big and small PTTH) display different mechanisms of action. However, it is also possible that both larval and pupal PG are activated by the PTTHs via cAMP, but that stage specific differences exist with regard to the amount of cAMP produced for a given level of activation. Data from similar studies using highly purified big PTTH favor the latter explanation. Day 3 larval PG incubated for 20 min with big PTTH exhibited a 50-fold increase in intracellular cAMP. Under identical conditions, pupal glands exhibited only a 50% increase. However, in the presence of a phosphodiesterase inhibitor, cAMP levels in pupal glands increased 16-fold. These results indicate the presence of a hormone-sensitive adenylate cyclase in both larval and pupal PG. Big PTTH is therefore likely to act through a cAMP second messenger at both stages. However, there appears to be a developmental difference in the degree to which cAMP accumulates following PTTH stimulation in pupal vs larval PG, as well as a difference in the amount of the cyclic nucleotide needed to elicit steroidogenesis. The effects of small PTTH on the cAMP levels are currently being investigated, as are other components of the system, e.g., adenylate cyclase and protein kinases, both enzyme classes being present at relatively high specific activities in the PG of *Manduca*.

3. Neuroendocrine Regulation of the Corpora Allata

It has long been postulated that specific neural factors from the brain modulate the production of JH by the CA [23], particularly during diapause and reproduction [6, 24]. Since the CA receive direct innervation from the brain by axons of both neurosecretory [25, 26] and ordinary neurons [27, 28], it is conceivable that the CA are controlled by nervous regulation, by neurohormones, or by a combination of both. In the case of neuroendocrine control, the neurohormones involved would appear to be delivered via a paracrine or semiprivate (directed) pathway. However, since a neurohemal organ, the corpus cardiacum (CC), is interposed between the brain and the CA, it is equally plausible that the putative neurohormones would be delivered via the hemolymph (neurohumorally).

Because of these several possible regulatory arrangements, the interpretation of studies on the control of the glands has become problematic. For example, severence of the nerves to the CA in situ can result in either an increase or decrease in gland activity [6] depending on the insect species, but it is not possible to determine whether strictly nervous control or paracrine neuroendocrine control has been interrupted. If such technical limitations to the study of CA regulation are to be overcome, alternative methods must be devised that will permit the direct assessment of only one of the various possible types of control mechanisms.

In Lepidoptera, the regulation of CA activity appears to involve additional levels of complexity, since during larval-pupal development the glands synthesize at least three JH homologs (JH I, II, and III) and perhaps four [29]. Preliminary results suggest that the synthesis of each homolog could be regulated independently and simultaneously [30]. Furthermore, neuroendocrine control may involve both allatotropins (stimulating factors) and allatohibins (inhibiting factors) [23, 30-33], each acting on the CA via an endocrine or paracrine pathway. Because of the prospective complexity of this regulatory scheme, elucidation of the control of the CA in lepidopterans will require new approaches for the direct demonstration of individual mechanisms. An in vitro approach has recently been taken by both Ferenz and Diehl [34] and our group [35], our system being modeled on that developed for our studies on PTTH (see Sect. 2). This approach has already led to the tentative identification of two cerebral neuropeptide regulators of CA activity, and this section summarizes those studies.

3.1 JH III Allatotropin

3.1.1 In Vitro CA Assay

The successful establishment of an in vitro CA assay for putative allatotropins and allatohibins required: (1) selection of an appropriate test animal *Manduca sexta* (see Sect. 2.1); (2) use of incubation conditions for the CA that would support gland activity in vitro and reflect CA activity in situ [36]; (3) the demonstration that the in vitro assay was sufficiently versatile to permit considerable manipulation without affecting its ability to demonstrate activation; and (4) confirmation that the assay was accurate and reproducible and, if possible, would permit the simultaneous assessment of the synthesis of the different JH homologs. The assay that essentially fulfills all of these criteria involves incubation of CA in 0.21 ml of Marks 19AB or Grace's medium plus 0.1% BSA, both at pH 6.5, in polyethylene glycol coated glass vessels at 25 °C, in the dark, in air, and high humidity without agitation [37]. The synthesis of two homologs, JH I and III, in these incubations are determined simultaneously by RIAs specific for each [38, 39]. These RIA methods have proven to be accurate since their cross-reactivities have been precisely defined. In addition, they are rapid and easy to perform, with the advantage of enabling the concurrent monitoring of the synthesis of more than one homolog in replicate aliquots of CA incubation medium [40].

Since the basal synthesis of JH I and III by CA can vary considerably between animals of the same stage and between experiments, a paired gland protocol was adopted (see Sect. 2.1). This was possible since members of a pair of CA exhibit comparable basal rates of JH I and III synthesis in vitro [38, 41]. Levels of JH synthesis by *Manduca* larval CA are generally low, so in order to obtain detectable amounts of JH synthesis within a relatively brief incubation period, a pair of right glands was used as experimentals and the contralateral left glands as controls. Activation of JH synthesis by the experimental CA in response to putative allatotropins in vitro is expressed as an A_r and is defined as the experimental synthesis divided by the control synthesis. Thus, an in vitro assay was available with which the existence of neuroendocrine regulators of the CA of *Manduca* could be probed.

3.1.2 Evidence for the Existence of a JH III Allatotropin

The successful implementation of an in vitro protocol for investigating neurohormonal regulation of CA required the selection of appropriate glands to be activated or inhibited and appropriate brains to serve as a source of regulatory factors. The choice of glands and brains was in part empirically determined from the results of a developmental study of CA activity during larval, larval-pupal, and early pupal development, in which the biosynthetic activity of isolated glands was compared to that of glands incubated as a complex with the brain-CC [36]. This comparison plus the results of brain-CA coincubation experiments [30] identified stages of development at which the brain appeared to specifically stimulate the synthesis of JH III and inhibit the synthesis of JH I, thus suggesting the possible existence of an allatotropin for JH III and an allatohibin for JH I. Since other brain factors have not yet been identified, the subsequent discussion will concern only these two moieties.

High levels of JH III allatotropic activity were found in brains of day 0 pupae [30], and day 7 last (fifth) instar larval glands (V/7 CA) responded with the highest levels of JH III synthesis [41]. Thus, these CA and brains were used to investigate the apparent allatotropic factor (ATF) for this JH homolog.

If a JH III allatotropin existed in the day 0 pupal brain, the CA should be activated by the brain to synthesize JH III in a dose-dependent manner. An initial attempt to demonstrate this activation employed a whole brain plus CA co-incubation protocol, and while a good dose response of JH III synthesis was not obtained, an A_{max} of 3–4 was frequently observed. To overcome the apparently inconsistent release of the ATF from whole brains in vitro, an attempt was made to demonstrate a dose response of activation with an extract of the pupal brain. The method of extracting the brains proved critical to the success of this study because the procedure not only had to minimize degradation of the ATF, but also had to efficiently extract the factor while interfacing directly with the in vitro CA assay. The extraction method ultimately adopted involved homogenization in Grace's medium at 4 °C followed by centrifugation (8,000 × g) of the homogenate to remove large particulate matter. The resulting supernatant was then centrifuged at 143,000 × g to obtain a final postmicrosomal supernatant which was termed the "brain extract" and was the source of the JH III ATF.

Activation of an endocrine gland by a peptide neurohormone is usually an acute response requiring only a short period of exposure to a small amount of the neurohormone. Thus, the duration of the in vitro CA assay for the JH III ATF was critical for its successful implementation. The assay time had to be sufficiently long to permit observable activation, but not so long that the activated gland became substrate depleted, resulting in a decreased rate of synthesis and an apparently low A_r. From several time course analyses of JH III synthesis by CA in vitro, as well as preliminary time courses of ATF-activated synthesis, an incubation time of 2 h was selected as optimal for the assays.

A dose response of JH III synthesis by V/7 CA, typical of hormonal activation, was generated with the day 0 pupal brain extract, indicating the presence of a JH III ATF. JH I synthesis did not change, suggesting that the ATF was homolog specific. The A_{max} for JH III synthesis was approx. 3, the same as that achieved with whole brains; and the A_{50} for the dose response was 0.05 brain equivalents, which suggested either that large amounts of the ATF were present in the brain or that the assay was highly sensitive. This dose response analysis thus demonstrated directly for the first time the specific activation of synthesis of a single JH homolog by a cerebral factor with the properties of a neurohormone.

3.1.3 Chemistry of JH III ATF

Characterization and purification of the JH III ATF first required an assessment of its stability and a determination of its chemical nature. This moiety proved to be stable with repeated freezing and thawing for at least 1 month, but activity was lost after heating at 100 °C for only 15 s. The ATF was identified as a protein by precipitation with ammonium sulfate (40%–60%) and its sensitivity to proteolysis by trypsin, which, under the assay conditions employed, destroyed more than 95% of the ATF activity.

Gel filtration of a day 0 pupal brain extract yielded an M_r for the ATF of approx. 40,000. The ATF has been rechromatographed with several different gel and buffer systems (including high salt to minimize aggregate formation) and the same M_r was obtained under all conditions. The large size of the ATF was unexpected, but fits the data on its heat lability and its precipitation by ammonium sulfate at only 40% saturation. IEF of the post-gel filtration ATF revealed a single pI of 5.5. The ATF is currently being purified with the above methods plus ion exchange chromatography, with the intention of generating antibodies to the purified factor.

3.1.4 Tissue Distribution of ATF Activity

Although the above information suggested that the JH III ATF acted as a neurohormone, the possibility still existed that its effect on the CA in vitro was not specific and represented an artifact of the system. This was tested by assessing activation of the CA by extracts of tissues other than the brain with the dose response protocol. Extracts of brain, CC, subesophageal plus prothoracic ganglia, the first three abdominal ganglia, and muscle were tested for their ability to activate the CA in vitro. Of these, only neural, brain, CC, and abdominal ganglia tissue exhibited ATF activity. The CC and abdominal ganglia had 1.4% and 23%, respectively, of the activity in a single brain. The activity in the brain was further localized to the lateral regions of the protocerebrum, which suggested that the lateral neurosecretory cells might be the source of the ATF. The exact cerebral location of the JH III ATF is currently being probed by methods similar to those used in our investigations identifying the prothoracicotropes [2]. Titers of the JH III ATF in the brain during postembryonic development are also being determined to see if fluctuations occur, and if so, whether they correlate with the critical times at which JH affects development. The establishment of such a functional correlation would argue for a physiological role for the factor.

In summary, the above data strongly support the existence of an ATF for JH III. However, they can provide no information about the role of this factor in the regulation of CA activity in situ. The problem in assigning a physiological role to an allatotropin for JH III in the control of *Manduca* larval-pupal development is that a discrete endocrinological function for JH III, or a metabolite of the hormone, has not been defined. If this is accomplished, the argument that the JH III ATF has a physiological function in situ will be bolstered substantially. Even if the ATF should ultimately be defined as a neurohormone, the question of the route by which the moiety reaches the CA to exert its effect cannot be determined with the present experimental design. Thus, the use of antibodies to the ATF provides the only practical approach toward determination of the neurohumoral or paracrine mode of delivery to the CA.

3.2 JH I Allatohibin

Evidence for the inhibition of CA activity in *Manduca* indicates that it is exerted by both neuroendocrine [32] and nervous [42] mechanisms. The data that exist to support the involvement of inhibitory neurohormones (allatohibins) in CA regulation are fragmentary and have been obtained for the most part by indirect methods. Our developmental study of the JH biosynthetic activity of isolated CA and brain-CC complexes suggested an inhibition of the CA by the brain, and this apparent inhibition correlated temporally with the fluctuations in the titer of JH in the hemolymph during the last larval instar [36]. In toto, these studies on *Manduca* indicate that inhibitory neuroendocrine mechanisms govern CA activity, but provide no hard information about their nature and possible role(s) in regulating insect development.

As with the work on PTTH and the JH III ATF, an in vitro assay method was developed to investigate the potential inhibitory regulation of the CA by an allatohibin(s). The demonstration of inhibition with an in vitro approach is conceptually different from a demonstration of activation because a decrease in gland activity in vitro could occur either in response to a neurohormone or as a consequence of a nonspecific, toxic effects on the CA by the sample (brain extract) being assayed. Furthermore, the dramatic effects of subtle differences in incubation conditions on CA biosynthetic activity [37] underscores the need for consistency in the routine use of an in vitro system for investigating the existence of allatohibins. Thus, any in vitro system for investigating these moieties has associated with it a variety of unique problems which must ultimately be addressed before it can be concluded that inhibition in vitro represents a specific response to a neurohormone. For example, one way of demonstrating that inhibition is not the result of a toxic effect would be to show that the inhibition obtained in vitro is reversible; this can be accomplished by comparing CA activity in the presence of an AHF to its activity following removal of the factor.

These potential problems with an in vitro investigation of CA inhibition necessitated an initially conservative approach to the problem. From the results of our developmental study of CA activity [36], day 0 larval CA appeared to be well-suited for the assay of putative allata inhibiting factors (AHF), since they were biosynthetically active and should demonstrate a maximal response to inhibition. The brains selected as the best possible source of an AHF were those from day 4 last instar larvae, an empirical choice based in part on the results of the developmental study.

Using the incubation protocol previously described, we initially co-incubated whole brains with CA and observed an apparent inhibition of JH I synthesis. Synthesis of JH III was not affected. Thus, it appeared that a JH I AHF was present in the day 4 brain; the fact that its effect was homolog-specific and was observed in whole organ co-incubations argued against the effect being due to a nonspecific and/or toxic factor.

Next, a dose response of CA inhibition was generated with a brain extract prepared as described (see Sect. 3.1.1). Incubation conditions were the same as those for the ATF assay, except that the incubation time of the assay was extended to 6 h to maximize synthesis by the control glands. As in the co-incubations, a specific inhibition of JH I synthesis by the CA was observed. The negative effect of the brain extract on gland activity demonstrated by the dose response of inhibition is expressed as an inhibition ratio (I_r). The dose (ED_{50}) of the JH I AHF required to half-maximally inhibit (I_{50}) the glands was 0.75 brain equivalents, a value substantially greater than the ED_{50} for the JH III ATF. The greater number of brain equivalents of JH I AHF needed to elicit the inhibition could reflect any one or all of a number of factors: (1) lower levels of this factor in the day 4 brain in comparison to levels of the JH III ATF in day 0 pupal brains; (2) different sensitivities of the test glands to the respective factors; (3) greater lability of the AHF during the extraction process (recently confirmed); and (4) suboptimal assay conditions for the demonstration of inhibition. Irrespective of the reason(s) for the large ED_{50} observed for the JH I AHF, further studies of assay conditions affecting the demonstration of AHF activity are required. Once conditions are optimized, the reversibility of the inhibition must then be demonstrated before it can be concluded that the observed effect represents the presence of an AHF for JH I in the brain of *Manduca*.

3.3 Summary

In vitro systems have been developed for investigating the possible contribution of neurohormones to the regulation of CA activity in *Manduca*. These systems provide the opportunity to test directly the effects of putative allatotropins and allatohibins on CA activity in both quantitative and qualitative terms. On the basis of initial studies using this in vitro approach, it appears that the CA of *Manduca* are controlled by cerebral neurohormones that either activate or inhibit the synthesis of individual JH homologs. These initial studies have opened a wide range of possible investigative paths which include the purification of the factors, identification of their site of synthesis, and the elucidation of their mechanism of action. Lastly, if these factors are indeed neurohormones, then other JH homolog-specific allatotropins and allatohibins may exist. The establishment of appropriate assays with which to investigate their existence and characteristics is currently being explored.

Acknowledgments. The authors thank Ms. Janet Hightower and Ms. Sheila King for their assistance in the preparation of this manuscript. This research was supported by grants from the National Institutes of Health (W.E. Bollenbacher, NS-18791 and AM-31642; L.I. Gilbert, AM-30118, and N.A. Granger, NS-18114) and the National Science Foundation (L.I. Gilbert, PCM-8116931). W. Smith is a postdoctoral fellow of NIH training grant NS-07166.

References

1. Granger NA, Bollenbacher WE (1981) In: Gilbert LI, Frieden E (ed) Metamorphosis: A problem in developmental biology, 2nd edn. Plenum, New York, p 105
2. Agui N, Granger NA, Gilbert LI, Bollenbacher WE (1979) Proc Natl Acad Sci USA 76:5694–5698
3. Bollenbacher WE, Smith SL, Wielgus JJ, Gilbert LI (1977) Nature (London) 268:660–663
4. Bollenbacher WE, Agui N, Granger NA, Gilbert LI (1979) Proc Natl Acad Sci USA 76:5148–5152
5. Gilbert LI, Goodman W (1981) In: Gilbert LI, Frieden E (eds) Metamorphosis: A problem in developmental biology, 2nd edn. Plenum, New York, p 139
6. deKort CAD, Granger NA (1981) Annu Rev Entomol 26:1–28
7. Bollenbacher WE, Granger NA (1984) In: Kerkut GA, Gilbert LI (eds) Comprehensive insect physiology, biochemistry, and pharmacology, vol VII. Endocrinology I. Pergamon, Oxford, in press
8. Steel CGH, Davey K (1984) In: Kerkut GA, Gilbert LI (eds) Comprehensive insect physiology, biochemistry and pharmacology, vol VIII. Endocrinology II. Pergamon, Oxford, in press
9. Bollenbacher WE, Agui N, Granger NA, Gilbert LI (1980) In: Kurstak E, Maramorosch K, Dubendorfer A (eds) Invertebrate systems in vitro. Elsevier/North Holland, Amsterdam, p 253
10. Bollenbacher WE, O'Brien MA, Katahira EJ, Gilbert LI (1983) Mol Cell Endocrinol 32:27–46
11. Agui N, Bollenbacher WE, Gilbert LI (1983) Experientia 39:984–988

12. Agui N, Bollenbacher WE, Granger NA, Gilbert LI (1980) Nature (London) 285:669–670
13. Bollenbacher WE, Gilbert LI (1981) In: Farner DS, Lederis K (eds) Neurosecretion: Molecules, cells, systems. Plenum, New York, p 361
14. Gibbs D, Riddiford LM (1977) J Exp Biol 66:255–266
15. Gilbert LI, Bollenbacher WE, Agui N, Granger NA, Sedlak BJ, Gibbs D, Buys CM (1981) Am Zool 21:641–653
16. Kingan TG (1981) Life Sci 28:2585–2594
17. Schimmer BP (1980) Adv Cyclic Nucleotide Res 13:181–214
18. Hall PF (1982) In: Conn PM (ed) Cellular regulation of secretion and release. Academic Press, London New York, p 195
19. Vedeckis W, Bollenbacher WE, Gilbert LI (1974) Zool Jahrb Physiol 78:440–448
20. Vedeckis W, Bollenbacher WE, Gilbert LI (1976) Mol Cell Endocrinol 5:81–88
21. Smith WA, Gilbert LI, Bollenbacher WE (1983) Am Zool 23:885
22. Smith WA, Gilbert LI, Bollenbacher WE (1984) Mol Cell Endocrinol (in press)
23. Scharrer B (1952) Biol Bull 102:261–272
24. Gilbert LI, Bollenbacher WE, Granger NA (1980) Annu Rev Physiol 42:493–510
25. Sedlak BJ (1981) Gen Comp Endocrinol 44:207–218
26. Arvy L, Gabe M (1952) Ann Sci Nat Zool 14:345–374
27. Cazal P (1948) Bull Biol Fr Belg Suppl 32:1–227
28. Scharrer B (1964) Z Zellforsch 62:125–148
29. Bergot BJ, Jamieson GC, Ratcliffe MR, Schooley DA (1980) Science 210:336–338
30. Granger NA, Bollenbacher WE, Gilbert LI (1981) In: Bhaskaran G, Friedman S, Rodriguez JG (eds) Current topics in insect endocrinology and nutrition. Plenum, New York, p 83
31. Sehnal F, Granger NA (1975) Biol Bull 148:106–116
32. Williams CM (1976) In: Gilbert LI (ed) The juvenile hormones. Plenum, New York, p 1
33. Bhaskaran G (1981) In: Bhaskaran G, Friedman S, Rodriguez JG (eds) Current topics in insect endocrinology and nutrition. Plenum, New York, p 53
34. Ferenz JH, Diehl I (1983) Z Naturforschung 38:856–858
35. Granger NA, Mitchell LJ, Bollenbacher WE (1984) In: Borkovec AB, Kelly TJ (eds) Insect neurochemistry and neurophysiology. Plenum, New York, pp 365–367
36. Granger NA, Niemiec SM, Gilbert LI, Bollenbacher WE (1982) Moll Cell Endocrinol 28:587–604
37. Granger NA, Mitchell LJ, Niemiec SM (1984) In: Kurstak E (ed) Invertebrate systems in vitro. Elsevier/North Holland, Amsterdam, in press
38. Granger NA, Bollenbacher WE, Vince R, Gilbert LI, Baehr JC, Dray F (1979) Mol Cell Endocrinol 16:1–17
39. Granger NA, Niemiec SM, Gilbert LI, Bollenbacher WE (1982) J Insect Physiol 28:385–391
40. Granger NA, Goodman WE (1983) Insect Biochem 13:333–340
41. Granger NA (unpublished observation)
42. Bhaskaran G, Jones G, Jones D (1980) Proc Natl Acad Sci USA 77:4407–4411

I.10 Isolation of an Allatotropic Factor in *Locusta migratoria* and Its Effect on Corpus Allatum Activity in Vitro

H.J. FERENZ[1]

1. Introduction

Insect corpora allata (CA) synthesize juvenile hormone (JH). This hormone regulates development and reproduction (for reviews see [1, 2]). In *Orthoptera,* such as *Locusta migratoria* ovarian growth and vitellogenesis are well known to be regulated by JH. The oocytes grow in synchrony and several batches of eggs are formed. These gonotrophic cycles are closely correlated with cyclic changes of CA activity [3, 4]. In young females JH biosynthesis increases through previtellogenesis and reaches a maximum during vitellogenesis. At the end of an oogenic cycle JH synthesis declines to low previtellogenic levels until the onset of the next oogenic cycle. Thus, in locusts the regulation of biologically active JH occurs rather at the level of synthesis than at the level of transport, sequestration, catabolism, or excretion [5].

Which physiological factors contribute to the rise and the decrease in rate of JH synthesis? There is general agreement that the function of the CA is coordinated by the brain, including nerve impulses and/or neurohormones, which reach the CA by direct innervation or via hemolymph. There are several possibilities how JH biosynthesis might be regulated: (a) regulation by activity or concentration of critical enzymes involved in JH synthesis; (b) regulation of release of stored JH – in general JH is not stored in the CA; (c) feedback inhibition by JH – while in some insect species a negative feed back could be observed there is no evidence for a direct effect of JH on CA activity in locusts; (d) neural regulation of JH biosynthesis – neural control of CA has been implicated in many insect species; in locusts its direct or indirect function in conjunction with a humoral control needs further investigation; (e) regulation by neurohormones (allatotropin, allatostatin) – investigations in various insect species including *Locusta migratoria* indicate a humoral control of CA activity. However, up to now no such hormone has been isolated and characterized; (f) regulation by factors not originating in the central nervous system – in insects producing successive batches of eggs an ovarian off-signal may be necessary upon completion of a gonotrophic cycle; in cockroaches ecdysterone has apparently such a function [6].

In the present study a humoral factor originating in the brain of *Locusta migratoria* is described which stimulates JH biosynthesis in isolated CA (Fig. 1). The allatotropic effect of this neurohormone is examined with an in vitro bioassay. Some of the proper-

1 Universität Oldenburg, FB 7, Postfach 2503, 2900 Oldenburg, FRG

Biosynthesis, Metabolism and Mode of Action of Invertebrate Hormones
(ed. by J. Hoffmann and M. Porchet)

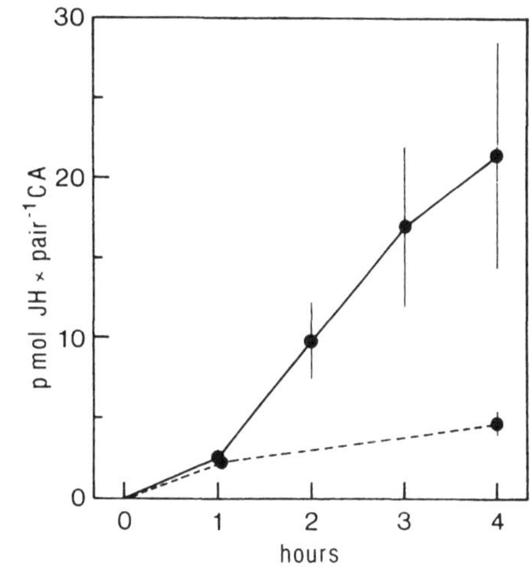

Fig. 1. The effect of CC extracts on JH biosynthesis by isolated CA. CC extracts were added after 1 h. Each *point* is the mean of five individual measurements; *vertical bars* show the SEM. The incubation medium contained the equivalent of 5 CC per incubation in 100 μl (–•–); control incubations contained no allatotropin extract (– – • – –)

ties of the hormone are also reported (Data included in this essay were recently published in part elsewhere [7].

2. Isolation and Bioassay of the Allatotropic Factor

Brains and CC were collected from female locusts within 48 h after emergence and from locusts with maturing ovaries (Table 1). The tissues were homogenized in 80%

Table 1. The effect of extracts of CC and brains from emerged and maturing females (equivalent of five organs added to each incubation), of female fat body, and flight muscles (equivalent of 22 mg tissue added to each incubation) and of farnesenic acid on JH biosynthesis by isolated CA

Material tested	JH synthesis[a] (pmol pair^{-1} CA)
CC extract, emerged	12.5 ± 4.0
Control	5.7 ± 1.8
CC extract, mature	29.0 ± 11.2
Control	3.6 ± 2.7
Brain extract, emerged	8.8 ± 6.1
Brain extract, mature	57.7 ± 11.1
Control	2.6 ± 0.3
Fatbody extract	4.5 ± 0.6
Flight muscle extract	4.3 ± 1.2
Control	7.0 ± 2.8
Farnesenic acid 100 μ*M*	274.3 ± 46.1
Control	13.1 ± 5.6

[a] Values are means of four to five individual incubations ± SEM performed for 4 h

methanol, centrifuged at 8,000 *g* and the supernatant subjected to chromatography on a Sephadex-G-10 column and eluted with 80% methanol. The resulting front peak was concentrated with nitrogen and lyophilized. Prior to incubation the extracts were redissolved in incubation medium and insoluble material was removed by centrifugation. Fat body and muscle tissue were extracted in the same way.

The JH biosynthesis in isolated CA was measured with the short-term radiochemical assay developed by Pratt and Tobe [8]. This assay utilizes the fact that the labelled methyl group of methyl-methionine is incorporated into the methylester group of JH. The resulting radiolabelled JH can be extracted and the rate of JH III synthesis measured. The culture conditions were as described previously [3]. CA pairs taken from females within 24 h after emergence are most suitable for testing extracts for allatotropic activity with a rather small individual variation.

3. Stimulation of CA Activity in Vitro

CA of 1-day-old adult female locusts have a very low JH biosynthetic activity. In general it is below 1.0 pmol per hour (see controls in Table 1), while vitellogenic females synthesize up to 120 pmol per h[3]. However, these young, little active glands are capable of biosynthesizing large amounts of JH – provided sufficient amounts of precursor are available. The addition of farnesenic acid (gift of Dr. Schooley, Zoecon Corp. Palo Alto) results in a very high rate of JH biosynthesis similar to that of maturing females (Table 1). Thus, rate limitation of the JH biosynthetic pathway occurs prior to the final two enzymatic steps in JH biosynthesis (methylation and epoxidation). Extracts prepared from brains and CC stimulate the JH biosynthesis of CA cultured in vitro significantly. A 10- to 20-fold stimulation can be observed with allatotropin extracts obtained from

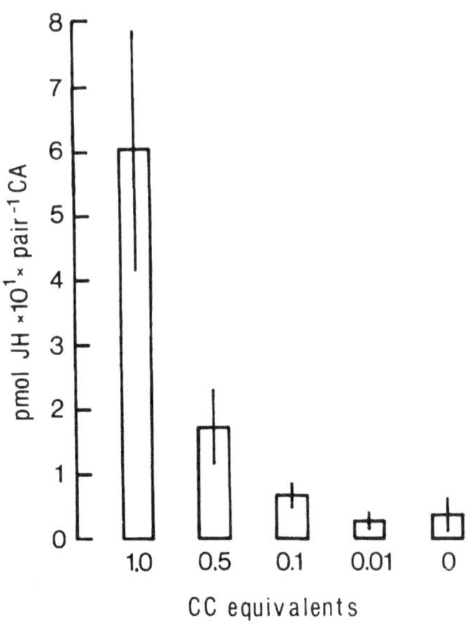

Fig. 2. Dose response of CA by extracts of CC from maturing females. Each *point* represents the mean of five incubations ± SEM; incubations were for 4 h

maturing female locusts. Extracts prepared from immature locusts are less effective. This difference in stimulation is in good agreement with the observed cyclic activity of the CA during the first gonotrophic cycle. The fact that both CC- and brain-extracts stimulate CA biosynthetic activity demonstrates that the allatotropin originates in the brain. Fat body and flight muscle extracts have no stimulating effect.

When added to unstimulated CA, CC extracts rapidly cause a significant increase in JH biosynthesis. The resulting elevated rate of JH synthesis remains high in the presence of the stimulating allatotropic factor for at least 3 h (Fig. 1). The stimulation by the allatotropic factor is dose dependent. In the experiment shown in Fig. 2 the addition of 1 CC equivalent caused a 15-fold higher CA activity compared with untreated controls. Addition of less CC-extract resulted in lower stimulation.

4. Properties of the Allatotropic Factor

The locust allatotropic factor can be identified as a small peptide. The described purification procedure takes advantage of the fact that the allatotropin readily dissolves in methanol. This neuropeptide is heat stable. It can be digested by pronase (Boehringer), but not by trypsin (Serva) (Table 2). Based on these first data a further purification of the allatotropic factor is in progress. A purified allatotropin is essential for additional immunological and biochemical studies.

Table 2. Assays of untreated, heated (100 °C, 5 min), and digested (enzyme concentrations 1 mg ml^{-1}, 37 °C for 2 h) allatotropin preparations

Material tested[a]	JH synthesis[b] (pmol $pair^{-1}$ CA)
Control with allatotropin	19.5 ± 4.8
Control without allatotropin	2.1 ± 0.7
Allatotropin heated	18.6 ± 7.0
Allatotropin trypsin digested	19.7 ± 8.3
Control with allatotropin	39.7 ± 16.8
Control without allatotropin	8.3 ± 4.1
Allatotropin pronase digested	10.7 ± 5.1

[a] Allatotropin extracts for the two separate experiments were prepared from CC
[b] Values are means of five incubations ± SEM performed for 4 h

5. Regulation of CA Activity

Certainly it is too premature to establish a concisive model of the regulation of locust CA biosynthetic activity. However, based on the present knowledge some conclusions can be drawn. Locust CA are able to biosynthesize JH III under in vitro conditions for several days (unpubl. data) [9]. Thus, CA are fully equipped to produce JH from the simple substances present in the incubation medium. A basal activity is possible without neural or neurohormonal influence. The present study demonstrates that a neurohormonal factor, an allatotropin, can stimulate the CA to biosynthetic activities above a basal level. This neurohormone originates in the brain, very probably in the pars intercerebralis (PI), and is released by the CC. Destruction of the PI suppresses CA activity in *Locusta* and thus oogenesis ceases [10-12].

Electrostimulation of the PI results in an enhancement of JH biosynthesis by the CA [13]. The allatotropin can be inactivated in vivo by injection of a crude antibody solution prepared against PI containing parts of the locust brain [14]. Thus, the allatotropin reaches the CA via hemolymph rather than via nervi corporis allati I. A suppression of CA activity at times when no or small amounts of JH are needed seems to be possible in two ways: (a) by neural inhibition of allatotropin release from the cells in the PI and the CC; (b) by direct neural inhibition of JH synthesis in the CA. Since the CA are solely innervated by lateral neurosecretory cells, these protocerebral cells might be responsible for inhibitory signals [15]. In addition these lateral neurosecretory cells are apparently synaptically coupled to the PI with the possibility to exert there a direct effect.

For the precise regulation of cyclic activity of CA during oogenesis signals from the maturing ovary are necessary. The nature of such signals in locusts is unknown. In cockroaches, however, ecdysterone has been shown to act as an allatostatin, possibly by acting through the central nervous system. It seems likely that the lateral neurosecretory cells in the brain are not only sensitive to such biochemical messages, but also to information about photoperiod, temperature, food quality, and others.

In summary, it is suggested that JH biosynthesis by the locust CA can be modulated by both a humoral factor, the described allatotropin, and neural signals. The biochemical approach of the present study should prove to be a helpful method for further elucidation of the regulation of CA activity in *Locusta migratoria*.

References

1. Engelmann F (1970) Physiology of insect reproduction. Pergamon, Oxford New York
2. Doane WW (1973) In: Counce C, Waddington H (eds) Developmental systems: Insects, vol I. Academic Press, London New York, p 291
3. Ferenz HJ, Kaufner I (1981) In: Pratt GE, Brooks GT (eds) Juvenile hormone biochemistry. Elsevier, Amsterdam, p 135
4. Girardie J, Tobe SS, Girardie A (1981) CR Acad Sci 293:443–446
5. De Kort CAD, Granger NA (1981) Annu Rev Entomol 26:1–28
6. Friedel T, Feyereisen R, Mundall EC, Tobe SS (1980) J Insect Physiol 26:665–670
7. Ferenz HJ, Diehl I (1983) Z Naturforsch 38C:856–858
8. Pratt GE, Tobe SS (1974) Life Sci 14:575–586
9. Baehr JC (1981) In: Pratt GE, Brooks GT (eds) Juvenile hormone biochemistry. Elsevier, Amsterdam, p 47
10. Goltzené F, Porte A (1978) Gen Comp Endocrinol 35:35–45
11. McCaffery AR (1976) J Insect Physiol 22:1081–1092
12. Lazarovici P, Pener MP (1978) Gen Comp Endocrinol 35:375–386
13. Tobe SS, Girardie J, Girardie A (1982) J Insect Physiol 28:867–871
14. Remboldt H, Eder I, Ulrich G (1981) Z Naturforsch 36C:466–469
15. Poras M, Baehr JC, Cassier P (1983) Int J Invertebr Reprod 6:111–122

I.11 Neurosecretion in Insects: Multisynthesis, Mode of Action, and Multiregulation

A. GIRARDIE, J. GIRARDIE, J. PROUX, C. RÉMY, and J. VIEILLEMARINGE[1]

1. Introduction

In insects, the products of neurosecretory cells (NSC) play crucial roles in coordinating growth, development, and reproduction. Sufficient morphological information has been now accumulated (histochemistry, histofluorescence, immunohistology, cytology) to reveal that neurosecretory systems contain various types of NSC. Recent investigations have provided evidence, as is the case in vertebrates, of the existence of several neurosecretory products in the same neuron. In some insects, experimental studies have determined the function of some NSC: prothoracicotropic, allatotropic, diuretic, myotropic activities. It is generally agreed that the mechanism of neurohormone action on the target tissue acts through a cAMP-mediated process. The activity of the NSC is influenced by several factors probably including a multineuroendocrine regulation as in other invertebrates.

In this review, we assess our recent findings in this field for two neurosecretory systems (median NSC of the brain and aldehydefuchsinophil NSC of the subesophageal ganglion) of the locust and, where possible, we extend this to other insects.

2. Neurosecretory Cells and Multisynthesis

2.1 Median Neurosecretory Cells of the Pars Intercerebralis

Two types of peptidergic median neurosecretory cells (M–NSC) of the pars intercerebralis can be found in insects: A–NSC which are basophilic after oxidation and B–NSC which remain acidophilic after oxidation. In the holometabola, M–NSC are large and scarce, while they are small and numerous in the hemimetabola. In *Locusta migratoria*, after Victoria blue-paraldehyde fuchsin staining, two categories of A–NSC are distinct [32]: the blue A_1–NSC containing proteins with thiol (–SH) and/or disulfur (S–S) groups which give strong acids after oxidation, and the pink A_2–NSC which contain proteins with carboxyl groups (–COH) giving weak acids.

1 Université de Bordeaux I, Laboratoire de Neuroendocrinologie, UA CNRS 683, Avenue des Facultés, 33405 Talence Cedex, France

Biosynthesis, Metabolism and Mode of Action of Invertebrate Hormones
(ed. by J. Hoffmann and M. Porchet)

Since 1977 [49] immunohistological exploration of insect nervous systems using antisera against vertebrate secretory products has allowed a new and more selective detection of neurosecretion materials especially in brain. Moreover, this approach has revealed NSC which were not visualized by classical staining methods. Neurosecretory products related to vertebrate peptides were detected in several areas of the brain and more particularly in the pars intercerebralis of seven species: *Locusta migratoria* with somatostatin-like neuropeptide [12] and substance P-like peptide [2] in A–NSC; *Bombyx mori* [64] with pancreatic polypeptide (PP), gastrin and insulin-like materials in A–NSC; *Eristalis aeneus* [17] with somatostatin, insulin, enkephalin, and endorphin-like materials; *Calliphora erythrocephala* [13–16] with insulin and α-endorphin-like materials in A–NSC, PP-like peptide in B–NSC and gastrin/cholecystokinin-like peptide in chromophob NSC; *Periplaneta americana* [19] with PP-like material; *Manduca sexta* [18] with somatostatin, insulin, substance P, enkephalin, β-endorphin, calcitonin, and vasoactive intestinal polypeptide (VIP)-like materials in A–NSC; *Ostrinia nubilalis* with insulin-like material in A_2–NSC (Lavenseau et al., unpubl. results). Although many peptides immunologically related to vertebrate peptides are present in the pars intercebralis, the coexistence of two or several of these materials has not up until now been detected in a single immunoreactive M–NSC. Nevertheless, in *Locusta migratoria* we have recently observed a great number of insulin-immunoreactive A median NSC (Vieillemaringe et al., unpubl. results). The great number of these immunoreactive cells compared to the great number of the somatostatin-immunoreactive A median NSC suggests the likelihood that the two immunoreactive materials could be present in the same A–NSC of the pars intercerebralis.

Median NSC of insects do not contain only peptides: the presence of catecholamines and indolamines has been demonstrated by the classical Falck and Hillarp histofluorescence technique (reviewed in [20, 36]). In African locust, the pars intercerebralis contains a few indolaminergic and numerous catecholaminergic cell bodies [37, 60]. Their processes are entirely visible right to their endings [60] in the corpora cardiaca [37, 38, 60]. Using the Falck and Hillarp technique and the paraldehyde fuchsin staining on alternate sections of freeze-dried brain included in paraffin, we have demonstrated that most of the catecholaminergic neurons are fuchsinophilic [61]. With a specific anti-dopamine serum recently prepared [24], we have observed that the greater part of the catecholaminergic M–NSC contains dopamine [62]. This amine is located in A_1 and A_2 peptidergic M–NSC. Thus, in *Locusta migratoria*, there is coexistence in the same neuron of amine and peptide materials. This observation is in agreement with recent results obtained in vertebrates and is inconsistent with the first part of Dale's principle "one neuron has the ability to synthesize only one transmitter".

In addition to amines and peptides, neurosecretory proteins have been characterized from M–NSC of some insects (reviewed in [25]). Recently, a neurosecretory protein from brain and corpora cardiaca of adult locusts has been isolated and purified [22]. This protein could present, both in size and amino acid composition, a similarity with the neurophysins of vertebrates. However, this putative insect neurophysin has a pHi more basic (6.3–6.5) than those of vertebrate neurophysins (4.45–4.70–4.82), and contains 8 cysteine residues instead of 14 for the neurophysins. Moreover, using several neurophysin-antisera, no immunoreactive deposits were found over the A_1 and A_2 median NSC of the locusts [48].

The proteins of delipidated nervous corpora cardiaca from adult locusts were studied in our laboratory after aqueous or alcoholic extraction (70% ethanol or methanol). They were analyzed by a combination of exclusion-diffusion chromatography on a high performance column (TSK 2000 SW) [30] or on a conventional column (Sephadex G-100), two dimensional polyacrylamide gel electrophoresis involving denaturing agents (sodium dodecyl sulfate alone or associated with β mercaptoethanol) in the second dimension, and isoelectric focusing in polyacrylamide gel. Three major acidic cystine-rich proteins were characterized:

- a trimer with an apparent molecular weight of 58,000 composed of 18,000 subunits, each subunit composed of polypeptidic chains of 9,000 held together by disulfide bonds; this trimer focuses at pH values 5.65 and 5.60;
- a monomer with an apparent molecular weight of 18,000 composed of two polypeptidic chains of 9,000 estimated molecular weight held together by disulfide bonds; this monomer has 5.05 for pHi;
- a dimer with an apparent molecular weight of 14,000 composed of 7,000 subunits, each subunit composed of two polypeptidic chains of 4,000 estimated molecular weight held together by disulfide bonds; the dimer focuses at three values of pH 4.29, 4.00, 3.92, and its yield is enhanced by the alcoholic extraction.

The trimer, the dimer, and the monomer are nearly undetectable in the proteinogram of the storage lobes of the corpora cardiaca 21 days after axotomy of the M–NSC by electrocoagulation of their tracti at decussation level in the brain. A large decrease of the fuchsinophilic material of the nervous corpora cardiaca has been demonstrated in this situation [31]. Thus, these three proteins, rich in cystine as are the fuchsinophilic materials and which decrease in parallel with the fuchsinophilic neurosecretory materials, are neurosecretory proteins of A median NSC. At the present time, we cannot define which type of A–NSC (A_1 or A_2 or both) synthesizes the trimer, the dimer, and the monomer. Nevertheless, the M–NSC of the locusts are able to synthesize multiple neurosecretory products: amines, peptides, and proteins. No clear analogy exists between these three proteins and the neurosecretory protein characterized by Friedel et al. [22]. On the one hand, the trimer evokes Friedel's neurosecretory protein with its high concentration in the nervous corpora cardiaca and its less acidic pHi. On the other hand, the monomer approaches Friedel's neurosecretory protein by its molecular weight. Therefore, the trimer, the dimer, and the monomer may be carrier proteins for locust neurohormones in the same way as neurophysins are the carrier proteins for vasopressin and oxytocin. They may also represent precursors of neurohormones or neurohormones themselves. Active neurosecretory products of comparable molecular weight have been characterized in other insects [27, 35].

2.2 Fuchsinophilic Neurosecretory Cells of the Subesophageal Ganglion

Another example of the coexistence of peptide and protein neurosecretory products in a single NSC is found in *Clitumnus* [49] and *Locusta* [50]. In these insects, the two A–NSC of the subesophageal ganglion synthesize a neuropeptide immunologically related to arginine-vasopressin (AVP) which is, as in vertebrates, associated to a carrier protein related to neurophysin. In *Locusta*, an immunohistological exploration of the entire central nervous system with anti-AVP and anti-neurophysin antisera showed immunoreactive tracts originating in the two A subesophageal perikarya and running forwards towards the brain and backwards towards the thoracic and abdominal cord. Collaterals branch off from these tracts and give arborizations in the root of each segmental nerve of thoracic and abdominal ganglia where the vasopressin-neurophysin-like neurosecretory products could be released into hemolymph [48]. Moreover, in embryo 2 days before eclosion, AVP and neurophysin immunoreactivities appear simultaneously in the two fuchsinophilic subesophageal NSC suggesting a common precursor for these two molecules [51].

2.2.1 Material Immunologically Related to Neurophysin in Locust

Crude acido-alcoholic extracts of subesophageal ganglia recognize anti-neurophysin antisera [4]. The displacement curve obtained parallels the reference curve produced under the same conditions by bovine neurophysin II. So, the neurophysin-like material could have some similarities with the neurophysin. Indeed, the neurophysin-like material is recovered by radio immunoassay in the 10,000 mol.wt. elution volume after Biogel P-10 filtration and focuses into two acidic pHi 4.5 and 4.75. Moreover, 95% of the 10,000 mol.wt. immunoreactive material was absorbed by antibovine neurophysin II antibodies during affinity chromatography on ammonium sulfate-precipitated antibovine neurophysin II antiserum immobilized on CH-Sepharose 4 B. Thus, extracts of subesophageal ganglia contain proteins possessing a molecular weight, isoelectric points, and immunochemical properties analogous to those of neurophysins.

2.2.2 Material Immunologically Related to Arginine-Vasopressin (AVP) in Locust

Crude acidic extracts of subesophageal ganglia bind to antibodies raised not only against neurophysin, but also against AVP. Displacement curves obtained from synthetic AVP and subesophageal extracts being parallel, an AVP radio immunoassay was used to research and quantify the AVP-like material in different tissues [44]. Thus, AVP-like material is present throughout the central nervous system from the brain to the abdominal ganglia with a maximum in the subesophageal ganglion. These results are in agreement with those obtained by the immunohistological technique, i.e., AVP-like neurosecretory perikarya in the subesophageal ganglion with distal processes from the optic lobes to the last abdominal ganglion. Moreover, radio immunoassay of AVP permitted the detection AVP-like material in the organs of diuresis, Malpighian tubules, and rectum as well as in the hemolymph. In addition, the organs of diuresis exhibited a degradative activity towards the AVP-like material. Purification of the AVP-like material by gel filtration (Sephadex G-50) then immunoabsorption (Sepharose 4B) yielded a purified extract which we used to determine some biochemical data [7].

Because of the parallel displacement curves of the AVP and AVP-like material it was likely that these two molecules shared a common antigenic determinant. The antigenic determinant of AVP recognized by the antibody of the radio immunoassay was determined to be the sequence of the C-terminal ending: –Arg(Lys)–Gly(NH_2) [8]. To confirm the presence of the same antigenic determinant in the AVP-like molecule,

AVP-like material was submitted to a tryptic hydrolysis followed by a glycinamide linkage. After trypsin treatment, 97% of the initial immunoreactivity was abolished and the glycinamide linkage restored 90% of the immunoreactivity. So, the C-terminal ending of the AVP-like material must be –Arg(Lys)–Gly(NH_2). To discriminate between Arg and Lys, a succinylation of the AVP-like molecule was performed. This treatment left the immunoreactivity unchanged. Thus, the antigenic determinant of the AVP-like molecule is similar to that of the AVP, i.e., –Arg–Gly(NH_2). The ability of the AVP-like factor to be radiolabelled with ^{125}I proved the presence of a tyrosine as in the AVP molecule.

However, the AVP-like molecule also exhibits some differences to AVP. The molecular weight, established both by gel filtration (Sephadex G-50) and gel permeation (TSK 2000 SW) with columns previously calibrated with peptidic and proteic markers, is close to 2,500 (AVP = 1,080). The pHi found by electrofocalization is neutral (7.5) in contrast to the basic (10.9) AVP pHi. This result is confirmed by an electrophoretic behavior different from that observed with AVP probably because of the presence of additional acidic residues, such as glutamic or aspartic acids.

The structural differences between AVP and AVP-like molecules were biologically confirmed. The AVP molecule has no effect in locust diuresis [45], while AVP-like peptide has only a very weak antidiuretic effect on mammalian diuresis (Burlet, pers. comm.).

The relationship between AVP and AVP-like peptides spreads to the biosynthetic process. As in the neurophysin-AVP pattern, the AVP-like molecule seems to be the product of high molecular weight precursors. The lack of immunoreactivity in the pellets of the subesophageal ganglia extracts [46] suggests that these putative precursors are not spontaneously immunoreactive with an anti-AVP antibody because the AVP-like sequence is likely enclosed inside the precursor forms. So, to reveal these putative forms with the radio immunoassay procedure, we performed a two step artificial maturation [7]:

- release of a truncated AVP-like C-terminal ending by tryptic hydrolysis;
- restauration of the complete C-terminal ending by amidation.

In this way, we obtained a truncated form terminated on one side by the AVP-like sequence. This molecule is recognized by the radio immunoassay for AVP. We therefore suggest that the protein treated as above is a precursor of AVP-like material. Thus, from homogenates of different nervous structures, we were able to quantify:

- free AVP-like forms from the supernatants;
- putative precursor forms from the pellets.

We observed that the distribution of the putative precursors in nervous tissues is generally similar to that of the free molecule. In contrast, control tissue (muscle) is devoid of both AVP-like and precursor forms.

In conclusion, the pattern of the subesophageal ganglion neurophysin-AVP-like material in locust has several similarities with the mammalian neurophysin-AVP pattern: a peptide neurohormone (AVP-like/AVP) associated to a neuroprotein (neurophysin-like/neurophysin) and synthetic processes involving common precursors as suggested above by immunohistology [51].

3. Neurosecretory Products: Physiological Activity and Mode of Action

We report only the physiological activity of AVP-like neurosecretory material of the subesophageal ganglion in *Locusta migratoria* [44, 45].

A possible diuretic function of the AVP-like material was suggested by establishing a relationship between the hemolymphatic AVP-like titer and the state of hydration of the locusts. This possible diuretic function of the AVP-like material was then confirmed and specified using Malpighian tubules in vivo and in vitro bioassays.

A purification of AVP-like peptide from subesophageal ganglia was performed by successive steps: acidic extraction, centrifugation, ultracentrifugation, gel filtration, immunoabsorption. Whatever the purification step, only the AVP-like material (characterized by means of radio immunoassay) stimulated diuresis.

The Malpighian tubules respond to the AVP-like diuretic factor in a dose-dependent way and the shape of the dose-response curve is characteristic of a hormonal response. The presence of AVP-like material in the hemolymph and the ability of the electrical stimulated subesophageal ganglion to release more AVP-like material into hemolymph provided additional evidence for the hormonal character of this compound. The diuretic action of the AVP-like neuropeptide seems to be mediated by cAMP [42] as is suggested in several insect neurohormones studies (see Table 1).

Table 1. cAMP: a possible mediator of insect neurohormones

Neurohormones	Target	Species	References
PTTH	Prothoracic glands	*Manduca*	Vedeckis et al. [59]
Diuretic H	Malpighian tubules	*Rhodnius*	Aston [1]
	Salivary glands	*Calliphora*	Berridge and Patel [3]
AVP-like H	Malp. tubules, rectum	*Locusta*	Proux and Cailla [42]
CTSH	Rectum	*Schistocerca*	Philipps et al. [41]
Hypertrehalosemic H	Fat body	*Periplaneta*	Gäde [23]
AKH	Fat body	*Schistocerca*	Spencer and Candy [53]
Eclosion H	Central nervous system	*Hyalophora*	Truman et al. [57]
Bursicon	Hemocytes	*Periplaneta*	Vandenberg and Mills [58]
	Epidermis	*Tenebrio*	Delachambre et al. [9]
Puparium tanning H	Tegument	*Sarcophaga*	Fraenkel et al. [21]
Melanizing H	Tegument	*Leucinia*	Matsumoto et al. [40]
Myotropic H	Hindgut	*Leucophaea*	Cook et al. [6]
Ovulation H	Ovary	*Glossina*	Delinger et al. [10]
EDNH	Ovary	*Aedes*	Shapiro [52]

4. Neurosecretory Cells and Multineuroendocrine Regulation

Dramatic perturbation of locust diuresis, resulting from localized destruction of different brain neurosecretory centers (reviewed in [47]) could suggest the involvement of these centers on the regulation of the two A diuretic subesophageal NSC.

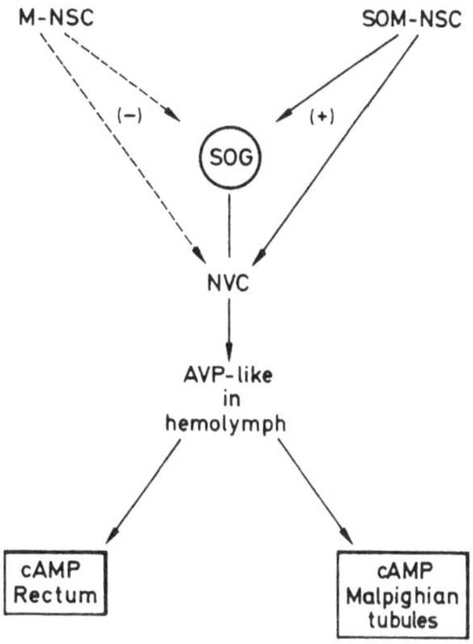

Fig. 1. AVP-like material in locust: multineuroendocrine regulation of the release, and diuretic mode of action. *M–NSC* median neurosecretory cells; *NVC* nervous ventral cord; *SOG* subesophageal ganglion; *SOM–NSC* subocellar median neurosecretory cells

The release of the AVP-like hormone into hemolymph (measured using radio immunoassay) is under the antagonistic control of two cerebral neurosecretory centers, the NSC of the pars intercerebralis and the subocellar NSC. The observed hemolymphatic AVP titer is high after destruction of the pars intercerebralis and very low after destruction of the subocellar NSC [46].

Electrical stimulation of the NSC, in contrast to destruction, produces opposite results [43]. The AVP-like titer increases when the subocellar NSC are stimulated, and decreases when the M–NSC of the pars intercerebralis are stimulated. These results suggest both diuretic and antidiuretic activities of the protocerebral NSC in insects. Thus, in view of these results, different neurosecretory centers (A subesophageal NSC, pars intercerebralis, subocellar NSC) up to now thought of as being separately involved in locust diuresis, can be gathered into a single regulatory system (Fig. 1).

In its original form, the neuroendocrine regulation of the diuretic subesophageal NSC appears to include stimulating (subocellar NSC) and inhibiting (M–NSC of the pars intercerebralis) mechanisms. In our opinion, it is the only example of a multineuroendocrine regulation of a neurosecretory center known in insects. On the other hand, such multineuroendocrine regulations of endocrine glands (prothoracic glands, corpora allata) are well known in some insects (Table 2).

Table 2. Insect neuroendocrine multiregulations

Effectors	Species	M–NSC	L–NSC	SOM–NSC	SOG	NVC	References
AVP-like NSC	Locusta	–		+			Proux and Girardie [43]
Prothoracic glands	Platysamia	+	+				Williams [63]
	Locusta	+	–				Carlisle and Ellis [5]
		+	+				Girardie [29]
	Galleria	+			+	–	Malá et al. [39]
	Manduca		+		–	–	Gibbs and Riddiford [26]
Corpora allata	Locusta	+–	+				Girardie [28]
		+	+				Tobe et al. [55]
	Schistocerca	+	+				Strong [54]
	Pyrrhocoris	+–					Hodková [33]
	Diploptera	+–					Tobe and Stay [56]
	Leptinotarsa	+–					De Kort and Granger [11]

L–NSC lateral neurosecretory cells; *M-NSC* median neurosecretory cells; *NVC* nervous ventral cord; *SOG* subesophageal ganglion; *SOM-NSC* subocellar neurosecretory cells

References

1. Aston RJ (1975) J Insect Physiol 21:1873–1877
2. Benedeczky I, Kiss JZ, Somogyi P (1982) Histochemistry 75:123–131
3. Berridge MJ, Patel NG (1968) Science 53:171–172
4. Camier M, Girardie J, Rémy C, Girardie A, Cohen P (1980) Biochem Biophys Res Commun 93:792–796
5. Carlisle DB, Ellis PE (1968) Nature (London) 220:706–770
6. Cook BJ, Holman GM, Marks EP (1975) J Insect Physiol 21:1807–1814
7. Cupo A, Proux J (1983) Neuropeptides 4:309–318
8. Cupo A, Rougon-Rapuzzi G, Delaage M (1981) Eur J Biochem 115:169–174
9. Delachambre J, Delbecque JP, Provansal A, De Reggi ML, Cailla H (1979) Experientia 35:701
10. Delinger DL, Chaudhury MFB, Dhadialla TS (1978) Experientia 34:1296–1297
11. De Kort CAD, Granger NA (1981) Annu Rev Entomol 26:1–28
12. Doerr-Schott J, Joly L, Dubois MP (1978) CR Acad Sci 286:93–95
13. Duve H, Thorpe A (1979) Cell Tissue Res 200:187–191
14. Duve H, Thorpe A (1980) Cell Tissue Res 210:101–109
15. Duve H, Thorpe A (1981) Gen Comp Endocrinol 43:381–391
16. Duve H, Thorpe A (1983) Cell Tissue Res 233:415–426
17. El Salhy M, Abou El Ela R, Falkmer S, Grimelius L, Wilander E (1980) Regul Peptides 1:187 to 204
18. El Salhy M, Falkmer S, Kramer KJ, Speirs RD (1983) Cell Tissue Res 232:245–317
19. Endo Y, Iwanaga T, Fujita T, Nishiitsutsuji-Uwo J (1982) Cell Tissue Res 227:1–9
20. Evans PD (1980) Adv Insect Physiol 15:317–474

21. Fraenkel G, Blechl A, Blechl J, Herman P, Selignan MI (1977) Proc Natl Acad Sci USA 74: 2182–2186
22. Friedel T, Loughton BG, Andrew RD (1980) Gen Comp Endocrinol 41:487–498
23. Gäde G (1977) Zool Jahrb Physiol 81:245–250
24. Geffard M, Kah O, Chambolle P, Le Moal M, Delaage M (1982) CR Acad Sci 295:797–802
25. Gersch M (1976) Actualités sur les hormones d'invertébrés. Publ Colloq CNRS 291:281–305
26. Gibbs D, Riddiford LM (1977) J Exp Biol 66:255–266
27. Gilbert LI, Bollenbacher WE, Agui N, Granger NA, Sedlak BJ, Gibbs D, Buys CM (1981) Am Zool 21:641–653
28. Girardie A (1965) CR Acad Sci 261:4876–4878
29. Girardie A (1974) Zool Jahrb Physiol 78:310–326
30. Girardie J, Allard M, Faddoul A, Neuzil E (1983) J Liq Chromatogr 6:2501–2512
31. Girardie J, Girardie A (1977) Arch Anat Microsc Morphol Exp 66:291–304
32. Girardie J, Rossi C (1978) CR Acad Sci 286:97–100
33. Hodková M (1979) Zool Jahrb Physiol 83:126–136
34. Hökfelt T, Johansson O. Ljungdahl Å, Lundberg JM, Schulzberg M (1976) Nature (London) 284:515–521
35. Ishizaki H, Ichikawa N (1967) Biol Bull 133:355–368
36. Klemm N (1976) Prog Neurobiol 7:99–169
37. Klemm N, Falck B (1978) Gen Comp Endocrinol 34:180–192
38. Lafon-Cazal M, Calas A, Bosc JJ (1973) J Microsc 17:223–226
39. Malá J, Granger NA, Sehnal F (1977) J Insect Physiol 23:309–316
40. Matsumoto S, Suzuki A, Yamamoto H, Ogura N, Ikemoto H (1979) Appl Entomol Zool 14: 159–163
41. Phillips JE, Mordue W, Meredith J, Spring J (1980) Can J Zool 58:1851–1860
42. Proux J, Cailla H (in preparation)
43. Proux J, Girardie A (1982) Neurosci Lett 33:73–77
44. Proux J, Rougon-Rapuzzi G (1980) Gen Comp Endocrinol 42:378–383
45. Proux J, Rougon G, Cupo A (1982) Gen Comp Endocrinol 47:449–457
46. Proux J, Rougon-Rapuzzi G, Rémy C (1980) J Physiol Paris 76:277–282
47. Raabe M (1982) Insect neurohormones. Plenum Press, New York London
48. Rémy C, Girardie J (1980) Gen Comp Endocrinol 40:27–35
49. Rémy C, Girardie J, Dubois MP (1977) CR Acad Sci 285:1495–1497
50. Rémy C, Girardie J, Dubois MP (1979) Gen Comp Endocrinol 37:93–100
51. Romeuf M, Rémy C (1984) Cell Tissue Res 236:289–292
52. Shapiro JP (1983) Insect Biochem 13:89–279
53. Spencer IM, Candy DJ (1976) Insect Biochem 6:289–296
54. Strong L (1965) J Insect Physiol 11:271–280
55. Tobe SS, Girardie J, Girardie A (1982) J Insect Physiol 28:867–871
56. Tobe SS, Stay B (1980) Gen Comp Endocrinol 40:89–98
57. Truman JW, Fallon AM, Wyatt GR (1976) Science 194:1432–1434
58. Vandenberg RD, Mills RR (1975) J Insect Physiol 21:221–229
59. Vedeckis WV, Bollenbacher WE, Gilbert LI (1976) Mol Cell Endocrinol 5:81–88
60. Vieillemaringe J, Cailley-Lescure H, Bensch C, Girardie J (1981) J Physiol Paris 77:989–995
61. Vieillemaringe J, Duris P, Bensch C, Girardie J (1982) Neurosci Lett 31:237–240
62. Vieillemaringe J, Duris P, Geffard M, Le Moal M, Delaage M, Bensch C, Girardie J (1984) Cell Tissue Res (in press)
63. Williams CM (1948) Growth Symp 12:61–74
64. Yui R, Fujita T, Ito S (1980) Biomed Res 1:42–46

I.12 Neuropeptides of Vertebrate Type in the Blowfly, *Calliphora*

A. THORPE and H. DUVE[1]

1. Introduction

Neuropeptides are the focal point of many present-day research programmes. In the vertebrates (and especially mammals) rapid progress has been made in the localisation and characterisation of a large number of neuropeptides and the list of these is continually being extended. Many of the substances are, in fact, rediscoveries of bioactive molecules that were originally found as hormones or neurotransmitters within the gastrointestinal tract (e.g. VIP). In some instances this situation is reversed and a substance previously categorised as belonging to nervous tissues only has been recognised as a gastrointestinal peptide (e.g. neurotensin).

In invertebrates too a number of neuropeptides have now been identified and studied extensively. In some instances the full amino acid sequence has been determined and the physiological significance elucidated. Several examples of this type of investigation are presented in this volume.

Invertebrate hormones are normally considered to be unrelated to those of vertebrates largely because the functions ascribed to them do not usually appear to have a place in vertebrate physiology. The eclosion hormone of insects, for example, would not be considered by many to figure prominently as a product of the genome of a mammal. Nevertheless, evolution has obviously produced the complex mammalian organism from simple metazoan origins and there is plentiful evidence for the almost immutable continuity of at least some of the DNA during this period. Thus, the calmodulin molecule of coelenterates differs by only three amino acid residues out of more than 120 from that of cattle [1].

Quite recently, an example of this type of conservation in evolution has come directly from the field of invertebrate endocrinology. Strikingly, the peptide concerned, the head activator peptide of *Hydra,* is one of the type just referrèd to above, that one would not perhaps expect to find in vertebrates. As reported in this volume, however, Schaller and Bodenmüller [2] have isolated a peptide with exactly the same sequence as that of *Hydra* from human and bovine hypothalami and rat intestines. The authors have suggested, furthermore, that during evolution this peptide may also have conserved its role in the control of growth or differentiation processes although the proof of this is not yet to hand.

---

1 School of Biological Sciences, Queen Mary College, University of London, Mile End Road, London E1 4NS, Great Britain

Biosynthesis, Metabolism and Mode of Action of Invertebrate Hormones
(ed. by J. Hoffmann and M. Porchet)

In our studies of the blowfly *Calliphora* we have examined the interesting problem of evolutionary conservation of neuropeptide structure and function from the opposite direction, by asking the question – do insects possess any of the known active peptides of vertebrates? Our methods have been twofold: (1) the immunocytochemical localisation of peptides using antisera raised against mammalian hormonal molecules or fragments of molecules and (2) the isolation and purification of peptides from extracts of *Calliphora* with continuous monitoring during the process by means of radioimmunoassay. Wherever possible, we have examined the functional significance of the peptides.

Initially, studies were directed especially to compare the hypoglycaemic and hyperglycaemic factors (hormones) of *Calliphora* with the corresponding hormones of vertebrates, insulin and glucagon and they were later extended to include another of the pancreatic hormones, pancreatic polypeptide (PP). The results of these studies have been documented in several papers and review articles [3–11]. The evidence, which will not be repeated here, is strongly in favour of peptides closely resembling those of mammals. Unfortunately, we still do not know the complete structure of any of these molecules although we hope this will not be too long delayed. In other insects [12–18] and in other invertebrates too [19–22] similar studies have been carried out with respect to the insulin-, glucagon-, and pancreatic polypeptide-like substances and there are many close similarities to the results of our own studies.

More recently we have been studying two of the newly discovered "mammalian" neuropeptides, those of the gastrin/cholecystokinin (CCK) type and those of the endorphin/enkephalin family. The results of these studies will be reviewed here.

2. Gastrin/CCK-Like Peptides

Sound evidence for the existence of the gastrointestinal peptides that we now know as gastrin and CCK was produced as early as 1905 and 1928, respectively. Endocrinologists were in a distinct minority in those days, however, as compared with neurophysiologists and this rich vein of biological material lay unprospected and unmined for many years. It was only in the 1960's that the two gut hormones were characterized and a further 10 years or so until Vanderhaeghen [23] provided the first evidence that the brain as well as the gut was a source of gastrin-like material. Many subsequent studies have shown that the two closely related peptides gastrin and CCK are indeed mammalian neuropeptides occurring in the brain [24–28] and elsewhere within the nervous system [29, 30]. Phylogenetic studies have revealed the presence of gastrin/CCK peptides in lower vertebrates [31] and since then, several investigators have looked for their existence in animals below this level of organisation [32, 33].

In *Calliphora*, for the immunocytochemical localization of the peptides, we have used a variety of antisera kindly made available to us by Professors G.J. Dockray of Liverpool University and J.F. Rehfeld of Copenhagen University and we have subsequently carried out collaborative biochemical studies with these key workers in the vertebrate field of gastrin/CCK peptides in attempts to characterise the *Calliphora* peptides.

The technique of immunocytochemistry (ICC) is most meaningful when region-specific antisera are used to build up a detailed structural analysis of the peptide under investigation. This technique has been very successfully used to help in the identification of the members of the gastrin/CCK families of peptides in vertebrates [34, 35] and we have followed this procedure in *Calliphora*. The antisera, all of polyclonal origin, were raised against the common C-terminal pentapeptide amide of gastrin and CCK, the NH_2 terminal regions of the two peptides and several against mid-core regions. The technique adopted was generally the peroxidase anti-peroxidase (PAP) procedure with aqueous Bouin's fluid as the fixative and diaminobenzidene (DAB) as the final substrate. Controls were those normally adopted for the ICC technique and included the use of absorption of antisera with gastrin/CCK peptides or peptide fragments. Mammalian tissues known to contain the respective peptides were treated in parallel in order to validate both technique and antiserum specificity.

The results of these studies show that material immunoreactive against the common COOH-terminal region of gastrin and CCK is widespread throughout the nervous system of *Calliphora*. Within the brain, immunoreactive neurones form a part of the median neurosecretory cell groups (MNC) and they also occur in several other well-defined areas [36] (Fig. 1A,B). The cells belong to one of two quite different size groupings (6–10 μm and 12–20 μm). The situation is further complicated by the results of further recent studies on the brain of *Calliphora* [37] which show that the precise number of immunoreactive cells and the quantity of material contained within them (as determined by the density of the immunoreaction product) is dependent upon age and diet. At the moment the data suggest that a greater amount of both cells and cellular content is correlated with increasing age and a high content of protein in the diet. The so-called giant neurones (35–45 μm) of the MNC demonstrate this correlation extremely well. In young sugar-fed flies they have only a very small amount of immunoreactive material, whereas in older protein-fed flies the reaction product fills the cells. The neuropile of the brain (and other ganglia) of older flies also shows a much greater amount of material than that of young (ca. 3-day-old) flies. Information of this type is difficult to interpret, but it seems clear that the differences are related to the function of the gastrin/CCK-like material. Whether or not large quantities of immunoreactive material correspond to an active or, as seems more likely, an inactive state of the gastrin/CCK neurones, the preponderance of material shown by older flies has made them ideal subjects in which to trace peptidergic pathways. This approach will, we feel, prove useful in the eventual solution to the problem of the functional significance of the peptides. Earlier [36] we have speculated that gastrin/CCK-like material may well be produced in the brain to act both locally within the brain itself and perhaps also to act at a distance from the brain as a metabolic neurohormone released from long neurosecretory axons. We now know that pathways allowing for such a functional interpretation do indeed occur in *Calliphora*. In a collaborative study with Dr. N.J. Strausfeld of the neurobiology unit of the European Molecular Biology Laboratory, Heidelberg, we have conducted a series of experiments in which cobalt retrograde filling of neurones is combined with normal ICC [37]. The former technique permits the complete axonal pathway of particular neurones to be elucidated and, superimposed upon this, we can specify by means of ICC, something of the precise chemical nature of the peptidergic material (in this case gastrin/CCK-like). By means of these combined techniques we have shown that one or more of the dorso-lateral neurosecretory cells of the brain, immunoreactive to gastrin/CCK COOH-terminal specific antisera, makes contact with

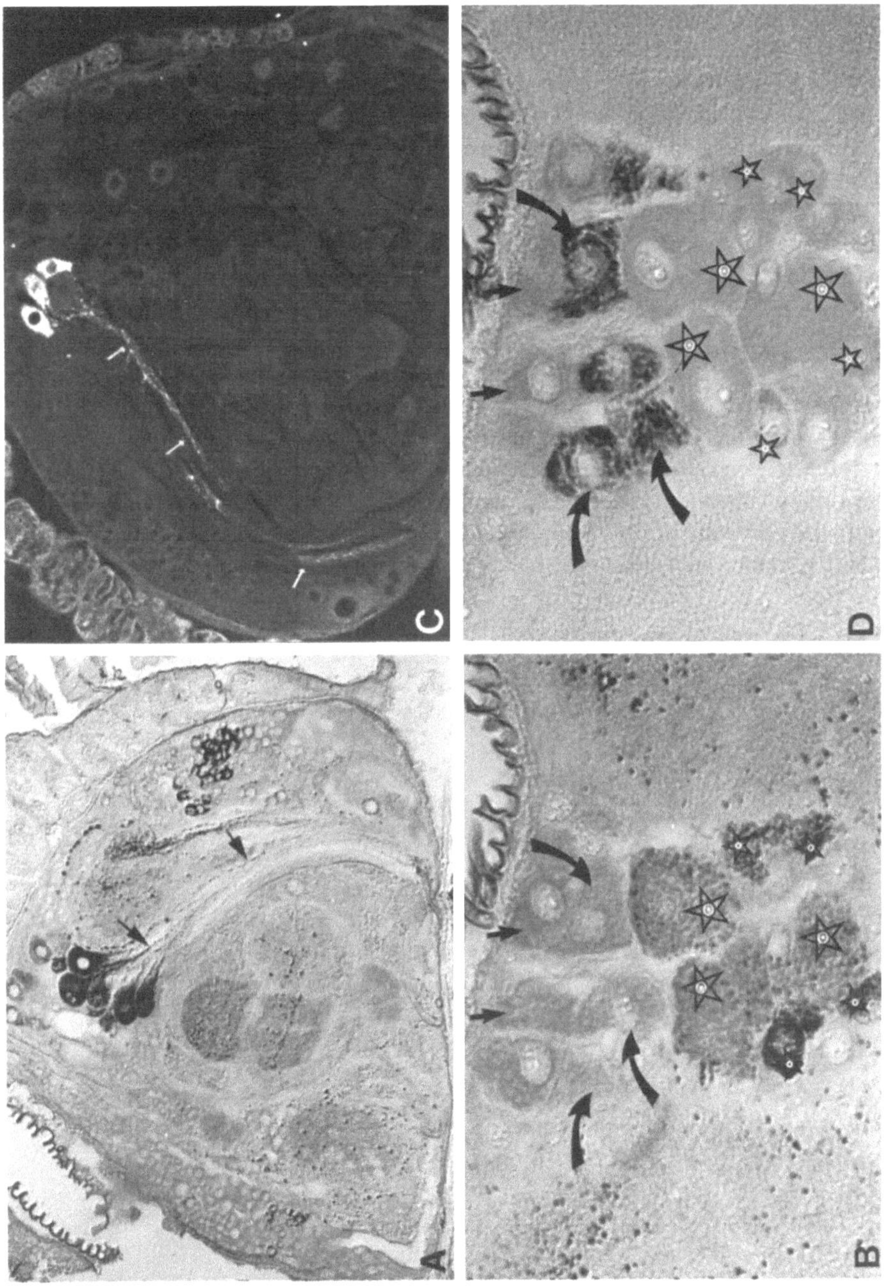

Fig. 1 A–D. Immunocytochemical preparations of the brain of *Calliphora* in sagittal plane (**A, C**) and frontal plane (**B, D**). (**A, B, D** PAP procedure; **C** immunofluorescent procedure.) **A, B** show gastrin/CCK-like and **C, D** α-endorphin-like immunoreactive neurones and fibres in the median neurosecretory cells and median bundle (*arrows*, **A, C**). **B, D** are adjacent sections to show comparisons of the different cell types: ☆ gastrin/CCK cells: ↷ α-endorphin cells: ↓ cells nonimmunoreactive to either antiserum. **A** × 265; **C** × 280; **B, D** × 650

the central body. This region is believed to be a coordination centre in *Calliphora*, the equivalent of a "higher brain centre" of vertebrates. This being so, it would suggest that the gastrin/CCK-like peptidergic material is acting in either a neurotransmitter or a neuromodulatory role. On the other hand, a second type of gastrin/CCK pathway, revealed in a series of backfillings from the cardiac-recurrent nerve is seen to commence in the MNC region and terminate in the corpus cardiacum or beyond. The corpus cardiacum, well-known as a neurohaemal organ, would provide exactly the outlet for any gastrin/CCK-like material acting in a purely metabolic (hormonal) role. At the present time, the exact functional significance of the material in the life of *Calliphora* is a matter for speculation only and will not be seriously entered into here.

For the gastrin/CCK-like material of *Calliphora*, as with other peptides in this [5, 6] and other insect species [18, 38], it is impressive to observe both the widespread nature and the symmetrical arrangement of the immunoreactive cells and fibres. This feature is particularly well demonstrated by a study we have carried out on the gastrin/CCK-like peptides within the fused thoracic ganglionic mass of *Calliphora* [39]. For each of the three thoracic ganglia per se there are two pairs of cells ventrally, one on each side of the longitudinal midline (Fig. 2). The fused abdominal ganglia possess a similar arrangement, with five pairs of cells ventrally and six to seven pairs dorsally, closely adjacent to the neuropile of this region. In the same manner as for the brain, the neuropile

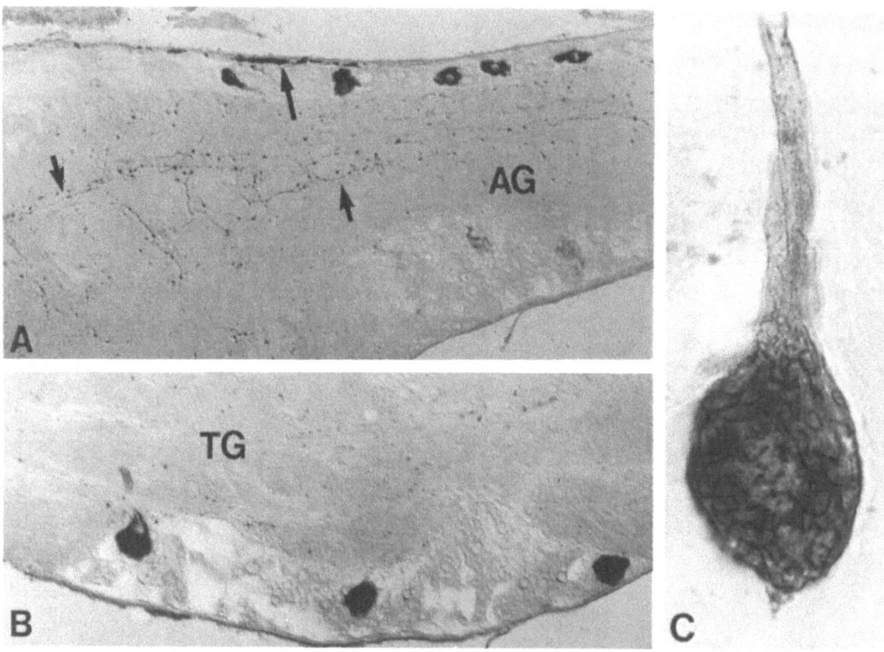

Fig. 2 A–C. Immunocytochemical preparations of the thoracic ganglion of *Calliphora* (gastrin/CCK-like immunoreactivity; PAP procedure). **A** Detail of region of fused abdominal ganglia showing immunoreactive cells in dorsal rind. Note material in dorsal sheath and extensive pathways of immunoreactive fibres in neuropile (*arrows*). **B** Detail of thoracic ganglion showing symmetrical arrangement of cells associated with each of the three component ganglia. **C** Detail of one cell. **A** x 330; **B** x 320; **C** x 1,600

of the thoracic ganglionic mass of older flies contains significant amounts of immunoreactive material and even in purely ICC preparations it is possible to trace elaborate gastrin/CCK pathways seemingly running in an anterior/posterior direction (Fig. 2A). We have also observed material leaving the neuropile of the suboesophageal ganglion, apparently joining up with that in the thoracic ganglion via the cervical connective. (It could be, of course, that the material is passing in the opposite direction, from the thoracic ganglion to the head.) Finally, gastrin/CCK immunoreactive material is observed in several of the abdominal nerves leaving the fused thoracic and abdominal ganglionic mass and also in considerable quantity in a possibly new neurohaemal region in the dorsal sheath of the ganglion [6]. Direct peptidergic innervation is suggested by the former and distribution of the material to the haemolymph by the latter.

In the corpus cardiacum, gastrin/CCK COOH-terminal specific material is present in the intrinsic neurosecretory neurones as well as within the neuropile of this organ. It is also present within the nerve that leaves the CC dorso-posteriorly (variously known as the cardiac or aortic nerve, or posterior extension of the cardiac-recurrent nerve). Immunoreactive material is certainly not present in the hypercerebral ganglion neither has it been observed in the corpus allatum nor in the nerve fibres running ventrally from the CC in the direction of the crop and crop duct.

3. Biochemical Studies on the Gastrin/CCK-Like Peptides of *Calliphora*

Results from ICC provide evidence of a gastrin/CCK COOH-terminal specific material widespread within the nervous system of *Calliphora*. Thus, whilst the results just described were all obtained with COOH-terminal specific antisera, negative results were seen with all types of NH_2 terminally-directed antisera. More recently we have obtained some evidence for immunoreactivity against certain mid-core CCK antisera, but these studies require amplification and verification.

In a further attempt to identify the material we have collaborated with Prof. G.J. Dockray of Liverpool University in isolation and purification studies of extracts of the heads of *Calliphora* [40]. The technique adopted was essentially that used in vertebrate studies [41] and consisted of extraction of the material in boiling water followed by immunoaffinity purification according to the method developed by Dockray et al. [25] for the purification of CCK8 from rat brain. This technique is extremely useful in removing possible non-specific effects in the subsequent RIA. The antibody used in this study (L112) was one of the COOH-terminal antisera used successfully in ICC. The IgG fraction of the antiserum was conjugated to Sepharose beads after which the reconstituted freeze dried head extract was applied to the column. Elution of the absorbed gastrin/CCK immunoreactive material bound to the antibody-Sepharose complex was carried out with 15% formic acid. The immunoaffinity concentrates were lyophilised, reconstituted in 0.05 *M* ammonium bicarbonate and fractionated at 4 °C on Sephadex G-50 with 0.02 *M* sodium barbitone as the eluent. Radioimmunoassay of the column eluates was carried out with several antisera having the same range of specifities as described for the ICC studies. With appropriate tracers and standards it could be shown that a peptide present in the extracts behaved in the manner of the common

COOH-terminal fragments of gastrin and CCK. Thus, the material reacted with three C-terminal-specific antisera, but it did not react at all with antisera specific for either intact G17 or its N-terminus. Values for the concentration of immunoreactive material were approx. 2 pmol/g of heads used in the initial extract when using the antiserum L112. The material caused an inhibition of the binding of tracer (^{125}I-labelled desulphated CCK8) to the antiserum L112 in parallel to the inhibition shown by the synthetic sulphated CCK8 used as a standard (Fig. 3 A). Other C-terminus antisera also gave parallelism in the RIA although they showed up to a 20-fold lower concentration of the immunoreactive material.

Fractionation of the material on Sephadex G-50 produced an elution pattern not in agreement with either the G17 or CCK8 standards (Fig. 3 B). This, of course, suggests a lack of correspondence (in absolute terms) with the structures of either of these two peptides and one must assume that the *Calliphora* material is different in some respect from known mammalian forms of either gastrin or CCK. Further studies aimed at characterizing the peptides more completely are currently in progress.

4. Opioid-Like Peptides

The discovery of the opioid peptides in vertebrates in 1975 [42] was shortly followed by a report stating that similar substances were not present in invertebrates [43]. Since then, however, there have been several observations pointing to the contrary. In insects, evidence of opiate-like substances has been based on immunocytochemical studies [44, 45] and also on receptor binding studies [46, 47]. In *Calliphora* we have screened a variety of vertebrate opioid peptide antisera and have demonstrated neuropeptides that are α-endorphin and enkephalin immunoreactive. The α-endorphin-like material contrasts sharply with that which is gastrin/CCK immunoreactive. Thus, it is restricted to a few (12–14) of the median neurosecretory cells of the brain (Fig. 1 C,D) and it is not present in any other cells elsewhere in the nervous system. From the MNC, the α-endorphin-like material may be followed in axons of the median bundle that lead into the corpus cardiacum. The axons carrying the immunoreactive material appear to pass through the CC in the cardiac-recurrent (aortic) nerve, dorsal to the proventriculus in the general direction of the abdomen. Unlike the gastrin/CCK-like peptides, the α-endorphin-like material is not found elsewhere within the neuropile of the brain or thoracic ganglion.

The particular antisera used in this study (raised against synthetic α-endorphin conjugated to thyroglobulin) have been shown in the enzyme-linked immunosorbent assay (ELISA) to be highly specific for this peptide, the 61–76 sequence of a precursor molecule β-lipotrophin (β-LPH). The antigenic determinant is believed to reside in the region of the C-terminus, but this has not yet been fully established.

The opioid peptides of vertebrates are proving to be a complex array of substances formed from at least two series of precursor molecules [48]. The sequence of α-endorphin is contained within the 61–91 sequence of β-endorphin which, in turn, is part of the larger 1–91 sequence of β-LPH. Interestingly enough, although the 61–65 sequence of met-enkephalin appears in the larger molecule of β-endorphin (and, of course, α-endorphin

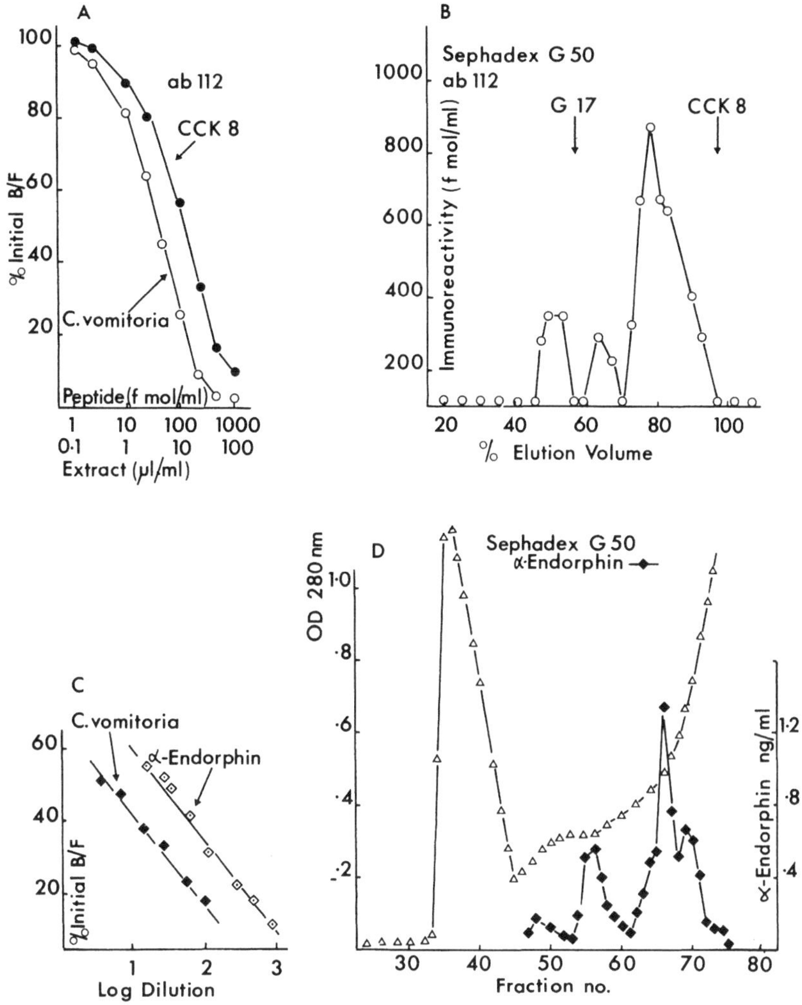

Fig. 3 A–D. Details of stages in the purification of the gastrin/CCK-like (**A, B**) and α-endorphin-like (**C, D**) peptides of *Calliphora*. **A** Radioimmunoassay dilution curve of immunoaffinity-purified boiling water extract of *Calliphora* heads compared with CCK8 standard. (Antibody L112; ^{125}I-labelled desulphated CCK8. **B** Sephadex G-50 elution profile of the immunoaffinity-purified extract assayed against antibody L112. The markers for gastrin G17 and CCK8 are shown. (Redrawn from Ref. [40]). **C** Radioimmunoassay dilution curve of Sephadex G-50 1% formic acid-eluted α-endorphin immunoreactive material compared with synthetic α-endorphin standard (Antibody αEI (7), ^{125}I-labelled synthetic α-endorphin. **D** Sephadex G-50 1% formic acid elution profile of 0.2 *N* HCl extract of *Calliphora* showing the α-endorphin-like immunoreactive peaks. [The met-enkephalin-immunoreactive material (not shown) follows immediately after the α-endorphin peak]

too) these are not thought of as likely precursors because of the adjacent Thr^{66} - Ser^{67} sequence which is a poor cleavage site. Furthermore, in mammalian studies, met-enkephalin and β-endorphin are found to be distributed in different regions of the brain and it is now believed that the enkephalins have precursors with either an Arg–Phe, Arg–Lys or Arg–Arg sequence immediately adjacent to either the Met or Leu residue.

These interesting problems have a bearing on our findings in *Calliphora*. Here too, we find Met- and Leu-enkephalin immunoreactive material in neurones situated in regions of the brain and thoracic ganglion distinct from those containing α-endorphin [49]. Our results with β-endorphin are inconclusive and require further studies with different antisera from those available to us at the moment.

5. Biochemical Studies on the α-Endorphin-Like Peptides

In order to characterise the α-endorphin-like peptides of *Calliphora* we have recently begun a series of extraction and purification studies. The procedure we have adopted is a modified version of that used in vertebrate opioid peptide studies with the initial extraction taking place in 0.2 *N* HCl. Despite a favourable ratio (1:10) of fly material (bodies/heads/or whole flies) to extraction fluid, the method presents severe problems. These mainly arise from the eye pigments of *Calliphora* and from the large amounts of fat present within the fat bodies. Centrifugation alone does little to clarify the extracts, but with a succession of filtration (especially by means of Hyflow) and centrifugation stages it is possible to obtain a sufficiently clear extract to apply to a G-50 Superfine column. The eluted fractions (1% formic acid) from such a column show clear, if broad, separate peaks of both α-endorphin- and met-enkephalin-immunoreactive materials (Fig. 3D). The radioimmunoassays by which both peaks can be identified are specific for the two peptides as judged by parallelism with the standards (Fig. 3C) and by the lack of cross-reactivity of the antisera with other of the known opioid peptides. The α-endorphin immunoreactive material has been further purified by high pressure liquid chromatography. In a solvent system comprising 40%–60% acetonitrile with 0.01% TFA, the *Calliphora* peptide is seen to have a retention time approximately equal to that of synthetic α-endorphine. Similarly, when this purified material is subjected to polyacrylamide gel electrophoresis the Rf value corresponds almost precisely with that of the synthetic peptide. These results certainly suggest a close similarity of the invertebrate and vertebrate α-endorphin peptides, but the identity of the *Calliphora* material will only be determined when sufficient quantities are obtained for amino acid composition and sequence studies. The identification of the met-enkephalin material of *Calliphora* has not progressed as far as that of α-endorphin, but the only problems to be overcome are those associated with the known difficulties of radioimmunoassaying this peptide.

6. Concluding Comments

The results of the immunochemical studies on *Calliphora* presented in this brief review suggest that at least two biologically active peptides previously considered as "vertebrate" may also occur at a position much lower down in the phylogenetic scale. Comparable findings for these and other peptides in other invertebrates are being obtained by laboratories elsewhere.

Despite the fact that we have perfectly good examples to illustrate the common occurrence of certain molecules such as respiratory pigments and enzymes in both vertebrates and invertebrates, nevertheless there are several difficulties in extending this idea to encompass biologically active molecules of the type discussed here. The main cause for concern and doubt is that apart from one or two studies that have taken the identification process as far as amino acid composition [8, 50], the occurrence of vertebrate-type peptides in invertebrates has to be judged almost entirely upon immunological evidence. Frequently, immunocytochemical studies are used as the sole means of identification and whilst results of such investigations are undoubtedly useful as a first step in the characterisation process, it is essential to follow this up by extracting and purifying the peptides. The combined use of ICC and RIA gives a greater degree of assurance that material closely similar to the vertebrate peptide(s) is indeed present in the tissues. The ultimate proof of common identity, of course, rests upon the determination of the amino acid sequence of the particular peptide and so far there are no instances where this has been successfully completed[2], largely because invertebrate material is usually in short supply and difficult to work with.

The other major difficulty in accepting the idea of vertebrate-type peptides in invertebrates relates to their functional significance. For certain of the key metabolic hormones of vertebrates such as insulin and glucagon, it is not conceptually difficult to transfer their known functions across to invertebrates, particularly when they are organisms such as *Calliphora*, known to have a high dependency upon carbohydrates as a fuel source. The greater difficulty is presented by neuropeptides such as those of the gastrin/CCK and endorphin series, the function of which, even in vertebrates, is at the present time uncertain. It may eventually be possible to relate a proven vertebrate function to these peptides in invertebrates, but perhaps we have to think in a totally different direction and invoke completely new ideas on functions. This is similar to the immediate problem in vertebrate neuropeptide studies where the well-known functions of a peptide such as CCK within the gut prove something of a mental handicap to thoughts on its function as a neuropeptide within the brain.

In our studies of the gastrin/CCK peptides of *Calliphora* we have suggested that there may be at least two quite separate and different roles associated with the two identified pathways. Correspondingly, we may, of course, be dealing with two different peptides derived from a common precursor in much the same manner as seen in the vertebrates.

The future prospects for studies of the vertebrate-type peptides in *Calliphora* and other invertebrates will depend very much on the continuing advances being made in protein and peptide separation techniques. In particular, developments in HPLC will continue to play a major role in this work. On the other hand, it may prove impossible to obtain pure peptide in sufficient quantity for sequence analysis despite the great advances in micro-sequencing which have been made recently. In this case, the newer advances in DNA technology and particularly in the use of vertebrate cDNA probes to look for the corresponding invertebrate mRNA appropriate for a specific peptide will be of the utmost value.

---

2 See, however, the identification of opioid peptides in molluscs, referred to on p. 23

Finally, the physiological approach is one that cannot be neglected for much longer. Ultimately it has to be proved that the identified peptides have a specific role in the life of the invertebrate concerned whether it be as hormone, neurohormone, neurotransmitter or neuromodulator, and experiments designed to elucidate such roles are required most urgently.

Acknowledgments. We gratefully acknowledge the support of the Science and Engineering Research Council of Great Britain for research grants GR/B88136 and GR/C20789 to A.T.

We would also like to record our sincere thanks to our collaborators in various phases of the studies reviewed here: Prof. G.J. Dockray, Physiology Laboratory, Liverpool University, U.K.; Prof. J.F. Rehfeld, Department or Clinical Chemistry, Rigshospitalet, Copenhagen University, Denmark; Dr. N.J. Strausfeld, European Molecular Biology Laboratory, Heidelberg, West Germany, and Mr. Alan Scott, School of Biological Sciences, Queen Mary College, London, U.K. We also thank Prof. Dockray, Prof. Rehfeld, and Dr. Sue Jackson, Pituitary Hormone Laboratory, St. Bartholomew's Hospital, London, U.K. for gifts of antisera and peptides without which these studies could not have been undertaken. We are extremely grateful to Mrs. Lorna A. Wheatley for typing the manuscript.

References

1. Iida Y (1982) Bull Chem Soc Jpn 55:2638–2684
2. Bodenmüller H, Schaller HC (1981) Nature (London) 198:685
3. Duve H, Thorpe A (1979) Cell Tissue Res 200:187–191
4. Duve H, Thorpe A, Lazarus NR (1979) Biochem J 184:221–227
5. Duve H, Thorpe A (1980) Cell Tissue Res 210:101–109
6. Duve H, Thorpe A (1982) Cell Tissue Res 227:67–77
7. Duve H, Thorpe A, Neville R, Lazarus NR (1981) Biochem J 197:767–770
8. Duve H, Thorpe A, Lazarus NR, Lowry PJ (1982) Biochem J 201:429–432
9. Thorpe A, Duve H (1984) Mol Physiol 5:235–260
10. Duve H, Thorpe A (1984) In: Bořkovec AB, Kelly TJ (eds) Insect neurophysiology and neurochemistry. Plenum Press, New York, pp 171–195
11. Thorpe A, Duve H (1984) In: Bořkovec AB, Kelly TJ (eds) Insect neurophysiology and neurochemistry. Plenum Press, New York, pp 197–222
12. El-Salhy M, Abou-El-Ela R, Falkmer S, Grimelius L, Wilander E (1980) Regul Peptides 1:187 to 204
13. El-Salhy M, Falkmer S, Kramer KJ, Spiers RD (1983) Cell Tissue Res 232:295–317
14. Kramer KJ (1979) In: Miller TA (ed) Experimental entomology: Insect neurohormones. Springer, Berlin Heidelberg New York
15. Tager HS, Markese J, Spiers RD, Kramer KJ (1975) Nature (London) 254:707–708
16. Tager HS, Markese J, Kramer KJ, Spiers RD, Childs CN (1976) Biochem J 156:515–520
17. Yui R, Fujita T, Ito S (1980) Biomed Res 1:42–46
18. Endo Y, Iwanaga T, Fujita T, Nishiitsutsuji-Uwo J (1982) Cell Tissue Res 227:1–9
19. LeRoith D, Lesniak MA, Roth J (1981) Diabetes 30:70–76
20. Sanders B (1983) Gen Comp Endocrinol 50:366–328
21. Schot LPC, Boer HH, Swaab DF, Van Noorden S (1981) Cell Tissue Res 216:273–291
22. Sundler F, Håkanson T, Alumets J, Walles B (1977) Brain Res Bull 2:61–65
23. Vanderhaeghen JJ, Signeau JC, Gepts W (1975) Nature (London) 257:604–605
24. Dockray GJ (1976) Nature (London) 264:568–570
25. Dockray GJ (1980) Brain Res 188:155–165
26. Rehfeld JF (1978a) J Biol Chem 253:4016–4021
27. Rehfeld JF (1978b) J Biol Chem 253:4022–4030
28. Rehfeld JF (1978) Nature (London) 271:771–773

29. Dockray GJ, Vaillant C, Hutchinson JB (1981) In: Grossman MI, Brazier MAB, Lechago J (eds) Cellular basis of chemical messengers in the digestive system. Academic Press, London New York, pp 215–230
30. Larsson L-I, Rehfeld JF (1979) Brain Res 165:201–218
31. Vigna SR (1979) Gen Comp Endocrinol 39:512–520
32. Grimmelikhuijzen CJP, Sundler F, Rehfeld JF (1980) Histochemistry 69:61–68
33. Thorndyke MC (1982) Regul Peptides 3:281–288
34. Larsson L-I, Rehfeld JF (1977) J Histochem Cytochem 25:1317–1321
35. Larsson L-I (1980) In: Jerzy Glass GB (ed) Gastrointestinal hormones. Raven Press, New York, pp 53–70
36. Duve H, Thorpe A (1981) Gen Comp Endocrinol 43:381–391
37. Duve H, Thorpe A, Strausfeld NJ (1983) J Neurocytol 12:847–861
38. Bishop CA, O'Shea M (1982) J Comp Neurol 207:223–238
39. Duve H, Thorpe A (1984) Cell Tissues Res (in press)
40. Dockray GJ, Duve H, Thorpe A (1981) Gen Comp Endocrinol 45:491–496
41. Dockray GJ (1979) Fed Proc 38:2295–2301
42. Hughes J, Smith TW, Kosterlitz HW, Fothergill LA, Morgan BA, Morris HR (1975) Nature (London) 258:577–579
43. Simantov R, Goodman R, Aposhian D, Snyder SH (1979) Brain Res 111:204–211
44. Rémy Ch, Dubois MP (1981) Cell Tissue Res 218:271–278
45. Rémy Ch, Girardie J, Dubois MP (1978) CR Acad Sci Ser D 286:651–653
46. Stefano GB, Scharrer B (1981) Brain Res 225:107–114
47. Stefano GB, Scharrer B, Assanah P (1982) Brain Res 253:205–212
48. Imura H, Nakai Y (1981) Annu Rev Physiol 43:265–278
49. Duve H, Thorpe A (1983) Cell Tissue Res 233:415–426
50. Kramer KJ (1983) In: Comprehensive insect physiology, biochemistry and pharmacology, vol VII. Endocrinology, chap 10. Pergamon Press, New York

I.13 Peptides from the Corpora Cardiaca and CNS of Locusts

W. MORDUE and P.J. MORGAN[1]

1. Introduction

It is somewhat surprising to realise that our knowledge of the chemistry of insect neurohormones is still rudimentary, it is only for adipokinetic hormone (AKH) and proctolin that structures are known [1, 2]. There are many reasons for this paucity of information and lack of progress [3, 4]. However, it is important to try and assess why so little progress has been made with the chemistry of peptides from insects whilst in recent years substantial progress has been made in other invertebrate groups - especially molluscs, crustacea and coelenterates. Much is known of the involvement of peptidergic materials as putative hormones and regulators in the control of developmental, physiological, and metabolic processes in insects. However, more rapid progress in our understanding of the endocrinology of peptides will be possible only when more is known of their structural chemistry. Over the past few years vast improvements in the techniques of chromatography and amino acid analysis have been made, exemplified by HPLC and microsequencing methods. These methods have already had a major impact upon insect neurohormone research. However, the main obstacle which impedes progress in the characterization of insect peptides is the bioassay; until more routine and reliable methods are devised our knowledge of the structural chemistry of insect peptides will progress only slowly. This is an area in which we are concentrating our research efforts. Accurate estimation of circulatory titres, perfection of RIA methods and cataloguing of chromatographic indices is possible only when quantitative and routine chemical methods are established. Receptor binding and degradation in vivo studies are also impossible until radio-labelled peptides of high specific activity are available. The only radio-labelled invertebrate peptides available until recently have been produced using radioiodination; however the large size of the iodine moiety relative to the mass of a small peptide limits the usefulness of such compounds in cellular investigations and binding studies.

This paper is concerned with two hormones: diuretic hormone and adipokinetic hormone which are both released from the corpora cardiaca of locusts. There is a considerable amount of information known concerning the characterisation and modes of action of these hormones. This material in itself is of special interest but both peptides in their own right and in their relationships with similar peptides from other insects

1 Department of Zoology, University of Aberdeen, Tillydrone Avenue, Aberdeen AB9 2TN, Great Britain

Biosynthesis, Metabolism and Mode of Action of Invertebrate Hormones
(ed. by J. Hoffmann and M. Porchet)

highlight many of the problems currently facing the insect-peptide endocrinologist. These problems encompass the establishment of endocrine, modulatory or transmitter functions: pharmacological and physiological effects; antigenic determinants and immuno-chemistry; and affinities in chemical composition.

2. Diuretic Hormone (DH)

In locusts the corpora cardiaca comprise two more or less discrete lobes. The ventral storage lobes contain material which originates predominantly, but not exclusively, from the cerebral neurosecretory cells. Storage lobe extracts are known to contain a number of biologically active factors [5]. Such a factor is diuretic hormone (DH). This neurosecretory peptide exerts a key role in the regulation of water balance. Its existence has been known for some time but its structure is still unknown. This has in part been due to the lability of the hormone; but in *Locusta* is probably also due to the small amounts of available starting material, compared with a peptide such as adipokinetic hormone (AKH) [6]. Moreover, it is still difficult to establish with certainty that the isolated material has been recovered in its full biologically active form. Serious problems have been encountered also with loss of hormonal activity by absorption onto surfaces and to increasing instability during purification.

2.1 Isolation

In our initial attempts to isolate DH from methanol extracts of storage lobes reversed-phase HPLC was employed [7]. These experiments revealed that locust DH chromatographs as at least two separate peaks of activity, suggesting that locust DH might exist in at least two distinct molecular forms. This separation of DH into two peaks of activity has been confirmed using a variety of columns. It was clear from these separations that methanol extracts of storage lobes are complex mixtures and that a chromatographic step prior to reversed-phase HPLC would be needed. It was decided to use High Performance Size-Exclusion Chromatography (HPSEC) using a TSK 2000 SW column, with a predicted fractionation range of 500 to 70,000 for proteins/peptides. Using a 30 cm column eluted in 0.1% trifluoroacetic acid (TFA) at a flow rate of 1 ml/min, fractions of 1 ml were collected at 1 min intervals and lyophilized prior to bioassay, (for bioassay method, see [6]. DH eluted between 10-12 min after injection (i.e. fractions 11 and 12). Accurate molecular weight estimations are not generally possible using HPSEC columns; however by calibrating the column with a number of marker molecules, it was found that diuretic hormone eluted close to AKH, oxytocin and vasopressin, confirming previous estimates of a molecular mass of $< 2{,}000$, and suggesting a molecular mass close to 1,000.

Each fraction produced from the separation of corpora cardiaca extracts on the TSK column were subsequently chromatographed on a Spherisorb OD5-2 column using the linear gradient system described previously [4, 7], it was possible to predict which peaks corresponded to DH. These results showed that the peaks corresponding to DH I and II occurred in the TSK fractions which had eluted between 10 and 12 min, as predicted by the bioassay data.

However following this two stage separation, the peaks containing predicted DH activity were not resolved sufficiently to allow the collection of uncontaminated hormone for amino acid analysis. Therefore the gradient system was modified as follows to improve the resolution: a slower linear gradient was employed running from 15% to 50% acetronitrile over 50 min; also the buffer strength was reduced to 0.025 *M* ammonium acetate, allowing the U.V. wavelength to be lowered from 230 to 220 nm. The absorbance profile and the elution conditions employed for the separation of a TSK fraction II are shown in Fig. 1. The regions of predicted DH I and DH II activity are shown by solid horizontal bars. A peak corresponding to DH I can be observed, but no peak for DH II. It is not clear from these separations whether the predicted DH I peak is due to a single compound. However, the DH I peak could not be resolved into further peaks using even a slower stepped gradient, suggesting that the absorbance peak corresponding to DH I might be due to only a single compound.

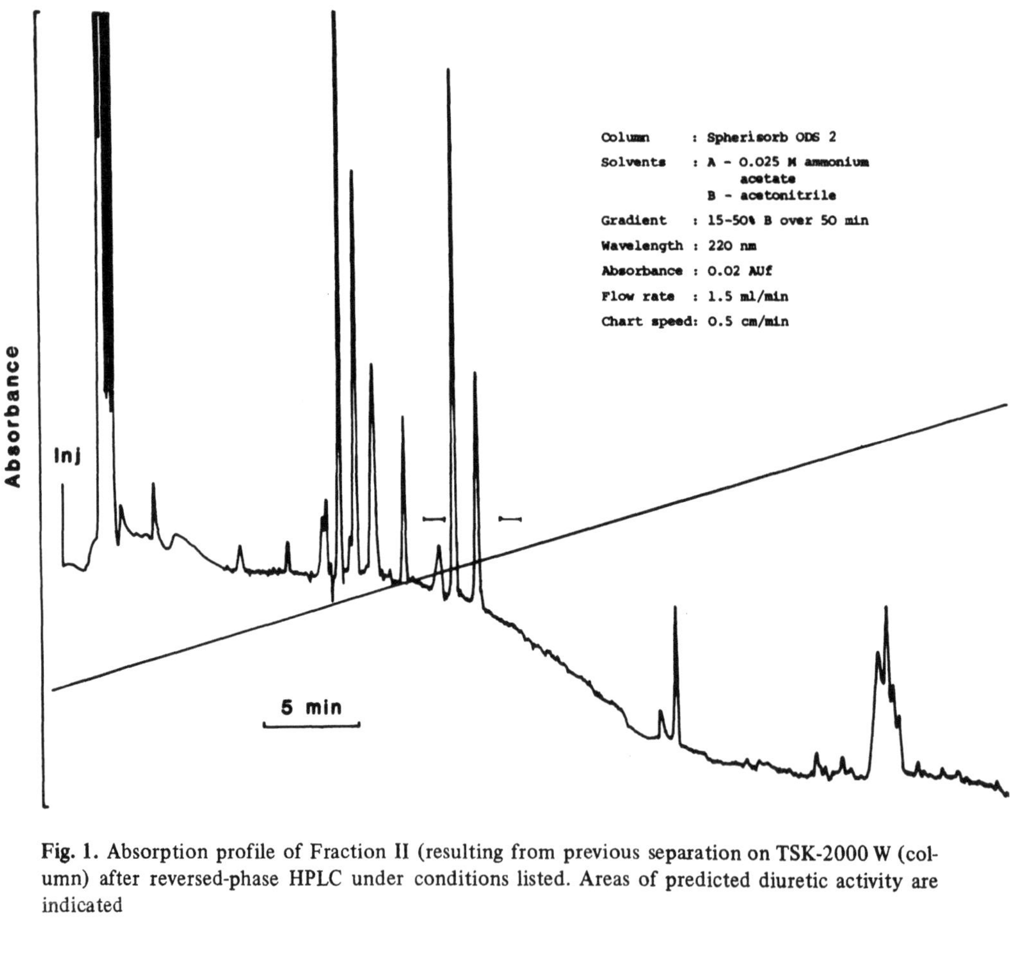

Fig. 1. Absorption profile of Fraction II (resulting from previous separation on TSK-2000 W (column) after reversed-phase HPLC under conditions listed. Areas of predicted diuretic activity are indicated

2.2 Bioassay

In this work a particular bioassay procedure [6] has been used more or less exclusively to identify DH activity present in methanol extracts or separated extracts from HPLC or HPSEC. The assay used is the ability of the test compound to increase fluid secretion by the Malpighian tubules. Such an assay is entirely consistent with the known physiological roles of the hormone. It is essential that the biological events which form the basis of a bioassay are related to the function of the hormone in vivo. At present other bioassay methods are also being investigated, e.g. the activation of adenylate cyclase in Malpighian tubules and their cell membranes. Once activity is detected it is hoped that chromatographic indices may be used to identify peaks with expected biological activity.

2.3 Comparison with Other Diuretic Peptides

Reversed-phase HPLC has demonstrated consistently two peaks of diuretic activity which are distinctly separated from one another [4, 7]. The peptides were separated from methanol extracts of storage lobes and therefore each peak contains a peptide of relatively small molecular mass. The peptides are therefore unlike the "high" and "low" molecular weight forms on DH that have been isolated from the thoracic ganglia of *Rhodnius* [8].

Moreover, these two diuretic peptides are isolated from the corpora cardiaca, and are chemically dissimilar to either vasopressin or the vasopressin-like diuretic factor identified in the sub-oesophageal ganglion (SOG) by Proux and colleagues [9, 10]. It has been demonstrated [4, 7] that arg-vasopressin, lys-vasopressin, vasotocin, and isotocin each chromatograph with distinctly shorter retention times than either DH I or DH II. In addition arg-vasopressin is inactive when assayed either dye elimination by locust tubules [10] or by fluid secretion [11].

Oxytocin, however, chromatographs almost identically with DH. Oxytoxin has been tested in a variety of ways but even concentrations as high as 10^{-5} *M* it has no diuretic effect on locust Malpighian tubules. Moreover Proux and Rougon-Rapuzzi [9] reported that the radioimmunoassay for vasopressin did not cross-react with oxytocin. In a recent study, it has been shown that the molecular mass of the arg-vasopressin-like molecule is ca. 2,500, twice that of arg-vasopressin (M.W. 1,080); furthermore, the two molecules have different isoelectric points and electrophoretic mobilities [12]. Estimates for the molecular mass of DH eluted from the size-exclusion column (TSK 2000 W) predict a size of ca. 1,000.

From these comparisons and from our data it seems unwarranted to describe the diuretic peptide isolated by Cupo and Proux [12] as vasopressin-like. It shows only some common antigenic determinants at the C-terminus with arg-vasopressin but differs substantially in many other chemical and physiological properties. Therefore from chromatographic and biological data we conclude that DH is not arg-vasopressin and from biological data that it is not oxytocin. The immunocytochemical and other evidence suggesting a similarity between mammalian vasopressin and DH may have to be re-examined.

2.4 Distribution of DH Within the CNS

Different regions of the CNS-brain, SOG, thoracic, and dominal ganglia have been prepared as aqueous extracts and tested for diuretic activity. The brain, SOG, and thoracic ganglia were all found to contain extractable diuretic activity [4, 6]; the terminal abdominal ganglion also has detectable activity. These results confirm the findings of Proux et al. [10] that the SOG of *Locusta* contains significant amounts of extractable DH. In addition, there is a striking similarity between the dose-response curves obtained in both studies. The results are similar to the findings for *Carausius morosus* [13]; in which similar amounts of diuretic activity were found in the various parts of the nervous system tested.

The nature of the diuretic activity in each ganglion was assessed to established whether it was due to the same factors in all ganglia or to different diuretic compounds. Brain, SOG, and thoracic ganglia were extracted in methanol and subsequently delipidated with dichloromethane. These extracts were then chromatographed on a Waters μ-Bondapak reversed-phase column using the method of Morgan and Mordue [6] and for each tissue a peak corresponding to DH I, possessing biological activity, was eluted. This result strongly supports the view that the diuretic activity extractable from the ventral nerve cord is chemically similar to DH I extractable from corpora cardiaca.

3. Adipokinetic Hormone (AKH)

AKH is a blocked decapeptide of the following structure: Gln–Leu–Asn–Phe–Thr–Pro–Asn–Trp–Gly–ThrNH_2. AKH is synthesised in the glandular lobes of the corpora cardiaca of locusts and is released in response to flight stimuli. Its particular actions are to mobilise lipid and to effect changes in metabolism which enable long term flight to occur [14].

3.1 Isolation and Characterisation of AKH

This peptide is characterised fully and is available in synthetic form commercially. A very full account of the methods employed in its successful isolation and characterisation is given by Stone and Mordue [15]. More recently we have been using AKH to standardise separation methods using both HPSEC and HPLC. We have shown that under HPSEC conditions given in [4] that the adipokinetic activity of glandular lobes cochromatographs exactly with synthetic AKH. The recovery of biological activity at this step is more than 90% and this preliminary purification is effected within some 11 to 12 min. Methanolic extracts of glandular lobes or lyophilised fractions from the TSK column have been subjected to reversed-phase HPLC. A typical elution profile is seen in Fig. 2; two peaks of adipokinetic activity have been detected. As indicated peak a is AKH and we predict that peak b is AKH II [16] because it possesses substantial adipokinetic activity.

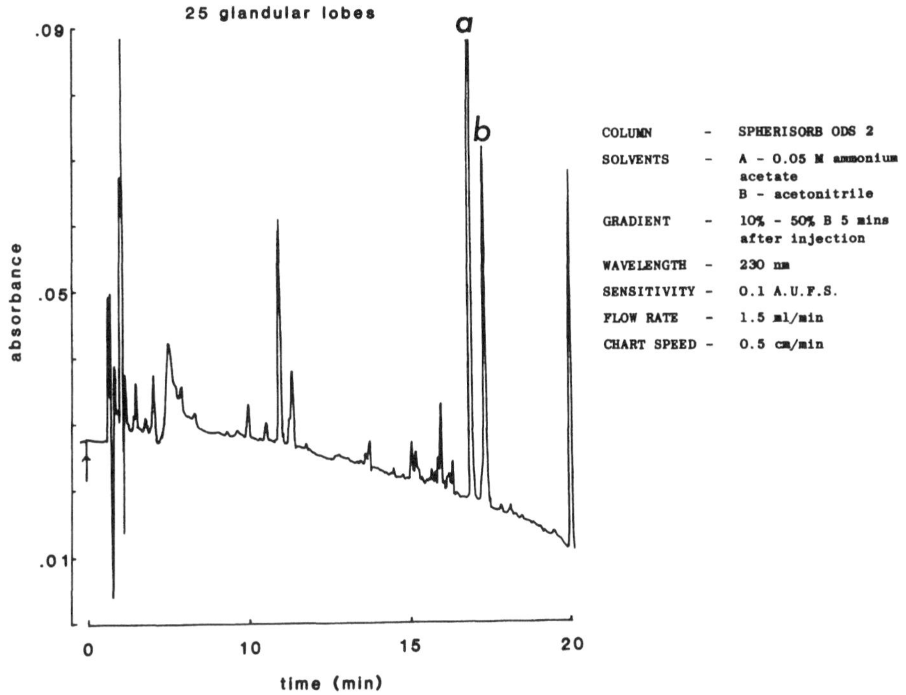

Fig. 2. Absorption profile of a methanol extract of 25 glandular lobes of *Locusta* after reversed-phase HPLC under conditions listed. Peaks a and b possess substantial adipokinetic activity: *a* AKH I, *b* AKH II

3.2 AKH and Related Peptides

It is likely that AKH belongs to a group of peptides which exert an important control over intermediary metabolism – especially those aspects concerned with the mobilisation of substrates. The presence of hyperglycaemic peptides has been reported in most orders of insects. AKH is able to induce hyperglycaemia in insects other than locusts and it has been of interest to investigate whether the well documented hyperglycaemic factors present in *Periplaneta* and other insects are related in structure to AKH. For no clear reasons this hyperglycaemic factor has eluded full characterisation in spite of the considerable efforts of a number of workers. *Carausius* possesses a corpus cardiacum factor which is strongly hyperglycaemic in roaches and hyperlipaemic in locusts [17]; this factor is also not yet characterised [18]. Characterisation using reversed-phase HPLC has shown that the factor is very similar to AKH [19]. It is of interest that a whole range of metabolic effects of such corpus cardiacum extracts can be exhibited. In the Colorado beetle and stick insects the actions of corpus cardiacum extracts upon metabolism are more difficult to detect. For example in the beetle such extracts affect proline metabolism; proline is the primary energy source for flight in this beetle and significant changes in haemolymph lipid and carbohydrate levels are not detectable [20]. This emphasises the need for correct bioassays; perhaps the assay often employed for AKH-related peptides may be inappropriate to the normal physiological role.

Table 1. Amino acid compositions of a number of AKH-related peptides

	AKH I [1]	AKH II [16]	Neurohormone [21] D	M I [22] Roach	M II [22] Roach	Roach [23]	Roach [24]	Carausius [19]
Ala						√	√	√
Asn (p)	2	1	2	2	1	√	√	√
Gln	1	1	1	1	1	√	√	√
Gly	1	1				√	√	√
Leu	1	1			1	√	√	√
Phe	1	1		1	1	√		√
Pro	1		1	1	1	√	√	?
Ser		1	1	1		√	√	√
Thr	2	1	1		2			√
Trp	1	1		1	1	?	?	?
Val			1	1		√	√	√

Precise numbers of residues are given if known: peptides of uncertain composition may well be mixtures

The similarities in amino acid composition of a number of peptides related to AKH but with apparently different metabolic and physiological actions are summarised in Table 1.

4. Summary

AKH and DH are critical importance to locusts for the control of metabolism and homeostasis. From the wide-ranging occurrence of peptides with adipokinetic activity and many similarities in amino acid composition it is likely that AKH-related peptides are common to many insect groups, e.g. locusts, grasshoppers, cockroaches, stick insects, moths, butterflies, beetles, flies, and honeybees. Further details on composition and roles are awaited. For the locust this is especially so for AKH II which may be a hormone or may be a metabolic byproduct of the production, release or degradation of AKH I. With respect to DH although the phenomenon of diuresis is common to many insects and under hormonal control it is likely that the hormones from different insects will prove to be very different in chemical nature.

Acknowledgment. The authors would like to gratefully acknowledge the support of AFRC, who funded this work.

References

1. Stone JV, Mordue W, Batley KE, Morris HR (1976) Nature 263:207–211
2. Starrat AN, Brown BE (1975) Life Sci 17:1253–1256
3. Mordue W, Stone JV (1980) Insect Biochem 10:229–239
4. Mordue W, Morgan PJ (1984) In: Borkovec E (ed) Insect neurochemistry and neurophysiology. Plenum, New York (in press)

5. Goldsworthy GJ, Mordue W (1974) J Endocrinol 60:529–558
6. Morgan PJ, Mordue W (1984) Physiol Entomol. In press
7. Morgan PJ, Mordue W (1983) Comp Biochem Physiol 75B(1):75–80
8. Aston RJ, White AF (1974) J Insect Physiol 20:1673–1682
9. Proux J, Rougon-Rapuzzi G (1980) Gen Comp Endocrinol 42:378–383
10. Proux J, Rougon G, Cupo A (1982) Gen Comp Endocrinol 47:449–457
11. Mordue W (unpublished observations)
12. Cupo A, Proux J (1983) Neuropeptides 3:309–318
13. Pilcher DEM (1970) J Exp Biol 52:653–665
14. Mwangi RW, Goldsworthy GJ (1977) J Comp Physiol 114:177–190
15. Stone JV, Mordue W (1980) In: Miller TA (ed) Neurohormonal techniques in insects. Springer-Verlag, New York, p 31–80
16. Carlsen J, Herman WS, Christensen M, Josefsson L (1979) Insect Biochem 9:497–501
17. Gade G (1979) Physiol Entomol 2:131–134
18. Mordue W (1981) In: Farner DS, Lederis K (eds) Neurosecretion molecules, cells, systems. Plenum Press, New York, p 391–401
19. Mordue W, Gade G (unpublished observations)
20. Weeda E (1981) J Insect Physiol 27:411–417
21. Baumann E, Gersch M (1982) Insect Biochem 12:7–14
22. Witten J, Schaffer M, O'Shea M (1983) Neuroscience Abs 25:1283
23. Jones J (1978) PhD Thesis, Univ of London
24. Traina ME, Bellino M, Serpietri L, Mass A, Frontali N (1978) J Insect Physiol 22:323–329

I.14 Adipokinetic Hormones in Locusts

G.J. GOLDSWORTHY and C.H. WHEELER[1]

1. Introduction

The actions of adipokinetic hormones are more diverse than their utilitarian name suggests. We will review here the known actions of these hormones in adult locusts to emphasise their importance in the integrated utilisation of metabolic fuels during flight. Other hormones are involved in the overall development of flight ability and control of flight metabolism [1], but we will restrict our attention to the adipokinetic peptides of locusts. Much work carried out in this area, especially during earlier years, necessarily used crude extracts of CC (corpora cardiaca) or their constituent glandular lobes as sources of hormone. It is now clear, however, that at least two adipokinetic peptides are present in such extracts. To distinguish between the two known peptides, we will use the abbreviation AKH–I for the peptide characterised by Mordue and his colleagues (this volume), and AKH–II for compound–II of Carlsen et al. [2]. Where no particular peptide is specified, or where tissue extracts have been used as a hormone source, we will use AKH as a general abbreviation for "adipokinetic hormone(s)".

2. Chemical Nature, Location, and Potency of Adipokinetic Hormones

2.1 Adipokinetic Hormone–I

AKH–I is a blocked decapeptide with the structure Glu–Leu–Asn–Phe–Thr–Pro–Asn–Trp–Gly–Thr–NH_2 [3,4]. It appears to be the most important of the adipokinetic peptides in the CC both in terms of its amount, and of its biological potency. Adipokinetic activity in locust CC was first demonstrated simultaneously by Beenakkers [5], and Mayer and Candy [6]. Subsequent investigations of the nature and action of AKH were facilitated by the development of a simple quantitative bioassay [7] and showed that the majority of biological activity is located in the glandular lobes of the CC. Recently, antibodies raised to a synthetic analogue, (Tyr_1)–AKH–I, have been used to identify apparent AKH-like immunoreactive material in the locust cerebral neuroendocrine system [8]. Although immunoreactive material is located mainly in the glandular lobes of the CC, it occurs also in groups of cells in the brain and suboesophageal ganglion

1 Department of Zoology, The University, Hull HU6 7RX, Great Britain

Biosynthesis, Metabolism and Mode of Action of Invertebrate Hormones
(ed. by J. Hoffmann and M. Porchet)

and in tracts in the circumoesophageal connectives, brain and storage lobes of the CC. The antiserum cross-reacts with authentic AKH–I, so it is not surprising that immunoreactivity is found in the glandular lobe cells but, because the antigen against which the antibodies were raised [8], is biologically inactive (the L-pyroglutamic acid N-terminal is an absolute requirement for adipokinetic activity [9], it is unknown how much of the authentic AKH–I molecule is recognised by the antiserum. The existence of immunoreactivity merely demonstrates groups of (related?) peptidergic neurones and the presence of biologically active AKH-like peptides cannot be stated with confidence. It would be interesting to screen the antiserum with a range of AKH–I analogues such as that used by Stone et al. [9] to determine the immunoreactive site(s). Even in the glandular lobes, much immunoreactive material may be due to biologically inactive molecules, or even to similar antigenic sites on totally different molecules. The presence of two adipokinetic peptides in CC confounds this situation further.

2.2 Adipokinetic Hormone–II

AKH–II is apparently an octapeptide [2] of similar composition to prawn RPCH (red pigment concentrating hormone [10]). Like AKH–I, AKH–II is confined largely to the glandular lobes, and appears to be released during flight [11, 12]. The peptide causes hypertrehalosaemia in head-ligated locusts [13], and it has been suggested [12, 13] that it may contribute to carbohydrate homeostasis during flight. Glycogen phosphorylase increases in activity within 5 min of flight initiation [14] but, while AKH–I is a potent activator of this enzyme [15], AKH–II appears to be relatively inactive. We have examined the relative efficacies of AKH–I and II in activating locust fat body glycogen phosphorylase [16]. Fifteen minutes after saline injection, phosphorylase activity (% of the total in the active form) is 17.5 ± 3.9% (Mean ± SE; n = 6) and injection of 0.02 pairs of glandular lobe equivalents of AKH–II does not increase this activity significantly (29.9 ± 4.32%, n = 6). A significant increase ($P < 0.01$) to 39.2 ± 3.2% (n = 6) is, however, obtained with a dose of 0.08 equivalents of AKH–II (this is the dose giving maximum lipid elevation in our locusts). With AKH–I, as little as 0.005 pairs of glandular lobe equivalents brings about a significant ($P < 0.001$) increase in phosphorylase activity to 49.6 ± 2.1% (n = 6), whereas injection of 0.02 equivalents of AKH–I is more than sufficient to cause maximum activation (61.9 ± 3.9%; n = 6). From these data it appears that, even allowing for differences in the amounts present [2] in the CC, AKH–II is considerably less potent in activating glycogen phosphorylase than is AKH–I.

AKH–II may be responsible for some lipid mobilisation during flight. Although this peptide is present in the CC in lesser amounts (ca. 20%) than AKH–I, it may have a similar lipolytic potency on a molar basis to AKH–I [2]. At first sight this appears unlikely if AKH–II is an octapeptide similar to RPCH; according to Stone and colleagues [9], only the decapeptide molecule, AKH–I, has full adipokinetic activity and AKH–II would be expected to have, at best, an activity of only 20% that of AKH–I at equimolar concentrations.

We have separated AKH–I and II in 6% butanol on Sephadex LH20 to assay the adipokinetic activity of natural AKH–I and II, and make comparisons with synthetic AKH–I and RPCH [17]. Additionally, we have measured the relative potencies of AKH–I and II in bringing about the characteristic changes in locust haemolymph lipoproteins [18]. In our *Locusta*, methanolic extracts of glandular lobes and natural AKH–I give maximum lipid mobilisation at a dose of 0.02 pairs of CC equivalents. With AKH–II, however, a lipid increase which is only 60% of the maximum AKH–II response requires as much as 0.06 equivalents; increasing this further to 0.2 equivalents does not increase the response. This is not due to differences in the timing of the responses to different hormones because the time courses for lipid mobilisation are identical. At optimum doses, therefore, AKH–II is only 60% as effective as AKH–I in eliciting maximum lipid mobilisation in locusts. Interestingly, in our locusts synthetic RPCH is as effective as AKH–I in stimulating lipid mobilisation when sufficient peptide is injected; a maximum response requires 20–30 pmol of synthetic RPCH, but only 2 pmol of synthetic AKH–I [17]. Injection of AKH–II does cause the formation of A^+ lipoprotein, as judged by both gel filtration and direct measurement using a heparin precipitation method [18–20]. The amounts formed are, however, significantly less than when using AKH–I, even when very high doses of AKH–II are used [17].

We estimate from preliminary analysis of the amino acid composition [21] that the molecular weight of AKH–II is ca. 906. If AKH–II is present in the CC at between 10% and 20% of the amount of AKH–I (ca. 5–10 ng estimated from reversed phase HPLC [22]), then we can tentatively compare their lipid mobilising activities on a molar basis (Fig. 1); they appear to have surprisingly similar activities at low doses, but at high doses, AKH–II cannot elicit as great a response as AKH–I. Thus, in defining structure/activity relationships for peptides such as AKH–I and II and their analogues, we advocate that a complete dose response curve for each peptide should be obtained before firm conclusions can be made concerning their relative activities. The response curves do not suggest a poor fit of AKH–II for AKH–I receptors. Does AKH–II act on the fat body via a population of receptors distinct from that for AKH–I?

3. Release of Adipokinetic Peptides During Flight

Release of adipokinetic hormones from the glandular lobes of the CC is under the control of secretomotor centres in lateral areas of the protocerebrum [23], and there is evidence for this control being octopaminergic [24, 25]. Severance of both the NCC–I and II is necessary to prevent lipid mobilisation during flight [26], but stimulation of the NCC–I in vitro does not release AKH [27]. Stimulation of only the NCC–II does release adipokinetic activity, and this is enhanced by simultaneous stimulation of the NCC–I. The NCC–I may, therefore, exert a modulatory influence on the release of AKH during flight [23, 27].

It has been suggested on numerous occasions that AKH release is caused by declining levels of haemolymph, trehalose during flight (see Goldsworthy [1]). Although the concentration of haemolymph carbohydrates can modulate release of AKH [28–30], it seems unlikely that their decreasing concentration in the haemolymph during flight is directly responsible for release of the hormones; injection of sucrose, which is not oxidised by the flight muscles, does not prevent the normal decrease in haemolymph trehalose during flight, but does delay the release of AKH. Clearly, more work on the supposed cerebral secretomotor centres is required to determine the causal stimuli for AKH release.

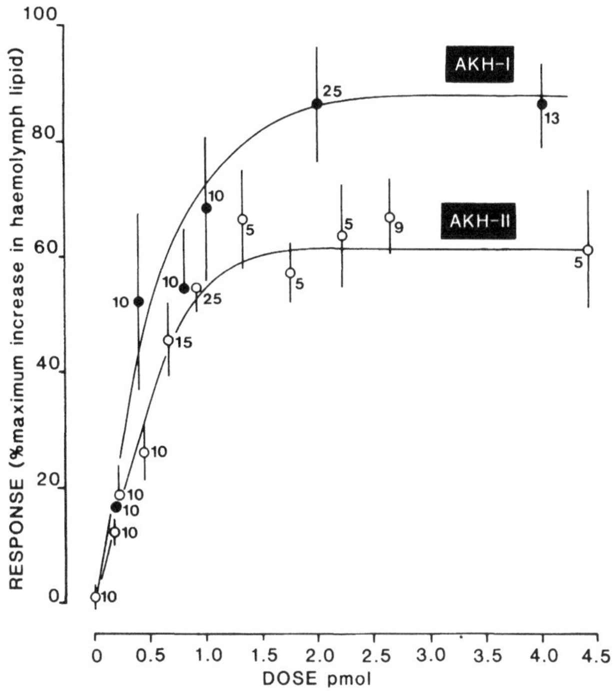

Fig. 1. Dose response curves for natural AKH–I and AKH–II from methanolic extracts of glandular lobes separated [2] on LH20 and assayed [7] by injection into adult male *Locusta*. The responses are the changes in total lipid concentration expressed as percentages (mean ± SE, n) of the increases induced by injection of 10 μl of a methanolic extract of glandular lobes containing 0.02 pairs of CC equivalents; in our *Locusta* this dose induces a maximal response which during the period of this study was ca. 25 μg/μl. The estimated recovery of biological activity from LH20 was > 95% for each hormone, and reversed phase HPLC analysis [22] suggests that there is on average ca. 95 ng of AKH–I and ca. 11 ng of AKH–II in methanolic extracts of one pair of glandular lobes

Orchard and Lange [12] propose that adipokinetic peptides are not released during the first 15 min of flight. They believe that lipid mobilisation during flight is biphasic, with the increase in haemolymph lipid during the first 15 min being effected by octopamine, and the subsequent more prolonged phase being due to AKH–I and II release. However, the magnitude and time course of lipid mobilisation in response to only 2 min flight [31] appears to be greater and more prolonged than that caused by octopamine [32]. In addition, it is not clear whether the concentration of octopamine in the haemolymph of flying locusts is sufficient to account for the observed rates of lipid mobilisation within the first 15 min of flight. Indeed, evidence for biphasic release of lipid is based solely on an apparent plateau in the increasing concentration of haemolymph lipid in flying locusts [12]. Previous authors have assumed an uninterrupted increase in haemolymph lipid concentration during flight [17, 31, 33–35] although, in common with the study of Orchard and Lange [12], it is arguable that some individual values for haemolymph lipid concentration do not appear to fall on a smooth curve. The variations in the timing of such apparent plateaus in lipid titre (15–30 min [33],

30–60 min [35], and 10–20 min [12]) suggest that they are random, and may be caused by differences in rates of mobilisation/utilisation of lipid. Such individual variation can give a false impression of lipid mobilisation [17]; observations of the rates of lipid mobilisation of individual locusts during flight may be necessary to demonstrate trends not noticeable when data from several animals are grouped.

We believe that the release of AKH occurs soon after flight initiation; 2 min flight on unpowered roundabouts is sufficient to elicit a sustained increase in haemolymph lipid concentration for over 2 h [31]. A rapid release of AKH is also suggested by Rademakers and Beenakkers [36] who measure a significant increase in the number of exocytotic profiles in glandular lobe tissue after only 5 min flight. However, when release is assessed by bioassay of haemolymph from flown locusts, adipokinetic activity is not measurable until between 10 or 15 min after the commencement of flight [12, 33]. This inability to measure AKH titres in the very early minutes of flight could be due to the methodology. In those studies where hormone titre is estimated by bioassay of haemolymph extracts [6, 12, 29, 33, 34] only small volumes of haemolymph are extracted. The amounts of peptide(s) recovered are sufficient to indicate hormone titre at the peak of release, but may be insufficient to determine lower hormone titres accurately during the early stages of flight. On the other hand, even at the peak of AKH–II titre, such small increases in lipid concentration are measured in bioassay animals (as little as 4–5 μg/μl) [12], that the precise quantification of AKH–II titre is difficult. Nevertheless, Orchard and Lange [11, 12] have demonstrated the release of both AKH–I and AKH–II during flight; earlier work [33] utilised extracts of haemolymph which contained both AKH–I and II.

Nevertheless, the estimated titre of adipokinetic activity in the haemolymph of flown locusts shows a rapid increase between 15 and 30 min of flight peaking at 30 min, but decreasing to about half by 90 min [12, 33]. There is some variation in the amounts of AKH recovered from the haemolymph at 30 min; from ca. 1 pmol/animal [29, 33, 34] to as much as ca. 9 pmol/animal (calculated from Orchard and Lange [12]). However, even the lower estimates would be adequate to achieve the rates of lipid mobilisation required to support lipid-powered flight; 1 pmol of AKH causes lipid mobilisation at rates of ca. 115 μg/min when injected into suitable resting adult locusts [37]. After 30 min of continuous flight, however, lipid is utilised only at between 57 [38] and 85 μg/min [31, 39]. According to Cheeseman and Goldsworthy [33], the AKH titre at 90 min is ca. 0.46 pmol/animal which would elicit lipid mobilisation at rates of ca. 83 μg/min [37]. That mobilisation and utilisation of lipid are balanced during the later stages of continuous flight is suggested by the steady state plateau of haemolymph lipid concentration [39, 40].

Why is there an apparent excess of hormone release during the first 30 min of flight? Are rapid rates of lipid mobilisation essential during the early stages of flight to ensure that high concentrations of haemolymph diacylglycerol saturate lipoprotein lipases [41] in the flight muscles? High rates of lipid mobilisation at the start of flight will ensure the formation of A^+ lipoprotein, the preferred substrate for flight muscles [42, 43], but high titres of AKH may be necessary to achieve its actions in the flight muscle [42, 44–46].

4. Mode of Action of Adipokinetic Peptides

Adipokinetic hormones act on the fat body to increase release into the haemolymph of stereospecific diacylglycerols [47–49] derived from stored triacylglycerols. It is generally assumed [1] that AKH stimulates lipolysis in the fat body via a mechanism involving phosphorylation of a lipase by a cAMP-dependent protein kinase. Crude and semi-purified extracts of CC increase the levels of cAMP in the fat body both in vivo [50–52] and in vitro [53]. Recently, crude extracts of glandular lobes of CC have been shown to activate a fat body cAMP-dependent protein kinase [54]. In addition, diacylglycerol lipases are activated by cAMP and cGMP in vitro [54]. The effect of either AKH–I and or II on these lipases has not been investigated, but RPCH activates locust fat body protein kinase in vitro, and monoacylglycerol transferase and an unspecified lipase in vivo [54]. Interestingly, monoacylglycerol transferase is not activated by cyclic nucleotides in vitro [54].

The second major site of action of AKH is at the flight muscles where it is thought to inhibit trehalose utilisation indirectly by increasing the oxidation of lipids [42, 44, 45, 46]. The evidence for a direct action of AKH on the flight-muscles has recently been discussed [1] and will not be dealt with further here. The exact mechanism of any proposed action of AKH on flight muscle metabolism remains, however, open to question.

Robinson and Goldsworthy [42] found that AKH only affects flight muscle metabolism when A^+ lipoprotein [18, 19] is supplied as substrate: they suggest that both the quality and quantity of lipoproteins are important in supplying diacylglycerols to the flight muscles. Corroborative evidence for this view is obtained from recent studies on lipoprotein lipases [43]. These enzymes (which hydrolyse lipoprotein-bound neutral lipids [41] are present on the plasma membrane of flight muscles [43] and are able to hydrolyse lipids associated with lipoproteins A^+ and A*yellow* [18, 19]. For each lipoprotein the rate of hydrolysis is dependent on the concentration of lipid. However, even when A*yellow* is used at twice its normal haemolymph concentration, its diacylglycerols are hydrolysed at only 25% of the rate for those bound to A^+ (at its normal haemolymph concentration in flying locusts [40, 43]. This substrate specificity of flight muscle lipoprotein lipase is part of a mechanism which ensures that glycolytic flux in the presence of AKH is inhibited maximally [42] when sufficient quantities of a suitable alternative substrate (lipoprotein A^+) are present in the haemolymph.

5. Adipokinetic Hormone and Lipoprotein Metabolism

In 1977, Mwangi and Goldsworthy [18] proposed an apparently unique model concerning lipoprotein-protein interaction during AKH-stimulated lipid transport in the locust. Subsequent work from this laboratory and from that of Prof. Beenakkers in Utrecht has given tacit support to the model. Thus, the model states that during AKH action, the resting lipoprotein, A*yellow* and the C_L-proteins combine together with extra diacylglycerol from the fat body to form a new lipoprotein, A^+. The C_L-proteins appear, at least in part, to be bound reversibly to A^+, because they can be pulled off the lipoprotein during electrophoresis, and bound ^{3}H-labelled C_L-proteins are displaced from A^+ in vitro by a large excess of cold C_L-protein [55]. This and other evidence for such a model has been reviewed recently [1].

Many problems relating to the A^+ lipoprotein remain; we know something of its particulate nature [1, 56, 57], its hydrated density [56], and its degradation and utilisation of constituent lipids by the flight muscles [41–43], but there are at least two important areas of uncertainty. We still have no clear idea of the detailed mechanism of A^+ formation (whether, for example, a proportion of the C_L-proteins play a structural role within the A^+ particles), and the dynamics of A^+ formation present an enigma. This latter point has recently been a focus of our attention.

The formation of A^+ has been studied in detail using both a gel filtration separation technique and a rapid quantitative precipitations method using heparin and EDTA [18–20]. This latter technique shows that lipoprotein-protein interactions in the haemolymph begin within the first 5 min of AKH action, but that the titre of A*yellow* does not change significantly until ca. 15 or 20 min after AKH injection. These observations are not artefacts of the methodology, because they are corroborated by parallel studies using the slower, less sensitive, and more laborious technique of gel filtration of haemolymph lipoproteins. When ^{3}H-labelled A*yellow* is injected into locusts, the radiolabel remains associated with the A*yellow* peak for about 20 min after AKH–I injection. Subsequently, a broadening of the leading edge of the peak occurs and moves forward gradually as a shoulder (that is, increasing in its molecular weight) until by 30 min there is a discernible peak of A^+ (Fig. 2 A). In contrast, in a parallel experiment using ^{3}H–C_L-proteins, a small amount of radioactivity associates with the leading edge of the A peak [18] of resting haemolymph, but immediately after AKH–I injection, this "moves forward" and increases in amount so that by 90 min, the majority of the total label is associated with the A^+ peak (Fig. 2 C). Changes in radioactivity derived from ^{14}C-labelled triacylglycerol stored in the fat body follow an almost identical pattern to that of the radiolabelled C_L-proteins; there is, however, a rapid but transient increase in label associated with A*yellow* [18, 20], coincident with a rapid increase in label associated with species of gradually increasing molecular weight. Within 90 min of AKH–I injection, there is a single peak of ^{14}C-radiolabeled associated with lipoprotein A^+ (Fig. 2 B). Both of these experimental approaches suggest that A*yellow* is not involved in the early lipoprotein-protein interactions associated with AKH–I action.

These observations show clearly the dynamic nature of A^+ formation and argue against the view that A^+ is merely a high molecular weight aggregate of lipoproteins and proteins formed by a clotting or coagulation process [58, 59]; A^+ forms in an ordered and gradual manner by processes which are reversible. This latter point is illustrated by the fact that radiolabelled C_L-proteins and A*yellow* return to their native protein peaks within 24 h of hormone injection (Fig. 2). Clearly, the original model for A^+ formation [18] is too simplistic. It appears now that a group of haemolymph lipoproteins, somewhat intermediate between A*yellow* and A^+ and present in small quantities, is involved in early events during AKH action. Locust fat body gives up lipids only to suitable haemolymph lipoprotein acceptors [60] and, therefore, the increases in lipid loading and in C_L-proteins binding, within minutes of AKH injection [20], suggest that there is a lipoprotein acceptor other than A*yellow* responsible for

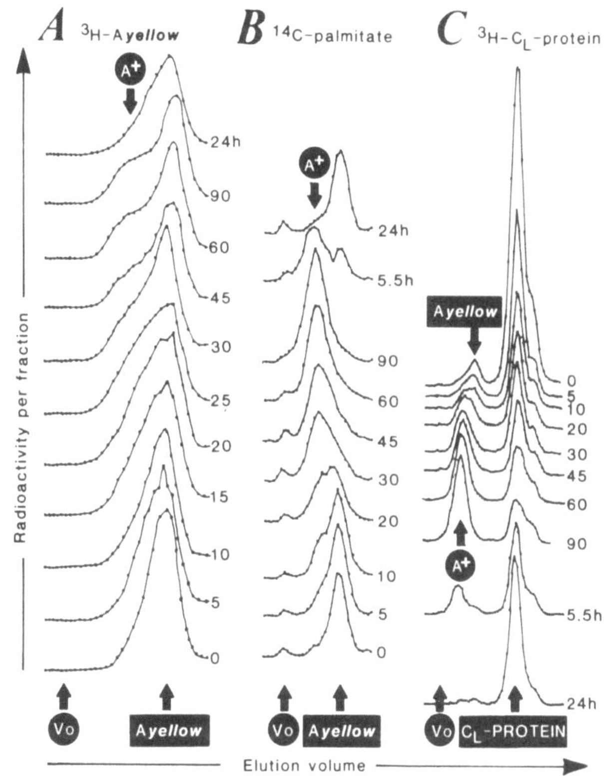

Fig. 2. Separation on AcA 22 of radioactivity (corrected for quenching) from pooled serial samples of ammonium sulphate-treated haemolymph [18] from groups of five adult male *Locusta* at different times after injection of ca. 2 pmol of natural AKH–I. Times after injection are in minutes except where indicated otherwise. In *A* and *C*, respectively, chromatographically purified A*yellow* lipoprotein and C_L-proteins were radiolabelled using *N*-succinimidyl [2,3–^{3}H] propionate [55]. The radiolabelled proteins were injected in tracer amounts into the assay locusts, and allowed to equilibrate for 30 min before hormone injection. In *B*, locusts were injected with 1 μCi of ^{14}C-palmitate (in an emulsion with albumin) [18] 4 h before hormone injection. Because of limitations of space, in *A* and *B* the radioactivity is indicated only in fractions associated with the lipoproteins peaks. In each case a small amount of activity was present in the salt fraction, but this did not change with hormone injection. The points at which the major protein and lipoproteins elute are indicated for each group of separations

this uptake. Electrophoretic separations of whole haemolymph from resting locusts reveal small quantities of lipoproteins migrating behind A*yellow* which increases their protein and lipid staining within the first minutes of AKH action [20]. We believe it is these lipoproteins which load lipid and bind C_L-proteins during the early stages of AKH action before the involvement of A*yellow*. A general scheme for the actions of AKH are summarised in Fig. 3.

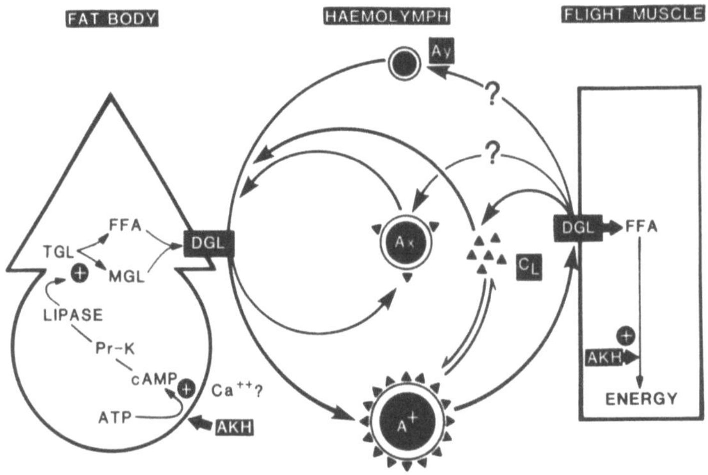

Fig. 3. Lipid mobilisation and the formation of lipoprotein A^+ during flight in *Locusta*. When the AKH response declines (e.g. after flight cessation), A*yellow* and C_L-proteins are slowly liberated from A^+; it is uncertain whether lipoprotein Ax is a necessary intermediate in these events, but it is certainly involved in the very early actions of AKH. For details see text. *Ay* A*yellow* lipoprotein; *Ax* an intermediate lipoprotein, Ax; A^+ lipoprotein A^+; *TGL* triacylglycerol; *DGL* diacylglycerol; *MGL* monoacylglycerol; *FFA* free fatty acid; *Pr–K* cyclic–AMP–dependent protein kinase

Acknowledgments. This work was supported by grants from The Royal Society and the S.E.R.C.

References

1. Goldsworthy GJ (1984) Adv Insect Physiol 17:149–204
2. Carlsen J, Herman WS, Christensen M, Josefsson L (1979) Insect Biochem 9:497–501
3. Stone JV, Mordue W, Batley KE, Morris HR (1976) Nature (London) 263:207–211
4. Broomfield CE, Hardy PM (1977) Tetrahedron Lett 25:2201–2204
5. Beenakkers AMT (1969) Gen Comp Endocrinol 13:Abstr 12
6. Mayer RJ, Candy DJ (1969) J Insect Physiol 15:611–620
7. Goldsworthy GJ, Mordue W, Guthkelch J (1972) Gen Comp Endocrinol 18:545–551
8. Schooneveld H, Tesser GI, Veenstra JA, Romberg-Privee HM (1983) Cell Tissue Res 230:67–76
9. Stone JV, Mordue W, Broomfield CE, Hardy PM (1978) Eur J Biochem 89:195–202
10. Fernlund P, Josefsson L (1972) Science 177:173–175
11. Orchard I, Lange AB (1983) J Insect Physiol 29:425–429
12. Orchard I, Lange AB (1983) J Insect Physiol 29:639–642
13. Loughton BG, Orchard I (1981) J Insect Physiol 27:383–385
14. Van Marrewijk WJA, Van Den Broek ATM, Beenakkers AMT (1980) Insect Biochem 10:675 to 679
15. Gade G (1981) J Insect Physiol 27:155–161
16. Goldsworthy GJ, Gade G, Ziegler R (unpublished)
17. Goldsworthy GJ, Mallison K (in preparation)
18. Mwangi RW, Goldsworthy GJ (1977) J Comp Physiol B 114:177–190
19. Mwangi RW, Goldsworthy GJ (1978) In: Gaillard PJ, Boer HH (eds) Comparative endocrinology. Elsevier, Amsterdam, pp 459–463
20. Wheeler CH, Goldsworthy GJ (1983) J Insect Physiol 29:339–347
21. Gade G, Goldsworthy GJ, Kegel G, Keller R (1984) Hoppe-Seyler's Z Physiol Chem 365: 393–398

22. Goldsworthy GJ, Gade G (unpublished)
23. Rademakers LHPM (1977) Cell Tissue Res 184:381–395
24. David JC, Lafon-Cazal M (1979) Comp Biochem Physiol 64C:161–164
25. Orchard I, Loughton BG (1981) J Neurobiol 12:143–153
26. Goldsworthy GJ, Johnson RA, Mordue W (1972) J Comp Physiol 79:85–96
27. Orchard I, Loughton BG (1981) Comp Biochem Physiol 68A:25–30
28. Houben NMD, Beenakkers AMT (1973) J Endocrinol 57:liv–lv
29. Cheeseman P, Jutsum AR, Goldsworthy GJ (1976) Physiol Entomol 1:115–121
30. Van der Horst DJ, Van Doorn JM, Beenakkers AMT (1979) Insect Biochem 9:627–635
31. Jutsum AR, Goldsworthy GJ (1976) J Insect Physiol 22:243–249
32. Orchard I, Loughton BG, Webb RA (1981) Gen Comp Endocrinol 45:175–180
33. Cheeseman P, Goldsworthy GJ (1979) Gen Comp Endocrinol 37:35–43
34. Houben NMD (1976) Regulatie van substraattransport tijdens vlucht in de sprinkhaan *Locusta migratoria*. PhD thesis, Univ Utrecht
35. Van der Horst DJ, Houben NMD, Beenakkers AMT (1980) J Insect Physiol 26:441–448
36. Rademakers LHPM, Beenakkers AMT (1977) Cell Tissue Res 180:155
37. Mwangi RW, Goldsworthy GJ (1977) Physiol Entomol 2:37–42
38. Van der Horst DJ, Baljet AMC, Beenakkers AMT, Van Handel E (1978) Insect Biochem 8: 369–373
39. Spencer IM, Candy DJ (1974) Biochem Soc Trans 2:1093–1096
40. Mwangi RW, Goldsworthy GJ (1981) J Insect Physiol 27:47–50
41. Wheeler CH, Van der Horst DJ, Beenakkers AMT (1984) Insect Biochem (in press)
42. Robinson NL, Goldsworthy GJ (1977) J Insect Physiol 23:9–16
43. Wheeler CH, Goldsworthy GJ (in preparation)
44. Robinson NL, Goldsworthy GJ (1974) J Comp Physiol 89:369–377
45. Robinson NL, Goldsworthy GJ (1976) J Insect Physiol 22:1559–1564
46. Robinson NL, Goldsworthy GJ (1977) J Insect Physiol 23:153–158
47. Tietz A, Weintraub H (1980) Insect Biochem 10:61–63
48. Tietz A, Weintraub H, Peled Y (1975) Biochim Biophys Acta 388:165–170
49. Lok CM, Van der Horst DJ (1980) Biochim Biophys Acta 618:80–87
50. Gade G, Holwerda DA (1976) Insect Biochem 6:535–541
51. Gade G, Beenakkers AMT (1977) Gen Comp Endocrinol 32:481–487
52. Gade G (1979) Gen Comp Endocrinol 37:122–130
53. Spencer IM, Candy DJ (1976) Insect Biochem 6:289–296
54. Pines M, Tietz A, Weintraub H, Applebaum SW, Josefsson L (1981) Gen Comp Endocrinol 43:427–431
55. Wheeler CH, Goldsworthy GJ (1983) J Insect Physiol 29:349–354
56. Goldsworthy GJ, Wheeler CH (1982) Gen Comp Endocrinol 46:380
57. Wheeler CH, Mundy JE, Goldsworthy GJ (1984) J Comp Physiol B (in press)
58. Brehélin M (1979) Comp Biochem Physiol 62B:329–334
59. Chino H, Kitazawa K (1981) J Lipid Res 22:1042–1052
60. Tietz A (1962) J Lipid Res 3:421–426

I.15 Ecdysteroids Regulate the Release and Action of Eclosion Hormone in the Moth *Manduca sexta*

J.W. TRUMAN[1]

1. Introduction

The regulation of hormone secretion often involves a complex interaction between neuroendocrine and endocrine centers. The classic example of such an interaction is between the vertebrate anterior pituitary and hypothalamus in which the cells of the former are regulated by peptides and amines that are released from neurosecretory cells in the latter [1,2]. These influences, however, are not unidirectional since the hormones released from endocrine glands may also act back on the central nervous system (CNS) to influence neurosecretory activity [2]. In insects, the existence of neurosecretory cells that release tropic hormones which then affect the activity of endocrine glands is well established. The best known example is the role of brain neurosecretory cells in regulating the secretory activity of the prothoracic glands [3]. There is now also a growing interest in insect endocrinology in how endocrine glands may in turn regulate neurosecretory centers in the CNS. One such example is the recent report that circulating ecdysteroids regulate the release of a myotropic peptide from the brain of female *Rhodnius* [4]. This paper will focus on another example in which ecdysteroids may control neuroendocrine function; in this case it involves the hormones involved in bringing a molt to its successful completion.

2. Background to Ecdysis

The process of molting in insects is terminated with the shedding of the remains of the old cuticle, a process termed ecdysis. This event is accompanied by abrupt behavioral and physiological changes as the insect goes from a state of relative quiescence, which was maintained during the molt, to an active feeding or reproducing stage [5]. These changes are most marked in the case of the metamorphic molts of holometabolous insects, especially in the transition from the pupal stage, which is essentially immobile, to the active adult. Similar changes in behavior are also seen in earlier larval molts, but typically they are not as dramatic as those seen during metamorphosis. The alterations in behavioral state seen at ecdysis are associated with a period of intense endocrine activity. Circulating titers of ecdysteroids and juvenile hormone often show peaks or

1 Department of Zoology, University of Washington, Seattle, WA 98195, USA

Biosynthesis, Metabolism and Mode of Action of Invertebrate Hormones
(ed. by J. Hoffmann and M. Porchet)

abrupt rises at ecdysis (e.g., [6, 7]). Also a number of peptide hormones are released at this time [8, 9]. Indeed, from an endocrine standpoint the ecdyses are probably the most active times in the life of most insects.

Eclosion hormone (EH) is a peptide hormone that has a central role in regulating ecdysis [10]. Although this hormone has only conclusively been shown to exist in Lepidoptera, there is evidence that it is probably found widely in most, if not all, insect groups [11-13]. The principal target for EH is the CNS and the peptide has been shown to activate directly the motor patterns used for the ecdysis behaviors [14, 15]. Eclosion hormone also has other effects on the CNS. Most of these involve the long-term activation of stage-specific behaviors, the best known of which are various simple reflexes that become functional at ecdysis (e.g., [16]). CNS centers that control neuroendocrine activity may also be affected by the peptide. For example, the release of the peptide bursicon after adult ecdysis in *Manduca* can occur only after the nervous system of the insect is exposed to EH [8].

The eclosion hormone system in insects is also of interest with respect to the precision and predictability of hormone release. The peptide is secreted in an episodic manner with 60%-70% of the stored activity being released during each episode [13]. These release episodes are confined to the terminal stages of each molt and there is no indication that significant amounts of eclosion hormone are secreted at any other times in the insect's life [13]. In the case of larval [17] and pupal molts [18] of *Manduca*, the timing of peptide secretion can be accurately predicted within ± 30 min by reference to external changes in the molting insect. For adult ecdysis, the timing of EH release is also tightly correlated with the developmental progression of the insect, but in addition a circadian system restricts the release of the hormone to a specific temporal "gate" during the day [19].

Eclosion hormone is produced and stored in two discrete regions of the nervous system and is released from these regions in a stage-specific manner [13]. The circadian gating of adult eclosion comes about through the action of a "clock" located in the moth brain [20] and it acts through an EH system that is also located in the brain. The earlier larval and pupal ecdyses are not directly gated, but rather occur at a specific time of development irrespective of time of day [21]. These earlier ecdyses are triggered by EH released from sites in the ventral ganglia rather than the brain [13]. Thus, release of EH from both sites appears dependent on the development state of the insect, but the brain site has the added influence of a circadian gating system. A primary focus of this essay is on the cues used by the EH centers in the brain and ventral ganglia to assess the developmental condition of the insect as the molt progresses to completion.

3. Regulation of Eclosion Hormone Release

As indicated above, eclosion hormone is released only once at the end of each molt. When molting insects are pretreated with EH so as to induce a precocious ecdysis these insects do not then release their own endogenous store of peptide [19]. Therefore, one action of EH is to turn off its own release. However, even though the EH release system is turned off after an ecdysis, it is obvious that it must later be reactivated in time for the next one. The appearance of ecdysteroids at the start of the next molt presumably is important for this reactivation. But, the major question remains as to how the release of EH is so precisely correlated with the subsequent completion of the molt.

A key to understanding the factors that regulate eclosion hormone release was provided by a series of experiments by Slama [22] on the ecdysis of *Tenebrio*. He showed that ecdysteroid injections delayed ecdysis in a dose-dependant fashion. These treatments were effective up until about 12 h before the scheduled time of ecdysis, but the insects became insensitive to treatment with the steroid shortly thereafter.

In response to these results on *Tenebrio*, we undertook an examination of the role of ecdysteroids in regulating the timing of ecdysis in *Manduca* [23]. The ecdyses of both the pupal and adult stages of this moth could be delayed in a dose-dependant fashion by treatment with 20-hydroxyecdysone (20-HE) (Fig. 1). In the case of pupal ecdysis, the 20-HE injections resulted in a smooth delay in ecdysis with increasing dosages delaying the behavior to progressively later times. For adult ecdysis, increasing dosages also resulted in later delays, but the delays were discontinuous with the insects being pushed to the end of one circadian gate and then abruptly jumping to the next gate almost 24 h later.

The ability of ecdysteroid treatments to delay ecdysis was examined in detail for the adult molt [23]. Injections of 20-HE were effective in delaying ecdysis up until the night preceeding the normal time of ecdysis. By midway through that night (approx. 12 and 14 h before the average times of EH release and adult ecdysis, respectively), 50% of the insects were committed to ecdyse on schedule and could no longer be delayed by the steroid treatment. At this time in development the endogenous ecdysteroid titer had dropped to a level of about 150 ng 20-HE equivalents/ml. Single injections of 500 ng 20-HE/g given at a time about 10 h earlier caused a significant delay in ecdysis.

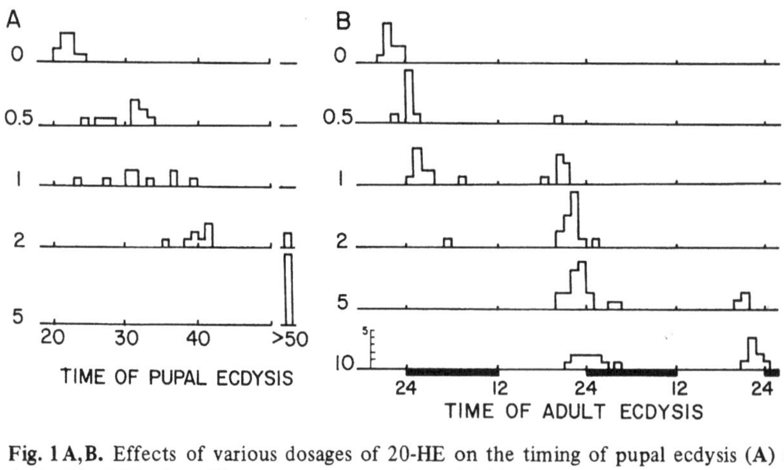

Fig. 1A,B. Effects of various dosages of 20-HE on the timing of pupal ecdysis (**A**) and adult ecdysis (**B**) of *Manduca*. Pharate pupae were injected 10 h after the appearance of a stage marker (the formation of the "dorsal bars") and times of ecdysis are given in hours relative to that marker. Insects in the > 50 group did not successfully ecdyse. Pharate adults were injected at the middle of the 17th day of development (about 22 h before the start of the graph). *Black bars* represent darkness. Amounts of 20-HE injected (μg/g) are given to the left of each distribution. Data redrawn from [23]

Infusion of ecdysteroids starting at the same time at a rate of 30 ng 20-HE/g/h was also effective in delaying ecdysis. Thus, the dosages of steroid required to effect the timing of ecdysis were within a physiological range.

This ability of ecdysteroids to delay ecdysis arises at least in part through a delay in the release of eclosion hormone. This conclusion was based on studies utilizing pharate adult *Manduca* that were fitted with a heart catheter through which blood samples could be withdrawn at various times surrounding the normal time of ecdysis on the 18th (last) day of development. Insects injected with saline on day 17 and then subsequently monitored for the appearance of EH in their blood showed an abrupt appearance of EH with the arrival of their normal ecdysis gate in the evening of day 18. By contrast, insects pretreated with 5 μg 20-HE/g at the same time failed to release EH during their normal gate. Rather they appeared to delay release for about 24 h until their circadian gate reopened on the following day.

In the case of pupal ecdysis it was not possible to implant catheters and remove successive blood samples from individuals. Therefore, EH secretion was examined in an indirect fashion. Normally, EH release into the blood at pupal ecdysis is accompanied by a measurable depletion of about 60% of the hormonal activity stored in the ventral nerve cord [13]. Consequently, we measured the amount of EH activity contained in the ventral nerve cord in control and 20-HE treated insects of the same age at a time after the ecdysis of all of the controls. The ecdysteroid treated group still showed levels of stored EH that were equivalent to those found in prepupae, indicating that significant release of hormone had not taken place.

The fact that EH release in both the pupal and adult stages is inhibited by high levels of ecdysteroid is important since these two ecdyses are controlled by anatomically distinct endocrine centers [13]. Therefore, both centers appear to be ecdysteroid sensitive. From the pattern of ecdysis delay (Fig. 1) seen for the two stages, it appears that the brain and ventral ganglion centers respond to ecdysteroids in different ways. The ecdysteroid titer seems to be the sole factor regulating secretion by the ventral ganglion cells, whereas the brain center is responsive to both a gating input and the steroid levels.

A second feature of note concerning the relationship of the ecdysteroids to EH release is that the response is not immediate. Rather in pupae 8 h elapse between the time that the insect is committed to ecdyse (and hence cannot be effected by exogenous steroid) and the subsequent secretion of EH. This interval in adults is 12 h. The events that are responsible for these rather long latencies are unknown.

4. Action of Ecdysteroids in Coordinating the End of the Molt

The successful completion of a molt requires the close coordination of numerous physiological and developmental processes. Indeed, during the last few days of adult development in *Manduca* a series of events, including pigmentation of the adult scales, digestion of the pupal endocuticle, resorption of the molting fluid, and degeneration of selected neurons and muscles occur in a precise order according to a highly stereotyped schedule [24].

The possibility that these late developmental events might be actively regulated was indicated by the results of isolating abdomens midway through adult development. This treatment caused an acceleration of the remainder of development such that the overall process was completed in 16 days rather than the normal 18 [24]. Thus, it appeared that the anterior end of the animal normally produced something that slowed the pace of development and abdomen isolation separated the abdomen from the source of this factor and, hence, this part of the insect then developed more rapidly.

That this factor was hormonal and related to prothoracic gland activity were shown by using *Manduca* that had received implants of prothoracic glands into their abdomens early in adult development [24]. When the abdomens of these insects were later isolated, they showed a normal developmental rate rather than the accelerated rate shown by abdomens without implants. Thus, the prothoracic gland implant appeared to release a factor that retarded the progression of the late stages of the adult molt.

The involvement of the prothoracic glands in controlling the developmental rates suggested that the ecdysteroids were the agents of interest. This hypothesis was directly tested by injecting various dosages of ecdysteroid into intact insects or isolated abdomens that were late in adult development [24]. In both cases treatment caused a delay in development that was proportional to the dosage of steroid administered. More telling information as to the role of ecdysteroids was obtained by infusing a constant rate of 20-HE into developing animals and then monitoring the progression of development by reference to external markers of the developing moth visible through the pupal cuticle (Fig. 2). Infusion of 0.09 μg 20-HE/h resulted in a progressive slowing and the eventual arrest of development. With higher rates of 20-HE administration the slowing

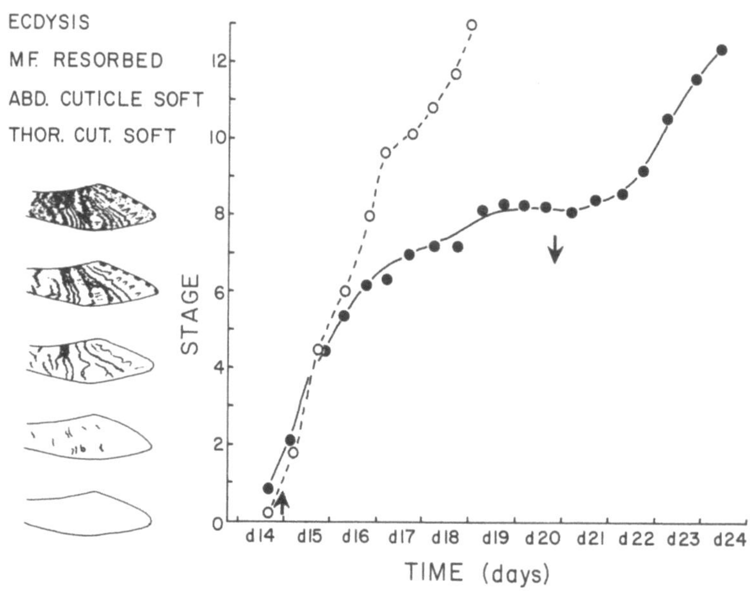

Fig. 2. Progression of the late stages of adult development in control (*open circles*) and 20-HE infused (*filled circles*) *Manduca*. Some aspects of the staging system, as described in [24], are represented to the *left*. Drawings of the wings show the progression of pigmentation used to denote stages 0, 2, 4, 6, and 8, respectively; *MF* molting fluid. Upward- and downward-pointing *arrows* show the start and end of infusion, respectively. Dosage was 0.18 μg 20-HE/h. Modified from [24]

was more pronounced and the developmental arrest occurred sooner. When the infusions were terminated, the insects usually required about a day to rid themselves of the effects of the ecdysteroids and then completed the remainder of development. The levels of ecdysteroid infused into the insect to bring about a slowing of development were within the levels found in the blood of the insect at this time. Thus, the ability of ecdysteroids to slow development appears to be a physiological, rather than a pharmacological effect of the steroid.

Another piece of evidence implicating the ecdysteroid titer in the regulation of development rates came from comparing the steroid levels late in the adult molt with the effects of abdomen isolation during the same period [24]. Abdomen isolation during the last 24 h of adult development had no effect on the rate of subsequent development. During this period, the ecdysteroid titer underwent a pronounced decay and abdomen isolation apparently did not significantly hasten this decline. This terminal decline, however, was immediately preceded by an approx. 12 h plateau in the levels of circulating ecdysteroids. Abdomen isolation prior to the plateau resulted in both an early decline in blood ecdysteroid titer and an advancement of the timing of the subsequent late developmental events.

This role of ecdysteroids in retarding development at the end of the molt is in marked contrast to its classical function of promoting differentiation such as is seen at the start of adult development [25]. This switch in the developmental role of ecdysteroids from stimulatory to inhibitory occurs about one-third of the way through development, just after the endogenous ecdysteroid titer has started to decline [24].

The inhibitory action of the ecdysteroids allows these hormones to maintain developmental synchrony among the various tissues during the late stages of the adult molt. Two features of the action of ecdysteroids are important in this respect. One has already been discussed above and is that at any particular time the rate of development was inversely proportional to the ecdysteroid concentration. The second feature is that as development proceeds, tissues became progressively more sensitive to the inhibitory effects of a particular dosage of steroid [24]. The potential importance of these two relationships can be appreciated when considering a condition during development in which two tissues are slightly out of synchrony with each other. Since both are exposed to the same ecdysteroid titer, both would be somewhat inhibited, but the more advanced tissue would be more affected, and, hence, slowed more than would be the less advanced tissue. This differential action would eventually bring the tissues back into synchrony.

One test of the hypothesis that the ecdysteroid titer maintains tissue coordination during development came from the examination of neuronal death in isolated abdomens of *Manduca* [26]. Certain abdominal motoneurons show a stereotyped degeneration that occurs after adult ecdysis and that is triggered by the low ecdysteroid levels at the time of ecdysis. A group of six dorsal motoneurons (the D–IV cells) was especially favorable for study. In intact individuals the cells within the group died at approximately the same time. However, when abdomen were isolated on day 15, the cells showed precocious degeneration and the synchrony of death among the members of this group was markedly disrupted. Thus, interference with the ecdysteroid decline resulted in a weakening of the coordination that was seen normally in the fate of these cells.

5. Ecdysteroid Regulation of Eclosion Hormone Action

The acquisition of sensitivity to eclosion hormone occurs late in development. In *Manduca* behavioral responsiveness to EH appears after the digestion of the pupal endocuticle is completed and is approximately coincident with the resorption of the molting fluid. As with the other events discussed above, the onset of this sensitivity is regulated by the declining ecdysteroid titer [23]. For example, when developing *Manduca* were treated with ecdysteroid on day 17 and then subsequently challenged with a supramaximal dosage of EH at various times thereafter, the insects showed a delay in the onset of EH responsiveness that was similar to the overall delay in development evoked by the steroid. Experiments on prepupae similarly showed that a decline in steroid levels was essential for the acquisition of EH responsiveness for the pupal molt.

It should be noted that the respective roles of ecdysteroids and eclosion hormone in coordinating events at the end of development may differ between species. An example of this variation is seen in the regulation of the degeneration of the abdominal intersegmental muscles after adult ecdysis. In *Manduca* the death of the muscles is triggered by the declining ecdysteroid titer and the presence or absence of EH is inconsequential in so far as the fate of these muscles is concerned [24]. By contrast, in the silkmoth *Antheraea polyphemus* EH is the signal that brings about the rapid death of these same muscles [27]. In the latter case ecdysteroids also have a role in the muscle death in that the cells must see the steroid decline before they are competent to respond to the peptide signal. It should be noted that other hormones besides the ecdysteroids and EH may be involved in some of these processes. For example, in the fly *Calliphora* the postecdysial muscle death can be delayed for variable periods after ecdysis until bursicon is released to trigger postecdysial tanning and the muscle degeneration [28].

6. Conclusions

The above sections show that the ecdysteroids and eclosion hormone are intimately involved in coordinating events at the end of the molt. The declining ecdysteroid titer plays a central role by modulating the rate by which cells play out their developmental programs. The inhibitory action of ecdysteroids coupled with the growing sensitivity of tissues to this inhibition as development progresses then insures that synchrony is maintained in these differentiating tissues. Presumably, this type of coordination is especially important in the case of extremely long molts such as occur during the pupal-adult transition of many holometabolous insects.

In both the pupal and adult ecdyses, release of EH does not occur if the ecdysteroids are maintained at a high level. At this time we have not shown that the steroids act directly on the CNS centers that control EH release. Indeed, the ecdysteroid effect could be indirect and the neuroendocrine centers could be cued by the developmental status of some peripheral structures. However, since implanted brains release EH on the appropriate day of development as well as the proper time of day [20], it is obvious that if such peripheral cues exist, they must be blood-borne. At this time we favor the hypothesis that the effects are due to a direct action of the steroids on the centers that control EH release.

Thus, the precise timing of EH release to coincide with the completion of development at the end of the molt is due to the two actions of the ecdysteroids (Fig. 3). The close linkage of developmental rate with the ecdysteroid titer insures that under normal conditions the insects have finished development and have achieved competence to respond to EH by the time that the steroid levels have reached a relatively low level. These low levels also allow the secretion of EH itself. This relationship then insures that when EH is finally released, the insect will be ready to respond to it.

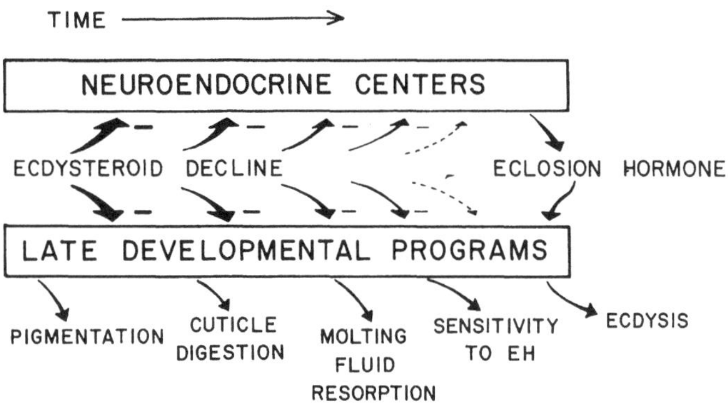

Fig. 3. Schematic representation of the endocrine interactions responsible for the successful termination of a molt in *Manduca*

References

1. Scharrer E, Scharrer B (1963) Neuroendocrinology. Columbia Univ Press, New York
2. Pang PKT, Epple A (eds) (1980) Evolution of vertebrate endocrine systems. Texas Tech Press, Lubbock
3. Williams CM (1952) Biol Bull 103:120–138
4. Ruegg RP, Kriger FL, Davey KG, Steel CGH (1981) Int J Invert Reprod 3:356–361
5. Reynolds SE (1980) Adv Insect Physiol 15:475–595
6. Dean RL, Bollenbacher WE, Locke M, Smith SL, Gilbert LI (1980) J Insect Physiol 26:267–280
7. Fain MJ, Riddiford LM (1975) Biol Bull 149:506–521
8. Truman JW (1981) Am Zool 21:655–661
9. Tublitz NJ, Truman JW (1983) Soc Neurosci Abstr 9:19
10. Truman JW (1980) In: Locke M, Smith DS (eds) Insect biology in the future. Academic Press, London New York, pp 385–401
11. Carlson JR, Bentley D (1977) Science 195:1006–1008
12. Miller PL, Mills PS (1976) In: Spencer-Davies P (ed) Perspectives in experimental biology, vol I, Zoology. Pergamon Press, Oxford, pp 199–208
13. Truman JW, Taghert PH, Copenhaver PF, Tublitz NJ, Schwartz LM (1981) Nature (London) 291:70–71
14. Truman JW (1978) J Exp Biol 74:151–174
15. Truman JW, Weeks JC (1983) Symp Soc Exp Biol 37:223–241
16. Levine RB, Truman JW (1983) Brain Res 279:335–338
17. Copenhaver PF, Truman JW (1982) J Insect Physiol 28:695–701
18. Truman JW, Taghert PH, Reynolds SE (1980) J Exp Biol 88:327–337

19. Truman JW (1978) In: Gaillard PJ, Boer HH (eds) Comparative endocrinology. Elsevier/North Holland, Amsterdam, pp 123–148
20. Truman JW, Riddiford LM (1970) Science 167:1624–1626
21. Truman JW (1972) J Exp Biol 57:805–820
22. Slama K (1980) Acta Entomol Bohemoslov 77:145–168
23. Truman JW, Rountree DB, Reiss SE, Schwartz LM (1983) J Insect Physiol 29:895–900
24. Schwartz LM, Truman JW (1983) Dev Biol 99:103–114
25. Williams CM (1968) Biol Bull 134:344–355
26. Truman JW, Schwartz LM (1984) J Neurosci 4:274–280
27. Schwartz LM, Truman JW (1982) Science 215:1420–1421
28. Cottrell CB (1962) J Exp Biol 39:431–448

I.16 Studies on Crustacean Eyestalk Neuropeptides by Use of High Performance Liquid Chromatography

R. KELLER and G. KEGEL [1]

1. Introduction

The neurosecretory system in the eyestalk of decapod crustaceans, comprising groups of perikarya and a conspicuous, "classical" neurohemal organ, the sinus gland, is widely believed to be the most important component of the endocrine system of this animal group. Largely on the basis of deficiency and replacement experiments, a considerable number of active "factors", regulating a large variety of physiological processes, have been postulated to be synthesized, stored, and released in the eyestalk (reviews [1, 2]). It is not known how many individual active substances are responsible for the many biological effects of eyestalk or sinus gland extracts. Despite many attempts at isolation and characterization, thus far only two eyestalk neuropeptides have been completely elucidated, the red pigment concentrating hormone (RPCH), a chromatophorotropin from *Pandalus borealis* [3] and, from the same species, one form of the distal retinal pigment hormone (light adapting), a neuropeptide regulating light adaptation of the compound eye [4] that also possesses melanophore dispersing activity. It may therefore also be called melanophore dispersing hormone (MDH, reviews [1, 2]). The next best known neuropeptide is the crustacean hyperglycemic hormone (CHH), a peptide of larger molecular size which has thus far been characterized only in terms of amino acid composition [5]. Apart from these achievements, progress has been slow due to (a) lack of convenient and reliable bioassays and (b) the shortcomings of conventional purification methods in the isolation of minute amounts of neuropeptides from small amounts of starting material. During the last years, high performance liquid chromatography has proved to be the method of choice for isolation of peptides, both in semimicro- or microanalytical and in preparative applications.

In this paper, some results from analytical as well as preparative HPLC separation of extracts from sinus glands (SG) will be reported. Apart from the well-known speed and sensitivity of the HPLC, it offers some special advantages in the present case. The SG is a highly specialized neurohemal tissue that, as shown by electron microscopy, consists almost entirely of axon endings which are densely filled with neurosecretory granules, i.e., the substances of interest are present in a highly concentrated form. Due to its resolving power, HPLC of such not too complex mixture (i.e., a mixture in which neuropeptides are present in high proportions over other, generally present cell constit-

1 Institut für Zoophysiologie der Rheinischen Friedrich-Wilhelms-Universität, Endenicher Allee 11–13, 5300 Bonn, FRG

Biosynthesis, Metabolism and Mode of Action of Invertebrate Hormones (ed. by J. Hoffmann and M. Porchet)

uents) can be expected to yield a chromatographic pattern in which individual neuropeptides can be immediately recognized. They can then be tested for biological activity. In some cases, substances may be pure enough after one or two HPLC steps to be analyzed directly by amino acid analysis or even to be sequenced. Thus, this method will undoubtedly prove useful for studies aimed at the elucidation of the hormone inventory of the sinus gland. Another possible application of HPLC is the study of molecular heterogeneity of certain hormones. Occurrence in different molecular variants of one biological activity in one species has, for example, been described for the melanophore dispersing hormone [1, 6], one form of which appears to be identical to the distal retinal pigment hormone (DRPH, light adapting) mentioned above. On the other hand, species or systematic group-related structural differences have been reported for CHH [1, 2, 7]. Differences in the relative amounts of individual neuropeptides and quantitative changes in relation to physiological conditions or seasons may also be analyzed.

2. Analytical HPLC of SG Extracts from Carcinus, Eriocheir, and Orconectes

To explore the potential of HPLC for analytical separation and characterization of SG neuropeptides from decapods, we have subjected SG extracts to reversed-phase chromatography on a Waters μBondapak Phenyl column. The three well-known hormonal activities were located in the chromatograms either by bioassay (CHH, MDH; in *Orconectes* RPCH) or by comparison of the retention time with that of synthetic RCPH. Some bioactive peaks were analyzed by amino acid analysis. The chromatograms shown have been obtained by directly injecting freshly prepared acetic acid (2 N) extracts from SG. By this procedure, generation of artifacts due to oxidation and/or proteolysis is supposed to be kept at a minimum. Pretreatment of samples, for example, by SepPak (Waters) separation was shown to result in losses affecting some peptides more than others. It was therefore avoided to obtain chromatograms that reflect the actual quantitative proportions of the different neuropeptides more faithfully.

In Fig. 1 a separation of SG extract from *Carcinus maenas* is shown. The hyperglycemic hormone (CHH) activity is found to be associated almost exclusively with a single, late eluting symmetrical peak. The size of the peak indicates that CHH is one of the most abundant neuropeptides in the SG, a result that agrees well with earlier findings obtained with other techniques [8]. The small peak directly preceding the main CHH peak also contains some activity. We do not know at present whether this represents a naturally occurring molecular variant or whether this peak arises as an artifact during extraction (for example, by rapid oxidation of Met residues). Amino acid analysis yields a composition that is almost identical to the composition established earlier with CHH of proven purity (Table 1); thus, it is obvious that highly pure material is obtained by a simple HPLC step carried out with crude SG extract. That there is essentially only one highly predominant form of CHH is also supported by radioimmunological analysis of all the fractions of the chromatogram [13]. In contrast to CHH, melanophore-dispersing activity (MDH) is associated with a larger number of fractions and it is note-

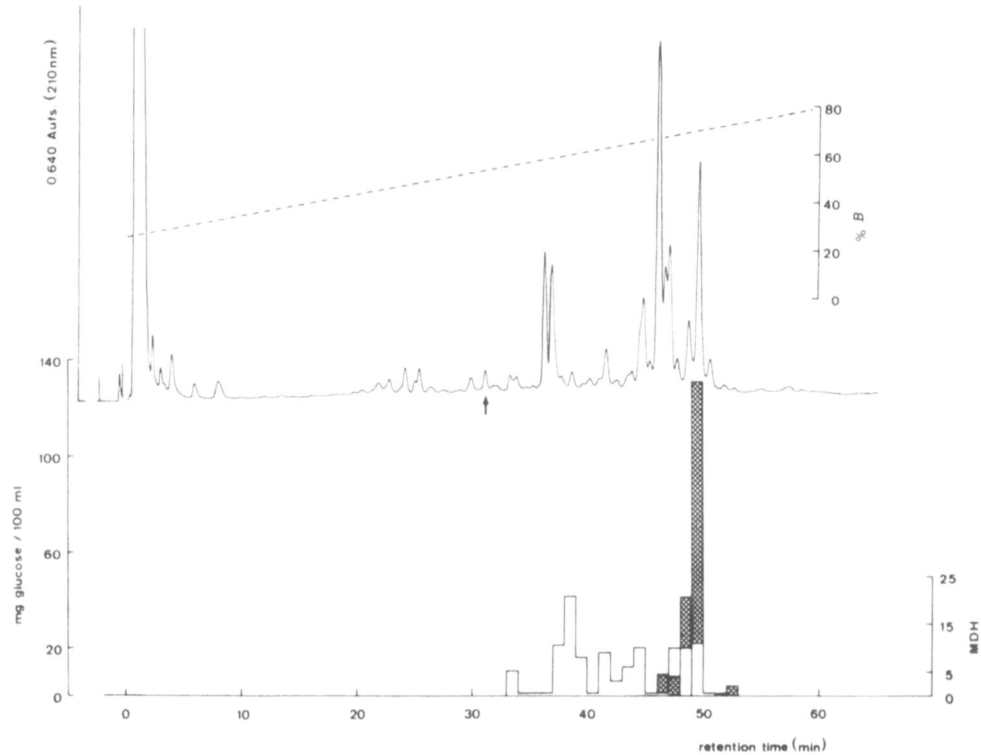

Fig. 1. HPLC of an acetic acid (2 N) extract from approx. 30 sinus glands of *Carcinus maenas* on a Waters μBondapak Phenyl column (30 × 0.39 cm) with gradient elution at 30 °C. Solvent A: 0.11% TFA; solvent B: 0.10% TFA; 60% acetonitril. The flow rate was 0.9 ml/min. Fractions were collected at 1 min intervals and hyperglycemic activity (*checkered bars*) was assayed according to [5]; MDH activity (*open bars*) according to [8]. *Arrow*: peak appearing at the retention time of synthetic RPCH

worthy that there is no clear association with one of the larger peaks. Thus, it appears to occur in several molecular variants which are all present in rather small amounts. The nature of the active fractions is currently under investigation.

Red pigment concentrating hormone was not located by bioassay, but a small though distinct peak is seen at an retention time corresponding to that of synthetic RPCH. Amino acid analysis of this peak is strongly suggestive of a substance closely resembling synthetic RPCH, but three additional amino acids not occurring in RPCH were found, indicating that this peak can not be considered completely pure. It is evident, however, that RPCH is a minor component of the SG neuropeptide inventory.

Figure 2 shows a chromatogram of SG extract from *Eriocheir sinensis* obtained under identical conditions. Again, CHH activity is found associated with a distinct, late eluting large peak and a small one directly preceding the latter. In this species, quantitative predominance of CHH is even more striking than in *Carcinus*. Preliminary amino acid analysis of the large peak shows a composition characteristic of other CHHs *(Carcinus, Orconectes)*, both in terms of proportions of individual residues and total number of residues. This indicates that the hormone is already highly, though not completely pure.

Table 1. Amino acid composition of sinus gland neuropeptides

	Hyperglycemic neuropeptides (CHH)				Neuropeptides with melanophore dispersing activity *(Eriocheir)*	
Amino acid	Number of residues *Orconectes*[a]	*Carcinus*[a]	*Porcellio*[b]	Molar ratio *Eriocheir*	Molar ratios Peak 7	Peak 8
Asx	8	9	7	9.70	1.89	1.55
Thr	2	2	2	–	–	–
Ser	3	4	2	5.40	1.84	1.72
Glx	6	5	5	4.83	2.79	3.18
Pro	1	1	1	–	1.32	1.37
Gly	2	1	2	2.00	4.50	4.52
Ala	3	4	4	3.75	3.59	3.58
1/2 Cys	4	4	4	n.d.	–	–
Val	5	4	3	4.00	1.54	1.36
Met	1	3	2	2.31	0.58	0.57
Ile	3	1	1	1.71	–	–
Leu	5	5	4	6.27	3.38	3.47
Tyr	5	4	2	3.96	–	–
Phe	3	2	3	3.50	0.63	0.62
Trp	n.d.	1	n.d.	n.d.	n.d.	n.d.
His	0	1	1–2	5.63	0.77	0.79
Lys	3	2	3	2.70	–	–
Arg	4	4	4	5.59	3.00	3.00
Total	58	57	50–52		28–29	

[a] Data taken from [5]
[b] Data taken from [12]

As in *Carcinus,* MDH activity is found in several fractions, none of which is clearly associated with a major peak. The pattern is somewhat different from that seen after separation of SG extract from *Carcinus* which indicates a certain degree of species specificity. The active fractions were combined and larger amounts of material were rechromatographed with a different solvent gradient (Fig. 3). MDH activity was now found associated with several distinct peaks. Two of them (peaks 7 and 8) were analyzed by amino acid analysis (Table 1) and were found to be very similar in composition. Although it is not quite clear whether the substances are homogeneous, it is interesting that Thr which is a constituent of DRPH [4] is not found and that, on the other hand, amino acids not found in DRPH are present. The Met content is lower than in DRPH. This may be taken as a first preliminary indication that there may be heretofore unknown neuropeptides with melanophore dispersing activity that differ from DRPH. The precise nature of these peptides is currently under investigation in our laboratory. It has to be considered that oxidation of Met residues may give rise to an artificial heterogeneity. With regard to peaks 7 and 8, however, the amino acid analysis did not show Met sulfoxide or sulfone. Thus, at least these two peptides may be considered native. Like in *Carcinus* material, RCPH was not directly assayed by bioassay, but a peak ap-

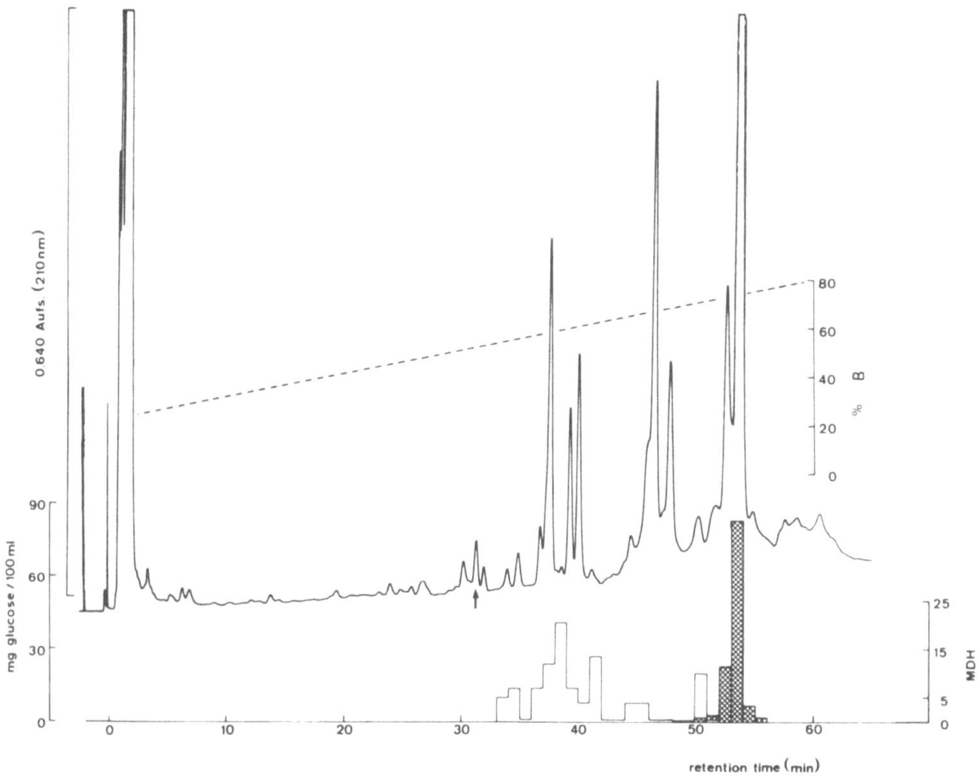

Fig. 2. HPLC of an extract from approx. 63 SG from *Eriocheir sinensis*. Conditions were identical to those described under Fig. 1. *Arrow*: peak appearing at the retention time of synthetic RPCH

pears at the retention time of authentic RPCH. Amino acid analysis of this peak indicates the presence of a RPCH-like substance. Again, RCPH is present in relatively small amounts.

In addition to the preparations from brachyuran sources, SG extract from an astacuran species, *Orconectes limosus*, was analyzed under identical conditions (Fig. 4). Throughout the chromatogram, MDH activity was found to be very low and was therefore not assayed in detail. Neuropeptides possessing this activity are obviously present in lower relative amounts compared to the *Carcinus* and *Eriocheir* material. On the other hand, a peak with a retention time of synthetic RPCH is very prominent. In this case, the material was analyzed by bioassay utilizing the well-documented [9] adipokinetic activity of RPCH in the locust, and a high level of activity was found. Amino acid analysis gave the expected values, but in addition, Arg and Val. It is, therefore, not quite clear to what extent this peak represents pure RPCH or an RPCH-like substance with an identical retention time. The results do, however, suggest that RCPH may be present in surprisingly large amounts in the SG of this astacuran species and the question remains whether this has any physiological significance. With regard to CHH, the pattern is very similar to that obtained with the *Carcinus* and *Eriocheir* materials. Again, there is only one zone of activity, corresponding to the largest peak in

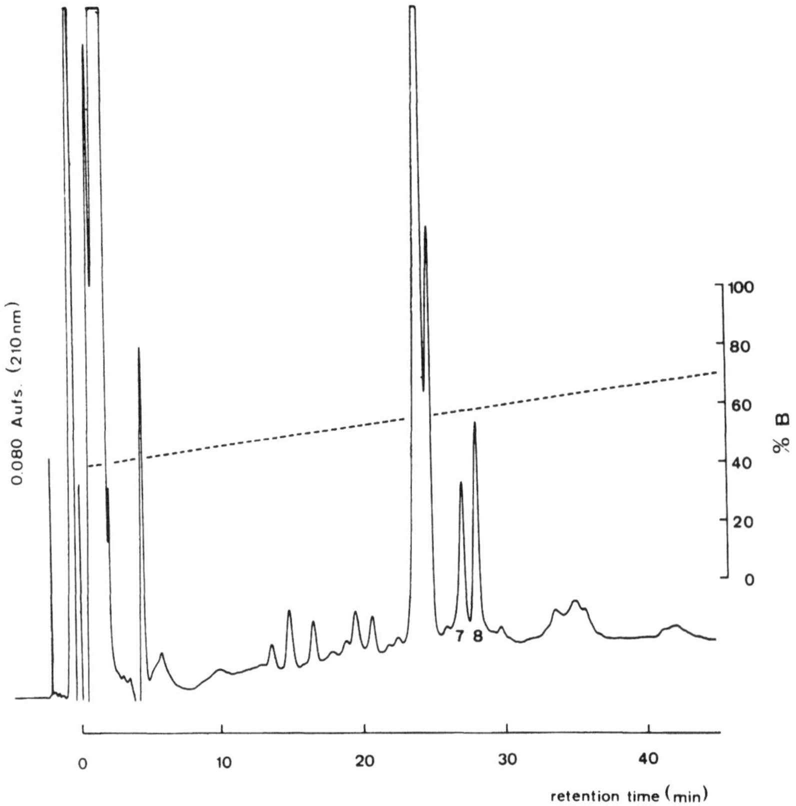

Fig. 3. Rechromatography of combined MDH-containing fractions from *Eriocheir* SG on a Waters μBondapak Phenyl column with a shallower gradient. MDH activity was found in 6 peaks. Peak 7 and 8 were analyzed by amino acid analysis (see Table 1 and text). Solvent rates as described in Fig. 1

the chromatogram. The characteristic small peak preceding the large one is also present and there is, in this case, evidence of another component eluting in the descending flank of the large peak. The nature of this heterogeneity is unknown at present.

As it can be seen in Figs. 1, 2, and 4, there are differences among the three species with regard to the retention times of the CHH peaks. This demonstrates that HPLC is a suitable method to study the species specificity of CHH for which evidence has been provided previously in other studies [5, 7]. To demonstrate this species specificity more clearly, material from CHH peaks from all three species was mixed and rechromatographed. All three CHHs are clearly separated (Fig. 5). The higher retention time of

Fig. 4. HPLC of an extract from approx. 75 SG of *Orconectes limosus.* Conditions identical to those described under Fig. 1. Adipokinetic activity was assayed according to [9]

Fig. 5. Rechromatography of a mixture of CHH peaks from three species under conditions described under Fig. 1. All three neuropeptides are clearly separated. *First peak: Carcinus* CHH; *second Orconectes* CHH; *Eriocheir* CHH third

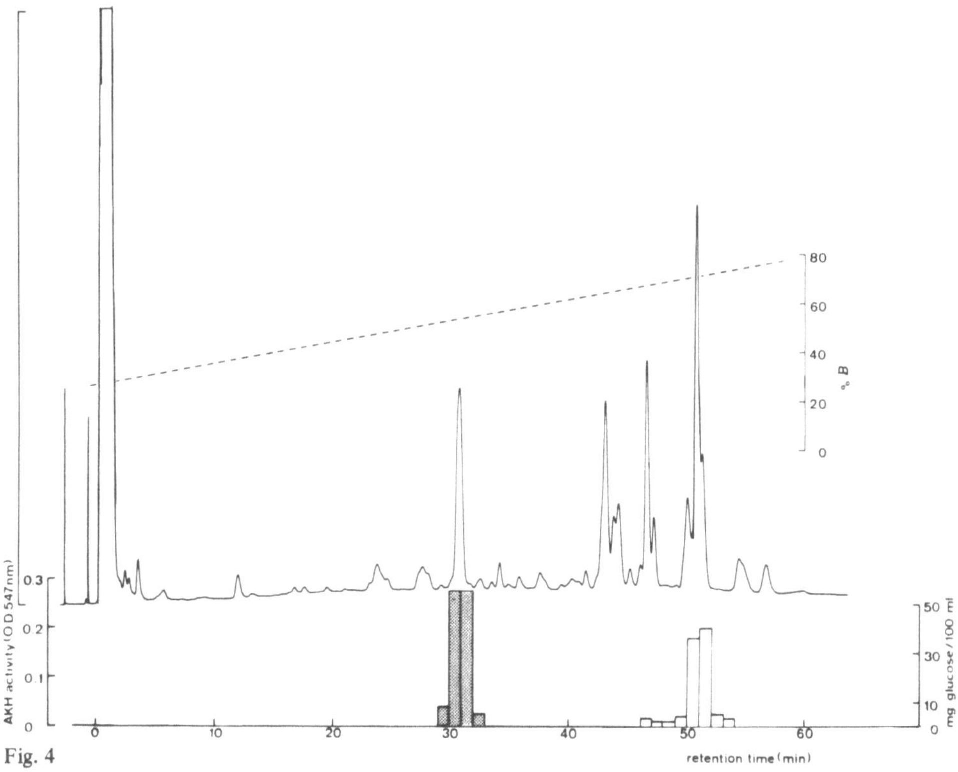
0.640 Aufs (210nm)
AKH activity (OD 547nm)
0.3
0.2
0.1
0
% B
80
60
40
20
0
mg glucose / 100 ml
50
30
10
0
0
10
20
30
40
50
60
retention time (min)

Fig. 4

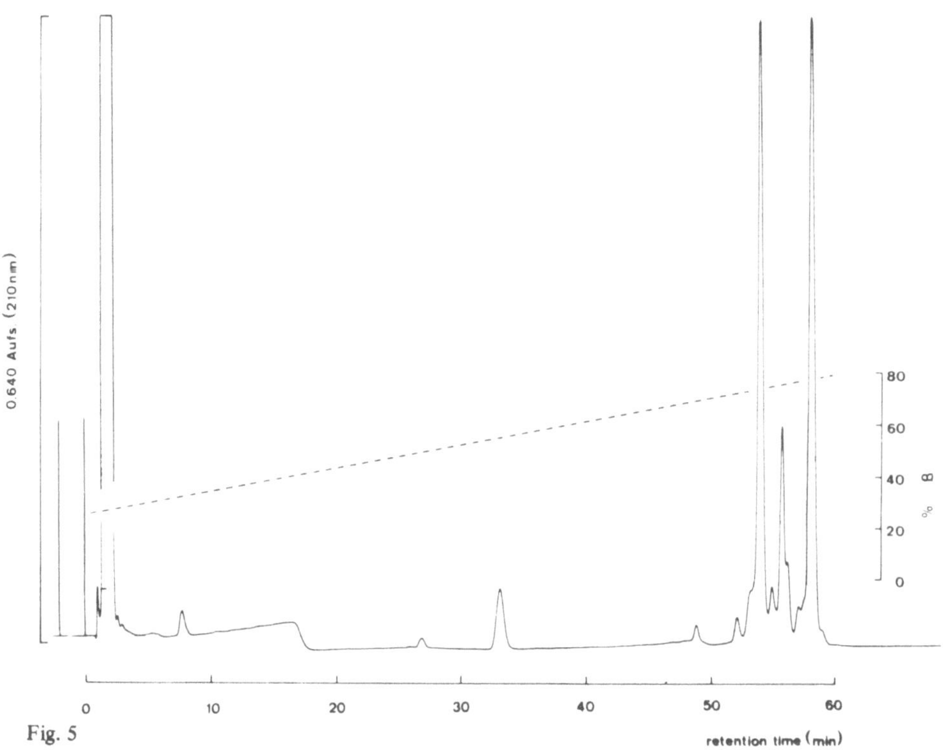
0.640 Aufs (210nm)
% B
80
60
40
20
0
0
10
20
30
40
50
60
retention time (min)

Fig. 5

Orconectes CHH as compared to that of *Carcinus* CHH may be explained by the higher content of hydrophobic residues (Table 1). The amino acid composition of *Eriocheir* CHH, which exhibits the highest retention time, is only preliminary, but the results appear to indicate an even higher amount of hydrophobic residues. Since only one hydrolysis time (22 h) was employed, values for Val, Ile, and Leu may actually be higher.

3. The Hyperglycemic Hormone of the Isopod, *Porcellio dilatatus*

Studies on the hyperglycemic neuropeptide have thus far been carried out almost entirely in decapod species. There are only a few reports suggesting the existence of a hyperglycemic principle in eyestalk or SG of other crustaceans, e.g., stomatopods [10] and isopods [11], but these factors have not been characterized. Especially in view of the molecular heterogeneity of CHHs among decapod species, it seemed interesting to investigate whether this neuropeptide from a species of a taxonomically distant group exhibits differences that exceed those found between different decapod CHHs or whether it resembles a "decapod-type CHH" closely. We have, therefore, isolated the CHH from SG of the isopod, *Porcellio dilatatus* and have determined its amino acid composition [12] which is shown in Table 1 together with those of decapod CHHs for comparison. This study was also carried out to explore the utility of HPLC in a preparative procedure for which only small amounts of starting material were available. In this case, gel filtration was used as a first step. This was chosen not only to achieve a first clean-up, but also to obtain information concerning the molecular size of the molecule in comparison to a decapod CHH. The chromatogram (Fig. 6) shows the separation of SG extract from *Porcellio* on a Sephadex G-50 column. As a molecular weight marker, a very small amount of ^{125}I-labelled *Carcinus* CHH, much too low to interfere in the bioassays and in amino acid analysis, was added to the extract prior to chromatography. The biological activity is associated with a rather marked peak with an elution volume between that of the excluded protein and the completely included low molecular weight material. The retention is almost identical to that of the radioactive marker, indicating close correspondence of the molecular weights. The active fractions were combined and the material was subjected to reversed phase HPLC (Fig. 7). Only two major peaks were seen, and hyperglycemic activity was found to be associated with the late eluting second peak. End group analysis by the dansyl-chloride procedure as well as the results of several amino acid analyses with subnanomol amounts showed that the material was highly pure. The amino acid composition (Table 1) does not indicate that the differences between the isopod CHH and decapod CHHs are greater than the differences between decapod CHHs from different species. It seems, however, that the number of residues may be somewhat lower. It should be noted, however, that *Porcellio* CHH did not show any hyperglycemic effect in *Uca pugilator*, whereas *Carcinus* CHH is fully active in this species. Reversely, *Carcinus* CHH is completely inactive in *Porcellio* [12]. Thus, although the hormones are undoubtedly homologous, it appears that enough changes, not revealed by amino acid composition alone, have been introduced during crustacean evolution to render, presumably, the isopod hormone unrecognizable by brachyuran target tissue receptors and vice versa.

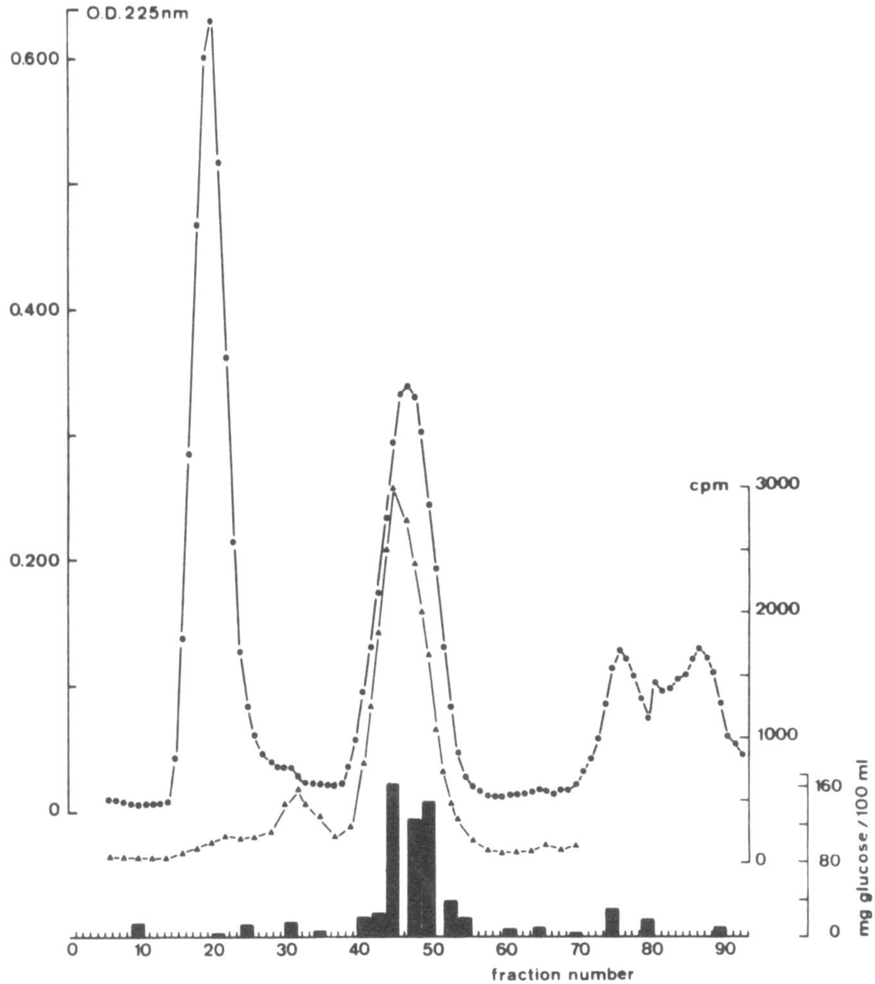

Fig. 6. Preparative gel chromatography, on Sephadex G-50 superfine, of an extract from approx. 2,000 SG of *Porcellio dilatatus*. For details see [12]. *Black bars:* hyperglycemic activity; *filled triangles:* radioactivity from 2.8 ng sample of ^{125}I-*Carcinus* CHH added to the extract prior to chromatography (Martin et al. 1983)

4. Conclusions

The first application of HPLC to separation and characterization of neuropeptides from the crustacean sinus gland which has been reported here, in our opinion demonstrates the great potential of the technique in the special case. The fact that the neuropeptides are present in the SG in highly enriched form together with high resolution and sensitivity of HPLC appears to permit, in many cases, a clear correlation of biological activities with distinct, albeit sometimes small peaks. These peaks may represent ma-

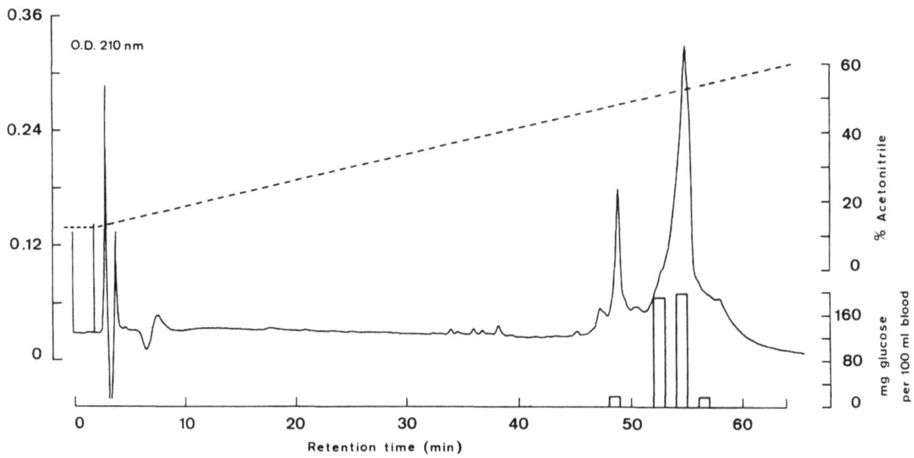

Fig. 7. HPLC combined active fractions from Fig. 6 on a Waters μBondapak C 18 column. Solvent A: 0.11% TFA; solvent B: 0.10% TFA, 60% acetonitril, flow rate 0.9 ml/min. For further details see [12] (Martin et al. 1983)

terial that is sometimes pure enough for at least preliminary amino acid analysis or even for sequencing.

Therefore, this method should prove useful in further work aimed at elucidation of the inventory of neuropeptides synthesized, stored, and released in the crustacean eyestalk.

Acknowledgments. The work was supported by the Deutsche Forschungsgemeinschaft (Ke 206/7). The excellent technical assistance of Barbara Reichwein is gratefully acknowledged and we thank Dr. G. Gäde for carrying out the bioassay of RPCH.

References

1. Kleinholz LH, Keller R (1979) In: Barrington EJW (ed) Hormones and evolution, vol I. Academic Press, London New York, pp 159–214
2. Keller R (1983) In: Gupta AP (ed) Neurohemal organs of arthropods. Thomas, Springfield, pp 118–148
3. Fernlund P, Josefsson L (1972) Science 177:173–174
4. Fernlund P (1976) Biochim Biophys Acta 439:19–25
5. Keller R (1981) J Comp Physiol 141:445–450
6. Fernlund P (1971) Biochim Biophys Acta 237:519–529
7. Leuven RSEW, Jaros PP, VanHerp F, Keller R (1982) Gen Comp Endocrinol 46:288–296
8. Keller R (1977) J Comp Physiol 122:359–373
9. Mordue W, Stone JV (1976) Nature (London) 264:287–289
10. Keller R (1969) Z Vergl Physiol 63:137–145
11. Gersch M, Eibisch H (1976) Zool Jahrb Physiol 80:267–273
12. Martin G, Keller R, Kegel G, Besse G, Jaros PP (in press) Gen Comp Endocrinol
13. Keller R (in preparation)

I.17 Present State of Knowledge on the Chemical Nature of the Androgenic Hormone in Higher Crustaceans

P. JUCHAULT, J.J. LEGRAND, and J. MAISSIAT[1]

1. Introduction

Studies carried out over the last 30 years have established the basic biological functions of the hormone secreted by the androgenic gland of higher crustaceans [1, 2], but have not yet been able to reveal the chemical nature of this hormone.

As early as 1960, investigations showed that some similarities existed between this hormone and the sexual hormones of vertebrates. An inhibitory effect of important doses of androsterone, testosterone, or methyl-testosterone on the ovarian growth was evidenced in prawns [3] and in the female crab *Ocypoda platytarsis* [4]. Transformation of ovary into a "testis" with spermatogonia and spermatocytes was observed after five injections of testosterone-propionate in doses of 4 to 5 mg. However, the external sexual characteristics were not modified, as opposed to the results observed after injection of an extract of androgenic glands. In *Ocypoda*, it was claimed that the sterol extract of the androgenic glands had the same masculinizing effects as the graft of an androgenic gland [5]. Similar results were obtained in immature female crabs of *Callinectes sapidus* injected with 0.3 mg testosterone per day for 30 subsequent days [6]. In this case, the ovarian structure was also affected, but the external sexual characteristics were not modified. On the other hand, the injection of testosterone-propionate and testosterone acetate into the prawn *Parapenaeopsis hardwickii*, induced hypertrophy and hyperplasia of the androgenic gland and an increase in the diameter of testicular follicles [7]. As opposed to this, injections of steroids (sterandryl-25,testosterone-propionate, methyl-17α-testosterone) did not elicit any masculinizing effect in the amphipod *Orchestia gammarella* [1, 8] and in the isopods *Porcellio dilatatus* and *Helleria brevicornis* [9]. Biochemical studies carried out in decapods pointed to the presence of steroids or enzymes of steroidogenesis in the region which included the androgenic gland and the adjacent part of the vas deferens. In *Carcinus maenas* the presence of a steroidic substance was monitored [10] and in other crustaceans, several steroidal enzymes were observed: 17β-hydroxysteroid dehydrogenase (HSD) in *Homarus americanus* [11], 3β-HSD in *Macrobrachium rosenbergii* [12], 17β- and 20α-HSD, C-21- and C-17β-hydroxylase, C-17- and C-20-lyases in *Callinectes sapidus* [6]. However, it was reported that the steroid metabolism detected in the male gonad of *Carcinus maenas* is extraglandular and occurs mainly in the testis and the vas deferens [13].

1 Laboratoire de Biologie animale, E.R.A. CNRS N° 070230 "Physiologie et Génétique des Crustacés", 40, Avenue du Recteur Pineau, 86022 Poitiers Cédex, France

Biosynthesis, Metabolism and Mode of Action of Invertebrate Hormones
(ed. by J. Hoffmann and M. Porchet)

In addition to these observations, a lipidic substance containing farnesylacetone and hexahydrofarnesylacetone [14, 15] was extracted from hemolymph and androgenic glands of the decapod *Carcinus maenas*. These substances inhibit vitellogenesis in the amphipod *Orchestia gammarella* and induce a male-type accumulation of carotenoid pigments in the antenna of females of *Talitrus saltator*. However, when 2 μl of an extract which corresponded to the androgenic glands of 450 crabs were injected every 2 days during 2 months, spermatogenesis and differentiation of the external male characteristics were not induced in the females of either *Orchestia* nor *Talitrus*.

Ultrastructural studies of the androgenic gland or normal and experimental animals have been carried out in various species, concomitantly to these biochemical studies. Most authors have suggested that the secretion of the androgenic gland is not of a steroid type, but is more likely a protein or a peptide [16–19]. This opinion is mainly based on the presence in active glands, of an abundant rough surfaced endoplasmic reticulum and of Golgi complexes which elaborate vesicles with dense cores. No signs of storage of the androgenic hormone have been observed in these studies.

Recently, a proteic substance exhibiting an androgenic function was extracted from the male reproductive organs of the isopod *Armadillidium vulgare* [20]. Independently from these results, our group has shown that the androgenic glands of intersexed males of *A. vulgare* contained a water soluble substance which is able to induce in the female the differentiation of all the external male characteristics [21]. We wish to report here on our recent data regarding the substance extracted from the androgenic glands of *A. vulgare*.

2. Material and Methods

2.1 Material

The androgenic substance can be extracted from entire male gonads [20] or from the androgenic glands or even from the hemolymph of normal males (Juchault, unpubl.). However, the best material for hormone extraction is the androgenic gland of males experimentally changed into intersexed. Some populations of *A. vulgare* present thelygenic lines in which the females lodge feminizing bacteria. When experimentally inserted into a normal male, these bacteria induce the differentiation of a female physiology [1]. The androgenic gland shows a marked hypertrophy and produces male hormone, although it does not seem to release it into the hemolymph. The various extraction procedures reported hereafter have been carried out on androgenic glands of males changed into intersexed 150–200 days prior to the extraction.

2.2 Extraction and Purification

Various extraction procedures were performed which included the separation of ethanol, ether, butanol, methanol, and aqueous fractions of these, only the aqueous fraction was found to be active. One of these procedures consisted in homogenizing the androgenic glands with a sonicator in distilled water. After centrifugation of the acetone-aqueous extract, hexane was added to the supernatant. The androgenic activity was not present in the hexane phase. When androgenic glands were plunged into freezing water and centrifuged, an androgenic activity was also found in the aqueous supernatant. The purification procedures consisted in: filtration on Sephadex G 75 or Biogel P_2 with water or 0.005 *M* ammonium acetate (pH 7) as eluents; thin-layer chromatography on silica plates (Whatman), in ethanol-water (80 vol:20 vol). HPLC separations were performed with a μBondapak C_{18} stainless steel column or a radial compression column; the eluents were: water, methanol-water (80 vol:20 vol; 60 vol:40 vol; 50 vol: 50 vol + 0.1% trifluoroacetic acid); elution gradient systems were (solvent A: water + 0.1% TFA; solvent B: 60% acetonitrile-water + 0.1% TFA).

2.3 Biological Test

The biological test of the extracts consisted in the injection into the body cavity of a young *A. vulgare* female (20–25 mg body weight) of 0.5 μl of an aqueous solution of the extract to be tested. Each animal received one injection; the effect was monitored at the time of molting.

3. Results

3.1 Active Extract

Regardless of the mode of extraction, the lipidic and steroidic fractions were found to be inactive. However, all injections of water soluble substances induced a male differentiation of the endopodites of the first two pairs of pleopods and even sometimes the appearance of spine brushes on the anterior pereiopods. A small genital apophyse and a deferent vas at the end of the ovary appeared occasionally. The transformation of the ovary into a testis has recently been obtained following several injections of extracts [22]. This sort of transformation is usually observed in young females which have received a graft of androgenic glands. If the molt appears within 6 or 7 days following the injection, females (in premolt at the time of the injection) are not masculinized. However, masculinization always occurs between the 9th and the 26th day and, more sporadically, between the 26th and the 36th. After this period, the injection is inefficient [21]. The androgenic unity of *A. vulgare* can be defined as the minimal quantity of substance able to inducing the differentiation of external male characters in a young female (of 20–30 mg body weight). One single androgenic gland of a male transformed into an intersexed (150 mg body weight) 5–6 months earlier yields a quantity of androgenic substance corresponding to 30 androgenic unities. The level observed in androgenic glands of normal males, is six times lower [20]. The most purified fraction obtained by Katakura et al. [20] can masculinize a young female with a dose containing 0.18 μg of proteins. On our side, we have been able to extract from the hemolymph of normal males a low quantity of active substance (about one androgenic unit per 35 μl of hemolymph).

3.2 Effect of Temperature and Proteolytic Enzymes

The androgenic activity persists after heating of the aqueous extract at 90 °C for 15 min. In addition, when the lyophilized extract is heated to 125 °C for 1 h, the activity is not lost. An intersexed androgenic gland kept at a temperature of 150 °C for 50 min is still able to masculinize a female, but the activity is lost when the glands are heated to 170 °C. The substance extracted by Katakura et al. [20] seems to be less thermostable since the activity is destroyed by heating for 5 min at 90 °C. The action of proteolytic enzyme is noteworthy [20, 23]: 1% pepsin (37 °C, 1 h treatment) does not inactivate the androgenic substance, but 0.1% pronase, 0.2% α-chymotrypsin and 0.2% trypsin, completely inactivate the extracts.

3.3 Specificity

The effect of the substance extracted from the androgenic glands of *A. vulgare* was investigated in five species of Oniscoid isopods among which three belong to the family of *Armadillidiidae* and two to that of *Porcellionidae*. Only females of the genus *Armadillidium* (*A. nasatum* and *A. maculatum*) were masculinized by this substance. However, it had no effect in *Eluma purpurascens* (primitive *Armadillidiidae*) and in *Porcellionidae*: *Porcellio scaber* and *Porcellio dilatatus*. These results are similar to those concerning the effect of crossed grafts on androgenic gland between these different species [24].

3.4 Purification and Molecular Weight

The different attempts at purification carried out by Katakura et al. [20] and ourselves have not yet led to the isolation of a pure product. We have noticed, with the techniques of filtration on Sephadex as well as with thin-layer chromatography or HPLC, that the androgenic activity was divided among different fractions of variable elution characteristics. This may suggest that the androgenic substance is linked to proteins of the androgenic gland (or even of the gonad, in the extractions of Katakura et al. [20]). Such a process could explain the differences in the estimated molecular weight between Katakura et al. [20] and ourselves. With the techniques of gel filtration of different marker proteins, Katakura et al. [20] recorded a molecular weight of about 16,000, while by techniques of dialysis through cellulose membrane, we estimate the molecular weight to be between 2,000 and 8,000. In fact, in the present state of the purification, it is premature to estimate the molecular weight of the androgenic substance, as we are not yet dealing with a pure product.

4. Discussion

The results obtained in decapods are not yet very precise. With regards to the masculinizing effect of the steroid hormones of vertebrates, we must mention that no external male differentiation was obtained and that the effect on the ovary never led to the differentiation of spermatozoa, but only to that of cells considered as spermatogonia and spermatocysts. Besides, no effect on the gonoduct (deferent vas) seems to have been recorded. It seems reasonable to say, in the present state of the published results, that several injections of testosterone can inhibit the ovarian function of decapods. This inhibition could correspond to a pharmacological effect. The data on the steroid metabolism of the androgenic gland are dubious as the positive results apparently concern the deferent vas and not the androgenic gland [13]. In addition, the existence of a steroid metabolism in the androgenic gland is not proof for the steroid nature of the male hormone. However, the results obtained with extracts of androgenic glands and hemolymph show that the lipidic fraction, and more precisely, hexahydrofarnesylacetone and farnesylacetone, exert some functions that several authors consider to be specific for the androgenic gland [14, 15]. In order to explain the difference in the results between decapods and isopods, given that it is most unlikely that the chemical nature of the male hormone basically varies from one order to another, we may assume that the androgenic gland secretes several hormones which would act specifically on different aspects of the sexual physiology. We must also notice that the activity test of a substance extracted from a decapod was carried out on amphipods, and this could explain the lack of masculinizing effect of the aqueous extract, with regard to the negative results obtained in *Porcellio* females injected with male hormone of *Armadillidium vulgare*. On the contrary, the results obtained both by Katakura et al. [20] and ourselves allow us to assert that in the isopod *A. vulgare*, the androgenic glands of normal males and males transformed into intersexed contain an hydrosoluble substance showing all the different functions of the androgenic hormone: differentiation of the external male characteristics and differentiation of the gonad into testis. It seems quite reasonable to consider that this product corresponds to the male hormone secreted by the androgenic gland. This substance acts at a low concentration, it is relatively thermostable, but is destroyed by proteolytic enzymes. Its effect is limited to species of the *Armadillidium* genus. The lipidic and steroid extracts do not have any masculinizing effect. These various results all plead in favor of a proteinic nature of the androgenic hormone.

References

1. Legrand JJ, Juchault P (1972) In: Wolff E (ed) Hormones et différenciation sexuelle chez les invertébrés. Gordon and Breach, Paris London New York, pp 179–218
2. Charniaux-Cotton H (1972) In: Wolff E (ed) Hormones et différenciation sexuelle chez les invertébrés. Gordon and Breach, Paris London New York, pp 127–178
3. Carlisle DB (1960) Mem Soc Endocrinol 7:9–16
4. Sarojini S (1963) Curr Sci India 32:411–412
5. Sarojini S (1964) Curr Sci India 33:55–56
6. Tcholokian AK, Eik-Nes KB (1971) Gen Comp Endocrinol 17:115–124
7. Veith WJ, Malecha SR (1983) S Afr J Sci 79:84–85
8. Charniaux-Cotton H (1962) Gen Comp Endocrinol 1:241–247
9. Juchault P (1966) These Doct Sc Nat, Univ Poitiers
10. Godbillon G, Balesdent ML (1974) CR Acad Sci 278:2823–2826
11. Gilgan MW, Idler DR (1967) Gen Comp Endocrinol 9:319–324
12. Nagabhushanam R, Kulkarni GK (1981) Aquaculture 23:19–27

13. Blanchet MF, Ozon R, Meusy JJ (1972) Comp Biochem Physiol 41B:251–261
14. Berreur-Bonnenfant J, Meusy JJ, Ferezou JP, Devys M, Quesneau-Thierry A, Barbier M (1973) CR Acad Sci 277:971–974
15. Ferezou JP, Berreur-Bonnenfant J, Tekitek A, Rojas M, Barbier M, Suchy M, Wipf W, Meusy JJ (1977) In: Marine natural products chemistry. Faulkner and Fenical, Ed., pp 361–366
16. King DS (1964) Gen Comp Endocrinol 4:533–544
17. Meusy JJ (1965) Zool Jahrb Physiol 71:608–623
18. Malo N (1970) CR Acad Sci 270:2843–2845
19. Malo N, Juchault P (1970) CR Acad Sci 271:230–232
20. Katakura Y, Fujimaki Y, Unno K (1975) Annot Zool Jpn 48:203–209
21. Juchault P, Maissiat J, Legrand JJ (1978) CR Acad Sci 286:73–76
22. Katakura Y, Hasegawa Y (1983) Gen and Comp Endocrinol 48:57–62
23. Juchault P, Maissiat J, Legrand JJ (1979) Bull Soc Sci Nat Tunisie 14:43–56
24. Juchault P, Legrand JJ (1968) Gen Comp Endocrinol 36:175–186

Part II Ecdysteroids

Introduction

P. KARLSON and J. HOFFMANN

Ecdysteroids were a major subject of interest at this CNRS Symposium. Fifteen plenary lectures, numerous poster communications plus a round table were devoted to these molecules. At the Lille meeting 8 years ago, ecdysteroids were primarily thought of as the "molting hormones" of Arthropods, whereas we now know that they probably control essential events of development and reproduction in most protostomian phyla. At Lille, two communications treated the mode of action of ecdysteroids at the molecular level, whereas in Strasbourg half of the communications were concerned with this aspect. Ten different ecdysteroids were known in arthropods at the previous meeting: more than 20 others have been isolated since. Significant advances have been made during the period between the two meetings; however, in spite of the intensive studies in numerous universities and institutions over the world, the reader of the following pages will sense that major questions in ecdysteroid biosynthesis, metabolism and mode of action remain almost totally unsolved. We hope that the articles presented in the second part of this volume will stimulate research on these numerous "black boxes" in our current knowledge.

We present the reader with a brief introduction to the various articles in this section and include essential points discussed at the round table which followed the plenary lectures.

The first two contributions deal with ecdysteroids in ovaries and embryos of orthopteran insects. This topic was already the subject of a preliminary report by the Strasbourg group at the Lille meeting and has since attracted considerable interest, as illustrated by the continuously increasing literature devoted to ovarian and egg ecdysteroids in many insects species. M. Lagueux and co-workers summarize the results which they have obtained in *Locusta* since the Lille meeting; they also review the studies on ecdysteroids in reproductively competent female insects. In dipterans at least, these molecules are thought to play an essential role in the control of vitellogenesis (see references in the article of Lagueux et al.) whereas in *Locusta* and *Schistocerca* (as in Lepidopterans) an obvious endocrine relationship exists between the maturing ovary and embryonic development. This relationship is mediated via surprisingly large quantities of ecdysteroid conjugates which are stored in the oöcyte and later made available to the embryo. Over the last 2 years, a large list of free and conjugated ecdysteroids have been isolated and identified in ovaries and eggs of both *Schistocerca* and *Locusta*. The results are summarized in these first two contributions. For the majority of the structure the results obtained in *Schistocerca* and *Locusta* are superposable, but a difference was obtained for the predominant maternal ecdysone conjugates which were reported in *Schistocerca* to be 22-phosphate esters and for which in *Locusta* an

Biosynthesis, Metabolism and Mode of Action of Invertebrate Hormones
(ed. by J. Hoffmann and M. Porchet)

adenosine moiety was detected on the 22-phosphate group (adenosine-monophosphate esters of 2-deoxyecdysone and ecdysone). The presence or absence of an adenosine moiety on the phosphate became a matter of lively discussion at the round table. G. Ourisson, in whose laboratory the identification of the complex ester had been performed, summarized the chemical evidence for its structure. Dr. Rees suggested that 22-phosphates and adenosine could have been cocrystallized, to which G. Ourisson objected that such a process could not account for the covalent binding observed between the ribose and the phosphate. Alternatively, H.H. Rees wondered whether the AMP-conjugate could not represent a minor form, the 22-phosphate being the major maternal conjugate. This possibility was left open, as well as the problem of the stability of the complex conjugates during the extraction procedure.

The contribution of Rees and Isaac also reviews current knowledge on the biosynthetic pathways of ecdysone. The ovarian system has proved to be an excellent model for investigating the sequence of the biosynthetic events leading from cholesterol to ecdysone, although the early steps in the pathway remain somewhat enigmatic. There is evidence from studies on prothoracic glands and ovaries of various insect species that more than one sequence of hydroxylations may exist, involving either complete formation of the nucleus before side-chain hydroxylations or synthesis of 2-deoxyecdysteroids. In the round table discussions, Bollenbacher pointed out that in contradistinction to ovaries, the other major ecdysone producing tissue in insects, i.e., the prothoracic glands, are apparently unable to hydroxylate precursors in the 14α-position. Prothoracic glands of *Manduca sexta* easily convert under in vitro conditions precursors which are already hydroxylated in C–14α, but are unable to transform cholesterol into ecdysone. Addition of lipoproteins to the incubation medium considerably enhances ecdysone secretion by excised prothoracic glands of *Manduca*, as monitored by RIA measurements, and the question was raised again if prothoracic glands use not cholesterol, but 14α-hydroxysteroids provided by lipoproteins as precursors for ecdysone biosynthesis rather than cholesterol. Clearly, this crucial problem should be the object of intense research efforts in the next years.

The contribution of R. Lafont and J. Koolman is a very complete and precious survey of our present knowledge on ecdysone metabolism. A long list of metabolites derived from ecdysone is presented which includes, in addition to 20-hydroxyecdysone, such molecules as 3-dehydroecdysteroids, 3-epiecdysteroids, 26-hydroxyecdysteroids and the recently described ecdysonoic acids, C28-ecdysteroids and both polar and nonpolar conjugates. The presence of stearate, oleate, and palmitate esters of 20-hydroxyecdysone in ticks, where they represent major metabolites, is stricking. As the nonpolar esters are rather resistant to *Helix pomatia* hydrolytic enzymes, they may have escaped detection in many investigations on ecdysone metabolism so far. R. Lafont and J. Koolman also present a survey on the enzymes involved in ecdysone metabolism and their tissue localization before adressing the question of the physiological significance of the various pathways.

The two authors rightly stress the spectacular advances made in the field of ecdysone metabolism since the Lille meeting. This is illustrated by the number of ecdysteroids isolated and identified in arthropods which is now more than 30, as compared to 10, 8 years ago. In this field, as in others, the introduction of refined analytical techniques has been extremely rewarding.

During the round table discussion, it was emphasized that no conjugates containing glycosidic bonds have yet been isolated, with the exception of a report by Heinrich and Hoffmeister (a study which should be repeated, according to Karlson). It is well known that insects can produce β-glycosides of a variety of compounds which include protocatechuic acid and N-acetyldopamine, but apparently ecdysteroids are not conjugated by such reactions. Similarly, the apparent absence of conjugation of ecdysteroids by glucuronic acids is unexpected. It was also mentioned at the round table that sulfates had been tentatively identified by the double-labeling technique. However, isolation in pure form from biological material and identification is still lacking. It was stressed that digestion of naturally occurring conjugates with commercially available enzyme preparations cannot be accepted as proof of structure, as most commercial enzymes are contaminated with other hydrolases, so that a "sulfatase" could presumably also hydrolyse phosphate esters.

In Lille, the only report on the biochemistry of an enzyme system of ecdysone metabolism was that by J. Koolman on ecdysone-3-oxidase. Since that meeting the studies have been extended primarily to the ecdysone-20-monooxygenase system to which the contribution of the Beltsville group is entirely devoted. This key enzyme which converts ecdysone to 20-hydroxyecdysone outside the prothoracic glands is perplexing in as far as its subcellular localization is concerned. Indeed, in some insects and tissues it is a mitochondrial enzyme, in others it is microsomal and in still others, like *Manduca* midgut, it is both mitochondrial and microsomal. In their contribution, G.F. Weyrich and associates critically review the data available on mitochondrial and microsomal ecdysone-20-monooxygenase; they conclude that much work is necessary to gain a better understanding on the mechanism of ecdysone 20-hydroxylation and its regulation. Furthermore, as pointed out by these authors, hydroxylation at C–20 is only one among five hydroxylation steps required to convert cholesterol to 20-hydroxyecdysone, and we have no information at all regarding the other four enzyme systems. The synthesis of high specific activity tritiated 2-deoxyecdysone which was reported at the meeting by Hetru and associates, and the projected synthesis of the other precursor molecules, will hopefully help to overcome this gap in the near future.

The very first question which was raised at the round table concerned the concept of "active ecdysteroid". It is a general belief that, although the steroid secreted by the prothoracic gland is essentially ecdysone, the molting hormone acting on the target cell is 20-hydroxyecdysone. This belief is essentially based on in vitro experiments with dipteran cells, tissues or organs in which 20-hydroxyecdysone was found to be active at concentrations down to two magnitudes lower than those required for ecdysone. Could ecdysone, or even 2-deoxyecdysone in the embryo model, be hormones per se? Although this did not seem to be the opinion of the majority of the round table discussants, the contribution by P. Porcheron and associates presents interesting in vitro data which are obviously best interpreted by admitting that ecdysone and 20-hydroxyecdysone have separate and complementary functions in the control of the integumental events in orthopteran insects; ecdysone controls at relatively low concentrations the cell changes observed in postexuvial integuments prior to apolysis while higher concentrations of 20-hydroxyecdysone induce apolysis, epidermal protein synthesis and epicuticle and procuticle depositions. The in vitro data match the respective fluctuations of ecdysone and 20-hydroxyecdysone which we ourselves reported 8 years ago and which are corroborated by P. Porcheron and his coworkers. This report on separate functions of ecdysone and 20-hydroxyecdysone is not isolated in the literature and a careful reinvestigation in other species might prove to be rewarding. Most intriguing is the observation presented in Porcheron's paper that incubated integuments release ecdysteroids into the medium in amounts which are in excess of those which are found to be present (with conventional techniques) in the integuments at the time of excision.

The release of ecdysteroids by epidermal cells is also a central issue in the following paper by J. Delachambre and his group. These authors conclusively show that in *Tenebrio*, isolated abdomens in vivo, and epidermal explants in vitro, release ecdysteroids, the intensity of the release being directly dependent on the intensity of the mitosis in the epidermis. The rise in ecdysteroid titres in pupae (in which prothoracic glands have degenerated) does not appear to be the cause of the mitoses but to be their consequence, which leads J. Delachambre to the apparently paradoxical interpretation that epidermal cells are both secretory and target cells, a conclusion which is close to that of P. Porcheron. The careful investigation of the cell cycle of *Tenebrio* epidermis in larvae and pupae, its correlation to the fluctuation of hormone titres and the results obtained by administration of exogenous hormones leave little doubt in the view of J. Delachambre and associates that in *Tenebrio* pupae ecdysteroids act as homeostatic factors (as proposed by Delbecque and Slama in 1980) which behave as cell to cell signals triggering the cell cycle arrest.

This article introduces a series of important contributions on the mode of action of ecdysteroids. The first of the series is the paper by K.-D. Spindler and co-workers devoted essentially to the mechanism of uptake of ecdysteroids and the presence and characterization of ecdysteroid receptors. The original investigations of the paper concern crustaceans but comparison with data available from insect receptor studies are also included. In the round table discussions the problem of ecdysteroid receptors in general was a matter of debate. It is beyond any doubt that various insect or crustacean tissues contain proteins capable of binding radioactive ligands such as ecdysone, 20-hydroxyecdysone and ponasterone A. It is certainly important to characterize these proteins further. It was felt however that they should be called receptors only when the interaction with ecdysone mediates the biological activity, i.e. induction of transcription of certain genes.

In her contribution on the hormonal control of sequential gene expression in insect epidermis, L.M. Riddiford provides an overall picture on how ecdysteroids and juvenile hormones can regulate protein synthesis during intermolt-molt cycles. When ecdysteroid titres rise at the end of the intermolt, the epidermis responds by shutting off the synthesis of all major intermolt products and readies itself to produce the new cuticle of the next stage. In the presence of juvenile hormones, this turning-off is temporary and the same products are made during a subsequent molt; however if juvenile hormones are absent, the turning off is permanent and the epidermal cells switch to their next polymorphic state during the following intermolt. L.M. Riddiford shows that juvenile hormones also play an important role during larval intermolts to maintain ongoing synthesis: when the JH titres decline, the low intermolt ecdysteroid levels can cause changes in synthesis which result in changes of the structure and properties of the larval cuticle. The regulation by juvenile hormones of these two actions of ecdysteroids remains to be elucidated in molecular terms.

One area in which there has been steady progress since the Lille meeting is the molecular analysis of ecdysteroid action in *Drosophila*, and a complete session of the present Symposium was organized on this theme. Presentations covered the majority of model systems, notably salivary glands, fat body, imaginal discs and cell lines. The authors have collaborated in preparing their contributions to this volume so as to present a comprehensive review and current assessment of these studies. The historical background leading to the choice of system, the early morphological observations, the later biochimical analysis and the state of recombinant DNA studies are described in turn and parallels between the systems are discussed.

All of these studies will eventually benefit from the P element transformation system (Rubin and Spradling 1982) which provides a functional assay for DNA sequences. This system enables cloned DNA molecules to be reintroduced into the chromosomes of flies and the activity of the sequence to be determined in subsequent generations. With P element transformation, insect systems are currently the most successful model for the dissection of chromosome regulatory sequences surrounding known gene coding sequences. G. Richards and J.A. Lepesant introduce the system and summarise findings on the major intermoult ecdysteroid regulated transcripts of larval salivary glands and fat bodies respectively, which they are using to analyze stage and tissue specific gene regulation. The authors are enthusiastic about the future of such studies in *Drosophila* and predict that, thanks to the P element transformation system, the DNA sequences involved in ecdysteroid regulation will soon be characterized.

O. Pongs takes up the developmental analysis with studies on the ecdysteroid-induced proteins and mRNAs of these tissues just prior to pupariation and relates them to classical puffing studies and the work of his group with indirect-immunofluorescence which localized ecdysteroids to salivary gland puffing sites. He reviews the data to be accommodated in any model of puff regulation and proposes a modified model to explain the findings derived from the immunofluorescence studies. He also reports preliminary findings on the transcriptional activity of one of the early ecdysteroid induced puff loci analyzed by cloned DNA genomic sequences.

L. Cherbas, J.W. Fristrom, and J.D. O'Connor have taken this opportunity to make a critical comparison between the results obtained from imaginal disc studies and those using Kc cell lines. Because of the striking difference in response to ecdysteroids in the pupal period between larval and presumptive adult tissues one might expect that different mechanisms may have evolved in the two classes of tissue. Pursuing this idea, the authors propose that similarities in the properties of the discs and the cell lines may reflect that the Kc cell lines have an imaginal origin. Whilst there may be differences in response in larval and adult tissues, there are many common strategies that rely on the same hormonal signals to elicit ordered gene responses. They emphasized important inhibitory effects of physiological levels of ecdysteroids in their systems, which supports the view derived from salivary gland studies that ecdysteroid titres may be used not only to induce new gene activity but also to repress existing activities or to delay later programmes of differentiation by regulating the competence of cells to respond to the hormone. The authors' recent finding also reveal something of the complexity of *Drosophila* ecdysteroid-induced gene transcription as, in both the case of dopa decarboxylase and the Kc cell line EIP28/29, multiple transcripts are detected from a unique gene sequence.

The Kc cell line work is also treated in the contribution of Dastugue's and Best-Belpomme's groups, which develops similar themes regarding their experiences with ecdysteroid-regulated enzymes and structural proteins. In the latter area the study of the cloned actin 5 C gene promises to be a useful model system for cell-line transformation work. Their observations on the hormonal regulation of transcription of members of repetitive gene families may also break important new ground in our studies of insect mobile sequences. They pose an interesting question with the observation that mitochondrial enzyme activity is sensitive to hormonal treatments as one must wonder whether this reflects that mitochondria are an intracellular target for transcriptional regulation by the hormone or whether indirect mechanisms are involved.

The session on ecdysteroids at the present meeting ended with a series of short communications on ecdysteroids in non-arthropod invertebrates. It has been established in recent years that ecdysteroids are present in coelenterates, annelids, plathelminths, nemathelminths, rotifera, and molluscs. It is not certain whether in all these groups ecdysteroids are used as hormonal messengers. In the hirudinean annelid *Hirudo medicinalis* it is proposed on the basis of correlations between ecdysteroid titres and integumental cycles that ecdysteroids control the process of cuticle replacement ("molting"). In the polychete annelid *Perinereis cultrifera*, ecdysteroid injections most significantly affect gametogenesis as explained in the collaborative contribution by Porchet, Sauber, and the Strasbourg group. Of considerable interest to the audience were the reports on the presence of ecdysteroids in parasite worms, namely in *Schistosoma*. P. Nirdé and associates demonstrate in their contribution to this volume that ecdysteroids synthesized by parasites appear in the blood and the urine of the infected host, which opens novel approaches to the early detection of parasitic diseases and also to the control of the efficiency of therapy in man. H. Rees and co-workers report the isolation and identification of ecdysteroids in a variety of trematode, cestode, and nematode species and J. Koolman presents a short survey on ecdysteroids in helminths.

At the round table discussion following these papers, P. Karlson insisted on the evolutionary interest of the presence of ecdysteroids in practically all protostomes investigated so far. He presented a possible route of evolution of the steroid hormone system and insisted that vertebrate steroid hormones proceed during their biosynthesis via 20,22-hydroxylated intermediates, which are evocative of the concomitant presence of hydroxyl-groups in C–20 and C–22 in 20-hydroxyecdysone. P. Karlson postulated that the pathway of biosynthesis in vertebrates repeats the evolution: thus ecdysteroids are primitive steroid hormones and vertebrate steroid hormones have evolved therefrom or from a common precursor (see also Hoppe, Seyler's, Z. Physiol. Chem., 364, 1067 to 1087, 1983).

II.1 Ecdysteroids in Ovaries and Embryos of *Locusta migratoria*

M. LAGUEUX[1], J.A. HOFFMANN[1], F. GOLTZENÉ[1], C. KAPPLER[1], G. TSOUPRAS[1], C. HETRU[2], and B. LUU[2]

1. Introduction

At the first International CNRS Symposium on Biosynthesis, Metabolism and Mode of Action of Invertebrate Hormones in Lille in September 1975, we reported the presence of ecdysteroid-immunoreactive molecules in vitellogenic ovaries of adult females of *Locusta migratoria* [1]. These molecules were active in the *Calliphora* bioassay, but it was only 1 year later that we could assess by physicochemical methods their identify as ecdysteroids [2, 3].

The presence of ecdysteroids in vitellogenic ovaries of *Locusta* was perplexing and prompted us to undertake a series of investigations with the main aims of understanding the origin and the roles of the ovarian ecdysteroids in this species. At this second CNRS meeting on the same topic, we will critically review the information which we have obtained since 1975 in this species and compare with data presented for other insects. The reader is also referred to the extensive literature survey in the recent review articles of Hagedorn [4, 5] and Hoffmann and Lagueux [6].

Our 1975 studies were concomitant to independent investigations by Hagedorn and associates [7-11] on the control of vitellogenesis in the mosquito *Aedes aegypti*. Their results first established that ovaries participate via the secretion of a Vitellogenin Stimulating Hormone (VSH) in the control of vitellogenin synthesis in the fat body [7] and second, that ovaries secrete ecdysone [11]. Hagedorn's proposal that VSH was identical to ecdysone led to a considerable number of studies aimed at establishing that ovarian ecdysone stimulates vitellogenin synthesis in mosquitoes and in other Dipteran species.

The proposed control of vitellogenesis by a steroid hormone in insects was appealing, but its extension to our laboratory insect *Locusta* seemed unrealistic for a variety of reasons, among which our observation that ecdysteroid synthesis in adult females of *Locusta* occurred well after the beginning of vitellogenin synthesis [3]. On the other hand, this synthesis has been conclusively demonstrated to be under the control of juvenile hormone in this species (see Wyatt, this vol.). Although it was obvious that ecdysteroids are present in the blood of vitellogenic females of *Locusta*, it became ap-

1 ERA CNRS 118 Endocrinologie et Immunologie des Insectes, Laboratoire de Biologie, Université Louis Pasteur, 12 rue de l'Université, 67000 Strasbourg, France

2 LA CNRS 31 Chimie Organique des Substances Naturelles, Université Louis Pasteur, 5 rue Blaise Pascal, 67000 Strasbourg, France

Biosynthesis, Metabolism and Mode of Action of Invertebrate Hormones
(ed. by J. Hoffmann and M. Porchet)

parent in 1977 that the bulk (95%–98%) of the ovarian ecdysteroids accumulate inside the oocytes and are present in the newly laid eggs [3, 12]. This observation led us to consider that the ovarian ecdysteroids might play a role in embryogenesis, and possibly in the control of the early synthesis of embryonic cuticles.

The hypothesis that embryonic cuticulogenesis is directed by ecdysteroids reopened a field of investigations initiated by Jones in 1956 [13, 14] on the control of embryonic molting in insects (review in [6]). A series of studies conclusively demonstrated that in contrast to larval molting, embryonic molting can occur in the absence of the brain and prothoracic glands (or their functional equivalents [15–17]). However, the possibility that embryonic molting might be controlled by ecdysteroids provided by other sources than the prothoracic glands was not eliminated in these studies and, indeed, Boohar and Bucklin reported as early as 1963 [18] that yolk extracts of *Melanoplus* showed molting hormone activity when assayed in the *Calliphora* bioassay.

Independently from the studies on *Locusta*, a series of investigations on ecdysteroids in eggs and egg-laying females of *Bombyx* by Ohnishi and co-workers [19–23] led in 1977 to the proposal that ovarian ecdysteroids played a role in the control of embryonic diapause.

In the following pages we will first analyze the data available on ovarian ecdysteroids, namely on their biosynthesis. We will then turn our attention to the distribution of these molecules in the blood and oocytes of the vitellogenic females. The last and major part of the paper will be devoted to the fate of ovarian ecdysteroids in eggs during embryonic development.

2. Ovarian Ecdysteroids

In the course of our studies it became rapidly evident that the ecdysteroids of *Locusta* present in vitellogenic ovaries were synthesized by the ovaries themselves [3]. This result was in agreement with data presented for mosquito ovaries [11] and has since been corroborated in a large variety of insect species [24–28]. Dissection of the various components of the ovaries indicated that the follicle cell epithelium was the precise site of this synthesis [3, 29]. A similar result has recently been obtained in *Nauphoeta cinerea* by Lanzrein and associates [28]. The excised follicle cell epithelium is inevitably contamined by fragments of the oolemma and adherent yolk. When it became clear that poorly immunoreactive ecdysteroid esters are bound to vitellin, as we will see below, the fear arose that what had been interpreted as a synthesis of ecdysteroids (in terms of increase in total immunoreactivity) by excised follicular epithelia, might represent the hydrolysis of the poorly immunoreactive esters in the contaminating yolk to strongly immunoreactive free ecdysteroids. To eliminate this objection, Goltzené (unpublished results) has recently prepared yolk-free cell suspensions of follicle cells by collagenase treatment of oocytes (Fig. 1). As illustrated in Fig. 2, these isolated follicle cells secrete ecdysteroids into the medium: the secretory kinetics generated in this experiment closely resemble some of those presented earlier by Goltzené et al. [29] when crude follicle cell epithelia ("ghosts") were incubated. In Fig. 3, it is apparent that the isolated follicle cells contain at the onset of incubation ecdysone and 2-deoxy-

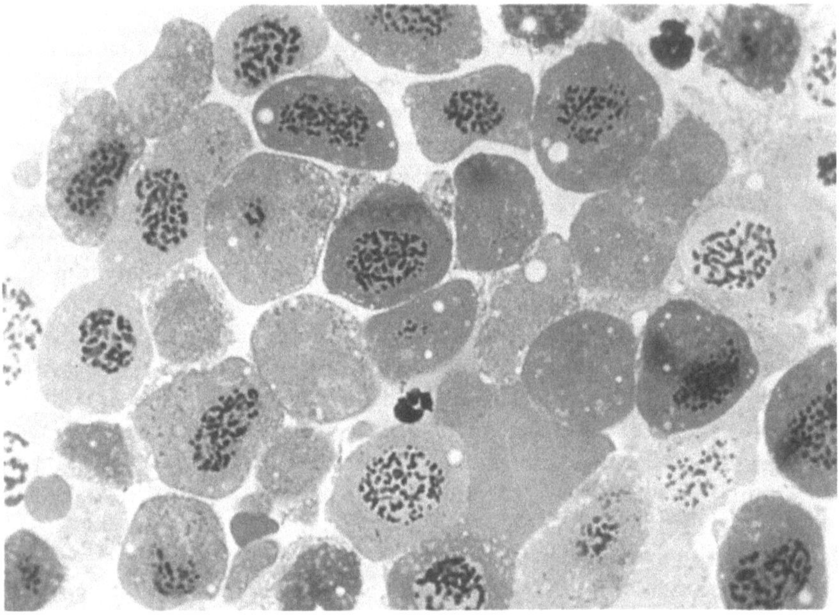

Fig. 1. Isolated follicle cells from vitellogenic ovaries of *Locusta migratoria*. Oocytes with their follicular epithelium were excised from ovaries and treated for 10 min with 0.1% collagenase in Landureau's medium at 32 °C under shaking. The isolated cells were repeatedly washed in medium to remove the vitellus. Semithin section after glutaraldehyde-osmium tetroxide fixation and araldite embedding (x 720)

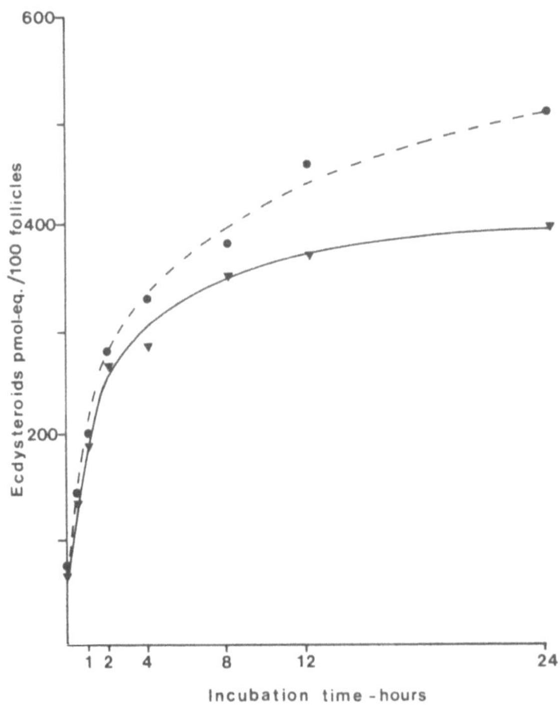

Fig. 2. Secretory kinetics of ecdysteroids by excised follicle cell sheaths from ovaries of *Locusta migratoria* (●) and by collagenase-treated isolated follicular cells (▼) (technique as in Fig. 1). Mean oocytes length 5.2 mm (oocyte length at ovulation: 6.5). Each *data point* represents the amount of ecdysteroids secreted into the medium by 100 follicle cell sheath-equivalents during the time of incubation. RIA measurements are expressed in pmol-equivalents of synthetic ecdysone

Fig. 3. Ecdysteroids present in the collagenase-treated isolated follicular cells before incubation (*Oh*) and in the cells and medium after 8 h of incubation. Conjugated ecdysone ▨; conjugated 2-deoxyecdysone ▤, ecdysone ■; 2-deoxyecdysone ▥. The ecdysteroids were separated by thin-layer chromatography and high pressure liquid chromatography. RIA measurements are expressed in pmol-equivalents of synthetic ecdysone for 100 follicle cell sheath-equivalents (mean oocyte length: 5.7 mm); the RIA measurements for conjugated ecdysone and 2-deoxyecdysone are given after hydrolysis of these conjugates by *Helix pomatia* enzymes

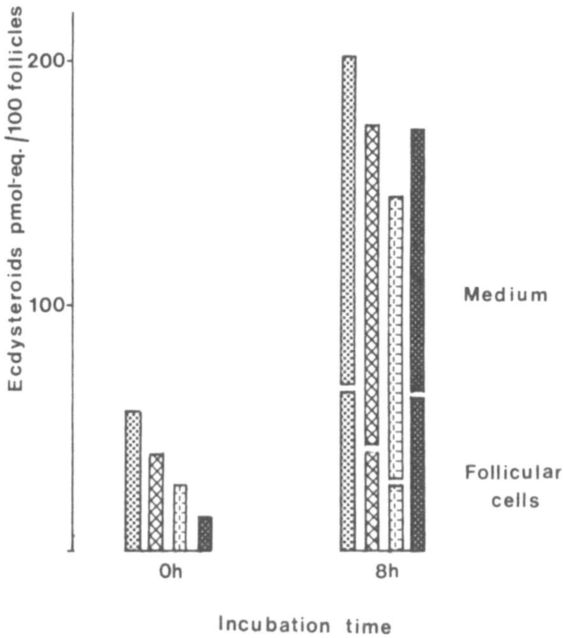

ecdysone both as free and as polar compounds (the latter being hydrolyzable by *Helix* enzymes) and that over a 8 h period of incubation a net increase in the level of these four types of ecdysteroids has occurred. This substantiates that follicle cells are actually capable of synthesizing 2-deoxyecdysone and ecdysone. It also shows that this synthesis is not restricted to free ecdysteroids, a finding which is corroborated by the observation of Kappler and Hetru (unpublished results) that isolated follicle cells convert tritiated 2-deoxyecdysone and ecdysone to polar metabolites. In the study of Zhu et al. [28] follicle cell epithelia have been shown to convert ecdysone to 20-hydroxyecdysone.

3. Hemolymph vs Intra-Oocytic Ecdysteroids

As noted above, ecdysteroids are present in the blood of vitellogenic females of *Locusta*, although their concentrations remain one to two magnitudes lower than those which are recorded during larval development [3, 30]. The blood ecdysteroids in maturing females are a complex mixture of free and conjugated 2-deoxyecdysone, ecdysone, 20-hydroxyecdysone, and unidentified immunoreactive molecules (Goltzené, unpublished results). We are unable so far to propose any function for the blood ecdysteroids in the adult female of *Locusta*, which of course does not mean that they have none. It should be stressed here that ovarian ecdysteroids have been implicated in a variety of functions in the reproductively competent females of many insect species. We have already referred to the work of Hagedorn and associates who conclude that in the mosquito ovarian ecdysteroids play a role in stimulating vitellogenin synthesis. This proposal has been challenged for mosquitoes (see review by Fuchs and Kang [31] and Hagedorn [4, 5]). On the other hand, several laboratories have recently provided evidence that ecdysteroids could be implicated in the control of vitellogenesis in other Dipteran insects, namely, in *Calliphora vicina* [32, 33], *Sarcophaga bullata* [34–36], and *Drosophila melanogaster* [37–44]. For the sake of completeness, we will mention that in mosquitoes additional functions have been attributed to ovarian ecdysteroids, such as stimulation of Dopa-decarboxylase activity in the egg [45] and separation of the incipient follicles from the germarium [46]. In *Rhodnius*, ovarian ecdysteroids might control the release of a myotropic ovulation hormone [47–49], while in cockroaches it is suggested that they play a role in the inactivation of the *corpora allata* at a stage when oocyte development ceases (*Nauphaeta cinerea* [50, 51], *Diploptera punctata* [52–54]).

It may be premature at this time to draw firm conclusions as to the roles which ecdysteroids of ovarian origin exert in the female insects.

The reason why our efforts have been primarily oriented towards elucidating a role of ovarian ecdysteroids in embryogenesis is simply that in *Locusta* up to 98% of the ecdysteroids present in the adult females are accumulated in the oocytes and later recovered from the newly laid eggs. Transfer of ovarian ecdysteroids to eggs has been documented in a variety of other insect species (e.g., *Bombyx mori* [20, 22, 23, 55–57], *Galleria mellonella* [24, 58], *Blaberus craniifer* [59], *Macrotermes bellicosus* and *M. subhyalinus* [60, 61], *Thermobia domestica* [62, 63], *Acheta domestica* [64], *Schistocerca gregaria* [65, 66], *Xyleborus ferrugineus* [67].

We have no data regarding the process by which ecdysteroids produced in the follicle cells are taken up into the oocytes. We have nevertheless information on the mode of sequestration of ecdysteroids inside the oocytes of *Locusta*: a combination of techniques, such as dextran-coated charcoal absorption of free ligands, gel filtration chromatography, and immunoprecipitation, conclusively demonstrates that the ecdysteroids are tightly bound to the major yolk protein vitellin, from which they dissociate only slowly at low temperature [68]. These tightly bound molecules make up 95% of the ecdysteroids in the oocytes at ovulation when they reach a concentration of 150 μM in this insect species. They are polar compounds which can be hydrolyzed to free 2-deoxyecdysone and ecdysone by enzymatic hydrolysis [69, 70]. As opposed to this, free ecdysone (approximate concentration at ovulation: 1 μM) is not tightly bound to proteins in the oocyte as shown by dextran-coated charcoal absorption or gel filtration. A special mention should be made here of the presence in oocytes (and subsequently in newly laid eggs) of 2, 14, 22, 25-tetradeoxyecdysone (3β-hydroxy-5β-cholest-7-en-6-one) and 2,22,25-trideoxyecdysone (3β,14α-dihydroxy-5β-cholest-7-en-6-one), which have been shown to be ecdysone precursors in *Locusta* ovaries [71]. The concentration of these compounds is in the micromolar range in the newly laid eggs [72], i.e., of the order of magnitude of that of free ecdysone.

4. Embryonic Ecdysteroids

There is no indication that quantitative or qualitative changes exist between the ecdysteroids present in oocytes at ovulation and those found in newly laid eggs. These eggs contain several unconjugated deoxyecdysteroids and ecdysone in the micromolar range and conjugated 2-deoxyecdysone and ecdysone at a concentration which is roughly two magnitudes higher. The fact that ecdysteroids are to a large extent conjugated in ovaries and/or in newly laid eggs was first reported by Ohnishi and associates in *Bombyx* [20]: these authors proposed that they represented "storage forms" yielding free molecules during embryogenesis. These observations were confirmed in *Schistocerca gregaria* [65, 66, 73-75] and later in *Galleria mellonella* [58] and *Locusta migratoria* [69, 70, 76]. Conjugation of ovarian ecdysteroids may, however, not be universal among insects, as in several species no indication for such a process has been found with technologies similar to those used in the studies cited above (cf. for instance *Leucophaea maderae* [77], *Nauphoeta cinerea* [78], *Clitumnus extradentatus* [79], and *Calliphora erythrocephala* [80]).

In 1982 and 1983 Rees and associates and our group reported the isolation and identification of a series of embryonic conjugates in *Schistocerca* [81-84] and in *Locusta* [85-88]. In both species, the conjugates of ovarian origin were found to be phosphate esters linked to 2-deoxyecdysone or ecdysone (or to 20-hydroxyecdysone which is only a minor ovarian ecdysteroid in *Locusta*) at the C–22 hydroxyl group. A discrepancy appeared, however, in the results in as far as in *Schistocerca* the conjugates were identified as 22-phosphates, whereas in *Locusta* evidence was obtained that an adenosine moiety was covalently linked to the C–22 phosphate of 2-deoxyecdysone in the newly laid eggs. That 22-adenosinemonophosphoric esters of 2-deoxyecdysone and ecdysone can be isolated from newly laid eggs of *Locusta* seems beyond any doubt, as far as the chemical evidence goes[3]. However, these compounds may not be predominant over the 22-phosphoric esters as we initially believed on the basis of RIA measurements of unhydrolyzed and hydrolyzed maternal conjugates. The problems raised by the extraction techniques, the relative instability of complex phosphate esters, and the cross-reactivity of the various conjugates in the RIA will probably only be overcome with pure reference compounds; we have now started the synthesis of these compounds. In any event, the reader can safely keep in mind that both in *Locusta* and *Schistocerca* the ovarian ecdysteroids are conjugated at their C–22 position to phosphate esters; whether or not some, a few, or most of these esters bear an adenosine moiety in these species is a matter of debate.

With the aim of understanding the roles of the huge amounts of ovarian ecdysteroids in embryogenesis of *Locusta* we have undertaken a few years ago a thorough study of the fate of these molecules from egg-laying till hatching [69, 70]. Some relevant information is summarized in Fig. 4.

---

3 Additional evidence for the structure of AMP conjugates of ecdysteroids has been obtained since the CNRS meeting by fast atom bombardment mass spectrometry by Hetru, Fraisse, Tabet, and Luu who observed in the spectra the presence of the quasi-molecular ions of the intact maternal conjugates (C.R. Acad. Sc. Paris, June 1984, in press)

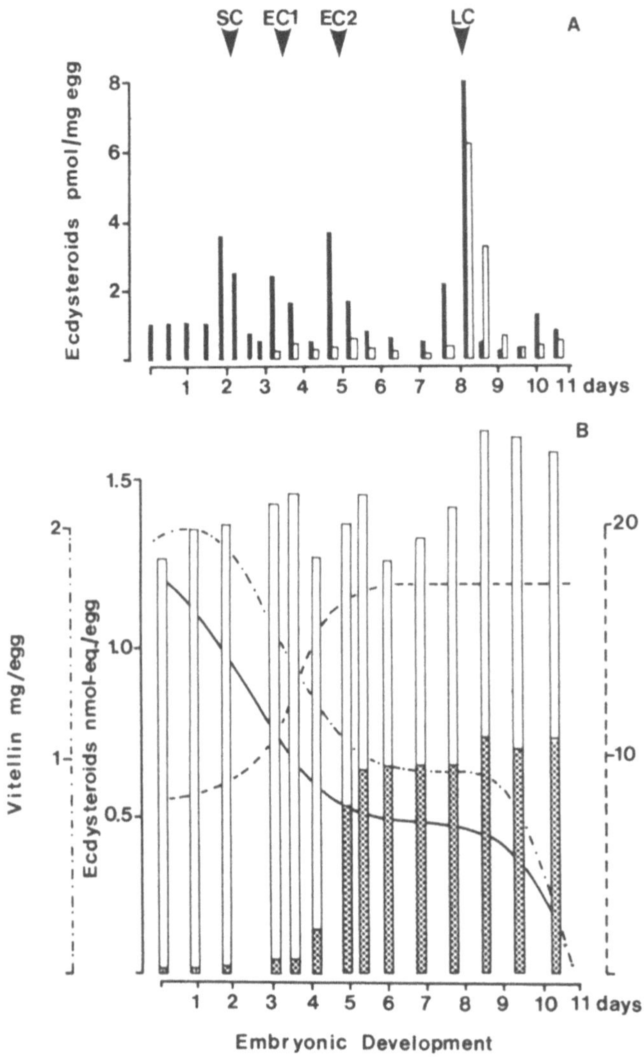

Fig. 4 A, B. Fluctuations of free and conjugated ecdysteroids in eggs during embryonic development of *Locusta migratoria*. **A** Concentrations of free ecdysone ▪ and 20-hydroxyecdysone □; RIA measurements expressed in pmol-equivalents of ecdysone and 20-hydroxyecdysone per mg of eggs (Modified from Lagueux et al. [12]). **B** Total ecdysteroid immunoreactivity of egg extracts before □ and after ▩ hydrolysis by *Helix pomatia* enzyme mixture; RIA measurements expressed in nmol-equivalents of synthetic ecdysone per eggs (Lagueux et al. [69] and Sall et al. [70]). *Curves 1 to 3* give the fluctuations of the following parameters during embryogenesis: *1* vitellin content per egg (– . –) measured by rocket electrophoresis (result obtained in collaboration with Emmerich, Darmstadt); *2* titer of vittelin-bound ecdysteroids (—) expressed in nmol equivalents per egg; ecdysteroids were measured after enzymatic hydrolysis (same scale as for total ecdysteroids) (for details, see Lagueux et al. [69]); *3* fresh weight of the egg (– –). *Arrows:* onset of deposition of the different cuticles: *SC* serosal cuticle; *EC1* and *EC2* first and second embryonic cuticles; *LC* first larval cuticle

First of all it must be stressed that this sort of undertaking has many pitfalls. Basically, the fate of the various ecdysteroids can be determined by measuring with a radioimmunoassay their respective titer fluctuations after appropriate separation by chromatographical techniques. In this case, a major drawback are the different cross-reactivities of the various free and esterified ecdysteroids. In addition, this technique makes of course no difference between ecdysteroids of maternal origin and those possibly synthesized de novo by the embryo. An alternative method is the injection of labelled cholesterol into the female: all ovarian ecdysteroids take up the label, although with a drastic isotopic dilution, and the fate of maternal ecdysteroids can be studied during embryogenesis by following the tritiated ecdysteroids after appropriate separations. But again, as tritiated cholesterol is also transferred to the eggs in this case, a de novo synthesis of ecdysteroids during embryogenesis – if existent – will occur at the expense of labelled cholesterol and can, therefore, not be accurately distinguished from the maternal hormone supply.

With these restrictions kept in mind, let us now consider the titers of free (unconjugated) ecdysone, 2-deoxyecdysone, 20-hydroxyecdysone. In Fig. 4B it is apparent that four concentration peaks of free ecdysone, ranging from 4 to 8 μM, occur throughout embryogenesis. In this species, 20-hydroxyecdysone is a minor ecdysteroid during early embryogenesis; only after blastokinesis, when the Malpighian tubules and the fat body are differentiated, the embryo significantly converts ecdysone to 20-hydroxyecdysone. Consequently, the two postblastokinetic concentration peaks of ecdysone are followed by peaks of 20-hydroxyecdysone [12]. We have been unable to monitor any clear-cut variations in the level of free 2-deoxyecdysone which fluctuates only slightly in the range of 0.3–0.6 μM.

The four concentration peaks of free ecdysone and/or 20-hydroxyecdysone have been correlated on a precise time basis to the onset of four successive depositions of cuticles; the serosal cuticle and three embryonic cuticles [12]. Similar information has been obtained in a variety of insect species [59, 77–80, 89, 90] which leaves no doubt that during embryogenesis, ecdysone, and/or 20-hydroxyecdysone control cuticle depositions, as these hormones do during postembryonic development.

The evident question raised by the observation of these peaks concerns the mechanism of rise and fall in titer concentration. Our initial belief was that the peak observed at the time of serosal cuticle deposition resulted from the hydroxylation of free precursor ecdysteroids, such as 2,14,22,25-tetradeoxyecdysone, 2,22,25-trideoxyecdysone, and 2-deoxyecdysone present in the newly laid eggs at (or slightly under) the micromolar range [72].

Although this hypothesis is still valid, it has not yet received any experimental support. On the other hand, the discovery of the remarkably high levels of ecdysteroid conjugates in newly laid eggs ("storage forms" according to Ohnishi and co-workers [20]) has favored the hypothesis that the peaks of free hormone result from the hydrolysis of these molecules. During embryogenesis in *Locusta* eggs, two major families of ecdysteroid conjugates are observed: (1) the original maternal conjugates of 2-deoxyecdysone, which are tightly bound to vitellin and are poorly immunoreactive in our RIA [91]; (2) a variety of conjugated metabolites of 2-deoxyecdysone and ecdysone, which are not tightly bound to proteins and which appear and accumulate in the eggs as embryogenesis proceeds [69, 70]; the immunoreactivity of these conjugates is stronger

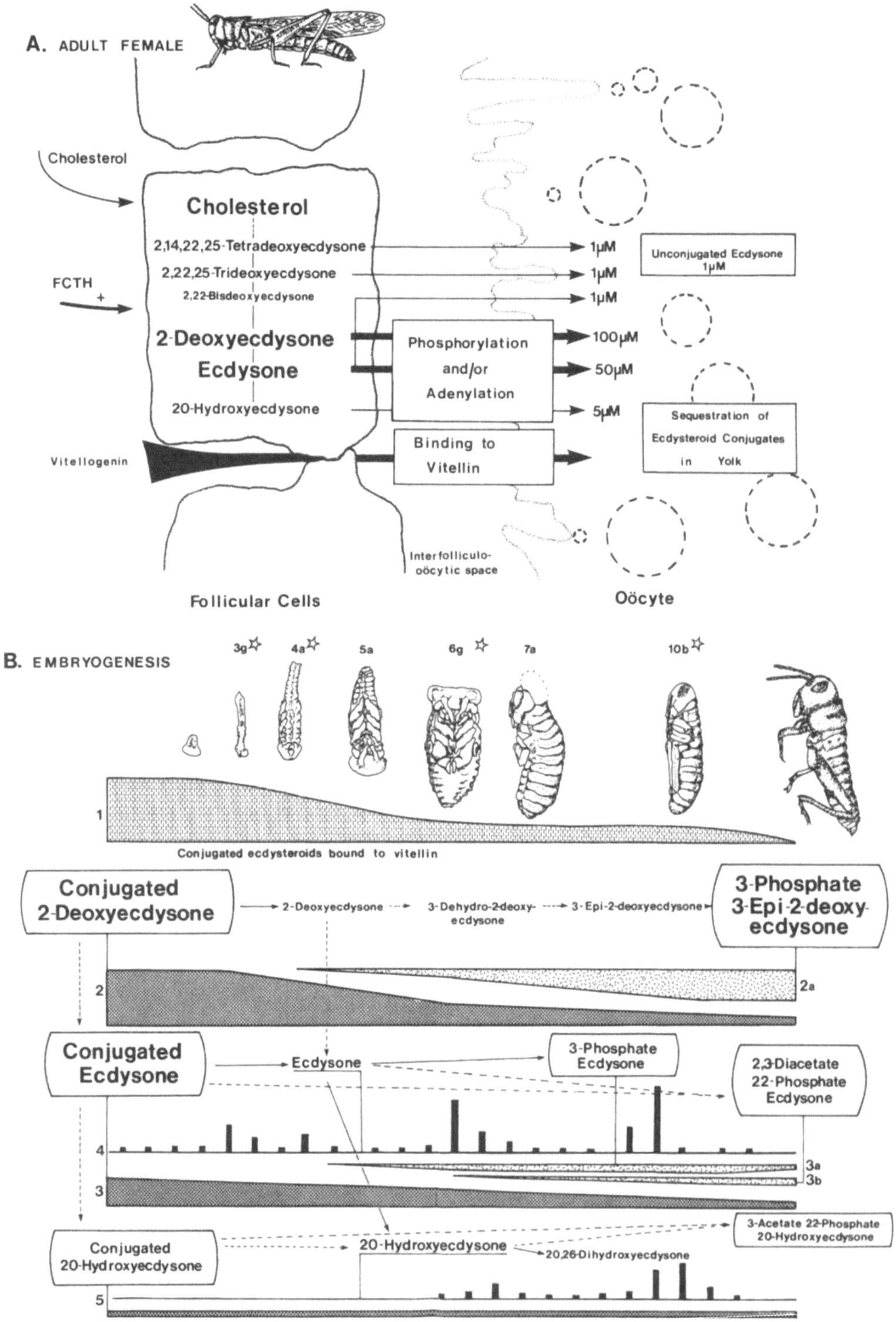
A. ADULT FEMALE
Cholesterol
FCTH
+
Cholesterol
2,14,22,25-Tetradeoxyecdysone
2,22,25-Trideoxyecdysone
2,22-Bisdeoxyecdysone
2-Deoxyecdysone
Ecdysone
20-Hydroxyecdysone
Phosphorylation
and/or
Adenylation
Binding to
Vitellin
Vitellogenin
1μM
1μM
1μM
100μM
50μM
5μM
Unconjugated Ecdysone
1μM
Sequestration of
Ecdysteroid Conjugates
in Yolk
Interfolliculo-
oöcytic space
Follicular Cells
Oöcyte
B. EMBRYOGENESIS
3g
4a
5a
6g
7a
10b
1
Conjugated ecdysteroids bound to vitellin
Conjugated
2-Deoxyecdysone
2-Deoxyecdysone
3-Dehydro-2-deoxy-
ecdysone
3-Epi-2-deoxyecdysone
3-Phosphate
3-Epi-2-deoxy-
ecdysone
2
2a
Conjugated
Ecdysone
Ecdysone
3-Phosphate
Ecdysone
2,3-Diacetate
22-Phosphate
Ecdysone
4
3a
3b
3
Conjugated
20-Hydroxyecdysone
20-Hydroxyecdysone
20,26-Dihydroxyecdysone
3-Acetate 22-Phosphate
20-Hydroxyecdysone
5

Fig. 5 A, B

than that of the maternal conjugates. A significant number of these molecules have now been isolated and identified in *Locusta* [85–88] (for *Schistocerca*, see Rees and co-workers, this Vol.).

Most of the transformation of the maternal ecdysteroids must occur between days 3 and 5 of embryonic development of *Locusta* (total duration: 11 days): as shown in Fig. 4B, when RIA measurements are performed on egg extracts prior to hydrolysis of the conjugates, a considerable increase in immunoreactivity is monitored between days 3 and 5; one is tempted to consider this increase as an indication for a strong ecdysteroid synthesis in the eggs. However, when the same extracts are subjected to hydrolysis before RIA, it becomes apparent that the total number of ecdysteroid

---

Fig. 5A, B. Ecdysteroids in ovaries and embryos of *Locusta*: an endocrine relationship between the mother and the embryo.

A Follicle cells convert cholesterol via a series of known deoxyecdysteroids to 2-deoxyecdysone (major compound), ecdysone and 20-hydroxyecdysone (minor compound in this system and insect). Evidence exists that this synthesis is under the control of a neurohormone (follicle cell tropic hormone, FCTH) produced by the neurosecretory cells of the brain and released by the *corpora cardiaca* [92]. The ecdysteroids are converted to esters by phosphorylation and/or adenylation of their C–22 hydroxyl function and these esters bind to vitellogenin or vitellin. The site of conjugation and binding to the proteins, as well as the mechanism of uptake, are under investigation. Very large amounts of ecdysteroid esters appear eventually to be sequestered in the yolk prior to ovulation; unconjugated ecdysone and several deoxyecdysteroids are present in the oocytes at a concentration roughly two magnitudes lower than those of conjugated 2-deoxyecdysone; the unconjugated forms are not tightly bound to vitellin.

B A survey of the fate of maternal ecdysteroids in eggs of *Locusta* during embryogenesis. The stages of embryonic development are symbolized by the drawings in the upper part of the figure, where the *numbers* refer to the staging table of Maltête [93] and the *stars* indicate the periods of cuticulogenesis. The absolute amounts of ecdysteroids bound to vitellin in the egg are symbolized by *line 1* which shows a continuous decrease (concomitant to the degradation of vitellin) with a steeper part between stages *3* and *6* (cf. also Fig. 4B). The fates of conjugated 2-deoxyecdysone and conjugated ecdysone are symbolized by *lines 2* and *3*, the height of the lines being proportional to the concentrations of the various compounds in the eggs. 2-Deoxyecdysone conjugates (*line 2*) gradually decrease in concentration. 3-Phosphate 3-epi-2-deoxyecdysone is the major 2-deoxyecdysone metabolite accumulating in the eggs (*line 2a*); its formation occurs certainly via 3-dehydro-2-deoxyecdysone and 3-epi-2-deoxyecdysone which are present in *Locusta* eggs. 2-Deoxyecdysone does not appear to be massively converted to ecdysone, although C–2 hydroxylation can occur in *Locusta* eggs [94]. Conjugated ecdysone (*line 3*) decreases similarly during embryogenesis; the main metabolites which appear are 3-phosphate of ecdysone (*line 3a*), 2,3-diacetate-22-phosphate of ecdysone (*line 3b*) (formation via acetylation of maternal 22-phosphates or acetylation and phosphorylation of free ecdysone?) and 20-hydroxyecdysone. Free ecdysone (*line 4*) shows four concentration peaks in the eggs which coincide with the four periods of cuticle deposition (cf. Fig. 4A). Conjugated 20-hydroxyecdysone (*line 5*) does not show clear-cut variations in concentration in this species. Free 20-hydroxyecdysone (*line 5*) shows two concentration peaks coincident or slightly following the third and fourth peak of free ecdysone (*line 4*). These peaks can result from hydrolysis of maternal 20-hydroxyecdysone conjugates or/and from hydroxylation of free ecdysone. 20-Hydroxyecdysone is metabolized to 20,26-dihydroxyecdysone; 3-acetate 22-phosphate of 20-hydroxyecdysone also appears in aging eggs, as well as other minor metabolites not listed in the figure.

The metabolism of the maternal conjugated ecdysteroids to a series of other molecules during embryogenesis seems to be a continuous process, but which apparently is accelerated between stages *3* and *6*, around blastokinesis. Theoretically, the absolute amounts of maternal ecdysteroid conjugates could account for the peaks of free ecdysone and 20-hydroxyecdysone; we do not yet know whether and when the embryo becomes capable of de novo synthesis of ecdysone

molecules has not significantly changed during this period. Indeed, the drastic increase in immunoreactivity between days 3 and 5, that is, around blastokinesis, reflects the transformation of the poorly immunoreactive maternal conjugates to more immunoreactive ecdysteroids. It is significant that this change occurs at the end of the first and major phase of degradation of vitellin, as illustrated in Fig. 4B. The release of the conjugates from vitellin is probably an essential event favoring the concomitant or subsequent hydrolysis of the ecdysteroid esters to the free hormones. It is assumed that the free hormones exert their biological role, among which as already mentioned, is the control of serosal and embryonic cuticle depositions. Unfortunately we have no information at present on any other role which free ecdysteroids might play in embryonic development. It is hoped that the establishment of embryonic cell cultures will help to overcome this ignorance. The fall in titer of free ecdysteroids is achieved via conjugation, the predominant reactions in *Locusta* being acetylation and phosphorylation. A series of metabolites of 2-deoxyecdysone, ecdysone, and 20-hydroxyecdysone have been isolated and identified in postblastokinetic and in preeclosion eggs of *Locusta* [86-88]. These compounds are listed in Fig. 5 (for *Schistocerca*, see contribution by Rees and Isaac) (Fig. 5, see p. 176).

As indicated in Fig. 5B, a large proportion of the endogenous 2-deoxyecdysone is converted to the 3-epimer which in turn is phosphorylated. The titres of free 2-deoxyecdysone did not exhibit clearent variations in our investigations and the role of 2-deoxyecdysone in embryogenesis of *Locusta* remains somewhat enigmatic.

2-Deoxyecdysone and ecdysone conjugates are found with the residual vitellin molecules till the end of embryogenesis. It is, therefore, a theoretical possibility that the later concentration peaks of free ecdysone and 20-hydroxyecdysone in the eggs (Fig. 4A) are of maternal origin. However, after blastokinesis, the embryonic prothoracic glands are fully differentiated and might take in charge de novo synthesis of hormone. Unfortunately we lack firm information on the time at which the embryo becomes independent from the maternal hormone supply.

Eight years after our initial report on the presence of ecdysteroid-immunoreactive and *Calliphora*-positive compounds in ovaries of vitellogenic females of *Locusta*, a wealth of information is available which allows the provisional conclusion that the majority of the ovarian ecdysteroids are in fact destined to play a role in embryogenesis in this species. Whereas a certain number of questions have received firm answers, many other problems remain open and new, and even controversial questions have arisen. Figure 5 summarizes that state-of-the-art for ecdysteroids in this biological model in *Locusta*.

Hopefully CNRS will fund a third meeting on Biosynthesis, Metabolism and Mode of Action of Invertebrate Hormones in a few years and by then we hope to have an answer to some of the unsolved questions.

References

1. Hoffmann JA, Lagueux M, Hirn M, De Reggi M, Goltzené F, Feyereisen R (1976) Actualités sur les Hormones d'Invertébrés. Colloq Int CNRS, pp 359–365
2. Luu B, Lagueux M, Hirn M, Hoffmann JA (1976) CR Acad Sci 283:1081–1084
3. Lagueux M, Hirn M, Hoffmann JA (1977) J Insect Physiol 23:109–119
4. Hagedorn H (1983) In: Downer RGH, Laufer H (eds) Endocrinology of insects, vol I. Liss, New York, pp 271–304
5. Hagedorn H (1984) In: Kerkut GA, Gilbert LI (eds) Insect physiology, biochemistry, pharmacology. Pergamon Press, Oxford New York, in press
6. Hoffmann JA, Lagueux M (1984) In: Kerkut GA, Gilbert LI (eds) Insect physiology, biochemistry, pharmacology. Pergamon Press, Oxford New York (in press)
7. Fallon AM, Hagedorn H (1972) Am Zool 12:697
8. Hagedorn H, Fallon AM (1973) Nature (London) 244:103–105
9. Hagedorn H, Fallon AM, Laufer H (1973) Dev Biol 31:285–294
10. Fallon AM, Hagedorn H, Wyatt GR, Laufer H (1974) J Insect Physiol 20:1815–1823
11. Hagedorn H, O'Connor JD, Fuchs MS, Sage B, Schlaeger DA, Bohm MK (1975) Proc Natl Acad Sci USA 72:3255–3259
12. Lagueux M, Hetru C, Goltzené F, Kappler C, Hoffmann JA (1979) J Insect Physiol 25:709–723
13. Jones BM (1956) J Exp Biol 33:174–185
14. Jones BM (1956) J Exp Biol 33:685–696
15. Mueller NS (1963) Dev Biol 8:222–240
16. Takami T (1963) J Exp Biol 40:735–739
17. Sbrenna-Micciarelli A, Sbrenna G (1972) J Insect Physiol 18:1027–1037
18. Boohar R, Bueklin DH (1963) Am Zool 3:496
19. Ohnishi E, Ohtaki T, Fukuda S (1971) Proc Jpn Acad 47:413–415
20. Mizuno T, Ohnishi E (1975) Dev Growth Differ 17:219–225
21. Hanaoka K, Ohnishi E (1974) J Insect Physiol 20:1375–2384
22. Chatani F, Ohnishi E (1976) Growth Differ 18:481–484
23. Ohnishi E, Chatani F (1977) Dev Growth Differ 19:67–70
24. Bollenbacher WE, Zvenko H, Kumaran AK, Gilbert LI (1978) Gen Comp Endocrinol 34:169 to 179
25. Nijhout MM, Koeppe JK (1978) Am Zool 18:626
26. Masler EP, Fuchs MS, Sage B, O'Connor JD (1980) Gen Comp Endocrinol 41:250–259
27. Rubenstein EC, Kelly TS, Schwartz MB, Woods CW (1982) J Exp Zool 223:305–308
28. Zhu XX, Gfeller H, Lanzrein B (1983) J Insect Physiol 29:225–235
29. Goltzené F, Lagueux M, Charlet M, Hoffmann JA (1978) Z Physiol Chem 359:1427–1434
30. Hirn M, Hetru C, Lagueux M, Hoffmann JA (1979) J Insect Physiol 25:255–261
31. Fuchs MS, Kang SH (1981) Insect Biochem 11:627–633
32. Thomsen E, Hansen BL, Hansen GN, Jensen PV (1980) Cell Tissue Res 208:445–455
33. Jensen PV, Hansen BL, Hansen GN, Thompsen E (1981) Insect Biochem 11:129–135
34. Huybrechts R, Loof A De (1977) J Insect Physiol 23:1359–1362
35. Huybrechts R, Loof A De (1981) Int J Invertebr Reprod 3:157–168
36. Briers T, Loof A De (1980) Int J Invertebr Reprod 2:363–372
37. Postlethwait JH, Handler AM (1979) J Insect Physiol 25:455–460
38. Jowett T, Postlethwait JH (1980) Dev Biol 80:225–234
39. Postlethwait JH, Laugé G, Handler AM (1980) Gen Comp Endocrinol 40:385–391
40. Postlethwait JH, Bownes M, Jowett T (1980) Dev Biol 79:379–387
41. Bownes M (1982) J Insect Physiol 28:317–328
42. Shirk PD, Minoo P, Postlethwait JH (1983) Proc Natl Acad Sci USA 80:186–190
43. Bownes M, Hames B (1978) FEBS Lett 96:327–330
44. Postlethwait JH, Shirk PD (1981) Am Zool 21:687–700
45. Schlaeger DA, Fuchs MS (1974) Dev Biol 38:209–219
46. Beckemeyer E, Lea O (1980) Nature (London) 282:92–94
47. Ruegg RP, Kriger FL, Davey KG, Steel CGH (1981) Int J Invertebr Reprod 3:357–361
48. Ruegg RP, Orchard I, Davey KG (1982) J Insect Physiol 28:243–248
49. Orchard I, Ruegg RP, Davey KG (1983) J Insect Physiol 29:387–391

50. Lanzrein B, Gentinetta V, Fehn R, Lüscher M (1981) Gen Comp Endocrinol 36:339–345
51. Lanzrein B, Wilhelm R, Gentinetta V (1981) In: Proc Int Conf Regul Insect Dev Behav, Wroclaw, Poland, pp 523–534
52. Stay B, Friedel T, Tobe SS, Mundall EC (1980) Sciences 207:898–900
53. Friedel T, Feyereisen R, Mundall EC, Tobe SS (1980) J Insect Physiol 26:665–670
54. Stay B, Tobe SS (1981) Am Zool 21:663–674
55. Ohnishi E, Mizuno E, Chatani F, Ikekawa N, Sakurai S (1977) Science 197:66–67
56. Ohnishi E, Mizuno E, Ikekawa N, Awata N (1977) J Insect Physiol 23:317–319
57. Ohnishi E, Mizuno E, Ikekawa N, Ikeda T (1981) Insect Biochem 11:155–159
58. Hsiao TH, Hsiao C (1979) J Insect Physiol 25:45–52
59. Bullière D, Bullière F, De Reggi M (1979) Wilhelm Roux's Arch 186:103–114
60. Bordereau C, Hirn M, Delbecque JP, De Reggi M (1976) CR Acad Sci 282:885–888
61. Delbecque JP, Lanzrein B, Bordereau C, Imboden H, Hirn M, O'Connor JD, Noirot C, Lüscher M (1978) Gen Comp Endocrinol 36:40–47
62. Bitsch J, Rojo de la Paz A, Mathelin J, Delbecque JP, Delachambre J (1979) CR Acad Sci 289:865–868
63. Rojo de la Paz A, Delbecque JP, Bitsch J, Delachambre J (1983) J Insect Physiol 29:323–329
64. Rennuci M, Strambi A (1981) CR Acad Sci 293:825–830
65. Dinan LN, Rees HH (1981) J Insect Physiol 27:51–58
66. Dinan LN, Rees HH (1981) Insect Biochem 11:255–265
67. Rao KDP, Norris DM, Chu HM (1983) Insect Biochem 12:531–536
68. Lagueux M, Harry P, Hoffmann JA (1981) Mol Cell Endocrinol 24:325–338
69. Lagueux M, Sall C, Hoffmann JA (1981) Am Zool 21:715–726
70. Sall C, Tsoupras G, Kappler C, Lagueux M, Zachary D, Luu B, Hoffmann JA (1983) J Insect Physiol 29:491–507
71. Hetru C, Kappler C, Hoffmann JA, Nearn R, Luu B, Horn D (1982) Mol Cell Endocrinol 26: 51–80
72. Hetru C (1981) Thèse Doct Etat, Univ Louis Pasteur, Strasbourg
73. Gande AR, Morgan ED (1979) J Insect Physiol 25:289–293
74. Rees HH, Davies TG, Dinan LN, William JS, Lockley JS, Goodwin TW (1980) In: Hoffmann JA (ed) Progress in ecdysone research. Elsevier/North Holland Biomed Press, Amsterdam New York, pp 125–137
75. Scalia S, Morgan ED (1982) J Insect Physiol 28:647–654
76. Hoffmann JA, Lagueux M, Hetru C, Charlet M, Goltzené F (1980) In: Hoffmann JA (ed) Progress in ecdysone research. Elsevier/North Holland Biomed Press, Amsterdam New York, pp 431–465
77. Matz G (1980) CR Acad Sci 291:501–504
78. Imboden H, Lanzrein B (1982) J Insect Physiol 28:37–46
79. Cavalin M, Fournier B (1981) J Insect Physiol 27:527–534
80. Sami L (1982) Thèse de 3e cycle, Univ Bordeaux I, France
81. Isaac RE, Rees HH, Goodwin TW (1981) JCS Chem Commun 594–595
82. Isaac RE, Rees HH, Goodwin TW (1981) JCS Chem Commun 418–420
83. Isaac RE, Rose ME, Rees HH, Goodwin TW (1982) JCS Chem Commun 249–251
84. Isaac RE, Rose ME, Rees HH, Goodwin TW (1983) Biochem J 213:533–541
85. Tsoupras G, Hetru C, Luu B, Lagueux M, Constantin E, Hoffmann JA (1982) Tetrahedron Lett 23:2045–2048
86. Tsoupras G, Luu B, Hoffmann JA (1982) Steroids 40:551–560
87. Tsoupras G, Hetru C, Luu B, Constantin E, Lagueux M, Hoffmann JA (1983) Tetrahedron 39:1789–1796
88. Tsoupras G, Luu B, Hoffmann JA (1983) Science 220:507–509
89. Dorn A, Romer F (1976) Cell Tissue Res 171:331–350
90. Mizuno T, Watanabe K, Ohnishi E (1981) J Insect Physiol 23:543–552
91. De Reggi M, Hirn M, Delaage M (1975) Biochem Biophys Res Commun 66:1307–1315
92. Charlet M, Goltzené F, Hoffmann JA (1979) J Insect Physiol 25:463–466
93. Maltête F (1962) Thèse de 3e cycle, Université de Bordeaux, France
94. Kabbouh M, Walcher E, Kappler C, Hetru C, Lagueux M, Nakatani Y (1983) Poster communication (this Symposium)

II.2 Biosynthesis of Ovarian Ecdysteroid Phosphates and Their Metabolic Fate During Embryogenesis in *Schistocerca gregaria*

H.H. REES and R.E. ISAAC[1]

1. Introduction

It is now well established that insect moulting hormones (ecdysteroids) not only occur in immature stages of insects, but also in ovaries of maturing adult females and in embryos (for review see [1] and the preceding chapter). That the ovaries are a site of ecdysteroid synthesis has been demonstrated in several insect species on the basis of the abolishment of hormone production in vivo as a result of ovariectomy and of an increase in immunoreactive ecdysteroids during incubation of ovaries in vitro. At least in maturing adult females of the migratory locust, *Locusta migratoria*, the follicle cells have been shown to be the site of ovarian ecdysteroid synthesis, the ecdysteroids being passed into the ooplasm of the oocytes [2]. The amounts of ecdysteroids generally detected in ovaries, in many species predominantly as polar conjugates, which are hydrolysable with hydrolase enzymes from *Helix pomatia*, are far greater than in larvae and pupae. In many, but not all species investigated, the bulk of the ovarian ecdysteroids is passed into the eggs [1].

This chapter primarily concerns the biosynthesis of ecdysteroids especially during oogenesis, the identification of the maternal polar ecdysteroid conjugates and their fate and possible function during embryogenesis in *Schistocerca gregaria*. Early work on this topic has been reviewed [3, 4].

2. Ovarian Ecdysteroids in *S. gregaria*

Initial investigation of ovarian ecdysteroids in *S. gregaria* revealed that they consisted predominantly of polar conjugate derivatives. Therefore, prior to determination of the ovarian ecdysteroid titres, the free and conjugated hormones were separated as groups on silicic acid columns. Since the cross-reactivity of the antiserum (Whitehead) used towards conjugated ecdysteroids was very low [5], in common with most other antisera raised against ecdysteroids, the ecdysteroid conjugate fraction was enzymically hydrolysed with a crude, so-called arylsulphatase preparation from *Helix pomatia* before radioimmunoassay of the ecdysteroids released. This illustrates the importance of

1 Department of Biochemistry, University of Liverpool, P.O. Box 147, Liverpool, L69 3BX, Great Britain

Biosynthesis, Metabolism and Mode of Action of Invertebrate Hormones
(ed. by J. Hoffmann and M. Porchet)

enzymic hydrolysis of ecdysteroid conjugate-containing fractions before radioimmunoassay. No doubt, significant amounts of such ecdysteroid derivatives have been overlooked in several species in the absence of enzymic hydrolysis in a number of early reports.

The titre of both conjugated and free ecdysteroids in the ovaries of *Schistocerca gregaria* increases sharply in the final stages of oogenesis, with approx. 95% of the total ecdysteroid being present as conjugates, hydrolysable with *Helix pomatia* hydrolases [5]. These results are in agreement with the demonstration by gas-liquid chromatographic analysis that ecdysteroids occur predominantly as conjugates in maturing *S. gregaria* adult females [6]. The maximum titre of the conjugated ecdysteroids is particularly high, being greater than 40 μg ecdysteroid/ovary [5].

Essentially all the ovarian ecdysteroids in *S. gregaria* are passed into the eggs [5]. The ecdysteroids released by enzymic hydrolysis of the conjugate fraction from newly laid eggs consist primarily of ecdysone (9) and 2-deoxyecdysone (14), with smaller amounts of 20-hydroxyecdysone (10) and 2-deoxy-20-hydroxyecdysone (16), as well as traces of 26-hydroxyecdysone (11) and 20,26-dihydroxyecdysone (12) [5-8] (see Fig. 1). We have utilised this comparatively large synthesis of ecdysteroids in maturing adult female *S. gregaria* as a system for investigation of aspects of ecdysteroid biosynthesis. In this procedure, [^{14}C] or [^{3}H] cholesterol (1) is injected into adult female *S. gregaria* at a developmental stage preceding the large increase in titre of ovarian ecdysteroids, and the labelled ecdysteroids (conjugated) isolated from the newly laid eggs [9]. This labelling procedure generally gives 0.4%-1% incorporation of radioactive cholesterol into ecdysteroids, which is far greater than the yield frequently obtained in larval systems (typically in the region of 0.01%-0.12%).

3. Biosynthesis of Ecdysteroids

Recent reviews of the biosynthesis of ecdysteroids are available [3, 4, 10, 11] and only a brief outline will be presented here. In general, insects cannot synthesize sterols de novo from small molecules and require a dietary source of these compounds [12]. For satisfactory growth, development and reproduction, most but not all insect species investigated require some C_{27} sterol such as cholesterol (1). Whereas carnivorous insects obtain cholesterol (1) directly from their food, most investigated phytophagous species and omnivorous ones can effect dealkylation of dietary plant C–24 alkyl sterols yielding C_{27} sterols. However, it is now apparent that certain phytophagous Heteroptera cannot dealkylate plant sterols and contain the C_{28} compound, makisterone A, as the major ecdysteroid [13, 14].

It is clear that during the formation of C_{27} ecdysteroids from cholesterol (1), at least some modification of the sterol nucleus precedes side chain hydroxylation (Fig. 1). 7-Dehydrocholesterol (2) is a possible early intermediate in this transformation and is, in fact, metabolised to ecdysteroids [15, 16]. However, whether or not it is converted into ecdysteroids directly or via cholesterol (1) is uncertain (see [11]). Formation of the Δ^7 bond of ecdysteroids in *Calliphora* involves removal of the 7β- and 8β-hydrogens of cholesterol [17], the same stereochemistry as in the formation of 7-dehydrocholes-

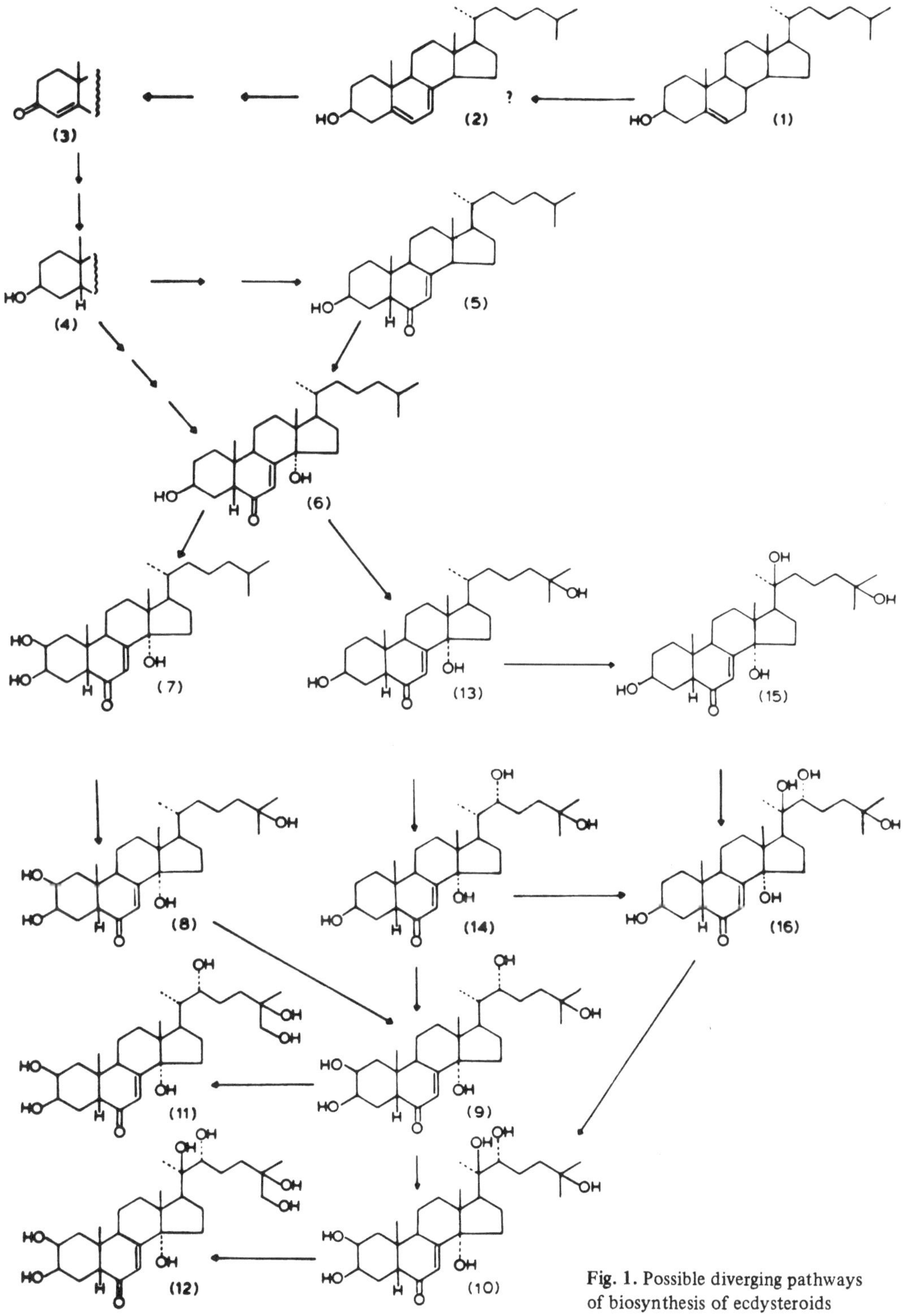

Fig. 1. Possible diverging pathways of biosynthesis of ecdysteroids

terol (2) in that species [18]. Elimination of the 3α- and 4β-hydrogens of cholesterol (1) during transformation into ecdysteroids has been interpreted in terms of the possible intermediacy of a 3-oxo-Δ^4 steroid (3) [19]. Reduction of the Δ^4 bond could then yield an A/B cis ring junction (i.e., 5β–H) [4, Fig. 1]. There is evidence from work on various species indicating that more than one sequence of hydroxylations may exist, involving either complete formation of the nucleus before side chain hydroxylation or 2-deoxyecdysteroids (see [11]). At least in ovaries, the occurrence and interconversion of 2-deoxyecdysteroids suggest that the route (5) → (6) → (13) → (14) → (9) may operate [11, 20]. The co-occurrence of 2-deoxyecdysone (14), ecdysone (9), 2-deoxy-20-hydroxyecdysone (16), and 20-hydroxyecdysone (10) in *Bombyx mori* ovaries [21] and primarily as conjugates in ovaries/eggs of *Schistocerca gregaria* [5–8] and *Locusta migratoria* (see preceding chapter) suggests that in these systems the terminal steps of biosynthesis of 20-hydroxyecdysone (10) may be branched with either the C–20 or C–2 hydroxylation reaction being possibly the final oxidation step (Fig. 1) [11]. The additional isolation of 2,22-dideoxy-20-hydroxyecdysone (15) from *Bombyx* ovaries may suggest another branch pathway in that system [21]. However, it is important to appreciate that the mere isolation of compounds does not necessarily implicate those compounds in a biosynthetic pathway. The mechanism of the C–2 and C–22 hydroxylations involves direct replacement of a hydrogen atom with retention of configuration at the respective carbon atoms, as observed in many vertebrate steroid hydroxylations [22, 23].

4. Structures of Major Maternal Ecdysteroid Conjugates in Newly Laid Eggs

The four major ecdysteroid conjugates (17–20) present in the newly laid eggs of *S. gregaria* were purified by reversed-phase and anion-exchange high-performance liquid chromatography, and the steroid moieties released by enzymic hydrolysis with the crude *Helix pomatia* "aryl sulphatase" preparation identified as ecdysone (9), 2-deoxyecdysone (14), 20-hydroxyecdysone (10), and 2-deoxy-20-hydroxyecdysone (16). The negative ion fast atom bombardment mass spectra of the intact conjugate samples showed prominent $[M-H]^-$ ions, which indicated molecular weights of 544, 528, 560, and 544 for (17), (18), (19), and (20), respectively. Possible structures consistent with the mass spectral data were the phosphate or sulphate esters of the parent ecdysteroids; the observation of the $[M-H]^-$ ions did not distinguish between these two possibilities. However, the prominent ions at m/z 79 and 97 in the negative ion spectra of all four compounds correspond to PO_3^- and $H_2PO_4^-$ ions, respectively. In contrast, sulphate esters of steroids (e.g., 20-hydroxyecdysone monosulphate) yield large ions at m/z 80 (SO_3^-) and 97 (HSO_4^-). Further corroboration of the presence of phosphate in conjugates (17) and (18) was also obtained [24]. Quantitative determination of inorganic phosphate in acid-hydrolysed samples of the conjugates showed a steroid:phosphate ratio of approx. 1:1 for all four compounds. These data demonstrated conclusively that the four ecdysteroid conjugates were phosphate derivatives. Fourier transform p.m.r. spectroscopy, showed that the only major differences between the spectra of the phosphate conjugates and those of the corresponding free ecdysteroids were the downfield shifts of the C–21 methyl [+ (0.03–0.07) ppm] and C–22 proton signals [+(0.59 to 0.74) ppm] in the conjugates, establishing that the phosphate group is attached to the steroid through an ester linkage at C–22. This was corroborated for (17–19) by the ^{13}C Fourier transform n.m.r. spectra, which also confirmed the absence of any carbon atoms in addition to those of the ecdysteroid moiety. The combined foregoing data demonstrate that the four major ecdysteroid conjugates isolated from newly laid eggs of *S. gregaria* were ecdysone 22-phosphate (17), 2-deoxyecdysone 22-phosphate (18), 20-hydroxyecdysone 22-phosphate (19), and 2-deoxy-20-hydroxyecdysone 22-phosphate (20) [24, 25].

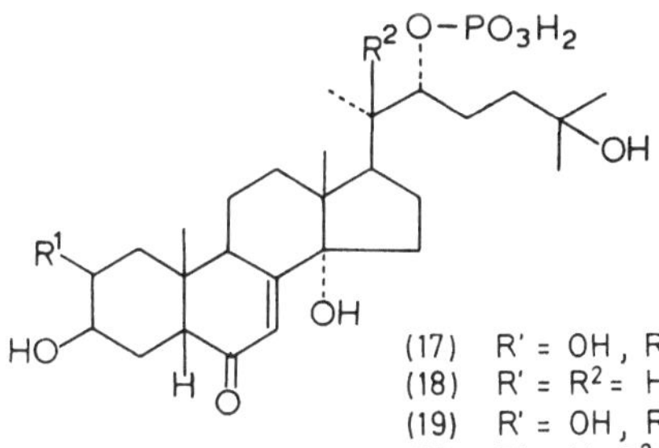

(17) R' = OH, R^2 = H
(18) R' = R^2 = H
(19) R' = OH, R^2 = OH
(20) R' = H, R^2 = OH

There have been numerous reports of the occurrence of highly polar ecdysteroid conjugates as metabolites of ecdysone and the tentative assignment of the conjugating moiety on the basis of hydrolysis of the conjugate with commercial enzyme preparations (for review see [26]). The fact that the ecdysteroid phosphates isolated from *S. gregaria* eggs were readily hydrolysed by an arylsulphatase preparation from *Helix pomatia* illustrates the danger of making any assumption regarding the structures of ecdysteroid conjugates on this basis. That the *Helix* enzyme preparation contains acid phosphatase activity has been demonstrated [24].

It is possible that the formation of the phosphate esters of ecdysteroids could promote the overall synthesis of ecdysteroids by facilitating the removal of free steroid from the site of synthesis in the ovaries. The exact stage in the ecdysteroid biosynthetic sequence at which phosphorylation occurs is uncertain. It must be a late step because the C–22 hydroxylation itself apparently occurs late in ecdysteroid biosynthesis (cf. Fig. 1). Whether ecdysteroid 22-phosphates can serve as substrates for later hydroxylation reactions is not clear.

In extracts of newly laid eggs of the related migratory locust, *Locusta migratoria*, a high proportion of the maternal ecdysteroid conjugates can be bound to the major yolk protein, vitellin, when it is precipitated by lowering the ionic strength of the medium [27]. In contrast, under the conditions we employed, only a small proportion (14% to 15% of the total) of the conjugates from *S. gregaria* are bound to vitellin, whereas 41% to 48% of the conjugates from *L. migratoria* are precipitated by the same procedure [28]. Although the reason for this different degree of binding of the conjugates from the two species to vitellin is unclear, it could possibly reflect a difference in the vitellin or a structural difference in the conjugates (see later) from both species. This precipitation with vitellin has been used as an initial step in the purification of the major ecdysteroid conjugates from newly laid eggs of *L. migratoria*. Interestingly, the two principal components were identified as the 22-adenosinemonophosphate ester of 2-deoxyecdysone [29, 30] and the 22-N^6-(isopentenyl)-adenosine monophosphate ester of ecdysone [31] (see preceding chapter). However, the net yields of the pure compounds isolated were apparently low on the basis of the amounts of immunoreactive ecdysteroids enzymically released from the unfractionated conjugates of newly laid eggs [32]. We have been unable, as yet, to detect such compounds in newly laid eggs of *S. gregaria*, nor during our preliminary attempt to isolate the authentic compounds for use as chromatographic standards from a small scale extraction of *L. migratoria* eggs [28]. In the latter case, the major conjugates isolated were also apparently the 22-phosphate esters of 2-deoxyecdysone and ecdysone. Our lack of success in isolating the adenosine monophosphate derivatives and the apparent low yields of isolation of these conjugates from *L. migratoria* eggs [29-32], may suggest that such compounds are either labile [33] or exist in much lower concentrations than was at first thought. The adenosine monophosphate ester of 2-deoxyecdysone readily undergoes solvolysis on heating in the presence of Tris buffer [29, 30], but little is known about the stability of the adenosine monophosphate conjugates under the usual extraction conditions.

5. Fate and Possible Function of Maternal Ecdysteroid 22-Phosphates During Embryogenesis in *S. gregaria*

As alluded to earlier, approx. 95% of the ecdysteroids in newly laid eggs of *S. gregaria* is in the form of polar conjugates (phosphates), the remainder being in the free form [5, 6, 25]. However, developing eggs of *S. gregaria* contain not only free ecdysteroids and polar hydrolysable conjugates, but also unhydrolysable polar ecdysteroid metabolites, the amounts of which increase in the later stages of embryogenesis (see Sect. 5.2). As in the case of newly laid eggs, the free ecdysteroids constitute a minor proportion of the free plus conjugated ecdysteroids throughout embryogenesis in eggs of *S. gregaria* [9, 28, 34, 35]. The titre of these free ecdysteroids in the developing eggs show distinct peaks during embryogenesis [34, 35]. In fact, studies on a number of insect species indicating an apparent correlation between peak titres of free ecdysteroids in developing eggs and embryonic membrane/cuticle formation imply a possible function for the hormones in controlling cycles of cuticulogenesis in embryos [35-42].

5.1 Enzymic Hydrolysis of Maternal Ecdysteroid 22-Phosphates During Embryogenesis

The possibility has long been considered that certain polar ecdysteroid conjugates might serve as inactive storage forms of ecdysteroids, with release of free, active hormone on enzymic hydrolysis at certain times during insect development [43]. It is significant that in the maternal ecdysteroid conjugates of *S. gregaria*, the C–22 hydroxy, which is essential for hormonal activity in ecdysteroids [44], is exclusively phosphorylated [25]. Thus, although the exact function of the maternal ecdysteroid 22-phosphates in the developing eggs is uncertain, they probably represent inactive storage forms of the hormone. Enzymic hydrolysis of the phosphates could yield free hormone at specific stages of embryogenesis and, thus, contribute to increasing titres of ecdysteroids. Of course, embryonic synthesis of ecdysteroids from cholesterol or precursors present in the ooplasm is not precluded.

To test whether the maternal ecdysteroid 22-phosphates could be hydrolysed and serve as a source of free hormone during embryogenesis in *S. gregaria*, [^{14}C] ecdysteroid conjugates consisting primarily of the 22-phosphates of ecdysone and 2-deoxyecdysone isolated from newly laid eggs were injected into day 3 and day 13 eggs, which were extracted after 4 $^1/_2$ h. (Under our incubation conditions, embryogenesis in *S. gregaria* takes 17–17 $^1/_2$ days and blastokinesis occurs on day 8.) The results indicated that at day 13, free ecdysteroids, primarily ecdysone (9), comprised 25% of the total recovered radioactivity, indicating that eggs at that stage of development possess an appropriate phosphatase enzyme for hydrolysis of the maternal ecdysteroid 22-phosphates [4, 45]. Even if the embryos do not synthesize ecdysteroid at that stage of development this level of enzymic hydrolysis should be sufficient to account for the free ecdysteroid titre at that time [34]. It was also apparent that 22% of the radioactivity recovered in the highly polar ecdysteroid fraction was refractory to hydrolysis with *Helix* hydrolases (cf. Sect. 5.2). In the case of the day 3 eggs, apparent hydrolysis of [^{14}C] ecdysteroid phosphates was minimal. However, this may partly reflect the insensitivity of detection owing to the low specific radioactivity of the initial substrate. The situation is further complicated by the ecdysteroid conjugating activity present during embryogenesis (see Sect. 5.2.2). Incubation in vitro of maternal [^{14}C] ecdysteroid conjugates with a homogenate of day 13 embryos also resulted in 10% hydrolysis of the substrate primarily to free ecdysone [45].

Having established that developing eggs of *S. gregaria* can hydrolyse ecdysone 22-phosphate (17), some properties of the phosphatase enzyme catalysing the reaction in day 15 eggs were investigated. In preliminary experiments using a day 15 embryo homogenate, the highest phosphatase activity towards ecdysone 22-phosphate (17) was observed at pH 4.0; there was very little enzymic activity at pH values above 5.0 [46]. Although this pH optimum is low compared to mammalian acid phosphatases [47], it is similar to such enzymes from insect tissues [48, 49]. When [1α, 2α-^{3}H] ecdysone 22-phosphate (17) was incubated at pH 4.0 with the embryo homogenate, [^{3}H] ecdysone identified by HPLC and mass spectrometry was the sole detectable product of the reaction at this pH [46]. In subsequent work, non-radioactive substrate was used and the ecdysone (9) product was quantified by HPLC, the eluent being monitored by UV absorbance.

It is interesting that a homogenate of yolk from day 15 eggs in pH 4.0 buffer also showed both acid phosphatase activity assayed by the method of Deloach and Mayer [50] using p-nitrophenyl phosphate as substrate and phosphatase activity towards ecdysone 22-phosphate [51]. However, whether the latter activity towards ecdysone 22-phosphate is also significant under the conditions (e.g. pH) prevailing in vivo is uncertain. All subsequent work on the hydrolysis of ecdysone 22-phosphate was carried out on embryos.

Determination of the subcellular distribution of the phosphatase activity towards ecdysone 22-phosphate (17) in embryo homogenate was carried out with the addition of Triton X-100 to the assay mixture in each case [46]. Relative specific activities of the enzyme (percentage total activity/percentage total protein) in subcellular fractions were: nuclear, 0.64; mitochondrial plus lysosomal, 3.0; microsomal, 0.99; cytosol, 0.60. Clearly, the mitochondrial-lysosomal fraction gives the highest relative specific activity. A similar subcellular distribution pattern was obtained when acid phosphatase activity was assayed using *p*-nitrophenyl phosphate as substrate. The latter phosphatase activity is often used as a lysosomal marker enzyme [50] and it is likely that the phosphatase catalysing hydrolysis of ecdysone 22-phosphate (17) is located in the lysosomal fraction. Therefore, unless otherwise stated, subsequent studies utilised a supernatant fraction from sonication and centrifugation of the mitochondrial-lysosomal fraction of day 15 embryos. Enzymic activity was linear for at least 1 h and up to 40 μg protein [51]. In common with lysosomal acid phosphatases from mammalian [52, 53] and insect tissues [49], the phosphatase hydrolysing ecdysone 22-phosphate is strongly inhibited by phosphate, fluoride, and L(+)tartrate, but much less affected by the sulphydryl reagent, iodoacetamide [51]. This evidence again is consistent with a lysosomal location of the enzyme. In contrast, mammalian cytosolic acid phosphatase is unaffected by tartrate or fluoride, but is readily inhibited by sulphydryl reagents [53].

Preliminary chromatography of a solubilised preparation of the mitochondrial-lysosomal fraction on Sephadex G-100 indicated that the apparent molecular weight of the phosphatase activity towards ecdysone 22-phosphate was approx. 70,000 and that it is apparently distinct from the majority of the acid phosphatase activity assayed with *p*-nitrophenyl phosphate as substrate [51]. However, these observations require substantiation. The tentative apparent molecular weight is similar to that of a lysosomal acid phosphatase from a tumorous blood cell line of *Drosophila* (60–65,000) [49], but is considerably lower than two lysosomal acid phosphatases of mammalian origin (> 200,000 and ~100,000) [52].

The apparent K_m value of the phosphatase towards ecdysone 22-phosphate was appreciably lower than those towards the 22-phosphates of 20-hydroxyecdysone and 2-deoxyecdysone (Table 1) [51]. In the latter case, the V_{max} was the highest; the reason for this is obscure. Of these three substrates, the concentration of ecdysone 22-phosphate (17) is the highest in *S. gregaria* eggs [25], so that hydrolysis of that compound would be expected to predominate.

Preliminary investigation of phosphatases activities in sonicated homogenates of *S. gregaria* embryos at different stages of development indicate that activity towards ecdysone 22-phosphate is already present at day 6 of embryogenesis and that this activity (20 pmol/h/embryo) increases approx. fivefold by day 14. However, the phosphatase activity towards *p*-nitrophenyl phosphate increases (from 2.2 nmol/h/embryo

Table 1. Kinetic parameters for the hydrolysis of ecdysteroid 22-phosphates by a phosphatase enzyme preparation from developing embryos of *S. gregaria*

Substrate	K_m (μM)	V_{max} (nmol/h/μg protein)
Ecdysone 22-phosphate (17)	305	0.19
20-Hydroxyecdysone 22-phosphate (19)	534	0.16
2-Deoxyecdysone 22-phosphate (18)	810	0.94

A solubilized preparation of the mitochondrial-lysosomal fraction from day 15 embryos was fractionated on Sephadex G-100 and incubated with various concentrations of the three substrates in turn. The ecdysteroid products of the reactions were quantified by HPLC

at day 6) by approx. 30-fold during the corresponding period. This is again another difference between the two enzymic activities.

In summary, the foregoing data suggest that ecdysone 22-phosphate (17) may serve as a storage form of ecdysone in *S. gregaria*. Uptake of the phosphate conjugate from the yolk and its subsequent hydrolysis is probably an important factor in determining ecdysteroid levels in the developing embryo.

5.2 Inactivation of Ecdysteroids During Embryogenesis

Having considered one possible source of free ecdysteroids during embroygenesis, we shall now consider their inactivation, another important factor in regulation of hormone levels.

5.2.1 Identification of Highly Polar Ecdysteroids in Developing *S. gregaria* Eggs

Initial information on the fate of the maternal ecdysteroid 22-phosphates of *S. gregaria* and the formation of possible ecdysteroid inactivation products during embryogenesis was obtained by allowing eggs labelled maternally from radioactive cholesterol to develop for varying periods of time [9]. Of course, it is impossible by this approach to differentiate between maternal ecdysteroids and any that might be synthesised by the embryo itself from cholesterol or ecdysteroid precursors which might also be incorporated into the eggs. Comparison of the radioactivity in the free and highly polar ecdysteroid fractions indicated, in agreement with results for analyses of mass of ecdysteroids in the two fractions, that the highly polar ecdysteroids predominate (mostly $> 80\%$) throughout embryogenesis [9,45]. It was also apparent that there was a marked increase during embryogenesis in the percentage of the highly polar ecdysteroid fraction that remained unhydrolysed by the *Helix* hydrolases (e.g. day 3 eggs, $< 5\%$; day 17 eggs, 31%) [9].

The ecdysteroids released by hydrolysis with the *Helix* enzymes of the [^{14}C] highly polar ecdysteroid fraction from eggs at various stages of embryogenesis were fractionated by TLC and by HPLC on a combination of reversed-phase [54] and APS-Hypersil [55] columns. It was clear that the pattern of radioactivity reflected the mass of each ecdysteroid present. The identities of the steroid components of [^{14}C] ecdysteroid conjugates formed during embryogenesis were established by comparison with mass of the compounds isolated and characterised from large-scale extractions of developed eggs. It was evident that the proportions of the individual conjugated [^{14}C] ecdysteroids did not change greatly up to approx. day 10 of embryogenesis, but thereafter some new ecdysteroid metabolites appeared in increasing quantities [45, 56].

Ecdysone-3-acetate (23) [57] could only be detected after approx. day 10, when it comprised 1%–6% of the conjugated ecdysteroids. This proportion increased to approx. 36% at day 17 around the time of hatching, whereas there was a marked decrease in the relative contribution of conjugated ecdysone [45, 56]. Although this was the first identification of ecdysone-3-acetate (23) as an ecdysone metabolite, there is little doubt that this compound has been detected in several studies, but erroneously assigned in many cases as 3-dehydroecdysone, which co-chromatographs on TLC with the acetate.

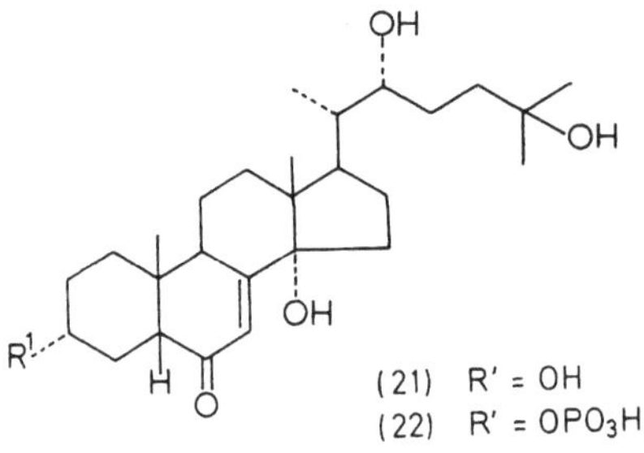

(21) R' = OH
(22) R' = OPO_3H_2

During the work-up and storage, ecdysone-3-acetate (23) readily undergoes acyl migration yielding ecdysone-2-acetate, as well as deacetylation to ecdysone [57]. Between day 6 and day 10 of embryogenesis, a conjugate of 3-epi-2-deoxyecdysone (21) [58] appeared, but unlike the conjugate of ecdysone-3-acetate (23), this compound was not a major ecdysteroid at any of the stages investigated. 2-Deoxy-20-hydroxyecdysone (16), 26-hydroxyecdysone (11), and 20,26-dihydroxyecdysone (12) were minor components (generally < 5% combined) of the conjugate fraction and did not change significantly during embryogenesis. Ecdysone (9), 2-deoxyecdysone (14), with smaller quantities of 20-hydroxyecdysone (10) were major components of the conjugate fraction throughout embryogenesis, except at the end when, in addition, conjugated ecdysone-3-acetate assumed greater significance [45].

Purification by HPLC and characterisation by physical techniques of the intact conjugates from developing eggs (day 15, 16) showed the presence of the 22-phosphate derivatives of ecdysone, 2-deoxyecdysone, 20-hydroxyecdysone, and 2-deoxy-20-hydroxyecdysone [58], which were also identified in newly laid eggs. However, in addition, the developing eggs also contained appreciable 3-acetylecdysone 2-phosphate (24) [59] and smaller amounts of 3-acetyl-20-hydroxyecdysone-2-phosphate (27), 3-epi-2-deoxyecdysone 3-phosphate (22), 3-acetylecdysone 22-phosphate (25), and 2-acetylecdysone 22-phosphate (26) [58]. The latter two compounds, after isolation are susceptible to non-enzymic acetyl migration and deacetylation to give mixtures of ecdysone 22-phosphate (17) and its 2- and 3-acetate derivatives (25 and 26). Similarly, 3-acetylecdysone 2-phosphate (24) readily formed ecdysone 2-phosphate on storage [59]. Owing to the instability of the acetyl, phosphate derivatives of ecdysteroids during

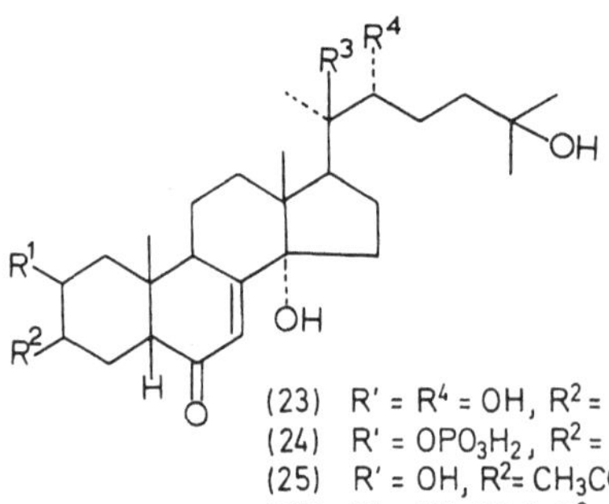

(23) R' = R^4 = OH, R^2 = $CH_3CO.O$, R^3 = H
(24) R' = OPO_3H_2, R^2 = $CH_3CO.O$, R^3 = H, R^4 = OH
(25) R' = OH, R^2 = $CH_3CO.O$, R^3 = H, R^4 = OPO_3H_2
(26) R' = $CH_3CO.O$, R^2 = OH, R^3 = H, R^4 = OPO_3H_2
(27) R' = OPO_3H_2, R^2 = $CH_3CO.O$, R^3 = R^4 = OH

isolation and storage, it is difficult to assess their true relative quantitative importance in the metabolism of egg ecdysteroids. The conjugates of 26-hydroxyecdysone (11) and 20,26-dihydroxyecdysone (12) have not been fully identified as yet in *S. gregaria* eggs.

3-Acetylecdysone 2-phosphate (24) and 3-acetyl-20-hydroxyecdysone 2-phosphate (27) are considered to be hormonally inactive, since modification of the A ring of the ecdysteroid nucleus reduces hormonal activity [44]. The observation that 3-acetylecdysone 2-phosphate (24) is a poor substrate [46], compared with ecdysone 22-phosphate (17), for the embryonic phosphatase enzyme preparation (Sect. 5.1) is consistent with an end-product role for this metabolite. Presumably, the addition of two distinct substituents on the ecdysteroid molecule limits any possible recycling of the ecdysteroid moiety to an active free hormone in the essentially closed system of the insect egg [59]. The detection of 3-acetylecdysone 2-phosphate (24) and 3-acetyl-20-hydroxyecdysone 2-phosphate (27) as [^{3}H] ecdysone metabolites in the faeces of fifth instar larvae of *S. gregaria* again suggests that they are end-products of metabolism [60].

Whether the acetates of ecdysteroid 22-phosphates can result from direct acetylation of maternal conjugate or the combined phosphorylation and acetylation of free hormone is uncertain. It is significant that the analogous compound, 3-acetyl-20-hydroxyecdysone 22-phosphate has been isolated in small amounts from developing eggs of *Locusta migratoria* [61] and as a metabolite by incubation in vitro of gut and Malpighian tubules from that species with 20-hydroxyecdysone [62].

3-Epi-2-deoxyecdysone 3-phosphate (22) was only a minor ecdysteroid component in developing eggs of *S. gregaria*, in contrast to those of *Locusta* where it is a major ecdysteroid [30]; the exact reason for this difference is uncertain. 3-Epiecdysteroids are formed from the corresponding 3β-hydroxy compounds via the 3-dehydro-derivatives [63]. Presumably, in developing *S. gregaria* eggs, free 2-deoxyecdysone (14) released by limited hydrolysis of the corresponding 22-phosphate is transformed via the 3-dehydro compound to the 3-epi-derivative, which undergoes phosphorylation at C–3. The co-occurrence of free 2-deoxyecdysone (14), 3-dehydro and 3-epi 2-deoxyecdysone (21) in late stage eggs of *Locusta migratoria* is in conformance with the intermodiary of the 3-dehydro compound in the epimerisation reaction [30].

As alluded to earlier, during embryogenesis in *S. gregaria*, there is increasing accumulation of unhydrolysable polar ecdysteroids in the eggs [4, 9]. These major compounds have been identified as ecdysonoic acid (28) and 20-hydroxyecdysonoic acid (29) [64]. Similarly, 20-hydroxyecdysonoic acid (29) has been characterised from *Spodoptera littoralis* pupae, where both compounds have also been detected as metabolites of [^{3}H] ecdysone [64]. Concurrently, 20-hydroxyecdysonoic acid (29) was identified in *P. bras-*

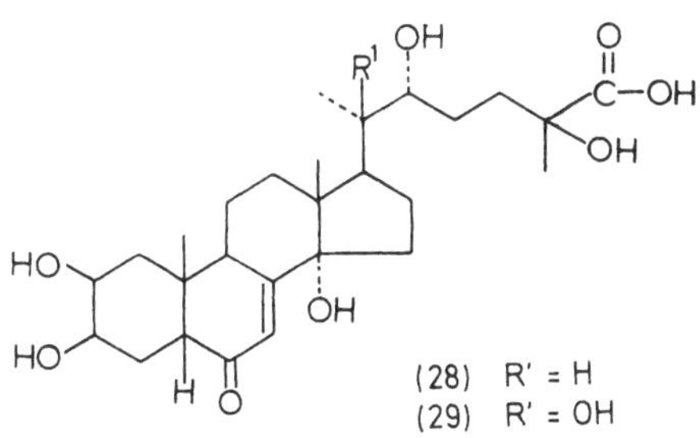

(28) R' = H
(29) R' = OH

sicae [65]. The acids are formed from the respective 26-hydroxy-ecdysteroids [64, 66], and although the latter transformations have not been demonstrated directly in developing eggs of *S. gregaria*, the 26-hydroxy-ecdysteroids have at least been identified in the conjugate fraction. Presumably the ecdysteroid-26-oic acids (28 and 29) represent inactivation products of ecdysone (9) and 20-hydroxyecdysone (10), respectively. It is significant that ecdysonoic acid (28) and 20-hydroxyecdysonoic acid (29) have been characterised as faecal metabolites of [^{3}H]ecdysone in fifth instar *S. gregaria* larvae [60]. In fact, ecdysteroid-26-oic acid formation is probably a widespread metabolic pathway in insects [65].

5.2.2 Metabolism of [^{3}H] Ecdysone in *S. gregaria* Embryos

Having identified likely ecdysteroid inactivation products in developing eggs of *S. gregaria*, it was necessary to establish the transformation of [^{3}H] ecdysone into such compounds. For this, [^{3}H]ecdysone was injected into day 13 embryos, which were extracted after 4 $^1/_2$ h. The free ecdysteroid fraction consisted almost entirely of unchanged [^{3}H] ecdysone with only a trace (~0.1%) of 20-hydroxyecdysone (10). The major metabolites were ecdysonoic acid (28) and 3-acetylecdysone 2-phosphate (24) with much less ecdysone 22-phosphate (17), 20-hydroxyecdysonoic acid (29), and possibly a phosphate ester of 3-acetyl 20-hydroxyecdysone [45]. However, it is important not to overlook the possibility that the metabolism of administered exogenous ecdysone may not accurately reflect that of endogenous hormone, where compartmentation and protein binding may be important.

6. Summary of Possible Metabolic Interrelationships of Ecdysteroid Derivatives During *S. gregaria* Embryogenesis

Based on the foregoing information, a possible metabolic scheme of events involved in regulating the hormone levels in developing eggs of *S. gregaria* is presented in Fig. 2. The embryonic phosphatase enzyme which hydrolyses maternal ecdysone 22-phosphate (17) can also act, though probably to a lesser extent, on the 22-phosphate esters of 2-deoxyecdysone (18) and 20-hydroxyecdysone (19). The small amount of 2-deoxyecdysone (14) produced could undergo epimerisation and reconjugation yielding 3-epi-2-deoxyecdysone 3-phosphate (22).

As we have seen, ecdysone can arise by hydrolysis of ecdysone 22-phosphate (17), but whether it also arises during embryogenesis in *Schistocerca* from synthesis de novo from cholesterol (1) or from metabolism of partly formed precursors of maternal origin is not yet established. In this context, it is significant that the hydroxylation of 2-deoxyecdysone (14) to ecdysone (9) has been demonstrated in embryos of *L. migratoria* (see preceding chapter). The detection of 20-hydroxy metabolites of [^{3}H] ecdysone, including traces of 20-hydroxyecdysone (10) itself in *S. gregaria* embryos, indicates that 20-hydroxylation of ecdysone (9) can occur, albeit to a limited extent, in that system. This is consistent with the occurrence of much smaller amounts of 20-hydroxy-

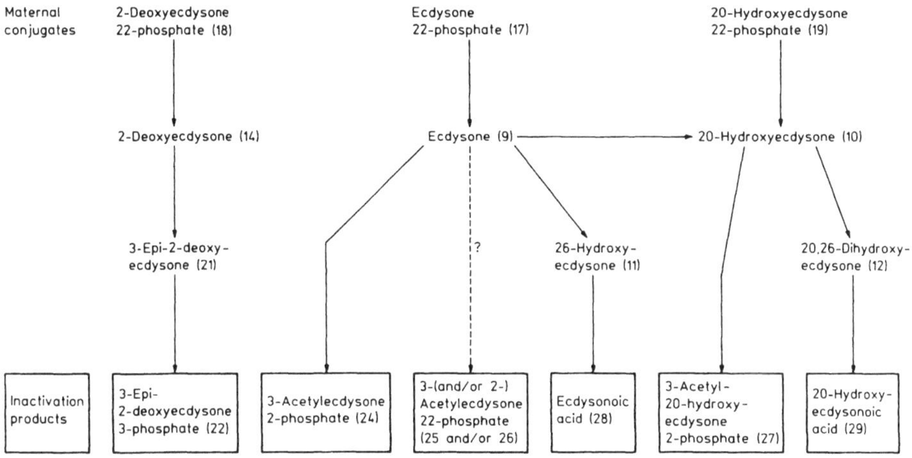

Fig. 2. Probable metabolic pathways of ecdysteroid 22-phosphates in developing eggs of *Schistocerca gregaria*

ecdysteroid derivatives than of the corresponding 20-deoxy compounds in developing eggs of *S. gregaria*.

Both ecdysone (9) and 20-hydroxyecdysone (10) can be inactivated by (a) conversion into the corresponding 26-oic acids (28 and 29) via the respective 26-hydroxyecdysteroids, and (b) formation of the corresponding 3-acetyl, 2-phosphate derivatives. The isolation of small amounts of 3-(and/or 2-)acetylecdysone 22-phosphate (25 and 26) from late stage eggs of *S. gregaria*, suggests that its formation from ecdysone (9) may constitute another inactivation route; formation of the latter compound by direct acetylation of maternal ecdysone 22-phosphate (17) cannot be discounted. However, quantitation of acetyl, phosphate derivatives of ecdysteroids is difficult due to their inherent instability. In fact, it is not clear whether the small amount of ecdysone 22-phosphate (17) detected as a metabolite of [^{3}H] ecdysone is formed directly or by breakdown during extraction/storage of 3-(2-) acetylecdysone 22-phosphate (25/26). There is little doubt that formation of both ecdysteroid acids and 3-acetylecdysteroid 2-phosphates are quantitatively important routes of hormone inactivation. At the end of embryogenesis in *S. gregaria*, ecdysteroid acids and the 2-phosphate ester of ecdysone-3-acetate (24) comprise 19% and 36% of the total ecdysteroids, respectively. This should readily account for the total decrease in ecdysteroid titres for the hormone peaks occurring during embryogenesis. It appears that a significant portion of the inactivation products from the developing embryos is passed into the yolk.

In larval and pupal stages of insects, 20-hydroxyecdysone (10) is generally regarded as the major active hormone, with ecdysone (9) being primarily a prohormone [67]. Should this tenet also hold for developing embryos, the formation of 20-hydroxyecdysonoic acid (29) and 3-acetyl-20-hydroxyecdysone 2-phosphate (27) in the closed system of the egg can be viewed as true inactivation routes responsible for decrease in hormonal activity. The detection of both 26-oic acid and 3-acetyl, 2-phosphate derivatives of ecdysone and 20-hydroxyecdysone as faecal metabolites of [^{3}H] ecdysone in fifth instar larvae of *S. gregaria* [60], indicates that late stage embryos already exhibit features of ecdysteroid metabolism common to fifth instar larvae.

Acknowledgments. We thank the Science and Engineering Research Council and the Agricultural Research Council for financial support and Dr. L.N. Dinan for his contribution to the earlier work and for valuable discussion.

References

1. Hoffmann JA, Lagueux M, Hetru C, Charlet M, Goltzené F (1980) In: Hoffmann JA (ed) Progress in ecdysone research. Elsevier/North-Holland Biomed Press, Amsterdam New York, p 431
2. Goltzené F, Lagueux M, Charlet M, Hoffmann JA (1978) Hoppe-Seyler's Z Physiol Chem 359: 1427–1434
3. Rees HH, Davies TG, Dinan LN, Lockley WJS, Goodwin TW (1980) In: Hoffmann JA (ed) Progress in ecdysone research. Elsevier/North-Holland Biomed Press, Amsterdam New York, p 125
4. Rees HH, Greenwood DR, Dinan LN, Isaac RE, Goodwin TW (1980) In: Sehnal F, Zabza A, Menn JJ, Cymborowski B (eds) Regulation of insect development and behaviour, part I. Wroclaw Tech Univ Press, Wroclaw, Poland, p 71
5. Dinan LN, Rees HH (1981) J Insect Physiol 27:51–58
6. Gande AR, Morgan ED, Wilson ID (1979) J Insect Physiol 25:669–675
7. Isaac RE, Rees HH, Goodwin TW (1981) J Chem Soc Chem Commun: 418–420
8. Isaac RE, Rees HH, Goodwin TW (unpubl. results)
9. Dinan LN, Rees HH (1981) Insect Biochem 11:255–265
10. Rees HH (1983) In: Porter JW, Spurgeon SL (eds) Biosynthesis of isoprenoid compounds, vol II. Wiley, New York, p 463
11. Rees HH (1984) In: Gilbert LJ (ed) Comprehensive insect physiology, biochemistry and pharmacology, vol VII. Pergamon Press, Oxford New York, in press
12. Clayton RB (1964) J Lipid Res 5:3–19
13. Svoboda JA, Dutky SR, Robbins WE, Kaplanis JN (1977) Lipids 12:318–321
14. Gibson JM, Majumder MSI, Mendis AHW, Rees HH (1983) Arch Insect Biochem Physiol 1: 105–120
15. Horn DHS, Middleton EJ, Thomson JA, Wilkie JS (1974) J Insect Physiol 20:2433–2445
16. Johnson P, Rees HH (1977) J Insect Physiol 23:1387–1392
17. Cook IF, Lloyd-Jones JG, Rees HH, Goodwin TW (1973) Biochem J 136:135–145
18. Johnson P, Cook IF, Rees HH, Goodwin TW (1975) Biochem J 152:303–311
19. Davies TG, Dinan LN, Lockley WJS, Rees HH, Goodwin TW (1981) Biochem J 194:53–62
20. Hetru C, Kappler C, Hoffmann JA, Nearn R, Bang L, Horn DHS (1982) Mol Cell Endocrinol 26:51–81
21. Ohnishi E, Mizuno T, Ikekawa N, Ikeda T (1981) Insect Biochem 11:155–159
22. Greenwood DR, Rees HH (1982) Biochem J 208:857–864
23. Greenwood DR, Dinan LN, Rees HH (1984) Biochem J 217:783–789
24. Isaac RE, Rose ME, Rees HH, Goodwin TW (1982) J Chem Soc Chem Commun 249–251
25. Isaac RE, Rose ME, Rees HH, Goodwin TW (1983) Biochem J 213:533–541
26. Koolman J (1982) Insect Biochem 12:225–250
27. Lagueux M, Harry P, Hoffmann JA (1981) Mol Cell Endocrinol 24:325–338
28. Isaac RE, Rees HH (unpubl. results)
29. Tsoupras G, Hetru C, Luu B, Lagueux M, Constantin E, Hoffmann JA (1982) Tetrahedron Lett 23:2045–2048
30. Tsoupras G, Hetru C, Luu B, Constantin E (1983) Tetrahedron 39:1789–1796
31. Tsoupras G, Luu B, Hoffmann JA (1983) Science 220:507–509
32. Lagueux M, Sall C, Hoffmann JA (1981) Am Zool 21:715–726
33. Hoffmann JA (1983) Round table discussion. CNRS Symp Invertebr Horm, Strasbourg
34. Dinan LN, Rees HH (unpubl. results)
35. Scalia S, Morgan ED (1982) J Insect Physiol 28:647–654
36. Dorn A, Romer F (1976) Cell Tissue Res 171:331–350

37. Bullière D, Bullière F, Mauchamp B, Lafont R, de Reggi M, Hirn M (1976) CR Acad Sci Ser D 283:647–650
38. Bullière D, Bullière F, de Reggi M (1979) Wilhelm Roux's Arch 186:103–114
39. Lagueux M, Hetru C, Goltzené F, Kappler C, Hoffmann JA (1979) J Insect Physiol 25:709 to 723
40. Matz G (1980) CR Acad Sci Ser D 291:501–504
41. Cavallin M, Fournier B (1981) J Insect Physiol 27:527–534
42. Imboden H, Lanzrein B (1982) J Insect Physiol 28:37–46
43. Willig A, Rees HH, Goodwin TW (1971) J Insect Physiol 17:2317–2326
44. Bergamasco R, Horn DHS (1980) In: Hoffmann JA (ed) Progress in ecdysone research. Elsevier/ North-Holland Biomed Press, Amsterdam New York, p 299
45. Isaac RE, Rees HH (1984) Insect Biochem (in press)
46. Isaac RE, Sweeney FP, Rees HH (1983) Biochem Soc Trans 11:379–380
47. Schmidt G (1955) Methods Enzymol 2:523–530
48. Gilbert LI, Huddleston CJ (1965) J Insect Physiol 11:177–194
49. Dinan L, Gläsener G, Emmerich H (1983) Insect Biochem 13:411–419
50. Deloach JR, Mayer RT (1979) Insect Biochem 9:653–659
51. Isaac RE, Sweeney FP, Rees HH (unpubl. results)
52. Lawrence GL, Van Etten RL (1981) Arch Biochem Biophys 206:122–131
53. Taga EW, Van Etten RL (1982) Arch Biochem Biophys 214:505–515
54. Isaac RE, Milner NP, Rees HH (1982) J Chromatogr 246:317–322
55. Dinan LN, Donnahey PL, Rees HH, Goodwin TW (1981) J Chromatogr 205:139–145
56. Isaac RE, Dinan LN, Rees HH, Goodwin TW (1983) In: Lofts B (ed) Comparative endocrinology. Hong Kong Univ Press (in press)
57. Isaac RE, Rees HH, Goodwin TW (1981) J Chem Soc Chem Commun 594–595
58. Isaac RE, Rees HH (1984) Biochem J 221:459–464
59. Isaac RE, Desmond HP, Rees HH (1984) Biochem J 217:239–243
60. Gibson JM, Isaac RE, Dinan LN, Rees HH (1984) Arch Insect Biochem Physiol (in press)
61. Tsoupras G, Luu B, Hoffmann JA (1982) Steroids 40:551–560
62. Tsoupras G, Luu B, Hetru C, Muller J-F, Hoffmann JA (1983) CR Acad Sci Ser III 296:77–80
63. Milner NP, Rees HH (1983) Pres CNRS Symp Invertebr Horm, Strasbourg
64. Isaac RE, Milner NP, Rees HH (1983) Biochem J 213:261–265
65. Lafont R, Blais C, Beydon P, Modde J-F, Enderle U, Koolman J (1983) Arch Insect Biochem Physiol 1:41–58
66. Beydon P, Claret J, Porcheron P, Lafont R (1981) Steroids 38:633–650
67. Gilbert LI, King DS (1973) In: Rockstein M (ed) The physiology of insecta, vol I. Academic Press, London New York, p 249

II.3 Ecdysone Metabolism

R. LAFONT[1] and J. KOOLMAN[2]

1. Introduction

Ecdysteroids represent a widespread family of steroids found in many invertebrate groups (zooecdysteroids) as well as in many plants (phytoecdysteroids) [1, 2]. They were first detected in insects [3], then in Crustacea and other Arthropods and, more recently, in Mollusca [4], various worms [5-7] and even in Coelenterata [8]. Their general occurrence in invertebrates may thus be assumed [9], but evidence that they play a hormonal role in every case is still lacking. In addition, it is not always known whether they are synthesized by the animals themselves or if they originate from the animal's diet.

Ecdysteroids differ strongly from vertebrate steroid hormones: (1) they have retained the entire carbon skeleton of the precursor molecule, which is usually cholesterol (Fig. 1) or a higher homologue of it; (2) they bear several hydroxyl groups that make them rather hydrophilic; and (3) they present a rather unusual configuration (5β-H = trans) of the A/B rings junction that otherwise is only found in bile acids.

The first ecdysteroid isolated was ecdysone (Fig. 1). This compound was isolated from *Bombyx mori* pupae [3], and it was demonstrated later to be the secretory product of molting glands [10, 11]. Ecdysone is further metabolized by several peripheral tissues in different ways. This metabolism has been investigated through the use of labelled hormones and the isolation/identification of several ecdysteroids from various biological materials.

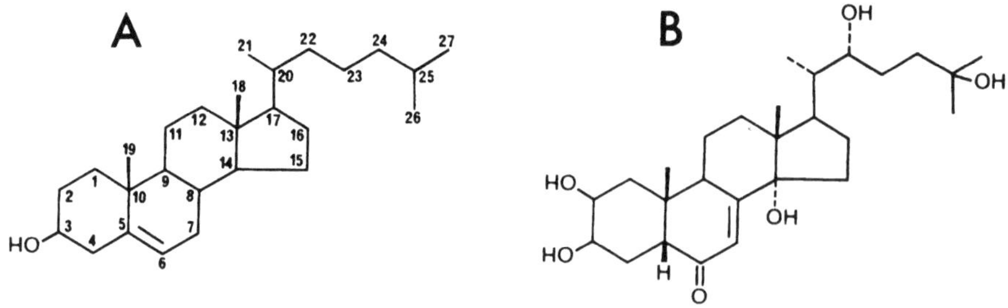

Fig. 1 A, B. Structural formulas of cholesterol (**A**) and ecdysone (**B**)

1 Ecole Normale Supérieure, Laboratoire de Zoologie, CNRS ERA 1048, 75005 Paris, France
2 Physiologisch-Chemisches Institut der Philipps-Universität, Marburg/Lahn, FRG

Biosynthesis, Metabolism and Mode of Action of Invertebrate Hormones
(ed. by J. Hoffmann and M. Porchet)

In this review we will give a short survey of the methods currently used for metabolic studies and ecdysteroid identification. We will describe the different pathways and the enzyme systems involved in these reactions. The last part of the review will be devoted to more physiological considerations concerning the functional significance of the different metabolites and the role of the different tissues in these conversions. Ecdysteroid metabolism has also been the subject of several recent review articles [12-15].

2. Methods Used for the Study of the Ecdysone Metabolism

Ecdysteroid metabolism has been studied by using two different and complementary strategies. The first one consists in the isolation of ecdysteroids from biological materials. A prerequisite for this strategy is that they can be detected by a quantitative method such as a bioassay or a radioimmunoassay. The second strategy involves the use of labelled hormones and the characterization of their metabolites. It is based on the assumption that the metabolism of injected or ingested compounds reflects the fate of endogenously synthesized ecdysteroids.

For the first way, sufficient amounts (in the order of mg) have to be obtained for the full characterization by physiochemical means (mass spectrometry, NMR). Large amounts of biological material (ranging from kg to ton!) are needed as starting material, except in the most favorable case of some insect eggs and embryos which contain up to 10^{-4} *M* ecdysteroids. In spite of the difficulties the isolation of new ecdysteroids is needed for their definite identification.

The second strategy uses labelled hormones which are introduced into biological systems either in vivo or in vitro. This often leads to the appearance of several metabolites that can be tentatively identified by comparison with available reference compounds and characterized by various reactions (Table 1). Ecdysone metabolism usually gives rise to polar derivatives and, in some cases, to additional less polar products. These compounds can be efficiently separated by various chromatographic procedures: thin-layer chromatography (TLC), liquid chromatography (LC), or high-performance liquid chromatography (HPLC). HPLC is at present the most widely used technique [24-26], especially in the reversed-phase mode. It represents a rapid and extremely powerful technique that provides the best available separations. The metabolites of ecdysone can be characterized by using a logical sequence of analytical steps (Fig. 2). Although the procedure involves only a limited number of experiments, it leads to a rather precise knowledge of the different metabolites.

The strategy has some important limitations that will be illustrated by two different examples. The first one concerns the conclusions that can be drawn from the comigration of two compounds in one HPLC system: although this technique is very efficient, one system is unable to resolve all known ecdysteroids, and comigration in at least two systems is needed for assessing compound identity. This can be performed by using a combination of reversed-phase and normal phase systems, or by using buffers at different pHs in the reversed phase mode (in the case of ionizable metabolites only). A second example concerns the use of commercial enzyme preparations for the hydrolysis of conjugates: they are often impure, and the fact that, e.g., a sulfatase preparation is able

Table 1. Techniques classically used for ecdysteroid metabolic studies

Techniques	Comments	References
Separation systems		
Thin-layer chromatography (TLC) on silica plates	Analytical/preparative	[16]
– Solvent chloroform/methanol (e.g., 80:20)	Non-polar metabolites	[17]
– Solvent ethyl acetate/ethanol/water (20:80:10)	Polar metabolites	[17]
– Two-dimensional TLC	Combines normal and reversed phases	[18]
– Two-dimensional TLC + fluorography	Complete metabolite patterns	[19]
High-performance thin-layer chromatography (HPTLC)	Improved separation	
– Normal phase		[20]
– Reversed phase	Improved recovery at elution	[21]
Liquid chromatography (LC) (low pressure)	Sample prepurification	
– Silicic acid		[22]
– Sephadex LH 20		[17]
– Ion exchangers	Polar conjugates	[23]
High-performance liquid chromatography (HPLC)		
– Normal silica or polar-bonded phases	Non-polar metabolites	[24, 25]
– Reversed-phases		[24]
Basic system (methanol/water mixtures)		[24]
Ion-suppression chromatography	Polar metabolites	[26]
Paired-ion chromatography	Polar metabolites	[26, 27]
– Ion exchangers	Polar metabolites	[27]
Gas-liquid chromatography (GLC)	Needs derivatization can be coupled with mass spectrometry (MS)	[26, 28–30]
High-voltage electrophoresis		[17, 31]
Double-labelling procedures	^{3}H Ecdysteroid + ...	
^{14}C Sugar (glucose) or UDP-sugar (with homogenates)	Glucoconjugates	[32, 33]
^{35}S Sulfate (or PAPS)	Sulfoconjugates	[17]
^{32}P Phosphate (or ATP)	Phosphoconjugates	[34, 35]
Hydrolysis experiments	Conjugates	
Enzymatic		
Use of crude *Helix pomatia* digestive juice or of semi-purified enzyme preparations (problems due to cross contamination of these preparations)	Use of some specific inhibitors can provide more specific information	[17, 36–38]
Chemical		
– Acid solvolysis (acetic acid in dioxane)	Sulfates (phosphates?)	[36]
– Alkaline hydrolysis (sodium hydroxyde)	Acyl esters	[39]
Derivative formation		
Acetylation		
– of a ^{3}H metabolite and an unlabeled reference	Identification	[40, 41]
– of a conjugate, followed by selective hydrolysis of the conjugate formed, and identification of the acetates formed	Determination of the conjugated position in a conjugate	[35, 42]
Acetonide formation	Characterizes cis-diols (C2–C3, C20–C22, C25–C26)	

Table 1 (continued)

Techniques	Comments	References
– Check for the presence of 3-epiecdysteroids		[16,41,43]
– Characterization of some conjugates (lack of reaction)		[44]
Methyl ester formation	For acidic compounds	
– Characterization of ecdysonoic acids (–COOH)		[45]
– of sulfate or phosphate esters (for MS analyses)	To be tested with ecdysteroids	
Reduction		
for 3-oxo compounds (3-dehydroecdysteroids)	With $NaBH_4$ (or enzymatic?)	[46]
Periodate reaction	Splits vicinal diols	[33]
Check for the presence of a free C20–C22 diol (with side chain labeled ecdysteroids)		
Physico-chemical analyses		
Infrared spectroscopy (IR)	Presence of carbonyl or carboxyl groups	[47]
Mass spectrometry		
– Electron impact (EI)	Maximal fragmentation	
– Chemical ionization (CI)	Gives large molecular ions except for ionized conjugates	[24]
– Fast-Atom Bombardment (FAB)	Gives the molecular ion for ionized conjugates	[34]
NMR spectroscopy		[15, 34, 48–51]
– ^{1}H–NMR	Methyl and proton signals	
– ^{13}C–NMR (visualization of all carbon atoms)	Needs mg amounts	
– ^{31}P–NMR	Presence of phosphorus in the molecule	

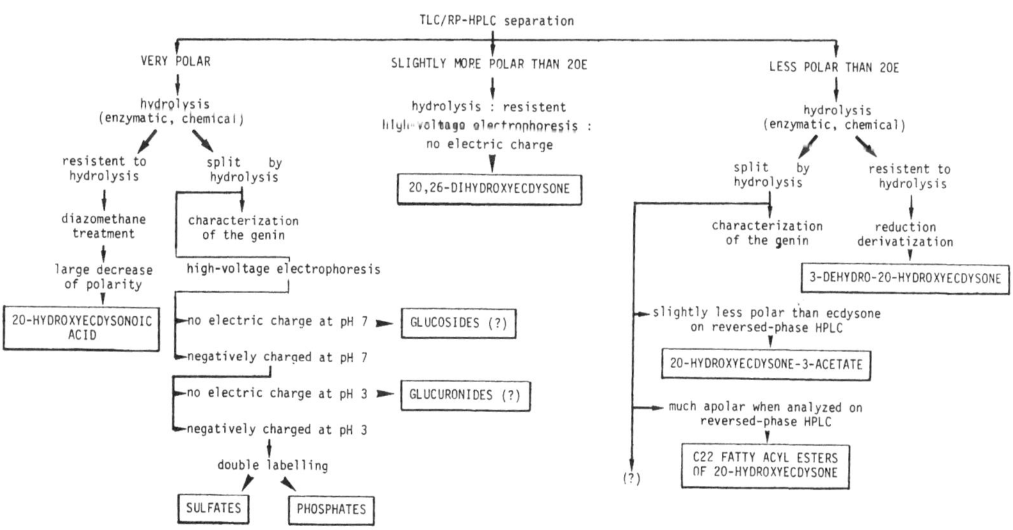

Fig. 2. General strategy for metabolite characterization using ^{3}H-20-hydroxyecdysone (20E) as starting material

to split a conjugate does not prove that it is a sulfoconjugate. In this case, only negative experiments can allow us to exclude some types of conjugates!

3. The Different Ecdysteroids and Their Metabolic Relationships

We have specified in Table 2 the ecdysteroids that were isolated from Arthropods as endogenous compounds or that were the result from in vivo or in vitro experiments. In the latter case, it is not always evident whether these ecdysteroids are true endogenous compounds. The formulas are given in Fig. 3. It might be worth mentioning that the isolation of metabolic end-products or storage forms is of course easier than the isolation of transient intermediates that are present in only low amounts (Table 2, p. 202 to 203; Fig. 3, p. 204 and 205).

3.1 20-Hydroxyecdysone and Its Potential Precursors

The major ecdysteroid found in nearly all species so far studied is 20-hydroxyecdysone (2). It usually arises from its immediate precursor, ecdysone (1). The entire biosynthetic scheme from cholesterol is not yet fully elucidated, but in several cases the final steps of hydroxylation are known [81, 82]. Several compounds lacking one or more hydroxyl groups have been isolated (3–6), but this area will not be discussed in this review, which is more concerned with the further conversion of ecdysone and 20-hydroxyecdysone. Conversion of ecdysone into 20-hydroxyecdysone is a general reaction occurring in all Arthropods so far analyzed. It also takes place in non-Arthropods since it was detected also in the molluscan *Cepaea nemoralis* [83]. According to developmental stage or species, 20-hydroxylation is more or less efficient (see Sect. 3.1), which allows the transient accumulation of ecdysone at some periods, e.g., in young pupae of Lepidoptera [84] and in embryos of Orthoptera [85].

The subsequent conversion of ecdysone and 20-hydroxyecdysone gives rise to parallel series of metabolites that are interconnected, as 20-hydroxylation can be performed on many different ecdysone metabolites.

3.2 26-Hydroxyecdysteroids and Ecdysonoic Acids

Ecdysone (1) and 20-hydroxyecdysone (2) are converted into the corresponding ecdysonoic acids [ecdysonoic acid (10) and 20-hydroxyecdysonoic acid (11)] via 26-hydroxylated metabolites [26-hydroxyecdysone (7) and 20,26-dihydroxyecdysone (8)]. The major compound accumulating is 20-hydroxyecdysonoic acid (11), the end-product of a pathway already found in many insect species, including Lepidoptera (*Pieris brassicae* [31, 45] and *Spodoptera littoralis* [63]), Diptera (*Calliphora vicina* [31, 44]), Orthoptera (*Locusta migratoria* [31, 35] and *Schistocerca gregaria* [63]), Phasmoptera (*Carausius morosus* [31]), and Coleoptera (*Tenebrio molitor* [52]) and also in Crustacea (*Carcinus maenas* [64]). The many reports on nonhydrolyzable polar metabolites in other species lead us to assume that this pathway is a general reaction in Arthropods.

Compounds lacking a 25-hydroxyl group are converted similarly into the corresponding acid: thus, ponasterone A (3) is converted via inokosterone (9) into 25-deoxy-20-hydroxyecdysonoic acid (12) [31, 64]. Generally, 26-hydroxyecdysteroids which are transient compounds do not accumulate. There are, however, two known exceptions to this rule: the eggs of *Manduca sexta*, which contain large amounts of 26-hydroxyecdysone (7) [60], and the crab *Gecarcinus lateralis*, which apparently forms 20,26-dihydroxyecdysone (8), but no ecdysonoic acids [86].

Both 26-hydroxyecdysteroids and ecdysonoic acids bear an asymmetric carbon atom at C25, and thus can theoretically exist as 25R and 25S isomers. This has indeed been observed to be the case for inokosterone (9) [50, 87], 26-hydroxyecdysone (7) [88], and 25-deoxy-20-hydroxyecdysonoic acid (12) [31]. So far no conjugates or other metabolites of ecdysonoic acids were found. Apparently they are end-products of ecdysone metabolism. Ecdysonoic acids are inactive in the *Calliphora* bioassay [31], thus they may be considered as major inactivation products of the molting hormone (see Sect. 5.2).

3.3 3-Dehydroecdysteroids and 3-Epiecdysteroids

3.3.1 3-Dehydroecdysteroids

In some insect species, such as the blowfly *Calliphora vicina*, the injection of ecdysone or 20-hydroxyecdysone leads to the formation of apolar metabolites. However, the products which were identified as 3-dehydroecdysone (13) and 3-dehydro-20-hydroxyecdysone (14) [65, 66] were detected in minor quantities only, and so far have not been isolated from untreated animals. Thus, their chemical identification was performed on compounds prepared from ecdysone and 20-hydroxyecdysone either enzymatically with an enzyme preparation from *Calliphora* or by chemical oxidation [66, 89]. Some evidence for their endogenous occurrence results from long-term labelling experiments with (^{3}H)-cholesterol in *Pieris brassicae* larvae in which significant amounts of labelled 3-dehydro-20-hydroxyecdysone (14) were observed [90]. Recently, 3-dehydro-2-deoxyecdysone (15) was tentatively identified as a minor metabolite in *Locusta* embryos [48]. This compound would be identical with silenosterone, a phytoecdysteroid already isolated from *Silene praemixta* (Caryophyllaceae) [91].

3-Dehydroecdysteroids comigrate on TLC with the corresponding 3-acetates (see Sect 3.5.1). For this reason the identification of 3-dehydroecdysteroids requires additional experiments, e.g., a combination of TLC and color reactions [89] or efficient HPLC systems. Consequently, reports on their occurrence that are merely based on TLC are questionable. It appears that the presently available data do not allow to conclude that 3-dehydroecdysteroids generally occur in untreated insects.

3.3.2 E-Epiecdysteroids

20-Hydroxyecdysone and most of its metabolites belong to the 3β series. These ecdysteroids, however, undergo an irreversible epimerization into 3α compounds in some species. Such a reaction leads to a parallel family of compounds comprising also 26-hy-

Table 2. Ecdysteroids isolated from arthropods or obtained from in vivo or in vitro experiments (formulas in Fig. 3)

Number	Name	Suggested abbreviation	Species	Stage[a]/Organ	References
	20-Hydroxyecdysone and potential precursors				
(1)	Ecdysone (= 2β,3β,14α,22R,25-pentahydroxy-5β-cholest-7-en-6-one)	E	*Bombyx mori*	P	[3]
(2)	20-Hydroxyecdysone	20E	*Bombyx mori*	P	[53, 54]
			Jasus lalandei	Total	[55]
(3)	Ponasterone A (25-deoxy-20-hydroxyecdysone)	PoA	*Callinectes sapidus*	E	[56]
(4)	2-Deoxyecdysone	2dE	*Bombyx mori*	Ovaries	[57]
(5)	2-Deoxy-20-hydroxyecdysone	2d20E	*Jasus lalandei*	Total	[58]
(6)	2,22-Dideoxy-20-hydroxyecdysone	222d20E	*Bombyx mori*	Ovaries	[59]
	26-Hydroxy compounds				
(7)	26-Hydroxyecdysone	26E	*Manduca sexta*	E	[60]
(8)	20,26-Dihydroxyecdysone	2026E	*Manduca sexta*	P	[61]
(9)	Inokosterone	I	*Callinectes sapidus*	Total	[62]
	Ecdysonoic acids				
(10)	Ecdysonoic acid (ecdysone 26-oic-acid)	EOIC	*Schistocerca gregaria*	E	[63]
(11)	20-Hydroxyecdysonoic acid	20EOIC	*Schistocerca gregaria*	E	[63]
			Pieris brassicae	P[b]	[31]
(12)	25-Deoxy-20-hydroxyecdysonoic acid	IOIC	*Pieris brassicae*	Imaginal disks[b]	[31]
			Carcinus maenas	Total[b]	[64]
	3-Dehydro compounds				
(13)	3-Dehydroecdysone	3DE	*Calliphora vicina*	Cytosol[b]	[65]
(14)	3-Dehydro-20-hydroxyecdysone	3D20E	*Calliphora vicina*	Cytosol[b]	[66]
(15)	3-Dehydro-2-deoxyecdysone (silenosterone)	3D2dE	*Locusta migratoria*	E	[48]
	3-Epi-ecdysteroids				
(16)	3-Epiecdysone (= 3α-ecdysone)	E′	*Manduca sexta*	P	[67]
(17)	3-Epi-20-hydroxyecdysone	20E′	*Manduca sexta*	P	[68]
(18)	3-Epi-2-deoxyecdysone	2dE′	*Schistocerca gregaria*	E	[69]
(19)	3-Epi-26-hydroxyecdysone	26E′	*Manduca sexta*	E	[41]

(20)	3-Epi-20,26-dihydroxyecdysone	2026E′	*Manduca sexta*	P	[67]
	Side chain cleavage product				
(21)	Poststerone	P	*Calliphora stygia*		[70]
	Acyl esters				
(22)	Ecdysone 3-acetate	E3A	*Schistocerca gregaria*	E	[71]
(23)	20-Hydroxyecdysone 3-acetate	20E3A	*Locusta migratoria*	A [b]	[35]
(24)	20-Hydroxyecdysone 22-acyl ester (-palmitate, -stearate, -oleate,...)	20E22-(acyl)	*Ornithodoros moubata*	N [b]	[72]
	Phosphate esters				
(25)	2-Deoxyecdysone 22-phosphate	2dE22P	*Schistocerca gregaria*	E	[34]
(26)	2-Deoxy-20-hydroxyecdysone 22-phosphate	2d20E22P	*Locusta migratoria*	E	[73]
			Schistocerca gregaria	E	[74]
(27)	Ecdysone 22-phosphate	E22P	*Schistocerca gregaria*	E	[34]
(28)	20-Hydroxyecdysone 22-phosphate	20E22P	*Schistocerca gregaria*	E	[74]
			Locusta migratoria	E	[73]
(29)	3-Epi-2-deoxyecdysone 3-phosphate	2dE′3P	*Locusta migratoria*	E	[75]
(30)	3-Epiecdysone 3-phosphate	E′3P	*Locusta migratoria*	E	[76]
	Nucleotides				
(31)	2-Deoxyecdysone 22-adenosinemonophosphate	2dE22AMP	*Locusta migratoria*	E	[75]
(32)	Ecdysone 22-adenosinemonophosphate	E22AMP	*Locusta migratoria*	E	[76]
(33)	Ecdysone 22-N^6-(isopentenyl)-adenosine-monophosphate	E22iAMP	*Locusta migratoria*	E	[77]
	Acetophosphates				
(34)	Ecdysone 3-acetate 2-phosphate	E3A2P	*Schistocerca gregaria*	E	[78]
(35)	20-Hydroxyecdysone 3-acetate 2-phosphate	20E3A2P	*Locusta migratoria*	A [b]	[35]
(36)	20-Hydroxyecdysone 3-acetate 22-phosphate	20E3A22P	*Locusta migratoria*	E	[73]
				L [b]	[79]
	C28 Ecdysteroids				
(37)	Makisterone A (24-methyl-20-hydroxyecdysone)	MaA	*Oncopeltus fasciatus*	E	[80]
(38)	Callinecdysone B	CaB	*Callinectes sapidus*	Total	[62]

[a] Abbreviations: *E* embryos; *L* larvae; *P* pupae; *N* nymphs; *A* adults
[b] Denotes that the compound was isolated from an in vitro or an in vivo experiment, not as an endogenous compound

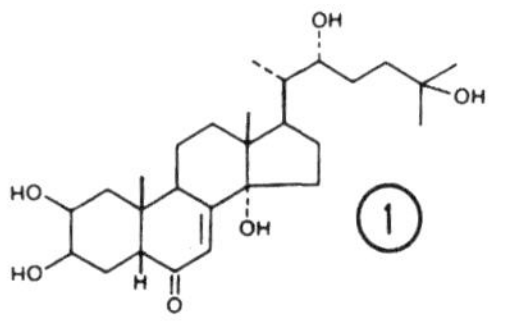
(1)
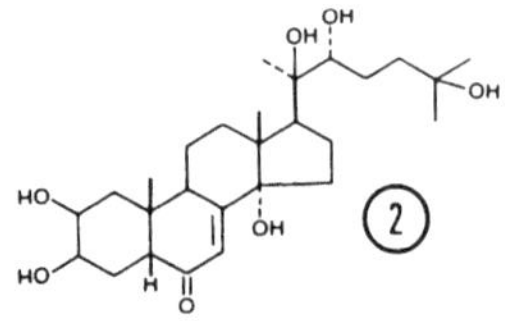
(2)
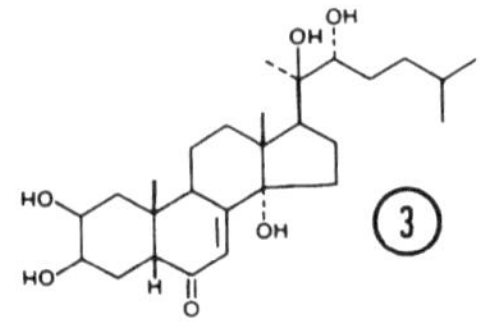
(3)
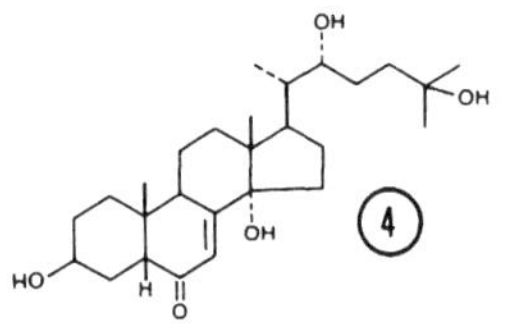
(4)
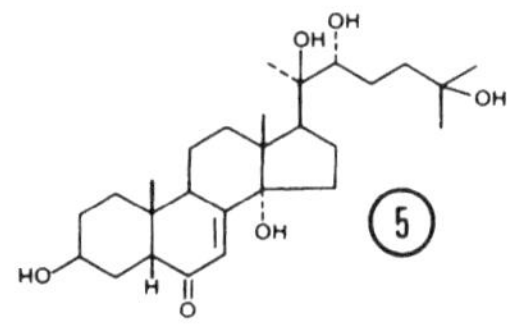
(5)
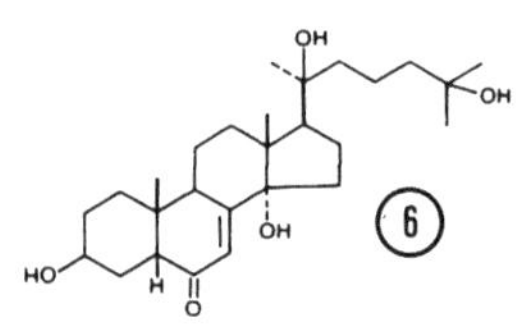
(6)
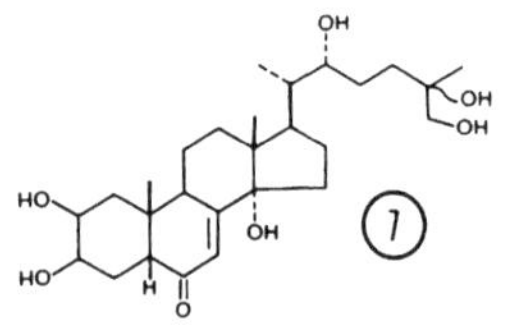
(7)
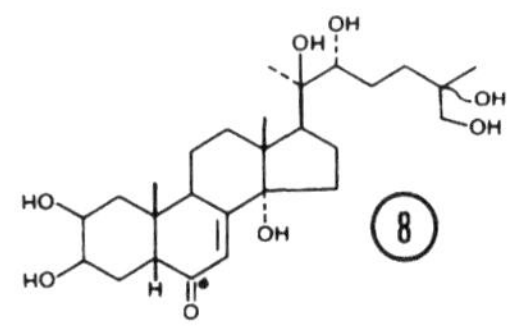
(8)
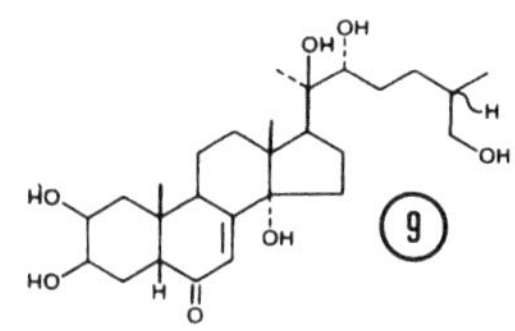
(9)
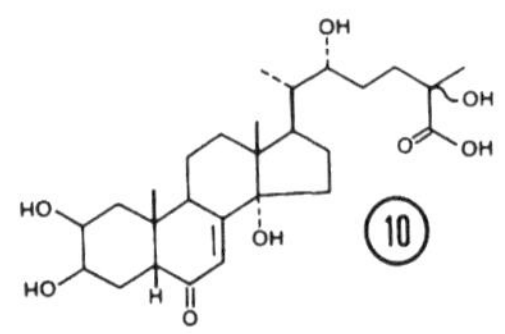
(10)
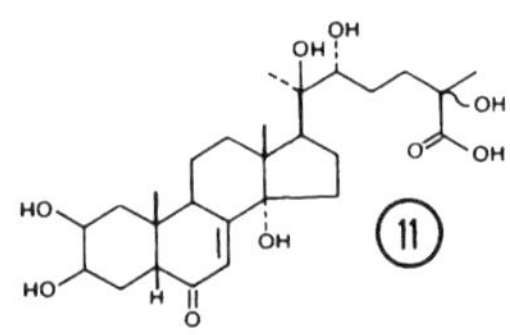
(11)
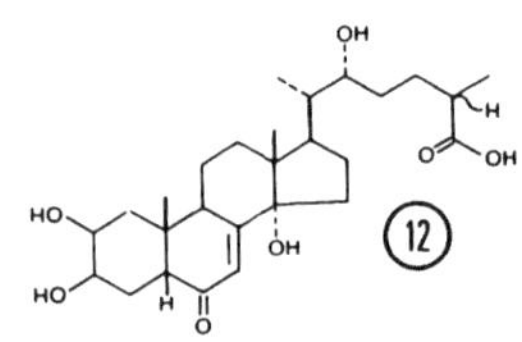
(12)
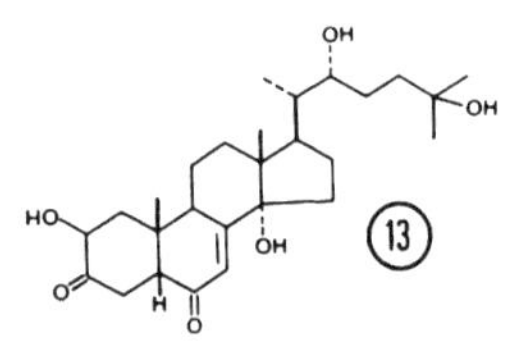
(13)
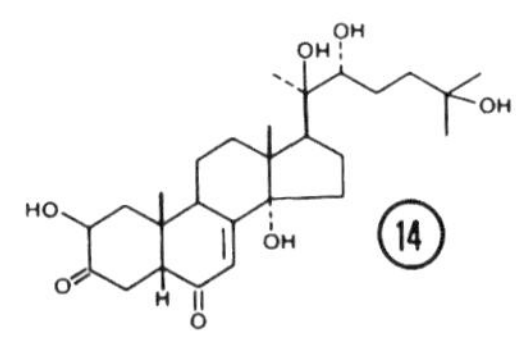
(14)
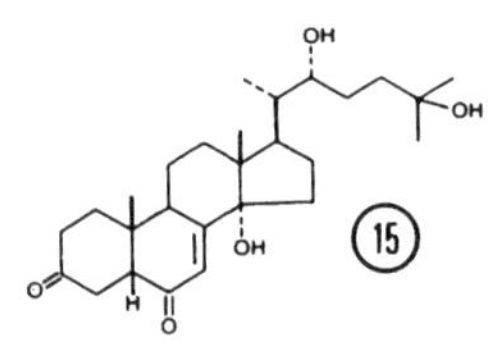
(15)
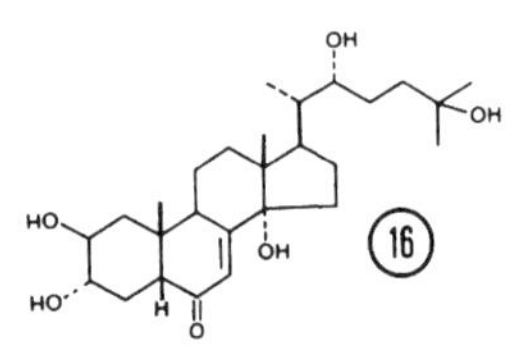
(16)
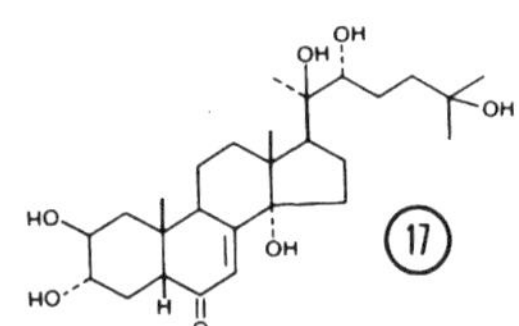
(17)
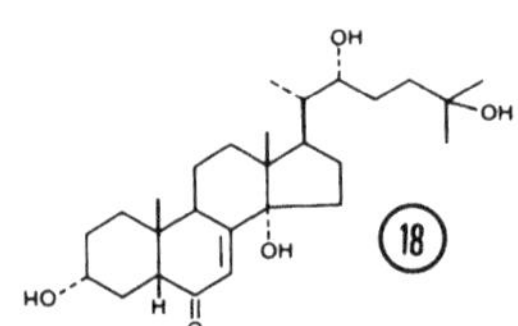
(18)
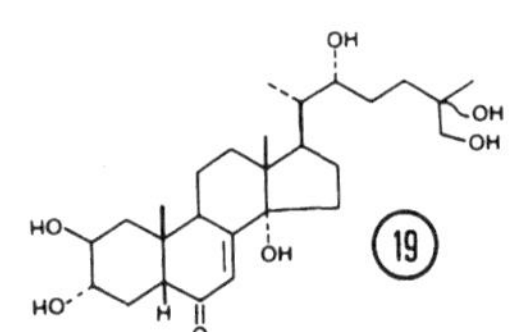
(19)
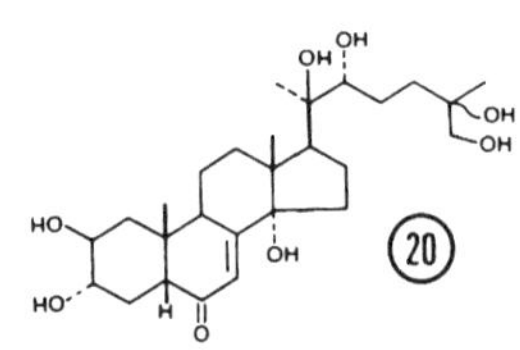
(20)
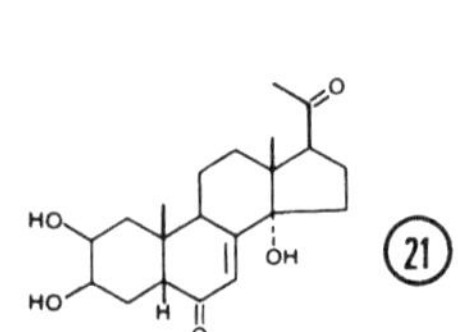
(21)

Fig. 3. Structural formulae of the presently isolated zooecdysteroids (see Table 2 for names)

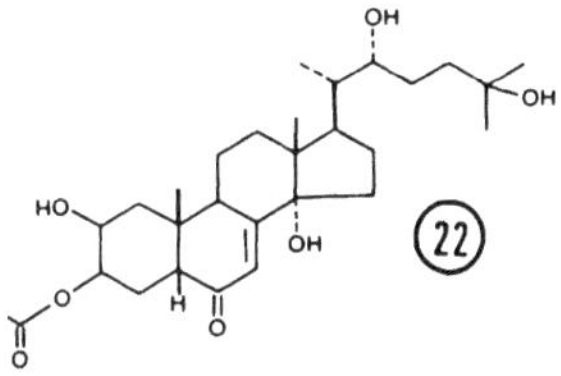
(22)

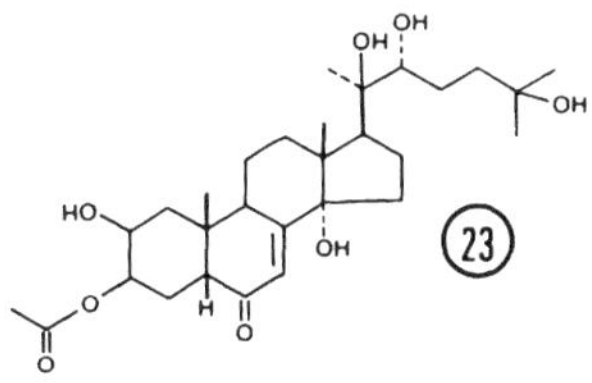
(23)

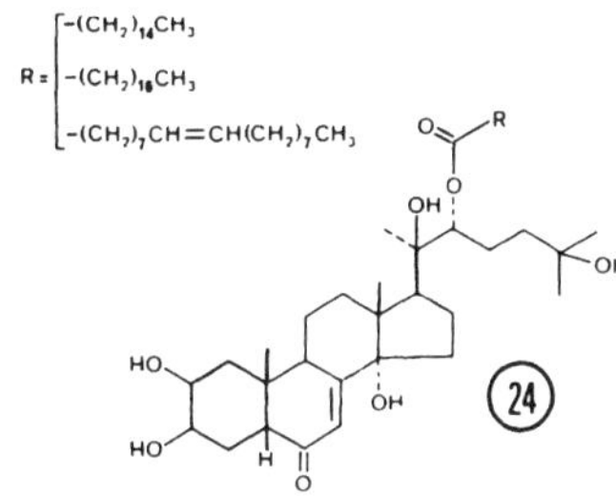

(24)

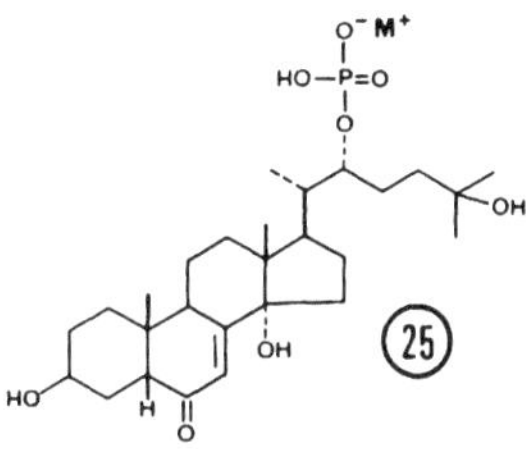
(25)

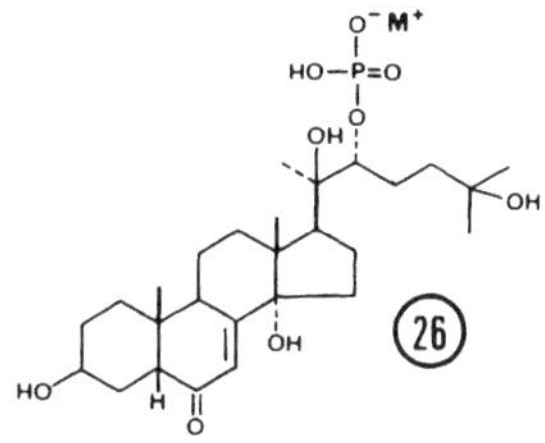
(26)

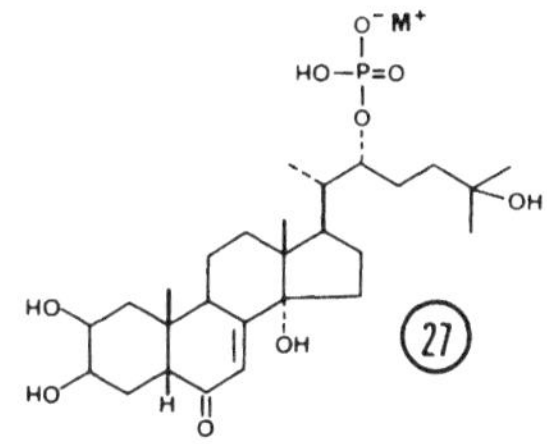
(27)

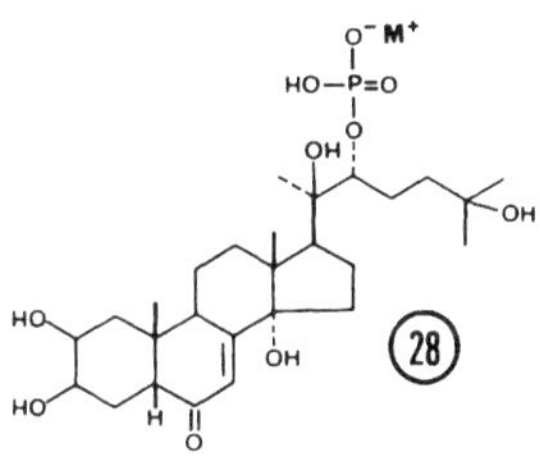
(28)

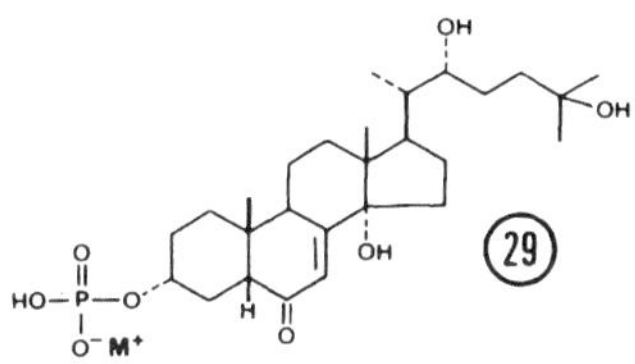
(29)

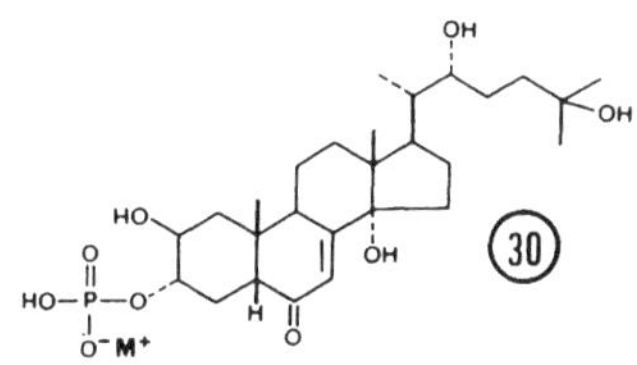
(30)

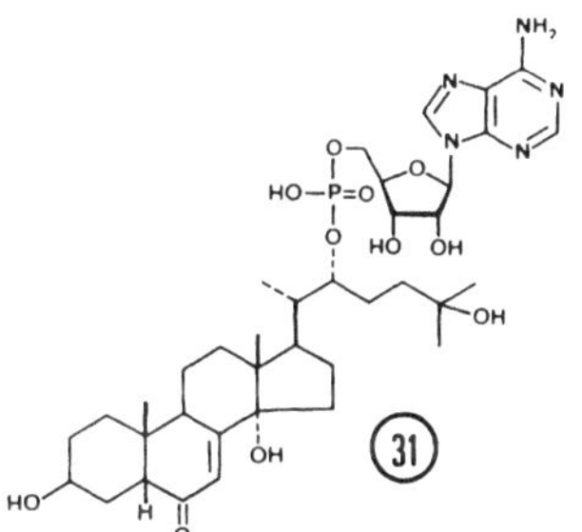
(31)

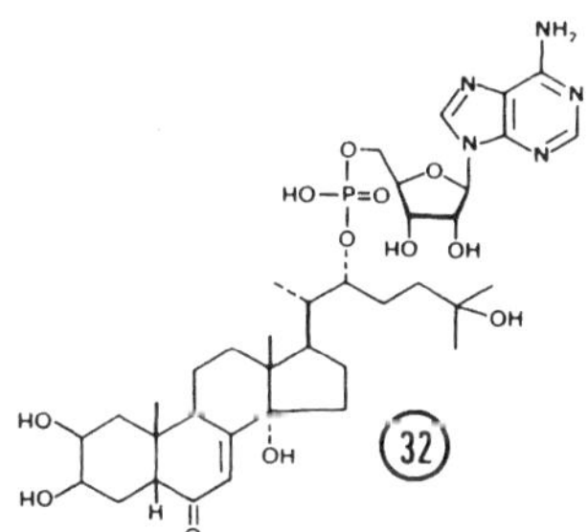
(32)

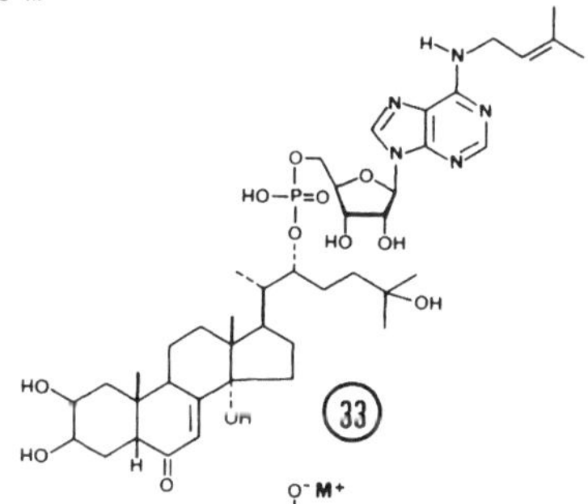
(33)

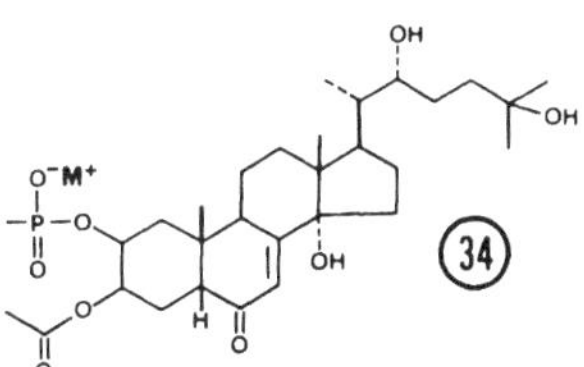
(34)

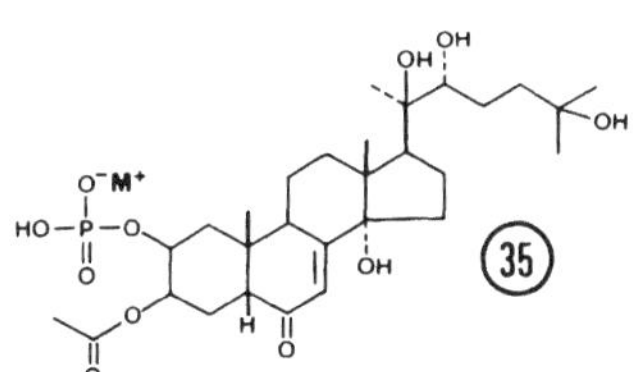
(35)

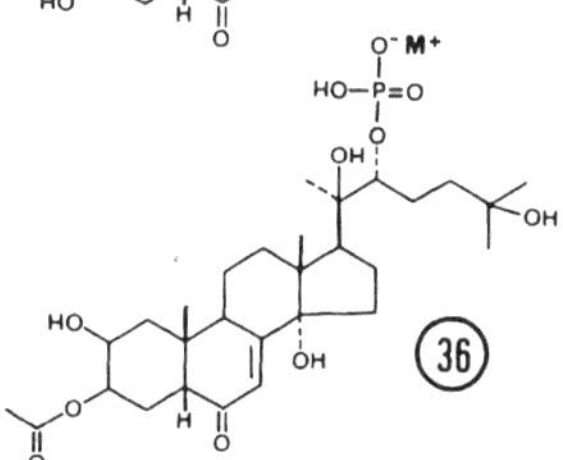
(36)

Table 3 (continued)

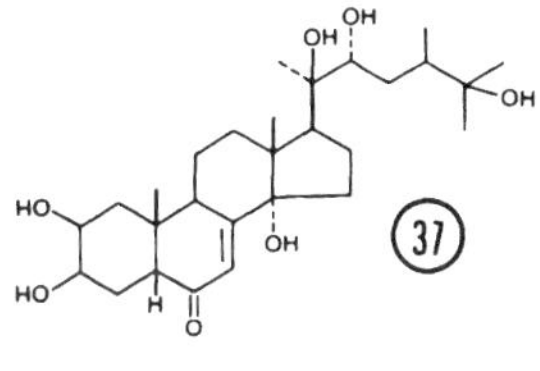
(37)

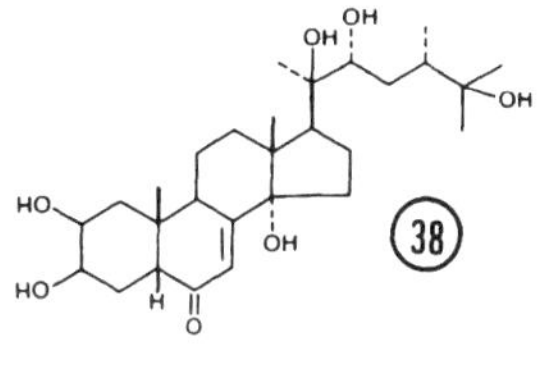
(38)

droxyecdysteroids and ecdysonoic acids [45]. So far 3-epiecdysone (16) [67], 3-epi-20-hydroxyecdysone (17) [68], 3-epi-26-hydroxyecdysone (19) [41], and 3-epi-20,26-dihydroxyecdysone (20) [67] have been isolated from pupae or embryos of *Manduca sexta*, and 3-epi-2-deoxyecdysone (18) from locust embryos [69]. There is also evidence for the formation of 3-epi-ecdysonoic acids in *Pieris brassicae* pharate pupae [45]. 3-Epiecdysteroids can exist in free or conjugated form (see Sect. 3.5). It is assumed that they arise through enzymatic reduction of 3-dehydroecdysteroids (see Sect. 4.3).

3.4 Side Chain Cleavage Product

Side chain scission between C20 and C22 is a classical reaction involved in the biosynthesis of vertebrate steroid hormones. In insects, however, it represents at best a minor reaction, leading to poststerone (21) [70, 87]. It is difficult to decide at present the generality of this assumption: in most studies of ecdysone metabolism, side chain labelled (23,24-^{3}H)-ecdysteroids were used. A potential side chain cleavage product would be devoid of the side chain and would, thus, escape detection. Attempts to demonstrate the formation of ^3H_2O from the catabolism of the side chain in larvae and pupae of the blowfly were, however, negative [92]. Similarly, long-term labeling experiments with nucleus-labelled cholesterol using *Pieris* pupae did not give any evidence for the formation of poststerone [84]. Recently, vertebrate-type C19 and C21 steroids were detected in the hemolymph of the flesh-fly *Sarcophaga bullata* [93, 94]. Indirect evidence suggests that they do not originate from the food of the flies. Whether or not these steroids derive from ecdysteroids by side chain cleavage remains to be clarified. In this respect, it would be of interest to investigate this problem with molluscs, which also contain both ecdysteroids [4, 181] and vertebrate-type steroids [95].

3.5 Conjugates

Although the presence of conjugates was suggested early [32, 96], and evidence for their presence was obtained through enzymatic hydrolysis experiments, the first isolation of a conjugate is only recent [71]. Up to now, only two classes of conjugates have been isolated, i.e., acyl esters and phosphate esters. The presence of many other classes has been conjectured (glucosides, glucuronides, sulfates), mainly on the basis of enzymatic hydrolysis (but see Sect. 2), but there is presently no unequivocal evidence for the presence of such conjugates.

The ecdysone molecule bears three secondary alcohol functions at C2, C3, and C22, respectively, and two tertiary ones at C14 and C25, which are much less reactive. Secondary alcohol functions are able to react and form esters with acids or ethers with sugars, and it appears logical to assume that conjugating mechanisms primarily involve these positions. The reactiveness of C2, C3, and C22 hydroxyl groups is different, as judged from chemical acetylation experiments [40] that gave the following sequence: C2 $\gg$ C22 $>$ C3. Thus, the C2 position is the most reactive one, and would be preferentially involved in nonspecific conjugating reactions.

3.5.1 Nonpolar Conjugates

The first conjugate isolated was an apolar conjugate, ecdysone 3-acetate (22) from *Schistocerca gregaria* embryos [71]. The same type of conjugate is also found for 20-hydroxyecdysone: 20-hydroxyecdysone 3-acetate (23). At the present time, 3-acetates have only been demonstrated in locusts, but it may be assumed that they occur in other insect groups also. Ticks (Arachnida) also form apolar conjugates [97, 98] that correspond to fatty acyl esters. Recently, 20-hydroxyecdysone esters (at C22) with stearic, oleic, and palmitic acids (24) were isolated as major metabolites [72]. It is important to note that such compounds are rather resistent to hydrolysis with *Helix pomatia* digestive juice and are better cleaved with (pig liver) esterases [39, 72].

3.5.2 Polar Conjugates

The more "classical" conjugates correspond to polar hydrolyzable metabolites, including a variety of compounds. All the identified compounds contain a phosphate group [compounds (25-36)].

Young locust embryos contain large amounts of conjugates that have been identified as the 22-phosphates of 2-deoxyecdysone (25), 2-deoxy-20-hydroxyecdysone (26), ecdysone (27), and 20-hydroxyecdysone (28). In *Locusta migratoria*, the occurrence of phosphodiesters with an additional adenosine moiety [compounds (31-33)] has been reported. These labile compounds would be transformed into the corresponding 22-phosphates early in embryonic development [99]. Conjugation always involves the C22 position in all these "maternal" (transferred by the mother to the egg) compounds. By contrast, conjugation mechanisms affect C2 and C3 positions during late embryonic and post-embryonic development. Phosphate ester formation involves the C2 position in ecdysteroids of the 3β series, and the C3 position in ecdysteroids of the 3α series, at least for the presently isolated conjugates, but any generalization would be premature. Once again, locust embryos have been the source of most of the isolated compounds.

C2-phosphates so far isolated bear an additional acetyl group in C3 position and are, therefore, complex conjugates: two such compounds have been isolated, the 3-acetoxy-2-phosphates of ecdysone (34) and of 20-hydroxyecdysone (35). They have been found both in old embryos, larvae, and adult of locusts [35, 78]. 3-Phosphates of 3-epiecdysteroids have so far been isolated from locusts (old embryos) only: they are the 3-phosphates of 3-epi-2-deoxyecdysone (29) and of 3-epi-ecdysone (30) [75, 76].

Evidence for the formation of phosphate esters in other groups has been obtained by double-labelling experiments ($^{3}H/^{32}P$) or by using selective hydrolysis procedures on conjugates. From these experiments, it can be concluded that phosphate-containing conjugates are formed in Lepidoptera (*Manduca sexta* [38], *Pieris brassicae* [100]) and in Diptera (*Calliphora vicina* [44, 100]). These compounds have not yet been fully characterized.

3.6 C28 Ecdysteroids

It is well established that insects are unable to synthesize the sterol nucleus [101, 102]. Instead they depend on dietary sources for their sterol supply. The major sterol source for the biosynthesis of ecdysteroids in insects is cholesterol. Phytophagous insects also use phytosterols which are C28 or C29 sterols with an additional methyl or ethyl group in the side chain [12, 103, 104]. The phytosterols in most cases are dealkylated to form cholesterol. In some specialized species the dealkylation for unknown reasons does not occur. Instead, the C28 sterols are converted into an analogue of the molting hormone, 24-methyl-20-hydroxyecdysone [makisterone A (37)] [80, 104, 105]. It is conceivable that these insects (bugs) have developed an interesting homologous pathway of ecdysteroid biosynthesis by slight modification of their enzymes. Concomitantly, the hormone receptors of these bugs changed their specificity (Table 5).

A C28 ecdysteroid [callinecdysone B (38)] has been isolated from a crustacean species *(Callinectes sapidus)* also [62]. It is likely that with time more structurally modified molting hormones will be found which will reflect how evolutionary constraints were overcome by biochemical adaptations of the organisms.

3.7 Metabolite Patterns

Animals at a given developmental stage display a more or less complex pattern of metabolites due to the possible simultaneous presence of several of the above-mentioned conversions. This will be exemplified by some typical HPLC chromatograms taken from experiments with various insects. The simplest ecdysteroid metabolite pattern is obtained in young pupae of *Pieris brassicae* (Fig. 4 A): the pathway leading to ecdysonoic acids in the only one present. The ecdysteroid pattern is rendered more complex in pharate pupae of the same species, in which additional epimerization occurs (Fig. 4 B). Other species contain both ecdysonoic acids and various conjugates, e.g., *Locusta migratoria* fifth instar nymphs (Fig. 4 C). Moreover, in some species metabolites are formed that do not correspond to presently identified compounds and, thus, do not fit into the above described pathways: the examples are from *Carausius morosus* (Fig. 4 D) and *Tenebrio molitor* (not shown) [52]. Clearly more pathways of the ecdysone metabolism remain to be discovered.

Figure 5 summarizes the different reactions of the ecdysone metabolism presently known in invertebrates. Any combination of these reactions is theoretically feasible, although a combination of all has not yet been observed: thus, conjugates of ecdysonoic acids are not known, nor 3-acetates of the 3-epimers. Stereospecificity at C3 appears to be important, and the same is true for vertebrates, in which isomers are converted to different conjugates:

Vertebrates	Insects (tentative)
3α (5β series) → glucuronides	3α → phosphates
3α (5α series)	
3β ((Δ^5 series) → sulfates	3β → acetates

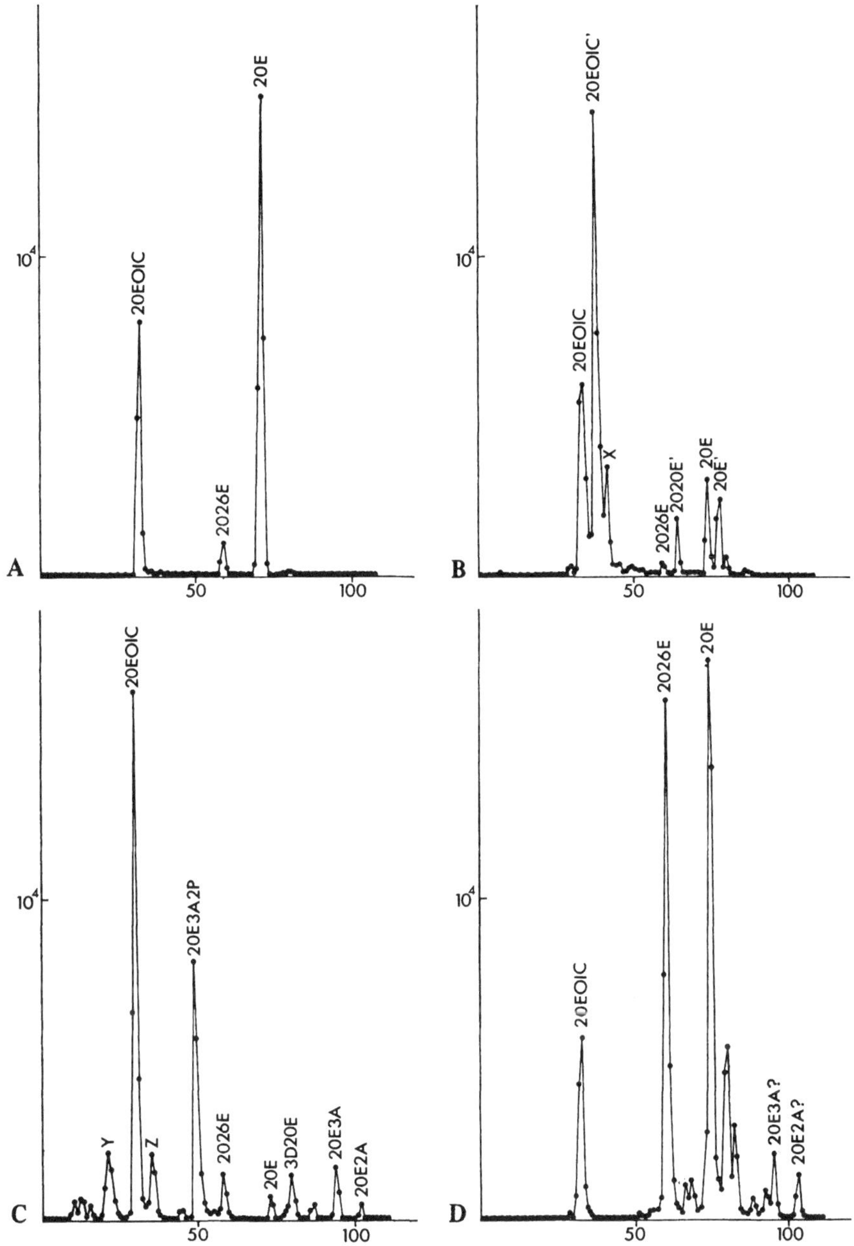

Fig. 4 A–D. HPLC separation of 20-hydroxyecdysone metabolites in various insect species. **A** *Pieris brassicae* pupae, **B** *Pieris brassicae* pharate pupae, **C** *Locusta migratoria* larvae (feces), **D** *Carausius morosus* larvae (feces). Operating conditions: column Ultrasphere-ODS (250 mm long, 4.6 mm i.d.) flow-rate 1 ml min^{-1}, fractions 0.4 ml, temperature: ambient. Elution with a linear gradient (8% to 40% in 60 min) of acetonitrile in a TRIS–$HClO_4$ buffer (20 m*M*, pH 7.5). *Abbreviations.* 20E: 20-hydroxyecdysone; 2026E: 20,26-dihydroxyecdysone; 20EOIC: 20-hydroxyecdysonoic acid; 20E′: 3-epi-20-hydroxyecdysone; 2026E′: 3-epi-20,26-dihydroxyecdysone; 20EOIC′: 3-epi-20-hydroxyecdysonoic acid; 3D20E: 3-dehydro-20-hydroxyecdysone; 20E3A: 20-hydroxyecdysone 3-acetate; 20E2A: 20-hydroxyecdysone 2-acetate. X: probably 3-epi-20-hydroxyecdysone 3-phosphate; Y: unidentified phosphate conjugate of 2026E; Z: unidentified phosphate conjugate of 20E

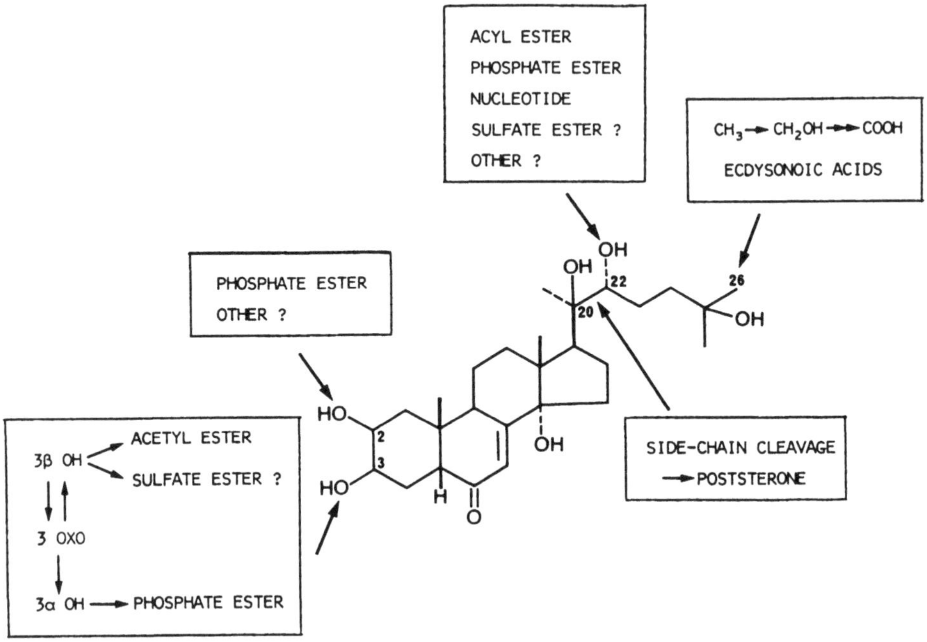

Fig. 5. Summary of the metabolic reactions that can affect the 20-hydroxyecdysone molecule. (?) means that the pathway has not been conclusively demonstrated

4. Enzyme Systems Involved in Ecdysteroid Metabolism

Although many metabolic conversions have been demonstrated to take place with ecdysteroids, the enzymes responsible have been characterized in a limited number of cases, and in a few species only (Table 3).

4.1 Ecdysone 20-Hydroxylase (Ecdysone 20-Monooxygenase)

This reaction is probably the best known one, and it represents an important step in the metabolism of ecdysteroids, as it converts ecdysone into 20-hydroxyecdysone, the metabolite which is believed to represent the active form of the molting hormone [122, 123] (see Sect. 4.2). The reaction is catalyzed by an enzyme system found in many tissues. Highest activities are detected in the fat body and the digestive tract. Remarkable is the dual localization of the system. It is found in mitochondria and the microsomal fraction, depending on the insect species and tissue (Table 3). Most studies have concerned the enzyme system from *Manduca sexta* fat body [106] and gut [38, 107] and from locust Malpighian tubules [13, 110, 111].

The enzyme system is a cytochrome P-450 containing monooxygenase [106, 111, 124]. For this reason ecdysone 20-hydroxylase is also called ecdysone 20-monooxygenase, although this terminology appears somewhat inaccurate, due to the fact that monooxygenases do not exclusively catalyze hydroxylation reactions. The enzyme

Table 3. Characteristics of the enzymes involved in ecdysone metabolism in insects[a]

Species/Stage	Tissue	Localization	Apparent K_m	References
Ecdysone 20-hydroxylase (EC 1.14.99.22)				
Manduca sexta larvae	Fat body	Mitochondria	1.6 10^{-7} *M*	[106]
Manduca sexta larvae	Midgut	Mitochondria	183 10^{-7} *M*	[107]
Manduca sexta larvae	Midgut	Mitochondria + microsomes	–	[38]
Pieris brassicae pupae	Imaginal discs	Mitochondria (?)	0.5 10^{-7} *M*	[108]
Calliphora vicina larvae	Total (in vivo)	–	4.5 10^{-7} *M*	[109]
Schistocerca gregaria larvae	Malpighian tubules	Mitochondria	7.1 10^{-7} *M*	[13, 110]
Locusta migratoria larvae	Malpighian tubules	Microsomes	2.7 10^{-7} *M*	[111]
Ecdysone oxidase (EC 1.1.3.16)				
Calliphora vicina larvae	Fat body	Cytosol	98.0 10^{-6} *M* (for E) 31.0 10^{-6} *M* (for 20E)	[112, 113]
Calliphora vicina larvae	Fat body, gut, ...	Cytosol	–	[42]
Musca domestica larvae	Total	Cytosol	–	[114]
Pieris brassicae	Imaginal disks, gut	Cytosol	–	[115]
Ecdysone 3-epimerase				
Manduca sexta larvae	Midgut	Cytosol	17.0 10^{-6} *M* (for E) 47.3 10^{-6} *M* (for 20E)	[104, 116, 117]
Musca domestica	Total	Cytosol	–	[114]
3-Dehydroecdysone reductase (3α-forming)[b]				
Schistocerca gregaria larvae	Various gut parts	Cytosol	–	[118]
Spodoptera littoralis larvae	Midgut	Cytosol	–	[119]
Pieris brassicae larvae	Gut	Cytosol	–	[115]
3-Dehydroecdysone reductase (3β-forming)[b]				
Schistocerca gregaria	Various tissues	Cytosol	–	[118]
Spodoptera littoralis	Midgut	Cytosol	–	[119]
Pieris brassicae	Imaginal disks, gut	Cytosol	–	[115]
Ecdysone sulfotransferase[b]				
Prodenia eridania larvae	Gut	Cytosol	–	[120]
Calliphora vicina	Fat body	Cytosol	–	[121]
Ecdysone Phosphotransferase[b]				
Manduca sexta larvae	Midgut	Cytosol	–	[38]

[a] This table does not take into account all the presently available data, in order to avoid redundant information
[b] Denotes that there is only preliminary evidence

system requires both oxygen and a reduced coenzyme (NADPH). The overall reaction can be written as follows:

$$\text{ecdysone} + \text{NADPH} + H^+ + O_2 \rightarrow \text{20-hydroxyecdysone} + \text{NADP}^+ + H_2O\ .$$

Substrate specificity is not restricted to ecdysone. Many other ecdysteroids can undergo 20-hydroxylation, e.g., 2-deoxyecdysone (4) probably also 2,22-dideoxyecdysone, as 2,22-dideoxy-20-hydroxyecdysone (6) has been isolated, 26-hydroxyecdysone (7), ecdysonoic acid (10), and 3-epiecdysone (16). The three last reactions have been observed in *Pieris brassicae* [45].

Ecdysone 20-monooxygenase activity varies during development. In *Locusta migratoria* Malpighian tubules, its activity has been measured throughout the last larval instar [36, 125], and it was observed that it was maximal at the time of the hemolymph 20-hydroxyecdysone peak. There was some (indirect) evidence that ecdysone could act as an inducer of the enzyme (substrate induction) [126]. It was also shown in the same species that juvenile hormone had no influence on the rate of conversion of injected ecdysone [127]. A different situation occurs in *Manduca sexta* last instar larvae [128]. In this species, enzyme activity was recorded both in fat body and midgut. It showed different profiles in these tissues that were both *out of phase* with 20-hydroxyecdysone levels [128]. Enzyme activities were maximal during the first half of the instar, i.e., several days before the hormonal peak, and strongly decreased at the time of the 20-hydroxyecdysone peak. Similarly, no correlation with 20-hydroxyecdysone levels was found in the case of *Calliphora vicina* [109] or *Schistocerca gregaria* [110]. It has been recently concluded from studies with *Manduca sexta* larvae that the increase of enzyme activity in the fat body is due to dietary factor(s), whereas the activity of the midgut enzyme is under endocrine cephalic control [128]. Thus, the important question of how the activity of ecdysone 20-hydroxylase is controlled is still controversial.

4.2 Ecdysone Oxidase

Ecdysone oxidase [89] catalyzes the irreversible oxidation of various ecdysteroids in the C3 position, according to:

$$\text{ecdysone} + O_2 \rightarrow \text{3-dehydroecdysone} + H_2O_2\ .$$

In vitro the hydrogen acceptor can be either oxygen or various dyes, like methylene blue [113]. Such a functioning is characteristic of "facultative oxidases" that are often flavoenzymes. However, this has not been proven for ecdysone oxidase. The enzyme has no equivalent in vertebrates: the 3α- and 3β-hydroxysteroid dehydrogenases (EC 1.1.1.50 and 1.1.1.51) from vertebrates (and microorganisms) involved in similar reactions are $NADP^+$ dependant and catalyze reversible reactions.

Ecdysone oxidase is a cytosolic enzyme [89]. It has been purified from *Calliphora vicina* pupae [113]. The enzyme has a relative molecular mass of 240,000, and an apparent K_m of 98.10^{-6} *M* for ecdysone. It can accept many ecdysteroids as substrates, but neither 22,25-dideoxyecdysone nor cholesterol [89, 113]. The enzyme from *Pieris brassicae* [115] (crude preparation) is also efficient with 26-hydroxyecdysteroids (7, 8), ponasterone A (3), and 2-deoxyecdysone (4). The steroid substrate of ecdysone oxidase is, thus, characterized by the presence of an unsaturated 6,7 ketone, a 22-hydroxyl group and a 3β-hydroxyl group: 3-epiecdysteroids are not substrates [115]. Ecdysone oxidase has been found in several insect species [129], and within an insect in many different tissues. In spite of the wide distribution and often high activities of ecdysone oxidase, 3-dehydroecdysteroids represent minor metabolites in vivo (see Sect. 3.3.1). This apparent contradiction which has already been pointed out [14] can now be explained, as described in the next section.

4.3 Ecdysone Epimerase and 3-Dehydroecdysone Reductases

The activity of an ecdysone epimerase [116] (also named ecdysone dehydrogenase/isomerase [117]) has been described in *Manduca sexta* midgut. The enzyme requires both oxygen and a reduced coenzyme [NAD(P)H]. It converts irreversibly ecdysone (3β-ecdysone) into its 3α isomer (3-epiecdysone). The epimerization reaction could involve two activities acting sequentially, i.e., an ecdysone oxidase (see Sect. 4.2) and a 3-dehydroecdysone reductase. However, in the *Manduca* system the formation of the postulated intermediate (3-dehydroecdysone) could not be observed [116]. In spite of this, the intermediary role of 3-dehydroecdysone appears a logical step by comparison with epimerization reactions in vertebrates. Other authors working with cytosolic preparations from Diptera [114] obtained some evidence for the simultaneous formation of 3-dehydroecdysone and 3-epiecdysone. This would support the two-step hypothesis, although in no case does it constitute a demonstration of the mechanism.

The problem was recently analyzed by incubating cytosolic preparations from various tissues of *Pieris brassicae* with tritiated 3-dehydroecdysone and a NADPH generating system [115]. The incubation resulted in a rapid reduction to 3-epiecdysone (3α form) *and* ecdysone (3β form). This finding supports the assumption that two reductases are present in this species. Parallel experiments with *Schistocerca gregaria* homogenates led to the same conclusion [118]. The experiments examplified in Fig. 6 and summarized in Fig. 7 indicate that: (1) many tissues contain a reductase which forms 3β-ecdysteroids – this would explain why 3-dehydroecdysteroids do not accumulate in vivo, even in insects that contain an active ecdysone oxidase and do not form 3-epimers; (2) some tissues contain an additional reductase which reduces irreversibly 3-dehydroecdysteroids into 3α-ecdysteroids – this enzyme seems to be restricted to the gut tissue, both in *Pieris* [115] and *Schistocerca* [118]. The most intriguing finding is the presence of both ecdysone oxidase and of a 3β-ecdysteroid forming reductase in several tissues [115]. Potentially these two enzymes can catalyze a futile cycle that produces noxious hydrogen peroxide at the expense of reduced nicotinic coenzymes. The cycle is characterized by the following equation:

$$NAD(P)H + H^+ + O_2 \rightarrow NAD(P)^+ + H_2O_2 ,$$

and its physiological significance is at present unclear. The occurrence of the combination of the two enzymes is likely in Orthoptera (*Schistocerca gregaria* [118] and *Locusta migratoria* [35]) and in Lepidoptera (*Pieris brassicae* [115] and *Spodoptera littoralis* [119]). Also in *Calliphora vicina* a rapid reduction of injected 3-dehydroecdysone occurred in vivo, but in this case the TLC system used for metabolite separation did not separate unambiguously ecdysone and its 3α-epimer [130]. Rapid reduction in vivo of injected 3-dehydro-20-hydroxyecdysone has been observed in *Locusta migratoria* [35]

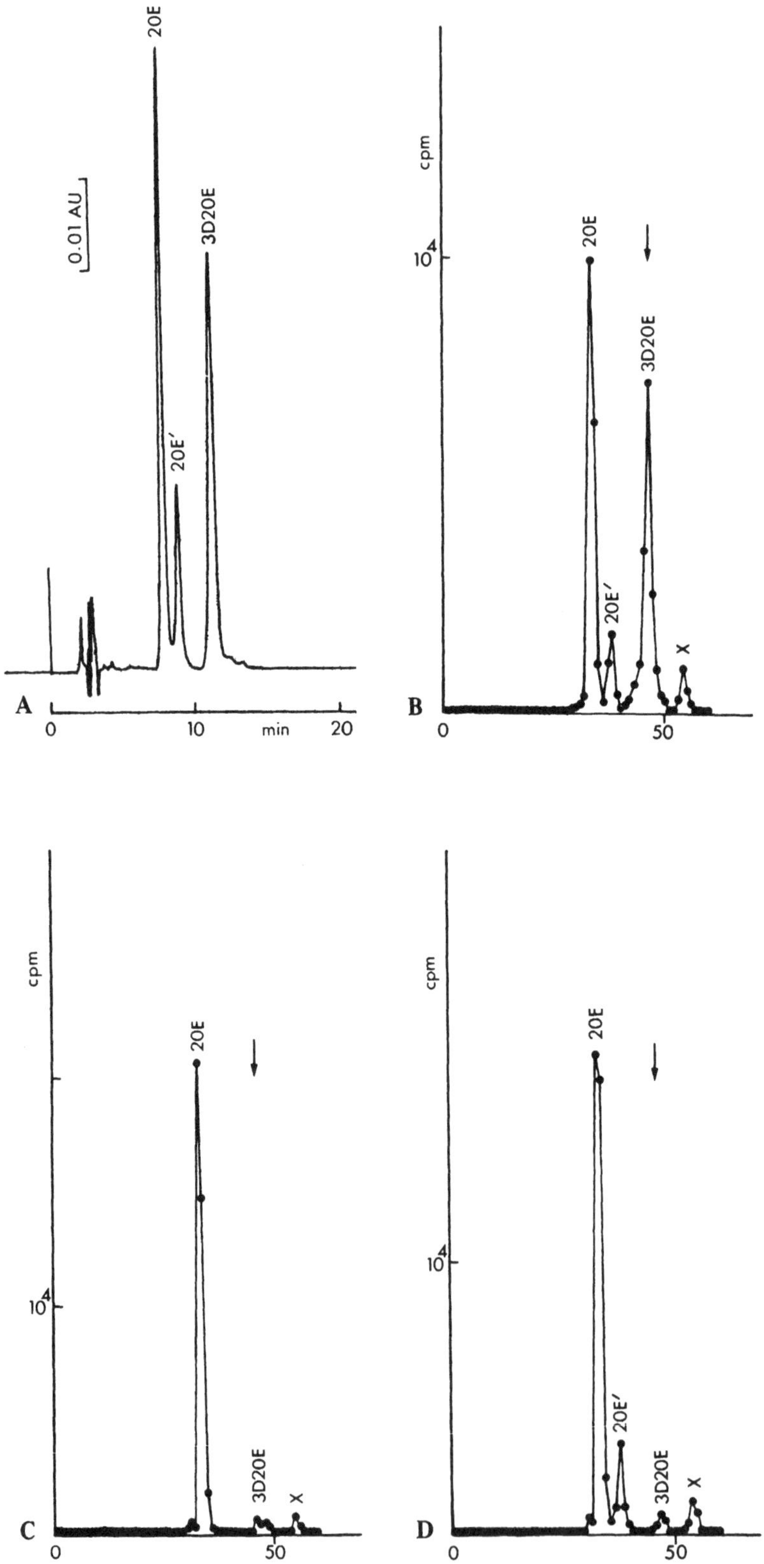
20E
0.01 AU
3D20E
20E′
A
0
10
min
20
cpm
10⁴
20E
3D20E
20E′
X
B
0
50
cpm
20E
10⁴
3D20E
X
C
0
50
cpm
20E
10⁴
20E′
3D20E
X
D
0
50

Fig. 6A–D

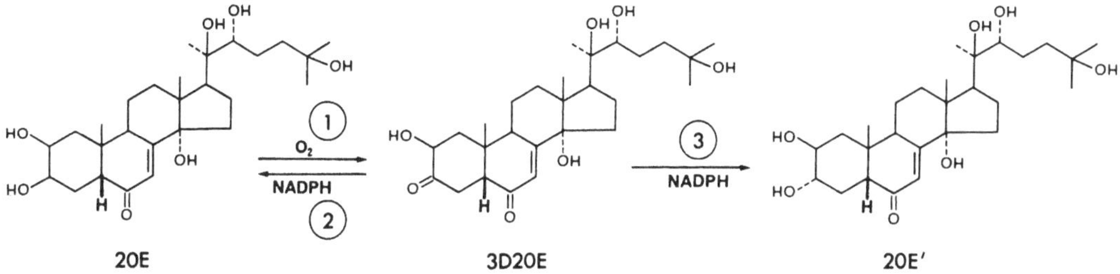

Fig. 7. Interconversion reactions between 20-hydroxyecdysone (20E), 3-dehydro-20-hydroxyecdysone (3D20E), and 3-epi-20-hydroxyecdysone (20E′). Enzymes: *1* ecdysone oxidase, *2* 3-dehydroecdysone reductase (3β-forming), *3* 3-dehydroecdysone reductase (3α-forming), *1* + *3* would correspond to the "epimerase" system

and *Pieris brassicae* [115]. The injected compound had almost entirely disappeared within 30 min following injection!

Clearly more experiments are needed to unravel the interrelationships between 3α-, 3β-, and 3-dehydro-ecdysteroids.

4.4 Enzymes Involved in the Formation of Ecdysonoic Acids

Although this pathway is general, nothing is known about the enzymes involved. In vitro experiments with homogenates or subcellular fractions from *Pieris brassicae* wing disks [108] or *Calliphora vicina* fat body [44] failed to demonstrate any 26-hydroxylation when using the same conditions as those used for 20-hydroxylation. This is surprising, because the same *intact* tissues are able to catalyze the reaction. It may well be that the enzyme system is not a hydroxylase, and that the reaction proceeds through an unsaturated intermediate. Such a mechanism could also explain the conversion of 20-hydroxyecdysone into inokosterone which takes place in *Calliphora vicina* fat body [43, 131].

Enzymes converting 26-hydroxyecdysteroids further into ecdysonoic acids are likewise still hypothetical [31]. Similar reactions leading to acidic steroids are also observed in vertebrates in the metabolism of vitamin D [132] and corticosteroids or progestins [133]. In this case, the responsible enzyme system is cytosolic or microsomal [134, 135].

4.5 Conjugating Enzymes

Ecdysteroid conjugates are formed enzymatically, but due to the only recent identification of ecdysteroid conjugates, the characterization of the enzymes responsible is

Fig. 6 A–D. Metabolism of 3-dehydro-20-hydroxyecdysone (3D20E) by cytosolic preparations from *Pieris brassicae*. (^{3}H)-3D20E was incubated with cytosol and a NADPH-generating system. HPLC conditions: column Ultrasphere-ODS, flow-rate 1 ml min^{-1}, solvent 19% acetonitrile, 0.4% acetic acid in water (isocratic), temperature +50 °C, fractions 0.3 ml. **A** Separation of reference ecdysteroids, **B** cytosol from whole pupae (6–7 day-old pupae), **C** cytosol from imaginal discs (6–7 day-old pupae), **D** cytosol from gut tissue (pharate pupae). Abbreviations as in Table 2. *X* unidentified metabolite

very preliminary. In fact, our best knowledge in this field results from studies of the enzymatic detoxication mechanisms of xenobiotics (e.g., pesticides) [136, 137]. In some of these studies the activity of these enzymes against ecdysteroids was also examined.

Fat body homogenates of *Calliphora vicina* and *Manduca sexta* apparently contain a *glucosyl-transferase* able to react with ecdysteroids [32, 33]. However, the enzyme does not seem to work with ecdysteroids in intact animals.

A cytosolic *sulfotransferase* has been reported to form sulfate esters in insects. It is found in the fat body of *Calliphora vicina* [121] and the gut of *Prodenia eridania* [120]. It works in vitro in the presence of ATP, Mg^{2+}, and sulfate. The sulfate donor is probably phosphoadenosine-phosphosulfate (PAPS) as with other sulfotransferases: the formation of PAPS has been demonstrated in insects [138].

Phosphate esters are widely distributed and we would expect that ATP is used as a phosphate donor. So far the presence of a *phosphotransferase* has been demonstrated in the Madagascar cockroach *(Gromphadorhina portentosa)* and the housefly *(Musca domestica)* [139]. Unfortunately the authors do not mention whether the enzyme works with ecdysteroids also. Recently it was shown that a midgut cytosol from *Manduca sexta* contained such an enzyme active with ecdysteroids as substrates [38].

Enzymes related with the formation of acyl esters have not been described so far.

Of course our present knowledge of conjugates is still preliminary and we may expect to find new types in the near future. In this respect, it is interesting to keep in mind that other conjugating mechanisms are already known in insects [136, 137]. These include conjugation with amino acids and with glutathione. Such reactions would transform ecdysteroids into very polar conjugates.

Finally, it must be remembered that also enzymes from the bacterial fauna of the gut might act on ecdysteroids. In vertebrates these enzymes are responsible of a number of reactions in the metabolism of bile acids and steroid hormones [140]. They include reductions, dehydroxylation of steroids, as well as hydrolysis of steroid conjugates [140]. Preliminary experiments with the gut content on *Locusta migratoria* [35] and *Calliphora vicina* [141] were essentially negative: neither ecdysone nor 20-hydroxyecdysone were metabolized. This does, however, not preclude the conversion of other steroids (e.g., of some conjugates).

5. Physiological Aspects of Ecdysone Metabolism

5.1 Tissue Localization of the Different Pathways

Ecdysone is the secretory product of molting glands, and it is not further metabolized by the gland themselves [33]. The various reactions described above take place in general or specific tissues, and this localization may vary according to species and/or developmental stage. Moreover, in the case of generally distributed reactions, a given tissue may have a predominant role due to a higher enzyme concentration/amount.

Ecdysone 20-monooxygenase is present in many tissues (Table 4). It can be detected in fat body, gut, Malpighian tubules, epidermis, or imaginal disks, but in every species

Table 4. Metabolism by arthropod tissues in vitro

Type of reaction	Tissue	Species	References
20-Hydroxylation	Malpighian tubules, fat body, gut hemocytes, gonads	*Locusta migratoria*	[142]
	Malpighian tubules, fat body, gut epidermis	*Locusta migratoria*	[35]
	Malpighian tubules, fat body	*Schistocerca gregaria*	[110]
	Leg regenerates	*Leucophaea maderae*	[143]
	Fat body	*Leucophaea maderae*	[144]
	Fat body, Malpighian tubules, midgut, integument	*Manduca sexta*	[33, 107]
	Fat body	*Calliphora vicina*	[43, 131]
	Imaginal wing disks	*Pieris brassicae*	[108]
	Gut (adult)	*Drosophila melanogaster*	[145]
	Epidermis, hepatopancreas, gut, testes, ovaries	*Carcinus maenas*	[146]
	Testes	*Pachygrapsus crassipes*	[147]
	Testes	*Cancer antennarius*	[148]
	Epidermis	*Astacus leptodactylus*	[149]
26-Hydroxylation (→ ecdysonoic acids)	Fat body, Malpighian tubules	*Manduca sexta*	[33]
	Fat body	*Leucophaea maderae*	[144]
	Fat body	*Calliphora vicina*	[43, 131]
	Imaginal wing disks	*Pieris brassicae*	[88]
3-Oxidation	Fat body, Malpighian tubules, gut, salivary glands, hemocytes	*Calliphora vicina*	[112, 113]
	Fat body, gut, hemocytes	*Locusta migratoria*	[36]
	Imaginal wing disks, gut	*Pieris brassicae*	[115]
3-Epimerization	Midgut	*Manduca sexta*	[116, 117]
	Gut	*Pieris brassicae*	[45]
3-Acetylation	Gastric ceca, midgut, rectum, fat body (adult only), epidermis (adult only)	*Locusta migratoria*	[35]
	Gut + Malpighian tubules	*Locusta migratoria*	[79]
22-Acylation	Ovaries	*Ornithodoros moubata*	[98]
Phosphorylation	Ovaries	*Locusta migratoria*	[150]
	Malpighian tubules (larvae) + fat body, epidermis, gut (adult)	*Locusta migratoria*	[35]
	Gut + Malpighian tubules	*Locusta migratoria*	[79]
	Midgut	*Manduca sexta*	[38]
Sulfatation	Gut (cytosol)	*Prodenia eridania*	[121]
	Fat body	*Calliphora vicina*	[122]
Glycosidation	Fat body	*Calliphora vicina*	[32]
	Fat body	*Manduca sexta*	[33]

This table has been restricted to be formation of identified metabolites. Experiments leading to non identified conjugates are not reported here

a tissue seems to have a predominant position. Locust Malpighian tubules are a major site of 20-hydroxylation [13, 35, 111]. On the other hand, fat body and gut are the major converting tissues in lepidopteran larvae [106, 107, 128]. One must, however, be cautious regarding the quantitative results obtained in most cases with homogenates that do not necessarily reflect the contribution of enzymes in situ.

The pathway to ecdysonoic acids via 26-hydroxyecdysteroids is also found in several tissues (Table 4). It was found in fat body, Malpighian tubules, and imaginal disks of Lepidoptera [35, 88]. It is, however, premature to decide whether these reactions occur in all target tissues.

Conjugation reactions take place essentially in the gut, Malpighian tubules, and fat body of animals (Table 4). They occur also in ovaries of reproductive females, but in this case the compounds formed seem different: in locusts, conjugation involves the C22 position in ovaries, whereas during post-embryonic development (or in the developing embryos) it involves rather C2 and C3.

An important problem is to determine whether all conjugating tissues of a given animal make the same conjugates or whether there is some tissue specificity. Presently available data are limited, but preliminary studies in *Locusta migratoria* [35] indicate that Malpighian tubules do not form the same conjugates as the gut, although both conjugates are phosphate esters.

A second important question is related to developmental changes affecting either the pathways or the tissues involved. In *Pieris brassicae*, it was observed (see Fig. 4A,B) that larvae form ecdysonoic acids, 3-epimers, and phosphate esters of the 3-epimers, whereas young pupae form only ecdysonoic acids and pharate adults form again conjugates [45, 84, 100]. On the other hand, in *Locusta* the pathways are identical in larvae and adults [35], and developmental changes are only quantitative: young fifth instar larvae excrete primarily ecdysonoic acids, whereas adults form predominantly acetophosphates [35]. Moreover, in vitro experiments with *Locusta* tissues have shown that the ability to form acetophosphates [compounds (34, 35)] is restricted to the gut in larvae, but that many other tissues (epidermis, fat body) become able to form these compounds in adults [35].

5.2 Biological Activity and Significance of the Different Metabolites

Ecdysone and its precursors or metabolites have been assayed for their biological activity, using in vivo or in vitro procedures. In vivo assays include the ability to induce molting (or pupation) or to break diapause, whereas in vitro assays include puffing assays on salivary gland polytene chromosomes, induction of morphological changes in *Drosophila* cell lines [168], ability to induce evagination or cuticle formation with imaginal discs [169, 170]. The results of these experiments are given in Table 5. Due to the recent isolation of conjugates, information about their biological activity is presently scarce. But first at all, we must discuss the significance of activity vs inactivity of ecdysteroids in bioassays.

Any active ecdysteroid can a priori be active either because it combines with the receptor, or because it is metabolized into an active molecule. For this reason, conclusions can be drawn only if the ecdysteroid is not metabolized during the bioassay. Metabolic conversion takes place in all in vivo assays, and this explains why ecdysone,

2-deoxyecdysone, 2-deoxy-20-hydroxyecdysone are as active as 20-hydroxyecdysone in vivo. On the other hand, these compounds show little activity (if any) in vitro assays. The same problem is encountered when using ecdysteroid conjugates that can be split by esterases or phosphatases.

Some metabolites are inactive both using in vivo or in vitro assays, and in that case it is considered that they are inactivation products. This is for instance the case for 3-epiecdysteroids (Table 5). Such a conclusion, however, would need to be substantiated by binding experiments. Actually, a metabolite can be inactive because it does not bind to the receptor, or because it does not enter the target cell. It is theoretically possible that a compound is inactive when externally applied, but active when present as an endogenous metabolite within the target cell. Moreover, a given metabolite may have a very specific function that cannot be evidenced by the presently used bioassays. All these facts lead us to be very cautious concerning the activity/inactivity of ecdysteroids.

Little data are presently available concerning the permeability of target cells towards ecdysteroids. In vertebrates, it is generally assumed that steroid hormones can freely enter target cells [171]. Ecdysteroids are more polar molecules, and concomitantly might have a different behavior. Thus, experiments performed with crayfish integument have provided some evidence for a carrier-mediated process [172, 173], which can be inhibited by sulfhydryl group inhibitors. In vitro experiments with *Pieris brassicae* imaginal disks have shown that they convert ecdysone into 26-hydroxyecdysone and ecdysonoic acid [88]. However, they do not metabolize externally applied 26-hydroxyecdysone, a compound which is unable to promote their differentiation [174]. In this case, it is probable that 26-hydroxyecdysone does not enter imaginal disc cells, and no definite conclusion can be drawn about its inefficiency as an endogenous metabolite.

The problem of ecdysteroid conjugates requires special attention. Presently available data lead us to distinguish two types of conjugates: *storage* conjugates and *excretory* conjugates. It is assumed that the first type of conjugates can be used as a source of free hormones further in development, whereas the second type represents waste products that are eliminated in feces. Both types of conjugates are found in locust embryos: C 22 phosphates and nucleotides represent a storage form, and acetates or acetophosphates, wastes. In this respect, it is noteworthy that locust embryos contain a phosphatase very active towards 22-phosphates, whereas acetophosphates are poor substrates [175]. In addition, experiments in *Sarcophaga peregrina* have given evidence that conjugates formed in pharate pupae could be split to free 20-hydroxyecdysone in pupae [176]. Thus, the existence of storage conjugates would not be restricted to embryos.

Excretion conjugates, on the other hand, are found in insect feces, and it is conceivable that the conjugation with a polar moiety leads to a more efficient excretion of the compound. This is, however, not a prerequisite for excretion, and insect feces contain also noticeable amounts of free ecdysteroids [177–179, 182]. Conjugation can be directed towards endogenous ecdysteroids or exogenous ones (phytoecdysteroids found in the animal's diet). This is for instance the case in *Locusta*. Ingested ecdysteroids undergo a 3-acetylation in gastric ceca, which is thought to inactivate them, and are thereafter converted to acetophosphates before being excreted [180].

Table 5. Biological activity of ecdysteroids

Number	Compound	Biological activity[a]	Type of bioassay[b]	References
20-Hydroxyecdysone and potential precursors				
[1]	Ecdysone (E)	7.5 ng	C	[3]
		20 ng	C	[151]
		5 ng	M	[152]
		5 ng	S	[153]
		5,000 ng	SM	[154]
		20E × 0.02–0.03	D	[155]
		20E × 0.002	DD	[170]
		20E × 0.006	Kc	[168]
[2]	20-Hydroxyecdysone (20E)	E × 1.25	C	[53]
		5 ng	C	[156]
		25,000 ng	CH	[157]
		3.5 ng	M	[158]
[3]	Ponasterone A (PoA)	E × ca. 1	C, S	[159]
		E × 3.2	S	[153]
		E × 0.5	C	[160]
		20E × 0.25	C	[156]
		E × 2	SM	[154]
		20E × 20	DD	[170]
		20E × 8	Kc	[168]
[4]	2-Deoxyecdysone (2dE)	20E × 1	C	[58, 161]
		20E × ~0.04	LD	[99]
		Inactive	DD	[170]
[5]	2-Deoxy-20-hydroxy-ecdysone (2d20E)	20E × 1	C	[58, 161]
		20E × 0.1	DD	[170]
		20E × 0.03	Kc	[168]
[6]	2,22-Dideoxy-20-hydroxy-ecdysone (222d20E)	E × 0.1	S	[59, 162]
26-Hydroxy compounds				
[7]	26-Hydroxyecdysone (26E)	Inactive	M	[41]
		(Corrects an earlier of hormonal activity)		[60]
[8]	20,26-Dihydroxyecdysone (2026E)	E × 0.07–0.10	M	[61]
		20E × 0.1	C	[163]
[9]	Inokosterone (I)	20E × 1	B	[160]
		E × 0.55	S	[153]
		E × 10	C	[160]
		20E × 0.07	DD	[170]
		20E × 0.01	Kc	[168]
Ecdysonoic acids				
[10]	Ecdysonoic acid (EOIC)	? (Probably inactive)		
[11]	20-Hydroxyecdysonoic acid (20EOIC)	? (Probably inactive)		
[12]	25-Deoxy-20-hydroxy-ecdysonoic acid (IOIC)	Inactive	C	[31, 44]

Table 5 (continued)

Number	Compound	Biological activity[a]	Type of bioassay[b]	References
3-Dehydro compounds				
[13]	3-Dehydroecdysone (3DE)	E × 0.1	C	[65]
		20E × 0.002–0.003	D	[155]
[14]	3-Dehydro-20-hydroxy-ecdysone (3D20E)	E × 0.07	C	[66]
		20E × 0.25–0.30	D	[155]
[15]	3-Dehydro-2-deoxyecdysone (3D2dE)	?		
3-Epi-ecdysteroids				
[16]	3-Epiecdysone (E′)	E × 0.1–0.2	M	[117]
		E × 0.1	M	[67]
[17]	3-Epi-20-hydroxyecdysone (20E′)	E × 0.07–0.10	M	[68]
		20E × 0.1	M	[67]
[18]	3-Epi-2-deoxyecdysone (2dE′)	Inactive	LD	[99]
[19]	3-Epi-26-hydroxyecdy-sone (26E′)	Inactive	M	[41]
[20]	3-Epi-20,26-dihydroxyecdysone (2026E′)	E × 0.003	M	[67]
Side chain cleavage product				
[21]	Poststerone (P)	Inactive	C	[163, 164]
		E × 0.25	S	[164]
Conjugates				
[25]	2-Deoxyecdysone 22-phosphate (2dE22P)	Inactive	LD	[99]
[29]	3-Epi-2-deoxyecdysone 3-phosphate (2dE′3P)	Inactive	LD	[99]
[31]	2-Deoxyecdysone 22-adenosine-monophosphate (2dE22AMP)	Inactive	LD	[99]
C28 Ecdysteroids				
[37]	Makisterone A (MaA)	E × 0.25–0.33	M	[80]
		20E × 1	C	[163]
		20E × 10	O	[165, 166]
		20E × 0.6	Kc	[168]
[38]	Callinecdysone B (CaB)	Active	CH	[167]

[a] Molting hormone activity is expressed either in absolute amounts equivalent to one biological unit or in multiples of molting hormone activity of ecdysone (*E*) or 20-hydroxyecdysone (20E)

[b] Abbreviations: *B Bombyx* bioassay; *C Calliphora* bioassay; *CH Chilo* bioassay; *D Drosophila* puffing assay; *DD Drosophila* imaginal discs evagination assay; *Kc* Kc cell line morphological assay; *LD Lucilia* imaginal discs evagination assay; *M Musca* bioassay; *O Oncopeltus* bioassay; *S Sarcophaga* bioassay; *SM Samia* bioassay

6. Conclusions

Our knowledge of ecdysone metabolism has experienced spectacular improvement during the last years, and several pathways have been elucidated. This is due to the finding of suitable biological materials (insect embryos) and to the development of new analytical techniques. Nevertheless, our present knowledge is very incomplete, and there is no doubt that many new metabolites are to be identified in the near future. In addition, metabolic pathways in invertebrates other than insects or Crustacea are presently unknown. The studies of ecdysone-metabolizing enzymes has been essentially restricted to ecdysone 20-hydroxylase and ecdysone oxidase, and this is obviously insufficient.

A better knowledge of these areas is a prerequisite for understanding the mechanism of action of some recently discovered inhibitors [104]. This could provide a new basis for controlling insect pests or parasitic worms, and merits our continuous efforts.

Acknowledgments. The authors wish to thank their colleagues, Drs. Philippe Beydon, Catherine Blais, Ulrike Enderle, Fabienne Lachaise, and Jean-Francois Modde for allowing them to mention their unpublished data.

This work was supported by C.N.R.S. (ERA 1048) and grants from the Deutsche Forschungsgemeinschaft (Ko 543).

References

1. Hetru C, Horn DHS (1980) In: Hoffmann JA (ed) Progress in ecdysone research. Elsevier/North Holland, Amsterdam, pp 13–28
2. Bergamasco R, Horn DHS (1983) In: Downer RGH, Laufer H (eds) Endocrinology of insects. Invertebrate endocrinology, vol I. Liss, New York, pp 627–656
3. Butenandt A, Karlson P (1954) Z Naturforsch 9b:389–391
4. Romer F (1979) Naturwissenschaften 66:471
5. Sauber F, Reuland M, Berchtold JP, Hetru C, Tsoupras G, Luu B, Moritz ME, Hoffmann JA (1983) CR Acad Sci Ser D 296:413–418
6. Mendis AHW, Rose ME, Rees HH, Goodwin TW (1983) Mol Biochem Parasitol 9:209–226
7. Nirde P, Torpier G, De Reggi ML, Capron A (1983) FEBS Lett 151:223–227
8. Sturaro A, Guerriero A, De Clauser R, Pietra F (1982) Experientia 38:1184–1185
9. Karlson P (1983) Hoppe-Seyler's Z Physiol Chem 364:1067–1087
10. Chino H, Sakurai S, Ohtaki T, Ikekawa N, Miyazaki H, Ishibashi M, Abuki H (1974) Science 183:529–530
11. King DS, Bollenbacher WE, Borst DW, Vedeckis WV, O'Connor JD, Ittycheriah PI, Gilbert LI (1974) Proc Natl Acad Sci USA 71:793–796
12. Svoboda JA, Thompson MJ, Robbins WE, Kaplanis JN (1978) Lipids 13:742–753
13. Rees HH, Greenwood DR, Dinan LN, Isaac RE, Goodwin TW (1981) In: Sehnal F, Zabza A, Menn JJ, Cymborowsky B (eds) Regulation of insect development and behaviour. Tech Univ Wroclaw Press, Wroclaw, Poland, pp 71–92
14. Koolman J (1982) Insect Biochem 12:225–251
15. Hoffmann JA, Hetru C (1983) In: Downer RGH, Laufer H (eds) Endocrinology of insects. Invertebrate endocrinology, vol I. Liss, New York, pp 65–88
16. Horn DHS (1971) In: Jacobson M, Crosby DG (eds) Naturally occurring insecticides. Marcel Dekker, New York, pp 333–459
17. Koolman J, Hoffmann JA, Karlson P (1973) Hoppe-Seyler's Z Physiol Chem 354:1043–1048
18. Wilson ID (1984) J Chromatogr (in press)
19. Hahn B, Koolman J (1983) Unpublished experiments
20. Briers T, De Loof A (1983) Insect Biochem 13:513–522

21. Wilson ID, Scalia S, Morgan ED (1981) J Chromatogr 212:211–219
22. Kaplanis JN, Thompson MJ, Yamamoto RT, Robbins WE, Louloudes SJ (1966) Steroids 8: 605–623
23. Hetru C (1981) Thesis, Strasbourg
24. Lafont R, Beydon P, Mauchamp P, Sommé-Martin G, Andrianjafintrimo M, Krien P (1981) In: Sehnal F, Zabza A, Menn JJ, Cymborowsky B (eds) Regulation of insect development and behaviour. Tech Univ Wroclaw Press, Wroclaw, Poland, pp 125–143
25. Dinan L, Donnahey PL, Rees HH, Goodwin TW (1981) J Chromatogr 205:139–145
26. Lafont R, Sommé-Martin G, Mauchamp B, Maume BF, Delbecque JP (1980) In: Progress in ecdysone research. Elsevier/North Holland, Amsterdam, pp 45–68
27. Isaac RE, Milner NP, Rees HH (1982) J Chromatogr 246:317–322
28. Katz M, Lensky Y (1970) Experientia 26:1043–1044
29. Morgan ED, Poole CF (1976) Adv Insect Physiol 12:17–62
30. Morgan ED, Wilson ID (1980) In: Hoffmann JA (ed) Progress in ecdysone research. Elsevier/ North Holland, Amsterdam, pp 29–43
31. Lafont R, Blais C, Beydon P, Modde JF, Enderle U, Koolman J (1983) Arch Insect Biochem Physiol 1:41–58
32. Heinrich G, Hoffmeister H (1970) Z Naturforsch 25b:358–361
33. King DS (1972) Gen Comp Endocrinol Suppl 3:221–227
34. Isaac RE, Rose ME, Rees HH, Goodwin TW (1982) Chem Commun:249–251
35. Modde JF (1983) Thesis, Paris
36. Feyereisen R (1979) Thesis, Strasbourg
37. Scalia S, Morgan ED (1982) J Insect Physiol 28:647–654
38. Weirich GF, Svoboda JA, Thompson MJ (1983) Commun Pres Int Symp Biosynth Metab Mode Action Invertebr Horm, Strasbourg, France
39. Wigglesworth K, Rees HH (1983) Poster Pres Int Symp Biosynth Metab Mode Action Invertebr Horm, Strasbourg, France
40. Galbraith MN, Horn DHS (1969) Aust J Chem 22:1045–1057
41. Kaplanis JN, Thompson MJ, Dutky SR, Robbins WE (1980) Steroids 40:551–560
42. Russel GB, Price GM (1977) Insect Biochem 7:197–202
43. Koolman J, Reum L, Karlson P (1979) Hoppe-Seyler's Z Physiol Chem 360:1351–1355
44. Enderle U (1983) Thesis, Marburg
45. Lafont R, Beydon P, Sommé-Martin G, Blais C (1980) Steroids 36:185–207
46. Dinan L, Rees HH (1978) Steroids 32:629–638
47. Karlson P, Sekeris CE (1966) Recent Prog Hormon Res 22:473–502
48. Tsoupras G, Hetru C, Luu B, Constantin E, Lagueux M, Hoffmann JA (1983) Tetrahedron 39:1789–1796
49. Blunt JW, Lane GA, Munro MHG, Russel GB (1979) Aust J Chem 32:779–782
50. Hikino H, Mohri K, Hikino Y, Arihara S, Takemoto T (1976) Tetrahedron 32:3015–3021
51. Russel GB, Greenwood DR, Lane GA, Blunt JW, Munro MGH (1981) Phytochemistry 20: 2407–2410
52. Delbecque JP, Lafont R (1983) Unpublished experiments
53. Hocks P, Wiechert R (1966) Tetrahedron Lett 33:2989–2993
54. Hoffmeister H, Grützmacher HF (1966) Tetrahedron Lett 33:4017–4023
55. Hampshire F, Horn DHS (1966) Chem Commun:37–38
56. McCarthy JP (1979) Steroids 34:799–806
57. Ohnishi E, Mizuno T, Chatani F, Ikekawa N, Sakurai S (1977) Science 197:66–70
58. Galbraith MN, Horn DHS, Middleton EJ, Hackney RJ (1968) Chem Commun:83–85
59. Ikekawa N, Ikeda T, Mizuno T, Ohnishi E, Sakurai S (1980) Chem Commun:448–449
60. Kaplanis JN, Robbins WE, Thompson MJ, Dutky SR (1973) Science 180:307–308
61. Thompson MJ, Kaplanis JN, Robbins WE, Yamamoto RT (1967) Chem Commun:650–653
62. Faux A, Horn DHS, Middleton EJ, Fales HM, Lowe ME (1969) Chem Commun:175–176
63. Isaac RE, Milner NP, Rees HH (1983) Biochem J 213:261–265
64. Lachaise F, Lafont R (1984) Steroids (in press)

65. Karlson P, Bugany H, Döpp H, Hoyer GA (1972) Hoppe-Seyler's Z Physiol Chem 363:1610 to 1614
66. Koolman J, Spindler KD (1977) Hoppe-Seyler's Z Physiol Chem 358:1339–1344
67. Kaplanis JN, Thompson MJ, Dutky SR, Robbins WE (1979) Steroids 34:334–345
68. Thompson MJ, Kaplanis JN, Robbins WE, Dutky SR, Nigg HN (1974) Steroids 24:359–366
69. Isaac RE, Rees HH, Goodwin TW (1981) Chem Commun:418–420
70. Galbraith MN, Horn DHS, Middleton EJ, Thompson JA, Siddall JB, Hafferl W (1969) Chem Commun:1134–1135
71. Isaac RE, Rees HH, Goodwin TW (1981) Chem Commun:594–595
72. Diehl PA, Connat JL, Girault JP, Lafont R (1984) submitted for publication
73. Tsoupras G, Luu B, Hoffmann JA (1982) Steroids 40:551–560
74. Isaac RE, Rose ME, Rees HH, Goodwin TW (1983) Biochem J 213:533–541
75. Tsoupras G, Hetru C, Luu B, Lagueux M, Constantin E, Hoffmann JA (1982) Tetrahedron Lett 23:2045–2048
76. Tsoupras G (1982) Thesis, Strasbourg
77. Tsoupras G, Luu B, Hoffmann JA (1983) Science 220:507–509
78. Isaac RE, Desmond HP, Rees HH (1984) Biochem J 217:239–243
79. Tsoupras G, Luu B, Hetru C, Muller JF, Hoffmann JA (1983) CR Acad Sci Ser D 296:77–80
80. Kaplanis JN, Dutky SR, Robbins WE, Thompson MJ, Lindquist EL, Horn DHS, Galbraith MN (1975) Science 190:681–682
81. Kaplanis JN, Thompson ML, Dutky SR, Robbins WE, Lindquist EL (1972) Steroids 20:105 to 120
82. Hetru C, Lagueux M, Luu B, Hoffmann JA (1978) Life Sci 22:2141–2154
83. Garcia M, Lafont R (in preparation)
84. Beydon P, Claret J, Porcheron P, Lafont R (1981) Steroids 38:633–650
85. Lagueux M, Hirn M, Hoffmann JA (1977) J Insect Physiol 23:109–119
86. McCarthy JF, Skinner DM (1979) Gen Comp Endocrinol 37:250–263
87. Hikino H, Ohizumi Y, Takemoto T (1975) Hoppe-Seyler's Z Physiol Chem 356:309–314
88. Blais C, Lafont R (1980) Wilhelm Roux' Arch 188:27–36
89. Koolman J (1984) Meth Enzymol (in press)
90. Beydon P, Lafont R (1983) Poster Pres Int Symp Biosynth Metab Mode Action Invertebr Horm, Strasbourg, France
91. Saatov Z, Usmanov BZ, Abubakirov NK (1979) Khim Prir Soedin 6:793–797
92. Koolman J (1979) Unpublished experiments
93. De Clerck D, Eechaute W, Leusen L, Diederick H, De Loof A (1983) Gen Comp Endocrinol 52:368–378
94. De Clerck D, Diederick H, De Loof A (1984) Gen Comp Endocrinol (in press)
95. Joosse J (1979) In: Barrington EJW (ed) Hormones and evolution. Academic Press, London New York, pp 119–157
96. Moriyama H, Nakanishi K, King DS, Okauchi T, Siddall JB, Hafferl W (1970) Gen Comp Endocrinol 15:80–87
97. Bouvier J, Diehl PA, Morici M (1982) Rev Suisse Zool 89:967–976
98. Connat JL, Diehl PA, Morici M (1984) Gen Comp Endocrinol (in press)
99. Sall C, Tsoupras G, Kappler C, Lagueux M, Zachary D, Luu B, Hoffmann JA (1983) J Insect Physiol 29:491–507
100. Beydon P, Modde JF, Lafont R (1982) Unpublished experiments
101. Clayton RB (1964) J Lipid Res 5:3–19
102. Clark AI, Bloch K (1959) J Biol Chem 234:2583–2587
103. Rees HH, Davies TG, Dinan LN, Lockley WJS, Goodwin TW (1980) In: Hoffmann JA (ed) Progress in ecdysone research. Elsevier/North Holland, Amsterdam, pp 163–186
104. Kaplanis JN, Weirich GF, Svoboda JA, Thompson MJ, Robbins WE (1980) In: Hoffmann JA (ed) Progress in ecdysone research. Elsevier/North Holland, Amsterdam, pp 163–186
105. Aldrich JH, Kelly TJ, Woods CW (1982) J Insect Physiol 28:857–861
106. Bollenbacher WE, Smith SL, Wielgus JJ, Gilbert LI (1977) Nature (London) 268:660–663
107. Nigg HN, Svoboda JA, Thompson MJ, Dutky SR, Kaplanis JN, Robbins WE (1976) Experientia 32:438–439

108. Blais C (1981) Personal communication
109. Young NL (1976) Insect Biochem 6:1–12
110. Johnson P, Rees HH (1977) Biochem J 168:512–520
111. Feyereisen R, Durst F (1978) Eur J Biochem 88:37–47
112. Koolman J, Karlson P (1975) Hoppe-Seyler's Z Physiol Chem 356:1131–1138
113. Koolman J, Karlson P (1978) Eur J Biochem 89:453–460
114. Yu SJ, Terriere LC (1977) Pestic Biochem Physiol 7:48–55
115. Blais C, Lafont R (1984) Hoppe-Seyler's Z Physiol Chem (in press)
116. Mayer RT, Durrant JL, Holman GM, Weirich GF, Svoboda JA (1979) Steroids 34:799–806
117. Nigg HN, Svoboda JA, Thompson MJ, Kaplanis JN, Dutky SR, Robbins WE (1974) Lipids 9: 971–974
118. Dinan LN (1980) Thesis, Liverpool
119. Milner NP, Rees HH (1983) Poster Pres Int Symp Biosynth Metab Mode Action Invertebr Horm, Strasbourg, France
120. Yang RSH, Wilkinson CF (1972) Biochem J 130:487–493
121. Enderle U, Koolman J (1981) Biochem Soc Trans 9:156p
122. Richards G (1981) Biol Rev 56:501–549
123. Koolman J, Spindler KD (1983) In: Downer RGH, Laufer H (eds) Endocrinology of insects. Invertebrate endocrinology, vol I. Liss, New York, pp 179–202
124. Feyereisen R (1977) CR Acad Sci Ser D 284:1831–1834
125. Feyereisen R, Durst F (1980) Mol Cell Endocrinol 20:157–169
126. Feyereisen R, Hoffmann JA (1977) J Insect Physiol 23:1175–1181
127. Feyereisen R, Joly L, Hoffmann JA (1976) CR Acad Sci Ser D 282:901–904
128. Smith SL, Bollenbacher WE, Gilbert LI (1983) Mol Cell Endocrinol 31:227–251
129. Koolman J (1978) Hoppe-Seyler's Z Physiol Chem 359:1315–1321
130. Karlson P, Koolman J (1973) Insect Biochem 3:409–417
131. Reum L (1981) Thesis, Marburg
132. Wichmann JE, DeLuca HF, Schnoes HK, Horst RL, Shepard RM, Jorgensen NA (1979) Biochemistry 18:4775–4780
133. Purkaystha AR, Martin KO, Goldberg A, Monder C (1982) J Steroid Biochem 17:51–59
134. Dey AC, Senciall IR (1978) J Steroid Biochem 9:1099–1104
135. Monder C, Purkaystha AR, Farhi RL, Han CA (1982) J Steroid Biochem 16:613–616
136. Yang RSH (1976) In: Wilkinson CF (ed) Insecticide biochemistry and physiology. Plenum Press, New York London, pp 177–228
137. Dauterman WC, Hodgson E (1978) In: Rockstein M (ed) Biochemistry of insects. Academic Press, London New York, pp 541–577
138. Isaac RE, Phua KK, Rees HH (1982) Biochem J 204:127–133
139. Yang RSH, Wilkinson CF (1973) Comp Biochem Physiol 46B:717–726
140. MacDonald IA, Bokkenheuser VD, Winter J, McLernon AM, Mosbach EH (1983) J Lipid Res 24:675–700
141. Koolman J (1983) Unpublished experiments
142. Feyereisen R, Lagueux M, Hoffmann JA (1976) Gen Comp Endocrinol 29:319–327
143. King DS, Marks EP (1974) Life Sci 15:147–154
144. Thompson MJ, Robbins WE, Dutky SR, Marks EP, Filipi PA, Finegold H (1978) Lipids 15: 719–733
145. Dübendorfer A, Maroy P (1983) Commun Pres VIth Ecdysone Workshop, Szeged, Hungary
146. Lachaise F, Feyereisen R (1976) CR Acad Sci Ser D 283:1445–1448
147. Chang ES, Sage BA, O'Connor JD (1976) Gen Comp Endocrinol 30:21–33
148. Chang ES, O'Connor JD (1978) Gen Comp Endocrinol 36:151–160
149. Daig K, Spindler KD (1980) In: Kurstak E, Maramorosch K, Dübendorfer A (eds) Invertebrate systems in vitro. Elsevier/North Holland, Amsterdam, pp 273–278
150. Hoffmann JA, Lagueux M, Hetru C, Charlet M, Goltzené F (1980) In: Hoffmann JA (ed) Progress in ecdysone research. Elsevier/North Holland, Amsterdam, pp 431–465
151. Karlson P, Shaaya E (1964) J Insect Physiol 10:797–804
152. Kaplanis JN, Tabor LA, Thompson MJ, Robbins WE, Shortino TJ (1966) Steroids 8:625–631
153. Ohtaki T, Milkman RD, Williams CM (1967) Proc Natl Acad Sci USA 58:981–984

154. Williams CM (1968) Biol Bull 134:344–355
155. Richards G (1978) J Insect Physiol 24:329–335
156. Hoffmeister H, Nakanishi K, Koreeda M, Hsu HY (1968) J Insect Physiol 14:53–54
157. Sato Y, Sakai M, Imai S, Fujioka S (1968) Appl Entomol Zool 3:49–51
158. Hsiao TH, Hsiao C (1979) J Insect Physiol 25:45–52
159. Nakanishi K, Koreeda M, Sasaki M, Chang ML, Hsu HY (1966) Chem Commun:915–917
160. Takemoto T, Ogawa S, Nishimoto N, Hoffmeister H (1967) Z Naturforsch 22b:681–682
161. Chong YK, Galbraith MN, Horn DHS (1970) Chem Commun:1217–1218
162. Mizuno T, Watanabe K, Ohnishi E (1981) Dev Growth Differ 23:543–552
163. Bergamasco R, Horn DHS (1980) In: Hoffmann JA (ed) Progress in ecdysone research. Elsevier/North Holland, Amsterdam, pp 299–324
164. Siddall JB, Horn DHS, Middleton EJ (1967) Chem Commun:899–900
165. Aldrich JR, Svoboda JA, Thompson MJ (1981) J Exp Zool 218:133–137
166. Kelly TJ, Woods CW, Redfern RE, Borcovek AB (1981) J Exp Zool 218:127–132
167. Imai S, Hori M, Fujioka S, Murata E, Goto M, Nakanishi K (1968) Tetrahedron Lett 36:3883–3890
168. Cherbas L, Yonge CD, Cherbas P, Williams CM (1980) Wilhelm Roux's Arch 189:1–15
169. Oberlander H (1974) Experientia 30:1409–1410
170. Fristrom JW, Yund MA (1976) In: Maramorosch K (ed) Invertebrate tissue culture. Research applications. Academic Press, London New York, pp 161–178
171. Westphal U (1983) J Steroid Biochem 19:1–15
172. Daig K, Spindler KD (1983) Mol Cell Endocrinol 31:93–104
173. Daig K, Spindler KD (1983) Mol Cell Endocrinol 32:73–79
174. Blais C (1981) Personal communication
175. Isaac RE, Sweeney FP, Rees HH (1983) Biochem Soc Trans 11:379–380
176. Moribayashi A, Ohtaki T (1980) Jpn J Med Sci Biol 33:189–201
177. Kaplanis JN, Thompson MJ, Robbins WE, Lindquist EL (1972) Steroids 20:621–625
178. Koolman J, Hoffmann JA, Dreyer M (1975) Experientia 31:247–249
179. Wilson ID, Morgan ED (1978) J Insect Physiol 24:751–756
180. Modde JF, Lafont R, Hoffmann JA (1984) Int J Invert Reprod Dev (in press)
181. Whitehead DL, Sellheyer K (1982) Experientia 38:1249–1251
182. Hoffmeister M, Rufen C, Ammon H (1965) Z Naturforsch 20b:130–133

II.4 Ecdysone 20-Monooxygenases*

G.F. WEIRICH, J.A. SVOBODA, and M.J. THOMPSON[1]

1. Introduction

Ecdysone 20-monooxygenase (ecdysone 20-hydroxylase; EC 1.14.99.20) is generally considered to be the key enzyme in the formation of the molting hormone 20-hydroxyecdysone from its precursor, ecdysone [1-3]. In this paper we will review in vitro studies of ecdysone 20-monooxygenases. The enzyme has been demonstrated in cell-free preparations from midgut [1, 4, 5] and fat body [2, 6, 7] of the tobacco hornworm, *Manduca sexta* (L.); Malpighian tubules, fat body [3, 8, 9], and gut [9] of the African migratory locust, *Locusta migratoria migratorioides* (R. & F.); Malpighian tubules and fat body of the desert locust, *Schistocerca gregaria* (Forskål) [10]; and testes of the crab, *Cancer antennarius* (Stimpson) [11]. Enzymes for the ecdysone 20-hydroxylation may also exist in other tissues as indicated by in vitro incubations of *M. sexta* Malpighian tubules and integument [2, 6, 12, 13].

2. Subcellular Localization

The subcellular localization of ecdysone 20-monooxygenase is not uniform. In *S. gregaria* Malpighian tubules [10] and *M. sexta* fat body [2, 6, 7], it is a mitochondrial enzyme; in *L. migratoria* [3, 8, 9], it is a component of the endoplasmic reticulum (microsomes); and in *M. sexta* midgut, it is present both as a mitochondrial [1, 4, 5] and as a microsomal [1, 5] enzyme system. Apparently, insects have developed two different enzyme systems for the 20-hydroxylation of ecdysone. We have no explanation at present for this duplication, and it is not known why some insect species and/or tissues utilize one of these systems to the exclusion of the other, or why both are utilized in the *M. sexta* midgut.

The typical mitochondrial monooxygenase system in mammals consists of three components: NADPH-ferredoxin reductase, ferredoxin (an iron-sulfur protein), and cytochrome P-450 (Fig. 1) [14]. The typical microsomal monooxygenase system, on the other hand, consists of only two components: NADPH-cytochrome P-450 reductase

---

* Dedicated to Professor Peter Karlson on the occasion of his 65th birthday

1 Insect Physiology Laboratory, Agricultural Research Service, USDA, Beltsville, MD 20705, USA

Biosynthesis, Metabolism and Mode of Action of Invertebrate Hormones
(ed. by J. Hoffmann and M. Porchet)

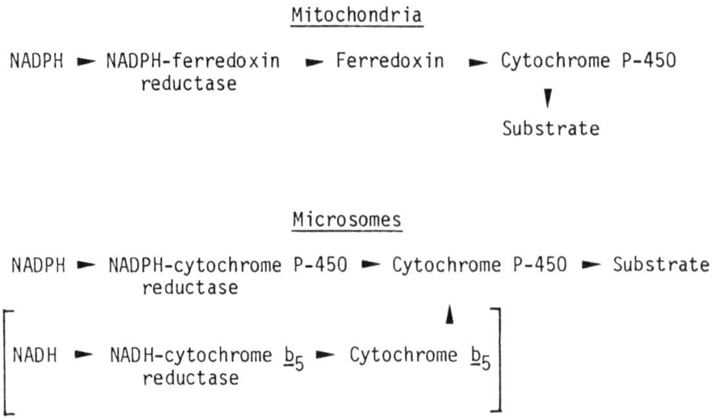

Fig. 1. Electron flow in mitochondrial and microsomal monooxygenase systems

and cytochrome P-450, with the possible involvement of NADH-cytochrome b_5 reductase and cytochrome b_5 in some cases [15, 16].

If we assume that insects have comparable enzyme systems, we have to conclude that the hornworm midgut is equipped with five different enzyme proteins employed for the ecdysone 20-hydroxylation, a task that could be accomplished with a minimum of two enzyme proteins. The picture is further complicated by the need for separate controls of the two ecdysone 20-monooxygenase systems.

These considerations prompted us to re-examine the subcellular distribution of ecdysone 20-monooxygenase in *M. sexta* midgut, and to determine the extent to which microsomes cosedimenting with mitochondria might contribute to the monooxygenase activity found in mitochondrial preparations. The specific ecdysone 20-monooxygenase activity of midgut microsomes is nearly twice that of mitochondria [1], and the tendency of microsomes to cosediment with mitochondria is documented in numerous reports dealing with mammalian as well as insect tissues.

Mitochondria of *M. sexta* midgut were washed three times, and after each washing, NADPH-cytochrome *c* reductase (a microsomal marker enzyme [17]) and ecdysone 20-monooxygenase activities were measured. At each step of the purification, the ecdysone 20-monooxygenase activity of the mitochondria far exceeded the activity accounted for by the microsomal contamination (Fig. 2). These results clearly demonstrated that the mitochondria have substantial ecdysone 20-monooxygenase activity of their own. Future studies concerned with developmental changes in activity and substrate specificity will, hopefully, shed some light on the function of these two monooxygenase systems.

3. Cosubstrates and Secondary Hydrogen Donors

NADPH is utilized as a cosubstrate by all ecdysone 20-monooxygenases investigated so far (Table 1). In mitochondria of *M. sexta* midgut, NADH, isocitrate, malate, and suc-

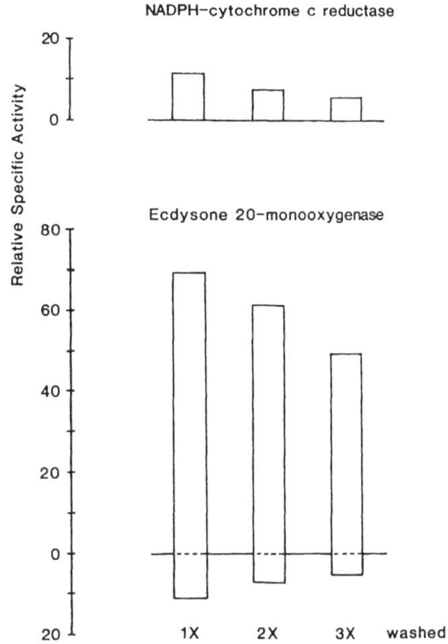

Fig. 2. NADPH-cytochrome c reductase and ecdysone 20-monooxygenase activities in midgut mitochondria of *M. sexta* after successive washings. Enzyme activities are expressed as relative specific activities [(spec. activity of mitochondrial preparation)/(spec. activity of microsomes) × 100]. NADPH-cytochrome c reductase (*upper panel*) is a microsomal marker enzyme and its relative specific activity is a measure of the microsomal contamination of the mitochondria. *Columns* representing the relative specific ecdysone 20-monooxygenase activities (*lower panel*) are arranged to show the contribution of the microsomes *below the zero line* and the genuine mitochondrial activities *above the zero line*

cinate can also sustain the ecdysone hydroxylation [2, 5], and in fat body mitochondria of *M. sexta*, NADH or malate plus $NADP^+$ can partially substitute for NADPH [6]. All of these compounds can generate NADPH through enzymatic pathways in the mitochondria [5].

Microsomal ecdysone 20-monooxygenases do not react with isocitrate, malate, succinate, or NADH [9, 18]. Microsomes of *L. migratoria* show a limited reaction with a combination of $NADP^+$ and NADH (±ATP), and also with $NADP^+$ plus glucose 6-phosphate, possibly through the involvement of an $NAD(P)^+$ transhydrogenase and a glucose-6-phosphate dehydrogenase, respectively [9]. NADH, though ineffective by itself, shows synergistic action with NADPH [9]. This synergism may be an indication for the involvement of NADH-cytochrome b_5 reductase and cytochrome b_5 in the ecdysone hydroxylation.

The requirement for molecular oxygen has been demonstrated for two mitochondrial [5, 10] and one of the microsomal [8, 9] ecdysone 20-monooxygenases.

4. Temperature, pH, and Kinetic Parameters

The enzyme systems differ only slightly in their pH and temperature optima, but show substantial differences in their Michaelis constants and maximum velocities (Table 1). It is not clear at present, whether the differences in the kinetic parameters reflect differences in the physiological functions and environments of the enzymes or differences in the sensitivities of the enzymes to experimental manipulations. Differences in the extent to which inhibiting or activating factors are removed during the isolation of mitochondria and microsomes can also affect the apparent kinetic parameters.

Table 1. Properties of ecdysone 20-monooxygenases[a]

	S. gregaria [10] Malpighian tubules Mitochondria	*L. migratoria* [8, 9] Malpighian tubules Microsomes	*M. sexta* Fat Body Mitochondria [2, 6, 7]	*M. sexta* Midgut Mitochondria [1, 4, 5, 18]
Cosubstrates, secondary hydrogen donors	NADPH[b]	NADPH, K_m 10.5 μM $NADP^+$ + NADH ± ATP $NADP^+$ + Glc-6-P	NADPH $NADP^+$ + malate NADH	NADPH NADH Isocitrate Malate Succinate
	O_2	O_2		O_2
pH Optimum	6.5	6.6	7.5	8.5
Temperature optimum	35 °C	40 °C	30 °C	30 °C
$Km_{ecdysone}$	12.5 μM	0.27 μM	0.16 μM	18.3 μM
$Vmax_{ecdysone}$	2.5 nmol/min/mg protein		5.4 pmol/min/mg protein	46.6 pmol/min/mg protein
Inhibitors				
Metyrapone	10 mM: 80% inhib.	I_{50} 4.0 μM	K_i 3.0– 4.0 μM	0.1 mM: 51% inhib.
Aminoglutethimide			I_{50} 10 –15 μM	
p-Chloromercuribenzoate		I_{50} 3.0 μM	I_{50} 0.3– 0.5 μM	
20-Hydroxyecdysone		K_i 0.75 μM	K_i 27.2 μM	
$NADP^+$		K_i 50.0 μM		
CO[c]	K 12.0 450 nm reversible	K 1.5 white light reversible	K 6.0 photochemical action spectrum, max. at 450 nm	K 1.0
CO difference spectrum showing 450 nm peak	+	+	+	+
Cytochrome P-450 content	0.91 nmol/mg protein	300 pmol/mg protein	10 pmol/mg protein	10–15 pmol/mg protein

[a] Open spaces indicate that no data are available
[b] No other hydrogen donors tested
[c] CO sensitivity is expressed by Warburg's partition coefficient $K = n/(1-n) \times [CO]/[O]$, where n is the ratio of the reaction rate with CO to that without CO

5. Inhibitors, CO Difference Spectra, and Cytochrome P-450 Content

Ecdysone 20-monooxygenases differ widely in their susceptibility to metyrapone (Table 1), a competitive inhibitor of steroid monooxygenases in vertebrates [19]. For *M. sexta* fat body mitochondria, this inhibition has also been shown to be competitive [6].

The effect of 20-hydroxyecdysone on the monooxygenase was determined for microsomes of *L. migratoria* Malpighian tubules [9] and fat body mitochondria of *M. sexta* [7]. Both enzyme systems showed competitive inhibition. A comparison of the K_m for ecdysone and the K_i for 20-hydroxyecdysone (Table 1) suggests that product inhibition may play a role in the regulation of ecdysone 20-hydroxylation in *L. migratoria* Malpighian tubules, but is unlikely to do so in *M. sexta* fat body.

All ecdysone 20-monooxygenases are inhibited by carbon monoxide and this inhibition is partially reversible by irradiation with white light or monochromatic light at 450 nm (Table 1). For fat body mitochondria of *M. sexta*, the photochemical action spectrum has been recorded; it showed a sharp peak at 450 nm and a lower broad peak around 420 nm [7]. The CO difference spectra of all enzymes show a 450 nm peak. However, the evaluation of mitochondrial spectra is complicated by strong interference from the cytochrome oxidase complex [5, 7, 10].

The cytochrome P-450 contents of the enzyme preparations differ widely (Table 1). Whereas mitochondria and microsomes would be expected to have quite different concentrations of P-450, the vast difference between *S. gregaria* mitochondria and *M. sexta* mitochondria requires experimental verification. Moreover, any cytochrome P-450 determination in mitochondria has to be considered as preliminary unless the microsomal contamination of the mitochondrial preparation is also determined and the P-450 value corrected accordingly. This has not been done in the studies reported so far.

As mentioned earlier, mitochondrial steroid monooxygenases in mammals contain non-heme iron proteins as an integral part of the electron transport pathway. The presence of such a non-heme iron protein has also been demonstrated in *M. sexta* fat body mitochondria by electron paramagnetic resonance spectroscopy [2].

6. Substrate Specificity

Mitochondria and microsomes are known to contain a variety of cytochrome P-450 species involved in various oxidations [15, 20, 21]. A definite determination of the substrate specificity of ecdysone 20-monooxygenase (or its specific component, cytochrome $P\text{-}450_{ecdysone}$) would, therefore, require pure cytochrome $P\text{-}450_{ecdysone}$, reconstituted to a functional monooxygenase system. Such experiments have not been undertaken so far.

Preliminary data on substrate specificity are available for *L. migratoria* microsomes [9] and *M. sexta* fat body mitochondria [2]. The microsomal enzyme system [9] does not react with 2β-acetoxy- or 3-dehydroecdysone, or with the 5α-epimer of ecdysone (A/B rings in *trans* configuration), nor does it hydroxylate 20-hydroxyecdysone or 3-dehydro-20-hydroxyecdysone at any other position.

Smith et al. [2] used a competition assay to obtain information on the substrate specificity of the mitochondrial monooxygenase in *M. sexta* fat body. Their results indicated comparable competition (and presumably enzyme-substrate binding) for ecdysone, 2-deoxyecdysone, and 22-deoxyecdysone. 2,25-Dideoxyecdysone was five times as effective as ecdysone; and 22,25-dideoxyecdysone, 2,22,25-trideoxyecdysone, 22-isoecdysone, and 22,25-dideoxy-5α-ecdysone (listed in order of decreasing effectiveness) were less effective than ecdysone.

If these results do indeed reflect different affinities of the active site of the cytochrome P-450$_{ecdysone}$ for the various substrates, they would indicate that the active site has a preference for an ecdysteroid with reduced polarity on the nucleus and on the side chain, and that the elimination of one hydroxyl group from either the nucleus or the side chain does not significantly alter the enzyme-substrate interaction. Further reduction of the number of hydroxyl groups, however, leads to a decrease in the enzyme-substrate binding, as does isomerization at C-22 or C-5.

Mammalian steroid hydroxylases (cytochromes P-450) can have a rather broad substrate specificity and may hydroxylate more than one position on the steroid substrate [19]. We do not know at present whether such broad-spectrum monooxygenases exist in insects. No hydroxylation other than C-20 has been reported for any of the cell-free enzyme systems.

7. Changes of Ecdysone 20-Monooxygenase Activity Related to Development

The activity of all ecdysone 20-monooxygenase systems changes during insect development [2-4, 10]. In *S. gregaria* [10], *L. migratoria* [3], and in the *M. sexta* midgut [4], these changes parallel the changes in the molting hormone titer. The *M. sexta* fat body monooxygenase, however, shows an inverse relationship to the ecdysone titer [2]. The activity of the *L. migratoria* enzyme systems (microsomes of Malpighian tubules, fat body, and gut) is enhanced by injection of ecdysone (into whole animals or isolated abdomens) and reduced by removal of the prothoracic glands [3, 22]. Thus, at least in some instances, the monooxygenase activity seems to be directly controlled by the concentration of its substrate, ecdysone, a mechanism that ensures the availability of adequate enzyme activity at a time of high demand for the ecdysone/20-hydroxyecdysone conversion.

8. Outlook

Much more work is necessary to achieve a complete understanding of the mechanisms and regulation of ecdysone 20-hydroxylation. Furthermore, the biosynthesis of ecdysone from cholesterol and the biosynthesis of 26-hydroxyecdysone require a series of other hydroxylation reactions, presumably also catalyzed by monooxygenase systems, about which we do not have any information at present.

References

1. Kaplanis JN, Weirich GF, Svoboda JA, Thompson MJ, Robbins WE (1980) In: Hoffmann JA (ed) Progress in ecdysone research. Elsevier/North-Holland, Amsterdam, pp 163–186
2. Smith SL, Bollenbacher WE, Gilbert LI (1980) In: Hoffmann JA (ed) Progress in ecdysone research. Elsevier/North-Holland, Amsterdam, pp 139–162
3. Feyereisen R, Durst F (1980) Mol Cell Endocrinol 20:157–169
4. Nigg HN, Svoboda JA, Thompson MJ, Dutky SR, Kaplanis JN, Robbins WE (1976) Experientia 32:438–439
5. Mayer RT, Svoboda JA, Weirich GF (1978) Hoppe-Seyler's Z Physiol Chem 359:1247–1257
6. Bollenbacher WE, Smith SL, Wielgus JJ, Gilbert LI (1977) Nature (London) 268:660–663
7. Smith SL, Bollenbacher WE, Cooper DY, Schleyer H, Wielgus JJ, Gilbert LI (1979) Mol Cell Endocrinol 15:111–133
8. Feyereisen R (1977) CR Acad Sci Ser D 284:1831–1834
9. Feyereisen R, Durst F (1978) Eur J Biochem 88:37–47
10. Johnson P, Rees HH (1977) Biochem J 168:513–520
11. Chang ES, O'Connor JD (1978) Gen Comp Endocrinol 36:151–160
12. King DS (1972) Am Zool 12:343–345
13. King DS (1972) Gen Comp Endocrinol Suppl 3:221–227
14. Lambeth JD, Seybert DW, Lancaster JR Jr, Salerno JC, Kamin H (1982) Mol Cell Biochem 45:13–31
15. Lu AYH (1976) Fed Proc 35:2460–2463
16. Sugiyama T, Miki N, Yamano T (1980) J Biochem 87:1457–1467
17. Weirich GF, Adams JR (1984) Arch Insect Biochem Physiol (in press)
18. Weirich GF (unpublished observations)
19. Sato H, Ashida N, Suhara K, Itagaki E, Takemori S, Katagiri M (1978) Arch Biochem Biophys 190:307–314
20. Suhara K, Gomi T, Sato H, Itagaki E, Takemori S, Katagiri M (1978) Arch Biochem Biophys 190:290–299
21. Thomas PE, Lu AYH, Ryan D, West SB, Kawalek J, Levin W (1976) Mol Pharmacol 12:746–758
22. Feyereisen R, Durst F (1980) In: Coon MJ, Conney AH, Estabrook RW, Gelboin HV, Gillette JR, O'Brien JP (eds) Microsomes, drug oxidations, and chemical carcinogenesis, vol I. Academic Press, London New York, pp 595–598

II.5 Ecdysteroids and Integuments in Locusts

P. PORCHERON, J.-P. CARUELLE, J.-C. BAEHR, and P. CASSIER[1]

1. Introduction

Since the previous Symposium on Invertebrate Hormones, our research has turned towards the study of the role which molting hormones play in the control of differentiation of the integument during the post-embryonic development of locusts.

The ultrastructural studies performed in this laboratory on the in vivo evolution of the larval integument of locusts during the molt cycle enables us to present the following sequence of events which take place within the tissues [1-5] (Fig. 1):

1. During the first stage of the molting cycle, post-exuvial endocuticular layers are laid down.
2. The second stage is characterized by cellular modifications of the epidermis and ends with apolysis.
3. During the third phase the epicuticle and the procuticular layers of the new cuticle are deposited.

We have correlated these various integumental events with the fluctuations of the blood titres of juvenile hormones and ecdysteroids [5-12].

The fluctuation of the ecdysteroid titres (Fig. 1) show that the first two stages of evolution of the integument (stages of deposition of post-exuvial endocuticular layers and of cell activation) occur in the presence of a low level of hormones (40 n*M*-400 n*M*) while the deposition of the new preexuvial cuticle corresponds to a period of significant increase in the level of circulating hormones (10 μM-20 μM).

Taking into account the relative specificity of the antibodies used in the assay, the quantitative variations observed could conceal qualitative variations. In fact, when ecdysone and 20-hydroxyecdysone are assayed separately, ecdysone is found to be proportionally more abundant during the first two stages of the cycle whereas 20-hydroxyecdysone is the major hormonal form during the high peak of circulating ecdysteroids at the time of the deposition of the new cuticle [9, 13-15].

Many investigations have been devoted to the action of molting hormones on morphogenesis and in vitro cuticle deposition by tissues of holometabolous insects [16, 17] while a few only concern heterometaboles [18-25].

1 ERA 620, Cytophysiologie des Arthropodes, Université Pierre et Marie Curie, 105, Boulevard Raspail, 75006 Paris, France

Biosynthesis, Metabolism and Mode of Action of Invertebrate Hormones
(ed. by J. Hoffmann and M. Porchet)

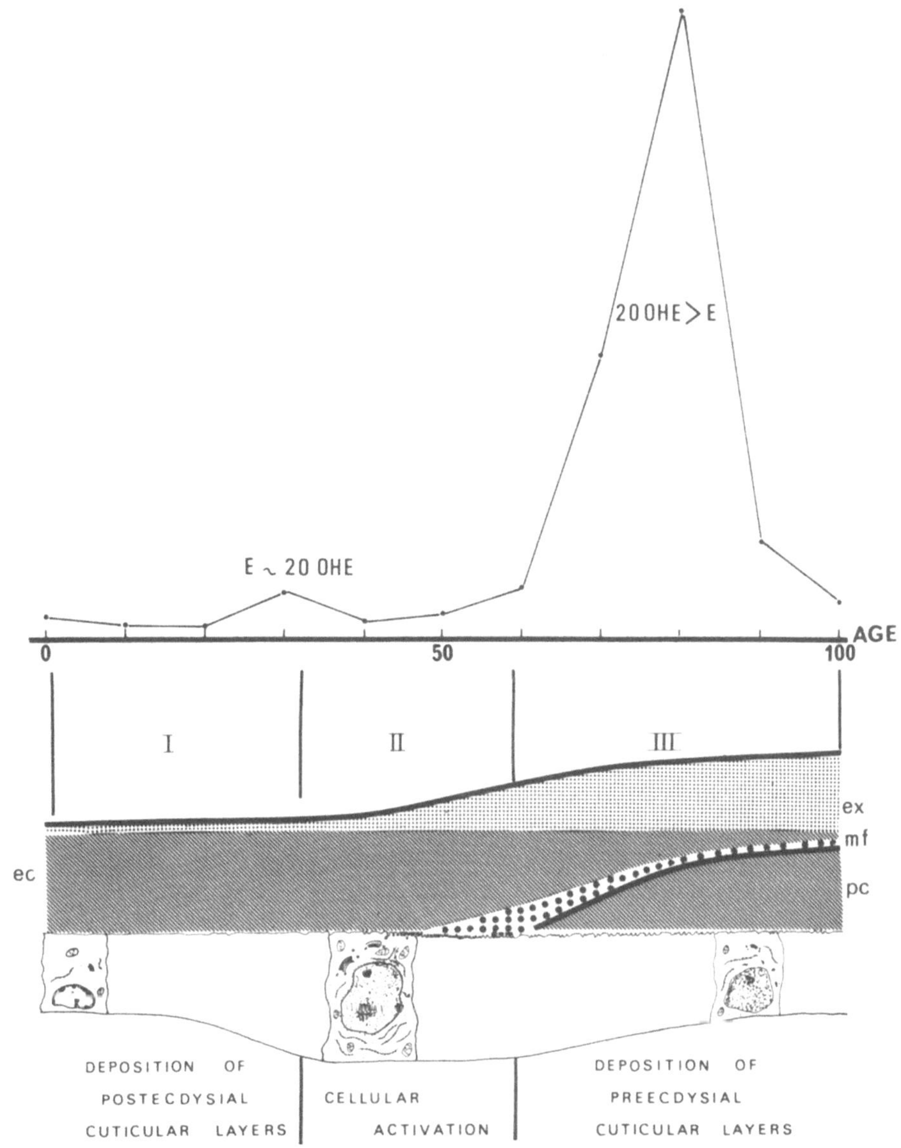

Fig. 1. Schematic representation of the relationship between ecdysteroids levels in the hemolymph and morphogenetic events in the integuments during molting cycle of locusts. Age is expressed as a percentage of total duration of larval instar. *E* ecdysone, *20 OHE* 20 hydroxyecdysone, *ec* endocuticule, *ex* exocuticule; *mf* molting fluid, *pc* procuticule

We have therefore decided to undertake a precise study on the roles exerted by ecdysteroids in the integumental molt-intermolt cycle; this study is based on the conjunction of ultrastructural techniques, organ culture and radioimmunoassay of hormones.

This paper constitutes the synthesis of a series of published papers and includes original data obtained in this field by our group. The conclusions offered herein will be discussed in the light of the data presented in literature.

2. In Vitro Studies on the Morphogenetic Role of Synthetic Ecdysteroids on the Integument of Locusts

2.1 Controls: the Fate of Explanted Integuments in Hormone-Free Medium

During our experiments, integumentary explants (abdominal tergites) were excised from young larvae of 4th instar *Schistocerca gregaria* or from larvae of 5th instar *Locusta migratoria* in their postexuvial period (stage 1, defined above). The integuments were cultured for 3 days at 30 °C in 1 ml of Landureau S20 medium [31], containing 2% of fetal calf serum. At the end of the culture, explants were withdrawn and fixed for electron microscopy study. The evolution of the integument was assessed taking into consideration two series of criteria corresponding to the implementing of the morphogenetical stages (stage 2–stage 3) described in vivo: importance of cytological alterations within the epidermal cells, possibilities of secretion and deposition of new cuticular structures (epicuticle–procuticle).

After 72 h of culture in a hormone-free medium, the integumentary structures had undergone few alterations (Fig. 2). The epidermis was not very thick and epidermal cells kept their quiescent aspects. The nuclei were parallel to the basal lamina, the cytoplasm was poor in organelles but rich in microcrystals. A dense endocuticular material was continuously secreted along with the deep layers of the old cuticle.

2.2 Induction of Cellular Activation by Addition of Ecdysone

After 3 days of culture in an ecdysone-supplemented medium, the epidermal cells revealed characteristic cytological changes (Fig. 3): increase in cell height, increase in nuclear volume and orientation of nuclei with respect to the apicobasal axis of the cells; proliferation of granular endoplasmic reticulum; activity of Golgi apparatus; chondriome development and intensification of the pinocytosis processes at the apex of the cell; glycogen accumulation in the infranuclear area; increase in number of mitoses, entailing more incorporation of radioactive thymidine in epidermal cell nuclei. These phenomena all occurred in the presence of ecdysone concentrations ranging from 2 n*M* to 200 n*M*.

When 20-hydroxyecdysone was added in similar concentrations, no comparable cell changes were observed.

The addition of Juvenile Hormone III to the culture media at concentrations ranging from 3 to 450 n*M* inhibited the action of ecdysone and prevented the changes of the epidermal cells. This inhibiting effect was not observed when Juvenile Hormone I was added in place of Juvenile Hormone III.

2.3 Induction of Secretion and Deposition of Cuticular Layers by Addition of 20-Hydroxyecdysone

When 20-hydroxyecdysone was added at a concentration of 200 n*M* to 2 μM, within 3 days of culture the integument underwent a true apolysis with detachment of the

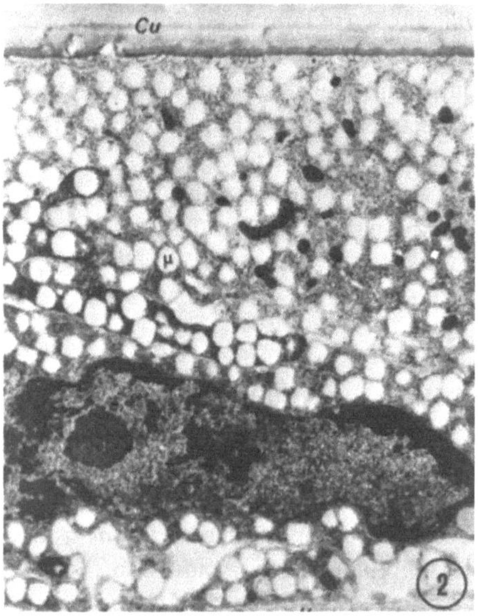

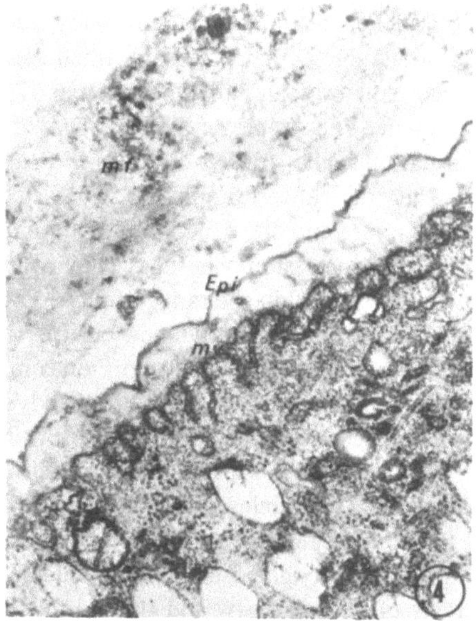

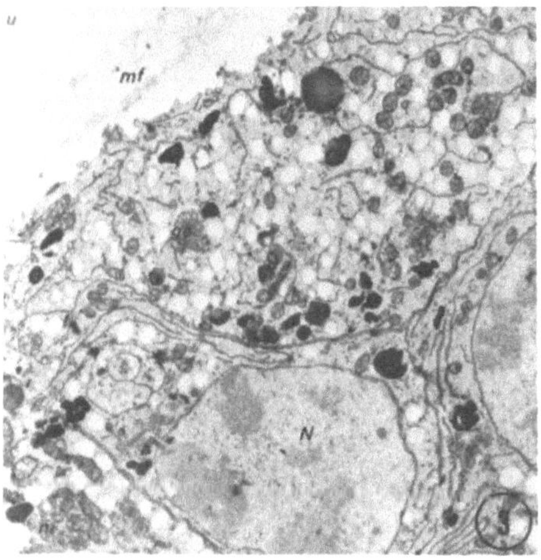

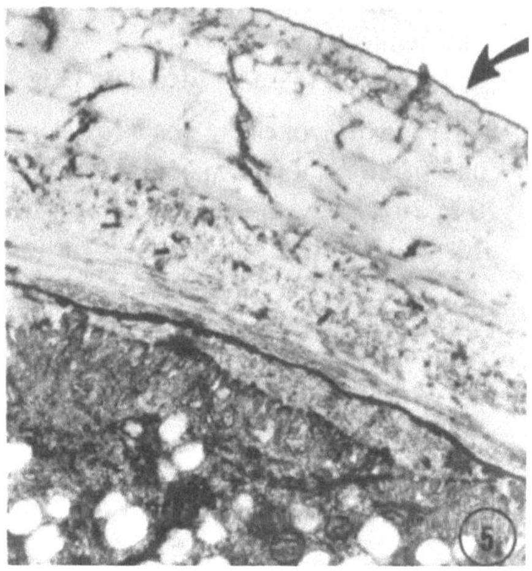

Fig. 2. 2-day-old 4th instar larva abdominal integuments of *Schistocerca gregaria* after 3 days of culture in a hormone-free medium. *Cu* old cuticle, *μ* microcrystals, *N* nucleus

Fig. 3. 2-day-old 4th instar larva abdominal integuments of *Schistocerca gregaria* after 3 days of culture is a medium containing 100 ng ml^{-1} of ecdysone. Epidermal cells are activated (*N* nucleus, *go* Golgi apparatus, *m* mitochondria) and scarce fibrous endocuticular material (*mf*) is secreted under old cuticle (*Cu*)

Fig. 4. 2-day-old 4th instar larva abdominal integuments of *Schistocerca gregaria* after 3 days of culture in a medium containing 100 ng ml^{-1} of 20-hydroxyecdysone. Microvilli (*mv*) appeared at the apex of epidermal cells and new epicuticule (*Epi*) is secreted

Fig. 5. 2-day-old 4th instar larva abdominal integuments of *Schistocerca gregaria* after 3 days of culture in a medium containing more than 1 μg ml^{-1} of 20 hydroxyecdysone. A complete new cuticle (*arrow*) and a second supernumerary epicuticule are deposited

old cuticle and development of many microvilli at the epidermal cell apex. In contact with the microvilli, a perfectly formed epicuticle was deposited (Fig. 4).

Subsequently, an increasing number of procuticular lamellae was deposited under this epicuticle. The deepest cuticular fibers of the procuticle did not show any orientation, a situation similar to that observed in the subcuticle in vivo.

When the concentrations of 20-hydroxyecdysone were increased beyond 2 μM the manufacture and the deposition of a complete and well organized cuticle was observed as well as a succession of cuticles which were reduced to the external epicuticle (Fig. 5).

Ecdysone within the range of physiological concentrations used (2 nM–2 μM) never induced the deposition of a new cuticle.

The presence of juvenile hormones in the medium altered the response of integuments to the action of 20-hydroxyecdysone: JHI increased the importance of secretions and cuticular depositions and lowered the threshold of response to 20-hydroxyecdysone, whereas JH III decreased the secretory activity; in its presence, the repeated depositions of cuticles were not observed.

Autoradiographic studies showed that 20-hydroxyecdysone promoted the incorporation of N-acetylglucosamine and leucine into the epidermal cells and into the deep layers of both the new and the old cuticles. Nevertheless, the new cuticle was, in this case, made up of extraepidermal proteins, the majority of which were taken up from the culture medium.

To summarize these in vitro data (Fig. 6), we may state that:

- Ecdysone provokes in postexuvial integuments cell changes prior to apolysis, at concentrations lower than 200 nM;
- 20-hydroxyecdysone induces apolysis, epidermal protein synthesis, epicuticle and procuticle depositions at concentrations ranging from 200 nM to 2 μM.

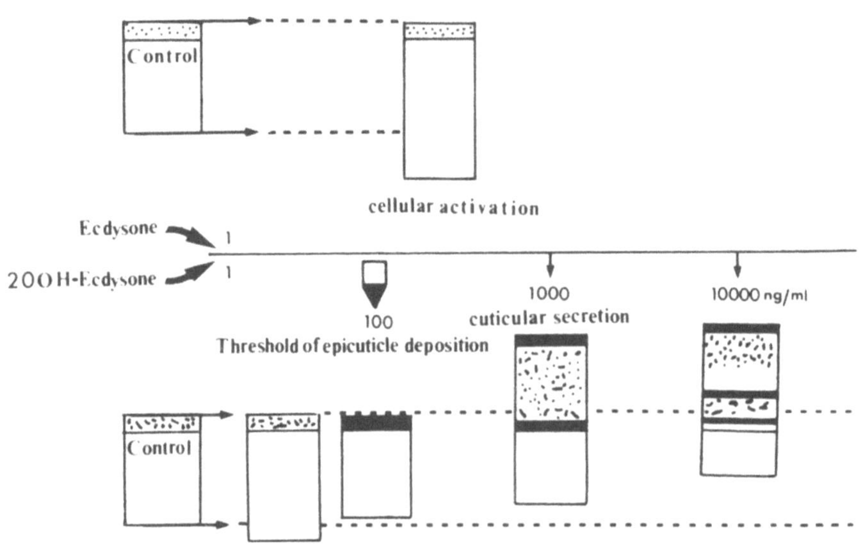

Fig. 6. Schematic representation of in vitro actions of ecdysone and 20-hydroxyecdysone on locusts' integuments. This diagram shows the effects of the hormones on cuticular secretion or cellular activation, and the range of their effective concentrations

The effects observed in vitro, as well as the hormonal concentrations used to elicit these effects, correspond to the sequence of events described in vivo (cf. Introduction) and under these conditions, it is possible to conclude that ecdysone and 20-hydroxyecdysone exert different roles in the morphogenesis of the integument.

This conclusion agrees with the hypothesis already put forward by Clever et al. in 1973 [32] regarding the specific actions of ecdysone and 20-hydroxyecdysone on the induction of puffs on the chromosomes of the salivary glands in *Chironomus tentans*.

The efficiency of 20-hydroxyecdysone in triggering apolysis and cuticular material deposition is generally accepted and was revealed in vitro by several authors [33-42].

It is believed that this hormone acts at a high concentration, allowing:

- on one hand, the implementing, in the form of a synthesis of specific proteins, of a program previously established within epidermal cells [43-45].
- on the other hand, the use of proteins of extraepidermal origin [5].

This action happens in vivo during the second half of the molting cycle after a critical period corresponding to a mitotic crisis [4, 25, 46].

Ecdysone would act at a low level by inducing cell changes necessary to the subsequent action of 20-hydroxyecdysone.

This specific role of ecdysone in epidermal differentiation has also been disclosed in *Tenebrio molitor* [17]. At the end of the first part of the molting cycle, the short and low increase in ecdysone rates probably brings about, in locusts, a commitment of epidermal cells, similar to that described in Lepidoptera and Diptera, characterized by the synthesis of specific messenger RNA [45, 47-51].

The role of juvenile hormones is more difficult to interpret in our experiments. However, in vitro, JH I seems to promote the action of ecdysteroids, while JH III apparently plays an antagonistic role.

3. In Vitro Release of Ecdysteroids by Integuments of Locusts

During the culture experiments described above we had noted that some integumentary explants evolved spontaneously, in vitro, without addition of exogenous hormone.

In fact, in the culture media which contained these integuments, ecdysteroids could be detected by radioimmunoassay at the end of the experiments.

3.1 Quantitative Data

The comparison between ecdysteroid contents of integuments of 4th larval instar *Schistocerca gregaria* at the time of the excision and in culture media after incubation of similar integuments for 24 h led to the following remarks [5, 9, 30].

- Some integuments were able to release quantities of ecdysteroids superior to those which could be disclosed in tissues upon withdrawing. The difference could reach a factor 100 for some explants from the 1st day.
- Integuments came from animals in their post-exuvial period or at the beginning of apolysis.

- Under our experimental conditions, only 20% of integuments withdrawn during these periods showed important capacities of ecdysteroid release.

From a research work on 5th stage larval integuments of *Locusta migratoria*, the same type of results as those previously obtained in *Schistocerca gregaria* is emerging (Fig. 7).

When the ratio between the quantity of ecdysteroids released by integuments after a 24-h culture and that which existed within tissues before incubation is plotted, one realizes that this ratio is maximum for integuments coming from donors at the beginning of the stage; it decreases during the whole period of cuticle stabilization and increases slightly again at the time when a high peak of ecdysteroids appears in the hemolymph.

These data raise the question of the biological relevance of this phenomenon observed under in vitro conditions. In particular, one may ask whether the integument can function as a source of hormones at certain periods of the postembryonic development, either by synthesizing ecdysteroids de novo or by recycling previously stored metabolites.

The absolute quantities of this release are nevertheless difficult to assess. As a matter of fact, whereas the values obtained by radioimmunoassay in extracts of integuments are relatively precise, since ecdysone and 20-hydroxyecdysone represent the majority of the immunoreactive products, this is not always the case for the values obtained from extracts of culture media, owing to the variety of released ecdysteroids.

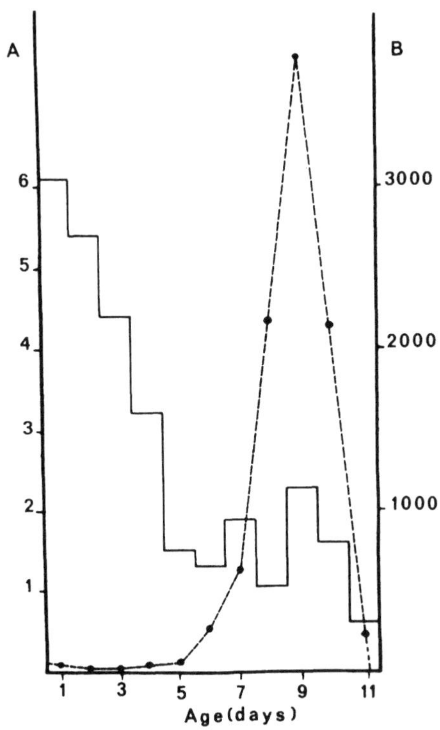

Fig. 7. Capacity of in vitro release of ecdysteroids by integuments of last larval instar of *Locusta migratoria* as a function of donor's age. *A* Ratio between the quantities of ecdysteroids present in the medium after 24 h culture of a piece of integument and in the integument at the time of withdrawing (*full line*). *B* Evolution of ecdysteroid titres of the donor's animals expressed in picograms per microliter of hemolymph (*dotted line*)

3.2 Qualitative Data

Medium in which integumentary fragments from either 3- or 9-day-old fifth instar larvae had been incubated were extracted and the ecdysteroids were separated by reverse phase HPLC followed by RIA of the collected fractions. As shown in Fig. 8, the elution profiles of the immunoreactive products are noticeably different.

In both cases, products comigrating with 20-hydroxyecdysone and ecdysone are present. They represent more than 80% of immunoreactive products in media containing integuments of 9-day-old animals, but less than 40% in media containing integuments of 3-day-old animals.

In the case of media containing integuments of 3-day-old animals, more immunoreactive compounds are observed which are either more or less polar than ecdysone.

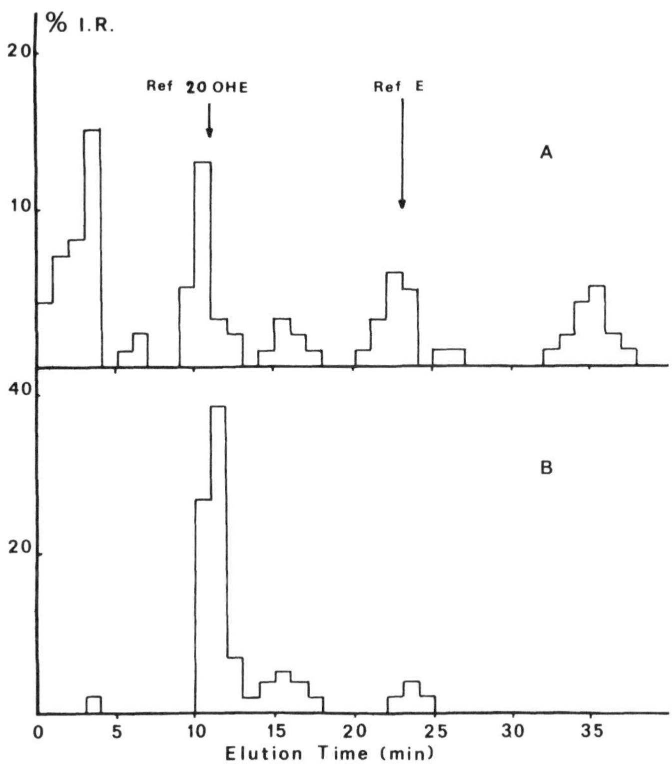

Fig. 8 A,B. Separation by reversed phase HPLC of immunoreactive products extracted from culture medium containing for 24 h integuments of fifth instar larvae of *Locusta migratoria*. **A** 3-day-old fifth instar larvae. **B** 9-day-old fifth instar larvae. Results of radioimmunoassay measurements of ecdysteroids in eluted fractions are expressed as percentage of total immunoreactivity of the extract (*% I.R.*). *Ref E* and *Ref 200HE* give migration of reference ecdysone and reference 20 hydroxyecdysone

3.3 Discussion

There exists in the literature a wealth of reports on the occurrence of concentration peaks of ecdysteroids in the absence of prothoracic glands [52–58], either after natural degeneration or surgical removal of these glands; in addition, a variety of isolated organs or tissues have been repeatedly claimed to synthesize ecdysteroids under in vitro conditions [59–64]. The production of ecdysteroids outside the prothoracic glands (or equivalent structures) appears therefore an appealing hypothesis.

Our observations show that larval integuments of locusts can release in vitro significant quantities of ecdysteroids. Since our antibodies bind preferentially ecdysone and 20-hydroxyecdysone, the absolute quantities of ecdysteroids released by integumentary explants could be underestimated, particularly when donors are in their preapolysis stage.

The purification and characterization of all ecdysteroids present in the integument and their exact titration is now in progress to define whether integumentary activity observed in vitro may account for significant ecdysteroid releases in vivo at some stages of the postembryonic development of insects.

Ecdysteroid production by integuments has also been observed in *Tenebrio molitor* [17, 65]. The common embryologic origin of epidermis and prothoracic glands should be recalled in this context. The situation of apterygote insects and of many Crustacea [66] is also of interest in this connection since they ensure their growth through moults in the absence of any gland anatomically isolated from the epidermis.

Ecdysteroid production could therefore be regarded as one of the primary properties of epidermal structures. Differentiated moulting glands on their behalf might play a role in the synchronization of the activity of various epidermal territories via the ecdysone they secrete, or through a molecule related to this secretion [67, 68].

4. Conclusion

Organ cultures of integumentary explants of Locusts allowed us to disclose specific roles for ecdysone and 20-hydroxyecdysone in integumentary morphogenesis. In addition the release of ecdysteroids by these explants raises a number of challenging questions.

This model can serve for further studies aimed at:

- On the one hand investigating the metabolic pathways of the molting hormone within a target tissue.
- On the other hand, the relevance of the in situ ecdysteroid metabolism and the role of the resulting metabolites on the biochemical differentiation of the integuments.

References

1. Cassier P, Joulie-Delorme C (1976) Arch Zool Exp Gen 117:95–116
2. Cassier P, Joulie-Delorme C (1976) Ann Sc Nat Zool 18:295–309
3. Cassier P, Joulie-Delorme C (1976) Insectes sociaux 23:179–198
4. Caruelle JP, Caruelle D (1978) CR Acad Sci Paris 286:339–341
5. Caruelle JP (1980) Thèse Doctorat d'Etat, Paris, p 140
6. Porcheron P, Cassier P, Dray F (1977) Bull Soc Zool Fr, Suppl 1:71–75
7. Baehr JC, Porcheron P, Papillon M, Dray F (1979) J Insect Physiol 25:415–421
8. Papillon M, Porcheron P, Baehr JC (1980) Experientia 36:419–422
9. Porcheron P (1979) Thèse Doctorat d'Etat, Paris, p 182
10. Porcheron P, Foucrier J, Gros Cl, Pradelles Ph, Cassier P, Dray F (1976) Febs Letters 61:159 to 162
11. Baehr JC, Pradelle Ph, Lebreux C, Cassier P, Dray F (1976) Febs Letters 69:123–128
12. Baehr JC, Pradelles Ph, Dray F (1978) Ann Biol Bioch Biophys 19:1827–1836
13. Morgan ED, Poole CF (1975) J Insect Physiol 22:885–889
14. Gande AR, Morgan ED, Wilson ID (1979) J Insect Physiol 25:669–675
15. Hirn H, Hetru C, Lagueux M, Hoffmann JA (1979) J Insect Physiol 25:255–261
16. Marks EP (1980) Ann Rev Entomol 25:73–101
17. Quennedey A, Quennedey B, Delbecque JP, Delachambre J (1983) Cell Tissue Res 232:493 to 511
18. Bullière F, Bullière D (1973) CR Acad Sc Paris 276:2533–2536
19. Bullière F, Bullière D (1975) CR Acad Sc Paris 281:925–928
20. Marks EP (1970) Gen Comp Endocrinol 15:289–302
21. Marks EP (1972) Biol Bull 142:293–501
22. Marks EP (1973) Gen Comp Endocrinol 21:472–477
23. Marks EP, Leopold RA (1971) Biol Bull 140:73–83
24. Marks EP, Sowa BA (1976) In: Hepburn HR (ed) The Insect integument. Elsevier, Amsterdam, pp 339–357
25. Miciarelli AS, Sbrenna G, Colombo G (1967) Experientia 23:64–66
26. Caruelle JP, Caruelle D (1978) CR Acad Sci Paris 286:969–972
27. Caruelle JP, Cassier P (1979) Arch Anat Microsc 68:127–138
28. Caruelle JP, Cassier P, Joulie-Delorme C (1978) Ann Sc Nat Zool 20:53–78
29. Caruelle JP, Caruelle D, Cassier P (1979) Ann Sc Nat Zool 13:49–56
30. Cassier P, Baehr JC, Caruelle JP, Porcheron P, Claret J (1980) In: Hoffmann JA (ed) Progress in ecdysone research. Elsevier, Amsterdam, pp 235–246
31. Landureau C, Grellet P (1972) CR Acad Sci Paris 274:1372–1375
32. Clever U, Clever I, Storbeck I, Young NL (1973) Develop Biol 31:47–60
33. Mandaron P (1970) Develop Biol 22:298–320
34. Mandaron P (1973) Develop Biol 31:101–113
35. Marks EP, Sowa BA (1976) In: Hepburn HR (ed) The Insect integument. Elsevier, Amsterdam, pp 339–357
36. Mitsui T, Riddiford LM (1976) Develop Biol 54:172–186
37. Oberlander H (1976) In vitro 12:225–235
38. Agui N (1977) J Insect Physiol 23:837–842
39. Fain MJ, Riddiford LM (1977) Wilhelm Roux' Archiv 181:285–307
40. Blais C, Lafont R (1980) Wilhelm Roux' Arch Develop Biol 188:27–36
41. Ryerse JS, Locke M (1978) J Insect Physiol 24:541–550
42. Wielgus JJ, Gilbert LI (1978) J Insect Physiol 24:629–637
43. Riddiford LM (1978) Gen Comp Endocrinol 34:438–446
44. Riddiford LM (1980) Ann Rev Physiol 42:511–528
45. Riddiford LM (1981) Amer Zool 21:751–762
46. Delachambre J, Besson MT, Connat JL, Delbecque JP (1980) In: Hoffmann JA (ed) Progress in ecdysone research. Elsevier, Amsterdam, pp 211–234

47. Shaaya (1976) Insect Biochem 6:553–559
48. Dean RL, Bollenbacher WE, Locke M, Smith SL, Gilbert LI (1980) J Insect Physiol 26:267 to 280
49. Riddiford LM, Chen AC, Graves BJ, Curtis AT (1981) Insect Biochem 11:121–127
50. Riddiford LM (1982) Develop Biol 31:101–113
51. Shaaya E, Levenbook L (1982) Insect Biochem 12:663–667
52. Srivastava US (1960) Experientia 16:445–446
53. Hsiao TH, Hsiao C, Dè Wilde J (1975) Nature 255:727–728
54. Delbecque JP, Delachambre J, Hirn M, De Reggi M (1978) Gen Comp Endocrinol 35:436–444
55. Glitho I, Delbecque JP, Delachambre J (1979) J Insect Physiol 25:187–191
56. Delbecque JP, Slama K (1980) Z Naturforsch 35:1066–1080
57. Brehelin M, Aubry R (1982) Bull Soc Zool Fr 1:21–32
58. Redfern CPF (1983) J Insect Physiol 29:65–71
59. Gersch M, Stürzebecher J (1971) Experientia 27:1475–1476
60. Nakanishi K, Moriyama H, Okauchi T, Fujioka S, Koreeda M (1972) Science 176:51–52
61. Romer F, Emmerich H, Novock J (1974) J Insect Physiol 20:1975–1987
62. Studinger G, Willig A (1975) J Insect Physiol 21:1793–1798
63. Gersch M, Eibisch H (1977) Experientia 33:468
64. Loeb MJ, Woods CW, Brandt EP, Bořkovec AB (1982) Science 218:896–898
65. Delbecque JP, Quennedey A, Besson MT, Delachambre J (1983) 2ème coll internat CNRS Biosynthese Metabolisme, mode d'action, hormones d'Invertébrés, Strasbourg
66. Spindler KD, Keller R, O'Connor JD (1980) In: Hoffmann JA (ed) Progress in ecdysone research. Elsevier, Amsterdam, pp 247–280
67. Joly L, Weins MJ, Hoffmann JA, Porte A (1973) Z Zellforsch 137:387–397
68. Charlet M, Hoffmann JA (1981) 6ème coll Physiol Insecte, Les Eyzies

II.6 Relationships Between Hormones and Epidermal Cell Cycles During the Metamorphosis of *Tenebrio molitor*

J. DELACHAMBRE, M.T. BESSON, A. QUENNEDEY, and J.P. DELBECQUE[1]

1. Introduction

The epidermis of holometabolous insects is a very attractive model for studies in developmental biology, particularly during metamorphosis: in a relatively short time indeed, epidermal cells change their program twice (larval-pupal and pupal-adult reprogramings), undergo several rounds of differentiative and proliferative mitoses, and successively elaborate completely different cuticles, leading to pupal and then adult structures.

Such complex and rapid changes proceed through a precise sequence which is undoubtedly under the control of hormones: the role of ecdysteroids in the initiation of cuticle deposition is generally admitted. The balance between juvenile hormones and ecdysteroids is also known to control the program changes [1].

Numerous studies have investigated the role of these hormones on DNA synthesis and mitoses in target tissues, but the various results so far obtained do not allow a precise and general interpretation [1, 2].

In the present paper, we have reinvestigated the effects of developmental hormones on the epidermal cells of *Tenebrio*, first with the aim to establish precisely the temporal correlations between hormonal and epidermal events. Moreover, we have undertaken experiments in order to evidence possible feed-back mechanisms on hormonal variations from target-tissues.

2. Chronology of Epidermal Cell Cycle and Hormonal Events

2.1 The Main Features of Epidermal Cell Cycle in *Tenebrio*

They have been investigated during metamorphosis in abdominal epidermis of sternites by measurements of cell density and mitotic index, by ^{3}H-thymidine incorporation, and by cytophotometry after Feulgen staining of whole mounts or of epidermal squashes [3].

In the last larval instar, most epidermal cells are blocked in G2 (4C DNA content) during the long period from larval ecdysis to larval-pupal apolysis (Fig. 1a). After apo-

1 Laboratoire de Zoologie, Université de Dijon, 21100 Dijon, France
(Equipe de Recherche Associée au C.N.R.S. no 231)

Biosynthesis, Metabolism and Mode of Action of Invertebrate Hormones
(ed. by J. Hoffmann and M. Porchet)

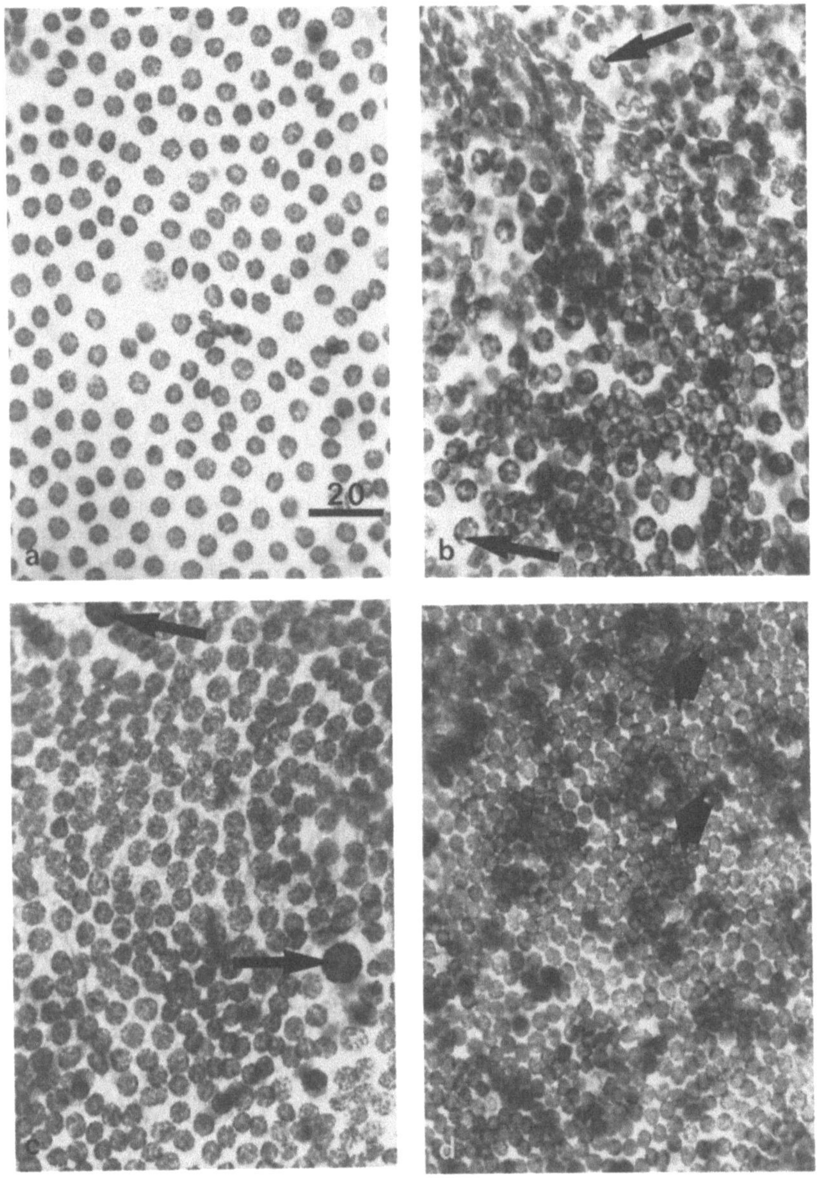

Fig. 1 a–d. Whole mounts of sternal epidermis after Feulgen staining. **a** Last larval instar before the mitotic period (Scale: 20 μm). **b** During the larval-pupal mitotic period numerous small G1 and G1–S nuclei are observable, together with polyploidizing oenocytes (*arrows*). **c** G2 nuclei of the pupal epidermis before the pupal-adult mitotic period; 32 C oenocytes are observable (*arrows*). **d** S and G2 nuclei at the end of the pupal-adult mitotic period. Note the high cell density and the formation of pit glands (*arrows*)

lysis, some epidermal cells become polyploid (8–32 C) and differentiate into oenocytes; others enter into mitosis (M), then go into G1 and duplicate their DNA (Fig. 1b). The peak of DNA synthesis for the normal epidermal cells occurs 2 days after that of oenocyte polyploidization. Two days before pupal ecdysis, most epidermal cells become G2 blocked (Fig. 1c) until the beginning of the next mitotic crisis (36 h after pupal ecdysis) This crisis, associated with DNA duplication, is two- to threefold stronger than the larval one and leads to a considerable increase of the cell density (Fig. 1d). Five days after pupal ecdysis, about one epidermal cell out of two degenerates and the other remains G2 blocked for the whole adult life.

Figure 2 illustrates schematically this peculiar cycle of *Tenebrio* epidermis, characterized by a long G2 block and a fast release into M, G1, and S phases during a short mitotic period. Several other studies have also emphasized the peculiar length of G2 phase during the differentiation of imaginal disks in *Ephestia* [4], *Calliphora* [5, 6], and *Drosophila* [7]. In the epidermis of other holometabolous insects, a substantial proportion of 4C and 8C nuclei has been observed during the last larval instar of *Manduca* [8] and *Bombyx* [9]. In *Galleria* [10], epidermal cells are also G2 blocked in a very similar way as in *Tenebrio*.

On the contrary, in *Schistocerca* [11] and in *Oncopeltus* [12] such long G2 phases have not been observed: though further investigations on other species are needed, it is thus possible that the cycles with long G2 phases are a characteristic of holometabolous insects.

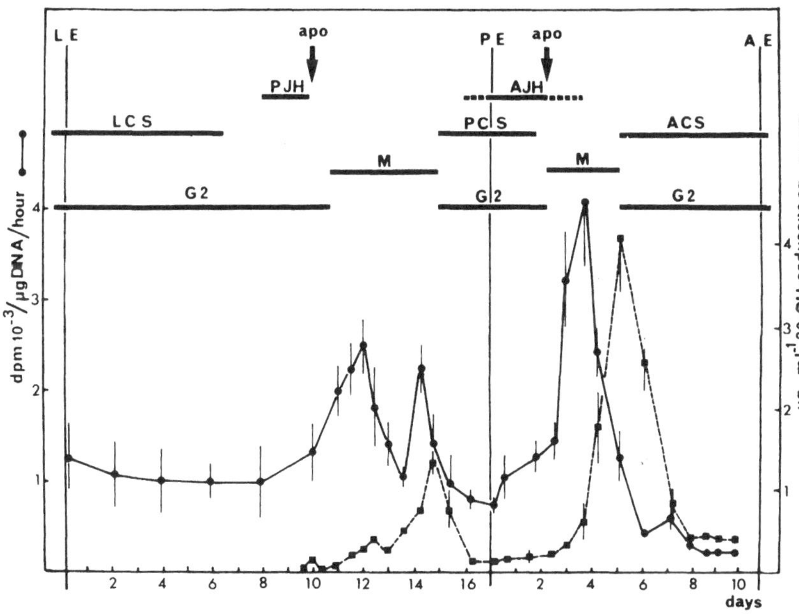

Fig. 2. Temporal correlations between the different cell events in the epidermis and hormonal variations during the last larval and pupal stages. *Points:* 3H-thymidine incorporation (± SEM). *Squares* and *dotted lines:* ecdysteroid titers in the hemolymph as measured by RIA. *Abbreviations: ACS* adult preecdysial cuticle secretion, *AE* adult ecdysis, *AJH* juvenoid sensitivity period for the adult program, *APO* (*arrows*) apolysis, *G2* G2 arrest periods, *LCS* larval postecdysial cuticle secretion, *LE* larval ecdysis, *M* mitoses, *PCS* pupal preecdysial cuticle secretion, *PE* pupal ecdysis, *PJH* juvenoid sensitivity for pupal program

2.2 The Correlations with Hormonal Events

From the above results, it is clear that the control of G2 arrest and G2 opening is the central piece of the control of metamorphosis. In order to see whether hormones are involved in these processes, temporal correlations have been evidenced during the metamorphosis of *Tenebrio* (Fig. 2): the G2 arrest coincides with the beginning of cuticle secretion and with a strong increase of ecdysteroid titers in last instar larvae as in pupae. The G2 opening, on the contrary, does not appear identical in the two successive stages: in the last larval instar it follows the larval-pupal apolysis, whereas in pupae it begins before the pupal-adult apolysis. In both cases, however, it occurs under low ecdysteroid titers. The DNA synthesis associated to the two mitotic crises also begins under low ecdysteroid titers and reaches a maximum about 36 h before the main ecdysteroid peaks.

Though the variations of juvenile hormone (JH) titers are unknown during the metamorphosis of *Tenebrio*, the effects of exogenous JH or juvenoids are well-known in this insect [13-15] and can be summarized as follows: (1) two periods of JH-sensitivity have been observed, the first one immediately before larval-pupal apolysis in the last larval instar and the second at the beginning of the pupal stage. It is likely that the endogenous JH titers are lowest during these two periods. (2) The periods of sensitivity occur before the expression of a new program and though they have not the same temporal importance, they occur close to the G2 opening, either before (in the case of the last larval instar) or during (in pupae).

From these correlations, it appears probable that close relationships could be evidenced between developmental hormones and cell cycles in *Tenebrio*. The pupal-adult transformation being the more intensive, further experimental investigations have been done essentially on this period.

3. Effects of Developmental Hormones on the Cell Cycle

3.1 Experimental Design

The aim of our study was to investigate the effects of juvenile hormone analogues and ecdysteroids immediately before, during, and after pupal ecdysis, i.e., from the G2 arrest in the last larval instar to the adult G2 arrest in the pupa. Juvenoids (mainly K–421d, given by Dr. K. Sláma and also ZR 515, Zoecon Corp.) were applied topically in 1 μl acetone at a dosage of 0.1 μg/animal. 20-Hydroxyecdysone was injected at a dose of 10 μg (dissolved in 3 μl saline) per animal. Hormonal effects were analyzed by cell density measurements and thymidine incorporations in abdominal sternal epidermis. For the present time, the methodical investigation of all experimental cases is not yet exhaustive and other doses and hormones (e.g., ecdysone) have to be tested in the future. However, the following results give clear information.

3.2 Effects of Juvenoids

Topical application of juvenoid at pupal ecdysis (or during the first day of pupal stage) induces a supernumerary pupal molt: in ca. 5 days a new pupal cuticle is visible and retracts under the old one (pseudoecdysis), but the animal does not succeed to ecdyse. The animals are, thus, unable to express the adult program. Applications during the 3–4 days *before* pupal ecdysis have exactly the same effect, though less intensive: in fact, the juvenoid being slowly eliminated, it is highly probable that remaining molecules exert their action during the sensitivity period, i.e., after pupal ecdysis. This is confirmed by the observation that the peak of juvenile hormone esterase activity occurs at the beginning of the pupal stage [16]. If animals are treated 2 or 3 days after the pupal ecdysis, they molt into pupal-adult intermediates [14, 15, 17].

Cell density measurements after treatment at pupal ecdysis (Fig. 3) show that adult mitoses and subsequent DNA synthesis (Fig. 4) are inhibited. When animals are treated before pupal ecdysis (provided it is after the sensitivity period for larval-pupal molt), only pupal-adult development is perturbed and not the end of the larval-pupal molt.

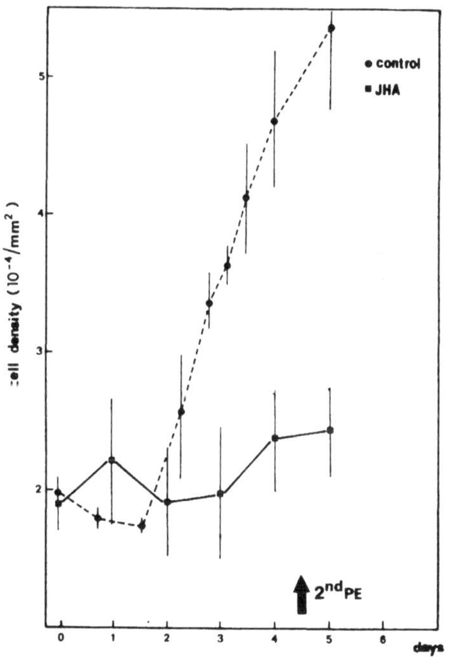

Fig. 3

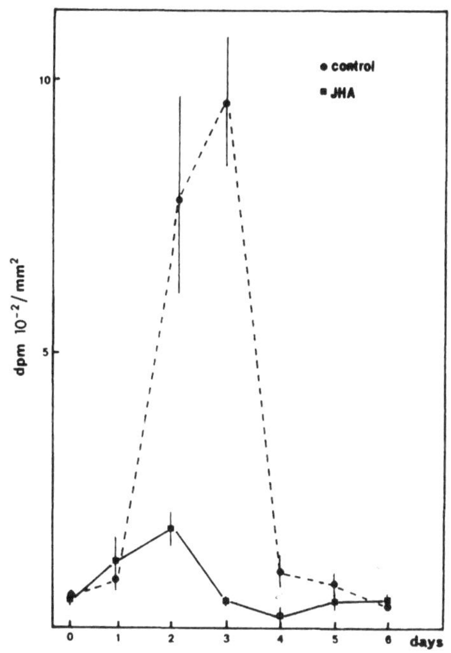

Fig. 4

Fig. 3. Evolution of the cell density in epidermis (*full line*) after topical application of the juvenoid K–421d at pupal ecdysis (day 0). Untreated controls in *dotted line*. *Arrow* indicates the supernumerary pupal (pseudo-)ecdysis in treated animals

Fig. 4. 3H-thymidine incorporation in epidermis treated at pupal ecdysis (day 0) in *full line*. The *points* of Figs. 3 and 4 represent the means (*vertical bars:* ± SEM) from four to six animals

3.3 Effects of Ecdysteroids

Unlike for juvenoids, injections of a high dose of 20-hydroxyecdysone have very different effects when injected immediately before or after the pupal ecdysis. At the end of the last larval instar, during the G2 arrest, 20-hydroxyecdysone generally delays pupal ecdysis, but does not induce mitoses or DNA synthesis (Fig. 5/1). Injections at or after pupal ecdysis, induce pupal-adult intermediate molts. Epidermal cells, which were G2 arrested at the time of injection, undergo a precocious cycle (Fig. 6) and DNA synthesis (Fig. 5/2, 3). The number of mitoses is smaller than in controls, suggesting that only one cycle occurs in presence of exogenous hormone. Cytophotometric analysis indicates that DNA contents of the daughter cells correspond to 4C and 2C–4C intermediates.

3.4 Interpretations

Our experiments suggest that hormones act on specific phases of the cell cycle. Both juvenoids and 20-hydroxyecdysone appear to have no specific effect at the end of the

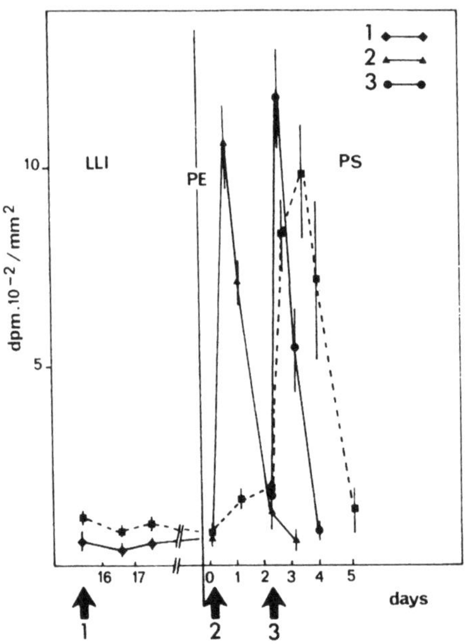

Fig. 5

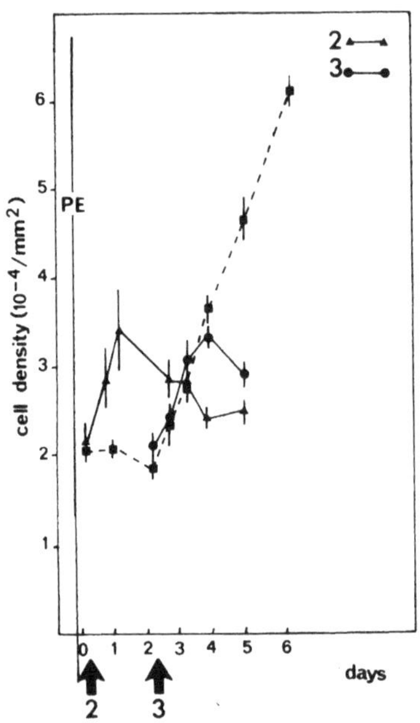

Fig. 6

Fig. 5. Effect of 20-hydroxyecdysone on 3H-thymidine incorporation. *Arrows* indicate time of injections: *1* 60 h before the pupal ecdysis, no induction of DNA synthesis is observed, *2* just after pupal ecdysis, induction of precocious DNA synthesis, *3* at the beginning of the normal DNA synthesis increase, faster but shorter induction of DNA synthesis than in controls (*dotted lines*), *LLI* last larval instar, *PE* pupal ecdysis, *PS* pupal stage

Fig. 6. Effects of injection of 20-hydroxyecdysone at pupal ecdysis (*arrow 2;* to be compared with Fig. 5) and at the beginning of the mitotic period (*arrow 3*) on the epidermal cell density. *Dotted line* controls. Measurements represent the mean (± SEM) from 3 to 12 animals

larval life, when epidermal cells are G2-arrested. However, after pupal ecdysis, these G2 cells become sensitive to hormones: juvenoids prevent them from changing their program and dividing; ecdysteroids, on the contrary, seem to quicken mitoses. It is, thus, likely that during G2, a maturation of cells occurs at pupal ecdysis, either autonomously or under the control of some ecdysial factor, which allows the G2-opening. This G2 opening or the subsequent G2–M transition can be inhibited by juvenile hormone analogues. This interpretation seems to be particularly well illustrated by the pupal-adult intermediates obtained after juvenoid applications at the end of the sensitivity period: the intensity of expression of adult program can be correlated with the intensity of mitotic crisis before JH application.

Once the epidermal cells are G2-opened, high doses of 20-hydroxyecdysone appear to fasten G2–M transitions, mitoses, and DNA synthesis. However, this rapid enhancement is also accompanied by a rapid block of mitoses and cells appear to be stopped during a G2 or a S–G2 state. Such an interpretation agrees with the results obtained on Kc cells which are arrested in G2 after 20-hydroxyecdysone action [18].

4. Influence of the Cell Cycle on Ecdysteroid Variations

The finding that the main effect of ecdysteroids is to arrest epidermal cells in G2 is consistent with the fact that the hormonal peak in the animals does not precede, but follows mitoses and DNA synthesis (see Fig. 2). Since this peak is consecutive to mitoses it cannot be regarded as being the cause of cell cycling. On the contrary, it can be hypothesized that mitoses could have a feedback effect on ecdysteroid release in order that high endogenous titers do not block cell cycling before sufficient differentiation and proliferation is reached. The following experiments have been done to verify this hypothesis.

4.1 Effects of Cell Cycle Inhibition on Ecdysteroid Levels in Vivo

Juvenoids which have been shown to be powerful inhibitors of cell cycles in *Tenebrio* also inhibit ecdysteroid release: when ZR 515 is applied before or at the beginning of the adult mitotic crisis, only a weak ecdysteroid peak is observable (Fig. 7); when applied later and later, the resulting ecdysteroid peak becomes closer and closer to the normal. It, thus, appears that there is a relation between the inhibition of mitoses by ZR 515 and the inhibition of ecdysteroid release.

As previously observed [14], the effects of juvenoids in *Tenebrio* pupae can be mimicked by inhibitors of the cell cycle. We have verified that such inhibitions also affect ecdysteroid concentrations. Cytosine arabinoside (Ara–C), when injected at the beginning of the mitotic period, arrests the epidermal cells during the S phase and affects the ecdysteroid titers (Fig. 8) leading to pupal-adult intermediates. Demecolcine, which definitively blocks the cells in M phase, completely prevents further mitoses and also completely inhibits the ecdysteroid secretion (Fig. 9) giving blocked pupae.

These experiments prove that reduction or inhibition of mitoses is followed by a similar reduction or inhibition of the ecdysteroid peak. They raise the question of the

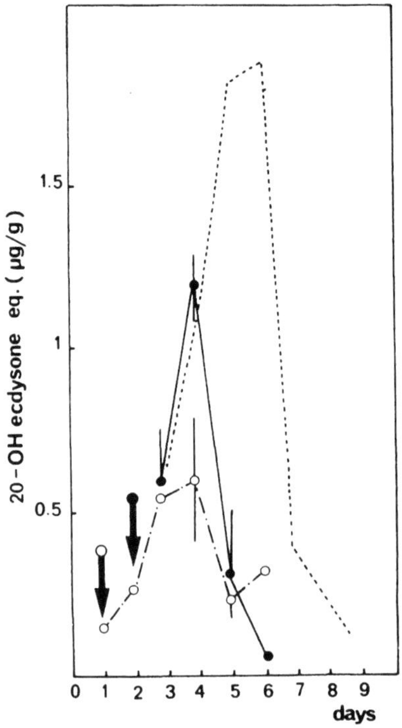

Fig. 7

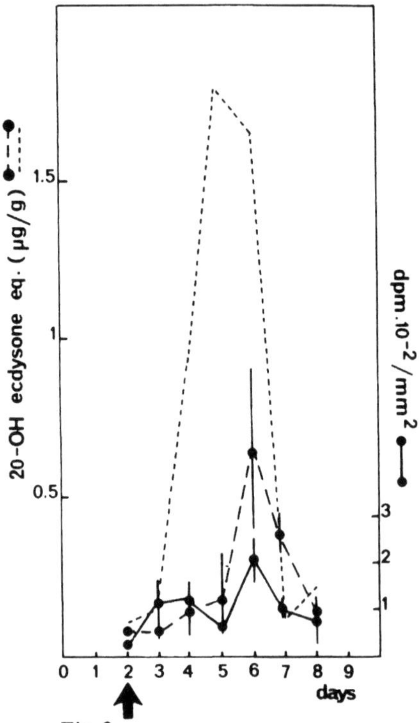

Fig. 8

Fig. 7. Effects of topical application of 0.5 µg ZR 515 on ecdysteroid contents in whole pupae. *White circles:* animals applicated 1 day after pupal ecdysis; *black points:* animals applicated at day 2; animals applicated later are identical to controls (*dotted lines*). *Points* represent the mean (± SEM) of at least six measurements with RIA

Fig. 8. Effects of Ara–C injections (200 µg/animal) at day 2 in pupae (*arrows*) on DNA synthesis (*full line*) and on ecdysteroid contents (*circles with dotted line*). Each *point* is the mean (± SEM) from three to five animals. *Small dots:* ecdysteroid contents in controls

nature of the link between mitoses and ecdysteroids, which leads to the problem of the origin of these hormones.

4.2 Origin of Ecdysteroids

In *Tenebrio* pupae, prothoracic glands are degenerated [19, 20] and it has been demonstrated by ligation experiments that isolated abdomens are capable to produce ecdysteroids in sufficient amounts for the secretion of adult cuticle [21]. As already observed [14], this autonomous development of isolated abdomens can be inhibited by juvenoids and other cell cycle inhibitors. We have verified that the treatments described for whole animals (cf. Sect. 3.1) similarly inhibit ecdysteroid production in isolated abdomens. Thus, juvenoids and cell cycle inhibitors do not inhibit ecdysteroid titers *via* brain and PGs, but *via* a peripheral organ. This organ responsible of ecdysteroid secretion is not located in a specific segment (as indicated by ligations) and has the same developmental timetable as the epidermis, as far as the cell cycle is concerned.

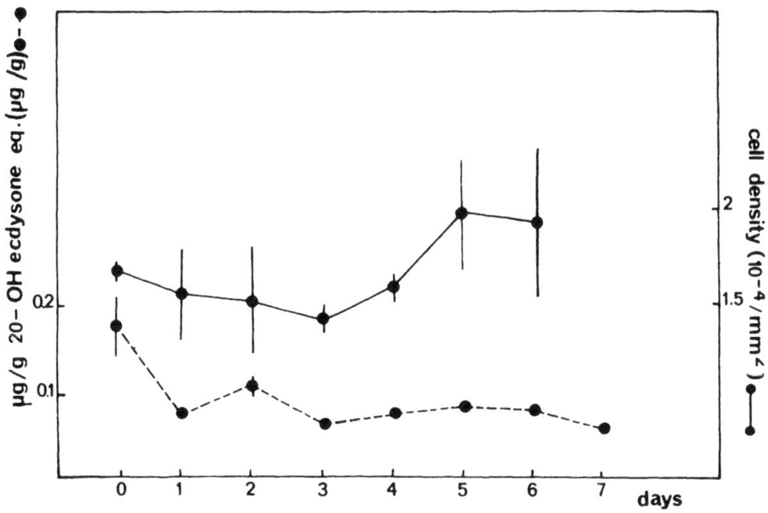

Fig. 9. Effect of demecolcine injection (10 µg/animal) at pupal ecdysis (day 0) on cell density (*full line*) and ecdysteroid contents (*dotted line*) in pupae. Each *point* represents the mean from three to six animals. *Vertical bars:* ± SEM (when absent SEM is inferior to the size of the point). See Fig. 8 for ecdysteroids in controls and Fig. 3 for their cell densities

The ability of epidermis to secrete ecdysteroids in vitro has already been observed [22]. We have verified in *Tenebrio* that various epidermal explants cultured in vitro are also able to secrete ecdysteroid-like molecules in the incubation medium and particularly ecdysone, as indicated by RIA after reverse phase HPLC (not shown). Organs which divide the most actively, such as wings, produce comparatively more ecdysteroids than other epidermal parts, such as abdominal sternites (Table 1) which confirms previous hypotheses [23, 24]. Moreover, the treatments by juvenoids or cell cycle inhibitors in vitro also reduce this ecdysteroid secretion, as shown in whole animals. This demonstrates that the production of ecdysteroids by epidermal cells is an active phenomenon clearly dependent on mitoses.

Table 1. Ecdysteroid production after a 5-day incubation of wings and sternites from 2-day-old pupae, cultured in different conditions: Normal Landureau medium; sterile water (giving controls for the initial RIA response); medium added with ZR 515 (50 µg/2 ml) and medium added with demecolcine (10 µg/2 ml). RIA results are expressed as picograms of ecdysone equivalents per mg of cultivated organ ± SEM (fresh weight of cultivated organs from 6 to 11 mg)

Conditions	Number of experiments	Wings	Sternites
Normal medium	37	2,300 ± 516	300 ± 72
Sterile water	6	197 ± 98	87 ± 18
Medium + ZR 515	6	934 ± 775	159 ± 113
Med. + demecolcine	6	958 ± 260	191 ± 63

In conclusion, our experiments suggest the following interpretations:

1. Epidermal cells, which are G2 arrested at the end of the last larval instar, undergo the G2 opening at pupal ecdysis, either autonomously or under the influence of some endocrine ecdysial factor, and become sensitive to both juvenoids and ecdysteroids.
2. Juvenoids inhibit either the adult G2–M transition or some step before.
3. The ecdysteroid peak in pupae occurs after mitoses, at the end of DNA synthesis and it appears probable that it blocks cell cycling by arresting epidermal cells in G2. This is confirmed by 20-hydroxyecdysone injections at high doses which, after enhancing temporarily mitoses and DNA synthesis, rapidly lead to a G2 arrest of epidermal cells.
4. The intensity of ecdysteroid release in vivo, but also in isolated abdomens and in epidermal explants in vitro appears to be directly dependent on the intensity of mitoses in the epidermis. Ecdysteroids appear to be a consequence and not the cause of mitoses. Such an interpretation is paradoxical since it implies the contradiction that epidermal cells are both secretory and target cells. However, due to the degeneration of PG and to the absence of control of the adult molt by the neuroendocrine system in *Tenebrio*, ecdysteroids seem to be homeostatic factors [24] and to behave as cell to cell signals triggering the cell cycle arrest.

References

1. Riddiford LM (1980) In: Hoffmann JA (ed) Progress in ecdysone research. Elsevier, Amsterdam, p 409
2. Kumaran AK (1981) In: Sehnal F, Zabra A, Menn JJ, Cymborowski B (eds) Regulation of insect development and behaviour. Tech Univ Wroclaw, Wroclaw, Poland, p 451
3. Besson-Lavoignet MT, Delachambre J (1981) Dev Biol 83:255–265
4. Löbbecke EA (1969) Wilhelm Roux' Arch Entwicklungsmech Org 162:1–18
5. Want JJL Van Der, Spreij E (1976) Neth J Zool 6:432–434
6. Egberts DJN (1979) Insect Biochem 9:89–93
7. Fain MJ, Stevens B (1982) Dev Biol 92:247–258
8. Dyer KA, Thornhill WB, Riddiford LM (1981) Dev Biol 84:425–431
9. Kato Y (1977) Zool Mag 86:250–253
10. Wolbert P, Kubbies M (1983) J Insect Physiol 29:787–794
11. Fontana F (1974) Caryologia 27:73–82
12. Lawrence PA (1968) J Cell Sci 3:391–404
13. Wigglesworth VB (1958) J Insect Physiol 2:73–84
14. Socha R, Sehnal F (1972) J Insect Physiol 18:317–337
15. Sláma K, Romaňuk M, Šorm F (1974) Insect hormones and bioanalogues. Springer, Berlin Heidelberg New York
16. Connat JL (1983) J Insect Physiol 29:515–521
17. Connat JL, Delbecque JP, Delachambre J (1984) J Insect Physiol (in press)
18. Stevens B, Alvarez CM, Bohman R, O'Connor JD (1980) Cell 22:675–682
19. Srivastava US (1960) Experientia 16:445–446
20. Glitho I, Delbecque JP, Delachambre J (1979) J Insect Physiol 25:187–191
21. Delbecque JP, Delachambre J, Hirn M, De Reggi M (1978) Gen Comp Endocrinol 35:436–444
22. Cassier P, Baehr JC, Caruelle JP, Porcheron P, Claret J (1980) In: Hoffmann JA (ed) Progress in ecdysone research. Elsevier, Amsterdam, p 235
23. Sláma K (1980) Acta Entomol Bohemoslov 77:145–168
24. Delbecque JP, Sláma K (1980) Z Naturforsch 35(c):1066–1080

II.7 On the Mode of Action of Ecdysteroids in Crustaceans

K.-D. SPINDLER[1], L. DINAN[2], and M. LONDERSHAUSEN[1]

1. Introduction

The "classical scheme" of the action of moulting hormones in crustaceans is depicted in Fig. 1. It closely resembles the situation in insects, but some details are different. For example, a stimulatory prothoracicotropic hormone is present in insects vs a moult inhibiting hormone in crustaceans. In order to elucidate the mechanism of action of ecdysteroids, crustaceans may be used with advantage, for at least certain steps of this complex sequence, not only because of the size they can reach, but also because the overall scheme is perhaps less complex than in insects. For example, there are no JH-ecdysteroid interrelationships as in insects, since JH has never been detected in crustaceans. The present paper will, therefore, focus on two areas where crustaceans have already been used with success, namely, the mechanism of uptake of moulting hormones and the presence and characterization of ecdysteroid receptors.

In recent years another aspect of ecdysteroids, apart from their role in moulting physiology, has attracted increasing interest, namely, the presence, chemical nature, and possible role of ecdysteroids during oogenesis and early development. Again, as in the moulting physiology, there are similarities between insects and crustaceans in this respect. A summary of the available data concerning the presence and significance of ecdysteroids during early development will be presented and some new findings will be added.

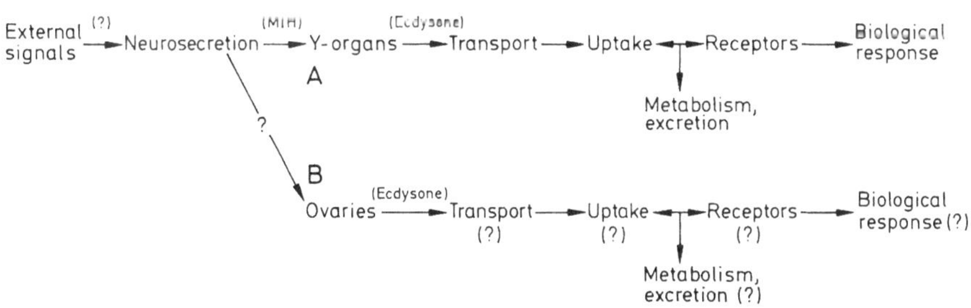

Fig. 1. Possible mechanism of action of ecdysteroids during postembryonic (*A*) and embryonic development (*B*) in crustaceans. ? = uncertain, *MIH* = moult-inhibiting hormone

1 Institut für Zoologie III, Universität Düsseldorf, Universitätsstraße 1, 4000 Düsseldorf 1, FRG
2 Department of Biological Sciences, University of Exeter, Exeter EX4 4QG, Great Britain

Biosynthesis, Metabolism and Mode of Action of Invertebrate Hormones
(ed. by J. Hoffmann and M. Porchet)

Other aspects of crustacean moulting physiology will not be dealt with, since they have already been treated in a separate review [1].

2. Materials and Methods

Most of our investigations were performed with the commercially available crayfish *Astacus leptodactylus* or *Orconectes limosus*. Rearing conditions and moulting stage determination of these animals have already been described [2]. For comparative purposes, the common shore crab, *Carcinus maenas* (kept at 12 °C in aerated artificial seawater), and the brine shrimp, *Artemia salina* (1 g dry cysts per 600 ml medium [0.4 *M* NaCl, 0.01 *M* $MgCl_2$, and 0.01 *M* $CaCl_2$], 30 °C, shaken on a rotary shaker), were used.

Ecdysteroids were extracted and separated according to [3] and quantified by a radioimmunoassay using the antiserum Anti-ICT-1 [4]. The uptake of moulting hormones was determined as already described [5], using an in vitro culture system [6]. Studies concerning the binding of tritiated moulting hormones to cytosol or nuclei were performed using the methods described [7–10].

3. Results and Discussion

3.1 Ecdysteroids During Oogenesis and Early Development of Crustaceans

In contrast to insects, where several species have been investigated and different functions of ecdysteroids during early development have been discussed (for reviews see [11, 12]), only three crabs and one amphipod have so far been studied with regard to

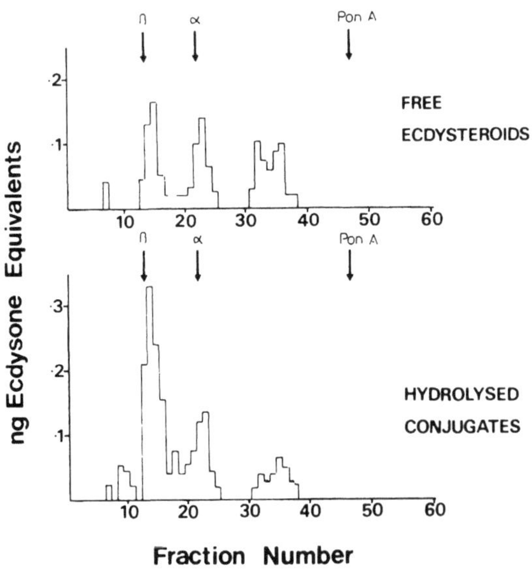

Fig. 2. HPLC/RIA of free ecdysteroid and hydrolysed conjugate fractions derived from *Astacus leptodactylus* eggs. The fractions were prepared according to [3] and aliquots were subjected to reverse-phase HPLC separation (Waters μ Bondapak C 18 column; solvent: MeOH: H_2O = 45:55, v/v; flow rate = 1 ml/min). Fractions of 30 s duration were collected, evaporated to dryness and analysed by RIA, using the antiserum ICT–1 [4]. α = ecdysone, β = 20-OH-ecdysone, *Pon A* = ponasterone A

the presence and possible function of ecdysteroids during oogenesis and early development in crustaceans. We are now able to add some knowledge about a macruran species, *Astacus leptodactylus* (Fig. 2) and about an anostracan species, *Artemia salina* (Fig. 3). These two species contain ecdysone, 20-OH-ecdysone, low and high polarity products, as were found in the three crabs. In *Artemia* there is evidence from HPLC studies for the presence of ponasterone A, as in *Carcinus* and *Callinectes* (Table 1). In *Artemia* conjugated and free ecdysteroids are already present in the dormant cysts and they also can be traced throughout postembryonic development up to the adult stage [13]. This last finding is also in favour of a maternal origin for the embryonic ecdysteroids. However, there is no direct proof for this in either crustacean species.

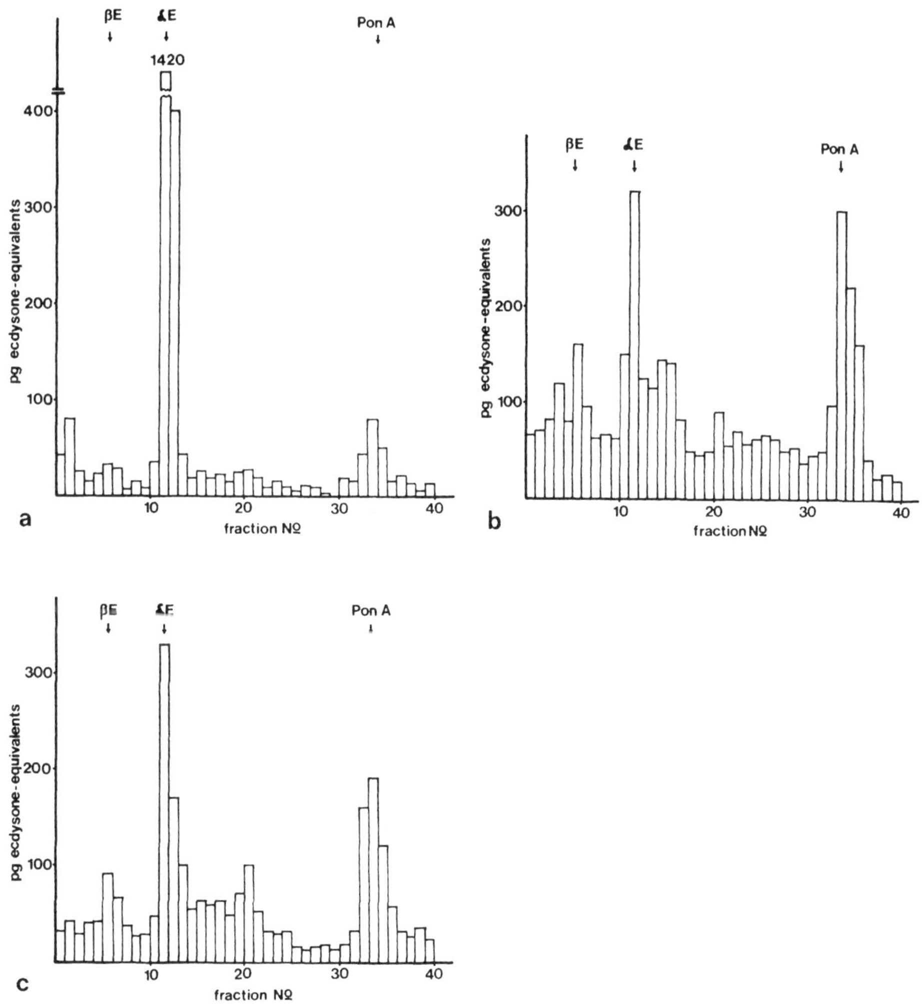

Fig. 3a–c. Separation of free ecdysteroids from dormant cysts (a), prenauplii (b), and prenauplii and nauplii (c). Extraction, separation and RIA according to Fig. 2. HPLC conditions: MeOH: H_2O = 4:6, v/v; flow rate = 1.3 ml/min. Fractions (100 drop) were collected and analysed by RIA, using the antiserum ICT–1

Table 1. Ecdysteroids during early development in crustaceans

Species	Remarks	References
Carcinus maenas Common shore crab	Maturing oocytes, ovary and hemolymph: free and conjugated ecdysone, 20-OH-ecdysone, Ponasterone A (TLC/RIA, HPLC; MS) The titer of ponasterone A shows two peaks, the second of which occurs just a few hours before the secretion of the first embryonic envelope	[18–21]
Acanthonyx lunulatus Spider crab	Ovary: Ecdysone, 20-OH-ecdysone, high-polarity products (HPP) Embryos: Ecdysone, 20-OH-ecdysone, HPP and low polarity products (TLC/RIA, RP–HPLC) Ecdysteroids are involved in oocyte maturation and in embryogenesis	[14]
Callinectes sapidus Blue crab	Embryos: Ecdysone, 20-OH-ecdysone, HPP and ponasterone A (RP–HPLC, RIA, MS) Total RIA-activity increases during maturation of the embryos and declines before hatching	[22, 23]
Astacus leptodactylus Crayfish	Eggs: Free and conjugated ecdysone, 20-OH-ecdysone and low polarity ecdysteroids (RP–HPLC/RIA). Male gonads: No activity (RIA)	see Fig. 2
Orchestia gammarellus	Ovaries, eggs, embryos: 20-OH-ecdysone, low polarity products (RP–HPLC, RIA) Vitellogenic and sexual resting females contain equivalent amounts of ecdysteroids	[24]
Artemia salina Brine shrimp	Pre-emergence stages: Free and conjugated ecdysone, 20-OH-ecdysone and ponasterone A (RP–HPLC/RIA)	see Fig. 3

MS = mass spectroscopy; RIA = radioimmunoassay; RP–HPLC = reversed phase high performance liquid chromatography; TLC = thin layer chromatography

One can so far only speculate as to the functions of ecdysteroids during oogenesis and early development in crustaceans. The most probable function is their involvement in the secretion of embryonic envelopes and in embryonic moulting. The high concentrations of ecdysteroids sometimes found during oocyte maturation cannot be clearly interpreted in terms of function. In one species there is a correlation between the rise in ecdysteroids and the onset of secondary vitellogenesis [14], but in other species ecdysteroids do not have an influence on vitellogenesis despite the fact that a low titer of ecdysteroids must be present at the beginning of vitellogenesis [15–17]. Identical ecdysteroid titer patterns cannot be expected in different crustaceans since the relation between moulting and reproduction is different throughout the crustacean phylum. For example, there is a terminal ecdysis in the spider crab with subsequent cycles of vitellogenesis and embryonic development, whereas ovarian activity in *Orchestia* is cyclic and closely synchronized with the moulting cycle.

In the few crustacean species which have been investigated so far, ecdysteroids have been found during ovarian development and during embryogenesis, but the important questions about the origin and chemical nature of the moulting hormones, their metabolism and especially their function during early development remain a task for future research.

3.2 Uptake

In vertebrates there is increasing evidence for extra-genomic effects of steroid hormones [25] and steroid hormone-membrane interactions. The mechanism of uptake of these compounds has also been investigated on several occasions [26, 27]. In contrast to vertebrates, our knowledge about these problems in invertebrates is very scarce. The available data have been summarized recently [28]. We have begun to study the mechanism of uptake of moulting hormones in detail, using crayfish integument in vitro. From our data (see Table 2) we conclude that, in addition to diffusion, an energy dependent, carrier-mediated process is involved in the uptake of moulting hormones by crayfish integument. Experiments with the integument from the common shore crab, *Carcinus maenas* yielded similar results (unpublished results). It is of interest that ouabain is able

Table 2. Characteristics of the uptake of moulting hormones by the integument of *Astacus leptodactylus*

Parameter	Effects	References
Temperature	Initial uptake is strictly temperature-dependent, $Q_{10} = 3$	[5]
Activation energy	45 kJ/mol	
Saturability	No linear relationship between hormone concentration and initial uptake; saturability between 2 and 4 × 10^{-9} *M*	
Hormone specificity	No pronounced differences between various ecdysteroids. A vertebrate steroid hormone is without any effect on the uptake system	
Metabolic inhibitors Antimycin A (0.1 m*M*) 2.4-Dinitrophenol (1 m*M*) Ouabain (0.1 m*M*)	Up to 60% inhibition of the uptake	[29, 30] and unpubl. results
Thiol-group Blocking agents p-Chloromercuriphenyl- -sulfonate (1 m*M*) N-Ethylmaleinimide (1 m*M*)	Up to 60% inhibition of the uptake. Simultaneous incubation of p-Chloromercuriphenyl-sulfonate with dithiotreitol prevents inhibition	

to inhibit the uptake of moulting hormones (at concentrations as low as 5×10^{-7} *M*), indicating the possible involvement of membrane-bound Na^+/K^+-ATPase (unpublished results).

The question arises whether the binding and recognition sites which are involved in uptake are identical with the already demonstrated binding sites in the cytosol and/or nuclei from the same species and tissues. There are several arguments against this: the binding and uptake is very rapid compared with the binding of the cytoplasmic and nuclear receptors, the membrane bound binding sites have a much less pronounced steroid-specificity and finally metabolic inhibitors partially inhibit the binding and uptake [29, 30], but do not interfere with the cytoplasmic receptor (unpublished results).

Whereas the evidence for the carrier-mediated transport of ecdysteroids into decapod integuments is increasing, the physiological significance of such binding sites still has to be elucidated.

3.3 Ecdysteroid-Binding Proteins

A prerequisite for the molecular action of steroid hormones is the presence of specific binding proteins (receptors) in the cytosol and nuclei of target tissues. Such receptors are characterized by high affinity for the steroid hormone, saturability, high ligand specificity and the ability to bind to chromatin and to initiate a specific biological response (for review see [31, 32]). In spite of the importance of steroid hormone receptors, ecdysteroid receptors have been unequivocally demonstrated in only two arthropod species, namely, the insect *Drosophila melanogaster* [33–35] and the crustacean *Orconectes limosus* [2, 7–10].

We, therefore, tried to ascertain whether ecdysteroid receptors exist in different crustacean groups and whether the distribution of ecdysteroid receptors within one animal is tissue specific. Table 3 gives a summary of the present knowledge of ecdy-

Table 3. Demonstration of ecdysteroid binding proteins in tissues of crustaceans

Species	Tissue	References
Orconectes limosus	Hypodermis, cytosol	[2, 7, 10]
	Hypodermis, nuclei	[8–10]
	Midgut gland, cytosol	[2]
	Midgut gland, nuclei	Unpubl. results
	Hindgut, cytosol	[2]
	Eggs, cytosol	Own unpubl. results
Astacus leptodactylus	Hypodermis, cytosol	
	Hypodermis, nuclei	
	Midgut gland, cytosol	
Carcinus maenas	Hypodermis, cytosol	Own unpubl. results
	Midgut gland, cytosol	
	Brain, cytosol	
Artemia salina	Dormant cysts, total	
	Nauplius, total	
	Adult, total	

steroid binding proteins in crustaceans, including the already published data on *Orconectes* and the so far unpublished work of our group with another species of crayfish, *Astacus leptodactylus*, with the common shore crab, *Carcinus maenas* and with the brine shrimp, *Artemia salina*. In the four species which we have investigated, ecdysteroid binding proteins are present in several, but not in all tissues. For example we were not able to detect ecdysteroid binding proteins in the haemolymph or the musculature.

Interestingly, competable high-affinity binding components have been found in the cytosol of eggs from *Orconectes limosus* and in the dormant cysts of *Artemia salina*. At the moment we cannot decide whether this is a specific binding of ecdysteroids to vitellin as has been described for *Locusta migratoria* [38], or whether there are receptors for ecdysteroids already present in the egg or in early embryonic stages, which are similar to those in later developmental stages. Nuclear binding components indistinguishable from the receptors in the imaginal discs have been described in *Drosophila melanogaster* eggs [39].

With regard to the affinities of ecdysteroid receptors in crustaceans (see Table 4), the following conclusions can be drawn:

1. the affinities of ecdysteroid receptors are very similar in the two crayfish species which have been investigated;
2. there are no pronounced differences in the affinities of cytoplasmic ecdysteroid receptors from different organs of *Orconectes limosus*;
3. the affinities of cytoplasmic and nuclear receptors from one organ are nearly identical;
4. the receptor always has the highest affinity for ponasterone A, followed by 20-OH-ecdysone and then by ecdysone. This last finding resembles the situation in *Drosophila melanogaster* [33, 34], but – with the exception of *Artemia salina* – the affinities of ecdysteroid receptors are higher in crustaceans than in insects which may reflect the generally lower circulating ecdysteroid titer in decapod crustaceans as compared to insects [1]. Concerning the K_D values, one has to keep in mind that – with the exception of *Drosophila* K_c cells – all other systems tested so far contain endogenous ecdysteroids. In the integument of *Astacus*, the endogenous ecdysteroid

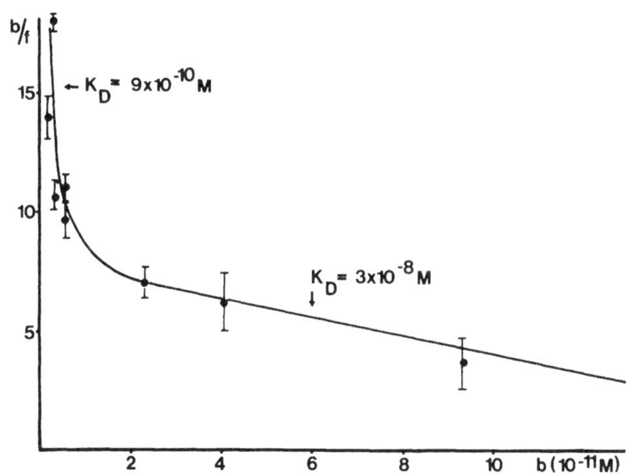

Fig. 4. Scatchard analysis of binding data from cytoplasmic ecdysteroid receptor from nauplii of *Artemia salina*, ^{3}H-ponasterone A being used as ligand. Only the values for the specific binding have been drawn

Table 4. Affinities of ecdysteroid receptors in crustaceans

Species	Organ	K_D[a] Value (*M*)						References
		Ecdysone	(n)	20-OH-ecdysone	(n)	Ponasterone A	(n)	
Orconectes limosus	Integument, cytosol	$2-7 \times 10^{-8}$	(5)	$1-3 \times 10^{-9}$	(5)	$3-6 \times 10^{-11}$	(4)	[2, 7, 10]
	Integument, nuclei	$4-7 \times 10^{-9}$	(3)	$1-2 \times 10^{-9}$	(2)	6×10^{-11}	(1)	[8, 10]
	Hindgut, cytosol	$2-3 \times 10^{-8}$	(4)	$1-2 \times 10^{-9}$	(4)	n.d.		[2]
	Midgut gland, cytosol	$2-3 \times 10^{-8}$	(3)	$1-2 \times 10^{-9}$	(3)	n.d.		[2]
Astacus leptodactylus	Integument, cytosol	n.d.		$1-2 \times 10^{-9}$	(3)	$2-6 \times 10^{-11}$	(4)	unpubl. results
	Integument, nuclei	n.d.		n.d.		$3-5 \times 10^{-11}$	(3)	unpubl. results
Artemia salina	Nauplii, cytosol	n.d.		n.d.		9×10^{-10} and 3×10^{-8}		see Fig. 4

[a] K_D = values were determined according to [36, 37]
(n) = number of determinations
n.d. = not determined

is partially bound to receptors (own published results). Since no reliable exchange assay exists at the present time, the determined K_D values may be incorrect.

4. Conclusions and Outlook

In the present paper we have focussed our attention on three different aspects of ecdysteroid action in crustaceans, namely, maternal-embryonic interrelationships (Sect. 3.1), extra-genomic effects (Sect. 3.2) and genomic aspects (Sect. 3.3). Although knowledge has begun to accumulate regarding these three aspects, fundamental questions on the mechanism of ecdysteroid action still remain and those deserve further investigation. Some examples of such unsolved problems are:

1. The site of synthesis of embryonic ecdysteroids, the transfer of maternal ecdysteroids into oocytes and embryos and the functional significance of ecdysteroids during oogenesis and early embryogenesis have to be elucidated. This is true not only for crustaceans, but at least partly for insects as well.
2. Besides the "classical" action of steroid hormones, extra-genomic effects have been described in vertebrates. Our findings on the mechanism of ecdysteroid entry in crayfish hypodermis may be an indication that such effects may also play a role in invertebrates.
3. Last but not least, the concept of steroid hormone receptors seems to be now well-established both in crustaceans and insects.

However, a great deal of work still has to be done: for example, purification of ecdysteroid receptors, immunological studies in order to demonstrate whether different domains exist as in the glucocorticoid receptor as well as a titration of ecdysteroid receptors during the moulting cycle. Finally, the physiological significance of the ecdysteroid receptors has to be demonstrated unequivocally either at the level of ecdysteroid-dependent protein synthesis or as a specific binding of hormone-receptor complex at ecdysteroid regulated genes. Protein synthesis during the moulting cycle and the possible hormonal regulation of this process is already under investigation in our laboratory.

Acknowledgments. Our own work was supported by grants from the Deutsche Forschungsgemeinschaft to K.D. Spindler. We gratefully acknowledge the excellent technical assistance of Manfred Schwab.

References

1. Spindler K-D, Keller R, O'Connor JD (1980) In: Hoffmann JA (ed) Progress in ecdysone research, vol VII. Elsevier/North-Holland Biomed Press, Amsterdam New York, pp 247–280
2. Kuppert P, Wilhelm S, Spindler K-D (1978) J Comp Physiol 128:95–100
3. Dinan L, Rees HH (1981) J Insect Physiol 27:51–58
4. Spindler K-D, Beckers C, Gröschel-Stewart U, Emmerich H (1978) Hoppe Seyler's Z Physiol Chem 359:1269–1275
5. Daig K, Spindler K-D (1983) Mol Cell Endocrinol 31:93–104
6. Daig K, Spindler K-D (1979) Z Naturforsch 34c:1243–1247
7. Londershausen M, Spindler K-D (1981) Mol Cell Endocrinol 24:253–265

8. Spindler-Barth M, Bassemir U, Kuppert P, Spindler K-D (1981) Z Naturforsch 36c:326–332
9. Kuppert P, Spindler K-D (1982) J Steroid Biochem 17:205–210
10. Londershausen M, Kuppert P, Spindler K-D (1982) Hoppe Seyler's Z Physiol Chem 363:797 to 802
11. Hagedorn HH (1980) Adv Invertebr Reprod:97–107
12. Hoffmann JA, Lagueux M, Hetru C, Charlet M, Goltzené F (1980) In: Hoffmann JA (ed) Progress in ecdysone research, vol VII. Elsevier/North-Holland Biomed Press, Amsterdam New York, pp 431–465
13. Funke B, Spindler K-D (in preparation)
14. Chaix J-C, de Reggi M (1982) Gen Comp Endocrinol 47:7–14
15. Charniaux-Cotton H (1980) In: Clark WH Jr, Adams TS (eds) Invertebrate reproduction. Elsevier/North-Holland, Amsterdam, pp 177–186
16. Meusy J-J (1980) Reprod Nutr Dev 20:1–21
17. Chang E. Personal communication
18. Lachaise F, Hoffmann JA (1977) CR Acad Sci Ser D 285:701–704
19. Lachaise F, Goudeau M, Hetru C, Kappler C, Hoffmann JA (1981) Hoppe Seyler's Z Physiol Chem 362:521–529
20. Lachaise F, Hoffmann JA (1982) Hoppe Seyler's Z Physiol Chem 363:1059–1067
21. Goudeau M, Lachaise F (1983) Tissue Cell 15:47–62
22. McCarthy JF, Skinner DM (1979) Dev Biol 69:627–633
23. McCarthy JF (1979) Steroids 34:799–806
24. Blanchet MF, Porcheron P, Dray F (1979) Int J Invertebr Reprod 1:133–139
25. Duval D, Durant S, Homo-Delarche (1983) Biochim Biophys Acta 737:409–442
26. Rao GS (1981) Mol Cell Endocrinol 21:97–108
27. Szego CM, Pietras PJ (1981) In: Litwack (ed) Biochemical actions of hormones, vol VIII. Academic Press, London New York , pp 307–463
28. Koolman J, Spindler K-D (1983) In: Downer R, Laufer H (eds) Insect endocrinology. Liss, New York, pp 179–201
29. Daig K, Spindler K-D (1980) In: Kurstak E, Maramorosch K, Dübendorfer A (eds) Invertebrate systems in vitro. Elsevier/North-Holland Biomed Press, Amsterdam New York, pp 273–278
30. Daig K, Spindler K-D (1983) Mol Cell Endocrinol 32:73–79
31. Yamamoto KR, Alberts BM (1976) Annu Rev Biochem 45:721–746
32. Higgins SJ, Gehring U (1978) Adv Cancer Res 28:313–397
33. O'Connor JD, Maroy P, Beckers C, Dennis R, Alvarez CM, Sage BA (1980) In: Roy A, Clark JH (eds) Gene regulation by steroid hormones. Springer, Berlin Heidelberg New York, pp 263 to 277
34. Fristrom JW (1981) In: Gilbert LJ, Frieden E (eds) Metamorphosis – a problem in developmental biology. Plenum Press, New York, pp 217–240
35. Schaltmann K, Pongs O (1982) Proc Natl Acad Sci USA 79:6–10
36. Scatchard G (1949) Ann NY Acad Sci 51:660–672
37. Chamness GC, McGuire W (1975) Steroids 26:538–542
38. Lagueux M, Harry P, Hoffmann JA (1981) Mol Cell Endocrinol 24:325–338
39. Osterbur DL, Yund MA (1982) J Cell Biochem 20:277–282

II.8 Hormonal Control of Sequential Gene Expression in Insect Epidermis

L.M. RIDDIFORD[1]

1. Introduction

The insect epidermis makes the overlying cuticle or exoskeleton which must change in structure to accommodate changing life styles as the insect grows, then undergoes metamorphosis. These cells also make various pigments or cuticular pigment precursors which again can differ according to the stage examined. Radically different proteins are often required and, thus, dramatic changes in gene expression must occur within the already differentiated epidermal cell.

Insects have opted for two different solutions to this problem of changing gene expression. One solution is seen in the Diptera in which the larval epidermis dies at metamorphosis and the imaginal discs and the histoblasts take over to produce the pupal and adult cuticle. In contrast, other insects have polymorphic epidermal cells, i.e., ones that sequentially make different types of cuticle and often different pigments to meet the animal's changing needs. In Hemimetabola these changes may be relatively simple, such as producing new bristles or setae or a different pigmentation as Wigglesworth [1] and Lawrence [2] have amply described. In the Holometabola, such as the Lepidoptera, this polymorphism involves marked changes in the type of cuticle produced – a flexible, constantly growing larval cuticle to allow for substantial increase in size, while maintaining mobility, then a relatively rigid pupal cuticle inside of which the adult forms, and finally the thin tanned adult cuticle covered by pigmented scales.

This sequential polymorphism of the insect epidermis is controlled primarily by the two hormones – ecdysone and its metabolite 20-hydroxyecdysone (20HE) and juvenile hormone (JH). The pioneering studies of Wigglesworth [3], Piepho [4], and Williams [5] suggested that when these polymorphic changes occurred during a molt, the level of JH present at the time of the ecdysteroid-induced molt determined the type of cuticle produced. Thus, in the Hemimetabola the presence of JH allowed another larval molt, whereas its absence signaled metamorphosis and a switching to adult-specific products. Since the Holometabola formed an intermediate pupal stage, the level of JH at the outset of metamorphosis was thought to be intermediate and, thus, to direct the cell's activities along the pupal line. More recently, the larval-pupal transformation in Lepidoptera has been shown to be controlled by two releases of ecdysteroid – the first in the absence of detectable JH in the hemolymph, the second in the presence of JH [6]. During the first release the cells lose their ability to make larval cuticle and pigment

1 Department of Zoology, University of Washington, Seattle, WA 98195, USA

Biosynthesis, Metabolism and Mode of Action of Invertebrate Hormones
(ed. by J. Hoffmann and M. Porchet)

(i.e., become pupally committed), but no pupal-specific products are made until the second ecdysteroid release causes pupal cuticle synthesis [7]. Importantly, however, the presence of JH is not necessary for the synthesis of pupal cuticle by the polymorphic epidermis at this time since cells cultured in vitro with only 20HE formed normal pupal cuticle [8].

Although the major polymorphic transitions are seen during metamorphosis, other polymorphic changes occur at larval molts [9, 10] and even during the intermolt period [11]. These changes may involve changes in pigmentation or in cuticular derivatives, such as setae or even more subtle changes in cuticular structure that are seen only at the cellular level. These relatively minor polymorphic changes are also regulated by ecdysteroid and JH, the level of JH directing the action of ecdysteroid on the genome.

The questions now of interest are: how does ecdysteroid act on the genome of a polymorphic cell and how does JH regulate the selection of a particular synthetic program? This paper will focus on our approach to answering these questions using the abdominal epidermis of the tobacco hornworm *Manduca sexta*.

2. Sequential Polymorphism in *Manduca* Epidermis

In addition to the transparent larval cuticle [12, 13], the larval epidermis of *Manduca* also makes a blue pigment insecticyanin [14], part of which is stored in apical granules and part of which is secreted into the hemolymph. During the intermolt period as the larva is growing, the larval endocuticle is continually deposited [13] and insecticyanin synthesis is high [11, 15] (Fig. 1). During the ensuing molt, syntheses of the endo-

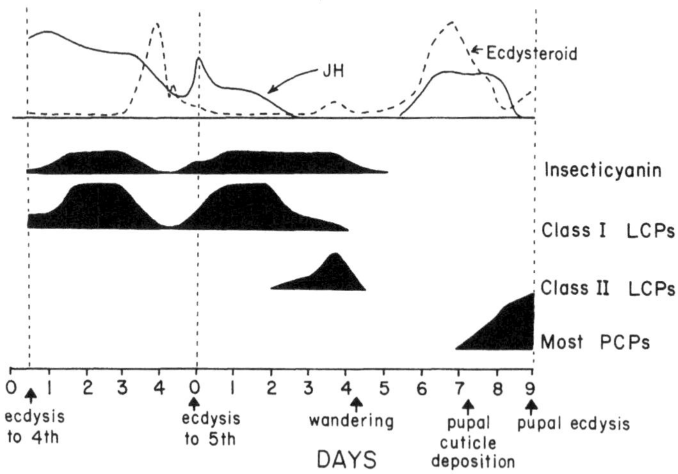

Fig. 1. Diagrammatic representation of the ecdysteroid [23] and juvenile hormone [26] titers and of the increases and decreases in syntheses of insecticyanin, the larval endocuticular proteins, and the pupal cuticular proteins and their respective mRNAs [11, 15, 17, 25] during the penultimate (4th) and final (5th) larval instars of *Manduca*. *JH* juvenile hormone; *LCP* larval endocuticular proteins; *PCP* pupal cuticular proteins

cuticular proteins and of insecticyanin shut off, only to begin again after ecdysis when the larva resumes feeding and growing [11, 16]. At the onset of metamorphosis when the larva ceases feeding and wanders off the food to find a pupation site, the synthesis of these larval-specific proteins again ceases, but in this instance the cessation is permanent [7, 11, 15, 17]. At this time the epidermis becomes committed to pupal differentiation [7, 18] which means that these larval-specific genes can no longer be expressed even if the cells are placed in a larval environment, such as that of the intermolt penultimate instar larva. Instead, during the subsequent larval molt of the host, only pupal cuticle is made by this epidermis. Except for those few proteins which appear to be common to larval and pupal endocuticle [17], synthesis of all larval-specific cuticular proteins and of insecticyanin is permanently repressed at this time. Consequently, the cell has switched to its second major polymorphic state, that of the pupa.

Polymorphic changes in the epidermis also occur within the larval stages. During a larval molt the absence of JH at the time of head capsule apolysis causes the deposition of premelanin granules into the newly forming cuticle [19]. The subsequent decline of the ecdysteroid titer leads to activation of these granules so that melanin [20] is formed just before ecdysis, resulting in a black cuticle. The absence of JH also causes ommochrome synthesis by the postecdysial epidermis and a reduction in intermolt insecticyanin synthesis [21].

Changing patterns of cellular syntheses can be seen during the larval intermolt as well. For example, during the final larval instar the lamellar structure of the endocuticle changes on the final day of feeding accompanied by a change in the proteins being synthesized by the epidermis and deposited into the cuticle [11, 13, 16]. This change is accompanied by an increase in cuticular stiffness [16] which apparently is in preparation for the time that the larva will burrow into the soil to form a pupation cell.

3. Hormonal Regulation of the Cessation of Gene Expression

Studies of the regulation of the synthesis of insecticyanin and the larval endocuticular proteins in *Manduca* epidermis showed that whenever the epidermis was exposed to ecdysteroid, either in vivo or in vitro, the synthesis of these proteins [11, 15, 22] declined due to a disappearance of their mRNAs [17]. During the larval molt, the translatable mRNAs for insecticyanin and most of the larval endocuticular proteins had declined to very low levels [11] within 12 h of prothoracicotropic hormone (PTTH) release and the initiation of the ecdysteroid rise [23]. 6 h later (at which time the ecdysteroid titer had risen to 2 μg/ml 20HE equivalents [19]), these RNAs were no longer present [11] and no larval endocuticular proteins were being synthesized [16]. Similarly, when the epidermis from day 2 fifth instar larva was incubated with 1.2 μg/ml 20HE, these mRNAs also declined [17] leading to the virtual cessation of synthesis of insecticyanin and the larval cuticular proteins by the end of the 24 h incubation period [22]. Decreased synthesis of many of these proteins was already evident by 6 h incubation. Therefore, one of the early actions of 20HE during the rising phase of the ecdysteroid titer is to act on the epidermis to turn off the synthesis of the major products it makes during the intermolt period. The cell then can marshall all of its resources to make the new cuticle. Importantly, this switching-off action of ecdysteroid occurs irrespective of whether the ensuing molt is to be larval or pupal.

In many insects [24] the deposition of the new cuticle begins at about the peak of the ecdysteroid titer. In the larval molt of *Manduca* the epicuticle is deposited during a 6 h period beginning 2 h after head cap apolysis [16]. As the ecdysteroid titer begins to decline [19], first the amorphous cuticle, then the endocuticle is deposited [16]. Since each type of cuticle has its own complement of proteins [16], sequential gene expression is required for this pre-ecdysial deposition. With the pre-ecdysial deposition of the endocuticle the intermolt larval endocuticular proteins begin to reappear and 24 h after ecdysis the full complement is again being expressed [11, 15-17]. Also, insecticyanin mRNA reappears shortly before ecdysis and attains its full expression during the intermolt [11, 17]. Thus, in the larval molt the rise of ecdysteroid turns off the genes for the major intermolt products; they then reappear as needed after the ecdysteroid titer falls.

Whether the cessation of ongoing mRNA and protein synthesis by ecdysteroid is temporary or permanent depends on the JH titer at the time that 20HE acts. If JH is above some threshold level in the tissues at the time that they are exposed to the 20HE, then the cessation of the synthesis is temporary as seen above in the larval molt. A decline in translatable mRNAs and in the synthesis of endocuticular proteins and insecticyanin in day 2 fifth instar larval epidermis in response to 1.2 μg/ml 20HE in vitro was nearly as marked when 1 μg/ml methoprene (a JH mimic) was present as when it was absent [17, 22]. Yet when it was present, the cells could again make both larval cuticle and insecticyanin when subsequently exposed to larval molting conditions. By contrast, when JH was absent, these genes were permanently turned off.

This permanent switching-off of genes by 20HE in the absence of JH does not occur instantaneously. Rather the mRNAs for the larval-specific endocuticular proteins and for insecticyanin disappear with differing time courses [17]. Preliminary studies with varying concentrations of 20HE in vitro [11, 25] show that some genes are repressed by a low intermolt concentration of ecdysteroid in the absence of JH, whereas others require exposure to a much higher concentration before ceasing activity. Just as each cell appears to change its commitment independently of its neighbors [18] so also the larval-specific genes may respond independently to the increasing ecdysteroid in the absence of JH.

4. Hormonal Regulation of Cuticular Polymorphism During the Intermolt

During the premetamorphic ecdysteroid rise in the absence of JH on day 3 of the final larval instar, the width of the cuticular lamellae decreases tenfold due to changes in cuticular protein synthesis [15, 16]. Some mRNAs disappear, while several new ones appear [17]. These mRNAs appear after the fall of JH on day 2, but the changes in protein synthesis and the consequent change in lamellar structure are not evident until after the beginning of the rise of ecdysteroid on day 3 (Fig. 1). When 10 μg methoprene was applied to larvae on day 1 before the decline in endogenous JH, the change in proteins and lamellar structure normally seen on day 3 were prevented [11, 16]. By contrast, the same dose of methoprene given on the afternoon of day 2, when the changes in mRNAs were first evident, was ineffective in preventing the changes in protein synthesis and cuticular structure on day 3. Both treatments delayed the onset of wandering to day 5 and prevented most, but not all, of the ecdysteroid rise on day 3. Further experiments showed that applications of methoprene up to about 4 h before the appearance of the new mRNAs (about 10 h before the appearance of the thin lamellae) were effective in preventing the change in cuticular structure [16].

These data suggested that the fall of JH somehow initiated the changes in cuticular protein gene expression. Yet when larvae were allatectomized on day 0, thin lamellae and their associated cuticular proteins were not seen until day 3 as in control larvae. Thus, in addition to the fall in JH, ecdysteroid also appeared to be necessary for the implementation of that change in terms of protein synthesis and secretion by the cell. In vitro studies [25] have recently shown that ecdysteroids are involved in the changes in both mRNAs and proteins. To simplify the discussion of this evidence, the endocuticular proteins present up to day 2 during the deposition of the thick lamellae are called Class I cuticular proteins and those that appear only during the deposition of the thin lamellae are called Class II cuticular proteins. Figure 2 illustrates the hormonal regulation of representative proteins of each of these classes.

Two representative Class I cuticular proteins which are no longer synthesized during the deposition of the thin lamellae have molecular weights of 28,000 and 22,500 on SDS gels [16]. These proteins are major synthetic components of the endocuticle from day 0 to day 2, but then abruptly shut off at the time of the beginning of the ecdysteroid rise on day 3 (Fig. 2). Surprisingly, when day 2 epidermis was cultured in vitro in the absence of hormones or in the presence of 1 μg/ml methoprene, the synthesis of these proteins nevertheless decreased, so that by 24 h none were being made [22] and their mRNAs were no longer translatable [17]. The addition of 20HE simply accelerated this decline. By contrast, when epidermis from day 1 larvae before the decline of JH was cultured under the same conditions, cessation of synthesis only occurred in the presence of 20HE [25] (Fig. 2). As little as 25 ng/ml 20HE (slightly above the intermolt level of about 15 ng/ml 20HE equivalents [25]) was sufficient to initiate this response. The simplest interpretation of these data is that after the decline of the JH titer to below some threshold level (which normally occurs during the night between days 1 and 2, see Fig. 1), the intermolt level of ecdysteroid acts on the epidermis to program the turning-off of these genes. This turning-off then occurs slowly in the presence of the intermolt level of ecdysteroid, but rapidly when the ecdysteroid titer begins to rise on the night between days 2 and 3.

A major Class II protein has a molecular weight of 27,000 and is comprised of at least four different charged forms when analyzed by 2-dimensional gel electrophoresis [16]. Whether these are different proteins or are all modifications of a single gene product is not known, so they will be referred to as the 27K complex. Part of the complex is synthesized in low amounts on the morning of day 2 after the JH titer has become undetectable, but the major syntheses and two new spots are correlated with the appearance of the thin lamellae on day 3 (Fig. 2b). When day 1 epidermis was cultured in the absence of hormones or in the presence of JH, none of the new spots appeared and synthesis remained at the day 1 level [16, 25]. Their appearance required exposure to ecdysteroid, the time course depending on the concentration of the hormone

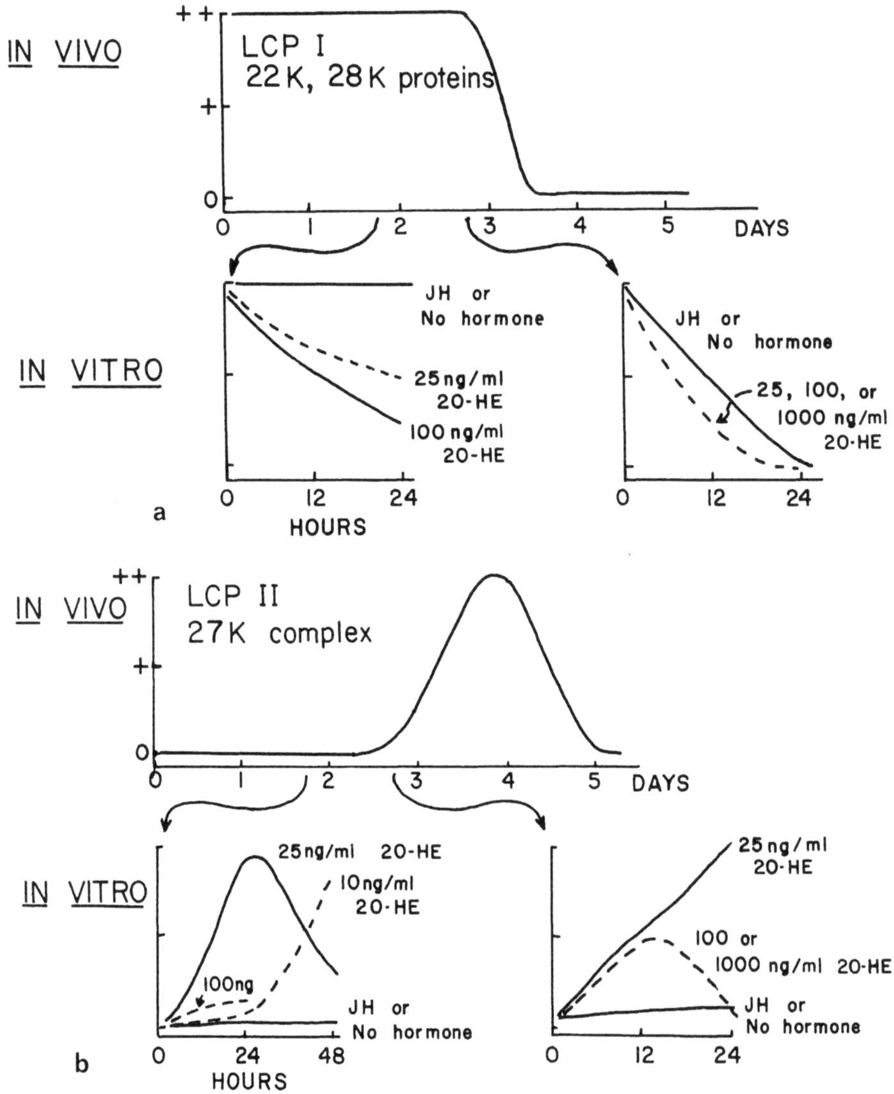

Fig. 2a,b. Diagrammatic representation of (**a**) the cessation of synthesis of two selected Class I larval endocuticular proteins and (**b**) the transient synthesis of the major Class II larval endocuticular protein complex in vivo (*top*) [15, 16] and in vitro (*bottom*) in response to various hormonal conditions after explanation at the indicated time [16, 22, 25]. *LCP* larval endocuticular protein; *JH* juvenile hormone (1 μg/ml methoprene); *20HE* 20-hydroxyecdysone

(Fig. 2b). Higher concentrations equivalent to those seen at the maximal titer in vivo on day 3 caused only a transient increase in synthesis. Day 2 epidermis explanted after the decline of JH showed similar responses to the various hormones in vitro (Fig. 2b). Apparently the low intermolt levels of ecdysteroid are sufficient in the absence of JH on day 2 to initiate the changes which become manifest as the ecdysteroid titer begins to rise early on day 3. Then the maximal titer later on the same day causes the irreversible shutdown of these Class II endocuticular genes as well of the Class I genes.

5. Conclusions

The polymorphic insect epidermal cell then is exquisitely sensitive to its hormonal environment. When ecdysteroid rises during a molt, the epidermis responds by shutting off syntheses of all major intermolt products, thus, readying itself to produce the new cuticle of the next stage. When JH is present, this turning-off is temporary and the same products are made during the subsequent intermolt. When JH is absent, the turning-off is permanent and the cell switches to its next major polymorphic state. Juvenile hormone is also important during the larval intermolt period to maintain on-going syntheses. When it declines, the low intermolt levels of ecdysteroid can cause changes in syntheses leading to changes in structure of the larval cuticle which may in turn cause changes in its properties.

How JH regulates these two actions of ecdysteroid on the polymorphic epidermal cell is not yet understood. Several points are clear [7]: (1) 20HE acts directly on the genome to initiate a cascading series of mRNA and protein syntheses, some of which require further direct intervention of ecdysteroid since during the larval molt its presence is necessary for 12 h to direct all the subsequent events; (2) JH must be present at the very outset of ecdysteroid action in order to prevent the polymorphic changes; (3) JH does not prevent the shutting-off of genes by ecdysteroid. Therefore, JH cannot prevent all of the primary, secondary, or later actions of ecdysteroid on the genome. Future studies using specific genes, such as the one for insecticyanin or one of the larval endocuticular genes, should help us begin to answer the question of how JH guides the action of ecdysteroid on the genome.

Acknowledgments. I thank Dr. James Truman for helpful discussions and a critical reading of this paper. The unpublished studies cited here were supported by grants from NIH (AI–12459) and NSF (PCM 80–11152).

References

1. Wigglesworth VB (1976) In: Wigglesworth VB (ed) Insects and the life of man. Chapman and Hall, London, pp 149–167
2. Lawrence PA (1969) Dev Biol 19:12–40
3. Wigglesworth VB (1940) J Exp Biol 17:201–222
4. Piepho H (1950) Biol Zentralbl 69:261–271
5. Williams CM (1961) Biol Bull 116:323–338
6. Riddiford LM (1980) In: Hoffmann JA (ed) Progress in ecdysone research. Elsevier/North Holland, Amsterdam, pp 409–430
7. Riddiford LM (1981) Am Zool 21:751–762
8. Mitsui T, Riddiford LM (1976) Dev Biol 54:172–186
9. Willis JH (1974) Annu Rev Entomol 19:97–115
10. Chihara CJ, Silvert DJ, Fristrom JW (1982) Dev Biol 89:379–388
11. Riddiford LM, Kiely ML, Wolfgang WJ (1983) In: Kurstak E, Maramorosch K, Oberlander H (eds) Invertebrate tissue culture: Developments and applications. Elsevier/North Holland (in press)

12. Sedlak BJ, Gilbert LI (1979) Tissue Cell 11:643–653
13. Wolfgang WJ, Riddiford LM (1981) Tissue Cell 13:757–772
14. Cherbas P (1973) PhD thesis, Harvard Univ, Cambridge
15. Kiely ML, Riddiford LM (1984a) Submitted
16. Wolfgang WJ (1984) PhD thesis, Univ Washington, Seattle
17. Riddiford LM (1982) Dev Biol 92:330–342
18. Riddiford LM (1978) Gen Comp Endocrinol 34:438–446
19. Curtis AT, Hori M, Green JM, Wolfgang WJ, Hiruma K, Riddiford LM (1984) J Insect Physiol (in press)
20. Hori M, Hiruma K, Riddiford LM (1984) Insect Biochem 14:267–274
21. Hori M, Riddiford LM (1982) J Comp Physiol 147B:1–9
22. Kiely ML, Riddiford LM (1984b) Submitted
23. Bollenbacher WE, Smith SL, Goodman W, Gilbert LI (1981) Gen Comp Endocrinol 44:302 to 306
24. Riddiford LM (1984) In: Kerkut G, Gilbert LI (eds) Comprehensive insect physiology, biochemistry, and pharmacology, vol VIII. Pergamon, Oxford (in press)
25. Riddiford LM (unpublished studies)
26. Sedlak BJ, Marchione L, Devorkin B, Davino R (1983) Gen Comp Endocrinol 52:291–310

II.9 Ecdysteroid Regulation of the Major Transcripts of *Drosophila melanogaster* Larval Salivary Glands and Fat Bodies

G. RICHARDS[1] and J.-A. LEPESANT[2]

1. Introduction

Although there is extensive literature concerned with the study of ecdysteroids and gene regulation in insects, it is a relatively simple task to trace the major developments that lead to the present encouraging position that promises so much for the near future. Beermann [1] proposed that the puffs of the polytene chromosomes of *Chironomous tentans*, which showed reproducible stage and tissue specific patterns in development, were sites of intense gene activity and, therefore, a suitable model system for understanding developmental gene regulation in eukaryotes. Clever and Karlson [2] found that some of these *C. tentans* puffs responded to injections of ecdysone and Clever [3] distinguished early and late classes of response to hormone.

Becker [4, 5] provided the link to *Drosophila* by describing developmental puffing patterns and in vivo responses to ligatures that isolated the ring gland from the larval salivary gland. This work was developed in *D. hydei* by Berendes [6] and in *D. melanogaster* by Ashburner [7]. The establishment of an in vitro technique for salivary gland culture and its subsequent use to dissect puffing patterns [8, 9] resulted in a detailed description of ecdysteroid responses in *Drosophila* larval and prepupal stages. Since the midsixties a parallel system has followed imaginal disk development (see Fristrom et al., this Vol.), and more recently the larval fat body has been the subject of such studies [10]. These studies have two aims; the first is to understand at the molecular level the means by which a steroid hormone regulates the activity of an individual gene, the second is to understand how complex gene sequences are regulated in the tissues of higher organisms. For this latter purpose there is perhaps no better system currently described than the developmental puffing patterns of *Drosophila* polytene tissues.

Despite the sophisticated techniques of molecular biology which have been added to the broad background of genetics in *Drosophila*, until very recently these studies have remained largely descriptive. However, the measurement of gene activity using cloned DNA sequences in cell free systems, in cells, and in whole organisms now enable an approach based on functional assays. We will conclude this review with some of the first results derived from P element gene transfer [11] of ecdysteroid regulated genes.

1 Laboratoire de Génétique Moléculaire des Eucaryotes du CNRS U.184 de Biologie Moléculaire et de Génie Génétique de l'INSERM 11, rue Humann 67085 Strasbourg Cédex, France
2 CNRS Institut de Recherche en Biologie Moléculaire, Université Paris VII, Paris, France

Biosynthesis, Metabolism and Mode of Action of Invertebrate Hormones
(ed. by J. Hoffmann and M. Porchet)

2. Salivary Gland Studies

2.1 Background

Although many *Drosophila* studies rely on salivary gland chromosome cytogenetic, either for the determination of mutant breakpoints and rearrangements, or more recently, for the in situ hybridization of cloned DNA sequences to their genomic homologues, relatively few studies of salivary gland developmental puffing patterns exist (see [12], for review). Nonetheless, the appeal of this material, which enables us to study genes in action by following changes in chromosome structure, is such that many laboratories are currently involved in research that derives essentially from these few studies. This enthusiasm results from the success of another polytene chromosome puffing phenomenon, that of heat shock [13]. Heat shock puffs were first described in *Drosophila melanogaster* by Ritossa [14] and are a rapid cellular response to increases in temperature. Upon heat shock there is an immediate induction of a few heat shock puffs and the regression of developmental puffs until the temperature change has been reversed or accomodated. Although we still know very little about the products of the heat shock genes, or their function, the idea of a coordinately induced set of genes of known chromosomal location has tempted many molecular biologists to work with *Drosophila* in the last decade. These studies have provided a solid base of techniques, designed to allow for the relatively small size of the fly, which are now proving invaluable for our studies of developmental gene activity.

In all, six broad classes of developmentally regulated puffs have been described, all of which respond in vitro to ecdysteroids. The first group consists of the third larval instar intermolt puffs of which the most prominent are now known to code for components of the glue secretion, the major product of the gland in the third instar. These puffs regress as the ecdysteroid titer rises some 8 h prior to pupariation and the second and third group of puffs, the early and late ecdysteroid induced puffs are activated [9]. The function of these puffs is unknown, but at this stage, the stored glue granules are mobilized and enter the gland lumen. While it is tempting to speculate that the major puffs' products are involved in this process, a word of caution is necessary as many of these regions are found puffed in other larval tissues at this time [15].

The fourth group of puffs, the mid-prepupal puffs [16], are interesting because their appearance depends upon a rapid drop of ecdysteroid titer following pupariation. The ability of high ecdysteroid titers to inhibit specific aspects of insect development is of increasing interest, and is distinct from the idea that steroid hormones act essentially by stimulating development. The induction of these puffs as the ecdysteroid titer drops is different from their later regression at the end of the prepupal stage. At this time the fifth and sixth group of puffs are induced, the early and late-late prepupal puffs [17]. Formally the mid-prepupal puffs and these latter groups behave identically to the first three groups prior to pupariation. The interest in the prepupal puffs results from the fact that while many puffs are common to both larval and prepupal induced sequences, others are specific to one or the other sequence. As the hormone and receptor are evidently present for both sequences, this specificity of response must depend on other regulatory molecules.

The molecular analysis of these complex puff sequences has started (see Pongs, this Vol.), but is undoubtedly a long project until a little more is known concerning the gene products and their role in salivary gland development. We will concentrate our attention on the present state of studies of glue genes, the best defined group of developmental puffs.

2.2 The Glue Genes

In outline, the salivary gland synthesizes a number of glue polypeptides, at least some of which are glycosylated post-translationally [18], during the second half of the third larval instar. These are first accumulated in the form of granules in the cells and are then mobilized and secreted into the gland lumen following the increase in ecdysteroid titer prior to pupariation. At pupariation the glue is pulsed through the mouthparts in a series of slow contractions and fixes the animal to its pupation site. The developmental regulation of the initiation of activity of the glue genes is not yet resolved, although data have been obtained by four approaches, puff activity, RNA analysis, protein synthesis, and ultrastructural and histochemical studies of glue granules. In a third larval instar starting 72 h after egg laying and ending at 120 h, at pupariation, puff studies detect activity at 80–90 h when the chromosomes reach a degree of polyteny and compactness for a reliable analysis [19]. This activity persists until the rise in ecdysteroid titer at 112–114 h. In Northern analysis of *sgs3* RNA, larvae from 96 to 110 h show the same levels of RNA as those found in active wandering larvae, between 110–115 h [20], and for *sgs4*, in a strain showing a third larval instar lasting from 84 to 140 h, RNA levels increase from 106 h onwards (equivalent to ca. 92 h in other studies) [21]. When ^{3}H proline incorporation was used to follow synthesis of the glue proteins [18] activity was detected in the period 106–120 h, rather later than that expected from ultrastructural studies of the accumulation of the distinct glue granules in the salivary gland cells or light microscope observations on PAS (Periodic Acid-Schiff reaction) stained glands, which detects the polysaccharide component of the glue glycoproteins (19 and see [22] for general review of salivary gland studies). Although attempts to define the initiation of synthesis have been made with the mutant *ecd'*, considerable caution must be used in the interpretation of such studies (see Sect. 3.1).

The known glue genes and their chromosome locations are given in Table 1. Before extensive cytogenetic correlations were made, two separate nomenclatures were in use, those derived from urea-acrylamide gel electrophoresis of the gland secretion proteins [23] and those from SDS-acrylamide gel electrophoresis [18]. As the genes are systematically named following the former, we shall refer to them as *sgs1*, etc. Even in this system there remain ambiguities concerning *sgs6, 7,* and *8*. The *sgs6* mapped by Velissariou and Ashburner [24] clearly does not correspond to the band 6 of Korge (ref. [26] his Fig. 4) and bands 7 and 8 of that figure are not necessarily produced by *sgs7* and *sgs8* of Crowley et al. [27]. It seems logical that designations derived by genetic or molecular analysis should take priority over those used to number bands in a particular gel electrophoresis system.

To date, molecular analysis has concentrated on *sgs3* and *sgs4* which are the most abundant components of the glue. This abundance facilitates the preparation of

Table 1. The chromosomal location of *D. melanogaster* salivary gland secretion genes

Gene	Cytogenetic locus	References
sgs1	25B	[24]
sgs3	68C	[23, 27, 28, 31]
sgs4	3C	[21, 23]
sgs5	–	[23]
sgs6	71CF	[25]
sgs7	68C	[27]
sgs8	68C	[27]

Note that band 2 of [23] is not considered an *sgs* gene, hence the absence of *sgs2* from this table

cDNA clones to the messenger RNAs extracted either from salivary glands [21, 28] or whole animals [20]. Both genomic sequences have been isolated and the organization of the genes described [29, 30]. Interestingly both genes have an internal repeat in the coding sequence of 15 bp *(sgs3)* or 21 bp *(sgs4)* which accounts for the variability in proteins previously detected and used in their cytogenetic localization to puffs 68C and 3C, respectively [23, 31]. The 68C region proved especially interesting as it contained not one, but three glue genes [27, 28, 30]. The analysis of mutant strains in the 3C region set upper limits of 16–19 kb on the functional unit for *sgs4*, one of the first indications of gene size in *Drosophila* [32]. Further experiments with mutant *sgs4* strains have shown that alterations some 4–500 pairs 5′ to the *sgs4* coding region result in the reduction or suppression of gene activity [29, 33]. In this region there are sites hypersensitive to DNase–I digestion in chromatin prepared from salivary gland nuclei, but not from embryonic nuclei [33, 34], suggesting, as in other systems, that DNase–I hypersensitive sites are correlated with gene activity. However, while the analysis of upstream mutants will reveal sequences that are necessary, but perhaps not sufficient, for normal gene activity and DNase hypersensitivity studies define regions of potential regulatory interest, both kinds of study may be considered preliminary now that the functional test of transformation is available (see Sect. 4.2).

3. Fat Body

3.1 Background

While much emphasis had been given to the developmentally regulated puffing patterns in *D. melanogaster* salivary glands, there was another fundamental aspect to these gene programs that required investigation. This was the question of the tissue specificity of sequences of gene activity and the generality of the salivary gland model. The larval fat body appeared a logical choice for a second experimental system being a major tissue implicated in general metabolism, somewhat distinct from the salivary gland with its specialized function of glue production. The problem was approached from three different directions, an extensive developmental and genetic analysis of the major larval serum proteins (LSPs) by Roberts and his co-workers (ref. [35] for review), a direct biochemical analysis of fat body protein and RNA populations [10] which was the basis of the molecular cloning of ecdysteroid regulated genes and finally a comparative cytogenetic study of polytene chromosomes in the fat body [36].

The Dipteran serum proteins are the major products of the last larval instar fat body which are secreted into the hemolymph during the instar and reabsorbed by the fat body at the end of the instar, where they are stored in the form of dense granules. They are often considered as a form of storage protein for metamorphosis and later adult development, although their exact physiological role is unknown [35]. Their interest is twofold, first their occurrence in different insects, e.g., Calliphorin and Lucilin, provides comparative material for regulatory studies, and secondly, in *Drosophila*, the existence of two larval serum proteins (LSP1 and LSP2), one of which, LSPI, consists of three subunits (α, β, γ) coded for by three distinct genomic sequences [37]. Therefore, the LSPs represent a coordinately regulated set of transcripts which may show regulatory DNA sequences in common. In *Drosophila* at least, there are two other major transcripts of the larval fat body, designated P1 and P6, respectively [10, 38]. The functions of these proteins are unknown, but they are both stage and tissue specific. The cytogenetic localization of the major fat body transcripts are given in Table 2.

As P1 and P6 were originally described in experiments using the *ecd'* strain [10], perhaps it is appropriate for us to discuss the mutant in this brief review. The strain *1(3)ecd'* was isolated as a temperature sensitive lethal larval mutant [39], which is viable at 20 °C, but which fails to pupariate at 29 °C. This mutant has proven valuable for the investigation of ecdysteroid regulated genes [10, 40, 41]. If mutant larvae, grown at the restrictive temperature, are transferred to ecdysteroid supplemented medium many form a tanned puparium within a few hours. We note here two points concerning the mutant, which is certainly more complex than a simple deficiency in ecdysone production [42]. Firstly, the spectrum of proteins of the nontreated larvae is broadly similar, but not identical to that found in normal midthird instar larvae and this difference probably increases in older mutant larvae. While these serve as a control for ecdysteroid fed mutants, often used as a plus-minus system for differential screening, it may not be an exhaustive procedure for the isolation of larval ecdysteroid regulated genes [41]. Secondly, while one undoubtedly isolates ecdysteroid regulated genes in such screens, it is not clear that the developmental status of all the tissues in the mutant larvae is identical, that is that all reach a competence comparable to that found just

Table 2. The chromosomal locations of the major larval fat body transcripts of *D. melanogaster*

Gene	Cytogenetic locus	References
P1	70D	[38, 49]
P6	30B	[38]
LSP1α	11A	[37, 38, 44, 46]
LSP1β	21DE	[37, 38, 44, 46]
LSP1γ	61A	[37, 38, 44, 46]
LSP2	68E	[31, 38]

prior to pupariation in the salivary gland. Thus, LSP1α, LSP1γ, LSP2, and P1 transcription is stimulated by feeding *ecd'* mutant larvae with ecdysteroids [38], but while the LSPs normally are induced just after the second/third larval molt [35, 38], P1 message first reaches appreciable levels in mid-instar [38]. Therefore, none of these transcripts in normal development are induced by the ecdysteroid peak immediately prior to pupariation, but perhaps by earlier third instar peaks (see [43] for review). These points should be borne in mind when one attempts to relate experiments with the *ecd'* mutant to normal developmental processes.

The puffing analysis of fat body chromosomes and ecdysteroid induced sequences in vivo and in vitro showed that the same general classes of response were found as in the salivary gland. However, some primary ecdysteroid puffs differed between the tissues (cf. Berendes [6]), and the in vitro system that had proven so useful in salivary gland studies did not give a faithful reproduction of normal developmental sequences with fat body chromosomes [15].

3.2 P1, P6, and the LSPs

The molecular analysis of these abundantly transcribed fat body genes is well advanced. Several genomic clones of the LSPs have been isolated [38, 44, 45] and analyzed by restriction mapping and heteroduplex analysis both for the localization of coding sequences and for estimates of sequence homologies between the LSP1 α, β, and γ subunits. Sequence homology has also been estimated by quantitative in situ hybridization to polytene chromosomes [46]. These studies support the view that the subunits arose by gene duplication and subsequent sequence divergence [35].

These LSP genomic clones have been used for transcriptional studies in vivo [38] and in vitro [47] and for studies of genome organization. In this context three independently regulated transcripts were detected within 12.5 kb of the LSP2 coding region [48], further evidence that *Drosophila* transcripts are more closely packed than was anticipated from classical genetic studies (see 12 for discussion). The LSP clones, and those coding for P1 and P6, have been used to study DNA sequence organization in different tissues [49], so as to test the hypothesis that development might be accompanied systematically by sequence rearrangement or amplification. While the *Drosophila* chorion genes are systematically amplified during oogenesis [50], this does not appear to be a general rule for the major transcripts of a polyploid tissue.

The structure and organization of the P1 gene has been investigated in detail. The P1 coding sequence is 3.5 kb long and is not interrupted by introns (noncoding intervening sequences) as determined by R loop and S1 nuclease mapping. The P1 mRNA start site has been located 20 bases downstream from a typical TATA box sequence and the translation initiation codon 157 base pairs downstream from the P1 mRNA start [51]. The localization and polarity of transcription of the P1 gene on the salivary gland cytogenetic map was determined by in situ hybridization of overlapping genomic clones encompassing an 85 kbp DNA segment [52]. Induction of the P1 RNA in in vitro cultured fat bodies of third instar larvae by 20-OH ecdysone was demonstrated recently. This induction is not inhibited by cycloheximide [47], a characteristic of a primary response to hormone.

4. Transformation

4.1 Principle and Purpose

Studies of the *Drosophila* genome have revealed a number of families of moderately repetitive dispersed elements (ref. [53] for review). While the properties of each family differ in respect to overall size, length, and orientation of terminal repeats and frequency in the genome, they all show variation in the sites of integration in different *Drosophila* strains and indeed are often polymorphic in both natural populations and inbred laboratory stocks. For this reason they are described as mobile elements, as their chromosomal position appears to change relatively frequently compared to unique gene sequences, whose positions are stable in evolution. While there has been much debate as to the significance of these sequences in the *Drosophila* genome, molecular biologists realized that they might provide suitable vectors for inserting cloned DNA sequences into the genome. The most successful vectors to date are those based on P elements, factors implicated in the phenomenon of hybrid dysgenesis, which occurs when crosses are made between P strain males and M strain females ([54, 55] for reviews). These crosses lead to high mutation rates, chromosomal rearrangements, male recombination, and varying degrees of sterility in the offspring, all of which may arise from the movement of P elements. The designation of a strain as P or M is made by observing the offspring of appropriate crosses to known standard P and M strains, as only P males crossed to M females give rise to the hybrid dysgenesis trait.

The complete P element has a sequence of 2.9 kb and a 31 base pair inverted repeat at each end [56]. Internally there are four open reading frames capable of coding for proteins, although the existence and function of these proteins is at present unknown. The analysis of P induced mutations showed that internally deleted P elements were inserted in chromosomes and Rubin and Spradling [11] constructed the first vector, p 6.1, by inserting such an element, with some flanking sequences from the *Drosophila white* locus, into a plasmid vector pBR322. The plasmid vector is required for the maintenance and multiplication of the *Drosophila* DNA in bacteria and for in vitro manipulation using recombinant DNA techniques. By digesting p 6.1 with the restriction enzyme XhoI it is possible to insert DNA sequences between the two P termini. This construct is then co-injected with a complete P factor, pπ25.1, to provide "transposase" activity, into an M strain embryo. In the majority of cases so far studied, the DNA sequences between the P termini are integrated as a single non-rearranged sequence in the fly chromosomes. Although the DNA is injected in the region where the germ line cells are forming, the frequency of integrations is somewhat variable. The integrations do not occur in all the germ line cells of the injected individual and the yield of transformants in the progeny of any one transformed individual may range from ca. 40% to less than 0.5%. Transformed flies yielding high levels of transformants may result from multiple insertions in more than one germinal cell.

This basic system allows two kinds of experiment. The first is the insertion of a wild type gene sequence into a mutant stock, as for example the original demonstration with XDH^+ which transformed *rosy* flies, having deep red eyes, to wild type [11, 57]. This system is practical wherever mutant or conditional mutants are available and can be highly selective where lethality can be rescued by gene transfer, e.g., *rosy* flies grown on purine, or alcohol dehydrogenase mutants grown on ethanol [58]. The second class of experiment requires a screening procedure of lines established from individual flies which may or may not have been transformed. For this kind of experiment it is essential that the transformation efficiency is reasonably high, $\geqslant 5\%$, and that the screening procedure is sufficiently simple so as to allow the processing of, for example, 50 sublines. Such a screen might be Southern, Northern, or in situ hybridization analysis, although it is evident that for successful Northern analysis (which detects RNA synthesis), the inserted sequence must be actively transcribed.

Because of these screening problems a second generation of transformation vectors are now in use. These have modifications to the plasmid vector to reduce its size, the insertion of a multiple cloning site (a "polylinker") which simplifies the insertion of genomic sequences with differing restriction site endpoints, and a selective marker gene, such as Xdh^+ or Adh^+ for the selection of transformants [59].

The transformation system provides a functional assay for DNA sequences. At its simplest it can test the hypothesis that a piece of DNA contains the wild type sequence of a given gene by a rescue of a mutant phenotype. The system may be extended to investigate the behavior of other pieces of DNA, e.g., putative origins of DNA replication, or to dissect in detail the regulatory sequences surrounding known gene coding sequences. It is in this latter area that we may expect the most rapid progress as the efficiency of the basic experiment has already been verified in a number of laboratories.

4.2 Sgs3 and P1 Experiments

We have described above two sets of ecdysteroid regulated genes, those of the larval fat body and those of the salivary gland. The genomic DNA sequences of these genes are present in all tissues, apparently without gross rearrangements, and at all developmental stages, yet each is limited in expression to a given stage and tissue. In addition we know that ecdysteroids are present in both tissues, as well as the receptors or binding proteins that are necessary for puff induction and which were presumably detected by the indirect immunofluorescence experiments with salivary gland nuclei [60], where ecdysteroids were photoactivated and cross-linked in situ to ecdysteroid puff sites. Where then does this specificity of activity lie? We can speculate that both trans and cis-acting elements may be involved. The former will include common elements, such as polymerase molecules, possibly co-factors that are tissue specific and interact with the hormone-receptor complex, and perhaps factors that alter the chromatin structure of a region in a cell or stage specific way. These trans factors will have their counterparts in the cis DNA sequences adjacent to, or within, the coding region. If developmental gene regulation is mediated by changes in chromatin structure, then it is apparent that our ideas of gene size derived from in vivo or in vitro assays of eukaryotic coding sequences inserted in plasmid DNA ([61] for review), and remaining extrachromosomal, may be insufficient.

The first step is, therefore, to define the cis acting sequences necessary for correct gene expression. For *sgs3* the approach chosen was to clone a natural variant (Formosa) which produces an mRNA some 300 base pairs shorter than most wild type stocks. This RNA results from a genomic segment that has a restriction map very similar to that found in Oregon R except for a deficiency in the repetitive region [30] of the *sgs3*

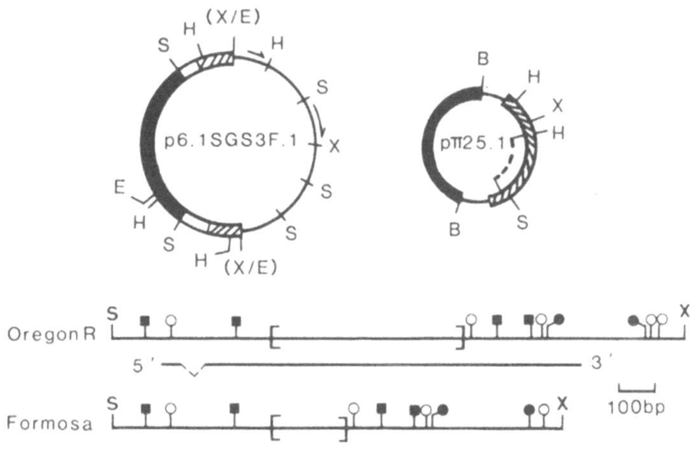

Fig. 1. The transformation plasmids p6.1 SGS3F.1, and pπ25.1. pBR322 vector sequences are shown by a *solid bar*, P element sequences by *hatching*. The extent of the deletion in the P sequence that was used to construct p6.1 is indicated on pπ25.1. The Formosa 68C genomic sequence was blunt end ligated into the unique XhoI site. The *sgs3* and *sgs7* coding regions are denoted by a *large* and *small arrow*, respectively. *Below*, the central SalI–XhoI segments of Oregon R and Formosa are compared by restriction analysis. The position of the Oregon R transcript is shown, including its short intron (*broken triangle*). *Brackets:* extent of the repeated sequences in the two strains. *E* Eco RI, *S* SalI, *H* HindIII, *X* XhoI, *B* BamHI. ▪ RsaI, ○ AluI, ● SacI. (Data from [11, 20, 30, 54] and Bourouis and Richards, unpubl.)

coding sequence (Fig. 1). We can distinguish this RNA from that of the longer alleles by Northern analysis. When this sequence is inserted into the transformation vector p6.1 and then into the fly chromosomes, it is expressed normally in two different chromosomal insertions [20] (Fig. 2). This limits the functional unit for the *sgs3* transcription unit to less than 7 kilobases of DNA, and experiments are currently in progress to further refine the limits of essential sequences.

For P1 the approach is somewhat different as neither natural variant alleles altering the size of the P1 RNA nor suitable deficiencies removing the P1 coding sequence are available at present. In order to circumvent this problem the P1 coding sequence can be replaced by the coding sequence of a gene naturally absent from the *Drosophila* genome, such as the *Escherichia coli gpt* gene coding for xanthine-guanine phosphoribosyl transferase. It has been shown recently that this gene, placed upstream from the SV40 small tumor antigen intervening sequence and termination and polyadenylation sites of SV40 early transcripts [62], can be expressed in a *Drosophila melanogaster* cell line in culture [63].

A 1.5 kb DNA fragment containing the P1 mRNA start site and the 5′ upstream sequences was fused with the *gpt*–SV40 hybrid gene. The resulting construct was inserted into the Carnegie 2 transformation vector [59] together with an 8.1 kb DNA fragment containing the XDH$^+$ gene (Fig. 3) and then into the fly chromosomes. The *gpt* gene is expressed in the fat body of third instar larvae from four independant transformed stocks [64]. This may indicate that sequences required for tissue specific expression of the P1 gene are contained within the 1.5 kb fragment linked to the *gpt* gene.

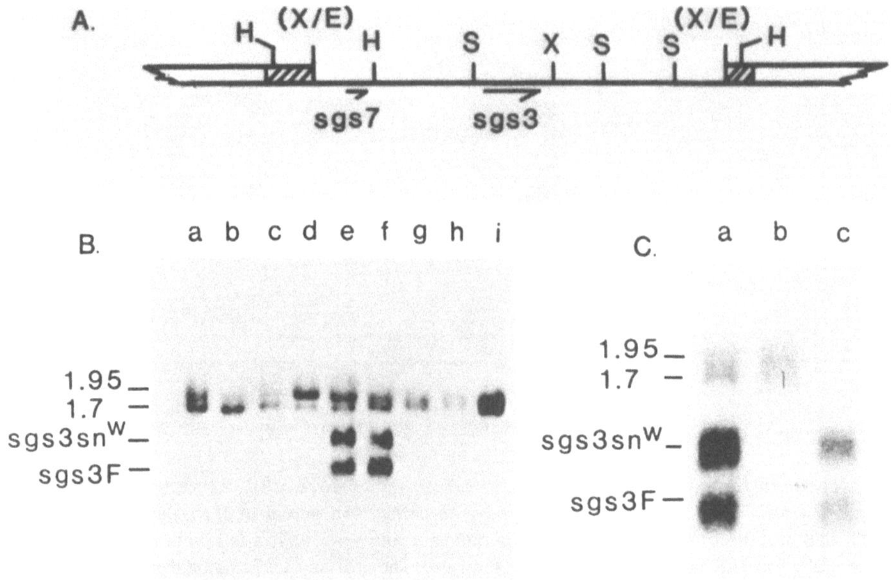

Fig. 2A–C. The inserted genomic segment and the activity of the *sgs3* gene. **A** The arrangement of the segment in the host chromosomes (*open bars*). **B** The developmental specificity of *sgs3* Formosa activity. Total RNA from *a* eggs, *b* 1st instar larvae, *c* 2nd instar larvae, *d* early 3rd instar larvae, *e* late feeding 3rd instar larvae, *f* late post-feeding 3rd instar larvae, *g* white prepupae, *h* pupae, *i* adults, probed with actin DNA (59), which reveals 1.95 and 1.7 kb transcripts at all stages, and *sgs3* DNA which detects both the sn^W host *sgs3* and the Formosa *sgs3* RNAs. Both glue genes are active only in late third instar larvae. **C** The tissue specificity of the *sgs3* genes. *a* RNA from one late third instar larva; *b, c* RNA from carcass minus salivary glands and salivary glands, respectively, of a late third instar larva. The actin transcripts are detected specifically in the carcass, while the *sgs3* transcripts are found only in the salivary gland. (Data from [20])

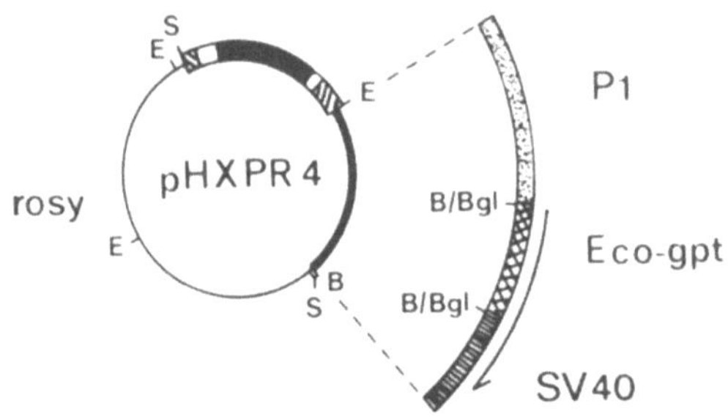

Fig. 3. The transformation plasmid pHXPR4. The chimeric P1-*Ecogpt*-SV40 gene was constructed as diagrammed on the expanded portion of the map; a 1.5 kb fragment (*stippled bar*) including the P1 mRNA start site and upstream sequences was ligated into the Bgl II (**Bgl**) site of a fragment derived from the pSV2gpt plasmid [62] and containing the Eco-gpt gene (*cross-hatching*) fused to the SV40 small *t* antigen intervening sequence and termination and polyadenylation sequences (*hatching*). The construct (*thick line*) was ligated into the EcoRI (*E*) and Bam HI (*B*) sites of the Carnegie 2 plasmid [59]. An 8.1 kb Sal I fragment containing the *rosy* gene [11] (*thin line*) was then ligated into the SalI (*S*) site of the plasmid. Puc8 vector sequences are shown by a *solid bar*, P element sequences by *hatching*. The *Ecogpt*-SV40 fused coding region is denoted by an *arrow*

5. Conclusions and Prospects

Ecdysteroid regulation in *Drosophila* is currently enjoying a stimulating phase of research thanks to the discovery of the P element transformation system [11]. We can reasonably hope that a number of laboratories, working on a range of ecdysteroid regulated genes, active in different tissues and at different developmental stages, will soon characterize the DNA sequences involved in that regulation. This possibility is important as *Drosophila* molecular studies had reached a somewhat stagnant phase, in common with many other model systems. After the initial excitement of cloning, mapping, and sequencing such genes, it became evident that in many respects, despite the sophisticated techniques employed, the results remained as descriptive as polytene chromosome puffing studies. Indeed, it is amusing to see that in other organisms the cycloheximide experimental protocols devised by Clever [3] (modified by the substitution of Northern hybridization for puff diameter) are now being used to distinguish direct or primary hormonal responses from indirect or secondary responses.

While we may hope for a rapid insight into the regulation of a set of related genes, such as the larval serum proteins, the glue genes, or the vitellogenins in adult females, the control of temporal puff sequences, or the genes contained therein, will remain a fascinating challenge for the future.

Acknowledgments. We thank our colleagues for permission to quote results from work in progress at Paris and Strasbourg and C. Kutschis, C. Werlé, and B. Boulay for assistance in the preparation of this chapter. J.A. Lepesant's research is supported by grants from the CNRS, the Ministère de l'Industrie et de la Recherche (50-9905) and the Fondation pour la Recherche Médicale Francaise. G. Richards is grateful to Prof. P. Chambon for support via grants from the CNRS (ATP 3582), the INSERM (PRC 124026), the Fondation pour la Recherche Médicale Francaise, and the Fondation Simone et Cino del Duca. G. Richards is a research fellow of the Université Louis Pasteur, Strasbourg.

References

1. Beermann W (1952) Chromosoma 5:139–198
2. Clever U, Karlson P (1960) Exp Cell Res 20:623–626
3. Clever U (1964) Science 146:794–795
4. Becker HJ (1959) Chromosoma 10:654–678
5. Becker HJ (1962) Chromosoma 13:341–384
6. Berendes HD (1967) Chromosoma 22:274–293
7. Ashburner M (1967) Chromosoma 21:398–428
8. Ashburner M (1971) Nature (London) New Biol 230:222–224
9. Ashburner M, Chihara C, Meltzer P, Richards G (1974) Cold Spring Harbor Symp Quant Biol 38:655–662
10. Lepesant J-A, Kejzlarova-Lepesant J, Garen A (1978) Proc Natl Acad Sci USA 75:5570–5574
11. Rubin GM, Spradling AC (1982) Science 218:348–353
12. Richards G (1984) In: Kerkut GA, Gilbert LI (eds) Comprehensive insect physiology, biochemistry and pharmacology, vol II. Postembryonic development. Pergamon Press, Oxford, UK
13. Schlesinger MJ, Ashburner M, Tissieres A (eds) (1982) Heat shock from bacteria to man. Cold Spring Harbor Laboratory, Cold Spring Harbor, New York
14. Ritossa FM (1963) Dros Inf Serv 37:122–123
15. Richards G (1982) Wilhelm Roux' Arch 191:103–111
16. Richards G (1976) Dev Biol 54:256–263

17. Richards G (1976) Dev Biol 54:264–275
18. Beckendorf SK, Kafatos FC (1976) Cell 9:365–373
19. Lane NJ, Carter YR, Ashburner M (1972) Wilhelm Roux Arch 169:216–238
20. Richards G, Cassab A, Bourouis M, Jarry B, Dissous C (1983) EMBO J 2:2137–2142
21. Muskavitch MAT, Hogness DS (1980) Proc Natl Acad Sci USA 77:7362–7366
22. Berendes HD, Ashburner M (1978) In: Ashburner M, Wright TRF (eds) The genetics and biology of drosophila, vol IIb. Academic Press, London New York
23. Korge G (1975) Proc Natl Acad Sci USA 72:4550–4554
24. Velissariou V, Ashburner M (1981) Chromosoma 84:173–185
25. Velissariou V, Ashburner M (1980) Chromosoma 77:13–27
26. Korge G (1981) Chromosoma 84:373–390
27. Crowley TE, Bond MW, Meyerowitz EM (1983) Mol Cell Biol 3:623–634
28. Meyerowitz EM, Hogness DS (1982) Cell 28:165–176
29. Muskavitch MAT, Hogness DS (1982) Cell 29:1041–1051
30. Garfinkel MD, Pruitt RE, Meyerowitz EM (1983) J Mol Biol 168:765–789
31. Akam ME, Roberts DB, Richards GP, Ashburner M (1978) Cell 13:215–225
32. McGinnis W, Farrell J Jr, Beckendorf SK (1980) Proc Natl Acad Sci USA 77:7367–7371
33. McGinnis W, Shermoen AW, Heemskerk J, Beckendorf SK (1983) Proc Natl Acad Sci USA 80:1063–1067
34. Shermoen AW, Beckendorf SK (1982) Cell 29:601–607
35. Roberts DB, Brock HW (1981) Experimentia 37:103–110
36. Richards G (1980) Chromosoma 79:241–250
37. Roberts DB, Evans-Roberts SM (1979) Genetics 93:663–679
38. Lepesant J-A, Levine M, Garen A, Lepesant-Kejzlarova J, Rat L, Somme-Martin G (1982) J Mol Appl Genet 1:371–383
39. Garen A, Kauvar L, Lepesant J-A (1977) Proc Natl Acad Sci USA 74:5099–5103
40. Kraminsky GP, Clark WC, Estelle MA, Gietz RD, Sage BA, O'Connor JD, Hodgetts RB (1980) Proc Natl Acad Sci USA 77:4175–4179
41. Poeting A, Koerwer W, Pongs O (1982) Chromosoma 87:89–102
42. Redfern CPF, Bownes M (1983) Mol Gen Genet 189:432–440
43. Richards G (1981) Mol Cell Endocrinol 21:134–142
44. Smith DF, McClelland A, White BN, Addison CF, Glover DM (1981) Cell 23:441–449
45. McClelland A, Smith DF, Glover DM (1981) J Mol Biol 153:257–272
46. Brock HW, Roberts DB (1981) Chromosoma 83:159–168
47. Nakanishi Y, Garen A (1983) Proc Natl Acad Sci USA 80:2971–2975
48. Yedvobnick B, Levine M (1982) Nature (London) 297:239–241
49. Levine M, Garen A, Lepesant J-A, Lepesant-Kejzlarova J (1981) Proc Natl Acad Sci USA 78: 2417–2421
50. Spradling AC (1981) Cell 27:193–201
51. Maschat F, Lepesant JA (in preparation)
52. Moreau J, Kejzlarova-Lepesant J, Brock H, Lepesant JA, Scherrer K. Submitted
53. Spradling AC, Rubin GM (1981) Annu Rev Genet 15:219–264
54. Kidwell MG, Kidwell JF, Sved JA (1977) Genetics 86:813–833
55. Bingham PM, Kidwell MG, Rubin GM (1982) Cell 29:995–1004
56. O'Hare K, Rubin GM (1983) Cell 34:25–35
57. Spradling AC, Rubin GM (1983) Cell 34:47–57
58. Goldberg DA, Posakony JW, Maniatis T (1983) Cell 34:59–73
59. Rubin GM, Spradling AC (1983) Nucleic Acids Res 11:6341–6351
60. Gronemeyer H, Pongs O (1980) Proc Natl Acad Sci USA 77:2108–2112
61. Manley JL (1983) Prog Nucleic Acid Res Mol Biol 30:195–243
62. Mulligan RC, Berg P (1981) Proc Natl Acad Sci USA 78:2072–2076
63. Sinclair JH, Sang JH, Burke JF, Ish-Horowicz D (1983) Nature (London) 306:198–200
64. Maschat F, Benes H, Lepesant JA (in preparation)

II.10 Ecdysteroid-Regulated Puffs and Genes in *Drosophila*

O. PONGS [1]

1. Introduction

Dramatic changes in the pattern of puffed chromosome loci occur toward the end of larval development of Diptera in response to the steroid hormone ecdysterone [1]. This hormonal stimulus induces a characteristic sequence of changes in puffing activity in the salivary gland chromosomes of *Drosophila melanogaster* [2]. Since it has been possible to reproduce the sequential ecdysone regulated puffing activities in vitro with cultured salivary glands [3], detailed analyses on this puffing sequence have been carried out [3-6]. Ecdysteroid induces three major changes in puffing activities in the larval/prepupal molting cycle. The puffs involved have been coined intermolt, early and late puffs [3].

2. Intermolt Puffs

Intermolt glands have a relatively simple puffing pattern described as puff stage 1 (PS1). Some of the puffs active at PS1 regress in response to ecdysteroid hormone. The puff at locus 68C, which codes for salivary glue proteins (Sgs) [7, 8], regress in culture also in the absence of hormone, but at a considerably faster rate in the presence of hormone [9]. Similarly, the puff at 25AC regresses more rapidly when PS1 glands are cultured in hormone containing medium [9]. The rate of regression of another prominent intermolt puff, 3C11, which codes for Sgs4-protein [10, 11], appears to be independent of hormone action, although the variability in the data did not allow a definitive conclusion. The effect of ecdysteroid on the regression of intermolt puffs is dose dependent, being half maximal at $\sim 5 \times 10^{-6}$ M hormone.

3. Early Puffs

Incubation of PS1 glands with ecdysteroid leads to a rapid induction of a small group of puffs within minutes, most notably at 2B5, 74EF, and 75B. These puffs are called

1 Lehrstuhl für Biochemie, Ruhr-Universität Bochum, Postfach, 4630 Bochum, FRG

Biosynthesis, Metabolism and Mode of Action of Invertebrate Hormones
(ed. by J. Hoffmann and M. Porchet)

the early puffs [3]. Early puffs are induced by ecdysteroid in a dose dependent manner [5, 9], yet threshold and maximum response differ by a factor of 500. Incubation in vitro of salivary glands in media containing cycloheximide or puromycin is without effect on puff induction at these early puff sites [12]. The action of ecdysteroid on early puff sites has been extensively characterized. There is a direct relation between ecdysone dose and puff size. As the ecdysone concentration is lowered, the time at which these puffs are active and attain the size maximal for the concentration being used becomes earlier and earlier: at 6×10^{-7} *M* ecdysteroid this time is 4 h; at 6×10^{-8} *M* ecdysteroid this time is 2 h; at 6×10^{-9} *M* ecdysteroid this time is 1 h. The regression of early puffs can be artificially induced by hormone withdrawal from the medium [13]. Having regressed as a consequence of ecdysone washout, the early puffs may be reinduced if the glands are reexposed to ecdysone. This regression, due to hormone withdrawal, is not inhibited by cycloheximide. However, the early puffs rapidly become refractory to reinduction after 2 h of hormone free culture. The development of insensitivity to ecdysone requires protein synthesis. Early puffs regress after 4 h in the continuous presence of hormone during normal development. This regression also requires protein synthesis and, consequently, is inhibited by cycloheximide. In summary, early puffs require ecdysteroid for their induction and their normal activity. Early puffs regress either upon hormone withdrawal or upon the action of factors, which depend on protein synthesis while early puffs are active.

Genetic data suggest that these factors are encoded in genes transcribed in early puffs. Walker and Ashburner [14] constructed genotypes which were either duplicate (three doses) or deficient (one dose) for the 74EF and 75B early puffs. The timing of puffing activity in these aneuploids was different from that in euploids. The puffs in the deficient larvae reach their maximum sizes after 4 h of salivary gland culture (rather than 3 h) and are active for a longer time than in salivary glands from euploid larvae. In the duplicate genotypes the early puffs are larger than in the euploids, reach their maximum activity 1 h earlier and are active for a shorter period of time. The early puffs show a compensation for altered gene dose by their activities. In three dose genotypes these puffs are active for a shorter time than in the usual two dose genotype. Vice versa, the puffs are active for a longer time in one dose genotypes; that is, the pattern of activities compensates for altered dose. These data are best understood if the activities of 74EF and 75B are controlled by the concentration of their own gene products in a feedback type mechanism. The detailed analysis of the gene products coded in these puffs will probably give a better understanding of their function.

A genetically well-studied early puff is the puff at 2B5 at the tip of the X-chromosome [15-18]. In the 2B5-region 42 mutations have been studied, which effect viability of third instar larvae. The mutations fall into six complementation groups [15], called dor, swi, and overlapping complex (o.c.c.). Dor is a larval lethal mapping to the right of 2B5, where the ecdysteroid controlled late larval puffing sequence is absent. In swi mutants early larval and early late puffs are normal (with a few exceptions), but all late larval and prepupal puffs are absent [18]. Interestingly, hormone withdrawal of in vitro incubated swi salivary glands causes premature appearance of late larval puffs. Product(s) of the o.c.c. complementation group are apparently of major importance for inducing the larval-prepupal puffing cycle in 3rd instar salivary glands. Larvae homozygous for the l(1)t435 mutation in the 2B5 band or carrying a deletion of this band fail to develop early puffs [16]. In vitro, ecdysteroid induces a puff at 2B5 to full size and puffs at 74EF and 75B to 30%–40% of maximum wild type size. All other puffs are absent. Intermolt puffs like 68C do not regress. This data suggests a key role for 2B5 in the onset of the larval puffing cycle [17].

4. Late Puffs

Late puffs appear upon culture of salivary glands with ecdysteroid only after a lag of 3 h or longer [3]. This induction is inhibited by cycloheximide [12]. The following late puffs have been studied in detail: 22C, 62E, 63E, 78D, 82F, with the exception of the 63E puff, which is induced to a low level when glands are cultured with inhibitors in the absence of hormone; the patterns of late puff activities are quite similar. In the continuous presence of hormone, late puffs reach maximum size between 8 and 10 h [4]. In contrast to early puffs, dose response of late puffs to ecdysteroid is steep and narrow, being maximal at 2×10^{-7} *M* ecdysteroid. Threshold and maximum response differ by a factor of 500 × for early puffs; and by a factor of only 4 × for late puffs. They are prematurely induced by hormone withdrawal from cultured glands [13, 18]. However, the size reached by the prematurely induced puffs is a function of the extent of puffing, during the initial exposure to ecdysteroid, of the early puff sites. Premature induction of late puffs is not inhibited by transferring glands from ecdysteroid into cycloheximide medium, but premature inductions of late puffs are reversed by reexposure to ecdysteroid. That is, ecdysteroid inhibits the premature induction of late puffs. Clever [19] first suggested that the early puffs might control the activity of the late puffs. This view is supported by the genetic data obtained with flies mutant at the 2 B locus or, respectively, aneuploid for the early puffs at 74EF and 75B. The basic idea is that late puffs respond as a consequence of early puff induction by ecdysteroid, not to the hormone itself. Puffing of late puff sites is controlled by factors of early puffs. Perhaps they are used in an inhibitory fashion for regression of early puffs and in a stimulatory fashion for activation of late puffs. This kind of self-regulation is the simplest model for explaining the data [5], though there are many alternatives possible.

5. Puffing Activities and Binding of Ecdysteroid to Polytene Chromosomes

A crucial test for this hypothesis was the correlation of sequential puffing activity between PS1 and PS11 with the sequential binding of ecdysteroid to puffed loci of the polytene chromosomes. This was accomplished by a combination of photoaffinity labeling with indirect immunofluorescence microscopy [20]. Ecdysteroid hormone can be photoactivated by irradiation at wavelengths of ~320 nm due to the α, β-unsaturated ketone moiety in its B-ring. Photoactivation induces a covalent bond of the hormone to a protein with the characteristics of a hormone receptor [21]. It can only be detected on polytene chromosomes after covalently binding the hormone to chromosomal sites. Photocross-linked 20-hydroxyecdysone was then localized by indirect immunofluorescence microscopy of squashed salivary glands. Ecdysteroid was detected at intermolt, early and late puffs with two notable exceptions [22]. Ecdysteroid was not detected at the intermolt puff 25AC, which regresses upon a rise in hormone concentration. Also, ecdysteroid was not detected at 23E, a site of early puff induction. In general, there was a good correlation between the sequential and temporal activity induced by 20-hydroxyecdysone and the binding of ecdysteroid hormone to these puffs. Ecdysteroid was detectable at early puffs as long as they were active. Also, it apparently accumulated at late puffs after induction. The observed accumulation of ecdysteroid at late puff sites after induction does not necessarily contradict the results of the "washout" experiments discussed above. On the other hand, it is difficult at the moment to fit all the data into one consistent picture.

6. Model for Ecdysteroid Action

I will first reemphasize the basic data, which are compatible with any model for explaining the temporal and sequential puffing activity and then, secondly, I will dwell on more controversial issues and will attempt to fit them in, where appropriate. There is now no doubt that puffs are the visual manifestation of active genetic units [23]. Generally, the terms puffing activity and gene activity are used synonymously. This is somewhat misleading, because there is ample evidence from recombinant DNA research that puffs contain more than one gene and that genes in a particular puff are not necessarily coordinately transcribed. Thus, looking at the same puff during development does not necessarily mean to look at the same gene activity. Furthermore, puff induction and transcription of gene(s) in a puff are certainly related, but yet separate events [24]. Therefore, terms such as sequential and temporal puffing activity are very complex and should not be used naively. The transcription of genes in early puffs (e.g., 2B5, 74EF, 75B) yields gene products (proteins, factors), which make the salivary gland competent to pursue the larval-prepupal puffing sequence [3-6]. The factors are not only necessary for the induction of the late puffs [14, 18], but also for the autoregulation (regression) of the early puffs as well as probably for the regression of some intermolt puffs (e.g., 25AC, 68C) [12]. The amount of these factors (gene dosage) determines the rate of temporal puffing. The more they are present, the faster early puffs regress and late puff activities commence. The discussion of the hormonal regulation of the synthesis of these factors and of their action is less straight forward. Clearly, early puffs are the initial targets of ecdysteroid hormone, which is absolutely necessary for inducing and maintaining the activity of early puffs. Also, ecdysteroid has a profound, although aberrant effect, on the rate of late puff appearance. Some, but not all!, late puffs are prematurely induced upon ecdysteroid withdrawal [13]. This is difficult to reconcile with the observed accumulation of ecdysteroid in late puff sites [22] as well as the different dose/response curves [4] for early and late puffs. As a "cascade" cannot explain premature induction of late puffs by hormone withdrawal [5, 13], the key event, I believe, is the autoregulation of early puff activity. Gene product(s) have to remove ecdysteroid-receptor complex from early puff sites. This could be accomplished by a modification of the hormone-receptor complex itself or of another component of the puffed locus. The former is more likely because a simple hormone withdrawal, which presumably induces a dissociation of hormone-receptor complex, leads to puff regression. Also, this hypothesis would explain why ecdysteroid is found in early puff stages at early puff sites and in late puff stages at late puff sites. Modification of the hormone-receptor complex (ER → ER*) would alter the affinity of hormone-receptor complex from early to late puff sites. And last but not least, the induction of late puffs might be generated by the cooperation between hormone-receptor and factors produced from early puffs, this being the basis of the narrow dose-response curves observed for late puffs. Figure 1 schematically illustrates this hypothesis.

The essence of the immunofluorescence data [22] is that ecdysteroid-receptor complex at first has a high affinity to early puff sites and no affinity to late puff sites; due to the activity of early puffs, ecdysteroid-receptor complex then has a high affinity to late puff sites and no affinity to early puff sites. As shown in the model, a modification of component(s) for ecdysteroid binding to early and to late puff sites is the simplest explanation for these observations. A slight variation of this model has been discussed by Belyaeva and Zhimulev [18], who proposed that a 2B5-product has a key role for the induction of other early puffs. That is that the puffing pattern is linearly extended for very early, early, early late, late puffs, and so on. This, however, does not influence the functional concept underlying the model of Fig. 1.

7. Prepupal/Pupal Puffs

The larval-prepupal puffing cycle is followed by a prepupal-pupal puffing cycle, which again is induced by ecdysteroid hormone [1]. Although many puffs reappear which have been active during the larval-prepupal puffing cycle, the temporal pattern and the sequence of puffing is substantially different (e.g., the 63E, 63F, 74EF/75B pattern is reversed) [6, 25]. This points to a different control circuit acting upon the prepupal-pupal puffing cycle. This notion is further corroborated by the observation that in o.c.c. mutants (t10) midprepupal and late prepupal puffing cycles are not disturbed in spite of disturbances of the larval puffing cycle [17]. Early puffs, such as 74EF and 75B, regress normally despite a high ecdysteroid titer. This has been explained above in terms of self-regulated accumulation of negatively acting factors (see Fig. 1). 74 EF and 75B are reinduced in the late pupal puffing period [25]. Apparently, the regression is a

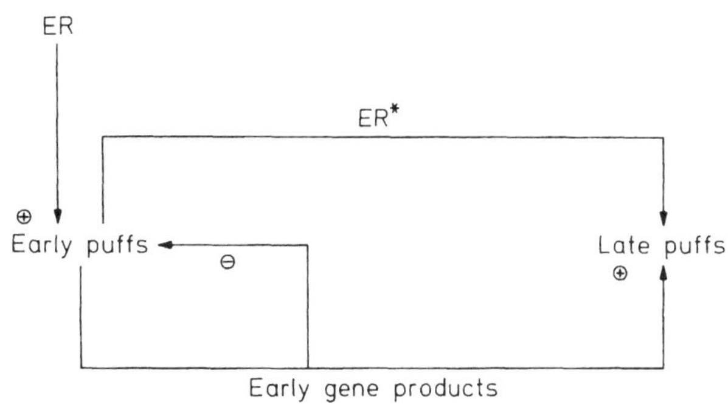

Fig. 1. Hypothetical model to explain the hormonal regulation of the temporal puffing sequence of polytene chromosomes of the 3rd instar larval salivary glands of *D. melanogaster*. *ER* ecdysteroid-receptor complex; *ER** modified ecdysteroid receptor complex

transient event, pointing to an instability of the inactivation mechanism of the early puffs in the larval-prepupal period. In vitro, a transient incubation of prepupal glands without ecdysteroid passes through the refractory stage. After a 3 h wash, puffs at 74EF and 75B are reinducible [6, 25]. Due to the lack of data, we do not know whether this reinduction leads to the production of products similar to the larval period. This is probably not the case, as temporal puffing patterns are not comparable. Many more puff sites have been described than discussed here [26]. The reader is referred to recent reviews [1, 27]. It should be remembered that puffs result from the accumulation of RNA and proteins at a band which is being transcribed. The number of chromosome sites visibly puffed is but a fraction, perhaps 50%, of the number of sites incorporating tritiated uridine into RNA as detected by autoradiography.

8. Puffing Patterns in Other Tissues

So far, I have discussed the role of ecdysteroid in regulating gene activity in the development of the larval salivary gland of *Drosophila melanogaster*. However, development not only requires a strict temporal and sequential activity of gene expression, but also tissue specificity, where each tissue with distinct functions obviously requires the activity of different genes. Since ecdysteroid has many different target organs in larvae, it is important to know whether the pattern of puffing activities, controlled by ecdysteroid in salivary glands, is tissue specific or whether it is a general phenomenon of the particular developmental stage under study. Based on early work by Berendes [28], it was believed that the initial response to ecdysteroids is the same in different insect tissues. This view was at first challenged by in situ hybridization experiments. Labeled RNA molecules, isolated from ecdysteroid treated salivary glands and imaginal disks hybridized to different chromosomal loci [29]. More recently, changes in puffing patterns in the fat body polytene chromosome 3L have been compared to those in the salivary gland [30]. Although this study included many of the prominent early, early-late, and late puffs, no data about the major early puff at 2B5 have been reported. When comparing puffs within the same subdivision of the 3L chromosome between fat body and salivary gland, three general situations exist: firstly, puffs are indistinguishable (e.g., 75B); secondly, puffs although similar in their location, may originate from adjacent bands (e.g., 62E) and thirdly, puffs are distinct (e.g., 68C). The changes in puffing pattern of fat body chromosomes are not as pronounced as those in salivary glands. The larval-prepupal period is marked by a rapid induction of 74EF and an increase in 75B. Other changes of the puffing pattern in salivary glands during this period are not reflected in the fat body. Also, 74EF and 75B remain fully active until pupariation, whereas, as discussed above, these early puffs in the salivary glands regress prior to pupariation [2]. Many puffs, which are typical for the larval-prepupal puffing pattern in salivary glands (e.g., 63E, 63F, 66B, 67B, 76D) [2], are not induced in fat body. These data indicate that the temporal puffing pattern is not only typical for a specific developmental stage, also different puffs are induced in different tissues at a particular developmental stage (e.g., 63F). Even if the same bands decondense to form a puff in two tissues there is no evidence to indicate that the same sequence is necessarily transcribed.

In fact, this situation will be discussed in molecular terms for the puff at 74EF, where it has been shown that this morphologically similar puff in salivary gland and in fat body produces different transcripts in the two tissues [31].

9. Ecdysteroid Induced Protein Synthesis

The complex sequence of changes in the puffing pattern observed in larval salivary glands is reflected in changes in the protein synthesis pattern [32, 33]. Patterns of protein synthesis in salivary glands of *D. melanogaster* larvae have been studied by pulse labeling the tissues with ^{35}S-methionine [32] or with ^{3}H-leucine [33]. As the former studies did not employ two-dimensional gel electrophoresis for analyzing labeled proteins, I will mainly discuss the latter. In broad terms, the two studies agree that the rate of synthesis of many proteins is either decreased or increased during larval-prepupal period; most notably the synthesis of two small proteins, which correspond to heat shock proteins hsp22 and hsp26 as judged by molecular weight, isoelectric point, and peptide map [34]. All in all, 26 proteins have been detected, of which the rate of synthesis is controlled by ecdysteroid in a puff stage specific manner. These proteins were compared to the proteins in fat body of which the rate of synthesis is under ecdysteroid control in vivo [35] and in vitro [33]. Only one protein was detected which had identical molecular weight and charge to a salivary gland protein (e.g., e13/f3 in [33]). This observation strongly supports the notion that ecdysteroid acts in salivary gland and in fat body of *Drosophila* larvae in a tissue specific manner.

10. Genes in Ecdysteroid Induced Early Puffs

Recent advances in cloning *Drosophila* genes by microdissection [36] allowed us to clones DNA of the 74EF puff. Fragments of the 74EF puff were microdissected from salivary gland squashes [31]. EcoRI-digested DNA of these fragments was cloned into λ-phage. Clones were screened with puff stage specific cDNA probes. One clone, which preferentially hybridized with PS4–cDNA, was used to isolate 70 kb of genomic DNA. This DNA hybridized in situ to locus 74F. Part of this DNA codes for two transcripts. One transcript was expressed in embryos and in 3rd instar larval salivary glands, the other transcript was not expressed in embryos, but in 3rd instar larval salivary glands *and* in fat body. Both transcripts are 5 kb apart from each other. As the average band contains 30 kb of DNA, it is very unlikely that both transcripts belong to genes, which are located in adjacent bands. They probably are within the same banded chromatin structure, which is puffed in salivary glands and in fat body chromosomes [30]. This suggests that adjacent genes within one puff can be transcribed differentially. The expression of these genes is apparently not controlled at the level of puffing, but at the level of selective activation. Though there are still alternative explanations possible for these data, they tentatively indicate that similarities in puffing between different tissues [30] does not necessarily mean similarities in gene expression or, respectively, in protein synthesis [33]. This observation makes it even more difficult to understand the molecular basis of temporal puffing activities controlled by ecdysteroid hormone.

Acknowledgments. The work of our laboratory described in this paper has been supported by grants of the Deutsche Forschungsgemeinschaft and the Land Nordrhein-Westfalen.

References

1. Ashburner M, Berendes HD (1978) In: Ashburner M, Wright TRF (eds) The genetics and biology of drosophila, vol IIb. Academic Press, London New York, p 316
2. Ashburner M (1972) In: Beermann W (ed) Results and problems in cell differentiation, vol IV. Springer, Berlin Heidelberg New York, p 101
3. Ashburner M (1972) Chromosoma 38:255–281
4. Ashburner M (1973) Dev Biol 35:47–61
5. Ashburner M, Chihara C, Meltzer P, Richards G (1974) Cold Spring Harbor Symp Quant Biol 38:655–662
6. Richards G (1976) Dev Biol 54:256–263
7. Muskavitch MA, Hogness D (1982) Cell 29:1041–1051
8. Akam ME, Roberts DB, Richards G, Ashburner M (1978) Cell 13:215–225
9. Ashburner M (1973) Dev Biol 35:47–61
10. Korge G (1977) Chromosoma 62:155–174
11. Beckendorf SK, Kafatos FC (1976) Cell 9:365–373
12. Ashburner M (1974) Dev Biol 39:141–157
13. Ashburner M (1976) Dev Biol 54:241–255
14. Walker VK, Ashburner M (1981) Cell 26:269–277
15. Belyaeva ES, Aizenzon MG, Semeshin VF, Kiss IL, Koczka K, Boritcheva EM, Gorelova TD, Zhimulev IF (1980) Chromosoma 81:281–306
16. Belyaeva ES, Vlassova IE, Biyasheva ZM, Kokpakov VT, Richards G, Zhimulev IF (1981) Chromosoma 84:207–219
17. Zhimulev IF, Vlassova IE, Belyaeva ES (1982) Chromosoma 85:659–672
18. Belyaeva ES, Zhimulev IF (1982) Chromosoma 86:251–263
19. Clever U (1966) Dev Biol 14:421–438
20. Gronemeyer H, Pongs O (1980) Proc Natl Acad Sci USA 74:2079–2083
21. Schaltmann K, Pongs O (1982) Proc Natl Acad Sci USA 79:6–10
22. Dworniczak B, Seidel R, Pongs O (1983) EMBO J 2:1323–1330
23. Beermann W (1972) In: Results and problems in cell differentiation, vol IV. Springer, Berlin Heidelberg New York, p 1
24. Berendes HD (1972) In: Results and problems in cell differentiation, vol IV. Springer, Berlin Heidelberg New York, p 181
25. Richards G (1976) Dev Biol 54:264–275
26. Belyaeva ES, Zhimulev IF (1976) Cell Differ 4:415–434
27. Zhimulev IF, Belyaeva ES, Semeshin VF (1981) CRC Crit Rev Biochem 11:303–340
28. Berendes HD (1967) Chromosoma 22:274–293
29. Bonner JJ, Pardue HL (1977) Cell 13:219–225
30. Richards G (1982) Wilhelm Roux' Arch 191:111–131
31. Möritz T, Edström JE, Pongs O (1984) EMBO J 3 (in press)
32. Zhimulev IF, Izquierdo ML, Lewis M, Ashburner M (1981) Wilhelm Roux' Arch 190:351–357
33. Poeting A, Koerwer W, Pongs O (1982) Chromosoma 87:89–102
34. Tekautschitz B, Pongs O (in preparation)
35. Lepesant JA, Kejzlarova-Lepesant J, Garen A (1978) Proc Natl Acad Sci USA 75:5570–5574
36. Scalenghe F, Turoco E, Edström JE, Pirotta V, Melli M (1981) Chromosoma 82:205–216

II.11 Pleiotropic Specific Responses Induced by Ecdysterone in the Cultured Kc Cells of *Drosophila melanogaster*

J.-L. COUDERC, M.-L. SOBRIER, G. GIRAUD, D. MICARD, and B. DASTUGUE[1]
M. ROPP, J.-L. BECKER, C. MAISONHAUTE, F. PERONNET, A.-M. COURGEON, and M. BEST-BELPOMME[2]

1. Introduction

For several years, we have developed a model for a better understanding of the mechanism of the ecdysteroid action at the cellular and molecular level: The *Drosophila melanogaster* cell lines established in vitro from fragments of 6-12 h old embryos [1-4].

Several reasons were in favor of this biological material. The first one was the sum of information concerning the genome and the biology of this insect. The cell lines grown in vitro allow a better control of several parameters impossible to deal with in vivo. Most of them have kept a diploid karyotype and it has been shown that the differentiating capacities of their genetic material had remained undamaged [5]. They can be cloned in homogeneous cellular populations. Some cellular clones have been shown to be ecdysteroid targets since they respond specifically to the hormone stimulus.

2. The Biological Material: *Drosophila* Cells in Culture

The cells of line Kc [1,2] are round (5-10 μm in diameter) with a voluminous nucleus. Line Kc has 85% diploid cells which have a XX karyotype with only one chromosome of the fourth pair. The length of the cell cycle was estimated to be about 20 h [6]. Isozymic studies suggested that these embryonic cells were of ectodermic origin [7]. Because we wanted to work with homogeneous cell populations, we isolated from line Kc numerous clones and subclones using a "feederlayer" technique giving a high plating efficiency [8-10]. Now a wide variety of clones are available. All these clones have a stable diploid karyotype. Line Kc 0% was adapted to grow without any fetal calf serum. It is aneuploid. All these cells were grown in monolayers at 23 °C.

1 Laboratoire de Biochimie, L.A. CNRS No 360, Faculté de Médecine, 28, Place H. Dunant, 63 001 Clermont-Ferrand Cedex, France
2 Laboratoire de Zoologie, E.R.A., CNRS, No 615, Université Pierre et Marie Curie, 75 230 Paris Cedex 05, France

Biosynthesis, Metabolism and Mode of Action of Invertebrate Hormones
(ed. by J. Hoffmann and M. Porchet)

3. The Effects of Ecdysterone on Morphological Modifications and Cell Division

Ecdysterone induces spectacular modifications of the morphology of the cells. The threshold steroid concentration is about 10 n*M*. The spherical cells become spindle-shaped and flatten. Then they elongate, emit long pseudopodia and aggregate [11]. The aggregates are formed from the progressive gathering of a few cells and then of neighboring cells which seem to be attracted by the existing clumps. At this moment, a drastic increase of cyclic GMP is shown while cyclic AMP remains constant [12]. Ecdysterone stops cell division within 2 days: the growth curve reaches a plateau which is concomitant with the induced elongation of the cells [13]. This arrest of the division is due to a blockage in G2 phase [14] and consequently the rate of incorporation of thymidine decreases [15]. The cells do not die, since during at least 4 days hydrolytic or proteolytic activities remain constant [4], the enzymatic inductions proceed during this period, and an increase of the rate of the cellular O_2 consumption without a decrease of the ATP cellular pool is observed.

The responses of the cells to ecdysteroids have been shown to be specific [16, 17]. One of the most powerful steroids is 20-hydroxyecdysone (ecdysterone). That is the reason why the effects of ecdysterone will be further described. We should emphasize that the range of steroid concentrations producing effects is usually lower than that measured in vivo [18-20].

4. Ecdysterone Induction of Enzymatic Activities

Activities of about 40 enzymes [4] have been studied before and after hormonal treatment. We did not detect any variation. For instance, the activities of proteases, aminopeptidases, total amino acyl tRNA synthetase, methionyl-tRNA synthetase, aspartate transcarbamylase, adenylate kinase, and glutathione peroxidase (indetectable) did not vary. These results led us (with other arguments, see above) to the conclusion that although the cell divisions were stopped by the hormone addition, this was not a "toxic" effect followed by a rapid death. The finding that acetylcholinesterase [21, 22] and β galactosidase [23] activities were ecdysterone inducible reinforce this statement and moreover showed that there was only a restricted number of functions which was modulated by the steroid. The advantage of enzymatic inductions was that they are quantifiable for following the ecdysterone action. For instance, the levels of enzymatic induction were ecdysterone concentration dependent (1-100 n*M*). However, if the ecdysterone concentrations were further increased (up to 0.5-1 μ*M*) the induced enzymatic levels now decreased [3]. This paradoxical effect can be compared with the lethal molts caused by experimental hyperecdysonism in insects. It was also observed [24] that polypodine B was more effective in inducing enzymes than ecdysterone, at low concentrations (1-10 n*M*). On the contrary, for higher concentrations (> 10 n*M*), ecdysterone was more efficient. This suggests that in *Drosophila* cells polypodyne B after reaching its optimal concentration could act as its own anti-hormone, perhaps in the same manner as anti-estrogens act through "anti-steroid binding sites" [25-28].

4.1 A Critical Period: The Maturation of the Cells

The *Drosophila* cell model has provided additional data on the mechanism of steroid responses from studies similar to those on the effect of the primary stimulation by estradiol in the "naive" ("incompetent") chick oviduct on a secondary stimulation after an intermediate withdrawal period [29–31]. This procedure adapted for *Drosophila* cells leads to analogous results [32]. We hypothesise that during primary stimulation a two-step mechanism is involved (1) a complex succession of cellular events which enables cells to respond phenotypically, this step involves the 'acquisition of competence' or 'maturation' of the cells, (2) a visible and measurable response to hormone. The experimental conditions have not allowed a detailed analysis of events during this phase of primary stimulation in the chick oviduct system.

In contrast, the in vitro *Drosophila* cell model offers us a good opportunity to resolve the primo-stimulation into the postulated two-step mechanism. For that, we mimicked the physiological situation such as it is known from the in vivo variation of the ecdysteroids level: the ecdysteroid peak corresponding to the molts in *Drosophila* is preceded by low levels. Thus, we "matured" the cells by a low and continuous hormonal concentration (1-5 n*M*). The maturation period is 2 days long. In a second step, without washing the cells, we stimulated them by the addition of higher ecdysterone concentrations (50-250 n*M*). With the maturation, the ecdysterone inductions, both for acetylcholinesterase and β galactosidase, are not only quicker, but higher when compared with those obtained without maturation. For instance, the initial rate of acetylcholinesterase induction is multiplied by six and detectable as early as 4 h after the addition of the stimulating hormone concentration. Moreover, the effect of the maturation on the acetylcholinesterase induction is abolished if during the maturation period a JH analog (altosid) is added [33]. Thus, the maturation period seems to be a critical period for the cells since during this "naive" period they become "mature" or "competent" even if their responses are not detectable. Competence is acquired during the primo-stimulation and is expressed during the secondary stimulation.

The molecular events which explain the competence acquisition during the maturation could involve a modification of the "chromatin structure". Another explanation could be that during the maturation period ecdysterone is metabolized to a more active compound. That is ruled out since HPLC studies of the metabolism of labelled ecdysone and ecdysterone demonstrated that the very large majority ($\geqslant$95%) of these two steroids was not metabolized in the cells [34].

4.2 Differentiation of the Cells and Aerobiosis

A working hypothesis is that the responses of the cells after the hormonal stimulation truly reflect the early step of a physiological differentiation induced by the steroid. For this differentiation we postulate that the cells need more energy; in other terms they have to cope with the energetic cost of this differentiation. If we keep in mind that (1) practically all the oxygen consumed by an aerobic cell is consumed by its mitochondria [35] and (2) that about 95% of the energy produced by an aerobic cell is produced by its mitochondria [36], we are thus led to study the oxygen consumption by *Drosophila* cells in the absence and in the presence of ecdysterone.

Using a Clark electrode, we observed that the rate of oxygen consumption (v) by one cell obeyed a Michaelis-like equation $v = V_M \frac{S}{K_m + S}$. This allowed the measurements of the kinetic parameters of the cellular oxygen consumption, i.e., the maximal rate of O_2 consumption (V_M) and the K_m for O_2 if S was the O_2 concentration dissolved in the medium. We found that the V_M value was 0.2 ± 0.05 fmol min^{-1} $cell^{-1}$ and the K_m is 4 ± 1 *μM* (25 °C). In the presence of ecdysterone the value of the K_m does not vary, whereas the value of the V_M is increased. It can be indeed multiplied by three. Assuming that 1 liter of oxygen corresponds to the biological production of 4.7×10^3 cal [35] we estimate the cellular energy production for a V_M of 0.2 fmol O_2 min^{-1} $cell^{-1}$, as 2.10^{-11} cal min^{-1} $cell^{-1}$, a value which corresponds to the synthesis of about 1.3 fmol ATP min^{-1} $cell^{-1}$. These values, in the presence of ecdysterone, have to be multiplied by the ratio between the V_M in the presence of the hormone and the V_M in the absence of the steroid. It is possible to strike a cellular energetic balance. Moreover, the measurements of the intracellular ATP pool show that it remains constant and even tends to increase on the fourth day after the hormone addition. These observations led us to assume that mitochondria could be an (in)direct intracellular target for the steroid. This hypothesis was reinforced when the increased cytochrome C oxidase activity was demonstrated in the hormone-treated cells, whereas cell sensitivity for cyanide was decreased.

However, we also know that oxygen is toxic for the cells, "all respiring organisms are caught in a cruel bind. The very oxygen which supports their lives is toxic to them and they survive precariously only by the virtue of elaborate defense mechanisms" [37]. Among these defense mechanisms are the superoxide dismutase and catalase activities [38, 39]. Observing that ecdysterone increases the maximal rate of O_2 cellular consumption, these two enzymatic activities were measured. Superoxide dismutase [40] and catalase activities [41] were increased during ecdysterone stimulation on the cells. Our present assumption is, thus, the following: ecdysterone leads the cells to a differentiated program. This costs more energy and the cells increase their oxygen consumption. If oxygen is the only energetic source for an aerobic cell, it is also toxic by producing dangerous reactive products of its reduction, i.e., O_2 and H_2O_2 [37-39]. To prevent the damages which could be caused by these two products the cells increase their superoxide dismutase (SOD) and catalase activities. In this case, it is expected that a sudden increase of the maximal rate of oxygen cellular consumption without the increase of SOD and catalase activities should provoke a quite different response. That seems to be the case since the rapid increase of the V_M obtained during the return to normoxia after a period of anoxia, but without increase of the SOD and catalase activities, is now accompanied by the induction of the heat shock proteins [42].

5. Ecdysterone Induces Specific Protein Changes in the Kc Cells

We observed the apparition of a class of soluble proteins after ecdysterone treatment. These proteins were relatively major ones since they were detectable by staining the electrophoretic gels after 24 h of 100 *nM* ecdysterone treatment and clearly accumulated after 4 days [16, 43]. This property allowed us to raise antibodies against them [44]. With these antibodies it was possible to show that (1) these proteins induced by the hormone in vitro were immunologically identical to those present in extracts of male and female adult flies, whereas they were not detectable in eggs, first and second instar larvae extracts and that (2) the increase of the cellular amount of these induced proteins was explained by an increase of their rate of synthesis.

More extensive experiments have been carried out on early changes in protein synthesis analyzed by ^{35}S methionine incorporation and autoradiography of two-dimensional gels [46]. We found that ecdysterone gives a rapid, very limited, and highly specific response in Kc 0% cells. It was shown that among the about 350 spots studied a limited number of those varied after ecdysterone treatment indicating that the rates of synthesis of about eight polypeptides were increased, whereas two, on the contrary, were decreased. The earliest effects were detected after 7 h of treatment. It has to be noted that two proteins whose synthesis level was increased by ecdysterone were identified as the two cytoplasmic actins [46] (actin II and III).

6. Ecdysterone Induces Accumulation and Polymerization of Actin in Kc Cells

Ecdysterone treatment of Kc cells induced an increase of actin synthesis as demonstrated by ^{35}S methionine incorporation. This increase of methionine incorporation was very important in the actin III form which is a short-term form and certainly the precursor of actin II [47, 48] in *Drosophila* cells. This increase of actin synthesis was correlated with an increase (X2) of the total actin content of Kc cells after ecdysterone treatment and an increase (X4) of polymerized actin [49] as measured by DNase I inhibition [50]. This result has also been independently found by another group [51], although they did not observe any change in the relative rate of actin synthesis nor in the relative rate of actin turnover. These variations of actin content and synthesis in Kc cells were correlated with the dramatic morphological changes, the aggregation, and the motility induced by ecdysterone. It is well-established that in many nonmuscle cells actin microfilaments are involved in many cellular functions, such as cytoskeletal support, pseudopod formation, and cell motility. Variations of the actin pools during transformation of embryonal carcinoma cells in fibroblast-like cells [52] during platelet activation [53] or during adipocytes differentiation [54] suggest that a reorganization of actin might be a marker of differentiation. So it may be possible to analyze the regulation of expression of actin and perhaps other constituants of microfilaments during morphological differentiation in a hormonally-controlled system, such as Kc cells.

7. Isolation of DNA Sequences Regulated by Ecdysterone in a *Drosophila* Cell Line

7.1 Differential Screening of a *Drosophila* Genomic λ Library

In order to obtain a wider field in the responses of the cells to the action of ecdysterone, we have screened a *Drosophila* λ library. This library was a *Drosophila melanogaster* randomly sheared DNA, cloned in λ Charon 4 using synthetic Eco RI linkers [55]. About four equivalent *Drosophila* genomes (i.e., 4.10^4 plaques) have been transferred onto duplicate nitrocellulose sheets. One filter was hybridized with poly A^+ RNA-^{32}P labelled, purified from cells grown for 30 h in the presence of 10^{-7} *M* ecdysterone, the second one with poly A^+ RNA-^{32}P labelled, extracted from untreated cells.

After autoradiography, we looked for spots in which the intensity was different on the two films. The corresponding areas were spread again twice more and tested with the two labelled RNA probes to obtain further purified specific recombinant sequences. At last, the DNA obtained from the screened recombinated phages were dot spotted and hybridized once more with the two RNA probes to ascertain the differential screening.

Thus, two kinds of *Drosophila* recombinant DNA sequences have been isolated [56]. (1) Sequences whose transcripts were increased by ecdysterone treatment: they will be referred to as "inducible" sequences. (2) Sequences whose transcripts were lowered by the hormone: they will be referred to as "repressed" sequences.

The terms induced and repressed did not include anything pertaining to the mechanism of the regulation of these ecdysterone-modulated sequences.

7.2 Characterization of These Sequences by in Situ Hybridization

All the isolated differential DNA sequences were ^{3}H-nick-translated and hybridized onto the giant chromosomes isolated from the salivary glands of the F1 progeny of the gt-1 X gt-11 cross, according to Bonner and Pardue [57]. All the inducible sequences have a single locus mapping, while all the repressed sequences hybridize to multiple loci distributed on each chromosome.

7.2.1 Inducible Sequences

Seven λ Dm containing inducible sequences were isolated [56]. These sequences are not clustered in the genome, but widespread on all the chromosomes. λ Dm A5A1 and λ Dm 1632 map at the same 5C locus, where one of the actin genes maps [58, 59]. λ Dm 43 is localized at the 37 A–B locus, a region of the second chromosome (II L) where the gene dopa decarboxylase (DDC) was previously described [60]. This gene is hormonally regulated in *Drosophila melanogaster* [61]. Moreover, Hirsh and Davidson [62] have isolated DDC probes from *Drosophila* embryos which lie within polytene chromosome band 37 C1–2. The comparison of the restriction maps of λ Dm 43 with the DDC probes are in good agreement.

The localization of λ Dm 531 at the 25 A–C locus is near a puff site described by Ashburner [63]. The restriction map of λ Dm 531 corresponds to the restriction map of a recombinant plasmid containing *Drosophila* genomic DNA bearing a collagen-like gene and hybridizing to the 25C locus on polytene chromosomes [64]. More precise identification of the induced gene on λ Dm 531 and λ Dm 43 is in course. The last three sequences: λ Dm 132, λ Dm 511, and λ Dm 951 are induced sequences whose in situ hybridization mapping is not in relation with known biochemical loci in *Drosophila melanogaster* and up to now cannot be assigned to a precise function.

7.2.2 "Repressed" Sequences

Six λ Dm clones (λ 261, 321, 1212, 1412, 1611, 1731) have been selected independently for a decreased hybridization with poly A^+ RNA of ecdysterone treated cells. In situ hybridization on the polytene chromosomes of the gt-1 × gt-x11 progeny reveals, for all of these λ Dm, multiple loci distributed on each chromosome and at the chromocenter (Becker, unpubl. data).

λ Dm 261 and λ Dm 321, two repressed sequences containing phages, were hybridized to polytene chromosomes of the two parental stocks of the larvae used in the original experiment. The two stocks, gt-1 and gtx11 show very different patterns of hybridization with a transposable element [65]. Although λ Dm 261 and λ Dm 321 each hybridize to multiple sites on chromosomes of both stocks, there is little, if any, overlap between the sites of hybridization of an element on one stock with its sites on the other stock. Both elements bind to the chromocenter on both stocks, but the lack of substructure in the chromocenter prohibits fine localization. Comparison of the sequence localization in the two stocks strongly support that these repeated sequences are transposable elements.

Restriction maps were done for λ Dm 261, 321, and 1611 (Couderc et al., unpubl. data). These restriction maps and cross-hybridization experiments show that these three phages overlap in part. This region of overlap contain repeated sequences that are actively transcribed in untreated Kc cells. Moreover, these repeated sequences hybridize to highly heterogenous poly A^+ RNA on Northern blots and the level of these RNAs decreases along ecdysterone treatment. Specific *Drosophila* probes are used for identification of these sequences.

8. Actin Gene Expression is Modulated by Ecdysterone in Kc Cells

8.1 Identification of the "Induced" Gene

Two overlapping λ Dm clones (λ Dm 1632 and λ Dm A5A1), hybridizing more with the RNA of ecdysterone-treated cells than with the RNA of untreated cells have been isolated [56] and more extensively studied [66]. They were chosen because they hybridize to abundant poly A^+ RNA and also map to the 5C locus – the restriction maps of these two λ Dm were deduced by digesting the chimeric phage DNA with Eco RI, Hind III, Sal I, Sst I, Bgl II, and Bam HI, and all combinations thereof. Comparison of the two maps suggests that they overlap by 8–7 kb. Moreover, this region of overlap corresponds to the maps of the 5C actin gene published elsewhere [67]. The position of transcribed sequences of λ Dm 1632 and λ Dm A5A1 were determined with labelled poly A^+ RNA of treated Kc 0% cells. There is only an actively transcribed sequence on the two recombinant phages λ Dm A5A1 and 1632 localized at the place of the 5C actin coding region previously described by Fyrberg et al. [67], suggesting that the ecdysterone inducible sequence that we have isolated is the 5C actin gene.

8.2 Expression of Actin Gene in Kc Cells

We isolated undegraded total cellular poly A^+ RNA from *Drosophila* Kc 0% cells. The RNA was subjected to size fractionation on denaturing agarose gels, transferred to nitrocellulose paper, and hybridized to the nick-translated actin coding-region. Figure 1 shows the results of such an experiment using poly A^+ RNA of cells treated with ecdysterone for 0, 4, 8, and 24 h. Two distinct heavy bands of actin-specific RNAs of 1.7 and 1.9 kb were detected. RNA prepared from control cells showed little hybridization to the actin probe, especially the lower band of 1.7 kb. RNA from ecdysterone-treated cells hybridized much more intensely to the probe, indicating that the level of the two actin specific RNAs increased proportionally to the time of treatment with ecdysterone at least during the 4–24 h period. We have not yet examined whether there was more than one mRNA in some of these bands. As a probe for Northern blots we

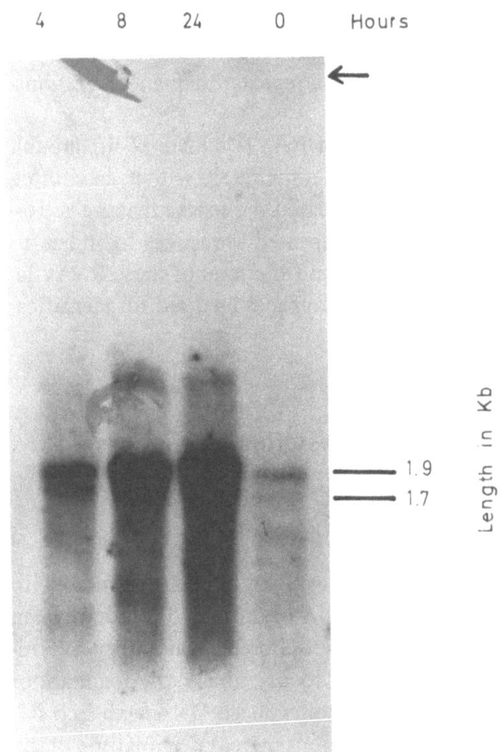

Fig. 1. Expression of ecdysterone-induced gene in Kc cells. Three μg of polyadenylated total cellular RNA was isolated from 4, 8, 24 h ecdysterone-treated cells and untreated cells (0) as indicated in figure, subjected to electrophoresis in denaturing glyoxal agarose gels, and transferred to nitrocellulose filters. RNA sizes are indicated in Kb. Blot was hybridized with nick-translated λDm 1632 sequences and subsequently autoradiographed

have also used the actin coding region of these genes that hybridizes to all actin genes and permits the detection of all actin mRNA. We observed the same pattern as that obtained with λ Dm 1632.

The intracellular distribution of actin mRNA was assayed by separating the nuclear and cytoplasmic contents of cultured cells. The extracted RNA was fractionated in poly A^+ and A^- fractions, separated by denaturing electrophoresis, transferred on nitrocellulose paper, and hybridized to actin coding region. Figure 2 shows that actin RNA was essentially poly A^+. Two bands of 1.7 and 1.9 kb were detected in poly A^+ and poly A^- fractions and were increased in 24 h ecdysterone-treated cells in cytoplasmic

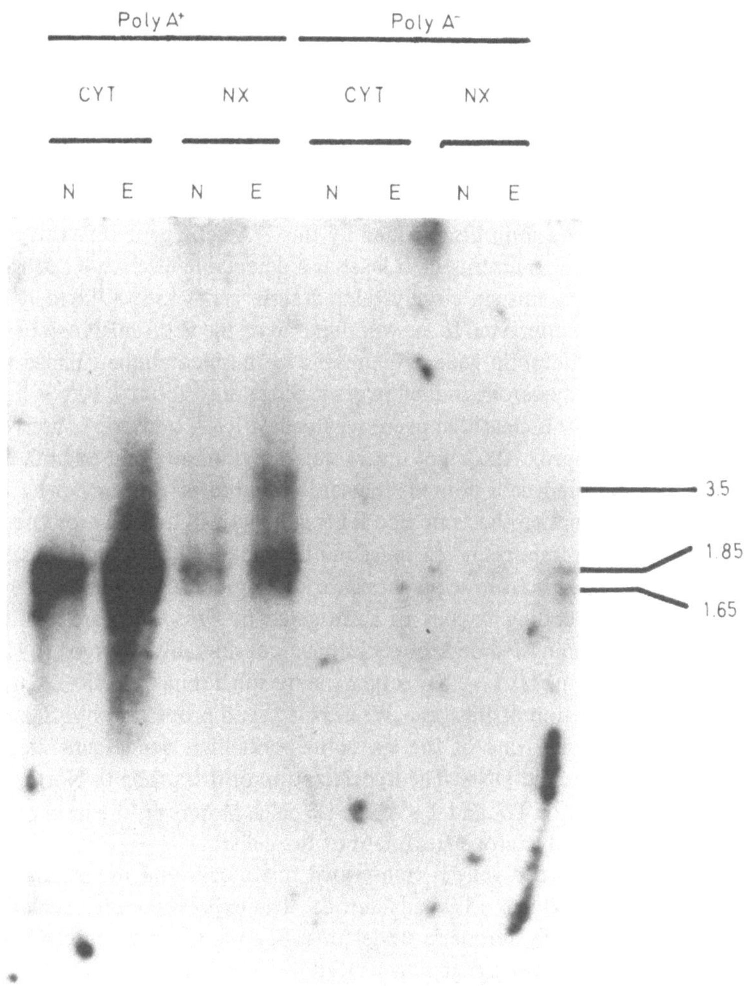

Fig. 2. 3 μg of poly A^+ cytoplasmic (*CYT*) and nuclear (*NX*) RNA and 20 μg of poly A^- cytoplasmic (*CYT*) and nuclear (*NX*) RNA were extracted from untreated (*N*) and 24 h ecdysterone-treated cells (*E*), subjected to electrophoresis in denaturing glyoxal agarose gels, and transferred to nitrocellulose filters. RNA sizes are indicated in Kb. Blot was hybridized with nick-translated actin coding region and subsequently autoradiographed

as well as in nuclear fractions. Moreover, in the nuclear poly A^+ fraction two other bands of about 3.5 kb appear essentially in ecdysterone-treated cells, but were also present in untreated ones (detected on overexposed autoradiograms). These RNAs which were more abundant in ecdysterone-treated cells, are the premessengers of actin mRNA since normal actin genes contain large introns [22]. The accumulation of actin premRNA in nuclei of ecdysterone-treated cells suggests that ecdysterone controls the level of actin mRNA at least in part at the transcriptional level.

The concentration of ecdysterone in the culture medium necessary to induce actin mRNA accumulation has also been investigated. The induction was maximal at 10 n*M* ecdysterone, while at 1 n*M* no induction of actin occurred. Half maximal induction required 5 n*M* of ecdysterone in Kc cells. This value may correspond to half maximal induction of the morphological response and the induction of acetylcholinesterase activity in Kc H cells that requires 10 n*M* and should be compared with the value of 100 n*M* giving half maximal puffing in organ culture [68].

Selection and in vitro translation of RNA homologous to λ Dm 1632 showed that the selected mRNA encoded the two forms of cytoplasmic actins [47] II and III and that no mRNA encoding muscle actin I was present either in untreated or in ecdysterone treated Kc cells. As actin I is encoded by the 57A actin gene [69] this result shows that this gene was never transcribed, at least at a detectable level, in Kc cells.

Drosophila contains six closely related actin genes [58, 59] widely dispersed in the genome. We were interested to see whether one of the actin mRNAs observed in Kc cells is encoded by the actin gene 5 C present in the recombinant phage that we isolated containing an ecdysterone inducible gene. Since only about 1.1 kb is needed to encode an actin protein, it is clear that even the smallest RNA transcript observed in our results must contain approx. 0.6 kb of untranslated regions at either or both ends of the transcript. Kc cultured cells contain the same number of actin genes by haploid genome and these genes are on the same Eco RI fragment as in Canton S or Oregon R flies. This suggests that no important rearrangements, under culture conditions, had occurred in the Kc cell line, at least in sequences surrounding the actin genes.

The protein coding regions of actin genes of *Drosophila* are highly conserved but 3′ and 5′ noncoding region sequences have considerably diverged [67]. So it is easy to isolate specific probes of the 5 C actin gene by subcloning in pBR 328 a transcribed, but untranslated portion of this gene. We used a 5′ end probe that hybridized in a Southern experiment to only one of the six actin containing bands generated by the Eco RI digestion of genomic DNA. The hybridization of this probe to Northern blots revealed again two RNAs of 1.7 and 1.9 kb in Kc cells. Moreover, the level of these two RNAs increased after ecdysterone treatment of the cells.

Thus, the 5 C actin gene is transcribed in Kc cells and the steady state level of 5 C actin RNA is modulated by ecdysterone. The expression and regulation of the other actin genes by ecdysterone is under investigation using specific probes and the Berk and Sharp S1 nuclease mapping procedure. Such data will yield insights about the molecular mechanism(s) governing actin gene expression in an hormonally-induced differentiation. Continued study of gene families whose members are expressed differentially will increase our understanding of the regulatory elements that control gene expression in a multigene family.

9. Summary and Perspectives

The different effects of ecdysteroids on the Kc cell lines from *Drosophila melanogaster* are described. These steroid specific responses include morphological modifications, induction of three enzymatic activities, and induction or suppression of a limited number of polypeptides. The question is raised whether these different aspects of ecdysterone action are controlled or not by a single biochemical mechanism. The cloning and the molecular study of *Drosophila* cell genes where expression is described as either induced or repressed by the hormone will help to answer this question. Ecdysteroid responsive genes have been previously cloned in the fat body and the salivary glands (see Chap. II.10). We now have clones of positive and negative ecdysterone responsive genes from a third cell type. It is possible to study comparatively the mechanisms of the two opposite effects due to a single steroid. Appropriate cultured cell transformation systems, i.e., vectors allowing the introduction [70] of cloned genes into cultured *Drosophila* cells, are now available and should permit functional analysis of the cloned ecdysterone-induced genes.

Acknowledgments. This work was supported by the Centre National de la Recherche Scientifique (ATP no. LT/DN 3651 and ATP no. LT/HL 3366) and by the Aide à la Recherche Universitaire, Biologie 1982. One of us (MLS) was supported by the Ligue Nationale contre le Cancer.

References

1. Echalier G, Ohanessian A (1969) CR Acad Sci 268:1771–1774
2. Echalier G, Ohanessian A (1970) In Vitro 6:162–172
3. Best-Belpomme M, Courgeon AM, Echalier G (1980) In: Hoffmann J (ed) Progress in ecdysone research. Elsevier/North Holland Biomed Press, Amsterdam New York, pp 379–392
4. Best-Belpomme M (1981) In: Sehnal F, Zabla A, Cymbroroswki B (eds) Regulation of insect development and behaviour. Wroclaw Tech Univ Press, Wroclaw, Poland, pp 715–722
5. Illmensee K (1976) In: Lawrence PA (ed) Insect development. Blackwell Sci Publ, Oxford, pp 76–96
6. Dolfini S, Courgeon AM, Tiepolo L (1970) Experientia 26:1020–1021
7. Debec A (1974) Wilhelm Roux' Arch 174:1–19
8. Courgeon AM (1975) Exp Cell Res 94:283–291
9. Courgeon AM, Best-Belpomme M (1976) In: Actualités sur les hormones d'invertébrés. Colloq Int CNRS 251:383–392
10. Richard-Molard C, Ohanessian A (1977) Wilhelm Roux' Arch 181:135–149
11. Courgeon AM (1972) Exp Cell Res 74:327–336
12. Courgeon AM, Cailla HL (1981) Exp Cell Res 133:15–22
13. Courgeon AM (1972) Nature (London) New Biol 238:250–251
14. Stevens B, Alvarez CM, Bohman R, O'Connor JD (1980) Cell 22:675–682
15. Rosset R (1978) Exp Cell Res 111:31–36
16. Best-Belpomme M, Courgeon AM (1976) CR Acad Sci 282:469–472
17. Cherbas L, Yonge CD, Cherbas P, Williams CM (1980) Wilhelm Roux' Arch 189:1–15
18. De Reggi ML, Hirn MH, Delaage MA (1975) Biochem Biophys Res Commun 66:1307–1315
19. Hodgetts RB, Sage B, O'Connor JD (1977) Dev Biol 60:310–317
20. Berreur P, Porcheron P, Berreur-Bonnenfant J, Simpson P (1979) J Exp Zool 210:347–352
21. Cherbas L, Cherbas P, Williams CM (1977) Science 197:275–277
22. Best-Belpomme M, Courgeon AM (1977) FEBS Lett 82:345–348
23. Best-Belpomme M, Courgeon AM, Rambach A (1978) Proc Natl Acad Sci USA 75:6102–6106

24. Ropp M (unpublished results)
25. Baulieu EE, Altger M, Best-Belpomme M, Corvol P, Courvalin JC, Mester J, Milgrom E, Robel P, Rochefort M, De Catalogne D (1975) Vitam Hrom 33:649–736
26. Sutherland R, San Foo M (1979) Biochem Biophys Res Commun 91:183–191
27. Sutherland R, Murphy L, San Foo M, Green M (1980) Nature (London) 288:213–215
28. Faye J, Lasserre B, Bayard F (1980) Biochem Biophys Res Commun 93:1225–1231
29. Oka T, Schimke RT (1969) J Cell Biol 43:123–137
30. Palmiter RD (1972) J Biol Chem 247:6450–6459
31. Palmiter RD (1972) J Biol Chem 247:6770–6580
32. Cherbas L, Cherbas P, Savakis C, Demitri G, Manteuffel-Cymbrorowska M, Yonge CD, Williams CM (1980) In: Kurstak E, Maramorosch K, Dübendorfer A (eds) Proc 5th Int Conf Invertebr Tissue Cult. Academic Press, London New York, pp 217–228
33. Best-Belpomme M, Courgeon AM (1980) Eur J Biochem 112:185–191
34. Lafont R, Best-Belpomme M (unpublished results)
35. Jöbsis FF (1964) In: Fenn, Rahn H (eds) Handbook of physiology, vol I, sect 3. Respiration. Am Phys Soc Wash DC, pp 63–124
36. Mitchell P (1979) Science 206:1148–1159
37. Fridovich I (1976) In: Pryor WA (ed) Free radicals in biology, vol I. Academic Press, London New York, pp 239–271
38. Fridovich I (1981) In: Gilbert D (ed) Oxygen and living processes. Springer Heidelberg Berlin New York, pp 250–272
39. Chance B, Gies H, Boveris A (1979) Physiol Rev 59:527–604
40. Ropp M, Best-Belpomme M (unpublished results)
41. Best-Belpomme M, Ropp M (1982) Eur J Biochem 121:349–355
42. Ropp M, Courgeon AM, Calvayrac R, Best-Belpomme M (1983) Can J Biochem Cell Biol 61: 456–461
43. Best-Belpomme M, Courgeon AM (1976) CR Acad Sci 283:155–158
44. Best-Belpomme M, Courgeon AM (1977) FEBS Lett 82:345–348
45. Cade-Treyer D, Munsch N (1980) FEBS Lett 117:19–22
46. Couderc JL, Dastugue B (1980) Biochem Biophys Res Commun 97:173–181
47. Horovitch SJ, Storti RV, Rich A, Pardue ML (1979) J Cell Biol 82:86–92
48. Berger E, Cox G (1979) J Cell Biol 81:680–683
49. Blikstad I, Markey F, Carlsson L, Persson T, Lindberg U (1978) Cell 15:935–943
50. Couderc JL, Cadic AL, Sobrier ML, Dastugue B (1982) Biochem Biophys Res Commun 107: 188–195
51. Berger EM, Cox G, Ireland R, Weber L (1981) J Insect Physiol 27:129–137
52. Markey F, Soriano L (1980) FEBS Lett 110:305–307
53. Pribluda V, Laub F, Rotman A (1981) Eur J Biochem 116:293–296
54. Spiegelman BM, Farder SR (1982) Cell 29:53–60
55. Maniatis T, Hardison RC, Lacy E, Lauer J, O'Connell C, Quon D, Sim GK, Efstratiadis A (1978) Cell 15:687–701
56. Couderc JL, Becker JL, Sobrier ML, Dastugue B, Best-Belpomme M, Lepesant JA, Pardue ML (1984) Chromosoma 89:338–342
57. Bonner JJ, Pardue ML (1976) Chromosoma 58:87–99
58. Tobin SL, Zulauf E, Sanchez F, Craig EA, McCarty BJ (1980) Cell 19:121–131
59. Fyrberg EA, Kindle Xl, Davidson N, Sodja A (1980) Cell 19:365–378
60. Hodgetts RB (1975) Genetics 79:45–54
61. Hodgetts RB (1980) Proc Natl Acad Sci USA 77:4175–4179
62. Hirsh J, Davidson N (1981) Mol Cell Biol 1:475–485
63. Ashburner M (1972) Chromosoma 38:255–281
64. Natzle JE, Monson JM, McCarthy BJ (1982) Nature (London) 296:368–371
65. Pardue ML, Dawid I (1981) Chromosoma 83:29–43
66. Couderc JL, Sobrier ML, Giraud G, Becker JL, Dastugue B (1983) J Mol Biol 164:419–430
67. Fryberg EA, Bond BJ, Hershey ND, Minter KS, Davidson N (1981) Cell 24:107–116
68. Richards G (1980) In: Hoffmann J (ed) Progress in ecdysone research. Elsevier/North Holland Biomed Press, Amsterdam New York, pp 363–378
69. Fyrberg EA, Mahaffey JW, Bond BJ, Davidson N (1983) Cell 33:115–123
70. Bourrouis M, Jarry B (1983) EMBO J 2:1099–1104

II.12 The Action of Ecdysone* in Imaginal Discs and Kc Cells of *Drosophila melanogaster*

L. CHERBAS [1], J.W. FRISTROM [2], and J.D. O'CONNOR [3]

1. Introduction

The imaginal discs and the cell line Kc from *Drosophila melanogaster* have been used in our laboratories as in vitro model systems for studying the mechanism of ecdysone action. The two systems are in many ways complementary. Both consist of imaginal cells which are induced to differentiate by treatment with the hormone. We shall attempt here to synthesize the results of independent work from our three laboratories using these model systems. We will not provide a comprehensive or unbiased review of the field. (We refer the reader to the papers by Richards and Lepesant [2] and by Best-Belpomme and Dastugue [3] in this volume for other points of view and additional *Drosophila* model systems.) Instead, we shall try to emphasize the experimental approaches for which these systems are particularly well-suited, and to point out similarities and differences in the details of their responses.

Several considerations led us to use these model systems. Both Kc cells and imaginal discs are isolated ecdysteroid-responsive tissues in which the hormone is neither synthesized nor metabolized. They thus provide the opportunity for precise experimental manipulation of the hormonal environment. Both tissue types are available in gram quantities, making biochemical studies – such as hormone receptor characterization – feasible. Finally, the use of *Drosophila melanogaster* provides a powerful combination of genetic and recombinant DNA techniques unavailable in other steroid-responsive organisms.

1.1 The Development in Vivo of *Drosophila* Imaginal Discs

The development of *Drosophila melanogaster* provides an excellent example of complete metamorphosis. The juvenile or larval tissues are extensively destroyed during metamorphosis and are replaced by imaginal (adult) tissues. The precursors of the imaginal

* In the nomenclature of Goodwin et al. [1], ecdysteroid is the generic name for compounds structurally related to ecdysone (α-ecdysone). We propose to employ ecdysone as the generic name for compounds with the appropriate biological activity, in analogy with estrogen or progestin. This usage accords with much of the literature

1 Department of Cellular and Developmental Biology, Harvard University, Cambridge, MA 02138, USA
2 Department of Genetics, University of California, Berkeley, CA 94720, USA
3 Department of Biology, University of California, Los Angeles, CA 90024, USA

Biosynthesis, Metabolism and Mode of Action of Invertebrate Hormones
(ed. by J. Hoffmann and M. Porchet)

tissues are carried in the larva as nests of diploid cells often surrounded by polytene larval tissue or as semi-autonomous organs. Examples of the former class of imaginal tissues are the abdominal histoblasts, nests of cells in the abdominal epidermis of the larva that ultimately form the abdominal epidermis of the adult [4]. Larval cells in the midgut surround islets of imaginal cells that will form the adult midgut [5]. Much of the adult nervous system is derived from pockets of neuroblasts which continue to divide through larval life [6]. Further examples are provided by the imaginal rings, groups of cells encircling tubular organs (e.g., foregut, ducts of the salivary glands) that form the adult versions of the organs following the degeneration of their larval counterparts [5]. Undoubtedly the best known adult precursors are the imaginal discs, discrete epithelial organs attached by stalks to the larval epidermis that form the majority of adult cuticular structures including all of the appendages of adult flies. Lying on the basal surface of the epithelium in each disc is a group of cells, the adepithelial cells that are precursors to adult musculature.

Imaginal discs arise as invaginations of the embryonic epidermis and are composed of 10-100 cells at the time of hatching and are already determined to develop into adult structures [7]. During larval development, imaginal discs grow by cell division to form the characteristic structures that are present at the end of larval development. At this stage the discs that give rise to each adult structure (e.g., leg, wing, eye, haltere) are readily recognized.

Larval development is terminated and metamorphosis begins in response to an increase in the titer of the steroid hormone, 20-hydroxyecdysone (20-HE). A major external sign of metamorphosis is the tanning of the larval cuticle to form a puparium. Within the puparium metamorphosis procedes. Most of the larval tissues degenerate (muscles, salivary glands, digestive tract, fat body). Much of the neural tissue survives, as do the imaginal precursors. The imaginal discs commence morphogenesis prior to puparium formation in response to the rise in hormone titer. By 6-8 h after puparium formation they form appendages and fuse together to give rise to the anterior shell, thorax, and head of the adult [5]. The discs also secrete a thin cuticle, the pupal cuticle, within 12 h after the formation of the puparium. The pupal cuticle has the typical structure of an insect cuticle, an outer epicuticle composed of waxes and proteins, and an inner, laminate procuticle composed of chitin and specific cuticle proteins [8-10]. Deposition of the procuticle begins 4-6 h after puparium formation.

The ecdysone-induced morphogenesis of imaginal discs to form appendages involves primarily short-distance cell movement and changes in cell shape, but no cell divisions [11, 12]. The formation of tissues from nests of imaginal histoblasts and imaginal rings involves extensive cell division with the imaginal cells invading and replacing their degenerating larval counterparts. These early, primarily morphogenetic events occur around and after puparium formation and much before the final differentiation of many adult structures.

The ecdysone-dependent differentiation of recognizable adult structures and functions occurs mostly during the mid-pupal period, starting about 36 h after the formation of the puparium. During this stage the characteristic adult bristles, hairs, and other sensilla form. The differentiation of adult bristles requires two cell divisions, yielding four cells. One of these cells differentiates into a sensory neuron that extends to connect with the adult nervous system [13]. Similar extensions originate from presumptive photoreceptor cells in eye imaginal discs, beginning as early as the middle of the third instar [14]. Thus, epithelial cells in imaginal discs develop not only into integument cells that secrete cuticle, but also into sensory nerve cells.

The separation in time of the major developmental events – cell division, determination, and differentiation – is a particularly useful characteristic of the imaginal discs from the point of view of the experimentalist. Determination occurs largely during embryogenesis and continues into the larval period. Cell division occurs during larval development. Morphogenesis (with the exception of the ommatidia in eye discs) occurs only after most cell division is completed. The formation of the pupal cuticle shortly after puparium formation is partially separated from morphogenesis. And the differentiation of adult cuticular structures occurs only afterwards, during the mid-pupal period. This temporal separation of the major developmental processes affords an opportunity to study each process free of the experimental complexities that ensue when such major developmental processes coincide.

1.2 The Cell Line Kc

Almost all of the *Drosophila melanogaster* cell lines have been established by similar procedures and share many characteristics. Our (L.C. and J.D.O.) laboratories have used the line Kc in most of our experiments; the choice of lines is primarily a result of historical circumstances and does not imply any special feature of this line. Indeed, much of what we have learned about Kc cells has since been shown to apply equally well to many other *Drosophila* cell lines [15–17].

The line Kc was established in 1968 from a primary culture of minced late-stage embryos [18]. Such primary cultures contain fully differentiated larval cell types. In addition, they contain a number of undifferentiated cells which continue to divide for several weeks; their appearance and continued division suggests that they may be imaginal cells of some sort (see Sect. 1.1). Some of these undifferentiated cells form hollow balls, which are capable, upon re-implantation into a larva, of metamorphosing into recognizable imaginal disc-derived structures [19–21]. After repeated subdivision of the hollow balls, a population of surface-adherent, dividing cells becomes apparent; these cells form the permanent line. The line cells are not capable of forming any identifiable structures upon implantation into a larva [22]. It is possible that they are derived from the balls of imaginal disc tissue, but since the primary cultures contain many other cell types, one cannot be sure of this. We suspect that they originate from some sort of imaginal cell type both because the only diploid dividing cells present in a late embryo are imaginal cells, brain cells, germ cells, and hemocytes [23] and because the cell lines continue to divide indefinitely until they are exposed to an ecdysteroid hormone, whereupon they differentiate, a characteristic of imaginal cells (see Sect. 1.1). The identification of Kc cells as imaginal cells is still a subject of some controversy. In particular, one of us (JDO) has argued that the lysis of some Kc cell clones in response to ecdysone (see Sect. 5.2.2) suggests a larval origin [24].

Kc cells offer two major advantages for studies of hormone action: homogeneity and quantity. Although the original Kc line was polyclonal, it has since been cloned in many laboratories, and the cloned cells seem to be quite stable in karyotype, growth characteristics, and hormone response. They have been adapted for growth in suspension and can be grown in spinner flasks; hence, the preparation of gram quantities of homogeneous cells is a simple matter.

The use of cell lines to study hormone action is, of course, subject to some conceptual problems. We do not know from which cell type the cells are derived, nor do we know how closely their differentiation resembles that of their homolog in vivo. We submit, however, that the Kc cells constitute an entirely suitable subject for studies of steroid action for two reasons. First, as we shall describe, the cells undergo an extensive response to the hormone, and the response is mediated by the same sort of receptor system as is used in more clearly defined cell types. Therefore, Kc cells should be considered to be a bona fide steroid-responsive tissue, and whatever we can learn about the mechanism of the ecdysone response of Kc cells enlarges our understanding of steroid hormone action. Second, insofar as Kc cells are an analog of normal *Drosophila* ecdysone-responsive cells, information derived from the more technically accessible cell line helps in the design and interpretation of experiments on imaginal discs and other tissues.

2. Ecdysteroid Levels Associated with Pupation in *Drosophila*

Because both of the culture systems with which we are concerned here are, or act as, imaginal cells which respond to ecdysone by undergoing differentiation, we shall digress briefly here to review the changes in level of ecdysteroids which normally accompany such differentiation in vivo. In particular, we are interested in the endocrine changes which bring about pupariation and pupation, the initial period of metamorphosis when most imaginal cells begin to develop adult characteristics. In *Drosophila*, as in other higher Diptera, the first metamorphic molt is preceded by pupariation, in which the larval cuticle becomes a tanned puparium. During pupation a pupal cuticle is laid down within the puparium. The two processes are separated by 6–8 h; both are under the control of ecdysone.

The levels of ecdysteroid increase about 6 h before puparium formation from approx. 0.02 μg/gm wet weight (about 4×10^{-8} *M*) to about 0.2–1.0 μg/gm wet weight (about 4×10^{-7} - 2×10^{-6} *M*). Hormone levels return to intermolt levels 2–3 after puparium formation. Some workers report another increase, to about 0.2 μg/g wet weight (about 4×10^{-7} *M*) about 12 h after the last larval molt. The principal ecdysteroid present at all times of elevated titer during embryogenesis, pupariation, and adult development is 20-HE ([25]; B. Sage, J.D. O'Connor, R.B. Hodgetts, unpubl. observations). (Since it is impossible to synchronize *Drosophila* populations accurately, these values probably represent averages of hormone titers from animals of slightly different stage. This problem has resulted in a large amount of variation between investigators and probably to an underestimation of peak values and an overestimation of troughs. The data given here are compiled from the results of several investigators; see Richards [26] for a review.)

3. The Ecdysteroid Response in Vitro

3.1 The Development of Imaginal Discs in Vitro

It has been known for many years that *Drosophila* imaginal discs can develop normally after they have been dissected from a larva and transplanted into the hemocele of a second larva [27, 28]. This phenomenon showed that normal disc development does not require normal in situ surroundings and suggested that imaginal discs might develop successfully in vitro in an appropriate medium. The first medium to allow the reproducible development of imaginal eye-antennal discs in vitro was developed by Schneider [29] and dependent on the co-culture of discs with brain and the associated prothoracic gland which was assumed to act as a source of ecdysone. The subsequent ready availability of ecdysteroids isolated from plant materials made possible the successful development of several culture systems for the in vitro development of imaginal discs.

3.1.1 Long-Term Culture (4–8 days)

The morphogenesis and differentiation of imaginal wing and leg discs in vitro was first accomplished by Mandaron [30–32]. He developed a culture medium that permitted the full development of discs over approx. 6–7 days. Discs were dissected from larvae grown axenically and placed in a hanging drop with continuous exposure to ecdysone. These discs formed appendages, deposited and apolysed pupal cuticles, and ultimately formed characteristic adult bristles, hairs, and other sensillae. Development of muscles derived from adepithelial cells also occurred and was revealed by occasional muscular contractions within some differentiated legs [32, 33]. Mandaron's results were duplicated, with some changes in the hormone regimen, by Fristrom et al. [34], using Robb's medium [35] and by Milner and Sang [36], using a medium developed by Shields and Sang [37]. The development of discs under these culture conditions requires significantly more time than in vivo development; this difference would appear to be caused by deficiencies in the culture medium or culture conditions. We shall elaborate upon this point later.

3.1.2 Short-Term Culture (up to 24 h)

Many of the developmental responses of imaginal discs to the rise in the level of 20-HE that initiates metamorphosis occur within 24 h in vitro. These responses, including morphogenesis (evagination) to form appendages, the fusion of the discs to form a continuous shell, and the secretion of the pupal cuticle, can be studied not only with discs dissected from axenically grown larvae [38, 39], but also with discs preparatively isolated from 500 to 1,000 g of larvae grown on corn meal medium seeded with yeast [40]. Such discs, obtained in gram quantities, not only permit analyses of the visible morphogenetic and differentiative events, but also, as we shall describe, allow molecular analyses of the mechanisms by which the steroid hormone mediates imaginal development. For these reasons, we (JWF) have concentrated upon the events taking place in short-term cultures of mass-isolated discs.

When discs are incubated under appropriate hormonal conditions (see below), the following events occur ([41]; see Fig. 1): Evagination occurs between 4 and 18 h after the addition of the hormone. Deposition of the epicuticle begins after the initiation of evagination and continues until the completion of evagination. Deposition of the chitin-containing procuticle begins at about 12 h and continues for at least 20 h. These events initiate at the same time as in vivo. However, once initiated, they may progress more slowly.

The hormone concentrations required to induce the various aspects of this differentiation vary. Evagination occurs at any 20-HE concentration above 10^{-7} *M* (we have used concentrations as high as 2×10^{-5}, but cuticle deposition is inhibited by high hormone concentrations [41]. Epicuticle formation - more precisely, the deposition of the cuticulin layer [42] - occurs only at concentrations of 20-HE between 10^{-7} and 2×10^{-6}. Procuticle deposition is prevented or delayed by concentrations of 20-HE above 10^{-7} *M*; deposition occurs when the discs are exposed to 10^{-6} *M* 20-HE for

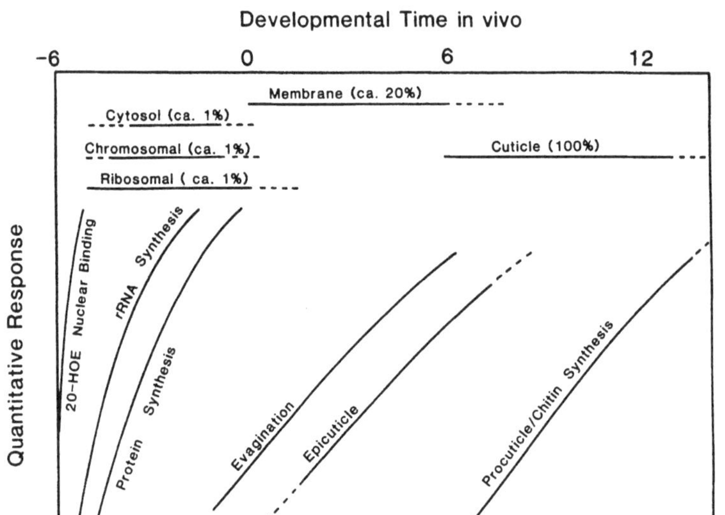

Fig. 1. Responses of imaginal discs to ecdysones in vitro. *20-HOE nuclear binding:* binding of 20-hydroxyecdysone to nuclear receptors. *rRNA synthesis:* relative increase in incorporation into rRNA and in RNA polymerase I activity. *Protein synthesis:* relative increase in protein synthesis. *Evagination:* increase in evagination index. *Epicuticle:* deposition of epicuticle. *Procuticle/chitin synthesis:* incorporation into chitin fraction. Times for epicuticle deposition are approximate and probably early. *Upper part:* changes in metabolic labeling of proteins in various subcellular fractions. *Numbers in parentheses:* approx. percent of labeled polypeptides on 1–D or 2–D gels that are altered by incubation with 20-hydroxyecdysone. *Solid lines:* times when changes have been detected. *Dashed lines:* uncertainty of presence of alterations at given times. For developmental time in vivo, 0 h, refers to time of puparium formation, –6 h to time when ecdysone titer is presumed to increase. Duration of culture in vitro is in hours

4-6 h, and then removed from the hormone [41]. (As we have described above, the full sequence of morphogenesis and cuticle deposition can take place during long-term culture in the continuous presence of 20-HE, but the process is much slower than in cultures from which the hormone is removed after 4-6 h. We interpret the slower course of development during continuous hormone exposure as a consequence of ecdysone induced delays in some of the later steps, including cuticle deposition. Consistent with this interpretation is the fact that the hormone concentrations used in successful long-term cultures are usually low, i.e., about 0.1 μg/ml [(2×10^{-7} *M*) 20-HE].)

Thus, different aspects of the differentiation require different hormone regimens. A consequence of this phenomenon is that in order to induce normal morphogenesis and cuticle deposition in vitro on a fairly normal time scale, one must provide a changing hormone concentration that closely parallels the changing concentration which accompanies development in vivo. It would appear then that the correct ordering of developmental events during pupariation in vivo is assured by the spectrum of sensitivities of developmental events to induction and inhibition by 20-HE [43]. The initial peak of 20-HE is sufficient to induce evagination and the later deposition of cuticle. Actual cuticle deposition occurs only when the titer of 20-HE drops; the deposition of epicuticle precedes that of procuticle because the former is inhibited only at very high 20-HE concentrations. In this way, a single peak of 20-HE causes a series of ecdysone-induced events to occur in a precise order.

3.2 Kc Cells

3.2.1 The Morphological Response to 20-HE

Kc cells undergo a dramatic morphological change during the first 24-48 h of exposure to physiological concentrations of ecdysone [44, 45]. The cells extend long, branching processes, often many times the length of the cell body. The processes have characteristic nodules that move rapidly back and forth along the processes. The morphological change is accompanied by a dramatic increase in cell mobility, and by a tendency to aggregate.

The extension of these characteristic processes occurs only in the presence of an active ecdysone. A maximal extension of processes requires at least 10^{-7} *M* 20-HE; concentrations as high as 10^{-4} *M* elicit the same response [45]. The morphological differentiation occurs only in the continuous presence of the hormone (LC, unpubl. observations); in this respect it differs from disc differentiation.

3.2.2 Proliferative Arrest

Kc cells divide continuously, with a doubling time of about 24 h, until they are exposed to ecdysone. The cells then undergo approximately one more doubling, and cease to divide further. The effects of ecdysone on division are complex. Our current understanding of these effects may be summarized as follows:

a) Time-lapse films of Kc cells show that it is not the case that the cells undergo a single differentiative division, followed by elongation. Most cells divide once in the presence of the hormone, but many divide twice or not at all. There is no temporal correlation between cell division and the extension of processes in individual cells [46].

b) All tested ecdysones elicit the same effects on division, including a transient stimulation at low concentrations and an inhibition at higher concentrations. When allowance is made for differences in relative activity (measured, for example, by the ability to induce the morphological response of Kc cells), all compounds tested give the same spectrum of responses [46–48]. An early report of a difference between the effects of ecdysone and 20-HE [49] turned out to be an artifact resulting from an unfortunate choice of concentrations.
c) Measurements of cloning efficiency in cells from which 20-HE is withdrawn after various times of treatment show that the commitment to proliferative arrest is very rapid, occurring in half the cells in less than an hour [46]. The arrest itself occurs after 12–18 h [46, 48].
d) The block to proliferation occurs in the G_2 portion of the cell cycle [48]. Cells enter G_2 at a normal rate, but fail to undergo mitosis. A similar G_2 arrest has been observed in a variety of imaginal discs [50, 51].
e) Our laboratories (JDO and LC) have obtained conflicting results about some aspects of the proliferative arrest. We do not know whether the differences in our observations reflect differences in experimental technique or differences in the properties of the Kc cell clones used in our laboratories. The G_2 block, measured by fluorescent flow cytometry, requires the continuous presence of ecdysone [52], but when ecdysone is withdrawn after a few hours the cloning efficiency of the cells is sharply reduced [46]. Cells in the continuous presence of 20-HE have been observed to reenter the cell cycle after about 100 h, and to continue to divide as an ecdysone-resistant, receptor-deficient population for several months; these conclusion were based principally on fluorescent flow cytometry measurements [48]. Yet simple cell counts failed to reveal any resumption of division for at least 300 h, and when cells were cloned in the continuous presence of 20-HE, only one cell in about 10^5 gives rise to a clone [46]. These disputed points are crucial to the description of ecdysone-induced proliferative arrest. Their resolution awaits further experimentation.

A cessation of division generally accompanies the 20-HE-induced differentiation of imaginal cells. The effect in Kc cells is complex and poorly understood. The effect in normal tissues is probably equally complex, but is even more poorly understood; study of effects on cell division in intact tissues is complicated by the effects of the injury involved in experimental manipulation. Hence, although it is very difficult to assess the relationship between the Kc cell proliferative response and that of other tissues, information obtained from Kc cells may give some suggestion of the phenomena occurring in vivo.

4. Ecdysone Receptors

In the target cells hormones interact with specific molecules called receptors that mediate the hormonal response. In vertebrates, steroid hormone receptors in hormonally naive target tissues are proteins located in the cytoplasm or the nucleus; when bound to the hormone, they are concentrated in the nucleus where they interact with the chromatin and, in some unknown manner, alter transcription of specific genes [53, 54].

Recently, specific acceptor sites in the DNA where steroid-receptor complexes bind have been identified for some steroid-responsive mammalian genes [55]; one such site appears to function as a steroid-dependent transcriptional enhancer element [56].

Several criteria may be used to distinguish hormone receptors, molecules that mediate target tissue responses, from hormone binding proteins that do not mediate specific responses. These criteria include a limited number of receptor molecules, a high binding constant, and a differential affinity for hormone analogs that parallels the differential activity of the analogs in eliciting characteristic target tissue responses. In addition, the equilibrium dissociation constant (K_d) for steroid hormones is generally close to the concentration of hormone required to elicit a half-maximal responses; this correlation is empirical, but there is no mechanistic requirement for it, and it does not hold true for all regulatory ligand-binding molecules (e.g., the *lac* repressor binds to IPTG with $K_d = 10^{-6}$ *M*, but 10^{-5} *M* IPTG is required to elicit a half-maximal response [57, 58]).

Ecdysteroid-binding proteins which meet all of these criteria have been identified in both imaginal discs and Kc cells [59, 60]. Their properties are summarized in Table 1.

Table 1. Properties of ecdysone receptors from *Drosophila* tissues [54, 55, 87]

Property	Discs	Kc cells	Embryos
K_D			
Ponasterone A	3×10^{-9} *M*	$3-4 \times 10^{-9}$ *M*	6×10^{-9} *M*
20-hydroxyecdysone	5×10^{-8} *M*	2×10^{-7} *M*	2×10^{-7} *M*
Ecdysone	7×10^{-5} *M*	3×10^{-6} *M*	5×10^{-5} *M*
Number of receptors per diploid cell	500–1,000	1,000–2,000	–
Subcellular localization	> 90% nuclear	40% nuclear	–
		60% cytosol	–
Sedimentation	5.95S	6S	6S
Molecular weight (gel sizing)	$\geqslant$ 100,000		

(We have also included data describing binding proteins from *Drosophila* embryos; these proteins have properties similar to those of the receptors isolated from discs and Kc cells, but they cannot be concluded to be receptors because no correlation has been made between the binding affinities and the abilities of various analogs to elicit target tissue responses.) The receptors from discs and Kc cell sources have similar properties; the differences in K_d and concentration are within the limitations of the experimental measurements. (The substantial differences in the measured K_d values reflect the large uncertainties in measurements of competition in crude extracts; this is especially a problem for compounds with low affinity for the receptor, such as ecdysone.)

Before hormone treatment, 90%–95% of imaginal disc receptors and 40% of Kc cell receptors are located in the nucleus [61, 62]; it is possible that the difference in subcellular localization reflects the exposure of the discs to 20-HE during larval molts. We think that this explanation is unlikely, since (a) three cell doublings take place between the last larval molt and the isolation of the discs [7], and (b) no translocation from the cytoplasm to the nucleus can be detected during treatment of Kc cells with the hormone [61]. The nuclear and cytoplasmic receptor components from Kc cells differ in that the cytoplasmic receptor sediments at 6 S when extracted from fresh cells but at 4 S when extracted from frozen cells, while the nuclear receptor extracted from both fresh and frozen cells sediment at 6 S [61]. Furthermore, the 4 S species has lowered affinity for DNA [63]. It is unclear whether the nuclear and cytoplasmic forms of receptor are different molecular species.

In general, the properties of the receptors suggest that the Kc cell response and the imaginal disc response to ecdysteroids are mechanistically similar, that ecdysteroids act in basically the same manner as vertebrate steroid hormones, and in particular, that their association with nuclei may reflect a role in regulating transcription of target genes [62, 64]. The proof that receptors mediate a steroid effect on transcription awaits direct experimental confirmation, but the properties of the receptors provide some assurance that studies of ecdysteroid action in imaginal discs and Kc cells will provide information which can be applied to vertebrate steroid hormones as well.

5. Molecular Description of the Response

5.1 Imaginal Discs

Imaginal discs in a variety of developmental stages have been the subject of molecular investigations [65-67]. The most detailed information describes the ecdysteroid response of mass-isolated third instar larval discs and, thus, provides a molecular description of the processes of evagination and cuticle deposition that we have referred to in Sect. 3.1. For early studies of general effects on DNA, RNA, and protein synthesis, we refer the reader to other reviews [68-70]. We shall describe here some approaches currently being used to identify specific macromolecules that may mediate particular developmental responses, such as evagination, epicuticle formation, and procuticle formation, and present preliminary results related to evagination and procuticle deposition.

5.1.1 Evagination

Evagination commences after 4-6 h of culture in the presence of an ecdysteroid, and is not inhibited by high concentrations of hormone. Hence, ecdysone-dependent synthesis of RNAs or proteins mediating evagination can be expected to begin within 4-6 h of hormone administration and to be insensitive to high hormone concentrations. Evagination results primarily from short-distance cell movement [11, 12] and, thus, depends on contractile machinery. The required changes in intercellular associations might depend on alterations in the properties or compositions of cell surface proteins. This seems particularly likely when one considers that the cells in disc epithelia are joined together through various junctions that must detach and reform during cell movements [71]. Thus, cell surface proteins with ecdysone-dependent synthesis also are likely candidates to mediate evagination. Rickoll and Fristrom [72] described extensive alterations in the metabolic labeling – presumably synthesis – of polypeptides isolated from a membrane vesicle fraction enriched for the cell surface marker Na^+/K^+-ATPase. No alterations in the ^{35}S-methionine labeling of polypeptides were found during the first 4 h after the addition of hormone. However, several new polypeptides were labeled during the period 4–8 h after hormone addition. Birr and Fristrom (unpubl. observations) have demonstrated that labeling of three major acidic, hormone-inducible, high molecular weight polypeptides (ca. 80,000) occurs even in extremely high hormonal concentrations (10^{-5} *M* ponasterone A, equivalent to 2×10^{-4} *M* 20-HE). Resistance to inhibition of synthesis in high hormonal concentrations rules out a controlling role for these polypeptides in cuticle formation, but these polypeptides remain candidates for mediating evagination.

Using the same criteria to identify putative evagination-mediating genes, Natzle, Hammonds, and Fristrom (unpubl. observations) have isolated a number of candidate clones from a library of *Drosophila* genomic sequences in λ. Each encodes a transcript that is localized in membrane-bound polysomes and, therefore, probably encodes a membrane or secreted protein. Accumulation of each transcript in discs is dependent on ecdysteroids and is insensitive to high concentrations of 20-HE. Transcripts homologous to one of the clones, λ7–14, appear within 2 h of the addition of 20-HE to the discs in vitro.

Both the protein synthetic pattern and the accumulation of transcripts that we have described parallel the induction of evagination by ecdysteroids. Proof that these products are causally involved in evagination awaits further studies.

5.1.2 Procuticle Deposition

In contrast to evagination, where the nature of molecules that mediate the process is unknown, many of the structural and functional components of insect procuticles have been identified. Procuticles contain chains of chitin, polymers of N-acetylglucosamine, to which specific proteins are bound covalently and noncovalently. Many of the components are cross-linked through metabolic products of the decarboxylation of DOPA by DOPA decarboxylase (DDC), an enzyme whose synthesis has long been known to be regulated by ecdysone during development [73, 74]. If the synthesis of the components of the procuticle parallels the hormonal dependencies for the formation of the procuticle, we can make some predictions about the regulation of synthesis of procuticle components. Specifically, synthesis of chitin, procuticle proteins, and DDC should occur only after discs are incubated with ecdysone for 4–8 h and the hormone is then removed from the culture medium.

This specific expectation has been met for several, but not all, procuticular components. The synthesis of chitin occurs at markedly elevated levels only after incubation of discs with a pulse of hormone [41] (Fig. 2). Comparable results have been obtained for the synthesis of low molecular weight (<25,000) pupal procuticle proteins [41]. However, the synthesis of high molecular weight pupal proteins (ca. 56,000 and 82,000) isolated in the procuticle fraction (but not necessarily components of the procuticle) is physiologically more complex. Current data suggest that synthesis of these proteins requires two pulses of 20-HE separated by a period of hormone withdrawal (Doctor,

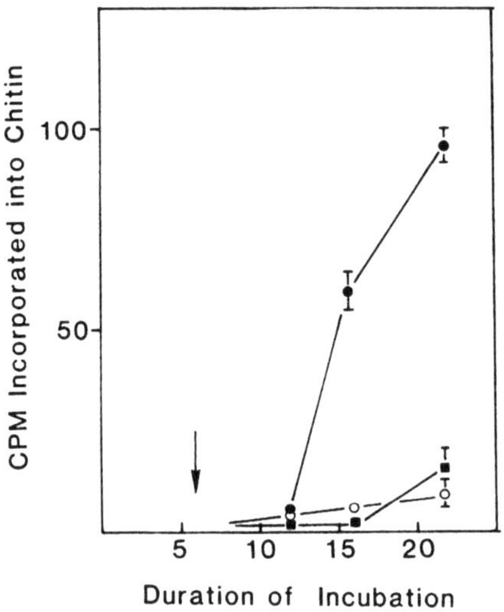

Fig. 2. Induction of chitin synthesis (procuticle deposition) in imaginal discs in vitro. *Arrow* indicates time of hormone removal (if applicable). Discs were incubated with 20-hydroxyecdysone (–○–); without 20-hydroxyecdysone (–■–); with and then without 20-hydroxyecdysone (–●–). Induction of DOPA-decarboxylase and synthesis of low molecular weight cuticle proteins follows the same time course as that of chitin synthesis

Fristrom, and Fristrom, unpubl. observations). The accumulation of DDC enzyme activity follows a time course like that for the synthesis of chitin. Analyses of DDC gene transcripts by hybridization to a gene-specific probe demonstrate that increases in gene transcripts follow the same time course (Clark, Doctor, Fristrom, and Hodgetts, unpubl. observations). Thus, the synthesis of known cuticular components closely parallels in time and hormonal physiology the actual formation of the procuticle. The requirement for readdition of an ecdysteroid for the accumulation of the 82,000 protein suggests a role for the second peak of 20-HE that apparently occurs in vivo 12 h after puparium formation.

We wish to draw particular attention to parallels between the hormonal responses of imaginal discs, as revealed through morphogenesis, differentiation, and the synthesis of specific macromolecules, and the hormonal responses of larval salivary glands, as revealed through the pattern of puffs in their polytene chromosomes (Table 2). Ashburner, Richards, and their colleagues [75, 76] have classified puffs into groups as a function of their time of appearance, sensitivity to ecdysone concentrations, and sensitivity to inhibitors of protein synthesis. Early puffs are believed to be direct responses to the binding of the steroid-receptor complex to the responding locus because they appear within minutes and are insensitive to inhibitors of protein synthesis. Late puffs are believed to be indirect responses to the hormone because they can appear even after the hormone is withdrawn. Furthermore, inhibitors of protein synthesis abolish the induction of the late puffs; this suggests that they are induced by an ecdysone-induced protein. Puffs that normally occur 6–10 h after puparium formation in vivo occur in vitro only if the hormone is administered and then withdrawn.

A series of apparent homologies between the imaginal disc response and the salivary gland response emerge from these observations. In particular, the salivary gland prepupal puffs and the synthesis of procuticle components by imaginal discs appear to be similar responses to hormone withdrawal and readdition. Although they occur at the same time, the relationship between epicuticle deposition and evagination-mediation components and salivary gland puff classes is unclear, but may well become apparent

Table 2. Ecdysone-induced responses in salivary glands and in imaginal discs

Stage and time	Salivary glands		Imaginal discs	
	Loci	Comments	Responses	Comments
Early −6 h[a]	2B5 23E 74EF 75B			
− 2 h			Evagination	Not inhibited by high hormone (> 10^{-5} *M* 20-hydroxyecdysone)
Late −2 h to +6 h	62E 63E 78D 82F	Earlier expression by removing hormone		
+2 h to +4 h			Epicuticle deposition	Inhibited or delayed by 2×10^{-6} *M* 20-hydroxy-ecdysone
Late prepupal +6 h to +8 h	75CD	Requires removal of 20-hydroxy-ecdysone below 10^{-8} *M*	Procuticle deposition	Requires removal of 20-hydroxyecdysone below 10^{-7} *M*
Late prepupal +10 h	93F	Requires readdition of 20-hydroxyecdy-sone	High molecular weight procuticle proteins	Requires readdition of 20-hydroxyecdysone

[a] Hours refers to time before or after puparium formation. The increase in titer of 20-hydroxyecdysone is assumed to occur at − 6 h. Times given are for initiation of puffs or responses

as we learn more of the imaginal disc system. It would appear that both tissues are taking advantage of the changing pattern of ecdysteroid titers in vivo to achieve a precise sequence of developmental events.

5.2 Kc Cells

A number of enzymatic changes have been observed during the ecdysone-induced differentiation of Kc cells. All are late effects, detectable only after 12 h or more of hormone treatment. There is insufficient information to infer any relationships between the various enzymatic inductions or to relate any of them to the morphological and proliferative changes we have already described; therefore, we will simply summarize them individually.

5.2.1 Acetylcholinesterase

The nerve-like morphology of 20-HE-treated Kc cells led Cherbas et al. [77] to test them for acetylcholinesterase (AChE) activity. AChE activity is undetectable in untreated cells, but after 24 h of hormone treatment it begins to rise. Within 48–72 h AChE activity reaches a maximum level corresponding to approx. 1% the specific activity of the *Drosophila* central nervous system.

Although the time course and dose-response curve for this induction are very similar to those of "late" salivary gland puffs (see Table 2), a number of properties distinguish AChE induction from a typical late puff [46]. The continuous presence of the hormone is required during the period of induction of AChE activity. Even after AChE activity has achieved its maximum level, ecdysone must be present to maintain that level. Although cells from which 20-HE is withdrawn after 6–12 h do not ever acquire detectable AChE activity, if the cells are primed by a short treatment and then reexposed to the hormone 24 h later, they produce AChE activity immediately, without the usual lag period. Dose-response curves for AChE induction by continuous hormone treatment and for the initial, priming effect are very sharp, as are the dose-response curves for late puffs. But the dose-response curves for maintenance of AChE activity once it has been induced and for AChE induction in previously primed cells are quite broad, resembling the curves for early salivary gland puffs (LC, unpublished observations). These observations suggest that ecdysteroids are required for more than one step in the process of AChE induction, and that under appropriate circumstances the production and maintenance of AChE activity may be a fairly direct effect of the hormone. Further molecular analysis of this process promises to be very fruitful.

5.2.2 β-Galactosidase

In *Drosophila*, β-galactosidase is a lysosomal enzyme. During the ecdysteroid response in some Kc sublines, cells accumulate large numbers of lysosomes [24], and in these sublines cell death is a prominent feature of the response. β-galactosidase activity is induced in some, but not all Kc sublines ([78]; LC, unpubl. observations); we suspect that the appearance of the enzyme and the lysosomes is correlated.

5.2.3 DOPA Decarboxylase (DDC)

Two Kc cell clones of about six tested respond to 20-HE with an induction of DDC activity [79]. DDC activity becomes detectable after 72–96 h in the continuous presence of 20-HE, but appears at about 48 h if the hormone is removed after 6 h. In this respect, DDC induction in Kc cells resembles that in imaginal discs.

The DDC activity found in ecdysone-treated Kc cells is distinguishable from that of late third instar epidermis both by its affinity for DEAE cellulose and by serological analysis. Preliminary observations indicate that DDC mRNA in Kc cells is larger than DDC mRNA in epidermis (R. Swiderwky, pers. comm.). Since there is only one known DDC gene in *Drosophila* [80], these results suggest that the DDC activity of Kc cells may be derived from a different pattern of transcription or posttranscriptional processing from that gene.

6. Early Aspects of the Response

In order to analyze at the molecular level the mechanism of ecdysteroid action, Cherbas and his colleagues have sought to identify, clone, and characterize genes which may be the analogs in Kc cells of the early puff sites in salivary glands. Two such genes have been cloned. These genes encode polypeptides whose rate of synthesis is rapidly increased by ecdysteroid treatment; they are called the ecdysteroid-inducible polypeptides (EIPs) [81]. The properties of their induction are very similar to those of early salivary gland puff induction [82] and it seems very likely that their induction is transcriptional. Our working hypothesis, therefore, is that these genes are examples of loci whose transcription is regulated in a direct way by the steroid-receptor complex.

Three EIPs have been identified by SDS-polyacrylamide gel electrophoresis; each is resolved into at least three isoelectric variants by two-dimensional electrophoresis [83]. cDNA clones for the EIPs were isolated from a library of Kc cell sequences and used to isolate genomic clones from a library in the λ vector Charon 4 [83, 84]. Using these cloned probes, we have learned that a single copy gene at chromosome locus 55B–D is responsible for all forms of EIP 40, and a single-copy gene at chromosome locus 71CD encodes all forms of EIP 28 and EIP 29 [83–85]. The charge variation in the EIPs 28 and 29 is due at least in part to posttranslational or cotranslational acetylations (A.J. Bieber and M.M.D. Koehler, pers. commun.). EIPs 28 and 29 are made from a single primary transcript; a variation in the splicing pattern leads to two mature mRNAs differing by 12 nucleotides of coding sequence [86, 87]. At least in the case of the EIPs 28 and 29, none of the posttranscriptional and posttranslational modifications are affected by ecdysone treatment.

We have examined the promoter region of the EIP 28/29 gene in some detail. It is clear that the effect of the hormone is to increase the activity of a previously active promoter, without causing any qualitative changes in the transcript it produces [86]. Experiments are now in progress to dissect the functional components of this promoter region in various transformation systems, in order to determine how this gene may interact with the ecdysone-receptor complex, and how the interaction may lead to an increased rate of transcription. Recent experiments with the glucocorticoid-responsive mouse mammary tumor virus transcription unit have led to the hypothesis that the receptor binds to a hormone-dependent transcriptional enhancer element [56]. The EIP 28/29 gene may provide an opportunity to determine whether the same mechanism applies to a cellular steroid-responsive gene.

7. Conclusions

The experiments described in this paper present a fragmentary picture of ecdysone action in imaginal discs and Kc cells. The similarities in the properties of the two systems are striking; this may simply reflect an imaginal origin of the cell line. In both systems, it is clear that ecdysone acts at a number of levels. There are events, such as the induction of the EIPs, which are induced quickly and apparently directly by the hormone-receptor complex. Other events, such as the induction of the high molecular weight procuticle proteins in the imaginal discs, and of AChE in Kc cells, require the presence of ecdysone, but do not occur for many hours. Thus, among the effects of ecdysone is the alteration of the target tissue's capacity to respond to the hormone. A clear description of the mechanism of ecdysone's action will require a combination of molecular, physiological, and genetic studies of these complex responses. We believe that the studies of imaginal discs and Kc cells outlined here constitute a strong beginning of such an analysis.

References

1. Goodwin RW, Horn DHS, Karlson P, Koolman J, Nakanishi K, Robbins WE, Siddall JB, Takemoto T (1978) Nature (London) 272:122
2. Richards GP, Lepesant J-A (this vol)
3. Best-Belpomme M, Dastugue B (this vol)
4. Madhavan MM, Schneiderman HA (1977) Roux' Arch 183:269–305
5. Bodenstein D (1950) In: Demerec M (ed) The biology of *Drosophila*. Wiley, New York, p 275
6. White K, Kankel DR (1978) Dev Biol 65:296–321
7. Nothiger R (1972) In: Ursprung H, Nothiger R (eds) The biology of imaginal disks. Springer, Berlin Heidelberg New York, pp 1–34
8. Poodry C, Schneiderman H (1970) Roux' Arch 168:1–9
9. Reed CT, Murphy C, Fristrom D (1975) Roux' Arch 188:163–177
10. Silvert D, Quesada L, Doctor J, King D, Fristrom J (1984) Biochemistry (in press)
11. Fristrom D, Fristrom J (1975) Dev Biol 43:1–23
12. Fristrom (1976) Dev Biol 54:163–171
13. Lees AD, Waddington CH (1942) Proc R Soc London Ser B 131:87–110
14. Perry MM (1968) J Morphol 124:227–248
15. Berger E, Ringler R, Alahiotis S, Frank M (1978) Dev Biol 62:498–511
16. Best-Belpomme M, Courgeon A-M (1977) FEBS Lett 82:345–347
17. Cherbas L, Cherbas P (1981) Adv Cell Cult 1:91–124
18. Echalier G, Ohanessian A (1969) CR Acad Sci 268:1771–1773
19. Schneider I (1972) J Embryol Exp Morphol 27:353–365
20. Dübendorfer A, Shields G, Sang JH (1975) J Embryol Exp Morphol 33:487–498
21. Dübendorfer A (1976) In: Kurstak E, Maramorosch K (eds) Invertebrate tissue culture. Academic Press, London New York, pp 151–159
22. Echalier G, Proust J (1973) In: Rehacek J, Blaskovic D, Hink WF (eds) Proc 3rd Int Colloq Invertebr Tissue Cult. Publ House Slovak Acad Sci, Bratislava
23. Schneider I, Blumenthal AB (1978) In: Ashburner M, Wright TRF (eds) The genetics and biology of *Drosophila*, vol IIa. Academic Press, London New York, pp 266–316
24. O'Connor JD, Maroy P, Beckers C, Dennis R, Alvarez CM, Sage BA (1980) In: Roy AK, Clark JH (eds) Gene regulation by steroid hormones. Springer, Berlin Heidelberg New York, pp 263 to 277
25. Hodgetts RB, Sage B, O'Connor JD (1977) Dev Biol 60:310–317
26. Richards G (1981) Mol Cell Endocrinol 21:181–197
27. Bodenstein D (1940) J Exp Zool 84:23–37
28. Ephrussi B (1942) Q Rev Biol 17:329
29. Schneider I (1964) J Exp Zool 156:91–104
30. Sengel P, Mandaron P (1969) CR Acad Sci 268:405–407
31. Mandaron P (1970) Dev Biol 22:298–320
32. Mandaron P (1971) Dev Biol 25:581–605
33. Milner M (1977) J Embryol Exp Morphol 37:105–117
34. Fristrom J, Logan WR, Murphy C (1973) Dev Biol 33:441–456
35. Robb JA (1969) J Cell Biol 41:876–884
36. Milner M, Sang JH (1974) Cell 3:141–143

37. Shields G, Sang JH (1970) J Embryol Exp Morphol 23:53–69
38. Mandaron P (1976) Roux' Arch 179:185–196
39. Milner M, Bleasby A (1984) Symp Br Soc Dev Biol (in press)
40. Eugene O, Yund MA, Fristrom JW (1979) Tissue Cult Assoc Man 5:1055–1062
41. Fristrom JW, Doctor JD, Fristrom D, Logan WR, Silvert D (1982) Dev Biol 91:337–350
42. Locke M (1966) J Morphol 118:461–494
43. Fristrom JW, Natzle J, Doctor J, Fristrom D (1984) Symp Br Soc Dev Biol (in press)
44. Courgeon A-M (1972) Exp Cell Res 74:327–336
45. Cherbas L, Yonge CD, Cherbas P, Williams CM (1980) Roux' Arch 189:1–15
46. Cherbas P, Cherbas L, Demetri G, Manteuffel-Cymborowska M, Savakis C, Yonge CD, Williams CM (1980) In: Roy AK, Clark JH (eds) Gene regulation by steroid hormones. Springer, Berlin Heidelberg New York, pp 278–305
47. Wyss C (1976) Experientia 32:1272–1274
48. Stevens B, Alvarez CM, Bohman R, O'Connor JD (1980) Cell 22:675–682
49. Courgeon A-M (1972) Nature (London) New Biol 238:250–251
50. Fain M, Stevens G (1982) Dev Biol 92:247–258
51. Graves BJ, Schubiger G (1982) Dev Biol 93:104–110
52. Stevens B, O'Connor JD (1983) In: Padilla GM (ed) Genetic expression in the cell cycle. Academic Press, London New York, pp 269–287
53. Yamamoto K, Alberts B (1976) Annu Rev Biochem 45:722–746
54. Anderson JE (1984) In: Goldberger RF, Yamamoto KR (eds) Biological regulation and developmental, vol IIIB. Plenum Press, New York, in press
55. Ross S, Ucker D, Yamamoto K (1981) Cell 27:257–266
56. Chandler VL, Maler BA, Yamamoto KR (1983) Cell 33:489–499
57. Riggs AD, Newby RF, Bourgeois S (1970) J Mol Biol 51:303–314
58. Barklay MD, Riggs AD, Jobe A, Bourgeois S (1975) Biochemistry 14:1700–1712
59. Maroy P, Dennis R, Beckers C, Sage B, O'Connor JD (1978) Proc Natl Acad Sci USA 75: 6035–6038
60. Yund MA, King DS, Fristrom JW (1978) Proc Natl Acad Sci USA 75:6039–6043
61. Sage BA, Tanis MA, O'Connor JD (1982) J Biol Chem 257:6373–6379
62. Hill RJ, Watt F, Yund MA, Fristrom JW (1982) Dev Biol 90:340–351
63. O'Connor JD (1984) In: Kerkut GA, Gilbert LI (eds) Comprehensive insect physiology, biochemistry and pharmacology. Pergamon Press, London (in press)
64. Gronemeyer H, Pongs O (1980) Proc Natl Acad Sci USA 77:2108–2112
65. Mandaron P (1976) Roux' Arch 179:169–183
66. Mitchell HK, Lipps LS, Tracy U (1977) Biochem Genet 15:575–587
67. Mitchell HK, Petersen NS (1981) Dev Biol 85:233–242
68. Fristrom JW (1972) In: Ursprung H, Nothiger R (eds) The biology of imaginal disks. Springer, Berlin Heidelberg New York, pp 109–154
69. Silvert D, Fristrom J (1980) Insect Biochem 10:341–355
70. Doctor J, Fristrom J (1984) In: Kerkut GA, Gilbert LI (eds) Comprehensive insect physiology, biochemistry and pharmacology, vol II. Pergamon Press, London (in press)
71. Fristrom D (1982) J Cell Biol 94:77–87
72. Rickoll W, Fristrom JW (1982) Dev Biol 95:275–287
73. Karlson P, Sekeris C (1962) Biochim Biophys Acta 63:489–495
74. Kraminsky G, Clark W, Estelle M, Gietz R, Sage B, O'Connor JD, Hodgetts R (1980) Proc Natl Acad Sci USA 77:4175–4179
75. Ashburner M, Chihara C, Meltzer P, Richards G (1974) Cold Spring Harbor Symp Quant Biol 38:655–662
76. Richards G (1976) Dev Biol 54:256–263
77. Cherbas P, Cherbas L, Williams CM (1977) Science 197:275–277
78. Best-Belpomme M, Courgeon AM, Rambach A (1978) Proc Natl Acad Sci USA 75:6102–6106
79. Spencer C, Stevens B, O'Connor JD, Hodgetts RB (1983) Can J Biochem 61:818–825
80. Hirsh J, Davidson N (1981) Mol Cell Biol 1:475–485
81. Savakis C, Demetri G, Cherbas P (1980) Cell 22:665–674

82. Cherbas P, Cherbas L, Savakis C, Koehler MMD (1981) Am Zool 21:743–750
83. Savakis C, Koehler MMD, Cherbas P (1984) EMBO J 3:235–243
84. Cherbas L, Cherbas P (in preparation)
85. Rebers J (1984) PhD thesis, Harvard University
86. Cherbas L, Schulz RA, Koehler MMD, Cherbas P (in preparation)
87. Schulz RA, Cherbas L, Cherbas P (1984) (in preparation)
88.. Osterbur DL, Yund MA (1982) J Cell Biochem 20:277–282

II.13 Ecdysteroids in Helminths

J. KOOLMAN[1], J. WALTER[1], and H. ZAHNER[2]

Helminths are medically and economically important parasites of many higher and lower animals. Several hundred million people, especially in countries of the third world, suffer from diseases caused by these parasites. Only a few representatives of the helminths will be mentioned here:

- the blood flukes, *Schistosoma mansoni, S. haematobium,* and *S. japonicum* causing bilharziosis;
- the hookworms, *Ancylostoma duodenale* and *Necator americanus*;
- the filariae, *Wucheria bancrofti* and *Brugia malayi* causing elephantiasis, *Onchocerca volvulus* causing the "river blindness", and the eyeworm *Loa Loa*;
- the large intestinal roundworm, *Asçaris lumbricoides*.

Helminths belong to different systematic categories of lower animals. They are found in two phyla: the nematodes in the phylum of Nemathelminthes; trematodes and cestodes in the phylum of Plathelminthes, both phyla belonging to the Protostomia. In a large group of the Protostomia, the arthropods, the hormone system of ecdysteroids plays an important role. However, according to the hypothesis by Karlson [1], other members of the Protostomia, such as nematodes, trematodes, and cestodes use ecdysteroids as hormones. These organisms should possess all the components of the hormone system: ecdysteroids should be synthesized in specialized cells or "glands" from which they are secreted and transported to target cells where they control various physiological processes. The hormone system would also include cells or organs that inactivate and remove ecdysteroids.

Ecdysteroids in fact have been detected in nematodes and trematodes but as yet in only a few species (Table 1). It appears that the conclusive demonstration of the occurrence of ecdysteroids in more species is obligatory before it can be generalized that ecdysteroids occur in helminths.

The presence of ecdysteroids in nematodes and trematodes cannot be taken as proof of their biosynthesis in these organisms without additional evidence. For example, some species develop during their juvenile stages in insects or molluscs (intermediate hosts) that are known to contain ecdysteroids which could have been taken up and concentrated by the parasites. Moreover, ecdysteroids are much more widespread in the biosphere than previously thought. The wide occurrence of ecdysteroids is due to the ability of many plants to synthesize and store phytoecdysteroids [7]. Thus, it is no surprise

1 Physiologisch-Chemisches Institut der Philipps-Universität Marburg, 3550 Marburg, FRG
2 Institut für Parasitologie der Justus-Liebig-Universität, 6300 Gießen, FRG

Biosynthesis, Metabolism and Mode of Action of Invertebrate Hormones
(ed. by J. Hoffmann and M. Porchet)

Table 1. Ecdysteroids detected in total extracts of Nemathelminthes and Plathelminthes

Species	Method of detection	Reference
Nematodes		
Haemonchus contortus	Bioassay	[2]
	Radioimmunoassay	[3]
Ascaris lumbricoides	Isolation + identification	[4]
Panagrellus redivivus	Bioassay	[2]
	Radioimmunoassay	[3]
Aphelenchus avenae	Radioimmunoassay	[3]
Trematodes		
Schistosoma mansoni	HPLC + radioimmunoassay	[5]
Fasciola hepatica	HPLC + radioimmunoassay	[6]

that even in the tissues of vertebrates uninfected with helminths ecdysteroids could be detected [5, 8].

The increase of ecdysteroid concentration (detected as immunoreactive material) during development of a helminth is a good indication of ecdysteroid biosynthesis, but of course, cannot be taken as a proof (cf. the discussion of the pitfalls of ecdysteroid analysis by radioimmunoassay by Sall et al. [9]). If the analyses are based on samples which were treated with perchloric acid prior to radioimmunoassay [5, 10, 11], then the interpretation of such data is even more difficult because acidic conditions may lead to the hydrolysis of potentially present ecdysteroid conjugates and, more importantly and known for a long time, cause decomposition of ecdysteroids [12, 13].

The unequivocal proof of the biosynthesis of ecdysteroids in helminths, however, is an essential part of the entire hypothesis. For this reason, we have attempted to demonstrate the conversion of radiolabelled cholesterol to ecdysteroids. Adult *Litomosoides carinii* that normally live in the pleural cavity of the American cotton rat were cultured in vitro in the presence of tritiated cholesterol. Unfortunately, the experiments with this nematode failed to show any synthesis of ecdysteroids, while the controls with insects were positive. Of course, there may be several explanations for the failure to prove ecdysteroid biosynthesis in this organism, e.g., use of adults, sex of adults, unsuitable incubation conditions. Therefore, the experiments have to be repeated and extended, in particular, to other stages and species.

Looking at the other end of the hormone system one may ask: what is known about the action of ecdysteroids in nematodes, trematodes, and cestodes?

During development the larvae of nematodes molt. The physiological and anatomical changes involved appear to resemble those in the molting process of insects and other arthropods [14]. Thus, it was postulated [15] that in nematodes ecdysteroids control the deposition of a new cuticle and shedding of the old. In fact, ecdysone showed developmental effects on a *Cephalobus* species [16] and stimulated the molting of last stage larvae to adults of *Nematospiroides dubius* [17]. Recently, the effects of ecdysteroids on the fine structure of the cuticle of *Litomosoides carinii* were studied [18]. The authors concluded that ecdysteroids are likely to be involved in the control of the molt in this species.

Trematodes do not molt like arthropods. Thus, ecdysteroids should control other functions in these organisms. Recently, it was speculated that ecdysteroids could be involved in membrane exuviation and vitellogenesis [6].

Additional information concerning the presumed role(s) of ecdysteroids in nematodes, trematodes, and cestodes may be expected from experiments with ecdysteroid antagonists, such as azadirachtin. Extracts from plant material containing azadirachtin were reported to be toxic to nematodes [19]. These extracts are in partical use in India to control the populations of plant parasitic soil nematodes.

In spite of the lack of evidence for ecdysteroid biosynthesis in helminths and the limited information available as to their possible role as hormone(s), it appears worthwhile to consider the potentials of ecdysteroids for practical purposes. Three ideas will be put forward and experiments reported to substantiate them.

1. The diagnosis of infections with helminths is often difficult. If helminths synthesize ecdysteroids they are also likely to inactivate and excrete them. Thus, it should be possible to detect an infection by helminths on the basis of the occurrence of ecdysteroids in the host's blood or urine.

The "multimammate rat", *Mastomys natalensis*, was used as an infection model to study whether infection with helminths leads to detectable concentrations of ecdysteroids in the blood of the host. The analyses reported in Table 2, first of all, show that uninfected *Mastomys* had no detectable ecdysteroids in their serum. In fact, the concentration of immunoreactive material was not always zero. But the concentration was below 1 nmol/l in each case. The ecdysteroid background in the sera probably originated from the food (see below). The second result from the analyses is that infection with a helminth species (either a nematode, trematode, or cestode) led in some, but not all, cases to an ecdysteroid concentration in the serum above the arbitrarily chosen level of 1 nmol/l. It is concluded that the increase in concentration of immunoreactive material in the serum of a mammalian host is most likely due to the excretion of ecdysteroids by the parasite. Detection of ecdysteroids, thus, may be taken as an indicator for infection by a helminth.

In insects ecdysteroid excretion is a function of their age and physiological stage [22, 23]. To see whether this is also the case with helminth species, the sera of *Mastomys* were tested at known intervals after infection (Table 3). Clearly, variations in the ecdysteroid concentration with the progress of development of the parasites were detected. This would explain why the analyses in Table 2 were not constantly negative or positive. In view of the rather large relative variation of the data obtained, which is due to the low concentrations measured, we will refrain from drawing any further conclusions.

Serum samples from uninfected *Mastomys* revealed a low, but variable background of ecdysteroids. The analysis of urine samples from these animals also led to the observation of ecdysteroids at low concentrations. To identify the origin of these parasite-unrelated ecdysteroids we tested the food and found significant amounts of immunoreactive material which we consider (pending further analyses) to be ecdysteroids (Table 4). Consequently, the ecdysteroid content of the food has to be taken into consideration when infections with helminths are to be detected by the presence of ecdysteroids.

2. Ecdysteroids and antiecdysteroids injected into the host should be able to interfere with the development and reproduction of the parasites. Therapy with such hor-

Table 2. Ecdysteroids in sera of helminth-infected *Mastomys natalensis* *

Species	Number of sera tested	
	Negative [a]	Positive [b]
Control: uninfected	15	0
Nematodes		
Litomosoides carinii	9	1 (1.0 n*M*)
Brugia pahangi	12	0
Brugia malayi	11	0
Dipetalonema viteae	11	1 (1.8 n*M*)
Trichinella spiralis	1	0
Angyostrongylus cantonensis	2	0
Ancylostoma caninum	2	0
Ascaris lumbricoides suum	3	0
Trematodes		
Schistosoma mansoni	4	1 (11.0 n*M*)
Opisthorchis felineus	2	1 (1.8 n*M*)
Cestodes		
Taenia crassiceps	0	3 (1.8; 2.8; 2.8 n*M*)
Hymenolepis fraterna	2	1 (1.4 n*M*)

* Blood samples were taken from *Mastomys* prior (= controls) and at several intervals after infection with various helminth species. The serum was stored (up to 3 months) at −20 °C until analysis of ecdysteroids by radioimmunoassay was done. Sample preparation as well as method of analysis were as described elsewhere [20]. The assays were performed in triplicate and the means are given here. Assays with a standard deviation above 20% were repeated. The antiserum used was DBL-1. Its characteristics have been reported [21]. The concentrations in brackets indicate radioimmunoassay positive material (expressed in equivalents of ecdysone). The lower limit of detection of ecdysone under the conditions used was 0.5 n*M*

a Negative: ecdysteroids were not detected by RIA, i.e., the concentration was below an arbitrarily chosen level of 1 nmol/l (25 pg ecdysone equivalents/assay)

b Positive: ecdysteroids were detected in concentrations above the limiting level of 1 nmol/l

mones and hormone analogues should have no side-effects on the vertebrate host because of their low toxicity in vertebrates.

Again *Mastomys natalensis* was used as host for the nematode, *Dipetalonema viteae*. In separate experiments ecdysone and 20-hydroxyecdysone were injected into the animals beginning at day 66 postinfection. The injections, in daily doses up to 20 mg/kg body weight of the host, were repeated for 4 days. No significant effects on the concentration of microfilariae in the blood or the number of adults in the host's body were observed when compared to controls. Perhaps injection at an earlier developmental stage is necessary for an effect on development and/or reproduction. While our results were negative, there are two reports in the literature more than 10 years old stating that the growth of *Caenorhabditis elegans*, *Panagrellus redivivus*, and *Haemonchus contortus* is inhibited by ecdysteroids [2, 24]. Experiments with the hormone analogue azadirachtin might also prove promising and will be done in the near future.

3. There is an active interplay between the serum proteins of the host and many parasites. Antibodies raised in the vertebrate host against ecdysteroids could perhaps

Table 3. Ecdysteroid concentration in the serum of helminth-infected *Mastomys natalensis*: time course

Species	Time elapsed after infection (days)	Concentration of radioimmunoassay positive material (nmol/l)
S. mansoni	31	5.4 (± 1.9; n = 3)
	51	1.7 (± 1.4; n = 3)
	81	3.1 (± 1.8; n = 3)
	115	3.3 (± 1.3; n = 3)
	210	3.7 (± 2.9; n = 3)
L. carinii	28	2.7 (± 2.2; n = 3)
	44	2.8 (± 1.5; n = 3)
	72	1.3 (± 1.4; n = 3)
	130	1.9 (± 2.2; n = 3)
	210	2.3 (± 2.0; n = 3)
	300	1.8 (± 1.6; n = 3)
	370	2.4 (± 1.9; n = 3)
B. pahangi	33	0.6 (± 1.0; n = 3)
	45	* (± 1.0; n = 3)
	65	1.0 (± 0.8; n = 3)
	170	2.8 (± 1.3; n = 3)
	330	*
	420	*
	500	*
B. malayi	50	*
	72	*
	88	*
	170	*
	270	*
	380	0.6 (± 1.4; n = 3)
	590	*
Controls		
Aqua bidest		*
Pooled serum from six uninfected individuals		*
Sera from single uninfected individuals		1.6 (± 2.8; n = 3)
		0.6 (± 1.1; n = 3)
		1.0 (± 1.0; n = 3)
		*
		1.2 (± 1.2; n = 3)
		*
		1.0 (± 0.9; n = 3)
		0.9 (± 0.8; n = 3)

Blood from six *Mastomys* fed with a special diet of low ecdysteroid content (no. 10, Table 4) was withdrawn at known intervals after infection. The blood was pooled and stored at −20 °C (up to 2 months) until analysis by radioimmunoassay (same conditions as in Table 2). This time the serum was split into three samples, each aliquot being analyzed in triplicate. (Means and standard deviations are given.) The lower limit of detection under the conditions used was 0.5 n*M* ecdysone

* Below limit of detection (= 0.5 n*M*)

Table 4. Ecdysteroids in the food of *Mastomys natalensis*

Number	Food	Concentration (ng/g)
1	Cod meal, type 66	500
2	Fish meal, type 64	93
3	Animal meal, protein content 55%	58
4	Pressed pellets, angular type	47
5	Sunflower kernels	29
6	Pressed pellets, round type, 9B	22
7	Sawdust	8.9
8	Oats	3.4
9	Shrimps	2.8
10	Special diet (low ecdysteroid content)	2.7
11	Wheat	1.5
12	Corn, whole meal	1.2
13	Barley	0.6

The ecdysteroid concentration was determined with radioimmunoassay (method see Table 2). The concentration is given as radioimmunoassay positive material in ng per g dry weight (= ppt) and is expressed in equivalents of ecdysone

perturb the hormone system of the parasite and could thus disrupt parasite development and reproduction.

To test this idea experimentally the animal model *Mastomys natalensis* as host infected with *Dipetalonema viteae* was used (Table 5). Immunized animals contained only half the number of microfilariae in their blood compared to nonimmunized controls. Furthermore, the embryogenesis of the microfilariae was effected. Pathologically altered microfilariae and embryos were detected in the uteri of dissected female worms. This would indicate that ecdysteroids are involved in the embryogenesis of nematodes. Further experiments will have to be performed to substantiate this assumption.

In conclusion: while direct evidence for the formation and hormonal function of ecdysteroids in nematodes, trematodes, and cestodes is still scarce, the experiments described here and in chapters of Nirde et al., Rees et al., Sauber et al. (this Vol.) are compatible with and support the hypothesis that ecdysteroids play a vital role as hormonal entities in helminths. It is our belief that the continuation of experiments to test this statement will eventually lead not only to proof of the hormonal function of ecdysteroids in helminths, but also to methods useful in the diagnosis and medical treatment of helminth infections.

Table 5. Immunisation of *Mastomys natalensis* against ecdysteroids

(A) Effects on Infection with *Dipetalonema viteae*

Animal no.	Number of adults found in subcutaneous tissue		Uterine embryos		Uterine microfilariae	
	Female	Male	Normal	Pathologically altered	Normal	Pathologically altered
1	6	7	++	++	+	+
2	2	7	++	++	+++	+
3[a]	0	3		Not applicable		
4	1	4	++	+	++	+
5[a]	0	1		Not applicable		
6	4	10	++	++	++	+
Controls (nonimmunized):						
1	5	10	+++	–	+++	–
2	7	5	++++	–	++++	–
3	10	8	++++	–	++++	–
4	6	9	++++	–	++++	–
5[a]	0	7		Not applicable		
6	2	9	++++	–	++++	–
7	1	7	++++	–	+++	–
8	2	3	++++	–	++++	–

[a] Microfilariae were detected in the blood, but no female adults were found in the connective tissue
\+ indicates the relative concentration of microfilariae and embryos

(B) Concentration of microfilariae in the blood of the host

Time elapsed after infection (days)	Concentration of filariae in six immunized *Mastomys* (μl^{-1})	in six unimminized controls (μl^{-1})
56	1.7 (± 1.2)	2.4 (± 1.8)
68	3.0 (± 2.0)	5.9 (± 3.6)
92	6.6 (± 5.4)	15.2 (± 13.3)
110	7.6 (± 5.4)	16.8 (± 7.5)

The synthesis of the ecdysone-albumin conjugate has been described elsewhere [20]. Coupling ratio was 5.4 (mol ecdysone/mol albumin). The steroid-protein conjugate was taken to immunize six individuals of *Mastomys*. Per animal 0.4 mg of the conjugate were injected dissolved in 0.5 ml aqua bidest and mixed with a twofold volume of complete Freund's adjuvant. The animals received four injections into the back. Booster injections with the same amount of antigen, but emulgated in Freund's incomplete adjuvant were applicated 21 and 50 days later. Before the animals were infected with *Dipetalonema viteae* formation of antibodies against ecdysteroids was controlled by analysis of the antibody titer. Equilibrium dialysis of diluted serum resulted in titers always better than 1:2,000

Acknowledgments. We thank Dr. Roger Dennis and Dr. Gerd Käuser for critical reading of the manuscript. The skillful technical assistance of Karin Schultheiss and Peter Möbus is gratefully acknowledged.

References

1. Karlson P (1983) Hoppe Seyler's Z Physiol Chem 364:1067–1087
2. Rogers WP (1973) Parasitology 67:105–113
3. Dennis RD (1977) Int J Parasitol 7:181–188
4. Horn DHS, Wilkie JS, Thomson JA (1974) Experientia 30:1109–1110
5. Torpier G, Hirn M, Nirde P, De Reggi M, Capron A (1982) Parasitology 84:123–130
6. Nirde P, Torpier G, De Reggi ML, Capron A (1983) FEBS Lett 151:223–227
7. Bergamasco R, Horn DHS (1983) In: Downer RGH, Laufer H (eds) Endocrinology of insects, invertebrate endocrinology, vol I. Liss, New York, pp 627–656
8. Walter J (1983) Unpublished
9. Sall C, Tsoupras G, Cappler C, Lagueux M, Zachary D, Luu B, Hoffmann JA (1983) J Insect Physiol 29:491–507
10. De Reggi ML, Hirn MH, Delaage MA (1975) Biochem Biophys Res Commun 66:1307–1315
11. Delaage MA, Hirn MH, De Reggi ML (1982) Methods Enzymol 84:350–358
12. Karlson P, Hoffmeister H, Hummel H, Hocks P, Spitteller G (1965) Chem Ber 98:2394–2402
13. Nakanishi K (1971) Pure Appl Chem 25:167–195
14. Davey KG, Kan SP (1967) Nature (London) 214:737–738
15. Dennis RD (1977) Int J Parasitol 7:171–179
16. Hitcho P, Thorson R (1971) J Parasitol 57:787–793
17. Dennis RD (1976) Comp Biochem Physiol 53:53–56
18. Spindler K-D, Mehlhorn (1983) Commun Symp Biosynth Metab Mode Action Invertebr Horm, Strasbourg, France, August 29–September 2
19. Saxena RC (1983) Commun IInd Int NEEM Conference. Rauisch-Holzhausen Castle, West Germany, May 25–27
20. Reum L, Koolman J (1979) Insect Biochem 9:135–142
21. Reum L, Haustein D, Koolman J (1981) Z Naturforsch 36c:790–797
22. Koolman J, Hoffmann JA, Dreyer M (1975) Experientia 31:247–248
23. Koolman J (1982) Insect Biochem 12:225–250
24. Dropkin VH, Lower WR, Acedo J (1971) J Nematol 3:349–355

II.14 Ecdysteroids in Schistosomes and Host-Parasite Relationship

P. NIRDÉ[1], G. TORPIER[1], A. CAPRON[1], M. DELAAGE[2], and M.L. de REGGI[2]

1. Introduction

Schistosomiasis is a major health problem in 73 tropical countries where the number of infected people is estimated to be about 300 million.

The parasitic worms which cause schistosomiasis (bilharziasis) in man or rodents belong to the genus *Schistosoma*, a somewhat unusual group of digenic trematodes. These digenea have complex life cycles involving intermediate hosts in addition to the definitive host in which they reach sexual maturity. The intermediate host must be a mollusc in which a phase of asexual larval multiplication of the parasite occurs. Schistosomes are the only trematodes in which sexes are separated in adult worms. The males are shorter than the narrowly elongate females. When mature, the females are carried by the males in a ventral groove of the body called gynaecophoric canal. During the mature stage, the females may lay several hundred eggs per day which cause irreversible sequelae (granuloma, liver fibrosis). Another characteristic of schistosomes is their adoption of the blood as their adult habitat.

While transformation of cercariae to schistosomula (first larval stages in infected host) and morphophysiological changes accompanying their subsequent developmental stages have been intensively investigated [1-3], the differentiation of the tegumental membrane complex has received little attention [4-6]. Freeze fracture studies of membrane differentiation of *Schistosoma mansoni* have allowed the observation of a membrane exuviation during the 10-20 day period of infection [7]. This phenomenon shows some similarity with the molting process of arthropods which is induced by ecdysteroids (namely, ecdysone and 20-hydroxyecdysone). After the first demonstration that platyhelminths produced ecdysteroids [7], further studies have shown that the two ecdysteroids are present at two crucial periods of the development in *S. mansoni*: (1) During the period extending from postinfection day 6-11 when a transition occurs between the juvenile worms and their exponential growth phase to adults and, (2) in adult males and females during sexual maturation and egg laying [8, 9]. The biological role of these ecdysteroids in the parasites, and the fate of these molecules in the infected host appear as challenging research topics. The aim of this present study is to demonstrate that ecdysteroids from parasites are released into the serum and the urine of infected vertebrates.

1 Centre d'immunologie et de biologie parasitaire, unité mixte, INSERM (U167), CNRS 624, Institut Pasteur, BP 245, 59019 Lille Cedex, France

2 Immunotech S.A., Case 915, 13288 Marseille Cedex 9, France

Biosynthesis, Metabolism and Mode of Action of Invertebrate Hormones
(ed. by J. Hoffmann and M. Porchet)

2. Material and Methods

Infestation: Rodents (hamster: *Mesocricetus auratus*, rat: Fisher) were infected by exposure to cercariae of *Schistosoma mansoni* as originally described by Smithers and Terry [10]. Monkeys *(Patas patas)*, aged 2 years and weighing 5 kg were infected 10 times at 2 week intervals by exposure to 500 cercariae of *Schistosoma mansoni*.

Ecdysteroid extracts: One ml of biological samples (sera or urine) were added to 3 ml methanol (Prolabo). The mixture was heated at 60 °C for 15 m and centrifuged at 800 × g for 15 m. The supernatant was dried under nitrogen flow. The residue was then dissolved in 0.1 *M* citrate buffer (pH 6.1) and assayed according to the RIA method described by De Reggi et al. (1975) [11].

Chemotherapy: 60 children, aged 12–14 (Niamey, NIGER) infected with *Schistosoma haematobium* were treated with a single dose of antihelminthic drugs. The first group of 20 patients received Oltipraz, the second (n = 20) received Oltipraz plus vitamin, and the last group (n = 20) received Praziquantel.

Electron impact mass spectroscopy: Urines of 20 people infected with *Schistosoma haematobium* were collected (600 ml). Urine sample was added to 1,200 ml of butanol. After centrifugation (1,000 g, 3 m) the butanol phase was dried under reduced pressure at 40 °C. The residue was then dissolved in 95% ethanol before being centrifuged (1,000 g, 3 m). The supernatant was dried under nitrogen and suspended in methanol prior to analysis by thin-layer chromatography (silicagel plates 20 × 20 cm, F 254, Merck) with chloroform/methanol as solvent (80v/80v). The plates were scraped by 1 cm bands and the products eluted with 95% ethanol. Immunoreactive samples were then analyzed by HPLC (reverse phase, methanol/water; 40v/60v), isocratic elution. Gas chromatography was performed on immunoreactive HPLC fractions after derivatization. For this, 100 μl of trimethylsilylimidazole (TSIM) were added. This solution was heated to 80 °C under argon, and separation of derivatized ecdysteroids from reagent was performed through HPLC on silica column with ethanol/hexane as solvents. After evaporation of the solvent under argon, acetone was added to the dried extract and 1–2 μl were injected into a gas chromatographic apparatus (Varian 1200, capillary glass column SE/30) at 290 °C. An LKB 9000 S electron impact mass spectrometer was coupled to the gas chromatographer. This spectrometer was equipped with an INCOSS computer system which allows the recording of one mass spectrum per second during the whole process of the chromatography.

3. Results and Discussion

Serum of infected rodents contains ecdysteroids. These molecules are detectable in the blood as early as 6 days after the infection, whereas no detection is possible in uninfected rodents or during the first 5 days of infection (Fig. 1). Moreover, the changes in hormone level are different between permissive (hamster) and nonpermissive hosts (rat). In the first case the RIA activity reaches 0.3 n*M* on day 12 postinfection and this value

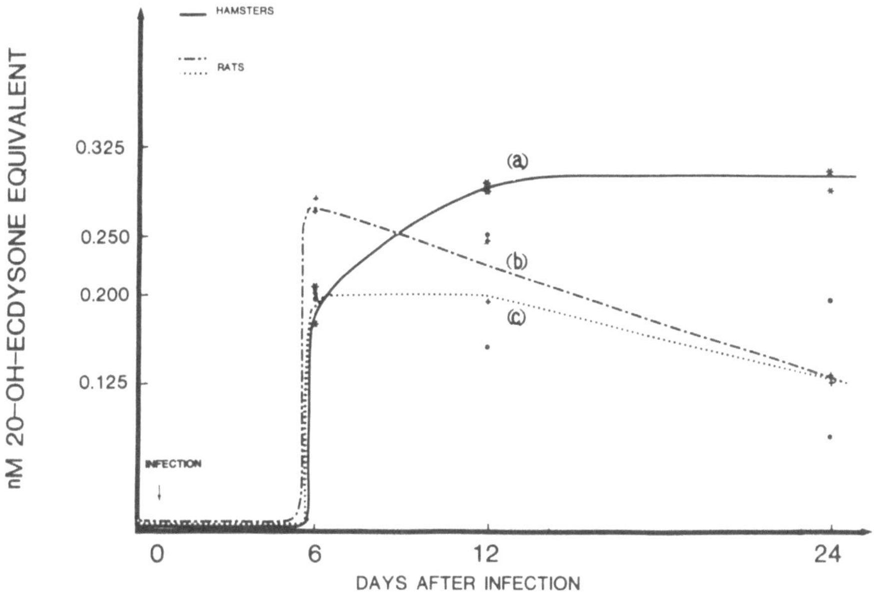

Fig. 1. Changes in ecdysteroid concentration in the blood of rodents (permissive and non permissive hosts) infected with *Schistosoma mansoni*. *a* Hamsters (permissive) infected with 2,000 cercariae. *b* Rats (nonpermissive) infected with 2,000 carcaria. *c* Rats infected with 500 cercariae. Each experimental point is the mean of four individual values. Bars are the SEM

is unchanged on day 24 (Fig. 1a). In contrast, in the nonpermissive hosts the ecdysteroid level reaches 0.3 n*M* on day 6 postinfection and, more importantly, it decreases thereafter, reaching 0.22 n*M* on day 12, and 0.13 n*M* on day 24 (Fig. 1b). Figure 1 also shows that the rise in ecdysteroid concentration which occurs on day 6 is significantly lower in rats which received a weaker challenge of 500 cercariae as compared to rats challenged with 2,000 cercariae (Fig. 1c). After day 12, the two groups of rats do not differ significantly with regards to ecdysteroid titers. We, therefore, conclude that the time course of ecdysteroid titer is correlated with the susceptibility or the innate resistance of the host to schistosome infection. Further analysis shows that ecdysteroids are also present in the blood of infected primates. Figure 2 shows that in *Patas patas* monkey, they reach 4 n*M* on day 218 postinfection and more than 8 n*M* on day 499 and thereafter. In contrast, ecdysteroid levels in an uninfected monkey were below 0.3 n*M* (Fig. 2c). It should be noted that *Schistosoma mansoni* eggs were found in the stools of monkey A during the duration of the experiments (10–180 eggs/g). Monkey B lacked eggs in the stools on day 499 and thereafter. Despite this difference in the form of the chronic model, ecdysteroid levels were very similar in the two animals. These important observations suggest that the concentration of ecdysteroids could be an early biological parameter of parasite infection and might be correlated with the viability of the parasites. To support this concept, ecdysteroid titers were determined in the urine of humans infected with *Schistosoma haematobium*. Figure 3 indicates that children have very high concentrations of insect molting hormones in their urine. The mean level observed was about 30 n*M* and the highest value found in three patients was as

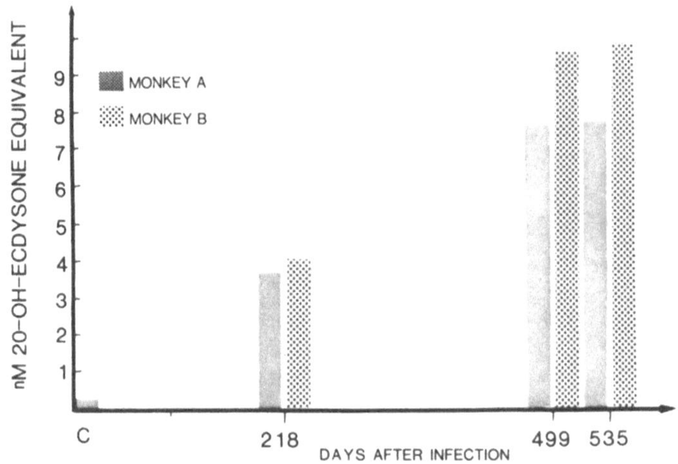

Fig. 2. Ecdysteroid titer in sera of monkeys infected with *Schistosoma mansoni*. *A* Chronic model of infected monkey with eggs in the stools (10–180 eggs g^{-1}). *B* Chronic model of monkey without eggs following day 499. *C* Noninfected monkey

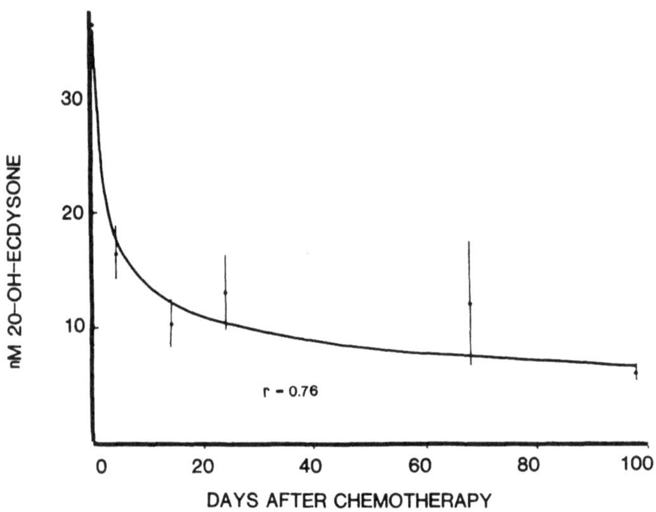

Fig. 3. Changes in ecdysteroid titer in urine from patients infected with *Schistosoma haematobium* before and after chemotherapy. (Do) Day of chemotherapy (single dose of drug). Each point is the mean of triplicates. Bars are the SEM

high as 180 n*M*. The figure also shows that the hormonal concentration decreased markedly after an antihelminthic chemotherapy. Four days after drug administration, the ecdysteroid titer fell to one-third, and then it slowly decreased reaching some 6 n*M* 100 days after treatment. We also have evidence that ecdysteroids from serum of patients infected with *Schistosoma mansoni* decreased after chemotherapy [8]. Chemical characterization of ecdysteroids found in urine of infected humans was achieved by GC and MS after trimethylsilylation (Fig. 4). Total ion recording and fragmentograph ion

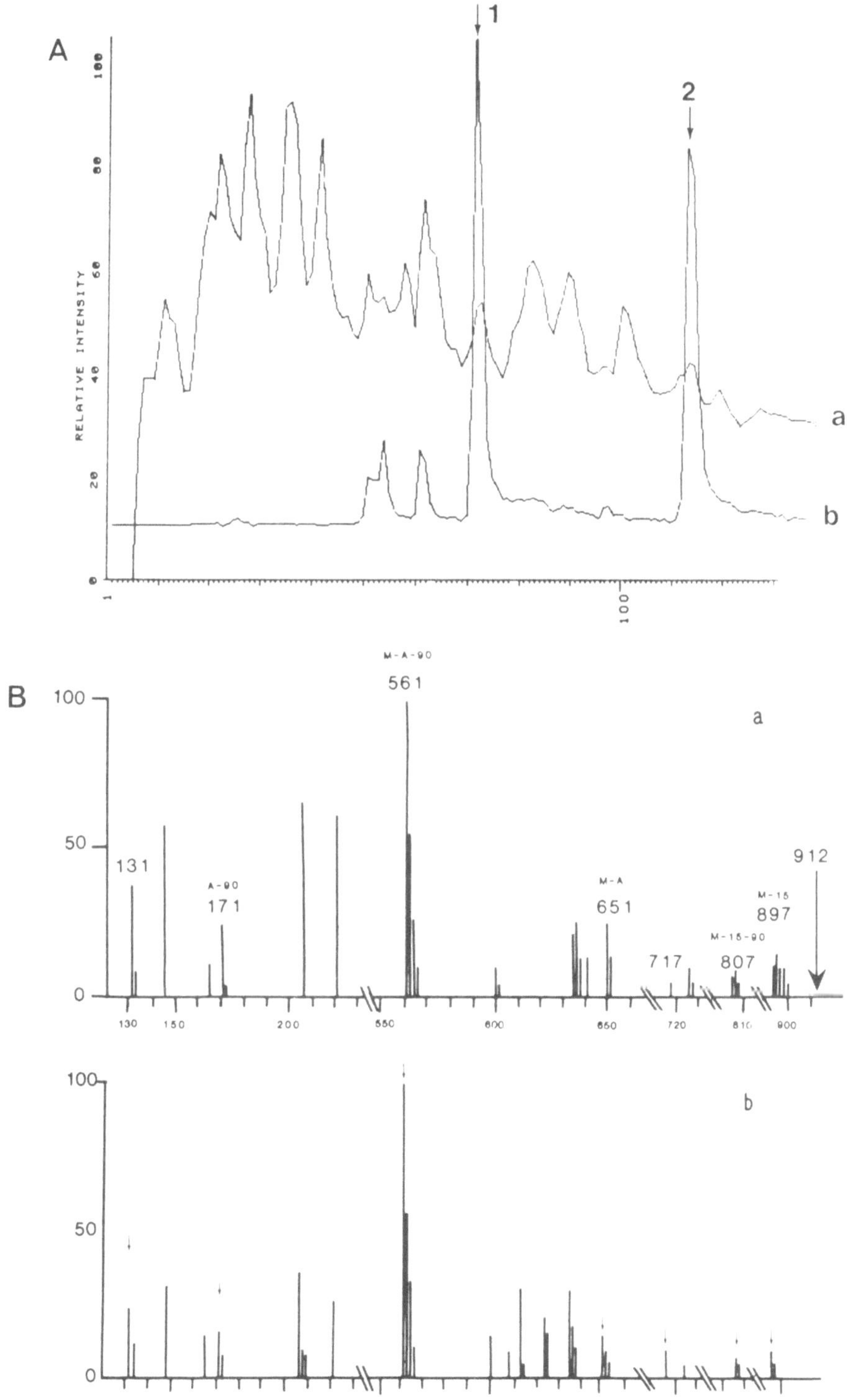

Fig. 4 A, B. Mass spectrum of hexa-trimethyl-silylated natural compounds present in urine of patients infected with *Schistosoma haematobium*. **A** Total ion chromatogram: *a* total ion recording; *b* recording of ion fragment at m/e 561. **B** Mass spectrum: *a* mass spectrum of hexatrimethylsilylate reference 20 hydroxy-ecdysone; *b* mass spectrum of compound yielding peak 1 in the fragmentogram of **A***b* (the compound yielding peak 2 has the same mass spectrum)

recording are shown in Fig. 4A. The total ion analysis (Fig. 4Aa) indicates that several molecules are present in urine sample. The fragmentograph recording of the main ion of 20-hydroxyecdysone at m/e 561 [12] (Fig. 4Ab) allows the detection of two major molecular forms. The first peak (vertical arrow 1) has the same retention time as standard hexatrimethylsilyl-20-hydroxyecdysone and the second peak (arrow 2) has the same retention time as standard hexatrimethylsilyl-5 epi, 20-hydroxyecdysone run under the same conditions (reference compounds prepared by Dr. Hetru [13]). The mass spectrum of these molecules is given in Fig. 4B. The spectrum of reference trimethylsilylated 20-hydroxyecdysone has the main ion at m/e 561 which corresponds to the loss of the side chain of the ecdysteroid after cleavage between C-20 and C-22 and the loss of a silanol in the case of fully silylated derivatives. This ion is highly characteristic of hexatrimethylsilylated 20-hydroxyecdysone. The fragments at m/e 131 and m/e 171 correspond to the fragmentation of the side chain of 20-hydroxyecdysone [14]. In addition, the fragment at m/e 897 (molecular ion minus 1 methyl) is also present. In conclusion, peaks 1 and 2 yield the same mass spectra (Fig. 4Bb) which are both similar to the mass spectrum of 20-hydroxyecdysone obtained under the same conditions. The difference in retention time is explained by the fact that peak 2 corresponds to the 5-α epimer of 20-hydroxyecdysone. These data clearly indicate that 20-hydroxyecdysteroids are the two main products. In contrast, in adult parasites from mesenteric vessels, ecdysone is more abundant than 20-hydroxyecdysone [9]. In addition to the roles that such a release might exert on the human endocrine or on immune systems, research on the effects of anti-ecdysteroid antibodies on the viability of schistosomes is of great interest. Preliminary results show that ecdysteroid antibodies significantly reduce the worm population in vitro as well as in vivo (unpubl.).

4. Conclusions

Our work demonstrates for the first time that the insect molting hormones 20-hydroxyecdysone present in schistosomes are released into the biological fluids of infected vertebrates. Ecdysteroids were also detected in the serum of rodents parasitized by the nematode *Dipetalonema vitae* (Nirdé 1982). From a clinical point of view, the presence of these hormones in the sera and in urine of infected humans is of practical interest because they can be used as a biological marker of parasitic infection, and, thus, palliate to three drawbacks of parasitological diagnosis which are: (1) no detection before the late egg-laying stage of the parasites; (2) decrease or absence of egg laying in chronic infection; (3) relative unreliability of the parasitological tests. Furthermore, besides the classical parasitological tests, no other biological tests allow an accurate determination of the efficiency of chemotherapy. Together with a new insight into host-parasite relationship (i.e., immunological system or endocrine system) these findings open novel approaches to the possible early detection of parasitic diseases, but also to the control of the efficiency of therapy in man.

Acknowledgments. The authors thank Dr. B. Sellin (CERMES, Niamey, NIGER) for the generous gift of serum and urine samples. The authors are indebted to Dr. J.A. Hoffmann for providing facilities in this work and to Dr. G. Teller for mass spectrometry. A grant by the Edna McConnell Clark Foundation, N.Y. is acknowledged. This work was also supported by INSERM (U 167) and CNRS (ERA 422).

References

1. Clegg JA (1965) Parasitology 16:133–147
2. Wilson RA, Draskau T, Miller P, Lawson JR (1978) Parasitology 77:57–73
3. Lawson JR, Wilson RA (1980) Parasitology 81:325–336
4. Hockley DJ, McLaren DJ (1973) Int J Parasitol 3:13–25
5. Hockley DJ, McLaren DJ, Ward BJ, Nermut MV (1975) Tissue Cell 7:485–496
6. McLaren DJ, Hockley DJ, Golgring OL, Hammond BJ (1978) Parasitology 76:327–348
7. Torpier G, Hirn M, Nirdé Ph, Reggi ML de, Capron A (1982) Parasitology 84:123–130
8. Nirdé Ph (1982) Thèse Doct Biochim, No 1010, Lille
9. Nirdé Ph, Torpier G, Capron A, Reggi ML de (1983) FEBS Lett 151(2):223–227
10. Smithers SR, Terry RJ (1965) Parasitology 55:695–700
11. De Reggi ML, Hirn M, Delaage MA (1975) Biochem Biophys Res Commun 66:1307–1315
12. Lafont R, Somme Martin G, Mauchamp B, Maume B, Delbecque JP (1980) In: Hoffmann JA (ed) Developments in endocrinology, vol VII (10). Elsevier, Amsterdam, pp 45–68
13. Hetru C, Nakatani Y, Luu B, Hoffmann JA (1983) Nouveau J Chim 7(10):587–591
14. Morgan ED, Woodbridge AP (1974) Prog Mass Spectrosc 9:102

II.15 The Occurrence and Possible Physiological Significance of Ecdysteroids During Nematode and Cestode Development

H.H. REES and A.H.W. MENDIS [1]

1. Introduction

This chapter will consider the occurrence, biological effects, and possible physiological functions of ecdysteroids in nematodes and, to a lesser extent, in cestodes. Reports of the occurrence of ecdysteroids outside the arthropod phylum are relatively few, a situation which may partly reflect the extremely low concentrations of ecdysteroids reported in such species and also the difficulty of obtaining sufficient biological material, particularly of parasitic species, for biochemical analysis.

"Hormones" may be defined as substances which are secreted internally by an endocrine organ and transported in the body fluids to other tissues, where they elicit, in minute quantities, physiological responses. For definitive establishment of the existence of a hormonal system in an organism, it is necessary to fulfill a number of criteria. The source of hormone should be identified, and the effects of extirpation (ablation) and implantation (replacement) of such tissue (organ) examined. Additionally, it is necessary to demonstrate that the hormone is released into the body fluid under defined physiological conditions, and that it elicits appropriate responses by interaction with specific receptors associated with its target cells. Ultimately the hormone should be purified and characterised chemically [1].

When considering the ecdysteroid endocrinology of nematodes and cestodes, it is convenient to use the moulting hormone system in insects as a basis, since there is much information available in the latter case. Obviously, this approach must be applied with caution, not only in the case of cestodes which do not moult, but also in the case of nematodes. However, aspects of the metabolic interrelationships of ecdysteroids, in particular, may well be expected to have parallel in the different phyla.

2. Nematodes

Development of nearly all nematodes proceeds by a series of four larval stages separated by moults and culminates, with the adult stage [2]. The moulting process consists of distinct steps and is apparently initiated by various stimuli leading to release of neurosecretory material from the central nervous system. This is followed by separation of

1 Department of Biochemistry, University of Liverpool, P.O. Box 147, Liverpool L69 3BX, Great Britain

Biosynthesis, Metabolism and Mode of Action of Invertebrate Hormones
(ed. by J. Hoffmann and M. Porchet)

the old cuticle from the hypodermis, and then the formation of the new cuticle between the hypodermis and the old cuticle. Finally, ecdysis (shedding) of the old cuticle leads to release of the next stage, either larva or adult. Although this moulting sequence is superficially similar to that in insects, the nature of the nematode cuticle, being non-chitinous, is quite different.

The infective stage in parastic nematodes is usually, but not invariably, the third stage larva (L3). In the infective stage of some parasitic nematodes, such as the trichostrongyles, the new cuticle is fully formed, but actual ecdysis is delayed and the old cuticle is retained to serve as a protective sheath. After infection the moulting process is resumed, apparently in response to host-related factors, and ecdysis or exsheathment occurs releasing the new larva, which continues development [3]. The proteolytic enzyme leucine aminopeptidase has been implicated [4] in this exsheathment process.

There is evidence for endocrine control of, at least aspects of, the molting and exsheathment processes in nematodes. A full treatment of this is beyond the scope of this chapter and is covered in excellent reviews [5-9].

2.1 Occurrence of Ecdysteroids

There have been several reports of the presence of insect moulting hormone (ecdysteroid)-like materials in nematodes. Since most of the early reports were based on bioassay, thin-layer chromatographic, or radioimmunoassay techniques, which generally lacked specificity, definitive establishment of the presence of ecdysteroids and their identification was not possible. Despite these shortcomings, the results *as reported* are summarised in Table 1. Due to the generally much lower concentrations of ecdysteroids in nematodes, compared with insects, and consequent paucity of material, complete characterization of the compounds by physical techniques was not feasible in any case. However, the recent characterization of ecdysone and 20-hydroxyecdysone as the major ecdysteroids in adult females and males of the filarial nematode, *Dirofilaria immitis*, by high-performance liquid chromatography in conjunction with radioimmunoassay (HPLC/RIA) on two different chromatographic systems and by gas-chromatography/mass spectrometry (selected ion monitoring) [GC/MS(SIM)] established for the first time the occurrence of ecdysteroids in a nematode species [18, 19]. Similarly, the utilization of the same techniques for analysis of the ecdysteroids from the hermaphrodite N2 strain of the free-living species, *Caenorhabditis elegans* [20], the swine parasite, *Ascaris suum* [16], and the mycophagous species, *Aphelencus avenae* [17], established the occurrence of ecdysone and 20-hydroxyecdysone as the major ecdysteroids in these nematodes; in the case of *Aphelencus*, identification was corroborated by fast atom bombardment mass spectrometry [17]. It was apparent that in these three species as well as in *D. immitis*, the ecdysteroids were present as both free compounds and as highly polar conjugates, hydrolysable with a crude enzyme preparation from the snail, *Helix pomatia*. The methodology employed in the other reports in Table 1 would not, in general, have allowed detection of such conjugates. Analogous polar conjugates from insects show low immunoreactivity [21] to several of the antisera commonly employed, and, consequently, necessitate enzymic hydrolysis to the free compounds before such assay.

Table 1. Reports of ecdysteroid-like substances in nematodes

Species	Developmental stage	Basis of identification	Compounds claimed[a]	Reference
Trichinella spiralis	Infective larvae	Bioassay on *T. spiralis* U.V., I.R., and mass spectra Co-chromatography on t.l.c.	Substance similar but not identical to 20-hydroxyecdysone and inokosterone	[10]
Phocanema depressum (probably *Porrocaecum* sp.)	Last stage larvae	Rf on t.l.c. U.V. and mass spectra Bioassay of crude extract on *Cynthia* moth	Substance similar to ecdysone	[11]
Haemonchus contortus	Larvae	Bioassay on *Calliphora stygia*	Ecdysone-like activity	[12]
	Larvae	Radioimmunoassay (RIA)	Ecdysone-like material	[13]
Ascaris lumbricoides (probably *Ascaris suum*)	Adult males and females	Bioassay on *Calliphora stygia* Co-chromatography of U.V. material with [^{3}H]20-hydroxy-ecdysone	Almost certainly 20-hydroxyecdysone	[14]
Ascaris suum	Adult females	HPLC/RIA	Ecdysone, 20-hydroxyecdysone	[15]
	Adult females	HPLC/RIA, gas chromatography/ mass spectrometry (selected ion monitoring) [GC/MS (SIM)]	Free and conjugated (polar) ecdysone and 20-hydroxyecdysone	[16]
Aphelencus avenae	Mixed stages	Radioimmunoassay	Ecdysone-like material	[13]
	Mixed stages	HPLC/RIA (2 systems) GC/MS (SIM) Fast atom bombardment MS	Free and conjugated (polar) ecdysone and 20-hydroxyecdysone	[17]
Panagrellus redivivus	Mixed stages	Radioimmunoassay	Ecdysone-like material	[13]
Dirofilaria immitis	Adult females Adult males	HPLC/RIA (2 systems) T.l.c./RIA GC/MS (SIM)	Free and conjugated (polar) ecdysone, 20-hydroxyecdysone, and possibly 20,26-dihydroxyecdysone	[18, 19]
Caenorhabditis elegans (N2 hermaphrodite strain)	Mixed stages	HPLC/RIA (2 systems) GC/MS (SIM)	Free and conjugated (polar) ecdysone, 20-hydroxyecdysone, and possibly 20,26-dihydroxyecdysone	[20]

[a] In most of the earlier reports, due to lack of specificity of the techniques, the results are not definitive but suggestive only

As in the case of insects, it is apparent that ecdysteroids occur both in larval stages and in adults of nematodes [15, 16, 18–20]. In adult females of many insect species, the ecdysteroids are present primarily as polar conjugates in the ovaries and are largely passed into the eggs, where they may serve as inactive storage forms of ecdysteroid, releasing free active hormone during embryogenesis by enzymic hydrolysis. During embryogenesis in several insect species there is apparent correlation between cycles of cuticulogenesis/membrane formation and peak titers of free ecdysteroids (see [22]). In fact, in females of *D. immitis* [16] and *Ascaris suum* [15, 16], a high proportion of the ecdysteroids is located in the reproductive tissue, which in the former species includes the intra-uterine developing eggs. Whether ecdysteroids are involved in control of cuticulogenesis during embryogenesis in nematodes is not clear. A pre-eclosion molt occurs in several plant-parasitic nematode species (for example, see [23]), a situation formerly believed to exist also in Ascarids, but recent observations on *Ascaris lumbricoides, A. suum, Toxocara canis,* and *Ascaridia galli* suggest the occurrence of two ecdyses before eclosion [24, 25].

It is interesting that comparable concentrations of ecdysteroids were observed in adult male and female *D. immitis* [19]; the exact significance of this is unclear. The situation in the insects, *Drosophila melanogaster* [26] and *Calliphora vicina* [27] is similar, the ecdysteroids in males of the latter species being primarily concentrated in the testes. In fact, testes of the tobacco budworm, *Heliothis virescens*, can apparently synthesis ecdysteroids [28].

A nematode tissue clearly responsible for synthesis of ecdysteroids has not been identified as yet, although laser ablation of particular regions is claimed to prevent certain events in the molting sequence (see [6]). By analogy with insects, nematode ecdysteroids should be formed from cholesterol (see [22]). However, as yet, the conversion of radioactive cholesterol into ecdysteroids in nematodes has not been reported; the low concentrations of nematode ecdysteroids renders demonstration of this difficult. In common with numerous other invertebrates, nematodes cannot synthesise sterols *de novo* (for references see [29, 30]), and must satisfy their sterol requirement from the diet or host.

2.2 Biological Effects of Ecdysteroids

The few investigations on the effects of exogenous ecdysteroids on nematodes are summarised *as reported* in Table 2. Obviously, no clear conclusions are justified from these data. Only in the case of *Nematospiroides* is a response reported to a very low concentration of ecdysteroid (2.2×10^{-11} *M* ecdysone) [35]. Most of the other effects recorded are in response to relatively high concentrations of steroid. However, the net concentration of ecdysteroid within the tissues is unknown, since the efficiency of uptake of the steroid cannot be assessed in these reports. Davey [6] argues that of the different steps in the molting process, only with respect to the control of ecdysis is there evidence for a dependency of such nematodes upon host hormones [37, 38], although respond to exogenous ecdysteroids indirectly via their own neurosecretory system and that the occurrence of these steroids is purely fortuitous.

Table 2. Biological effects *as reported* of exogenous ecdysteroids on nematodes

Test substance		Biological effect	Reference
Ecdysteroid	Concentration		
20-Hydroxyecdysone	1.1×10^{-7} *M*	Development of *Cephalobus* affected	[31]
20-Hydroxyecdysone	1×10^{-2} *M*	No effect on development of infective larvae of *Heterodera schachtii*	[32]
Ponasterone A Cyasterone Ecdysone Inokosterone 20-Hydroxyecdysone	0.01 to 200 µg/ml	All inactive in inhibition of growth of *Caenorhabditis elegans* and *Panagrellus redivivus*	[33]
Ponasterone A Ecdysone		No effect on several spp.	[34]
20-Hydroxyecdysone	5×10^{-5} *M*	Decreased the length of *Trichinella spiralis* larvae, but increased the number moulting	[10]
Ecdysone		Resulted in a "positive test for leucine aminopeptidase" in incubated excretory glands from *Phocanema depressum* (probably *Porrocaecum* sp.)	[11]
Ecdysone 20-Hydroxyecdysone	10^{-2} and 10^{-3} *M*	Inhibited growth of free-living stages of *Haemonchus contortus*	[12]
Ecdysone	2.2×10^{-11}	Stimulated adult moult of *Nematospiroides dubius*	[35]
20-Hydroxyecdysone	2×10^{-7} *M* to 2×10^{-4} *M*	No effect on moulting in vitro of L2 *Brugia pahangi*	[36]

Whether or not development of entomophilic nematodes is dependent on the molting hormones of the insect host is a pertinent question. Apparently, there is no definitive evidence for a dependency of such nematodes upon host hormones [37, 38], although the interpretation of some experiments may be complicated by the comparatively recent realisation of the occurrence and synthesis of ecdysteroids in adult insects.

3. Cestodes

It has been demonstrated that 20-hydroxyecdysone stimulates growth and asexual reproduction of *Mesocestoides corti* larvae both in vitro and in vivo [39]. Furthermore, thin-layer chromatographic fractions of lipid extracts from larvae of *M. corti* and adults of *Hymenolepis diminuta* stimulate increased growth of *M. corti* larvae in vitro and to some extent an increase in asexual reproduction [39]. In fact, on the basis of HPLC/RIA on two chromatographic systems and GC/MS (SIM), the occurrence of free, but not significant conjugated, ecdysone, 20-hydroxyecdysone, and possibly 20,26-dihydroxyecdysone has now been demonstrated in the sheep cestode, *Moniezia expansa* [40].

The ecdysteroid concentrations, albeit not at peak titers, were appreciably lower than those detected in adults of the filarial nematode, *Dirofilaria immitis* [18]. Radioimmunoassay of ecdysteroids in segments along part of the strobila of *M. expansa* suggested that the immunoreactive ecdysteroids were primarily concentrated in the anterior region [40]. However, GC/MS(SIM) analysis of the ecdysteroids in a strobilar segment containing the most mature proglottid suggested the presence of several unidentified putative ecdysteroid metabolites. It is possible that these metabolites have low immunoreactivity to the antiserum employed. These results suggested that there may be much metabolism of ecdysone and 20-hydroxyecdysone at some stage during proglottid maturation.

Preliminary results also indicate the presence of immunoreactive ecdysteroids in infective cysticercoids, patent adults, and eggs of the rat cestode, *Hymenolepis diminuta* [41].

The exact functions of ecdysteroids in cestodes are obscure. It may be significant that neurosecretory material first appears in neurosecretory cells in the scolex of *H. diminuta* adults just before the time that strobilization begins and is apparent in the axons about the time the first proglottid is shed [42]. Whether this connected to ecdysteroid production is unknown. Since the two cestode species investigated, *H. diminuta* [43] and *Echinococcus granulosus* [44], cannot effect cholesterol synthesis de novo, the expected cholesterol precursor of ecdysteroids in cestodes would presumably be derived from the host.

4. Possible Exploitation of the Ecdysteroid System for Control of Parasite/Pest Species

The demonstration that ecdysteroids occur in nematodes and cestodes as well as in trematodes (see Chap. II.14, this Vol.), coupled with the possibility that these compounds may be essential regulators (hormones) in the development of such species, raises the possible opportunity of exploiting this steroid system for control of parasite/pest species. This applies even to parasites of vertebrate animals and of man in view of the following distinctive structural features of ecdysteroids compared to vertebrate steroid hormones: the ecdysteroids possess a complete side chain, are polyhydroxylated, have an α,β-unsaturated oxo group in ring B (as opposed to ring A when this feature is present in vertebrate steroid hormones), and probably most important, an A/B cis ring junction, which results in a distinctly bent ring structure. Possible points of interference with the ecdysteroid system for exploitation in chemotherapy/control include, inhibition of (1) synthesis or (2) inactivation of the ecdysteroids, (3) the development of anti-hormones to interfere with hormone action, and (4) the application of potent ecdysteroid analogues to elevate hormonal activity at an inappropriate stage of parasite development. The possible development of immunological approaches to interfere with the ecdysteroid system is considered in Sect. II.14.

5. Conclusion

It is now clear that ecdysteroids occur in both nematodes and cestodes, albeit in comparatively low concentrations. However, the site(s) of synthesis of such steroids and their formation from cholesterol have not been demonstrated, as yet, in these species. Although biological effects of ecdysteroids on nematodes and cestodes have been reported, a particular requirement is the clear demonstration of a specific biological effect and the identification of target organs for these steroids.

Acknowledgments. Work from the authors' laboratory was supported by the Wellcome Trust and the Science and Engineering Research Council. We thank Dr. J.G. Mercer for valuable comments on the manuscript.

References

1. Barrington EJW (1979) In: Barrington EJW (ed) Hormones and evolution, vol I. Academic Press, New York, p vii
2. Rogers WP (1962) The nature of parasitism. Academic Press, New York
3. Davey KG (1972) In: Van den Bossche H (ed) Comparative biochemistry of parasites. Academic Press, New York, p 81
4. Rogers WP (1970) J Parasitol 50:329–348
5. Davey KG (1976a) In: Van den Bossche H (ed) Biochemistry of parasites and host-parasite relationships. Elsevier/North Holland Biomedical Press, Amsterdam, p 359
6. Davey KG (1976b) In: Croll NA (ed) The organization of nematodes. Academic Press, London, p 273
7. Willett JD (1980) In: Zuckerman BM (ed) Nematodes as biological models, vol I. Academic Press, New York, p 197
8. Davey KG (1982) In: Meerovitch E (ed) Aspects of parasitology. Institute of Parasitology, McGill University, Montreal, p 58
9. Rogers WP, Petronijevic T (1982) In: Symons LEA, Donald AD, Dineen JK (eds) Biology and control of endoparasites. Academic Press, Sydney, p 3
10. Hitcho PJ, Thorson RE (1971) J Parasitol 57:787–793
11. Rajulu GS, Kulasekarapandian S, Krishnan N (1972) Current Sci 41:67–68
12. Rogers WP (1973) Parasitol 67:105–113
13. Dennis RDW (1977) Int J Parasitol 7:181–188
14. Horn DHS, Wilkie JS, Thomson JA (1974) Experientia 30:1109–1110
15. Fleming MW (1983) American Soc of Parasitologists, December Meeting Abstracts, p 81
16. Cleator M, Delves CJ, Howells RE, Rees HH (1984) British Society for Parasitology Meeting
17. Mendis AHW, Rees HH (unpublished work)
18. Mendis AHW, Rose ME, Rees HH, Goodwin TW (1983) Mol Biochem Parasitol 9:209–226
19. Mendis AHW, Howells RE, Rees HH, Goodwin TW (1982) Parasitol 85:xxiii–xxiv
20. Mendis AHW, Keene NM, Rees HH, Goodwin TW (unpublished work)
21. Dinan LN, Rees HH (1981) J Insect Physiol 27:51–58
22. Rees HH, Isaac RE (1984) Chap II.2, this vol
23. Siddiqui IA, Taylor DP (1970) Nematologica 16:133–143
24. Araujo P (1972) Rev Inst Med Trop Sao Paulo 14:83–90
25. Araujo P, Bressan MCRV (1977) Annal Parasitol (Paris) 52:531–537
26. Hodgetts RB, Sage B, O'Connor JD (1977) Dev Biol 60:310–317
27. Koolman J, Scheller K, Bodenstein D (1979) Experientia 35:134–135
28. Loeb MJ, Woods CW, Brandt EP, Borkovec AB (1982) Science 218:896–898
29. Rees HH, Goodwin TW (1972) In: Geissman TA (ed) Biosynthesis (Special Periodical Reports), vol I. The Chemical Society, London, p 59

30. Comley JCW, Jaffe JJ (1981) J Parasitol 67:609–616
31. Webster JM, Craig S (1969) J Nematol 1:308
32. Johnson RN, Viglierchio DR (1970) Exp Parasitol 27:301–309
33. Dropkin VH, Lower WR, Acedo J (1971) J Nematol 3:349–355
34. Hansen EL, Buecher EJ (1971) Experientia 27:859–860
35. Dennis RD (1976) Comp Biochem Physiol 53A:53–56
36. Ogura N, Kobayashi M, Yamamoto H (1981) Dokkyo J Med Sci 8:74–82
37. Davey KG, Hominick WM (1973) Exp Parasitol 33:212–225
38. Gwadz RW, Spielman A (1974) J Parasitol 60:134–137
39. Kowalski JC, Thorson RE (1976) Int J Parasitol 6:327–331
40. Mendis AHW, Rees HH, Goodwin TW (1984) Mol Biochem Parasitol 10:123–138
41. Gibson JM, Rees HH, Arme C (unpublished results)
42. Davey KG, Brecklbridge WK (1967) Science 158:931–932
43. Frayha GJ, Fairbairn D (1969) Comp Biochem Physiol 28:1115–1124
44. Digenis GA, Thorson RE, Konyalian A (1970) J Pharmaceutical Sci 59:676–679

II.16 Ecdysteroids in Annelids

M. PORCHET[1], N. GAILLET[1], F. SAUBER[2], M. CHARLET[3], and J.A. HOFFMANN[3]

The epidermal cells of most annelid worms secrete a cuticle which consists of a mixture of proteins and sugars. The Hirudinea have been reported, more than a century ago [1], to molt, a process which is less conspicuous in other annelid worms, although it is regularly observed in species like the polychaete *Nereis diversicolor*.

We undertook a few years ago a study on the possible involvement of ecdysteroids, the molting hormone of arthropods, in the control of the epidermal cell cycles in annelid worms. In 1981, we reported the presence of ecdysteroids in the oligochaetes *Dendrobaena subrubicunda* and *Eisenia foetida* and the hirudinean *Erpobdella octoculata* and *Glossossiphonia complanata* [2]. Our results were based on the combination of chromatographical separations (silicagel thin-layer chromatography and reverse-phase high pressure liquid chromatography) and radioimmunoassay [3] and showed the presence of ecdysone, 20-hydroxyecdysone plus other as yet unidentified immunoreactive molecules. These data were corroborated in 1983 by our report [4] on the presence in the leech *Hirudo medicinalis* of 20-hydroxyecdysone, identified on the basis of combined gas chromatography-mass spectrometry after trimethylsilylation.

Of major interest was the observation that in the latter species, the whole body titer of ecdysteroids rised to peak values (in the range of 1 nmol per g body weight) at the time of apolysis of the old cuticle and synthesis of a new cuticle. Leeches having exuviated showed lower titers which were maintained during most of the intermolt period. These fluctuations obviously parallel those classically observed in arthropods during molt-intermolt cycles and suggest that this process is controlled in leeches by mechanisms which are basically similar to those of arthropods. These data have been published recently [4] and will, therefore, not be repeated here. Our investigations in *Hirudo* are now being extended to the elucidation of the biosynthetic origin and the mode of action of ecdysteroids in this species.

Concomitantly to these collaborative studies on leeches between Dr. Sauber's group in Luxembourg and the Strasbourg laboratory, a collaboration was initiated with the laboratory of Dr. Porchet to see whether ecdysteroids were implicated in the control of development and reproduction in *Perinereis cultrifera*, a marine polychaete which was introduced into research on endocrinology by Durchon in 1952. *Perinereis cultri-*

1 LA CNRS 148 Endocrinologie Comparée des Invertébrés, Laboratoire de Biologie Animale de l'Université des Sciences et Techniques de Lille, 59 Villeneuve d'Ascq, France

2 Laboratoire de Biologie des Cours Universitaires, Place Auguste Laurent, Luxembourg

3 ERA CNRS 118 Endocrinologie et Immunologie des Insectes, Laboratoire de Biologie Générale de l'Université Louis Pasteur, 12 rue de l'Université, 67 Strasbourg, France

Biosynthesis, Metabolism and Mode of Action of Invertebrate Hormones
(ed. by J. Hoffmann and M. Porchet)

fera is typically monotelic and shows one single reproductive cycle extending over 1 or 2 years and ending in a profound metamorphosis ("epitoky") which is concomitant to the final stages of maturation of the gametes. Durchon and co-workers established that gametogenesis and epitoky are under an inhibitory control by a cerebral neurohormone, called nereidin (see Durchon, this Vol., and [5, 6]. No evidence has been obtained so far for the existence of a gonadotropic hormone in this species.

A first series of RIA measurements [3] in various parts of the body of female *Perinereis* at different stages of oogenesis showed the presence of "ecdysteroids" in coelomic fluid, coelomocytes, muscles, and integuments and oocytes (Table 1). Interestingly, a high titer of immunoreactivity was evidenced in oocytes at the developmental stage of 130 μm, which corresponds to the end of the vitellogenic period and the onset of synthesis of glycoproteins characteristic for the cortical alveoli (review in [7]). Physicochemical characterization of these ecdysteroids is underway; the identity of ecdysone and 20-hydroxyecdysone was assessed recently by chromatographical characteristics in silicagel thin-layer chromatography and reverse phase high pressure liquid chromatography. For reasons essentially linked to the seasonal cycles of reproduction, our information on ecdysteroid levels in the late phases of gametogenesis and at the time of epitoky is scanty and will be completed in the near future.

Table 1. RIA measurements of ecdysteroid immunoreactivity[a] in *Perinereis cultrifera* at various stages of oogenesis

	Stages of oogenesis[b]			
Tissues	90 μm	130 μm	170 μm	200 μm
Oocytes	19.3	1,186.5	23.1	14.5
Coelomocytes	11.3	10.9	14.3	29.4
Muscle + epidermis fragments	13.3	19	22	13.4
Coelomic fluid	7	54.3	15.6	13.5

[a] In pmol ecdysone-equivalents per g of tissue or ml of fluid
[b] Stages are defined in relation to the diameter of the maturing oocytes (for details, see [7])

We have injected 20-hydroxyecdysone (2 nmol per worm every 2 days during 20 days) into the body cavities of females containing submature oocytes and monitored a spectacular acceleration of oocyte maturation which was accomplished in 3 weeks as compared to 3 months in controls (injection of saline) (Fig. 1). 20-Hydroxyecdysone shows also an effect on male gametogenesis: in the classical bioassay for the cerebral inhibitor hormone nereidin [8, 9] a brain-containing prostomium is cultured in the presence of a parapode from a young male containing spermatogonia. The nereidin secreted by the brain inhibits under these conditions spermatogenesis (which is scored as a positive result for the presence of nereidin in the bioassay). We have now observed that the addition of 10 *μM* 20-hydroxyecdysone to the culture medium opposes the inhibition by nereidin and allows complete spermatogenesis within 2 weeks, while in the absence of the ecdysteroid, spermatogenesis remains arrested under these experimental conditions.

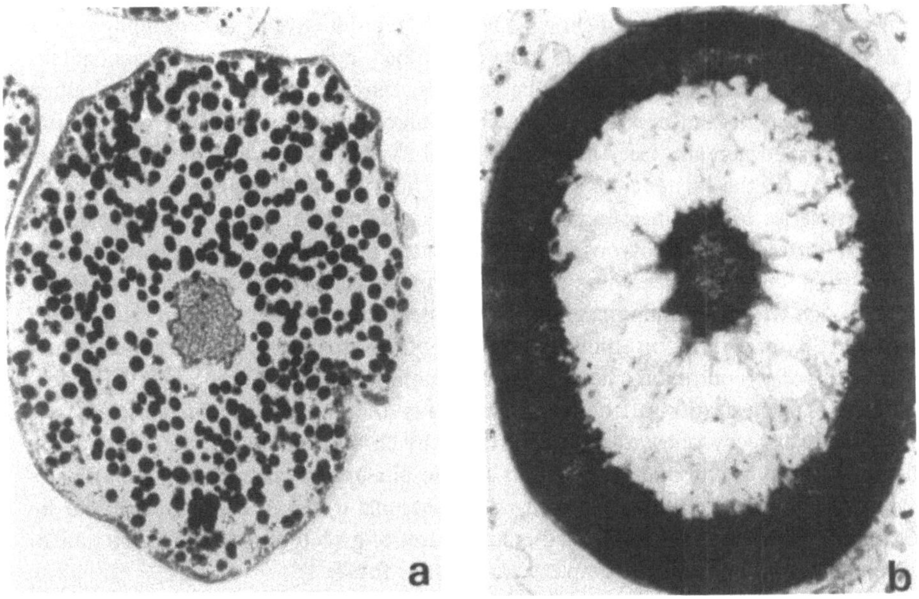

Fig. 1a,b. Physiological action of 20-hydroxyecdysone on maturation of oocytes in *Perinereis cultrifera*. a Control. Structure of submature oocyte (200 μm) after injection of seawater into coelomic cavity. No maturation is induced and the cytoplasmic inclusions [yolk (*black*) and cortical alveoli] remain dispersed in oocyte. **b** Injection of 20-hydroxyecdysone (2 nmol per worm every 2 days during 20 days). The characteristic structure of mature oocytes with segregation between ooplasmic inclusions have been induced by the hormone treatment. Note that the cortical alveoli have migrated towards the periphery and that the vitellus is in immediate vicinity of the nucleus

We are fully aware that these results are preliminary. However, a meeting like this Symposium is not restricted to the presentation of firm published data. The message which we wish to convey in this short note is: (1) annelid worms contain fluctuating concentrations of ecdysteroids; the presence of 20-hydroxyecdysone has been demonstrated in at least one case *(Hirudo)* by physicochemical methods; (2) the data available favor the idea that ecdysteroids are used as hormonal messengers in annelids; a role in gametogenesis and cuticulogenesis is proposed as a working hypothesis.

Finally, this study also shows that *Nereis* and *Hirudo* are excellent experimental animals for physiological studies on ecdysteroids in worms, which allow research beyond the level of titration or isolation of these molecules.

References

1. Leydig F (1849) Z Wiss Zool 1:103–134
2. Sauber F, Kappler C, Hoffmann JA (1981) Commun 5th Symp Eur Ecdysone, Berne
3. De Reggi ML, Hirn M, Delaage M (1975) Biochem Biophys Res Commun 66:1307–1315
4. Sauber F, Reuland M, Berchtold JP, Hertu C, Tsoupras G, Luu B, Moritz ME, Hoffmann JA (1983) CR Acad Sci 296:413–418
5. Porchet M, Spik G (1978) Comp Biochem Physiol 59B:175–181
6. Dhainaut A, Porchet M (1977) Biol Cell 28:233–240
7. Porchet M (1984) Fortschr Zool (in press)
8. Durchon M, Porchet M (1970) CR Acad Sci 270:1689–1691
9. Durchon M, Porchet M (1971) Gen Comp Endocrinol 16:555–565

Part III Juvenile Hormones

Introduction

L.I. GILBERT[1]

General Comments

The presence of steroid and peptide hormones in insects is certainly not surprising to students of evolution since all eukaryotic organisms probably use one or both classes of these molecules to control aspects of growth, reproduction, and homeostasis. Indeed, it is very likely that the steroid and peptide hormones of arthropods exert their effects in a manner similar or identical to the mechanisms elucidated for vertebrates, i.e., nuclear receptors for steroid hormones and membrane receptors and second messengers for peptide hormones. One may conclude that during the course of evolution, it has not been the regulatory molecules nor their basic mode of action that have changed drastically, but rather the uses for which they have been employed.

In contrast to the above concerning steroid hormones (ecdysteroids) and the various peptide hormones utilized by insects, the juvenile hormones (JH) are unique humoral agents, sesquiterpenes, with critical roles during postembryonic development (and perhaps embryonic development as well), and serve to regulate a variety of reproductive functions in the adult. There is a small, but dedicated group of investigators working on this capricious, but intriguing hormone, most of these individuals having contributed to the chapters in this section, or to the roundtable discussion.

There has been considerable progress since Janet, at the end of the nineteenth century, conferred the name corpus allatum (CA) on that now well-known structure attached to the insect brain, and Nabert indicated in the early twentieth century that the CA was probably an endocrine gland. It was not until the early and mid-1930s that V.B. Wigglesworth demonstrated by a series of decapitation studies on *Rhodnius* larvae that the CA was the source of an "inhibitory hormone" that controlled the change in form characteristic of metamorphosis. Confirmation of these studies occurred just a few years later with Bounhiol's elegant microsurgical analysis of larval-pupal metamorphosis in *Bombyx*, and then by many others in a variety of insects over the years. On the basis of parabiosis studies a few years later, Wigglesworth showed that this inhibitory hormone actually stimulated the synthesis of larval structures and proposed the term juvenile hormone (JH). However, since we are now aware of the multiple gonadotropic roles of JH in many adult insects, the term juvenile hormone may not be as appropriate as once thought. Over the years we have learned that JH controls a large variety of

1 Department of Biology, University of North Carolina at Chapel Hill, Chapel Hill, NC 27514, USA

Biosynthesis, Metabolism and Mode of Action of Invertebrate Hormones
(ed. by J. Hoffmann and M. Porchet)

physiological processes (e.g., diapause, behavior, reproductive processes, etc.) and exists in four homologous forms with JH III having the widest distribution among the Insecta.

The original pioneering experiments on the role of the CA, and therefore JH, were conducted by Wigglesworth, Piepho, Bounhiol, Pfeiffer, etc., while the elucidation of the chemical nature of JH is an outgrowth of the original studies of Williams in the mid-1950s followed by the chemical analyses emanating from the laboratories of Röller, Bowers, Meyer, and more recently, Schooley. Over the past 10 years increasing attention has been given to: the means by which the JH titer in the insect's hemolymph is regulated, entailing the quantification of JH titers in both larvae and adult females (Strambi, Schooley, Mauchamp); studies on JH synthesis using isolated CA (Pratt, Tobe, Lanzrein); investigations of JH degradation (DeKort, Goodman, Hammock); and experimental analyses of the neural and humoral control of CA activity (Tobe). At the same time other investigators have examined the mechanisms of action of JH utilizing information gained from investigations of the action of vertebrate steroid hormones, i.e., cytosol and nuclear receptors, etc. (Koeppe, Engelmann, Willis, Wyatt). Progress in the latter field has not been as rapid as we might wish, but the highly lipophilic molecule, JH, is difficult to handle in a quantitative manner and only a handful of laboratories are examining the effects of JH at the level of the cell or molecule. Indeed, the evidence gathered thus far suggests that JH may act at the nuclear and plasmalemma levels, depending on the system examined, and future studies could very well uncover multivalent effects in the same tissue. One possible problem in these studies is, with few exceptions, the use of unphysiological substances, i.e., JH mimics (insect growth regulators), at concentrations that are orders of magnitude greater than the circulating levels of the natural JH. These substances persist in the organism for abnormally long periods because of the high dose administered and because they are not prone to rapid degradation by endogenous enzymes, e.g., esterases. One of the critical problems facing students of JH in the future is to distinguish physiological from pharmacological effects and to extrapolate data from in vitro studies to the growth, development, and reproduction of the organism: a holistic appraisal will be required if the data are to be meaningful. One approach being taken by only a few laboratories and which could help alleviate the real possibility of spending years pursuing an artifactual situation is to monitor the in vitro regimen by conducting analogous studies in situ at regular intervals. After all, one of the subjects we understand least in the field of insect endocrinology is feedback effects of hormones on the various glands, i.e., endocrine integration, a phenomenon totally absent from almost all in vitro investigations.

This symposium dealt with some of the above noted material and the data presented will no doubt help to resolve some basic questions, albeit not in the near future. Before those papers are summarized, however, it should be mentioned that a roundtable discussion was held during this segment of the symposium and dealt almost exclusively with the measurements of JH in hemolymph and tissue extracts by a variety of methods. This is a most important subject since simulation of the in vivo situation in vitro requires knowledge of both the absolute JH titer and the nature of the JH homolog(s) present. However, even if these criteria are met, JH may be compartmentalized (see S.S. Tobe et al., this part) so that the hemolymph titer may not always reflect the concentration of JH at the target cell. In essence the roundtable participants concluded that physical-chemical analysis (i.e., mass spectrometry) was the ideal approach, but that since neither the equipment nor expertise are generally available, radioimmunoassay (RIA) could be of great value if used with precision and care. RIA is of particular value when studying CA regulation in vitro, as is the radiochemical assay of G.E. Pratt and S.S. Tobe when followed by HPLC in those situations where mixed homologs are secreted, but both tissue and hemolymph extracts must be purified in order to exclude non-JH entities that bind to the antibody with relatively high affinity. It was suggested that sample JH extracts containing known amounts of JH (1-4 homologs) be prepared by J. Koeppe and sent to several laboratories to determine the degree of correlation of data between different laboratories and different techniques. Unfortunately, the data are not yet available, although the goal of the project remains a noble one.

Comments on Section III

A recently developed RIA was described by C. Strambi in which the antibody recognizes the diol form of JH, the major advantage over the RIA of Baehr et al. being that after conversion to the diol the diols are easier to separate from lipid extracts than is intact JH. Indeed, the diols can be separated by TLC before RIA. Use of the RIA, combined in some cases with HPLC and mass spectrometry, revealed the presence of only JH III in *Labidura* and *Bombus*, and in the latter case, caste specific differences in JH titer. D.A. Schooley and F.C. Baker utilized mass spectrometry exclusively to examine the JH content of a wide variety of insects and showed that only in the Lepidoptera could JH 0, I, and II be identified. A few years ago, these investigators reported the presence of JH 0 in *Manduca* embryos and it was assumed that this homolog may have a unique function during embryogenesis. They now report the presence of JH 0 in male *H. cecropia* moths and although JH 0 comprises only 1%-2% of the total JH, JH 0 can no longer be considered a unique "embryonic hormone". JH III was the sole homolog in a wide variety of orders, including Diptera, Orthoptera, Dictyoptera, Hemiptera, Coleoptera, etc. Even in *Nauphoeta* where JH I and II were previously reported to be present, it is now agreed that JH III is the sole JH present. One rather amazing finding was the huge quantities of JH III present in the hemolymph of *Diploptera* adult females, up to 2 μg/ml. This is reflected in the exceedingly high activity of the CA in vitro reported by Tobe and his laboratory. Why moths and butterflies are unique in employing a diversity of JHs is conjectural at present, although there are exceptions even to this tenet in that larvae of *T.ni* yield JH II only.

As is well known, the JH titer is a resultant of synthesis by the CA and degradative activities, mainly by esterolytic cleavage. Data on *Leptinotarsa* (C.A.D. de Kort and M.A. Kahn) revealed the presence of a high molecular weight hemolymph protein that binds JH III with high affinity, presumably to protect the natural (10 R) enantiomer against the action of hemolymph general esterases. This is in contrast to the much smaller (28,000), high affinity binding protein known for several years to exist in the hemolymph of *Manduca sexta* (see Goodman, this part). W.G. Goodman's data revealed a large increase in the hemolymph titer of JH binding protein in female adult *Manduca*. This finding is as yet unexplained, but perhaps it may be involved with the transport of JH into the eggs and ovaries. One of the presumed reasons for a binding protein requirement is protection of JH from general esterase hydrolysis of the methyl ester, but when the JH titer must be lowered, as before commitment and metamorphosis, specific JH esterases are synthesized that degrade JH even in the presence of binding protein. B.D. Hammock and co-workers have studied the kinetics and regulation of these so-called JH esterases in a variety of Lepidoptera. Both nutrition and putative head factors appear to be important in JH esterase regulation in *T. ni* last instar larvae at the time of the first head critical period (pre-wandering), while JH may elicit JH esterase activity at the end of the instar (post-wandering). The presumed function of JH binding proteins and esterases are the subject of the contributions of C.A.D. de Kort, W.G. Goodman, and B.D. Hammock.

Although it has been known for more than 20 years that embryos can synthesize JH, it is only within the past few years that embryonic JH have been chemically characterized. The contribution by B. Lanzrein demonstrated that the embryonic CA of *Nauphoeta* synthesize JH III in vitro. To my knowledge this is the first direct demonstration that the embryonic CA is the source of embryonic JH. It is of interest that methyl farnesenate is also a product of the embryonic CA and may play a humoral role during embryogenesis. As in *Diploptera* female adults, the hemolymph of *Nauphoeta* embryos contain large quantities of JH III (in the μg/ml range).

Although a great deal of attention has been given to hemolymph titer regulation, JH in the hemolymph must make its way to target tissues (e.g., epidermis in larval development and fat body, ovaries, and accessory glands in female adults) and there elicit specific effects. The first event involves receptor recognition and putative JH receptors have been investigated in the fat body and ovaries of adult female *Leucophaea* by F. Engelmann and J.K. Koeppe, respectively. F. Engelmann's studies over several decades have shown that JH stimulates vitellogenin (Vg) mRNA synthesis in the fat body and that JH may also affect Vg translocation to the hemolymph. High affinity (K_d = 10^{-9} *M*) binding molecules appear to be present in both cytosol and nuclear preparations from fat body, whereas slightly lower affinity binding molecules (K_d = 10^{-8} *M*) were extracted from target tissues, nontarget tissues, and inanimate objects, such as urate crystals. The physiological significance of these lower affinity binders is obscure. Studying the ovary of *Leucophaea* and utilizing a photoaffinity label, J.K. Koeppe and co-workers have shown the presence of a high affinity, high molecular weight binding protein in this target tissue which has a higher affinity for JH III than JH I or JH 0. This is the first exception to the polarity rule developed on the basis of investigations by Goodman on the *Manduca* hemolymph binding protein. In the latter case, the least polar JH homolog binds with the greatest affinity (JH 0 > JH I > JH II > JH III), while with *Leucophaea* it is the converse, indicating different interactions between the hormone and binding protein in the two species. There is some data, but certainly not conclusive evidence, that these cellular binding molecules differ from those in the hemolymph. This is a critical distinction to be made if one is to be sure that putative receptors are being investigated rather than a hemolymph contaminant. However, utilizing the photoaffinity label in which binding is covalent, it should be possible to purify the cellular binding protein and utilize immunological techniques to distinguish cellular binding proteins from extracellular binding proteins. It must be noted that what Engelmann considers low affinity binding (K_d = 10^{-8} *M*) is considered to be high affinity binding by J.K. Koeppe for ovarian putative receptors and by G.R. Wyatt et al. for locust fat body putative receptors. The methodology used by these investigators must be critically compared.

Once it enters the target cell and is presumably bound to a receptor, the JH-receptor complex is postulated to act at the level of the gene. The studies of J.H. Willis and D.L. Cox on saturniid silkworms were designed to determine whether stage-specific gene sets are responsible for the sequential polymorphism occurring in the cuticle during metamorphosis. By examining the acidic cuticle proteins from different stages and two species of *Hyalophora* as well as from different body regions, they conclude that different, but homologous genes, may be involved in each species. Further, it appears that the body region influences gene expression more profoundly than developmental stage. This thoughtful and imaginative hypothesis provides the basis for further experimental analysis and may ultimately alter our concept of how JH exerts its effects.

Returning to JH as a gonadotropic agent, G.R. Wyatt and his colleagues summarized their studies of JH (analog) action on the fat body of adult female locusts. They report the first successful direct induction of Vg synthesis in fat body in vitro by a JH analog; until this report an indirect action was a possibility. However, until the experiment can be conducted successfully with physiological concentrations of the natural JH, the possibility of a pharmacological effect remains. G.R. Wyatt's group has also isolated two Vg genes in genomic DNA clones to be used as hybridization probes in future studies and they report the assay of fat body mRNA by hybridization. Of interest is the finding that primary induction of $mRNA_{Vg}$ by methoprene requires a long lag; in the case of secondary induction the lag period is markedly reduced. This is almost exactly analogous to the situation in the chick oviduct where ovalbumin synthesis is induced by progesterone. Binding proteins for JH I (not the natural homolog) are present in fat body cytosol, the highest affinity being $K_d = 1.7 \times 10^{-8}$ *M*. The cellular binding protein picture remains a problem since F. Engelmann's data (see above) suggest that binding with a K_d of 10^{-8} *M* in *Leucophaea* may be nonspecific, and in the case of the locust it is probable that the natural JH (JH III) has even less affinity for this cytosol binding molecule, i.e., $K_d > 10^{-8}$ *M*.

This was a well-organized and exciting symposium to which the JH section contributed considerable discussion and debate. JH is a truly insect-specific molecule in a sense, in contrast to peptides and ecdysteroids and, therefore, the conceptural framework for metabolic and mode of action studies cannot be borrowed in toto from the vertebrate literature. Progress in this field is not as rapid as one would hope, but the chapters in this section should convince the reader that there are few molecules to study that are as exhilarating in their prospects for flamboyant discoveries.

III.1 Radioimmunoassays of Juvenile Hormones. State of the Methods and Recent Data on Validation

C. STRAMBI[1], A. STRAMBI[1], M. de REGGI[2], and M. DELAAGE[2]

1. Introduction

The quantification of juvenile hormones is a major aim of invertebrate endocrinology. For a long time, the "galleria wax test" was the only way of quantifying juvenile hormone activity [1]. More recently, large efforts have been devoted to the development of mass spectography [2, 3] and radioimmunoassays [4, 5] (RIA). However, all approaches encountered difficulties because of the hydrophobic properties of JHs. It is only recently that reliable results were obtained. RIA is now applicable to any kind of biological material and can be used for routine assays; it is as accurate as ecdysteroids RIAs, provided that validated protocols are strictly followed.

2. Reagents: Antibodies and Tracers

Juvenile hormones are small molecules which become immunogenic only after conjugation to a protein. Similarly, the labelling by ^{125}iodine requires the conjugation of juvenile hormone to a tyrosine or an histidine residue. Both derivatizations have to be carried out on the same carbon, otherwise the iodinated derivative is not properly recognized by the antibodies. Juvenile hormones have two ways of conjugation, each of which defines a completely different radioimmunological system:

a) Carbon 1. After the early attempt by Lauer et al. [6], this particular method was subsequently explored with success by Baehr et al. [4, 7]. The methyl ester function was hydrolyzed under mild alkaline conditions. Then the carboxylic function was coupled to lysine residues on albumin using the classical mixed anhydride method. High titer (50% of the radioligand is bound by a serum diluted 1:100,000); antibodies were raised to JH I, JH II, and JH III immunogens, respectively. These are partially specific of the respective haptens so that they can potentially be used to quantify JH I and JH III (JH II cross-reactivity is 37% with JH I antiserum and 0.9% with JH III antiserum [4, 7]. The iodinated derivative was made by coupling the carboxylic function 1 to histamine, then iodinated. The radiolabelled derivative was purified by TLC.

1 CNRS, INP 6, BP 71, 13277 Marseille Cedex 9, France
2 IMMUNOTECH, Luminy, Case 915, 13288 Marseille Cedex 9, France

Biosynthesis, Metabolism and Mode of Action of Invertebrate Hormones
(ed. by J. Hoffmann and M. Porchet)

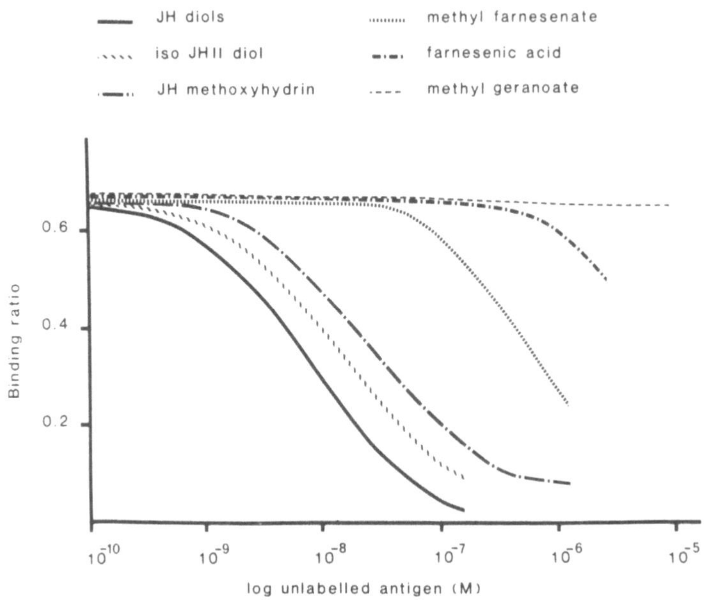

Fig. 1. Displacement curves of the radioligand by different compounds

b) Carbons 10–11. The epoxy function on these carbons provided another opportunity which was explored by our group [5]. After opening the epoxy under mild acidic conditions we synthesized a hemisuccinate which was then coupled to serum albumin.

The radioligand was obtained by coupling glycyl-tyrosine to JH I-diol-hemisuccinate, then iodinated. The presence of a carboxylic function on the iodinated derivatives has two advantages: better solubility in water and chromatographic behavior that is adjustable with pH. That is, at pH 4.5 the iodinated derivative readily separates from iodine and other contaminants on G25 Sephadex. The juvenile hormone is not damaged during iodination nor is the iodinated derivative damaged when it is diluted and frozen.

Antibodies raised against JHI diol exhibited a high titer (the serum diluted 1:2,000,000 binds 50% of the ligand) and identical affinity for JH I, JH II, JH III diols; JH 0 diol was 60-fold less immunoreactive (M. Cavallin, pers. comm.). Predictably, iso-JH II ethyl ester, which was closest to JH diols, was more immunoreactive than JH III methoxyhydrin. Cross-reactivity is less with molecules like farnesenic acid, which exhibit an acidic terminal group, or with shorter molecules like methyl-geranoate (Fig. 1).

3. Treatment of Biological Samples

Juvenile hormones are by far the most hydrophobic substances for which an RIA has been developed. The lack of appreciation and misunderstanding concerning this point is primarily responsible for the difficulties encountered by the first to use this method. It is essential to note that when dissolved in water, the hormones tend to stick to the vials, especially to plastic ones. Thus, only glass vials must be used. A less well-known, but equally important phenomenon is the adsorption of the hormones at the air-water interface (where up to 40% of the dissolved material could be trapped). Moreover, hormones from biological origin are extracted by solvents, together with other lipids. In aqueous solution, the latter are organized as micelles or as fatty microglobules and they can trap the hormones as well as the iodinated derivatives, preventing them from binding to the antibodies. Accordingly, juvenile hormone has to be completely separated from other lipids at the final stage of the assay. Again we have to consider two completely different cases:

1. Radioimmunoassay of unconverted juvenile hormones [4]. The chromatographic behavior of juvenile hormones is very similar to that of most other lipids (see Fig. 2). Thus, the purification of biological extracts cannot be accomplished unless one uses HPLC combined or not combined with other techniques. There is one notable exception, namely, in a culture medium where the level of free lipids is low enough to allow the assay of juvenile hormone to be made without prior purification [8].
2. In our radioimmunoassay, juvenile hormone is assayed after its conversion into diol. This procedure has two major advantages:
   - diols are more water soluble than JHs which facilitate the final recovery and the incubation with antibody;
   - the epoxy-diol conversion changes the chromatographic behavior of the molecule, thus, providing a simple and elegant separation of juvenile hormones from other lipids.

The general protocol presented in Table 1 gives access to JH I, II, and III determination for any biological material. This protocol is close to the one published in 1981 [5] except that an HPLC has been added. This step is intended to produce a better separation of the three diols from each other.

It is worth noting that the TLC step separates diols from all immunoreactive contaminants. Indeed HPLC eluates of hemolymph extracts did not exhibit any immunoreactivity except in fractions having the same retention time as JH-diol standards (Figs. 3 and 4). It should be emphasized that with the exception of Granger and Goodman's observations [9], the compounds extracted by organic solvents from TLC plates do not show any immunoreactivity in our test.

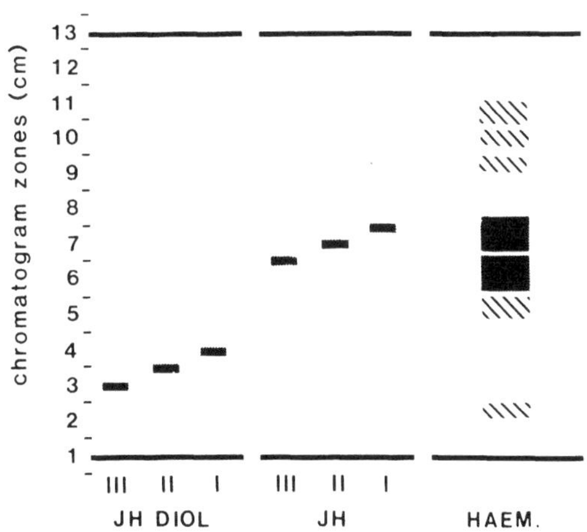

Fig. 2. TLC separation of JH and JH-diol standards and of a biological sample, displaying UV absorption. The biological sample is a haemolymph extract from *Acheta domesticus*, treated as described in Table 1. Note that the JH-diol areas are almost free of contaminants, most of which comigrate with JHs

Table 1. Treatment of biological samples (the quantities of biological material given as an example can be reduced)

1. Homogenize 100 mg of fresh tissue or 100 μl of biological fluid in 750 μl of methanol/water (70:30, v/v)
2. Add 1.5 ml of hexane; vortex for 1 min and centrifuge (1,000 × g; 2 min)
3. Save the upper phase; add 1 ml of hexane to the methanol/water, vortex and centrifuge
4. Save the lower phase for ecdysteroid assay
5. Pool the hexane extracts; add 1 ml of 0.01 N NaOH; vortex and centrifuge. Discard the lower phase
6. Repeat step 5 with 1 ml of water
7. Load the extract on a silicagel column (1 ml total volume) equilibrated in hexane. Discard the eluate
8. Rinse with 2 ml of hexane and then with 0.6 ml of dioxane; discard the eluates
9. Elute with 1.4 ml of dioxane; keep the eluate
10. Add 160 μl of 0.5 N H_2SO_4. Evaporate excess hexane; leave overnight (15 h) at 40 °C
11. Add 1.4 ml hexane and 1 ml of 0.01 N NaOH; vortex and centrifuge
12. Save the upper phase and add another 1.4 ml of hexane to the lower phase; vortex and centrifuge
13. Load the pooled hexane phases onto a silicagel plate
14. Develop with heptane/dioxane (70:30, v/v) up to 12 cm
15. Cut the areas corresponding to JH diols (C_{18} diol, R_f: 0.40; C_{17} diol, R_f: 0.35; C_{16} diol, R_f: 0.29)
16. Elute each area with 1 ml of acetone

Alternatively:

17a. Add 0.4 ml of citrate buffer (0.01 *M*, pH 6.2); lyophilize; add 0.4 ml of water to the lyophylisate and use this solution for the RIA

17b. Evaporate the acetone extract. Dissolve the residue in acetonitrile/water (50:50, v/v). Reverse-phase Bondapak C_{18} column (Waters). Chromatography performed with a Waters apparatus equipped with two M 6000 pumps and an M 720 system control: elution achieved with a linear gradient of acetonitrile/water 45% to 65% within 30 mn, and then a step gradient with 10 mn pure acetonitrile at flow rate of 1 ml/mn. To determine the RIA activity of HPLC fractions proceed as described in Step 17a

The protocol can be simplified in many cases, for example:

1. The HPLC step can be omitted when the biological material contain only one kind of JH.
2. Purification is no longer necessary when corpora allata (that produce only JH) are cultured in an artificial medium. In this case, conversion into diol is achieved merely by acidification with 1% acetic acid (G.E. Pratt, pers. comm.).

When the biological material is especially rich in lipids (eggs, whole body extracts) it is preferable to convert the JH into diol just after the first hexane extraction and then to do the chromatography on a silicagel column under slightly modified conditions. Since the diols contain fewer contaminants, the loading on TLC [10] is facilitated.

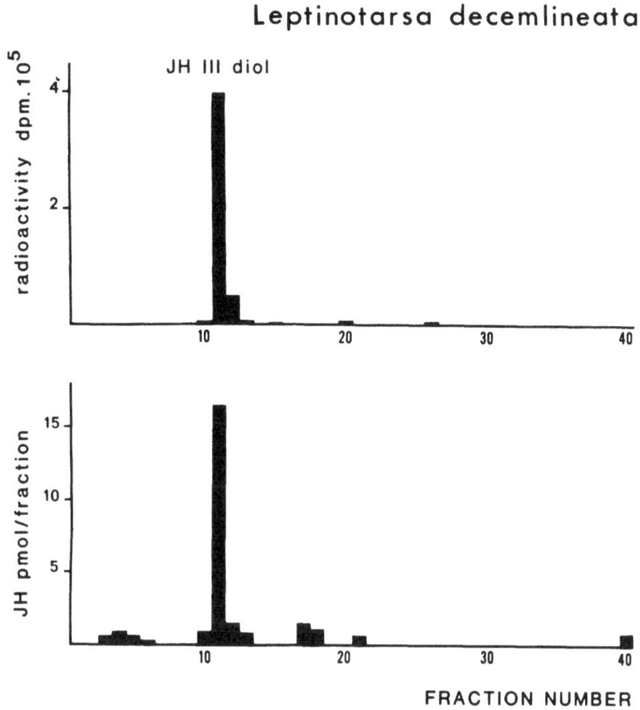

Fig. 3. RIA activity of HPLC fractions (*bottom*) and HPLC profile of [^{3}H] JH III diol standard (*top*). The biological sample is a haemolymph extract from *Leptinotarsa*. Comparison of the two profiles shows that the only immunoreactive compound comigrates with JH III diol standard

4. Validation of Juvenile Hormone Radioimmunoassay

The validation of JH radioimmunoassay depends on two conditions:

- the evaluation and the reproducibility of the overall yield of extraction,
- the identity of standards and extracted hormone when presented to antibody.

The overall yield is easily calculated by adding tritiated hormone in a biological sample [10]. The overall yield does not depend on the biological specimen since it reflects primarily the partition coefficient of JH and JH-diol between solvents and the epoxy-diol conversion yield.

The diol conversion is a crucial step that needs special comment. According to Mumby and Hammock [11], the presence of salts can partially inhibit conversion in a tetrahydrofuran acid medium. We have shown that this also holds in the case of dioxan extracts, when, after removing the protein pellets, diol conversion was achieved by adding sulfuric acid to the supernatant (Fig. 5). However, in our protocol the conversion is performed after hexane extraction, i.e., without any salts, so that the conversion yield is stable in the range of 80%–85% (Fig. 5). The microscale conversion into diol is reproducible provided that careful attention is paid to the percentage of water and of

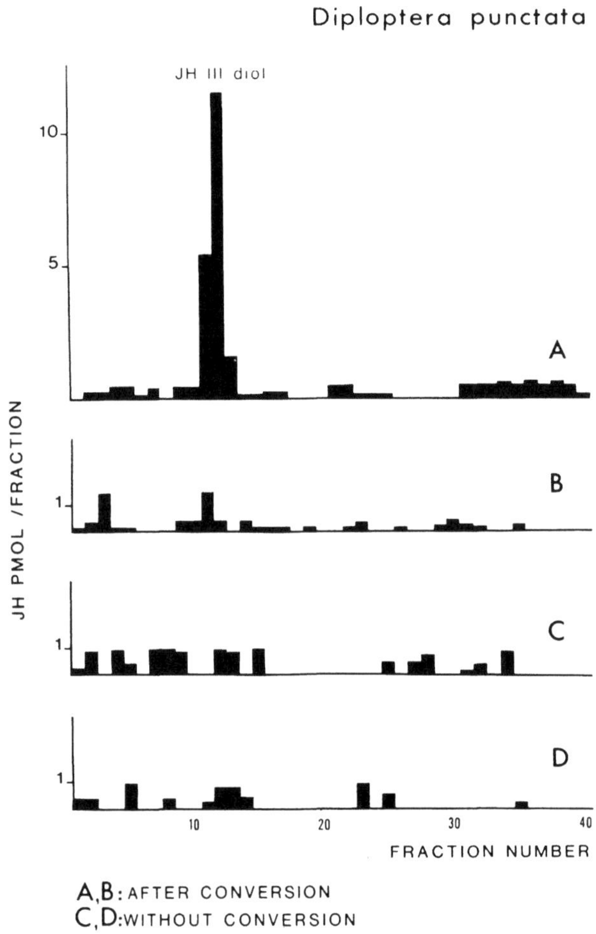

Fig. 4. HPLC analysis of TLC eluates from haemolymph of *Diploptera punctata*. The eluates of TLC areas corresponding to JH III diol (*A*) or to JH I diol (*B*) were loaded on a HPLC column (see Table 1) and the RIA activity of the 40 fractions was determined. *C* and *D* are identical to *A* and *B* respectively, except that diol conversion is omitted (in step 10, H_2SO_4 is replaced by water). The comparison of *A* and *B* shows that only the TLC area corresponding to JH III diol contained immunoreactive product. HPLC separation in *A* demonstrates that the only immunoreactive product in this area has the same retention time as JH III diol. *C* and *D* show that in the absence of diol derivatization no immunoreactive product (namely, no native diol) could be detected either in JH I-diol or in JH III-diol areas

acid (7.5% of 0.5 N sulfuric acid). Nevertheless, a control of the conversion is recommended for each new biological sample.

The second question concerns the immunoreactivity of the natural isomer versus that of synthetic JH. The natural isomer is 2-trans, 6-trans, and 10 R-11 S; but some synthetic hormones can be a racemic mixture of 10 R-11 S and 10 S-11 R. Some of us have shown that a racemic hapten (abscissic acid, a plant hormone) elicited anti-R and anti-S antibodies [12]. On the other hand, acidic hydrolysis retains the absolute

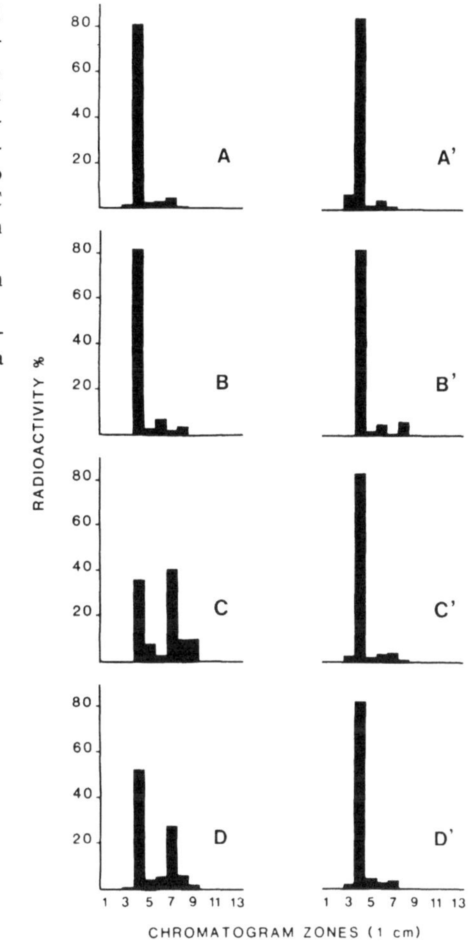

Fig. 5. Efficiency of diol conversion. [^{3}H] JH I was used as control or added to biological samples. *A,A'*: controls; *B,B'*: *Polistes* wasps pupae, whole bodies; *C,C'*: Tissue culture medium (199 Gibco); *D,D'*: *Acheta domesticus* haemolymph; *A–D*: extraction by dioxane; *A'–D'*: extraction by hexane/methanol (according to Table 1, steps 1 to 6). After conversion and TLC (Table 1, steps 9 to 14), the radioactivity pattern is determined.
JH I migration area: *zone 7*; JH I diol migration area: *zone 4*.
In several cases (examples *C* and *D*) the conversion into diol is partially prevented unless a hexane/methanol extraction is performed

configuration around C_{10} and C_{11} [13], thus, the JH diols from biological extracts have the same configuration as the natural JHs. In any case, the problem is less critical for our radioimmunoassay since the isomer concerns an area relatively less well recognized by our antibodies. Moreover, a displacement curve built from dilution of a highly active biological extract (from *Bombyx mori* embryos) has shown a strong parallelism with the JH-diol standard curve (Gharib and de Reggi, pers. comm.).

5. Conclusion

A great deal of data obtained from various insect species in all stages of development is required to assess the in vivo changes in JH levels. The RIA described in this report is a very convenient method for such routine investigations, in that it provides accurate and reliable results whatever biological material is being assayed (the method can be used even for fatty crude extracts). The user should be warned, however, that scrupulous attention must be paid to each step of the protocol, so that the hormones are carefully purified and that all the products are properly handled.

Another question is the comparison of results obtained by different techniques (biological tests, GC–MS, RIAs). Because of the large variety of species and physiological situations that are studied, direct comparisons can be done only with difficulty.

Nevertheless, in some species *(Diploptera, Leptinotarsa)* where combined gas chromatography and mass spectrometry revealed only the presence of JH III [14, 15], our RIA analysis of HPLC fractions also demonstrates that only JH III is present (Fig. 5). Moreover, quantitative data obtained from *Leptinotarsa decemlineata* by gas chromatography-mass spectrometry and RIA agree reasonably well [16].

Acknowledgments. We are grateful to Drs. G.E. Pratt and D.A. Schooley, to Drs. C.A.D. de Kort and S.S. Tobe who provided us with JH analogs and insect haemolymph respectively.

References

1. de Wilde J, Staal GB, de Kort CAD, de Loof A, Baard G (1968) K Nederl Acad Wet 71:321 to 326
2. Bergot BJ, Ratcliff M, Schooley DA (1981) J Chromatogr 204:231–244
3. Mauchamp B, Lafont R, Jourdain D (1979) J Insect Physiol 25:545–550
4. Baehr JC, Pradelles P, Lebreux C, Cassier P, Dray F (1976) FEBS Lett 69:123–128
5. Strambi C, Strambi A, de Reggi ML, Hirn MH, Delaage MA (1981) Eur J Biochem 118:401 to 406
6. Lauer RC, Salomon PH, Nakanishi K, Erlanger BP (1974) Experientia 30:558–560
7. Baehr JC, Pradelles P, Dray F (1979) Ann Biol Anim Biochem Biophys 19:1827–1836
8. Granger NA, Niemiec SM, Gilbert LI, Bollenbacher WE (1982) J Insect Physiol 28:385–391
9. Granger NA, Goodman WG (1983) Insect Biochem 13:333–340
10. Gharib B, de Reggi M, Connat JL, Chaix JC (1983) FEBS Lett 160:119–123
11. Mumby SM, Hammock BD (1979) J Agric Food Chem 27:1223–1228
12. Lepage E, Duval D, Bulard J, Delaage M (1984) J Immunol Methods (in press)
13. Peter MB, Stupp HP, Lentes KU (1983) 2nd Int CNRS Symp Biosynth Metab Mode Action Invertebr Horm, Strasbourg, 29–8/2–9/1983
14. Bergot BJ, Schooley DA, de Kort CAD (1981) Experientia 37:909–910
15. Tobe SS, Ruegg RP, Stay BA, Baker FC, Miller C, Schooley DA (1983) 2nd Int CNRS Symp Biosynth Metab Mode Action Invertebr Horm, Strasbourg, 29–8/2–9, 1983
16. de Kort CAD, Khan A (1983) 2nd Int CNRS Symp Biosynth Metab Mode Action Invertebr Horm, Strasbourg, 29–8/2–9, 1983

III.2 The Qualitative and Quantitative Determination of Juvenile Hormones by Mass Spectrometry

B. MAUCHAMP[1], M. ZANDER[2], and R. WOLFF[3]

1. Introduction

In order to understand the endocrinology regulation of insect morphogenesis and reproduction, an accurate and precise method of measuring juvenile hormone (JH) titers at different stages of the insect life cycle is necessary. This is not a simple matter due to the very low concentrations of these substances in insect tissue or fluids (10^{-7} to 10^{-9} *M*), and also to the simultaneous presence of large amounts of potentially interfering contaminants in the form of other endogenous organic substances.

Early approaches to the problem have involved bioassays, in which the response of some living target tissue to the JH(s) present in an extract was measured [1]. Such methods lack precision and cannot distinguish between different JHs present in a mixture containing more than one JH. A more recent development has been the use of radioimmunoassay [2, 3] – this method has a good sensitivity and some results have been published which are in reasonably good agreement with other (physicochemical) methods [4]. In a number of other cases, however, results have been obtained using radioimmunoassay in which the found levels of JH are unrealistically high [5], and a good reproducibility is not obtained. There appears to have been a lack of selectivity due to cross-reactions of the ligand with other substrates. Moreover, it is not possible to determine the chemical nature of the detected material.

Accordingly, a third method has arisen, using the physicochemical technique of mass spectrometry. The purpose of this chapter is to provide a review and commentary on those mass spectrometric methods currently available for the quantitative determination of JHs.

2. Experimental Approach

2.1 General Principles

A common feature of all the physicochemical methods, and perhaps the most important feature, is the use of a preliminary wet chemical cleanup in order to remove the more

1 Lab. Phytopharmacie, INRA, 78000 Versailles, France
2 Algonquin College, 200 Lees Ave., Ottawa, KIS OC5, Canada
3 Lab. Chimie Org. Substances Naturalles, 1 Rue Blaise Pascal, 67000 Strasbourg, France

Biosynthesis, Metabolism and Mode of Action of Invertebrate Hormones
(ed. by J. Hoffmann and M. Porchet)

grossly interfering substances from the initial biological extract. This being done, the analyst is then left with the problem of (a) separating the miniscule amounts of JHs from the remaining interfering substances, and (b) quantifying the JHs thus separated. In all cases the techniques used to achieve (a) involve one or more chromatographic separations, using either a gas chromatograph (GC) or high performance liquid chromatograph (HPLC), or a combination of the two. The components of the mixture are separated with respect to time in the chromatograph and the vapor phase eluate from the GC is passed directly into the mass spectrometer (MS). The organic molecules enter the ionization chamber where they are converted into ions which are then separated according to their mass [6] by a magnetic or electric analyzer. These ions are excited and relatively unstable species and, therefore, break up easily, giving rise to lower molecular weight charged fragments which are also separated and detected. The relative intensity of all ions observed is recorded as a function of their mass; the result is a mass spectrum of which a typical example is shown in Fig. 1.

This mass spectrum was obtained by using an ionization technique known as electron impact (EI). With this technique, the organic molecules from the GC enter the mass spectrometer in gaseous form and are bombarded with a high energy beam of electrons. A molecule (M) may have an electron struck from it, and becomes positively charged (M^+). Since the energy of the electron beam is high the M^+ ion formed is quite "excited", and in fact carries on it several times the energy required to break the chemical bonds therein. Consequently, the molecular ion M^+ is quite unstable in the ionization chamber and may get rid of this excess energy by breaking up into fragments. These fragments may still carry the positive charge, and if so, they will also be accelerated out of the ionization chamber and registered at the detector as independent fragment ions. In fact, most of the time the abundance of these fragments is much greater than that of the short-lived parent molecular ion M^+. This is evident in Fig. 1 – the high mass molecular ion of JH-3 at mass 266 is virtually invisible; the majority of fragments occur in the 50–100 mass range.

Electron impact mass spectra are very useful for structure determination and identification by comparison with reference spectra, but are less useful for quantitative analysis because the more characteristic high mass ions are usually of too low a relative

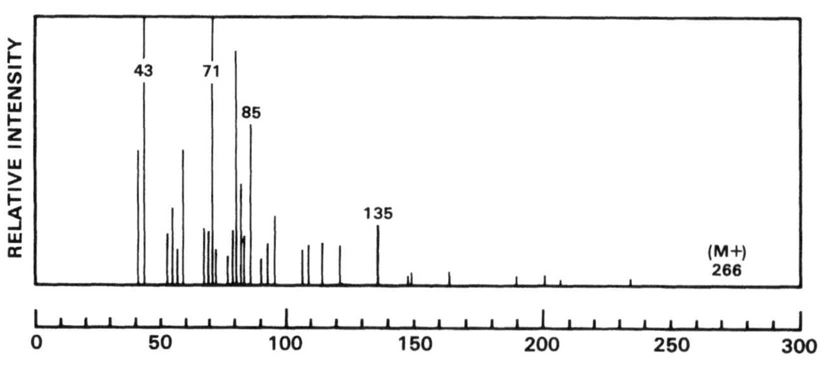

Fig. 1. EI mass spectrum of JH-3, showing the principal fragments seen in the 40–300 mass range

intensity, if not undetectable. One solution to this problem has been the use of derivatives.

The mass spectra in Fig. 2 (also of JH-3) were obtained using an alternative ionization technique, known as chemical ionization (CI). In this method the sample molecule is bombarded with a low velocity ionized reagent gas (instead of high velocity electrons), and becomes ionized by a proton transfer reaction. The energy transferred to the sample molecule is much less than with the EI technique, and consequently, the sample molecule is less prone to break into many fragments. A simpler spectrum is obtained, in which the high mass ions are usually highly characteristic, and the CI method, therefore, offers an advantage over EI for quantitative (as opposed to qualitative) analysis.

The reagent gas in CI is typically methane, isobutane, or ammonia. Figure 2 illustrates the CI mass spectrum of JH-3 using these three different gases.

It can be seen from the above examples that a mass spectrum gives a near unique "fingerprint" of a given organic molecule. When the mass spectrometer is coupled to

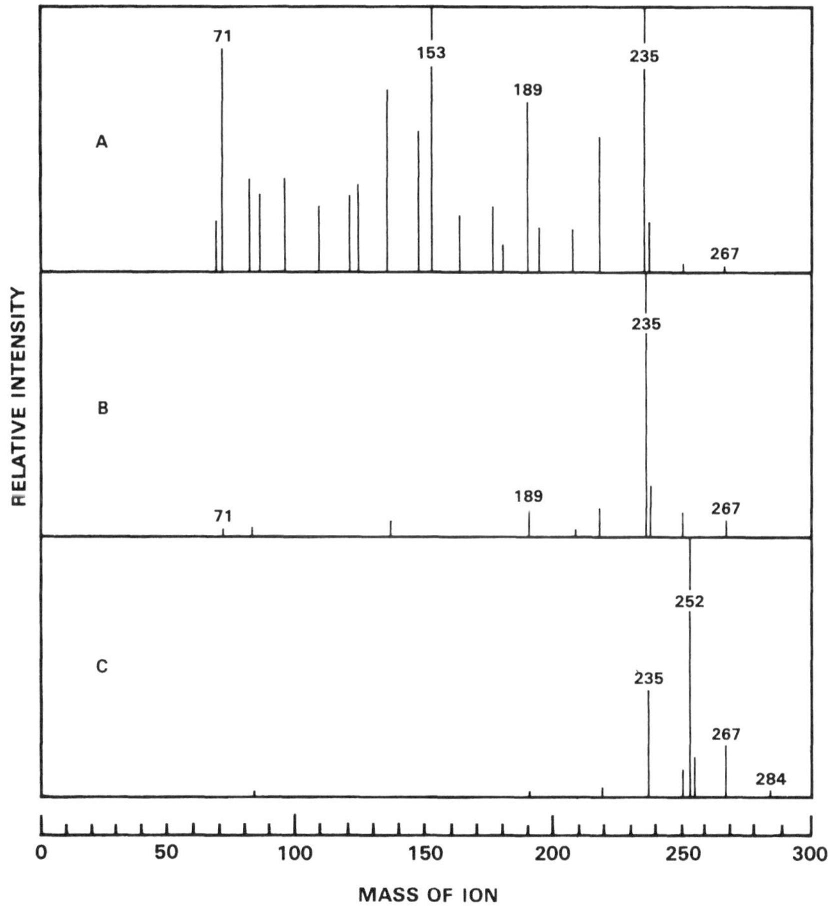

Fig. 2A–C. Ci mass spectra of JH-3, showing changes due to nature of the reagent gas. The ionizing gas is **A** methane, **B** isobutane, **C** ammonia

the GC, it is in effect being used as a specialized detector for the GC, the advantage over the traditional GC detectors being the structural information provided.

If the GC–MS is now linked to a computerized data collection system, the intensity of all ions produced in the mass spectrometer at given moments in time may be monitored and stored. This information may then be reconstructed in several ways. For example, a complete mass spectrum of any individual chromatographic peak may be retrieved, often permitting identification of the substance(s) giving rise to the peak. Alternatively, the summed intensity of all ions may be played back as a function of time, giving a chromatogram showing the sequential appearance of all substances emerging from the GC (this is called a total ion current chromatogram). In the case of a biological extract being analyzed for JHs, such a chromatogram may contain a very large number of peaks, due to the presence of many interfering contaminants, and in all probability these will obscure any peak(s) corresponding to the JHs. The solution to this problem is to select and record only a *few selected ions*, highly characteristic of the JHs, rather than all ions over the whole mass range. The result is a *selected ion mass chromatogram*. Such a chromatogram is far simpler than a total ion current chromatogram – the other contaminants are still present in the GC eluate, but are simply not recorded.

An example of a selected ion mass chromatogram of JH-3 is shown in Fig. 3. A single characteristic ion (that at mass 252 – cf. Fig. 2c) is monitored. Examination of this chromatogram shows numerous substances emerging from the GC having this ion, but only one with the GC retention time of 3 min 1 s, characteristic of JH-3 with this particular GC column. Further certainty of identification may be achieved by running additional selected ion mass chromatograms of the characteristic 235 and 267 mass

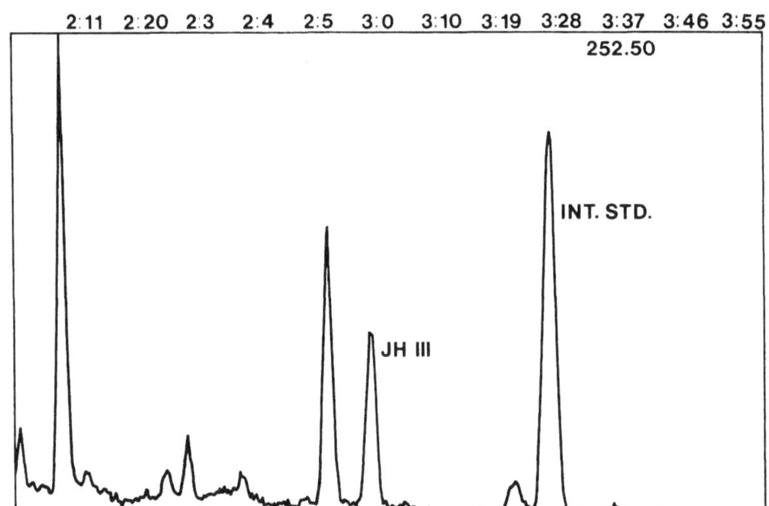

Fig. 3. Selected ion mass chromatogram of a GC eluate of an extract corresponding to 10 μl of *Locusta* hemolymph. The selected ion monitored is that having mass 252.30. Conditions are CI (ammonia)

ions, which should also show simultaneous peaks at the same retention time. Thus, the two parameters of (a) chromatographic retention time and (b) a mass spectrum containing selected ion peaks of a given intensity ratio, when taken together and compared with the retention time and mass spectrum of a known reference sample, can provide near unequivocal identification for an unknown substance. In this sense, the GC–MS method is unsurpassed in selectivity – the ability to distinguish a given substance from other interfering substances also present.

The selected ion monitoring technique may also be used to simultaneously analyze different JHs present in a mixture of JHs. Provided that the JHs are separated in the GC, a common ion (or ions) can be monitored. An example is shown in Fig. 4. Under EI conditions, the mass 95 ion is common to JH-3, JH-2, JH-1, and JH-0 – peaks at retention times corresponding to each of the four JHs are seen. The JH-0 can also be separately identified by the mass 128 ion, as this ion is specific to JH-0 only and is not produced by the other three JHs.

2.2 Quantification

Assuming that the analyst is now satisfied from the retention time and mass spectrum that a given peak emerging from the GC is indeed a JH, the next step is to quantify it. This is done by examining the intensity of the peaks in the selected ion mass chromatogram, following anywhere one to four selected ions. For simplicity of discussion we shall assume that only one (highly characteristic) ion is followed – the pros and cons of this will be discussed later. In the case of JH-3, using CI mass spectrometry with ammonia, this ion has mass 252.

There is a direct proportionality between the amount of JH-3 entering the mass spectrometer and the intensity of the mass 252 peak. It suffices to inject known amounts of JH-3 into the GC–MS, determine the MS response over a concentration range, and plot a calibration curve relating quantities injected to peak intensities. This calibration curve is then used for the analysis of unknowns potentially containing JH-3.

There are two ways of measuring the MS response for an ion of given mass:

1. Compare the ion current (i.e., peak area) of the characteristic mass 252 ion of JH-3 in the unknown with that of standard aliquots, injected separately. This is an absolute method, where absolute peak areas are compared. The principal disadvantages are:

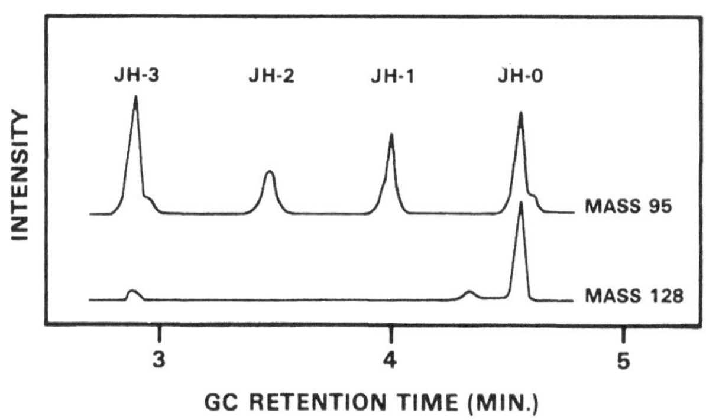

Fig. 4. Selected ion mass chromatogram of a GC eluate of a synthetic mixture of JH-3, JH-2, JH-1, and JH-0. The two ions selected have masses 95 and 128. Conditions are EI

a) the difficulties in injecting reproducible volumes in the 1–2 μl range;
b) variations in instrument sensitivity between two injections, leading to variations in peak areas; and
c) the fact that no account is taken of losses of JH during prior extraction and work-up. These losses can only be determined by adding a radioactive JH tracer to the initial extract, which is undesirable.

2. Use an internal standard. The internal standard is added to the biological sample right at the beginning of the extraction procedure. This immediately establishes a certain ratio between the compound to be measured (JH) and the internal standard (IS). Whatever losses may occur during the extraction and purification steps, this ratio will not change, and it is the *ratio* of these two compounds which we measure in the GC–MS. All of the disadvantages of method (1) are eliminated.

However, it should be noted that this reasoning applies only if the internal standard has physical and chemical properties very similar to the compound of interest (JH), so that during sample preparation there are negligible differences in extraction, derivatization, and chromatographic properties. Only in the final step in the analysis is the internal standard clearly distinguished from the compound of interest.

In the choice of the internal standard, there are two common possibilities:

a) *The stable isotope labelled analog.* This is the ideal choice. The labelled and unlabelled compounds can be distinguished in the MS. The labelled standard can be added in large excess and act as carrier for the unlabelled material, as it has identical chemical properties to the unlabelled material and is not even separated chromatographically in the GC [7]. Use of such a carrier requires firstly that it be synthesized, and have a high degree of isotopic purity. In addition, the synthesis of the labelled analog must be designed in such a way that the stable isotope occurs in the ion fragment being used in the analysis, for if the isotope is lost in another nonused fragment, the internal standard will not be distinguishable from the compound of interest in the mass spectrometer.
The advantage of the stable isotope internal standard is that of its carrier role and nonseparation from the compound of interest during extraction and workup. The disadvantage is the (sometimes lengthy) synthesis which may be required. A prior knowledge of the mass spectrum is obviously required so that the isotope is placed in the part of the molecule giving the fragment used.

b) *The structurally-related derivative.* In method (a) above the internal standard has the same GC retention time as the compound of interest, and is discriminated by having a different mass. In the alternative method, the internal standard and compound of interest have a common mass (ion), and are discriminated by having different GC retention times.
The internal standard is therefore, similar in physical and chemical properties to the compound of interest, but less so than the stable isotope labelled analog. After all, it has to be separable on the GC. This means that it may not provide adequate carrier protection to the compound of interest.

2.3 Sensitivity

Sensitivity may be described as the ability to discriminate signal from noise. There are a number of factors influencing sensitivity; a few of the more important ones are:

a) Chemical noise. This is due to the presence of other interfering substances. Chemical noise may be evaluated by running a procedural blank. If there are contributions to the selected ion mass chromatogram at the retention times corresponding to the JHs, it will be necessary to subtract these contributions from the sample spectrum. Obviously, it is best not to have any such contributions at all. The object of the initial sample cleanup is to reduce this background to an acceptable level. The lower the concentration of the hormone to be measured, the more critical this background becomes – what might be acceptable as background when measuring JHs in microgram amounts may completely drown any signal from JHs present in the picogram range.
b) The intensity of a peak in the mass spectrum of a compound, relative to the other ion peaks. Sensitivity is maximal if the most intense peak is used for the selected ion mass chromatogram (cf. for example the mass 252 peak of JH-3, Fig. 2c).
c) The time available for the computer to sit on a peak and integrate it. The longer this time is, the better the signal/noise ratio will be. For maximum sensitivity it is, therefore, best to monitor and integrate a *single* ion, and all the computer time can be dedicated to it. If by contrast we choose to monitor not just one iȯn, but say all ions of a mass between 200 and 350, the data analysis system will have much less time available to integrate the mass 252 ion. Consequently, the signal/noise ratio for a given mass (e.g., 252) is some 1,000 times poorer.

There is thus a trade-off between sensitivity and selectivity. The greater the number of ions followed in the selected ion mass chromatogram, the greater the certainty of the identity of the compound, but the price paid for this is a reduced sensitivity. In practice, therefore, a good compromise is to follow two or three masses.

3. Available Methods

A number of mass spectrometric methods have been described in the literature. The first, that of Bieber et al. [9] used a deuterium-labelled form of JH as both carrier and internal standard to estimate JH levels in the giant silkworm moth *Hyalophora cecropia*. Their initial sample cleanup involved a gel filtration on Sephadex LH-20, followed by two TLC purifications; GC–MS analysis of this extract with selected ion monitoring of two ions gave an observed level of JH in the male cecropia of 1.8 ± 0.2 μg per insect, and in the female 0.6 ± 0.1 μg.

In 1975 a paper by Lanzrein et al. [10] brought two important modifications – the use of HPLC in sample cleanup, and the use of the mass spectrometer in the CI mode. Their GC–MS analysis was qualitative only; complete mass spectra of 50–150 ng of JH-3 present in a biological sample were obtained.

In 1979–1981 three different groups published quantitative analytical methods for JH, using GC–MS:

Firstly, in 1979, Mauchamp, Lafont, and Jourdain described the measurement of JH-1 levels in *Pieris brassicae*, using CI (ammonia) mass spectrometry on the underivatized JH [11, 12]. Their sample cleanup involves sonication of a 3 ml hemolymph sample in hexane/methanol, passage of the extract through a small alumina column, then a purification by reversed phase HPLC. The resulting extract was injected into a 25-meter capillary glass column coated with SE-30, operating isothermally, coupled to the mass spectrometer. No internal standard was used, and quantification was done by measurement of the total ion current of three selected ions. A tritium-labelled JH-1 carrier was added to the initial extract in order to determine losses during extraction and workup; the amount of this exogenous added JH-1 must be subtracted from the experimentally obtained value. The stated detection limit for JH-1, using this method, is 100 pg/injection. In a later modification of this procedure [13] an internal standard (JH-3) is used for the quantitative analysis of JH-1, eliminating the need for a labelled carrier. The final paper in the series [14] describes a new technique whereby the HPLC is coupled directly to the mass spectrometer, omitting the GC separation stage. An additional prior cleanup step using reversed phase C-18 SepPak filtration is used, and the ethyl ester of iso-JH-2 is used as an internal standard. This novel technique has a good sensitivity (detection limit approaching 10 pg/injection with standard mixtures), but has encountered difficulties with biological samples because the omission of the powerful GC capillary separation step leads to increased chemical noise.

Shortly afterwards in 1980 and 1981 Rembold [15, 16] described protocols for the quantitative analysis of JH-0, JH-1, JH-2, and JH-3 in honeybees and locusts, using EI mass spectrometry on the derivatized JHs. A typical procedure involves methanolic extraction of the lyophilized insect, partition of the JH into hexane, delipidization on a normal phase silica SepPak cartridge, purification by normal phase HPLC, and reaction with a polyfluorinated reagent to give a derivative which is purified by another HPLC step. GC separation uses a 25-meter SE-30 coated capillary column, operating isothermally. Ethyl esters of JH-1 or JH-3, added to the initial extract, serve as internal standards. Selected ion monitoring of one or two ions gives a detection limit for biological samples of 10–50 pg/g of tissue.

In 1981 the group of Bergot and Schooley [17] described the quantitative determination of JH-0, JH-1, JH-2, and JH-3 in a range of different insect species, using GC–MS in the EI mode with selected ion monitoring. Like Rembold, they also include a derivative formation step (however, the derivative is different). A tritiated internal standard is used (the ethyl ester of iso-JH-2) and the semipurified extract is diluted with epoxy methyl geranoate as a carrier for the derivatization reaction (cf. [7]). The procedure requires extraction of hemolymph or whole insects with acetonitrile, partitioning into hexane, passage through a reversed phase C-18 SepPak cartridge, followed by passage through a small alumina column. After addition of the epoxy methyl geranoate carrier, reaction of the extract with deuterated methanolic trifluoroacetic acid gives a methoxyhydrin derivative which is then purified by HPLC. For the GC separation a packed column is used (3% OV-225), programed over a 190°–230 °C range. A detection limit of 10–40 pg/g of tissue is reported. Several applications of the method have been described subsequently [18, 19].

The most recent method, developed by the group of Temim et al. [20], is a modification of the Mauchamp et al. procedure for JH-1. This method has been used for the quantification of JH-3 in locust eggs and hemolymph. Individual hemolymph samples of as little as 25 μl have been successfully analyzed by this method. A sonication in methanol/hexane is followed by passage of the hexane extract through a reversed phase C-18 SepPak cartridge and then a basic alumina minicolumn. Further purification by reversed phase HPLC precedes injection into a 25-meter OV-1 capillary column coupled to a quadrupole mass spectrometer operating in the CI (ammonia) mode.

4. Discussion and Summary

In the case of the Bieber group, the choice of insect was fortuitous, as the levels of JH found are very high compared to those in other insects (e.g. *Locusta, Pieris*). The described cleanup would not have been adequate for the analysis of low nanogram or picogram quantities of JH – it is our experience that when working with these low levels, a purification by HPLC is always required to reduce chemical noise to an acceptable level. Consequently, the more recent methods, using newer technology, have a sensitivity at least 10^4 times greater than that of the Bieber method.

Of the methods developed since 1979, all four are of comparable sensitivity, with detection limits in the low picogram range. The paper of Bergot, Ratcliff and Schooley is noteworthy for the meticulous attention to detail and the rigorous emphasis on precision and accuracy. Their method has the advantage of being applicable to the quantification of all four JHs, and has been used for both hemolymph and embryos. Extraction yields are high and the label in the internal standard permits routine monitoring of these recoveries. The principal disadvantage is the requirement for a special internal standard and carrier, available only by multistage synthesis. In addition, the derivative-forming step makes the procedure more lengthy than those which measure the underivatized JH directly.

The method of Rembold is also applicable to the quantification of all four JHs. The internal standard is easily prepared and the one standard can be used for all four JHs. The principal disadvantage is the lengthiness of the procedure: not only is a two-stage derivatization required, but also *two* HPLC purifications. Consequently, the extraction yields are lower than for the other methods.

The method of the Mauchamp group is more rapid, as no derivatization is required. The internal standard is also easily prepared, and yields are good. The principal disadvantage is that the method is presently developed for the analysis of JH-1 only, although it should be possible to extend it to the other JHs. The lower degree of sample cleanup has led to difficulties in the case of insects having very low levels of JH-1.

The method of the Wolff group is also rapid, the internal standard easily prepared, and the yields good. Sample sizes analyzed are smaller than those reported by other groups. The principal disadvantage is that the method is developed for analysis of JH-3 only, although it should be possible to extend it to other JHs.

Three of the four methods use a quadrupole mass spectrometer; capillary columns are also used in three cases out of four. All methods require a HPLC cleanup step. The reversed phase operation we believe to be preferable as it permits rapid flushing of the column with a second solvent mixture to remove impurities which accumulate after repetitive runs - it is difficult to regenerate reproducible column activity with a normal phase column when treated in this way. Use of the SepPak cartridges is widely preferred as a quick and simple way of removing interfering lipids. The use of high purity solvents and very clean glassware to mimimize chemical noise is mandatory. For more detailed rationales for the individual steps used in each method the reader is referred to the original articles.

In summary, the methods presently available using GC–MS for the quantitative analysis of JHs offer a combined sensitivity and selectivity not available with the older methods. It should also be stated that these new methods do not come cheaply: a heavy investment in equipment (gas chromatograph, mass spectrometer, computer/data collection system, HPLC) is required, although such equipment can often be used on a shared basis with other departments. This disadvantage notwithstanding, it is our belief that the advantages offered by the GC–MS method are so substantial that it will become the method of choice for JH analysis.

References

1. Sláma K, Romanuk M, Sorm F (1974) Insect hormones and bioanalogues. Springer, Berlin Heidelberg New York
2. Baehr JC, Caruelle JP, Porcheron P, Cassier P (1981) In: Pratt GE, Brooks GT (eds) Juvenile hormone biochemistry. Elsevier, Amsterdam
3. Strambi C, Strambi A, De Reggi ML, Hirn MH, Delaage MA (1981) Eur J Biochem 118:401
4. Porcheron P, Papillon M, Baehr JC (1982) Experientia 38(8):970
5. See for example remarky by Bergot, Schooley, and de Kort (1981) in their article. In: Experientia 37(8):909
6. It is more correct to speak of 'mass-to-charge ratio', but since we are speaking of singly charged ions only, the term 'mass' is preferred here for simplicity
7. The use of a carrier is desirable because very small amounts of JH have a tendency to get 'lost' by irreversible adsorption to various substrates during workup procedures
8. See for example the article by Ende M, Spiteller G (1982) Contaminants in mass spectrometry. Mass Spectrom 1:29
9. Bieber MA, Sweeley CC, Faulkner DJ, Petersen MR (1972) Anal Biochem 47:264
10. Lanzrein B, Hashimoto M, Parmakovich V, Nakanishi K, Wilhelm R, Lüscher M (1975) Life Sci 16:1271
11. Mauchamp B, Lafont R, Jourdain D (1979) J Insect Physiol 25:545
12. Mauchamp B, Lafont R, Hardy M, Jourdain D (1979) Biomed Mass Spectrom 6(7):276
13. Mauchamp B, Lafont R, Pennetier JL, Doumas J (1981) In: Sehnal F, Zabra A, Menn JJ, Cymborowski B (eds) Regulation of insect development and behaviour, part 1. Tech Univ Wroclaw Press, Wroclaw, Poland
14. Mauchamp B, Lafont R, Krien P (1981) In: Pratt GE, Brooks GT (eds) Juvenile hormone biochemistry. Elsevier, Amsterdam
15. Rembold H, Hagenguth H, Rascher J (1980) Anal Biochem 101:356
16. Rembold H (1981) In: Pratt GE, Brooks GT (eds) Juvenile hormone biochemistry. Elsevier, Amsterdam
17. Bergot BJ, Ratcliff M, Schooley DA (1981) J Chromatogr 204:231
18. Bergot BJ, Baker FC, Cerf DC, Jamieson G, Schooley DA (1981) In: Pratt GE, Brooks GT (eds) Juvenile hormone biochemistry. Elsevier, Amsterdam
19. Edwards JP, Bergot BJ, Staal GB (1983) J Insect Physiol 29(1):83
20. Temim G, Zander M, Wolff R (in preparation)

III.3 Juvenile Hormones O, I, and II Exist Only in Lepidoptera

D.A. SCHOOLEY, F.C. BAKER, L.W. TSAI, C.A. MILLER, and G.C. JAMIESON[1]

1. Introduction

During the last 5 years there has been a growing amount of research on juvenile hormone titer determinations at physiological levels in a variety of insect species. These efforts have come from a number of research groups and have utilized a variety of methodologies, including radioimmunoassay (RIA) methods [1, 2], procedures using gas-liquid chromatography with electron capture detection [3-6] (GC/ECD), and methods utilizing gas chromatography or liquid chromatography coupled to mass spectrometric detection (GC/MS [7-10] or LC/MS [11]; with the mass spectrometer operated in the "selected ion monitoring" mode). These methods have supplanted procedures using bioassay for detection, as in many instances the sensitivity of bioassay procedures is insufficient, and quantification is difficult.

Despite these advances, our knowledge concerning JH titers has lagged behind that of ecdysteroid titers for a number of reasons. JH titers seem to be generally quite low (10^{-7}-10^{-9} *M*), and the JHs are similar in chemical and physical properties to the large amounts of lipids which are co-extracted by organic solvents. While ecdysteroid titers must also be measured with a small sample size, the physiological titers appear to be much higher [12] than for JH. Further, the relatively hydrophilic nature of the ecdysteroids makes removal of lipid from tissue extracts easy and apparently imparts a degree of antigenicity which is lacking in JH. In fact, a promising newer procedure for JH radioimmunoassay [1] incorporates a conversion of JH to the 10,11-diol derivative, which is reported to be more antigenic, and easier to separate from lipids than JH.

An unfortunate gap in our understanding of JH titers is the lack of reliable information on the qualitative identity of juvenile hormones in a variety of species. Such information is particularly important in light of the probability of greater application of RIA procedures for JH titer determination, as these methods appear less suited to qualitative identification of the JHs present in a particular species than are chromatographic methods.

In this paper, we review a number of recent studies conducted in these laboratories, many in collaboration with scientists in other institutions. The focus of these studies has been on qualitative and quantitative determination of juvenile hormones in certain insect species as a function of development, although the present discussion is centered

1 Biochemistry Department, Zoecon Corporation, 975 California Avenue, Palo Alto, CA 94304, USA

Biosynthesis, Metabolism and Mode of Action of Invertebrate Hormones
(ed. by J. Hoffmann and M. Porchet)

on the identification of the actual juvenile hormones present. Many of these studies were initiated with at least the partial aim of encompassing insect species from a diverse number of orders, so that eventually broad generalizations on JH identity could be made; other studies were undertaken to reinvestigate certain reports on JH identification in various species where we felt that stronger evidence was desirable. Much of the research reviewed here would not have been possible without the collaboration of scientists at other locations; we acknowledge our able colleagues at eight different institutions upon mention of the results in the text.

2. Materials and Methods

JH titers were determined essentially by the method of Bergot et al.; full experimental details are provided elsewhere [7]. Minor changes were made in the work-up and purification of the methoxyhydrin derivatives: after formation of the JH d_3-methoxyhydrin derivatives, samples were quenched (10 μl of 2 *M* $KHCO_3$), evaporated to a small volume (~50 μl) with a stream of N_2, diluted with distilled water (~50–100 μl), and extracted 3 × with pentane:ether (4:1). After addition of alkylphenylurea [7] marker compounds, the combined solvent extracts were purified by liquid chromatography (LC) using a 5 μm silica column [25 × 0.46 cm, LiChrosorb SI-100 or Spheri-5 (Brownlee Labs)] eluted with 3% ether in dichloromethane (mixture half-saturated with water).

Insect specimens were obtained from laboratory-reared colonies, using conditions reported in the collaborative studies cited in the results, with one exception. Pupae of *Attacus atlas* were field-collected (Dr. B. Morallo-Rejesus, University of the Philippines, Los Banos) and shipped by air to the USA. Cocoons were slit, pupae removed, and adults allowed to emerge in cages. Whole bodies of *Attacus* and other species (and hemolymph samples when possible) were extracted and processed according to our method [7].

3. Results and Discussion

3.1 Development of Methodology for JH Titer Determinations

The possibility of the existence of various juvenile hormones in different species arose following the identification [13] in 1968 of a second hormone, JH II, in *Hyalophora cecropia* soon after the identification [14] of JH I. Five years passed before the discovery of a new JH in a different species (JH III produced together with JH II in vitro by CC–CAs from *Manduca sexta* [15]). Following this report, a number of investigators studied the identification of JHs in a variety of species using in vitro methods pioneered by Judy et al., as well as in vivo methods. Particularly noteworthy was a 1974 report [16] by Trautmann et al. on in vivo identification of only JH III in six insect species selected from four nonlepidopterous orders. These identifications were based on capillary GC/MS analysis in which full spectra were obtained. This technique requires far more sample than the selected ion monitoring methods [7–10] in which only a few ions are measured. Titer determinations based on full spectra are in fact more rigorous, but the requirement for a 1 kg sample of insect tissue has prevented anyone from again utilizing this method!

The following year Lanzrein et al. [17] confirmed the detection of JH III in adult *Nauphoeta cinerea* [16], but reported that nymphs of *Nauphoeta* contained JH I and JH II in addition to JH III. This finding, based on a selected ion monitoring GC/MS technique, suggested that JH I and II might be important in morphogenesis, while JH III would more likely be the gonadotropin. This was virtually the first study on JH identity in larval insects. Subsequently, JH I, II, and/or III were reported to be present in a number of insects from various orders. It, thus, became widely assumed that JH I, II, and III coexisted in a variety of insects.

We have devoted several years to the development and refinement of a procedure for analysis of JH titers at physiological levels in samples of either whole bodies or hemolymph. Demonstration of the ability to detect picogram levels of JH or some derivative using gas chromatography with electron capture or mass spectrometric detection is, in fact, rather simple. However, utilizing such a detection scheme for biological samples requires development of an extraction and purification procedure which will remove interfering substances. These interferences are likely to vary from species to species, or even vary with development in a given species. Only the analysis of many samples from many species can verify the general validity of such a method. It is, in fact, this aspect of refining our method that proved difficult and time-consuming.

Figure 1 depicts our method of analysis [7]. A key feature of our procedure is the use of a tritium-labeled analog of JH as an internal standard, which is added to the insect tissue at time of extraction. This allows precise quantification and monitoring of any losses in the purification scheme for every sample. Extraction of insect tissue with acetonitrile removes JH, but extracts less lipid than a solvent, such as ether, due to the relatively low solubility of glycerides in acetonitrile. Partitioning the acetonitrile extract between pentane and water gives an easily evaporated pentane phase containing JH and a crude lipoidal extract which represents typically $\leqslant 3\%$ of the wet weight of a whole insect. (Male saturniid moths are exceptionally high in lipid. The yield of crude lipid from adult males of *Samia cynthia* is 6.5% with acetonitrile extraction, vs 36.7% with ether extraction [18].) Several extractions of the crude oily concentrate with acetonitrile removes all JHs, but leaves much of the poorly soluble triglycerides. Passage of the combined acetonitrile phases through a reversed-phase adsorption chromatography cartridge (C_{18} Sep-Pak) removes residual glycerides and sterols. After evaporating solvent, only one major lipid class remains, the free fatty acids. These can be conveniently removed by dissolving the residue in ether and percolating this through a short column of neutral alumina. These simple and efficient steps are far easier than classical methods for delipidating JHs, and could be used as well with procedures using RIA detection.

Following removal of major lipids, the samples are treated with acidic d_4-methanol to form the d_3-methoxyhydrin derivatives, worked up, and separated from residual neutral lipids by liquid chromatography using marker compounds to delineate the zone in which these derivatives elute. Final analysis is by gas chromatography-mass spectrometry using selected ion monitoring; the derivatives of the internal standard

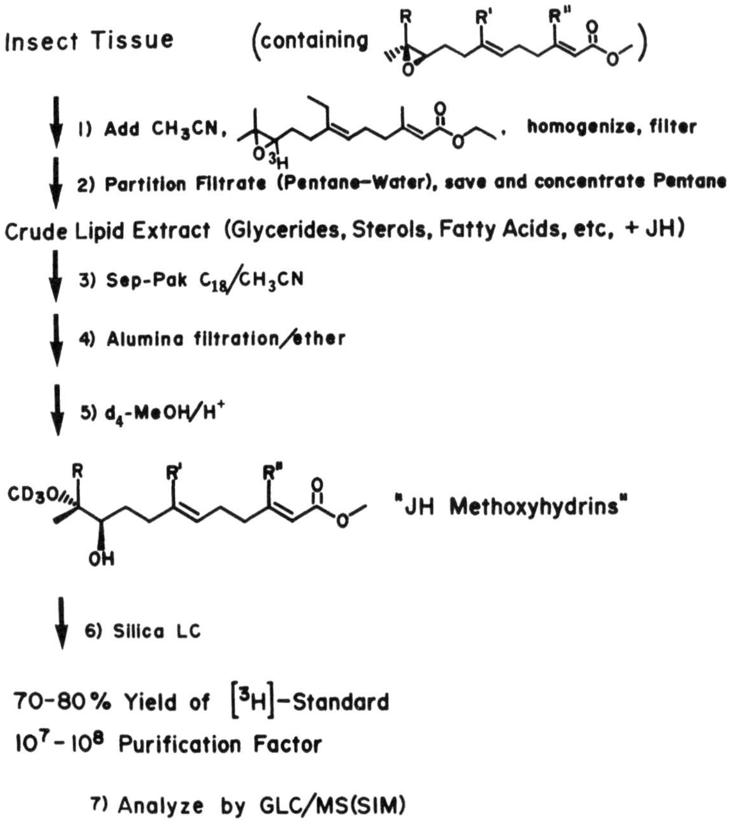

Fig. 1. Scheme for processing and analysis of insect tissues for determination of JH titer

and JH III are detected at m/z 76 and those of JH II, JH I, JH 0, and 4-MeJH I at m/z 90. A typical chromatogram is shown in Fig. 2 (Sect. 3.4). Frequently, samples need to be analyzed on two GC columns with different selectivities, especially when analyzing near the limit of detection of the procedure (~0.01 ng/g, or ~0.03 pmol/g).

3.2 Physiological Titer Determinations in Lepidoptera

During application of this method to analysis of juvenile hormones in *Manduca sexta* throughout the entire life cycle, we noticed that embryonic stages contained two apparent interferences on the m/z 90 trace, in amounts as high as JH I (1-5 ng/g). However, full mass spectra suggested these two materials were homologs of JH I, and indeed they were subsequently identified as such. JH 0 [19] and 4-methyl JH I (4-MeJH I) [20] are in fact isomers; the latter has been detected only in *M. sexta* eggs, whereas JH 0 has been detected in adult male *H. cecropia* (24 ng/g, 1.2% of total JH) [21] and as a minor constituent of the total JH in *M. sexta* larval hemolymph. While JH III is undetectable in *M. sexta* embryos and only a trace constituent (like JH 0) in *M. sexta* larval blood, JH III occurs in amounts comparable to JH II in adult female *M. sexta*. Thus, all five known juvenile hormones exist in various stages of *M. sexta*.

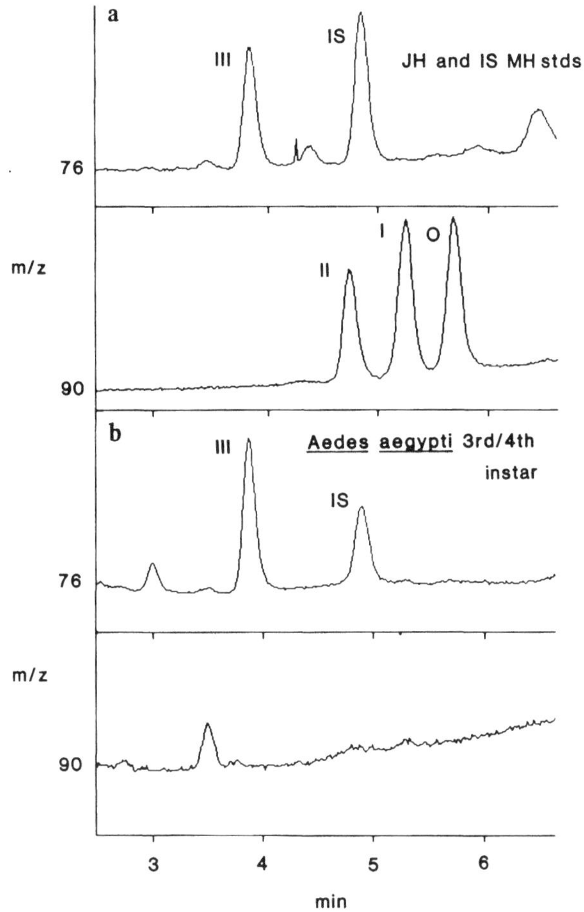

Fig. 2a,b. Selected ion chromatograms of a juvenile hormone and internal standard (*IS*) methoxyhydrin derivatives (*upper trace:* m/z 76: JH III and IS; *lower trace:* m/z 90: JH II, JH I, and JH 0); **b** analysis of a whole-body extract from *Aedes aegypti* 3rd/4th instars processed [7] as in Fig. 1. The observed level of JH III was 0.27 ng/g

We have reinvestigated a report [22] by Paguia et al. that a giant silk moth, *Attacus atlas*, contains JH bioactivity nearly comparable to *H. cecropia*, but that the chromatographic behavior of this bioactivity on TLC is similar to that of JH III. Our work was conducted in collaboration with Dr. B. Morallo-Rejesus, University of the Philippines, Los Banos, who supplied us with field-collected cocoons of *A. atlas*. Newly emerged adults were sexed and allowed to mature in cages. Analysis of whole body extracts revealed that the adult males contained ~4–17 ng/g of JH II and lesser amounts of JH I [23]. Females contained much lower levels of both hormones. While the sexual dimorphism in JH content is reminiscent of that found in the related saturniid moth *H. cecropia* [4], the overall JH levels are about 100 × lower than in *H. cecropia*, and the ratio of JH II/I of 14–40:1 is quite different (1:5–7 in *H. cecropia*). Using the values quoted by Paguia et al. for the specific biological activity of JH I and JH II, as well as the biological potency of their *A. atlas* extract, we calculate that the titer values we observe are in good agreement with their bioassay data [22] for *A. atlas*. Closer inspection of their biological potency data for the *H. cecropia* extract reveals it to be far lower than other literature reports [4, 18]. In summary, we find no evidence to support the possibility for the existence of a new JH in *A. atlas*.

We have completed a collaborative study with Drs. G. Jones and B.D. Hammock, University of California at Davis, on JH titers in the noctuid moth, *Trichoplusia ni*. Larvae of this moth contain JH II as the major detectable juvenile hormone [24] at levels lower than that found in *M. sexta*. Thus, *T. ni*, like all other lepidopterans studied to date, contains an homologous JH as the principal constituent, rather than JH III. In no case has *only* JH III been detected in *any* lepidopterous sample analyzed.

From a teleological standpoint, it appears likely to us that if the homologous JHs are found in nonlepidopterous insects, they should be found in other holometabolous insects. Accordingly, results of JH identification studies in Coleoptera and Diptera are of interest.

3.3 Identity of JH in Coleoptera

In the Coleoptera, Trautman et al. had reported the presence of only JH III in extracts of whole *Melolontha melolontha* [25] and *Tenebrio molitor* [16] (the latter data are supported by in vitro experiments [26] with cultured corpora allata). We reported in 1982 that larvae and adults of *Leptinotarsa decemlineata* also contain only JH III [27] (conducted in collaboration with Dr. C.A.D. de Kort, Agricultural University, Wageningen). Larvae and pupae of *L. decemlineata* had been reported to contain significant levels of JH I and JH II based on an RIA procedure [28]. However, we were unable to detect any juvenile hormone except for JH III. Recently, we investigated JH titers in adult *Diabrotica undecimpunctata*, and again detected only JH III (unpubl. data). Unfortunately, at this time *L. decemlineata* is the only coleopteran in which larval stages have been studied.

3.4 JH in Diptera

The Diptera constitute an order in which JH titers seem to be generally low, and reliable results have been slow to accumulate. Klages et al. reported that only JH III exists in larval and adult stages of *Drosophila hydei* [29]. In a recent study done together with Dr. H.H. Hagedorn and Mr. G. Wheelock, Cornell University, we have found that only JH III is present in larvae and adults of the mosquito *Aedes aegyptii* [30]. Furthermore, we established that the lower limit of detection for JH I and JH II was ~200–400 × lower than the highest JH III titer in adult *A. aegypti* of ~3 ng/g, yet the biological potency of JH III was only 10–25 × lower than JH I. From these circumstances, we concluded that it cannot be argued that trace levels of the more potent JH I may play a hormonal role in *A. aegypti* [30]. We have also reinvestigated some early JH titer studies that we conducted with *Musca domestica* and *Sarcophaga bullata*, using a method employing gas-liquid chromatography with electron capture detection [3]. We subsequently abandoned this procedure in favor of GC/MS [7] because of insufficient specificity inherent in the GC/ECD procedure. We had detected low levels of what appeared to be JH I, JH II, and JH III in both species of flies [31]. Recently, we have found that we cannot reproduce these results using the GC/MS method. In particular, JH I and JH II are totally absent from comparable stages of both *M. domestica* and *S. bullata*, and also from adults of *Calliphora vicina* (F.C. Baker and L.W. Tsai, unpubl.). Titers of JH III are close to the limit of detection and rather difficult to quantify reproducibly; it is, in fact, not possible currently to state for certain whether the small amount of material detected in these fly extracts is actually JH III or some impurity. Thus, at the moment the data from *D. hydei* [29] and *A. aegypti* [30] appear to be the only positive JH identification in Diptera, but clearly JH I and JH II are not detectable in *M. domestica, S. bullata,* and *C. vicina* using our GC/MS method.

3.5 JH in the Dictyoptera

There has been great interest in JH in the superorder Orthopteroidea, and of the order Dictyoptera in particular. The apparent co-existence of all three JHs in nymphs of the cockroach *Nauphoeta cinerea* [17] provided added credibility to subsequent studies claiming the existence of JH I and II in a variety of species. Accordingly, we have re-investigated JH titers in several stages of *N. cinerea* [32] in collaboration with Dr. B. Lanzrein, Universität Bern. Extracts of *N. cinerea* embryos, nymphs, and adult females were prepared in Bern and shipped to Palo Alto for analysis. In all samples, we can detect *only* JH III [32]. Embryos (not analyzed in the original study) contain extremely high (~1,000 ng/g) JH III titers in certain phases of development, larvae relatively low titers, and adult females intermediate titers (some additional details are provided in the article by Bürgin and Lanzrein, this vol.). The absence of JH I and II from nymphal stages is noteworthy. In the earlier study [17], quantification of JH levels had been achieved only by bioassay of zones separated on LC using a silica column, and GC/MS in the chemical ionization mode had been used for qualitative identification. As regards the substantial discrepancy between the current and earlier results, it appears that the most likely possibility is from contamination of *N. cinerea* extracts in the first study with synthetic standards of JH. The standards could have been introduced because of the practice of calibrating the LC column with nearly microgram quantities of JH I–III. The instrument used in the early study [17] was equipped with a septum injector device, which has a tendency to exhibit "memory" effects upon injection of blanks or small samples, following a larger standard sample.

Simultaneously, we have been conducting a collaborative study on JH identity and titers in another cockroach, *Diploptera punctata* [33] (with Drs. S.S. Tobe and R.P. Ruegg, University of Toronto, and Dr. B.A. Stay, University of Iowa). Adult females of this species contain exceptionally high JH III titers, ~2,000 ng/ml, in hemolymph at peak levels. Presence of such exceptionally high titers affords the opportunity of analyzing for traces of JH 0, I, or II which may be present. It is especially difficult to establish a low detection limit for one JH in a sample containing massive amounts of another JH (or internal standard) due to limitations in dynamic range of the peak detection and integration systems. Thus, we specially processed a sample of female *D. punctata* day 5 hemolymph. This sample, containing 0.87 ng/ml of internal standard, was separated on LC into a fraction containing JH III as well as a fraction containing the low level of internal standard plus any endogenous JH 0, JH I, and JH II. Additional internal standard was added to the JH III zone in an amount corresponding to 470 ng/ml. Each fraction was then converted to the methoxyhydrin derivative, purified, and analyzed as usual. Analyses revealed a JH III titer of ~880 ng/ml, but no JH 0, JH I, or JH II at a limit of detection of 0.02 ng/ml. If any of the higher JH homologs occur in *D. punctata*, they are present at levels ~5 × 10^4 lower than that of JH III. Previous studies [34] on the product of in vitro incubations of corpora allata of *Periplaneta americana* using GC analysis with a radioactivity monitor have also confirmed the absence of JH I and JH II at a detection limit of 10^{-5} times that of JH III. Other interesting aspects of JH titer and its regulation in *D. punctata*, which are outside the focus of this article can be found in a future journal report [33] or the chapter by Tobe (this vol.).

3.6 JH Identity in Orthoptera

Nymphs of *Locusta migratoria* were reported to contain substantial levels of JH I based on an RIA procedure [35]. We reinvestigated this finding in collaboration with Dr. C.A.D. de Kort, Agricultural University, Wageningen. We could not detect any JH other than JH III [36] in nymphs with stage and age nominally identical to that utilized in the RIA-based report. Moreover, the JH III titers we observed were 10–1,000 × *lower* than those that would have been required to cross-react with their JH I antibody to produce the data reported based on RIA [35]. In a more recent study done in collaboration with Dr. W. Loher, University of California, Berkeley, we studied [37] JH titers in nymphs and adults of the cricket, *Teleogryllus commodus*. Again, only JH III could be detected; levels in adults were substantially (20–30 ×) higher than in nymphs (maximum observed ~2 ng/g in the latter stages). Only a few other in vivo experiments have been conducted with Orthoptera (summarized in ref. [37]).

3.7 JH in Thysanura

We have also investigated a claim that mixed sexes of adult animals of the primitive insect *Thermobia domestica* contain JH I at unspecified levels [38]. We find [32] that *T. domestica* contain only JH III (0.4–1.4 ng/g in adult females, ~0.3 ng/g in nymphs 7-weeks-old), with a limit of detection of ~0.01 ng/g for JH I and JH II. It is difficult to disprove an assertion that JH I exists in this species, when no quantification was stated. However, the ~1 kg sample of *T. domestica* utilized should have contained about 1 μg of JH III, according to our data. This amount of JH should have been detectable, but its presence was not reported [38].

3.8 JH in Hemiptera

There has been a notable lack of information on JH in hemipterans, despite their extensive use in insect physiology. In general, insects of this order are rather sensitive to the effects of treatment with certain JH analogs, particularly chemicals of unusual structural type [e.g., a peptidic derivative of ethyl 3-aminobenzoate, the "paper factor" juvabione, and a 3-phenoxy terpenoid epoxide (see Henrick [39] for review)] which have much lesser or no effects on insects of other orders. Moreover, the anti-JH effects of the precocenes were discovered in a species of this order (initially on *Oncopeltus fasciatus*). From such circumstances, some researchers have suspected that Hemiptera have the potential of containing juvenile hormones different from those known. We have recently concluded a collaborative study [40] on the hemipteran *Megoura viciae* (suborder Homoptera), with Drs. J. Hardie and A.D. Lees, Imperial College at Silwood Park. Peak titer values were found in long-day reared adult aphids with a mean value of only ~0.15 ng/g of JH III. This constitutes the first JH identification in any hemipterous insect. It is not known whether JH III at these low levels fulfills all hormonal functions, or if a chemically different form could be present also. It is not surprising that there have been no other reports of JH identification in Hemiptera, if these low titer levels are representative of other species of that order.

4. Conclusions

The number of species whose juvenile hormones have been identified are quite small in comparison to the hundreds of thousands of known species of insects. However, we conclude that there is no solid evidence for the existence of the higher homologs of JH III (JH II, JH I, JH 0, and 4-MeJH I) in any order other than the Lepidoptera, and that all lepidopterans analyzed to date contain one of the higher homologous JHs as the principal form. The data presented here support the 1974 report [16] of Trautmann et al., who found only JH III in adult animals of six insect species spanning four orders. We have been able to use much smaller samples (100 X) than in their 1974 study, and yet still achieve a higher sensitivity. Consequently, we have generally attempted to analyze JH in both larval and adult stages, and on occasion also in embryonic stages. Moreover, we have identified JH III in three orders not investigated by Trautmann et al.: Diptera, Hemiptera, and Thysanura. Including a literature report [41] describing isolation of only JH III from an isopteran, *Macrotermes subhyalinus*, there are now eight orders from which select species have been shown to contain only JH III by GC/MS methods. The newest of several RIA techniques [1] for the juvenile hormones can detect equally well each JH, but is not "specific" for one JH over another, as claimed for an earlier procedure [2]. This inability to discriminate between the JHs using RIA in studying nonlepidopterous organisms would appear to be no disadvantage in our opinion, provided extracts are sufficiently cleaned-up.

There are enormous differences in what constitutes a high physiological titer of JH when comparing two insect species. For example, in *Megoura viciae* 0.15 ng/g (whole body) of JH III is a maximum value [40], as compared to 740 ng/g of JH III in *Diploptera punctata* [33]. In most insects analyzed to date, maximal physiological titers seem to fall within a range of 1-100 ng/g (or ng/ml), which is still a broad range, but narrower than the extremes of *M. viciae* vs *D. punctata*. The difference between maximal titers in the latter species of a factor of 5000 is difficult to rationalize at this stage of our understanding. Perhaps characterization of hormone receptors in these species will provide answers to this enigma.

In comparative bioassays of JH I, II, and III, the difference in biological potency has been found to vary from slight in Diptera [39] to substantial in most Lepidoptera [39], with JH I being most potent, JH III least, and JH II at an intermediate level. Certain exceptionally large differences have been reported, such as an injection assay with *Tenebrio molitor*, where JH I is 16,000 times as potent as JH III [42]. While only

JH III has been isolated [16] from *T. molitor*, the limit of detection was not sufficient to rule out the existence of biologically efficacious levels of JH I. Studies on relative biological potency of the natural JHs were initially responsible for a suspicion that JH I would most likely represent the most important and widespread of the known JHs.

We challenge the veracity of this view, which appears to still persist among many researchers. As we have made clear, there is in our opinion no clear-cut evidence for existence of JH I in nonlepidopterous orders. We have cited in vivo [33] and in vitro [34] studies in two cockroach species in which JH III has been shown to be present at a level $5-10 \times 10^4$ greater than any conceivable trace of other known JH. We also point out that synthetic JH analogs [such as methoprene (Zoecon) and fenoxycarb (Hoffman-LaRoche)] are known which generally possess biological activity orders of magnitude greater [39] than JH I, and arguments concerning relative bioactivity of the natural JHs should only be considered with this in mind.

Lastly, the question of JH identity within an order or species is certainly not a question of interest only in terms of structural chemistry. Recent studies on juvenile hormone biochemistry and physiology have provided examples where experiments conducted with exogenous JH I applied to a JH III-producing organism have resulted in incorrect conclusions regarding hormone transport and metabolism. For example, de Kort et al. [43] recently reexamined the question of JH binding proteins in hemolymph of *L. decemlineata*, an insect which produces only JH III. Earlier studies with [10-^{3}H]-JH I had shown no apparent high-affinity binding protein and rapid esteratic degradation of the applied JH I. When restudied with [10-^{3}H]JH III, a high affinity binding protein was detected which protected the 10*R* isomer of labeled JH III from degradation. Thus, the *L. decemlineata* hemolymph binding protein does not protect the hormone *analog*, JH I, as efficiently as its native hormone. Earlier, Peter et al. [44] studied the stereochemical interactions of JH I and JH III with hemolymph proteins of *Locusta migratoria*, a JH III producing species. From these studies they concluded that the binding protein of *L. migratoria* has highest affinity for the natural optical isomer of JH III. Moreover, they observed that exposure of racemic JH I and JH III to hemolymph of *L. migratoria* results in preferential metabolism of the *natural* isomer of JH I, but the *unnatural* isomer of JH III, due primarily to the exceptional protection afforded to (10*R*)-JH III by the binding protein.

Isolation of target tissue receptors for the juvenile hormones is a subject of much interest. Hopefully, considerations such as the above may assist in interpreting results of ongoing experiments in this field.

Acknowledgments. We again thank our many collaborators, who are cited in the text, for their cooperation and for allowing us to cite portions of the data we have gained together. We thank the National Science Foundation (USA) for partial financial support of this research (PCM 82-08665), and Dr. C.A. Henrick for helpful comments on the manuscript draft. One of us (DAS) thanks CNRS for partial support to attend the conference for presentation of this paper.

References

1. Strambi C, Strambi A, De Reggi ML, Hirn MH, Delaage MA (1981) Eur J Biochem 118:401–406
2. Baehr JC, Pradelles P, Dray F (1979) Ann Biol Anim Biochem Biophys 19:1827–1836
3. Bergot BJ, Schooley DA, Chippendale GM, Yin C-M (1976) Life Sci 18:811–820

4. Dahm KH, Bhaskaran G, Peter MG, Shirk PD, Seshan KR, Röller H (1981) In: Kloza M, Sehnal F, Zabza A, Menn JJ, Cymborowski B (eds) Int Conf Regul Insect Dev Behav. Wroclaw Tech Univ Press, Wroclaw, Poland, pp 183–198
5. Huibregste-Minderhoud L, Van Der Kerk-Van Hoof AC, Wijkens P, Biessels HWA, Salemink CA (1980) J Chromatogr 196:425–434
6. Hagenguth H, Rembold H (1979) J Chromatogr 170:175–184
7. Bergot BJ, Ratcliff M, Schooley DA (1981) J Chromatogr 204:231–244
8. Rembold H, Hagenguth H, Rascher J (1980) Anal Biochem 101:356–363
9. Mauchamp B, Lafont R, Hardy M, Jourdain D (1979) Biomed Mass Spectrom 6:276–281
10. Zander JM, Wolff R, Temim G, Kappler C, Hoffmann JA (1984) In: Hoffmann JA and Porchet M (eds) Colloq Int CNRS, Strasbourg, September 1983
11. Mauchamp B, Lafont R, Krien P (1981) In: Pratt GE, Brooks GT (eds) Juvenile hormone biochemistry. Elsevier/North-Holland, Amsterdam, pp 21–31
12. Hoffmann JA (1980) Progress in ecdysone research. Elsevier/North-Holland pp 29–44, 45–68, 69–82, 211–234, 409–430, 431–465
13. Meyer AS, Schneiderman HA, Hanzmann E, Ko JH (1968) Proc Natl Acad Sci USA 60:853 to 860
14. Röller H, Dahm KH, Sweeley CC, Trost BM (1967) Angew Chem Int Ed 6:179–180
15. Judy KJ, Schooley DA, Dunham LL, Hall MS, Bergot BJ, Siddall JB (1973) Proc Natl Acad Sci USA 70:1509–1513
16. Trautmann KH, Masner P, Schuler A, Suchý M, Wipf H-K (1974) Z Naturforsch 29c:757–759
17. Lanzrein B, Hashimoto M, Parmakovich V, Nakanishi K (1975) Life Sci 16:1271–1284
18. Gilbert LI, Schneiderman HA (1961) Gen Comp Endocrinol 1:453–472
19. Bergot BJ, Jamieson GC, Ratcliff MA, Schooley DA (1980) Science 210:336–338
20. Bergot BJ, Baker FC, Cerf DC, Jamieson G, Schooley DA (1981) In: Pratt GE, Brooks GT (eds) Juvenile hormone biochemistry. Elsevier, Amsterdam, pp 33–45
21. Bergot BJ, Jamieson GC, Ratcliff MA, Schooley DA (1981) In: Kloza M, Sehnal F, Zabza A, Menn JJ, Cymborowski B (eds) Int Conf Regul Insect Dev Behav. Wroclaw Tech Univ Press, Wroclaw, Poland, pp 173–182
22. Paguia P, Masner P, Trautmann K-H, Schuler A (1976) Experientia 32:122–123
23. Baker FC, Jamieson GC, Morallo-Rejesus B, Schooley DA (1984) To be published
24. Jones G, Hammock BD, Baker FC, Tsai LW, Miller CA, Schooley DA (1984) Manuscript in preparation
25. Trautmann KH, Schuler A, Suchý M, Wipf H-K (1974) Z Naturforsch 29c:161–168
26. Judy KJ, Schooley DA, Troetschler RG, Jennings RC, Bergot BJ, Hall MS (1975) Life Sci 16: 1059–1066
27. de Kort CAD, Bergot BJ, Schooley DA (1982) J Insect Physiol 28:471–474
28. Deleurance S, Baehr JC, Porcheron P, Cassier P (1979) CR Acad Sci 290:367–370
29. Klages G, Emmerich H, Rembold H, Hagenguth H (1981) In: Kloza M, Sehnal F, Zabza A, Menn JJ, Cymborowski B (eds) Int Conf Regul Insect Dev Behav. Wroclaw Tech Univ Press, Wroclaw, Poland, p 207
30. Baker FC, Hagedorn HH, Schooley DA, Wheelock G (1983) J Insect Physiol 29:465–470
31. Schooley DA, Judy KJ, Bergot BJ, Hall MS, Jennings RC (1976) In: Gilbert LI (ed) The juvenile hormones. Plenum Press, New York, pp 101–117
32. Baker FC, Lanzrein B, Miller CA, Tsai LW, Jamieson GC, Schooley DA (1984) Life Sci (in press)
33. Tobe SS, Ruegg RP, Stay BA, Baker FC, Miller CA, Schooley DA (1984) To be published
34. Hamnett AF, Pratt GE (1978) J Chromatogr 158:387–399
35. Baehr JC, Porcheron P, Papillon M, Dray F (1979) J Insect Physiol 25:415–421
36. Bergot BJ, Schooley DA, de Kort CAD (1981) Experientia 37:909–910
37. Loher W, Ruzo L, Baker FC, Miller CA, Schooley DA (1983) J Insect Physiol 29:585–589
38. Madhavan K, Ohta T, Bowers WS, Schneiderman HA (1981) Am Zool 21:733–735
39. Henrick CA (1982) In: Coats JR (ed) Insecticide mode of action. Academic Press, London New York, pp 315–402
40. Hardie J, Baker FC, Jamieson GC, Lees AD, Schooley DA (1984) Manuscript in preparation
41. Meyer DR, Lanzrein B, Lüscher M, Nakanishi K (1976) Experientia 32:773
42. Dahm KH, Bhaskaran G, Peter MG, Shirk PD, Seshan KR, Röller H (1976) In: Gilbert LI (ed) The juvenile hormones. Plenum Press, New York, pp 19–47
43. Kort CAD de, Peter MG, Koopmanschap AB (1983) Insect Biochem 13:481–487
44. Peter MG, Gunawan S, Gellisen G, Emmerich H (1979) Z Naturforsch 34c:588–598

III.4 Biochemical Aspects of Juvenile Hormone Biosynthesis in Hemimetabola

G.E. PRATT[1]

1. Introduction

Studies of the biochemistry and physiology of juvenile hormone biosynthesis in the corpora allata from adult female insects of the lower orders have formed an important part of the research effort of this laboratory during the last decade. As will become apparent below, this work has always benefitted enormously from the parallel works of my close colleagues in the fields of organic chemistry, reproductive physiology and molecular biology: these coordinated activities in related disciplines have provided many essential chemical tools for the biochemical research together with a broader physiological perspective from which to assess the results. It is the stipulated aim of this article to review our progress in this area from 1975 onwards, and to indicate briefly how this relates to other major developments in the field. The work is considered under four headings.

2. Corpus Allatum Activity in Vitro

A method had been devised by 1975 for measuring the endocrine activity of corpora allata immediately after removal from adult locusts and cockroaches, using short-term radiochemical incorporations from methyl-labelled methionine into juvenile hormone [1, 2]. This really came about because of our first finding [3] that large quantities of radio-labelled juvenile hormone III were produced by glands from adult female *Schistocerca gregaria*, irrespective of their physiological status, when a possible biosynthetic intermediate (farnesoic acid) was added to the incubation medium. Although such experiments are unphysiological, they did enable us to define quickly the kinetics of radio-incorporation from methionine into juvenile hormone and to begin the more difficult task of devising culture conditions which would afford good rates of synthesis under physiological control [4, 5]. Of course, we can not expect to find conditions in vitro which reflect exactly the rates of hormone biosynthesis in the animal, yet there has been sufficient confidence in the method for it to be used during the intervening years to assess the detailed regulation of the corpus allatum in a wide variety

1 Agricultural & Food Research Council, Group of Insect Chemistry and Physiology, University of Sussex, Brighton BN1 9RQ, Sussex, Great Britain

Biosynthesis, Metabolism and Mode of Action of Invertebrate Hormones
(ed. by J. Hoffmann and M. Porchet)

of physiological states and insect species; indeed, the largest part of this work has been carried out very successfully elsewhere.

Quantitative measurements of the rates of synthesis – and therefore release [1, 6] – of juvenile hormone have provided the necessary basis for a more exact description of the physiological regulation of the corpus allatum during adult female reproduction. In contrast to previously existing techniques of environmental or surgical manipulation followed by observation of the ultimate effects of hormonal disturbance, which admittedly laid the original foundation of corpus allatum physiology, direct measurements of hormone biosynthetic rate have afforded important quantitative data and enabled us to follow the precise timing of changes in the endocrine activity of the glands in different circumstances. In this laboratory we have concentrated on studying the effects of feeding [7], mating [8], and some surgical intervention [9–11] in one or other of *Periplaneta americana* and *Locusta migratoria*. In view of the wide range of reproductive strategies in insects [12], particularly with regard to the importance of inhibitory influences, it is to be expected that the exact profiles of corpus allatum modulation will be unique for each species. We even found (cf. [13]) that in the basic matter of functional symmetry between the left and right corpora allata of a single animal, there is a great difference between the nearly symmetrical *Periplaneta americana* and the highly asymmetrical *Schistocerca gregaria* [14] which may result from the dominance of nervous regulation in the latter species [15]. However in general, and at least qualitatively, these direct measurements of corpus allatum activity have not contradicted major existing physiological hypotheses, according to the literature.

Here, I wish to consider the findings according to the simple hypothesis that the corpus allatum is responsible for initiating and maintaining vitellogenesis in successive waves of oocytes in these species, i.e. that the profile of juvenile hormone synthesis may be the causal instrument in the regulation of their reproductive potential. The simplest case is found in species such as *Schistocerca* and *Locusta*, whose panoistic ovaries show complete inter-ovarian and intra-ovarian synchrony (apart from graded resorption), with no temporal overlap between successive waves of vitellogenesis and oviposition. Thus the minimal hypothesis was fulfilled by the early findings [2] that there was one "pulse" of juvenile hormone biosynthesis per vitellogenic cycle in *Schistocerca*. However, doubts were expressed as to the precision of the timing of peak hormone synthesis relative to follicular development in individual animals, and it was suggested that a very small peak of biosynthesis early in adult maturation might be responsible both for the initiation of the first wave of vitellogenesis and for inducing extra-follicular maturation of the reproductive tract and possibly the synthesis of vitellogenin in the fat body. Moreover, recent detailed studies of hormone synthesis and accessory gland activation in *Periplaneta* [8, 16–18] show that one of the major biochemical events of reproduction in females of this insect, the juvenile hormone-dependent synthesis of egg-case proteins in the collecterial gland, is associated with an elevation in the level of specific messenger RNAs, which is initiated prior to the first ovary-related major peak of hormone biosynthesis. Therefore, it seems increasingly probable that several important targets for juvenile hormone during female reproduction are not dependent upon the major "gonotrophic cycles" of corpus allatum activity which are so characteristic of insects with synchronous batch-wise development [2, 7, 19, 20].

We found that *Periplaneta* exhibits as many major peaks of juvenile hormone biosynthesis as it initiates consecutive waves of vitellogenesis, but here this means that each wave of oocytes experiences two cycles of hormone biosynthesis during its vitellogenic growth, because of the temporal overlap between terminal and pre-terminal oocytes in this species [8]. The follicles appear to be insensitive to this double pulse of hormone synthesis, because their growth proceeds without major perturbation during the period of approximately one day at mid-vitellogenesis when there is a more than fivefold reduction in the rate of hormone synthesis [8]. Apparently, whilst we can use the major pulses of corpus allatum to "count the vitellogenic cycles", the animal is not using them to exert rapid dynamic control over the rate of vitellogenesis, which proceeds unabated at maximum rate in fed mated animals. Of course, it is possible that the size of the hormone pool and its turnover time in the haemolymph, which have not been investigated in detail in this species, may exert a major damping on fluctuations in titer perceived by the target tissues, yet this would not help us understand how or why the number of major hormone pulses is numerically linked with the number of vitellogenic waves. Certainly this linkage is robust, for it persists in a modified form during the enforced trauma of starvation and re-feeding [7]. Nevertheless, it may be more than coincidence that the cessation of vitellogenesis during starvation, as evidenced by absence of growth and loss of patency in the follicular epithelium, only occurs when the rate of juvenile hormone synthesis has fallen below ca. 2 pmol/pair glands/h, i.e. below the normal minimum of the gonotrophic cycles. Furthermore, during the recovery after refeeding, the normal maximum rate of vitellogenesis has been reinstated by the time the activity of the corpora allata exceeds 5 pmol/pair gland/h, and the first recovery wave is nearly finished by the time the glands reach their maximal activity of ca. 30 pmol/gland pair/h [7]. According to this view only low levels of juvenile hormone biosynthesis are required to initiate maximum rates of vitellogenesis, and corpus allatum activities of greater than ca. 4 pmol/pair gland/h in this species must serve some other purpose, as yet unidentified.

If one pursues this hypothesis that the major cycles of hormone biosynthesis in the corpora allata are coordinated with successive waves of vitellogenesis in Orthoptera and Dictyoptera because the glands are subservient to the ovaries, possibly via the brain, then one may speculate that alternating positive and negative stimuli to the glands have a concerted effect on the amplitude of the allatal cycles. Positive stimuli would reflect a state of early vitellogenesis, negative stimuli would reflect impending chorionation or perhaps ovulation. The intensity of these stimuli could vary much with the reproductive strategy of the species, particularly with respect to the operation of phagatory and sexual autonomy and whether or not brood are carried after ovulation [12]. In many insects enforced virginity imposes a third ovarian state, that of arrested vitellogenesis. In *Rhodnius prolixus*, whose telotrophic ovaries show intra-ovarian asynchrony, vitellogenesis is arrested at all stages in the virgin: the follicular epithelia lose their patency and become refractory to juvenile hormone when their physiology is changed by the action of an abdominal anti-gonadotropin [21, 22], but no corpus allatum assay is available for this species. In *Periplaneta* virginity imposes a small delay on the entry into vitellogenesis of the first wave, which then proceeds normally to its mid-point (2.5 mm oocytes); at this time there is a rapid loss of patency in the follicle epithelia and a dramatic reduction in the growth rate of terminal and pre-terminal oocytes, which eventually enter complete arrest by age 25 days [8]. However, we found no such dramatic effect on the corpora allata, other than a failure to develope the normal cycles: the rate of juvenile hormone synthesis was "frozen" at an intermediate value of ca. 12 pmol/pair gland/h, then slowly declined the ensuing two weeks to a chronic minimum of ca. 4 pmol/pair gland/h. This rather "indecisive" response would be expected if the altered status of the follicles denied the brain-corpus allatum axis its usual positive stimulus but also completely prevented production of the normal "end of vitellogenesis" negative signal [19]. The result is a slow collapse to a level of hormone synthesis which we have already argued (vide supra) may be quite sufficient to maintain full reproductive function in all presently known tissues, provided they are receptive to the hormone. This suggestion is not inconsistent with our findings in surgically ovariectomised *Periplaneta* [9].

Thus, our attempts to thoroughly evaluate the quantitative data on rates of juvenile hormone biosynthesis in a few species under certain physiological and unphysiological conditions appear to raise as many questions as they help solve. Even seemingly straightforward experiments on the graded chemical allatectomy of *Locusta* with precocene [23], called to question the reproductive significance of the main peak of corpus allatum activity during the first vitellogenic cycle. We found, using a farnesoic acid stimulation assay to measure the maximum synthetic capability of the partially necrotic glands, that over 75% of their synthetic potential could be destroyed before there was any quantitative effect on the growth of the oocytes. Again, we were forced to suggest that so far as the ovaries are concerned there is "excess potential" in the corpus allatum, although we have no answer to the question of its physiological function. These ideas may not be novel, although they are difficult to reconcile with existing interpretations based upon physiological and hormone therapy experiments; moreover, they must be considered in the light of the now large body of quantitative biochemical data on corpus allatum activity from many other laboratories, a representation of which is beyond the scope of this article. What may be clear, is that quantitative assays for juvenile hormone biosynthesis will continue to play a vital role in elucidating the biochemistry of complex interactions between ovary, brain and corpus allatum.

3. Chemical Inhibition of the Corpus Allatum

Our finding that periods of maximum hormone synthesis by the corpora allata of adult female *Periplaneta* could be predicted fairly accurately by timing from ootheca formation [24], enabled us to devise a simple protocol with bi-laterally randomised glands to test for acute inhibitors of spontaneous hormone synthesis during a 3-h assay in vitro [25]. We examined examples from quite a wide range of organic chemical classes but only found interesting inhibitory activity in putative mixed function oxidase inhibitors containing methylenedioxyphenoxy or propinyloxy groupings. Subsequently, other laboratories have performed a more detailed survey of the activity of MFO inhibitors in a different cell-free preparation [26], or have used variations of the whole gland assay to identify two new types of inhibitor for some species: monacolin [27] and fluoromevalonate [28]. Meanwhile we have continued to use the whole gland assay to look at the activity of new methylenedioxyphenoxy and propinyloxy compounds; two of these have proved useful in exploring the mode of action of precocenes in other species [29, 30], but perhaps their most interesting outcome is the differential effect they appear to have on glandular levels of methyl farnesoate, the substrate for the epoxidase (vide infra).

Corpus allatum assays have also proved valuable in elucidating the biochemical mode of action of precocenes; these plant products were discovered elsewhere and heralded as possible lead compounds for a totally new type of insecticide [31]. They have the effect of selectively destroying the corpus allatum in those insects which are sensitive to them, and their action has been most studied in members of the Hemiptera and Orthoptera. Initially, the mode of action of precocenes was the subject of some confusion. We first reported that corpora allata of *Periplaneta*, a species which proved to be insensitive in vivo, are acutely and irreversibly inhibited by precocene in vitro, and some protection is afforded by co-administration of farnesoic acid, but the required concentrations of precocene (more than 1 m*M*) were unrealistically high [32]. Other laboratories explored the possibilities that precocenes might be bioactivated to reactive metabolites in peripheral tissues, or act on the corpus allatum via the brain, but the results were either negative or unconvincing [33–35]. Careful studies elsewhere on the course of shrinkage necrosis in the sensitive *Locusta* indicated that the inactivation was a rather slow process [36]. We therefore determined the circulating concentration of precocene-I during an effective dose to adult female *Locusta* [23], and showed that this was rapidly bioactivated by the corpora allata in vitro leading to massive alkylation of cellular molecules [30], which we suggest encompasses the instrument of cell death. Physicochemical studies on synthetic epoxide [37] have confirmed its exceptionally high reactivity, the solvent dependency of this reactivity, and its ability to scavenge nucleophiles [38, 39].

Our attempt to interpret the precocene effect, particularly with respect to structure-activity relationships, tissue and species specificity [40], has led to further work. We thought it would be helpful to determine the stereospecificity – if any – of precocene epoxidation in *Locusta* corpora allata, because it is well established that formation of the natural hormones is stereospecific 10R, which we have confirmed in *Locusta* [41]. We have analysed the cis and trans dihydrodiols of precocene-I, which are the main product of gland incubations in vitro, and found that the epoxidation is completely stereospecific 3R [42, 43]. By chance, this agrees with an earlier speculation [40]. The simplest interpretation of these stereospecificities is that both methyl farnesoate and precocene adopt respectively unique positions on the active site of the P-450 enzyme; moreover, it is interesting to see whether the observed stereochemistries would allow a high degree of iso-spatiality of these two rather different substrates. This is illustrated in Fig. 1, which shows how a particular configuration of the C_6–C_{12} portion of the flexible methyl farnesoate molecule can bring it into a high degree of congruence with the relatively rigid structure of precocene-I. The validity of this model is quite unproven at this time, but it does imply a number of structure-activity predictions which should be amenable to testing. The 6-methoxyl group of precocenes II and III should be iso-spatial with the 7-ethyl group of JH-I, and it is already known that the olefinic precursor of JH-I is a good, albeit unnatural, substrate for the allatal epoxidases in *Schistocerca* and *Periplaneta* [3, 44]. We have recently prepared a radio-labelled mixture of all

Fig. 1. Possible conformational similarity between methyl farnesoate (*left*) and precocene-I vis a vis the epoxidase, as deduced from the stereochemistry of their dihydrodiols. Drawn from photographs of Dreiding models: oxygen atoms are stippled. (After [43])

8 geometrical isomers of this substrate and found that the 6E geometry is more important than the 2E geometry for the microsomal enzyme from *Locusta* corpora allata (unpublished work from this laboratory): study of methyl farnesoate analogues with closed rings at the carboxy terminus might indicate the preferred configuration of this part of the molecule on the enzyme. We are currently investigating some other analogues and "hybrid-analogues", based upon geometric analogy.

However, prediction of substrate analogue activity becomes more complex when one considers the electronic structure in the region of the double bond; this cannot be ignored because it is this difference which confers the dramatic difference in reaction rate and mechanism between the resultant epoxides of methyl farnesoate and precocenes [40]. The alkoxy groups on the benzo ring of precocenes confer a strong electron induction away from the C_{3-4} centre, only partly reflected in the down-field shifts of their ^1H and ^{13}C NMR signals [45]. Since formation of the epoxide ring involves attack of the p-π electrons on positively polarised oxygen in the Fe^{3+} haem complex, one may suggest that the activation energy of the reaction would be reduced, and hence the reaction would be promoted, if the donor olefin could stabilise a formal positive charge in the transition state, relative to the ground state energy. Simple PMO theory [46] indicates that this would be available at C_4 in precocenes from the 7-alkoxy heteroatom, analogous to the stabilisation of the carbocation in solvolysis of the epoxide; this reasoning implies a predominantly C_3-O axis for the primary bond-forming reaction in epoxidation. Similarly, yet in contravention of the model analogy suggested by Fig. 1, a predominantly uni-centred attack from methyl farnesoate would be along the $C_{10}-O$ axis, with inductive stabilisation of a positive centre at C_{11}. The implication is that there may be a different geometry for the minimal energy reaction coordinate in the case of -I (e.g. tri-alkyl) and -E (e.g. *o*/*p* anisolyl) substituted double bonds. Unfortunately, these attempts at speculation seem to be presently unassailable by direct experimentation, although they would surely be improved or rejected by numerical computation of the energy surfaces for the electrophilic attack of oxygen on these differently substituted double bonds, but these are not yet available. In the meantime, and in an attempt to assess the importance of a balance between inductive and electromeric effects on the reactivity of the 3,4 double bond in precocenes, we have prepared the tritium-labelled positional isomer of precocene-I, 6-methoxy-2,2-dimethyl-benzo(b)-pyran (iso-precocene-I, see [39]), and examined its metabolism by isolated microsomes from *Locusta* corpora allata using a modification of our methyl farnesoate epoxidase assay [41] (unpublished results from this laboratory). The detection limit in these assays was ca. 3% the rates of dihydrodiol formation from precocenes-I and II at similar concentrations (10–15 μM), and we observed no significant enzymic formation of iso-precocene-I dihydrodiol. The simplest explanation of this result may be that whilst the shape of the iso-precocene molecule could be acceptable to the epoxidase, as previously suggested [40], there is an important need for powerful +I substituents to be balanced by –E effects, in order to facilitate enzymic epoxidation. Thus there may be a parallel importance of positional effects due to hetero-atoms in the π system of precocene analogues, for both enzymic formation of the 3,4-dihidroepoxide and then its high reactivity.

4. Studies on Individual Enzymes

Studies on isolated enzymes from the biosynthetic pathway are an essential complement to those on intact glands, principally for two reasons. First, they may lead to fully controlled assays for enzyme activities, which will be helpful in elucidating the mechanism of physiological regulation of juvenile hormone biosynthesis. Second, they allow investigation of substrate specificity and sensitivity to inhibition, free from the effects of limited access, intracellular compartmentalisation or product accumulation. Other laboratories have made important advances in the isolation and evaluation of enzymes from the early part of the pathway [47–50], whereas we have concentrated on the two final enzymes, namely the S-adenosylmethionine: farnesoic acid O-methyl transferase (EC 2.1.1) and the methyl farnesoate epoxidase [methyl farnesoate, reduced flavoprotein: oxygen oxidoreductase (10,11-epoxide forming) (EC 1.14.14)] [41], from corpora allata of mature female *Locusta*. Both enzymes appear to be present at high activity at all times in insects from the lower Orders [48, 51], although we have observed a major reduction in the overt epoxidase activity of intact *Periplaneta* glands after prolonged starvation, which could well be due to a reduction in enzyme level [7]. There are interesting differences between species in the relative overt activity of these two last enzymes, in intact glands [5, 52]. However, none of these quantitative aspects have yet been examined with defined cell-free assays.

The O-methyl transferase is a soluble enzyme capable of methylating a wide range of farnesoic acid analogues at high rates [53]. In all cases where the acid possessed a conjugated 2,3 double bond, the unnatural 2Z isomer was completely rejected by the enzyme; however, the enzyme shows little sign of stereoselectivity at the other terminus when the epoxide group is present [54], in marked contrast to the complete stereoselectivity for 10R juvenile hormone acid reported for the extra-corpora allata hormone from male *Hyalophora cecropia* [55]. Although the physiological substrate for this enzyme has not actually been determined in *Locusta* it would be rather suprising, from the results of precursor incorporation by glands of two species of cockroach [48, 54], if it were anything other than 2E, 6E farnesoic acid. Accordingly, we suppose that this enzyme plays no part in determining the isomer structure of the juvenile hormone released by glands from lower Insects. The situation in Lepidoptera seems to be unstudied, apart from the anomolous male *Hyalophora*.

The methyl farnesoate epoxidase is a microsomal enzyme whose known properties are completely consistent with it being a P-450 mixed function oxidase [41], although insufficient material has been available for spectral studies. The enzyme has a K_m for methyl farnesoate of 8 μM, which is close to the monomeric solubility of the ester in aqueous buffers, and a K_m for NADPH of ca. 2.6 μM; the Warburg partition coefficient for carbon monoxide inhibition (4.0) is unexceptional [41]. The enzyme is of interest to us because of its broad susceptibility to inhibition [26, 29, 30], which accounts for one of the specific IGR activities of piperonyl butoxide in a lepidopteran species [28], and because of its ability to convert precocenes to allatocidal epoxides (vide supra). Very little is known about the substrate specificity of this particular P-450 enzyme, apart from our finding of complete stereospecificity and regiospecificity towards both methyl farnesoate and precocene-I in *Locusta* [30, 41, 43]. From the point of view of agriculture, and the possible development of new pest control chemicals, it would be helpful if further investigation of the allatal epoxidase could include comparisons with the enzyme from lepidopteran source(s).

We have already stated that density gradient centrifugation is the preferred method for isolating clean microsomes from these small quantities of tissue [41]. I am indebted to Dr. D.A. Schooley for pointing out that this presently represents the highest degree of purification of any enzyme in the biosynthetic pathway, which must be a salutory thought for many juvenile hormone biochemists. In the course of our further studies we have refined the sucrose density method by adoption of non-linear gradients, which give tighter banding of the endoplasmic reticulum derivatives whilst retaining good separation from other cytoplasmic components. Figure 2 shows an application of this method in a typical experiment to determine the cytoplasmic distribution of bound precosyl moieties after incubation of *Locusta* corpora allata with radio-labelled precocene-I. A total of 20 pairs of corpora allata were incubated for 1 h with 20–25 μM (variously) of labelled precocene (32 mCi/mmol) exactly as described [30]. The glands were briefly washed in unlabelled culture medium, then homogenised, centrifuged and aliquots of the fractions analysed for the microsomal marker enzyme, exactly as described [41], except for the use of non-linear density gradients. As can be seen from Fig. 2, the distribution of precosyl residues which cannot be extracted by organic solvent nor discharged by brief boiling in water, falls into four areas. The distribution of radioactivity which is organic solvent extractable, but only after boiling, is qualitatively similar (not shown). Separate experiments have confirmed that only the microsomal band can convert precocene to its dihydrodiols [56], yet it is clear that most of the binding occurs to the soluble macromolecules at the top of the gradient. The identity of the two fastest-moving radioactive peaks is not certain. The fastest component bands at the same density as the mitochondrial marker succinate dehydrogenase from parallel control homogenates and cannot be due to contamination with nuclei (not shown);

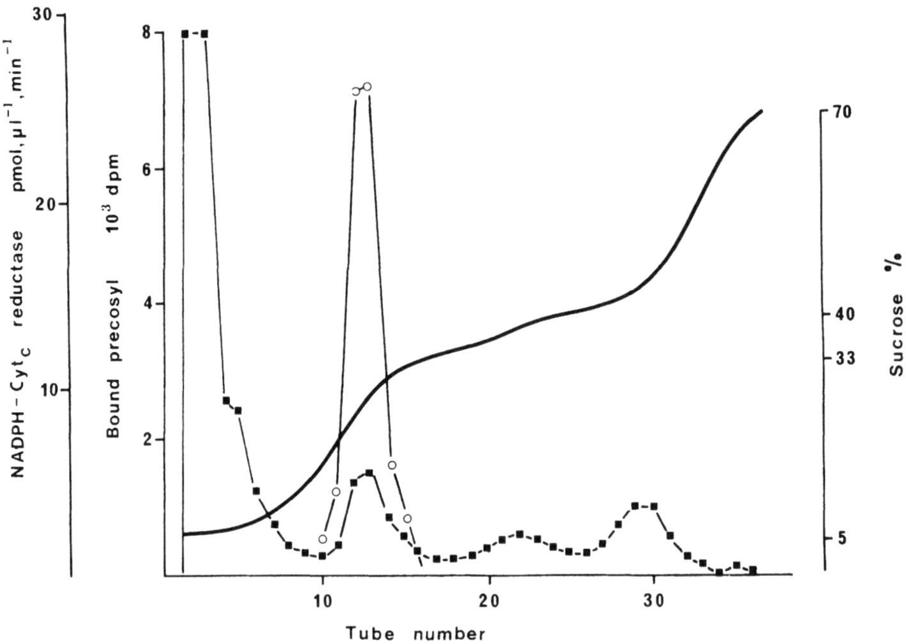

Fig. 2. Distribution of bound precosyl residues in the post-nuclear supernatant from *Locusta* corpora allata incubated with ^{3}H-precocene-I [30]. Buffered sucrose gradients were prepared as steps (0.7 ml 70%, 1.2 ml 40%, 1.2 ml 33%, 1.3 ml 5%) and allowed to stand at 5 °C for 3 h before application of 0.15 ml sample and centrifugation (3 h at 27.5 k r.p.m., MSE SS–50 swing-out). Aliquots of 8 drop fractions were co-precipitated with 2 mg bovine serum albumin plus 1 mg dextran sulfate prior to heat treatment, organic extraction and final dissolution [30]. Further details see text

therefore we strongly suspect that it represents labelling of corpus allatum mitochondria. The identity of the well resolved radioactive peak in between the microsomal and mitochondrial bands is quite unknown: it could represent stripped and free ribosomes, Golgi bodies, lysosomes or mitochondrial fragments. What is clear, from our failure to detect protein in these two fast peaks using standard micro-Biuret assay, is that there is a very high degree of specific binding to these cytoplasmic components. This can be compared with vertebrate studies on the bioactivation of benzo(a)pyrene, where a microsomal conversion to the reactive tetrahydrodihydroxyepoxide results in the highest specific labelling in the mitochondrial fraction [57]. Clearly, despite the very high reactivity of precocene dihydroepoxide towards water in ionically buffered aqueous model systems [38, 39], the complex milieu of the living cell seems to allow protected access of the microsomal product to many or all of the potential targets for the electrophilic carbocation [40]. Accordingly, it may be unwise for us to simplify the picture by concentrating further study on the labelling of microsomal components, even though the loss of membrane integrity does appear to be an important parameter of precocene-induced allatal necrosis [36].

5. Enzymic Balance and Interaction in the Pathway

The biosynthetic pathway for juvenile hormones has not really been investigated as a complete integrated whole in any insect, which is not suprising because of the considerable technical difficulties in making multiple biochemical measurements on those very small quantities of material which are usually available from experiments in vitro. Moreover, simple measurements of the overall rate of hormone biosynthesis have been quite adequate for many physiological purposes. However, it has proved possible to make some observations on the levels of biosynthetic intermediates in situ, by following radioactive incorporations from precursors such as acetate, mevalonate, farnesol, and farnesoic acid [5, 48, 52, 54]. It is recognised that there may be difficulties in using incorporation from early precursors to measure the pool size of biosynthetic intermediates, because of limited access or uncontrolled dilution from endogenous pools when glands are incubated in physiological media (cf. [58]). It may be neccessary to develope chemical methods, of appropriate sensitivity, in order to examine the relationships between major physiological changes in the flux through the pathway and the quasi-steady-state levels of earlier intermediates. These restrictions appear not to apply to measurement of methyl farnesoate levels, because incorporation from methyl-labelled methionine does not suffer from significant isotopic dilution [4, 5, 59].

We, and others, have found that methyl farnesoate levels fluctuate simply with the physiological flux through the pathway in several species [52, 59, 60] and unpublished results from this laboratory on *Tenebrio* and *Locusta*), although it is noteworthy that Lepidoptera have not yet been examined in this way. Moreover, it is well established by farnesoic acid replacement experiments, that the low levels of methyl farnesoate which are found at physiologically low rates of hormone synthesis are not caused by unavailability of the O-methyl transferase or its co-substrate S-adenosylmethionine, implying similarly low levels of farnesoic acid. Thus neither of the two final enzymes are responsible for physiological regulation of corpus allatum activity, although their levels might become rate-limiting at very high physiological rates in some species [48].

We had already found that methyl farnesoate levels could be made to „overflow", apparently because of saturation of the terminal epocidase, under conditions of farnesoic acid stimulation in glands from only some species of insect, notably *Periplaneta* and *Tenebrio* [5, 52], yet these very high levels (which are well in excess of the physical aqueous solubility) seem never to occur under even maximum physiological rates [52, 60]. We were also puzzled that several putatively specific inhibitors of mixed function oxidases could block juvenile hormone biosynthesis in vitro, without causing any accumulation of methyl farnesoate [25]. We tentatively speculated that high levels of methyl farnesoate might exert negative feed-back on the endogenous biosynthesis of farnesoic acid (at some unspecified earlier step) [52], but this was difficult to prove without reliable methods for measuring farnesoic acid levels in the glands. We are currently attempting to develope such methods, using HPLC [54], a novel internal standard and S.I.M. GC-mass spectrometry.

In the meantime, we have pursued our studies along these lines by examining the effects of a wider range of P-450 inhibitors (methylenedioxyphenoxy and propinyloxy compounds) on juvenile hormone biosynthesis by *Periplaneta* glands in vitro, and by making use of the fact that a wide variety of farnesoic acid analogues can be pushed into the system in parallel with endogenous farnesoic acid [4], because of the low substrate specificity of the O-methyl transferase in insects from lower orders [53, 61]. We have focussed particular attention on one compound [3,7-dimethyl-8-(3,4-methylene-dioxyphenoxy-) 2E,6E-octadienoic acid] which is a good substrate for the O-methyl transferase and, as its methyl ester, is an effective inhibitor of methyl farnesoate epoxidation (EC_{50} = 7 μM) in *Periplaneta* glands [60]. Thus, by using suitable separation systems, it is possible to measure simultaneously the rate of esterification of both farnesoic acid and the exogenous analogue, and hence investigate the interaction between the two. A full description of these experiments is beyond the scope of this article, and the results will be presented elsewhere. However, this is a good opportunity to summarise the findings and indicate how they have conditioned our current thoughts on the likely occurrence of negative feed-back loop(s) within the biosynthetic pathway.

Figure 3 illustrates some of the possible interactions between exogenous acids or P-450 inhibitors and the endogenous pathway; it is based solely upon measured incorporations from methionine into various esters, therefore that part of the pathway prior to farnesoic acid is treated as a "black box". It is to be expected that P-450 inhibitors will directly block the expoxidation of methyl farnesoate, and that exogenous farnesoic acid (or analogous thereof) may compete with endogenous farnesoic acid for the O-methyl transferase, but the point in question is whether they also augment negative feed-back in the endogenous pathway, as indicated by the question marks in Fig. 3. We have found [60] that the methyl octadienoate analogue described above can completely inhibit spontaneous juvenile hormone biosynthesis, with only a slight accumulation of methyl farnesoate, yet this powerful inhibition of farnesoic esterification can be completely relieved by simultaneous administration of farnesoic acid. We take this as prima facie evidence that this P-450 inhibitor (or methyl farnesoate analogue) blocks methyl farnesoate synthesis without inhibiting the O-methyl transferase itself or depriving it of its co-substrate S-adenosylmethionine, by default, we deduce a feed-back inhibition of farnesoic acid production. The question remains whether it is methyl

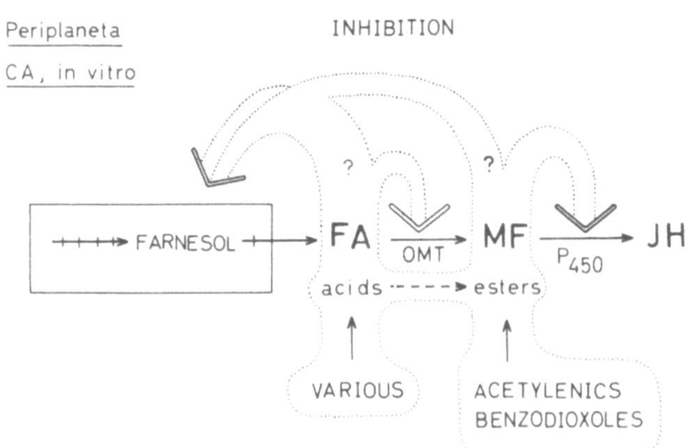

Fig. 3. Possible scheme for the inhibition of endogenous juvenile hormone (*JH*) biosynthesis by analogues of farnesoic acid (*FA*), and inhibitors of methyl farnesoate (*MF*) epoxidase. See text for interpretation

farnesoate, the analogue or some other factor which is mediating the proposed feed-back. There is also an important inhibition of methyl farnesoate synthesis (up to ca. 80%) when graded doses of farnesoic acid analogues are administered to spontaneously active glands, and some analogues are 50 times more effective as inhibitors than as alternative substrates for the O-methyl transferase. Again, it is not clear whether it is the endogenous intermediates or their chemical analogues which are responsible; possibly, there may be synergism. The *dènouement* of this pharmacological enigma, which certainly supports existing speculations of feed-back regulation in the biosynthetic pathway [48, 52], must come from more detailed studies of earlier enzymes in the pathway.

Acknowledgments. I am grateful to my colleagues in this laboratory, presently B.A.J. Alexander, G.T. Brooks, M. Drosopoulos, A.F. Hamnett, R.C. Jennings, D.W. Mace, A.P. Ottridge, R.N. Pau, M.B. Pople, K.M. Stott, and R.J. Weaver, whose work and encouragement, but not neccessarily opinion, has contributed much to this article. I also acknowledge the contributions from many colleagues elsewhere, whose signal discoveries in this area since 1975 could not be cited fully here, yet they have illumined many a previously darkened corner of the field.

References

1. Pratt GE, Tobe SS, Weaver RJ, Finney JR (1975) Gen Comp Endocrinol 26:478–484
2. Tobe SS, Pratt GE (1975) J Exp Biol 62:611–627
3. Pratt GE, Tobe SS (1974) Life Sciences 14:575–586
4. Pratt GE, Weaver RJ (1978) In: Gaillard PJ, Boer HH (eds) Comparative endocrinology. Elsevier/North Holland Biomedical Press, Amsterdam, pp 503–505
5. Weaver RJ, Pratt GE, Hamnett AF, Jennings RC (1980) Insect Biochemistry 10:245–254
6. Tobe SS, Pratt GE (1974) Nature 252:474–476
7. Weaver RJ, Pratt GE (1981) J Ins Physiol 27:75–83
8. Weaver RJ, Pratt GE (1977) Physiol Entomol 2:59–76
9. Weaver RJ (1981) Experientia 37:435–436
10. Weaver RJ (1984) J Insect Physiol (submitted)
11. Pratt GE, Pener MP (1983) J Ins Physiol 29:33–39
12. Engelmann F (1970) The physiology of Insect Reproduction. Pergamon Press, Oxford
13. Tobe SS (1977) Can J Zool 55:1509–1514
14. Weaver RJ (1979) Can J Zool 57:343–345
15. Tobe SS, Chapman CS, Pratt GE (1977) Nature 268:728–730
16. Weaver RJ (1981) In: Pratt GE, Brooks GT (eds) Juvenile hormone biochemistry. Elsevier/North Holland Biomiedical Press, Amsterdam, pp 271–283
17. Pau RN (1981) In: Pratt GE, Brooks GT (eds) Juvenile hormone biochemistry. Elsevier/North Holland Biomedical Press, Amsterdam, pp 285–293
18. Pau RN (1984) Biochem Biophys Acta (in press)
19. Stay B, Friedel T, Tobe SS, Mundall EC (1980) Science 207:898–900
20. Lanzrein B, Wilhelm R, Buschor J (1981) In: Pratt GE, Brooks GT (eds) Juvenile hormone biochemistry. Elsevier/North Holland Biomedical Press, Amsterdam, pp 147–160
21. Pratt GE, Davey KG (1972) J Exp Biol 56:223–237
22. Davey KG (1981) In: Pratt GE, Brooks GT (eds) Juvenile hormone biochemistry. Elsevier/North Holland Biomedical Press, Amsterdam, pp 233–240
23. Pratt GE, Jennings RC, Hamnett AF (1981) In: Sehnal F, Zabza A, Menn JJ (eds) Regulation of Insect development and behaviour, vol 2. Wroclaw Technical University Press, Wroclaw, Poland, pp 347–356
24. Weaver RJ, Pratt GE, Finney JR (1975) Experientia 31:597–598
25. Pratt GE, Finney JR (1977) In: McFarlane NR (ed) Crop protection agents: their biological evaluation. Academic Press, London New York, pp 113–132
26. Hammock BD, Mumby SM (1978) Pest Biochem Physiol 9:30–47

27. Edwards JP, Price NR (1983) Insect Biochem 13:185–189
28. Kramer SJ, Staal GB (1981) In: Pratt GE, Brooks GT (eds) Juvenile hormone biochemistry. Elsevier/North Holland Biomedical Press, Amsterdam, pp 425–437
29. Brooks GT, Hamnett AF, Jennings RC, Ottridge AP, Pratt GE (1979) In: Proc Brit Crop Prot Conf: Pests & Diseases, pp 273–279
30. Pratt GE, Jennings RC, Hamnett AF, Brooks GT (1980) Nature 284:320–323
31. Bowers WS, Ohta T, Cleere JS, Marsella PA (1976) Science 193:542–547
32. Pratt GE, Bowers WS (1977) Nature 246:548–550
33. Ohta T, Kuhr RJ, Bowers WS (1977) J Agric Food Chem 25:478–481
34. Bergot BJ, Judy KJ, Schooley DA, Tsai LW (1980) Pest Biochem Physiol 13:95–104
35. Unnithan GC, Nair KK, Kooman CJ (1978) Experientia 34:411
36. Schooneveld H (1979) Cell Tissue Res 203:25–33
37. Jennings RC, Ottridge AP (1979) Chem Commun 920–921
38. Hamnett AF, Ottridge AP, Pratt GE, Jennings RC, Stott KM (1981) Pesticide Science 12:245 to 254
39. Hamnett AF, Pratt GE, Ottridge AP, Alexander BAJ (1982) In: Hawk GL (ed) Biological/biomedical applications of liquid chromatography IV. Marcel Dekker, New York Basel, pp 111 to 133
40. Pratt GE (1983) In: Whitehead DL (ed) Natural products for innovative pest management. Pergamon Press, Oxford New York, pp 323–355
41. Feyereisen R, Pratt GE, Hamnett AF (1981) Eur J Biochem 118:231–238
42. Jennings RC (1982) Tetrahedron Letters 23:2693–2696
43. Hamnett AF, Pratt GE (1983) Life Sciences 32:2747–2753
44. Pratt GE, Hamnett AF, Finney JR, Weaver RJ (1978) Gen Comp Endocrinol 34:113
45. Ottridge AP, Jennings RC, Brooks GT (1981) In: Pratt GE, Brooks GT (eds) Juvenile hormone biochemistry. Elsevier/North Holland Biomedical Press, Amsterdam, pp 381–386
46. Dewar MJS, Dougherty RC (1975) The PMO theory of organic chemistry. Plenum Press, New York
47. Kramer SJ, Law JH (1980) Insect Biochem 10:569–575
48. Feyereisen R, Koener J, Tobe SS (1981) In: Pratt GE, Brooks GT (eds) Juvenile hormone biochemistry. Elsevier/North Holland Biomedical Press, Amsterdam, pp 81–92
49. Baker FC, Schooley DA (1981) Biochem Biophys Acta 664:356–372
50. Baker FC, Lee E, Bergot BJ, Schooley DA (1981) In: Pratt GE, Brooks GT (eds) Juvenile hormone biochemistry. Elsevier/North Holland Biomedical Press, Amsterdam, pp 67–80
51. Tobe SS, Pratt GE (1976) In: Gilbert LI (ed) The juvenile hormones. Plenum Press, New York, pp 147–163
52. Pratt GE, Tobe SS, Weaver RJ (1975) Experientia 31:120–122
53. Pratt GE, Stott KM, Brooks GT, Jennings RC, Hamnett AF, Alexander BAJ (1981) In: Pratt GE, Brooks GT (eds) Juvenile hormone biochemistry. Elsevier/North Holland Biomedical Press, Amsterdam, pp 107–121
54. Hamnett AF, Pratt GE, Stott KM, Jennings RC (1981) In: Pratt GE, Brooks GT (eds) Juvenile hormone biochemistry. Elsevier/North Holland Biomedical Press, Amsterdam, pp 93–105
55. Peter MG, Gunawan S, Emmerich H (1979) Experientia 35:1141
56. Pratt GE, Hamnett AF, Jennings RC, Alexander BAJ, Feyereisen R (1982) Gen Comp Endocrinol 46:378
57. Allen JA, Coombs MM (1980) Nature 287:244–245
58. Schooley DA, Judy KJ, Bergot BJ, Hall MS, Jennings RC (1976) In: Gilbert LI (ed) The juvenile hormones. Plenum Press, New York, pp 101–117
59. Feyereisen R, Friedel T, Tobe SS (1981) Insect Biochem 11:401–409
60. Pratt GE, Jennings RC, Weaver RJ (1984) Insect Biochemistry (submitted)
61. Hamnett AF, Pratt GE (1978) J Chromatog 158:387–399

III.5 Regulation of Juvenile Hormone Titer in the Adult Female Cockroach *Diploptera punctata*

S.S. TOBE[1], B.A. STAY[2], F.C. BAKER[3], and D.A. SCHOOLEY[3]

1. Introduction

It is now widely accepted that changes in the titers of both juvenile hormone(s) and ecdysteroids control, at least in part, the process of metamorphosis of most insect species [1]. Similarly, JH (juvenile hormone) appears to be an important regulator of reproduction in female insects of most species, with the possible exception of certain members of the Lepidoptera and Diptera [2]. JH may exert its effects at a number of levels, including synthesis of vitellogenin and uptake of vitellogenin by oocytes and synthesis of secretion by female accessory glands [3]. The original experiments suggesting a role for JH in reproduction involved the extirpation of the source of the JH, the corpora allata - following extirpation, egg production ceased; hormone replacement therapy using synthetic hormones or analogues was able to reverse this effect [2, 4]. Transplantation experiments as well as bioassay data have also strongly suggested the importance of JH in the regulation of oocyte growth and maturation. Nonetheless, extirpation and transplantation experiments simply demonstrated the necessity of the corpora allata for oocyte production and although these early results suggested that changes in corpus allatum activity were responsible for the cycles of reproduction observed in many species, with periods of high activity associated with rapid oocyte maturation, confirmation of this hypothesis awaited the development of techniques for measuring corpus allatum activity directly. Since the development of a radiochemical assay for JH biosynthesis [5-7], corpus allatum activity has been determined during the reproductive cycle of several species including *Schistocerca gregaria* [8], *Periplaneta americana* [9], *Diploptera punctata* [10], *Tenebrio molitor* [11], *Leptinotarsa decemlineata* [12], *Locusta migratoria* [13], and *Melanoplus sanguinipes* [14]. Unfortunately, these species show different degrees of regulation of their reproductive processes (some showing discrete gonotropic cycles and others producing oocytes continuously) and this may be related both to the degree and mechanisms of regulation of the corpora allata [15]. One of the most extensively studied species has been *D. punctata* and in this viviparous species, periods of both rapid basal oocytes growth and high corpus allatum activity are widely separated by periods of relative quiescence (gestation) with respect to both oocyte growth and corpus allatum activity [10, 15-18]. In this

1 Department of Zoology, University of Toronto, Toronto, Ontario M5S 1A1, Canada
2 Department of Zoology, University of Iowa, Iowa City, IA 52242, USA
3 Zoecon Corporation, 975 California Avenue, Palo Alto, CA 94304, USA

Biosynthesis, Metabolism and Mode of Action of Invertebrate Hormones
(ed. by J. Hoffmann and M. Porchet)

species, there is an excellent correlation between JH biosynthesis and basal oocyte length and in fact, oocyte length can be used as a predictor of corpus allatum activity. Nonetheless, it is ultimately the JH titer which controls oocyte growth. Although there is ample evidence to suggest that JH biosynthesis is an accurate indicator of JH titer, at least in this species [15], direct evidence is still required to demonstrate that JH biosynthesis parallels precisely JH titer. Another aspect of this same question concerns the factor which regulate JH titer. Our results suggest that rates of JH biosynthesis must be of major importance in the regulation of JH titer in adult female *D. punctata*.

2. JH III Titer in Diploptera

JH III titer [19], as determined by gas chromatography/mass spectrometry with selected ion monitoring [20], undergoes dramatic changes during adult life of *D. punctata*. Of particular interest to this paper are the values for the first gonotrophic cycle, spanning the time of emergence (0) to day 8 postemergence. Representative values are shown in Fig. 1 as are representative values for in vitro JH biosynthesis, as determined by the radiochemical assay [5-7] for the corresponding times.

Maximal JH III titer in the haemolymph occurs about mid-cycle, with values of up to 6 μM. This value is equivalent to a concentration of about 1,600 ng ml^{-1} JH III, a remarkably high concentration for a putative regulatory hormone. JH titer values are much lower at the beginning of the gonotrophic cycle and at the end of the cycle although even these levels are high in comparison to those observed in other species during the gonotrophic cycle (maximum levels in females of *L. migratoria* [21] - 41 ng ml^{-1}; *L. decemlineata* [22] - 88 ng ml^{-1}; *T. commodus* [23] - 64 ng ml^{-1}; and *Aedes aegypti* [24] - 1.7 ng g^{-1}).

Highest titer occurs during the period of rapid vitellogenesis of the basal oocytes whereas low titer occurs prior to the onset of vitellogenesis and during the period of chorion formation (Fig. 1). We have previously demonstrated a precise and predictable

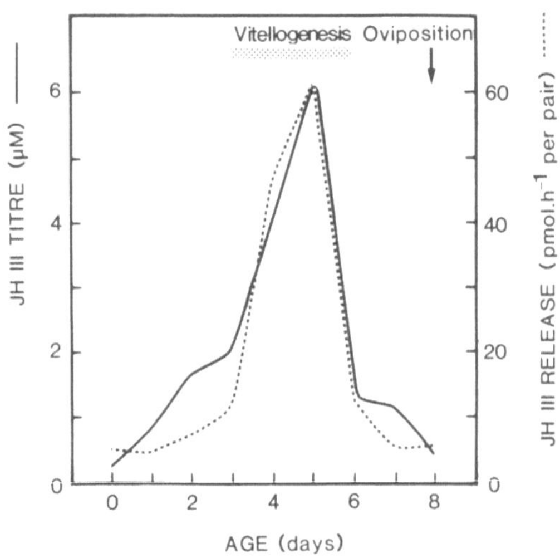

Fig. 1. Haemolymph JH III titer and JH III release in vitro in adult females of *D. punctata* during the first gonotrophic cycle. 0 represents the time of emergence and the arrow shows the time of oviposition. Vitellogenesis in basal oocytes occurs between days 3 and 6. Titer was determined on one group of animals and JH release on another group of animals. (Data from Tobe et al. [19])

correlation between length of basal oocytes and JH biosynthesis [15, 25] and the data presented in Fig. 1 suggest a similar relationship between oocyte length and JH III titer. The data also suggest that high levels of JH may be responsible for the rapid uptake of vitellogenin by basal oocytes during the period of intense growth. Nonetheless the data do not indicate that high JH titer is itself responsible for the *onset* of vitellogenesis [8, 15] because it is uncertain if JH titer increases prior to, at the same time or shortly after the onset of vitellogenesis.

The typical values for JH biosynthesis/release during the first gonotrophic cycle, shown in Fig. 1, vary somewhat between experiments, but the shape of the curve does not. Accordingly highest rates of biosynthesis are consistently observed on days 4 and 5. Rates may vary by up to a factor of 2 (range: 60-120 pmol h^{-1} per pair for maximal rates and 5-10 pmol h^{-1} per pair for minimal rates) but within any experimental series, rates show a proportional relation to one another, i.e., rates fall into the higher or lower range throughout the observation period. It should also be noted that although Fig. 1 shows data for JH release, these values are related in a precise linear fashion to JH biosynthesis in both this species and in other species [7, 10]. In *D. punctata*, values for rates of JH release for a 3 h incubation are 10-20% less than those of JH biosynthesis. We have previously defined JH release and JH biosynthesis [7, 10]; release represents the quantity of JH found in the incubation medium at the end of the incubation period whereas biosynthesis represents the sum of JH release and the quantity of JH within the corpora allata at the end of the incubation period.

JH titer closely parallels JH release/biosynthesis (Fig. 1). These data indicate conclusively that these two parameters are highly correlated and in fact, a plot of JH titer as a function of JH release (or biosynthesis) (Fig. 2) shows a very high correlation coefficient (r = 0.96; y intercept = 0.21 *μM*) (data from [19]). Thus JH release (or JH biosynthesis) is an excellent indicator of JH titer, at least in this species, during the first gonotrophic cycle. These data also suggest that JH titer is regulated largely by changes in rates of JH biosynthesis [15].

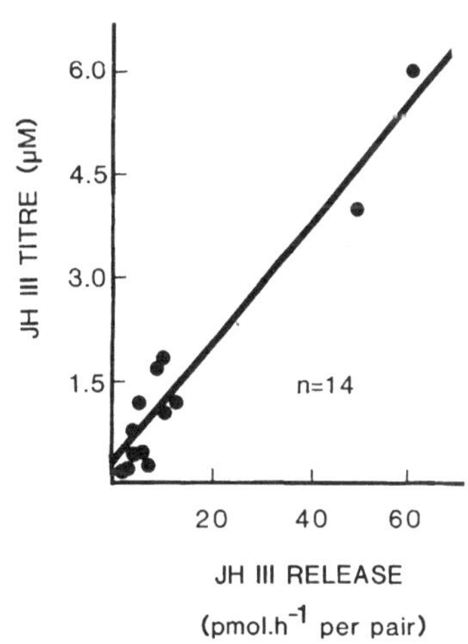

Fig. 2. Haemolymph JH III titer as a function of JH III release in female *D. punctata*. The y-intercept is 0.21 *μM* and the correlation coefficient, r = 0.96. (Data for this plot were taken from Tobe et al. [19])

3. Half-Life of Juvenile Hormone

We have previously observed that at least at certain times during the first gonotrophic cycle of *D. punctata*, there appears to exist an appreciable pool of JH outside of the haemolymph [19]. This conclusion was based largely on determinations of quantity of JH in both whole bodies and haemolymph. The existence of such a pool could have a dramatic effect on the apparent half-life of the hormone, particularly if the extra-haemolymph pool is in equilibrium with the haemolymph pool. Under these circumstances, degradation of the hormone in the haemolymph would result in the release of JH into the haemolymph from the extrahaemolymph pool and the half-life of the hormone would be considerably longer than expected.

Two methods have been used to estimate half-life of JH in *D. punctata*. The rate of disappearance of exogenous radiolabelled hormone following injection (in mineral oil) into the haemolymph has been determined and appears to be linear [26]. The half-life of the exogenous hormone in day 5 adult females appears to be 1.6 h and in view of the linear rate of disappearance, may be described by first-order kinetics. However, when the rate of disappearance of *endogenous* hormone following allatectomy was measured over a 16-h period, a very different pattern of disappearance emerged [19]. The initial rate of disappearance of the hormone is exponential (first 2–4 h) but thereafter declines appreciably [19]. Between 4 and 16 h after allatectomy, the rate of disappearance is non-exponential and much larger quantities of hormone are present than would have been predicted from the initial rate of disappearance (Fig. 3). The initial rate of disappearance provided an estimated half-life of the hormone of about 1–2 h, and this value is in good agreement with that obtained from experiments utilizing exogenous radiolabelled hormone (1.6 h) [26]. However, it is clear that these values underestimate appreciably the half-life of endogenous JH in *D. punctata*. Accordingly, great care must be exercised when estimating half-life of JH from experiments utilizing radiolabelled exogenous hormones.

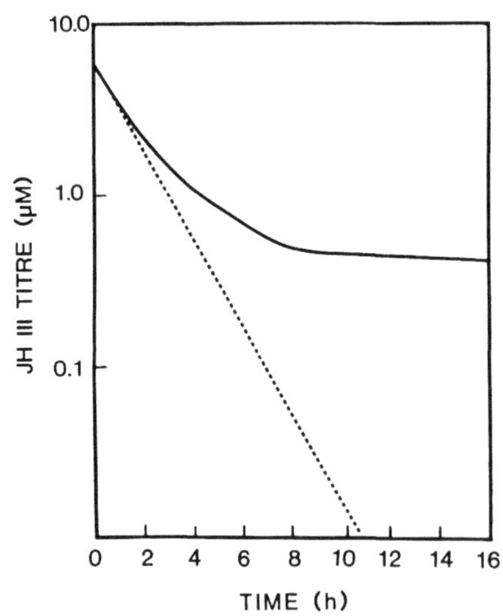

Fig. 3. Decay in endogenous haemolymph JH III titer following allatectomy of day 5 female *D. punctata* at time = 0 h. Initial half-life (*dotted line*) was derived directly from the time required for JH titer to decline to half that at t = 0 and was extrapolated thereafter. Apparent initial half-life is about 1.2 h. Actual measured JH titer is shown by the *solid line*. The predicted rate of decay in JH titer (*dotted line*), based on initial half-life, is considerably greater than that actually observed

The extended half-life value for JH which emerged from these studies has necessitated reinterpretation of data concerning effects of allatectomy on oocyte growth and theoretical rates of hormone biosynthesis. For example, several authors have used half-life values for JH to derive a value for the rate constant, assuming steady state conditions. This value was subsequently used to estimate the rate of hormone biosynthesis necessary to maintain a given hormone titer [27, 28]. Considerably higher rates of biosynthesis than those observed in vitro were predicted on the basis of these calculations; this subsequently led to the suggestion that rates of JH biosynthesis, as measured in vitro by the radiochemical assay, were less than those in vivo, probably because of deficiencies in the culture conditions. When similar calculations were done using the titer data from *D. punctata*, and an apparent half-life of 1.2 h, as expected, the calculated rates of hormone biosynthesis necessary to maintain the given titer were considerably greater than those actually observed [19]. However, when the longer observed half-life was taken into consideration (see Fig. 3), theoretical and actual values for rates of JH biosynthesis were similar (Tobe and Stay, in preparation). Thus it is likely that rates of JH biosynthesis in vitro are an accurate reflection of rates in vivo. However if half-life is underestimated, or if the rate of decay of hormone is assumed to be exponential, significant errors can be generated in the interpretation of data.

Calculations utilizing the rate constant to define rates of hormone biosynthesis or degradation usually assume steady-state conditions, that is V_S, the rate of hormone biosynthesis, is equal to the V_D, the rate of hormone degradation. However, it is clear from Fig. 1 that rates of hormone synthesis change dramatically during the first gonotrophic cycle as does the JH titer. Only at times when hormone titer remains constant could it be expected that $V_S = V_D$. However at times when the titer is increasing, for example between days 3 and 5, V_S should exceed V_D. Alternatively, it is possible that previously synthesized JH could be released from a storage pool during this period. We have previously demonstrated that there is only a small extrahaemolymph JH pool prior to day 3 and thus this pool cannot account for the large and rapid increase in haemolymph hormone titer between days 3 and 5 [19]. Accordingly, de novo hormone biosynthesis rather than reflux of previously synthesized hormone must be the principal factor causing the elevation in juvenile hormone titer. After day 5, the titer decreases sharply and this implies that V_D exceeds V_S. However, an additional or alternative mechanism for the rapid decline in JH titer is uptake of hormone by the tissues at this time and the subsequent release of the hormone from the pool at a rate which does not result in an appreciable elevation in hormone titer. Under these circumstances, degradation of hormone by esterolytic cleavage and subsequent excretion may occur, particularly if the JH is unprotected by association with either binding proteins or lipophilic pools, at a sufficiently high rate to prevent elevation in hormone titer. A significant percentage of the total JH in whole animals appears to be localized in pools outside of the haemolymph (up to 40%-50% [19]) on days 4 and 5, whereas by day 7, there is little if any hormone in the extrahaemolymph pool. Thus between day 5 and 7, there is probably an appreciable release of hormone from this pool into the haemolymph. Because haemolymph JH titer does not rise but rather declines during this period, the hormone must be undergoing degradation, either in the haemolymph or in the tissues. The uptake and subsequent release of hormone from the extrahaemolymph pool may therefore provide a mechanism for the animal to render the hormone susceptible to degradation by haemolymph esterases, possibly as a result of dissociation from protective binding proteins or lipophilic molecules.

4. Persistence of Juvenile Hormone Effects

Underestimates of half-life of JH can also cause difficulties with respect to interpretation of allatectomy data. This is particularly true when using oocyte growth as a biological indicator of JH titer as is often done in *D. punctata*. Accordingly we determined the time interval over which the effects of JH on oocyte growth could be observed following allatectomy. Groups of females were allatectomized on days 2, 3, 4, or 5 and basal oocyte length was measured on day 8 (the age at which oviposition occurs in normal females). Other females served as the unoperated controls and in groups of these animals, oocyte length was determined on each day between days 2 and 7. A final group of animals was used as the sham-operated controls and in this group, oocyte length was determined only on day 8. Following allatectomy of females, oocytes showed an increase in length only on days 2, 3, and 4, as shown in Fig. 4.

Figure 4 shows the basal oocyte lengths for normal females between days 2 and 7; the dynamics of oocyte growth observed are similar to those reported previously [10, 29]. Also shown in Fig. 4 are mean oocyte lengths on day 8 for females allatectomized on the day indicated on the abscissa. These values are higher than the control values and are indicative of the amount of oocyte growth following allatectomy. The time difference between these two curves represents the interval over which the effects of JH persisted; on day 2, it was 15.6 h, day 3, 15.6 h, and day 4, 19.2 h. The mean value was 16.8 h. Therefore, after allatectomy, the oocytes continued to increase in length for a period of about 17 h. Assuming that the oocytes responded rapidly to the presence or absence of JH, this value of 17 h must represent a minimum because there is probably of threshold concentration for hormone effect. This value is considerably greater than would be predicted on the basis of the observed *initial* half-life of JH and

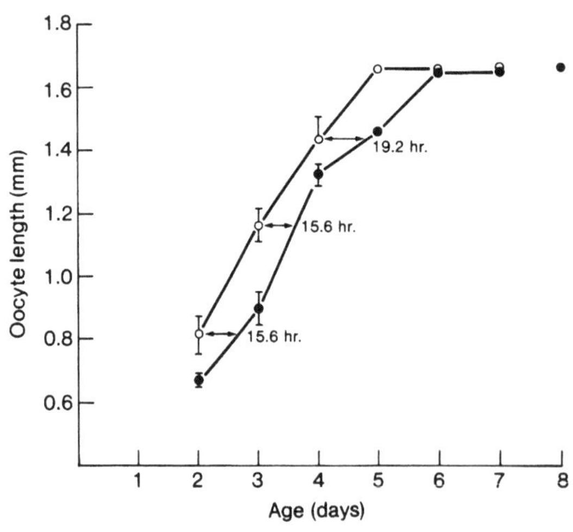

Fig. 4. Effect of allatectomy on basal oocyte growth in *D. punctata*. Vitellogenesis begins on day 3 (oocyte length about 0.80 mm) and mature oocytes attain a length of about 1.65 mm. Oviposition occurs on day 8 but oocyte length does not increase appreciably after 5–6. *Closed circles*: oocyte lengths in unoperated animals, on the days indicated. *Open circles*: oocyte lengths in animals, allatectomized on the indicated days, and measured on day 8. ⊗ represents the oocyte length of sham-operated animals, measured on day 8. The difference between the two curves, indicated by the *arrows*, represents the interval over which the effect of JH persists

supports our hypothesis regarding the existence of a pool of hormone outside the haemolymph which is in equilibrium with the haemolymph pool [19].

The persistence of JH effects provides evidence for the non-first-order decay in JH titer, which probably occurs as a result of efflux of JH from a sequestered pool. It is striking that the interval for persistence of the hormone effect is virtually identical on three different days studied (Fig. 4), particularly because oocyte growth in *D. punctata* appears to be dependent on "dose" or titer of JH [15, 30]. The half-life data on endogenous JH suggest that hormone titer approaches an equilibrium value (between 0.3 to 0.4 μM) about 16 h after allatectomy of the day 5 adults [19] (Fig. 3). This time is similar to the interval of persistence of JH effects following allatectomy (17 h). It is not surprising that the time for persistence of hormone effect is similar to the time required for the establishment of a "steady-state" condition with respect to JH titer following allatectomy, if it is assumed that the biological response (in this case, oocyte growth) exhibits a typical dose response and threshold for JH. If the threshold titer for biological effect is between 0.3–0.4 μM (see below), the interval for persistence of hormone effects would also be about 16 h.

The lowest JH titer observed to date in *D. punctata* is between 0.13 and 0.20 μM in last instar females and adult males [19] and it may be that haemolymph JH titer cannot be reduced below this level in normal animals, perhaps because of efflux of hormone from a sequestered pool. Even in animals with undetectable rates of hormone biosynthesis or allatectomized females, JH titer is appreciable (0.13 μM) [19]. Although this absolute quantity of JH is significant, when compared to many other species, this titer may represent essentially zero, i.e. the threshold JH titer in *D. punctata* for biological effect may be equivalent to a haemolymph JH titer of about 0.13 μM. This suggestion is supported by several observations on the occurrence of vitellogenin – in *D. punctata*, vitellogenin does not occur in normal males (JH III titer, 0.18 μM [19]), in female last instars (JH III titer, 0.13–0.22 μM [19]) or in newly emerged females (day 0) (JH III titer, 0.22 μM [19]) [31–33]. However, following denervation of corpora allata (a surgical procedure known to stimulate JH biosynthesis [34, 35]), implantation of supernumerary corpora allata, or treatment with the JH analogue ZR512, vitellogenin can often be detected in both males and in last instars [31–33]. Although the effect of these manipulations on JH titer is not known at present, it is unlikely that the increase in titer would be greater than twofold (0.26–0.44 μM), assuming that the JH titer is in fact proportional to the increase in rate of JH biosynthesis in vitro (as Fig. 2 would indicate). It is known that the increase in rates of biosynthesis after denervation are less than twofold in males and last instars [36, 37]. Similarly, implantation of supernumerary corpora allata results in overall rates of JH biosynthesis (by the host corpora allata plus implanted corpora allata) of slightly less than twice that of the host glands [29]. Thus, assuming proportionally between JH biosynthesis and titer, a twofold increase in biosynthesis would result in a similar increase in titer.

The line generated from data on JH titer as a function of biosynthesis (Fig. 2) also provides evidence for the existence of a threshold for JH action. The Y-intercept is 0.21 μM – thus, at a biosynthetic rate of zero, the JH titer is 0.21 μM [19]. Because there is an excellent correlation ($r = 0.96$) between these two parameters [19], the Y-intercept value is probably a good indicator of the JH titer expected after allatectomy and may approximate the minimum JH titer, encountered in *D. punctata*. Therefore, the threshold titer for JH effect in adult females may be about 0.2 μM.

5. Extrahaemolymph Pool of Juvenile Hormone

The results from whole-body and haemolymph JH titer determinations have led to the conclusion that there is a substantial pool of hormone outside the haemolymph at least at certain specific times during the gonotrophic cycle of *D. punctata* [19]. This conclusion is supported both by our observations on the extended half-life of the hormone following allatectomy and by the persistence of biological effects following allatectomy. However, the problem of the location(s) of this pool remains to be resolved. An obvious choice for the location of such a pool is the fat body, not only because of the large quantity of lipid material which this tissue contains (and accordingly, one would expect a substantial partitioning of JH into the lipid pools) but also because the fat body is a target tissue of JH and therefore must possess JH receptors. Similarly the ovary and particularly the developing vitellogenic oocytes might also be expected to sequester JH. If vitellogenin does in fact bind JH with high affinity in *D. punctata*, as is the case in *Leucophaea maderae* [38], this provides yet another route of entry for JH. Nonetheless, the fat body and ovary represent only a small percentage of the total body weight and although these tissues may sequester an appreciable quantity of JH, when expressed as a percentage of the tissue weight, the relatively small mass of tissue may be capable of sequestering only a small proportion of the total extrahaemolymph pool of JH. Measurement of JH content of these two tissues by physicochemical methods [20] has confirmed the presence of appreciable quantities of hormone within the tissues [19], particularly when expressed as a percentage of tissue weight. However, the total quantity of hormone in these tissues is only a small percentage of the total extrahaemolymph pool and thus we must continue to search for other sites of sequestration of JH. To date, our investigations have been inconclusive; however, *Diploptera* should prove to be a most useful species in which to investigate hormone sequestration, in view of the very large quantities of JH found in adult females of this species.

One final point with regard to the existence of the extrahaemolymph JH pool should be considered. All calculations for the quantity of hormone in the haemolymph were performed using values for haemolymph volume obtained by the radiolabelled inulin dilution technique [32]. Because JH and inulin are very different molecules with very different adsorption and partition characteristics, it is possible that the total inulin space is quite different from the total JH space. If the total JH space is appreciably larger than the inulin space, the calculated quantity of hormone in the haemolymph will be significantly underestimated and the size of the extrahaemolymph hormone pool overestimated. Koolman [39] has suggested a similar disparity in the ecdysteroid space when compared to inulin space.

6. Factors Regulating Juvenile Hormone Titer

The results summarized in this chapter and elsewhere [19] demonstrate an excellent correlation between JH titer and hormone release/biosynthesis. Thus regulation of the rate of biosynthesis of the hormone is an important factor in the overall regulation of JH titer. However, the role of JH-specific esterase in titer regulation remains to be clarified. The massive catabolic activity present in the haemolymph of most species which have been studied and the relative changes which occur in this activity both during development (see Hammock, this vol.) and during the reproductive cycle [40] argue strongly for a role for JH-catabolic enzymes in the regulation of JH titer. However, several points must be remembered. Firstly, the catabolic activity of the haemolymph far exceeds the in vitro rate of biosynthesis of JH (in *D. punctata*, by a factor of several thousandfold [26]) and accordingly, any free unprotected hormone could be expected to undergo degradation almost instantaneously. Secondly, virtually all studies on the JH-specific esterase have utilized the commercially available racemic [^{3}H]–JH rather than the natural enantiomer. Recent studies have suggested that in some species the esterase may not selectively degrade the natural enantiomer [41, 42], but clearly this must be established for each species. It will also be mandatory to determine the effect of the unnatural enantiomers on hydrolysis of the natural substrate. Thirdly, most studies have utilized crude diluted haemolymph preparations for study of the JH-specific esterase. If there are proteins in the haemolymph with high affinity for JH, as now seems likely in many species [38, 42], the presence of such molecules could affect appreciably the apparent rate of hydrolysis of the hormone substrate, especially in dilute solution. In addition, if the binding proteins exhibit stereoselectivity with respect to JH, the unnatural enantiomers could undergo rapid esterolysis whereas the natural hormone may not. Again the importance of using only the authentic natural enantiomer of JH for such studies is obvious.

Finally, the existence of JH-esterase inhibitors could affect profoundly the rate of degradation of the hormone. Engelmann [38] has reported that two different proteins in the haemolymph of *L. maderae*, which have a high affinity for JH ($K_D \cong 10^{-8}$ *M*), are also inhibitors of the esterase. Although the mode of inhibition of these proteins remains to be defined, it is possible that they function not only to prevent the substrate from reaching the active site of the esterase as a result of binding, but also inhibit the enzyme directly. In either case, inhibition of JH-specific esterases by binding components in the haemolymph could influence appreciably the rate of degradation in vitro, particularly in diluted haemolymph.

For the above reasons, it is likely that determinations of rates of JH esterolysis in vitro, presumably by JH-specific esterases, are inaccurate. Nonetheless, the enormous catabolic activity far exceeds the rates of JH biosynthesis and accordingly, the importance of JH esterases in the regulation of JH titer in *D. punctata* remain to be resolved. The esterase is capable of degrading JH at very high rates and accordingly, any free hormone in the haemolymph will be rapidly broken down. This suggests that most hormone in the animal is normally bound or sequestered. At times of declining hormone titer, the size of the sequestered pool decreases appreciably and this suggestion is supported by our observations on haemolymph vs whole body JH titer on day 7 [19]. However, we do not understand the nature of these sequestered pools or the means by which they are utilized. In addition, we have not identified definitively the sites of these pools. These areas, as well as reassessment of enantiomer-specific JH esterases and JH binding, will provide exciting research opportunities for the next few years and should permit a more complete understanding of the regulation of JH titer in insects.

References

1. Wigglesworth VB (1964) Advan Insect Physiol 2:247–336
2. Engelmann F (1970) The physiology of insect reproduction. Pergamon Press, Oxford
3. Engelmann F (1979) Advan Insect Physiol 14:49–108
4. Wigglesworth VB (1970) Insect hormones. Oliver and Boyd, Edinburgh
5. Pratt GE, Tobe SS (1974) Life Sci 14:575–586
6. Tobe SS, Pratt GE (1974) Biochem J 144:107–113
7. Pratt GE, Tobe SS, Weaver RJ, Finney JR (1975) Gen Comp Endocrinol 26:478–484
8. Tobe SS, Pratt GE (1975) J Exp Biol 62:611–627
9. Weaver RJ, Pratt GE, Finney JR (1975) Experientia 31:597–598
10. Tobe SS, Stay B (1977) Gen Comp Endocrinol 31:138–147
11. Weaver RJ, Pratt GE, Hamnett AF, Jennings RC (1980) Insect Biochem 10:245–254
12. Kramer SJ (1978) J Insect Physiol 24:461–464
13. Girardie J, Tobe SS, Girardie A (1981) CR Acad Sci Paris 293:443–446
14. McCaffery AR, McCaffery VA (1983) J Insect Physiol 29:259–266
15. Tobe SS (1980) In: Locke M, Smith DS (eds) Insect biology in the future. Academic Press, New York, pp 345–367
16. Engelmann F (1959) Biol Bull 116:406–419
17. Roth LM, Stay B (1961) J Insect Physiol 7:186–202
18. Stay B, Tobe SS (1981) Amer Zool 21:663–674
19. Tobe SS, Ruegg RP, Stay BA, Baker FC, Miller CA, Schooley DA (submitted)
20. Bergot BJ, Ratcliff M, Schooley DA (1981a) J Chromatogr 204:231–244
21. Bergot BJ, Schooley DA, de Kort CAD (1981b) Experientia 37:909–910
22. de Kort CAD, Bergot BJ, Schooley DA (1982) J Insect Physiol 28:471–474
23. Loher W, Ruzo L, Baker FC, Miller CA, Schooley DA (1983) J Insect Physiol 29:585–589
24. Baker FC, Hagedorn HH, Schooley DA (1983) J Insect Physiol 29:465–470
25. Feyereisen R, Friedel T, Tobe SS (1981) Insect Biochem 11:401–409
26. Rotin D, Tobe SS (1983) Can J Biochem 61:811–817
27. Fain MJ, Riddiford LM (1975) Biol Bull 149:506–521
28. de Kort CAD, Khan MA, Bergot BJ, Schooley DA (1981) In: Pratt GE, Brooks GT (eds) Juvenile hormone biochemistry. Elsevier/North Holland Biomedical Press, pp 125–134
29. Tobe SS, Stay B (1980) Gen Comp Endocrinol 40:89–98
30. Tobe SS, Stay B (1979) Nature 281:481–482
31. Mundall EC, Tobe SS, Stay B (1979) Nature 282:97–99
32. Mundall EC, Tobe SS, Stay B (1981) J Insect Physiol 27:821–827
33. Mundall EC, Szibbo CM, Tobe SS (1983) J Insect Physiol 29:201–207
34. Stay B, Tobe SS (1977) Gen Comp Endocrinol 33:531–540
35. Tobe SS, Stay B, Friedel T, Feyereisen R, Paulson C (1981) In: Pratt GE, Brooks GT (eds) Juvenile hormone biochemistry. Elsevier/North Holland Biomedical Press, pp 161–174
36. Szibbo CM, Tobe SS (1982) Gen Comp Endocrinol 46:533–540
37. Szibbo CM, Tobe SS (1983) Gen Comp Endocrinol 49:437–445
38. Engelmann F (1984) In: Hoffmann JA, Porchet M (eds) Biosynthesis, metabolism and mode of action of invertebrate hormones. Springer-Verlag, Berlin Heidelberg, pp 444–453
39. Koolman J (1981) In: Sehnal F, Zabza A, Menn JJ, Cymborowski B (eds) Regulation of insect development and behaviour. International Conference, Technological University Wroclaw Press, Wroclaw, Poland, pp 93–106
40. Rotin D, Feyereisen R, Koener J, Tobe SS (1982) Insect Biochem 12:263–268
41. Peter MG, Gunawan S, Gellissen G, Emmerich H (1979) Z Naturforsch 34:588–598
42. de Kort CAD, Peter MG, Koopmanschap AB (1983) Insect Biochem 13:481–487

III.6 Environmentally Controlled Changes in Juvenile Hormone Titers: The Colorado Potato Beetle as a Model System *

C.A.D. de KORT and M.A. KHAN[1]

For almost 3 decades, the Colorado potato beetle, *Leptinotarsa decemlineata*, has been used in our laboratory as a model insect for the study of the environmental control of diapause. Diapause in this beetle occurs in the adult stage and is mainly controlled by photoperiod [8].

When reared under short-day conditions (10L:14D) from the egg stage onwards, the newly emerged adults feed actively, but do not reproduce. After about 10 days they dig into the soil and enter diapause. If the newly emerged adults are transferred to long-day photoperiod (18L:6D), oviposition occurs on day 5, similar to animals which are reared under long-day from the egg stage. This confirms the sensitivity of the young adult to long-day exposure [8]. However, recently we observed that the III and IV larval instar are also highly sensitive to long-day because when these were exposed to long-day and then transferred back to short-day conditions, the subsequent adults obtained show normal oviposition and no diapause behavior during at least 21 days (Fig. 1). This suggests that long-day treatment is somehow imprinted into the brain and this information is retained throughout the pupal stage resulting finally in adults which are programed to behave as long-day beetles even when kept under short-day.

Classical endocrinological experiments have shown that photoperiod is translated into physiological and behavioral responses via the neuroendocrine system with the corpus allatum forming a crucial link in the chain of processes. Allatectomy of long-day beetles led similar effects as short-day treatment [6]. From such endocrinological studies it was hypothesized that the long-day treatment activates the brain neurosecretory system and hence the corpora allata, whereas under short-days such activation is lacking [7]. Although this hypothesis recognized the central role played by the corpora allata, the neuroendocrine control of reproduction and diapause appeared to be more complex. Implantation of two pairs of active CA into allatectomized long-day beetles completely restored reproduction and prevented the onset of diapause, but were ineffective when implanted into short-day beetles. To prevent the onset of short-day diapause, 10-12 pairs of active CA are needed [6]. Besides, large doses of JH or JH analogs were necessary to induce temporary oviposition in short-day beetles [12]. Diapause could only be broken permanently by transferring beetles to long-day conditions [20].

---

* Dedicated to the late Prof. Dr. J. de Wilde

1 Laboratory of Entomology, Agricultural University, P.O. Box 8031, 6700 EH Wageningen, The Netherlands

Biosynthesis, Metabolism and Mode of Action of Invertebrate Hormones
(ed. by J. Hoffmann and M. Porchet)

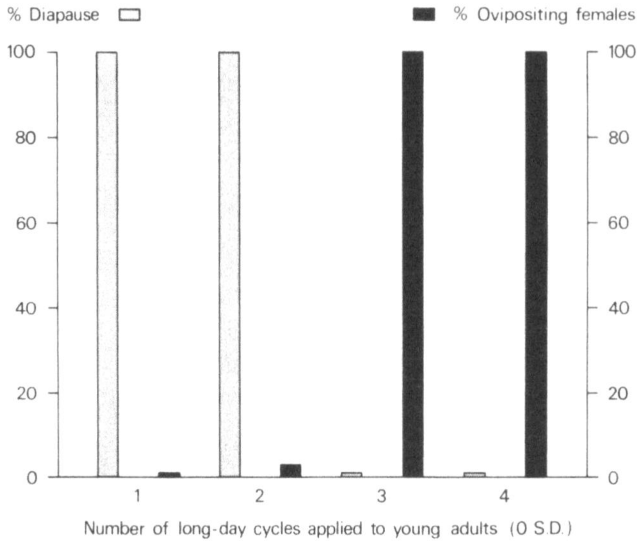

Fig. 1. The sensitivity of the egg and larval stages to long-day exposure. From the egg stage onwards until the end of the stage indicated, the experimental animals were kept under long-day photoperiod and then transferred to short-day conditions. The subsequent adults which emerged were observed for diapause or reproductive behavior (n = 60 for each observation)

These findings impelled us to investigate the factors involved in the regulation of the JH titer in more detail. Such studies require information about the in vivo concentration and nature of juvenile hormone in the hemolymph. We first published data on the JH titer in relation to the photoperiod using the Galleria bioassay in 1968 [9]. Later, in collaboration with our colleagues at Zoecon Corporation, California, our bioassay measurements were confirmed using the physicochemical method [1]. Under long-day conditions high titers of JH III are present in the hemolymph, whereas under short-day conditions small amounts of JH III could be detected just after adult emergence and these gradually decrease towards the onset of diapause. JH III is the only JH present in both adults and larvae [2]. We have now estimated the JH titer, using a recently developed RIA [21]. The results shown in Fig. 2 agree reasonably well with our earlier findings.

The steady state concentration of JH in the hemolymph is a dynamic equilibrium between synthesis and release of JH by the corpora allata and its uptake and degradation by the tissues. Hormone degradation occurs mainly by carboxylesterase in the Colorado potato beetle, and the highest breakdown capacity resides within the hemolymph [18]. The JH esterase activity in the hemolymph showed drastic changes in relation to photoperiodic treatment (Fig. 3). Under long-day conditions relatively low JH esterase activity was detected in the hemolymph, whereas in hemolymph from prediapause beetles very high levels of JH esterase activity were measured, with a peak on day 8 of the adult stage. In order to appreciate these findings, it is important to realize that in these studies JH esterase activity was measured under substrate saturated conditions [16]. This means that the activities given represent breakdown capacity rather than actual rates of breakdown of endogenous hormone. Comparative studies of hemo-

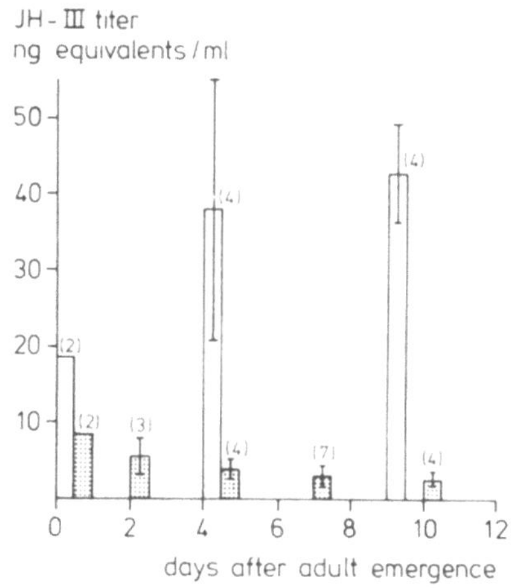

Fig. 2. The juvenile hormone (JH) titer in the hemolymph of adult females as determined by RIA under long-day (*open columns*) and short-day (*shaded columns*) conditions. The number of determinations are indicated in *parentheses*. *Bars* and *columns* represent means ± SD

lymph from different insect species revealed that the breakdown capacity of Colorado beetle hemolymph is extremely high [4]. If no protection of JH III occurred, most of it would be degraded within a few minutes. We, therefore, investigated the existence of a JH III binding protein which might offer some protection against the esterases in the hemolymph. In 1978 Kramer and de Kort [17] showed the presence of high molecular weight lipoproteins which bind JH I with low affinity. Since JH I is not the natural hormone in this beetle, these binding studies were reexamined.

Firstly, the kinetics of JH I and JH III degradation by concentrated hemolymph using physiological (10^{-7} *M*) instead of saturated concentrations were examined. It appeared that JH I was degraded much faster than JH III in contrast to the situation when saturated concentrations of hormone and very diluted hemolymph were used [16]. Moreover, the kinetics of JH III degradation did not follow a straight line when plotted logarithmically [5].

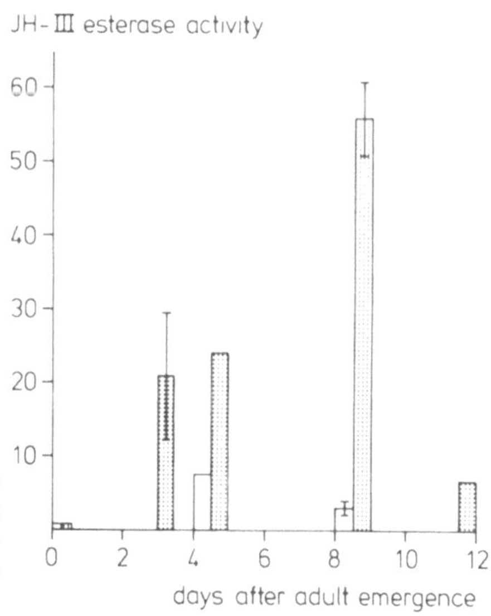

Fig. 3. The juvenile hormone III esterase activity (expressed in μmol/ml/h) in hemolymph of adult females under long-day (*open columns*) and short-day (*shaded columns*) photoperiod. The SD is indicated by *bars* only when n = 10

Since racemic mixtures of JH were used in these breakdown studies, it was essential to know whether the rate of degradation of the natural 10 R enantiomer differed from that of the unnatural 10 S enantiomer. To study this possibility the enantiomer composition of both undegraded JH and JH acid were analyzed after 50% of the hormone had been degraded. The analysis involved chemical derivatization and separation by HPLC and were carried out by Dr. Martin Peter (University of Bonn). The results revealed that undegraded hormone shows a clear excess of the natural 10 R enantiomer, whereas the JH acid fractions were mainly composed of the unnatural 10 S enantiomer [5]. This means that either the esterase preferentially degrade the unnatural 10 S enantiomer, or that the binding protein preferentially binds and protects the natural 10 R enantiomer. These alternatives were subsequently studied in more detail.

A known amount of hemolymph was incubated with physiological concentrations of labeled JH III to give about 50% binding of the label. After precipitation of the JH binding protein complex using polyethyleneglycol [15], both the bound and the free hormone containing fractions were separately extracted and subjected to diastereomer analysis by HPLC. The results, as illustrated in Table 1, show a clear excess of the natural

Table 1. Distribution of radioactivity between the natural (10 R) and unnatural (10 S) enantiomers in the bound and free hormone fractions, after incubation of 2.2×10^{-7} *M* JH III with paraoxon-treated hemolymph of the Colorado potato beetle. (Data from [5])

	Bound JH III		Free JH III	
% of racemate bound	% 10 R	% 10 S	% 10 R	% 10 S
56.5	72.3	27.7	14.9	86.1
58.9	73.5	26.5	15.9	84.1

10 R enantiomer in the bound hormone fraction, whereas the free hormone contains more of the unnatural 10 S enantiomer. This conclusively illustrates that the binding protein preferentially binds the natural enantiomer of JH III. To check binding specificity, competitive displacement experiments were carried out using racemic JH I, JH III, and JH III-diol. The results show that JH I does not compete as effectively as JH III for the JH III binding sites, whereas JH III-diol does not show any displacement (Fig. 4). From such displacement studies, the binding affinity of the binding protein for JH III can be calculated using a Scatchard plot, although such an analysis does not take into account enantiomer specificity. Nevertheless, such data have been published for several other insect species [3] and in all these cases racemic mixtures of hormone have been used. From such Scatchard plots we calculated a dissociation constant of $1.3 \pm 0.2 \times 10^{-7}$ *M* [5], which can be classified as a high affinity binding. Our results concerning the presence of JH III binding proteins in the hemolymph of the Colorado potato beetle can be summarized as follows:

1. Hemolymph of the Colorado potato beetle contains a large molecular weight lipoprotein which binds JH III with high affinity, a new class of JH binding proteins among insects.

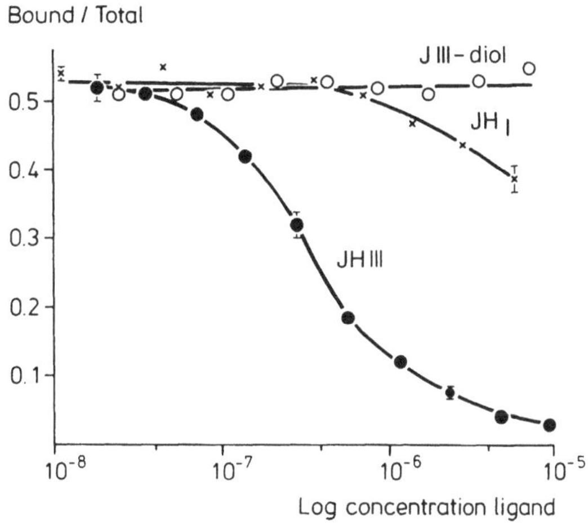

Fig. 4. Competitive displacement of (10-^{3}H)JH-III from hemolymph by JH III (•), JH-I (x), and JH-III-diol (○). Values represent the mean from four (JH-I and JH-III) or two experiments (JH-III-diol), respectively. *Bars* indicate the SD

2. The binding site has a higher affinity for JH III than for JH I, whereas JH III-acid or JH III-diol do not bind at all.
3. The protein preferentially binds the natural enantiomer of JH III and consequently, it protects the natural enantiomer of JH III more efficiently against degradation by JH-specific esterases.

Apparently, during evolution of insects, differences have evolved not only in the nature of the JHs, but also in the type of JH-binding proteins. In the Colorado beetle this protein apparently protects the natural hormone against degradation by highly active JH-specific esterases present in the hemolymph. This conclusion can be inferred from degradation kinetics [5]. If we assume that the slower part of the degradation curve is due to the breakdown of the natural JH III enantiomer, we arrive at a half-life of about 75 min for the natural hormone, which is about 3 times longer than previously published data from this laboratory [18] using racemic JH I.

Although JH esterases and binding proteins are essential components in the overall regulatory mechanism of the JH titer, they are not the only ones. Studies in our laboratory, using the in vitro radiochemical assay developed by Pratt and Tobe [19], have shown that a change in corpus allatum activity is the primary mechanism for control of the JH titer. Under long-day conditions the corpora allata were fully activated within 4 h after emergence and they remained active throughout adult life, although large individual variations can be observed (Fig. 5). Under short-day conditions the CA display significantly lower activities and they gradually became less active during prediapause, whereas during diapause no incorporation of (^{14}C)-methionine into JH III could be observed [10]. These results correlated well with JH titer measurements, although direct comparison in the same individuals is not possible.

The next question to be solved was: which factors are responsible for the activation or inactivation of the CA under the two photo-regimes?

These problems were studied with the aid of CA transplantation into experimental animals which were subsequently exposed to various external stimuli, such as starvation and changes in day length. The CA activity, measured in vitro by radiochemical assay,

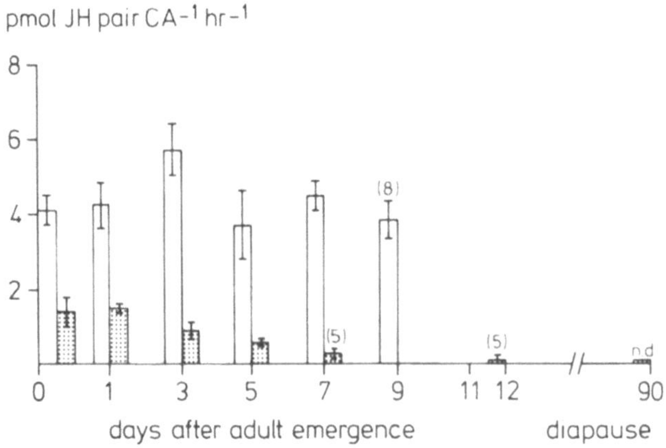

Fig. 5. The mean rates of JH biosynthesis by corpora allata from long-day (*open columns*) and short-day (*shaded columns*) females during the adult life cycle. Each *column* represents the mean of ten individual determinations unless otherwise mentioned in *parentheses*; *n.d.* = not detectable. *Bars* indicate SEM

of intact and denervated glands was then compared. In order to have some idea of the in vivo situation, juvenile hormone titers were also determined after some treatments. A few experiments to illustrate the main conclusions are described below.

To investigate the role of CA innervation under short-day conditions, 1-day-old short-day females were implanted with CA from similar donors and the activities of the implanted and host (intact) glands were followed under short-day conditions. The implants became temporarily activated, indicating that innervation plays an important role in restraining gland activity in the early prediapause period. After a few days, however, the activities of the implants declined suggesting involvement of humoral factors [13]. Thus, under short-day both neural and humoral factors suppress corpus allatum activity.

Implantation of active CA from long-day females into one-day-old short-day females resulted in a significant, but rather slow decrease in activity of the transplanted glands. The long-day CA implants never became completely inactivated even in hosts which had been in diapause for 20 days (Fig. 6). Thus, "humoral factors" alone are not very effective in restraining long-day corpora allata [13].

For the activation of CA, intact innervation is not essential, as illustrated by the following experiment. When 2-day-old short-day females were transferred to long-day, their CA became fully activated. Such an activation of intact CA did not occur if the beetles were simultaneously starved after transfer to long-day, although loose glands became completely activated under these circumstances (Fig. 7). Thus, under the influence of starvation, in contrast to the short-day treatment, the CA are restrained via the neural pathway, but the humoral milieu enhances CA activity [11, 13]. Such experiments also demonstrate that neural inhibitory signals take precedence over any humorally mediated "stimulatory" signals. The nature and source of these presumptive humoral factors remain to be established.

In order to locate the source of the neurally-mediated signals, it is essential to know the precise innervation of the CA. This was traced by using retrograde diffusion with

Fig. 6. The rate of JH biosynthesis of loose CA from 3-day-old, long-day females various periods after implantation into 1-day-old, short-day female hosts reared under short-day conditions (*shaded columns*). As controls, activities of intact CA of similar age from females maintained under long-day photoperiod are shown (*open columns*). *Columns* and *bars* represent means ± SEM

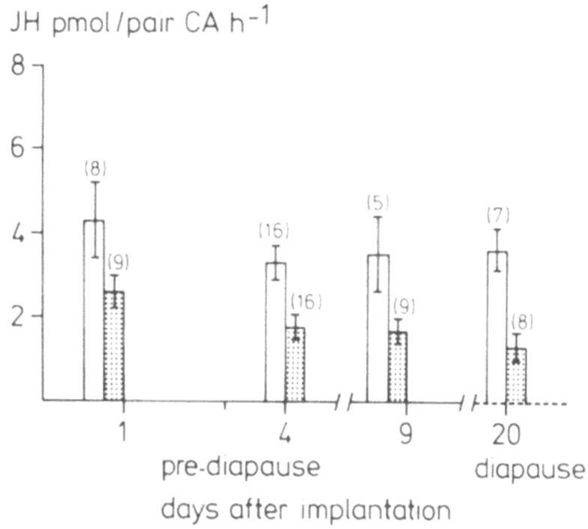

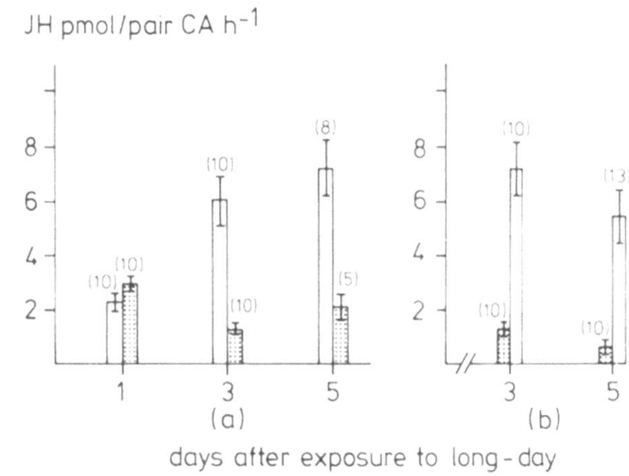

Fig. 7. a The effect on CA activity of transferring 2-day-old, short-day females to long-day photoperiod. After transfer the animals were either fed (*open columns*) or starved (*shaded columns*). b The activities of intact host (*shaded columns*) and implanted CA (*open columns*) after transfer of 2-day-old, short-day female hosts to long-day and simultaneous starvation. Both gland donors and recipients were 1-day-old, short-day females at the time of implantation

horseradish peroxidase. As shown in Fig. 8 the direct innervation of the CA is limited to a group of lateral neurosecretory cells in the ipsilateral protocerebral lobe of the brain. These cells display extensive arborizations in the dorsal region of the brain and may, therefore, be capable of receiving synaptic input from neurons in these areas and perhaps also from the median neurosecretory cells of the pars intercerebralis [14]. It is, therefore, possible that the neurally-mediated inhibitory signals are conveyed by the lateral neurosecretory cells, although the histological data remain to be confirmed by physiological studies.

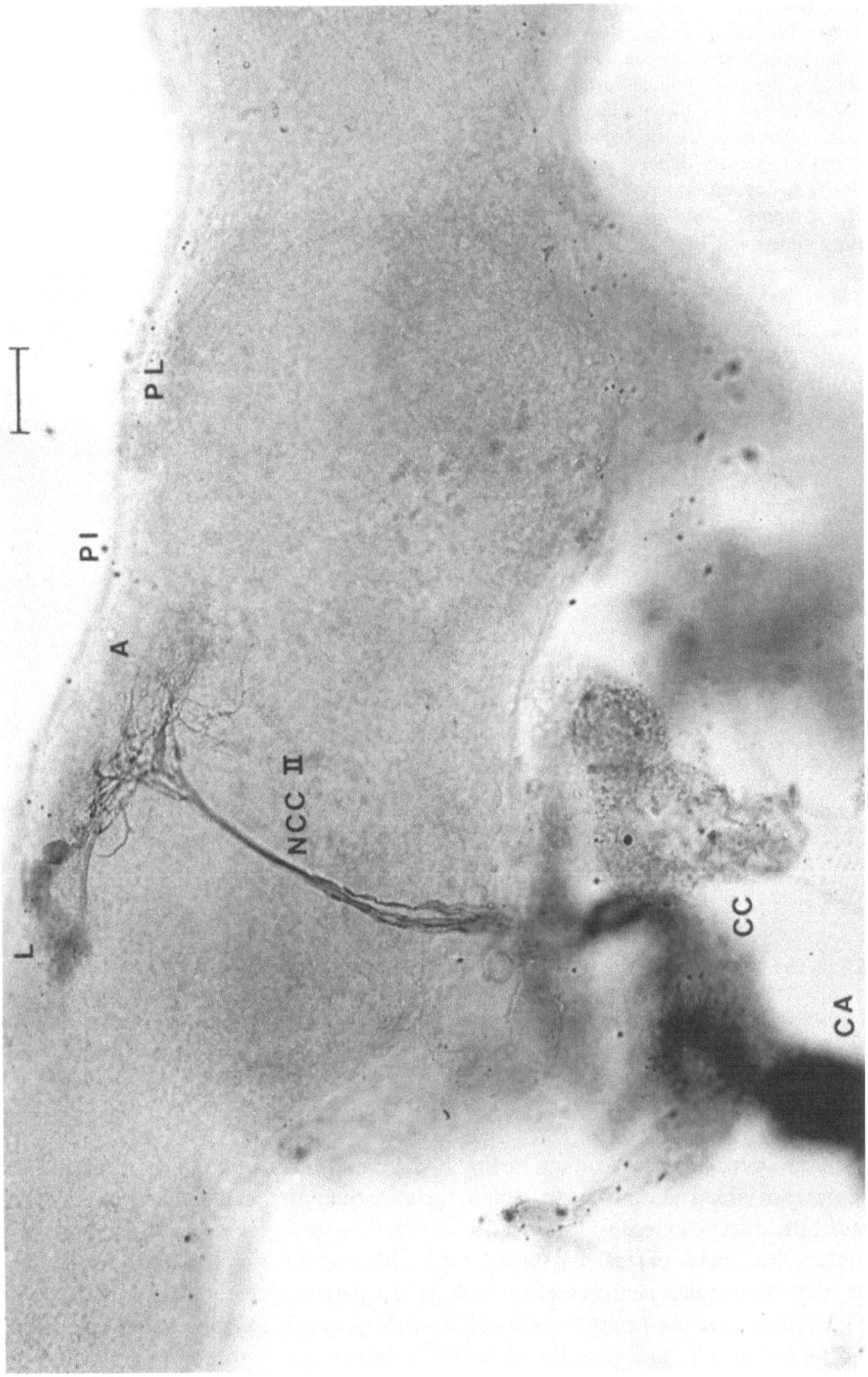

Fig. 8. A whole mount preparation of the brain after retrograde diffusion of horseradish peroxidase from the corpus allatum (*CA*). The innervation of the corpus allatum could be traced along the nervous corporis cardiacum (*NCC*) II to a group of lateral neurosecretory cells (*L*) in the ipsilateral protocerebral lobe (*PL*). Note the extensive arborizations (*A*) of the axons in the region adjacent to the pars intercerebralis (*PI*). The scale represents 50 μ; *CC* corpus cardiacum

References

1. Bergot BJ, Ratcliff MA, Schooley DA (1981) J Chromatogr 204:231–244
2. de Kort CAD, Bergot BJ, Schooley DA (1982) J Insect Physiol 28:471–474
3. de Kort CAD, Granger NA (1981) Annu Rev Entomol 26:1–28
4. de Kort CAD, Kramer SJ, Wieten M (1978) In: Gaillard PJ, Boer HH (eds) Comparative endocrinology. Elsevier/North-Holland Biomed Press, Amsterdam, pp 507–510
5. de Kort CAD, Peter MG, Koopmanschap AB (1983) Insect Biochem 13:481–487
6. de Wilde J, De Boer JA (1961) J Insect Physiol 6:152–161
7. de Wilde J, de Boer JA (1969) J Insect Physiol 15:661–675
8. de Wilde J, Duintjer CS, Mook L (1959) J Insect Physiol 3:75–80
9. de Wilde J, Staal GB, de Kort CAD, de Loof A, Baard G (1968) Proc K Ned Acad Wet (c) 71: 321–326
10. Khan MA, Doderer AB, Koopmanschap AB, de Kort CAD (1982) J Insect Physiol 28:279–284
11. Khan MA, Koopmanschap AB, de Kort CAD (1982) J Insect Physiol 28:791–796
12. Khan MA, Koopmanschap AB, de Kort CAD (1982) J Insect Physiol 28:995–1001
13. Khan MA, Koopmanschap AB, de Kort CAD (1983) Gen Comp Endocrinol 52:214–221
14. Khan MA, Romberg-Privee HM, Schooneveld H (1984) Gen Comp Endocrinol 55:66–73
15. Koeppe JK, Kovalick GE, Lapointe MC (1981) In: Pratt GE, Brooks GT (eds) Juvenile hormone biochemistry. Elsevier/North-Holland Biomed Press, Amsterdam, pp 215–231
16. Kramer SJ, de Kort CAD (1976) Life Sci 19:211–218
17. Kramer SJ, de Kort CAD (1978) Insect Biochem 8:87–92
18. Kramer SJ, Wieten M, de Kort CAD (1977) Insect Biochem 7:231–236
19. Pratt GE, Tobe SS (1974) Life Sci 14:575–586
20. Schooneveld H, Otazo Sanchez A, de Wilde J (1977) J Insect Physiol 23:689–696
21. Strambi C, Strambi A, de Reggi ML, Hirn MH, De Laage MA (1981) Eur J Biochem 118: 401–406

III.7 The Role of Juvenile Hormone Metabolism in the Metamorphosis of Selected Lepidoptera

B.D. HAMMOCK[1], Y.A.I. ABDEL-AAL[2], T. HANZLIK[1], D. JONES[3], G. JONES[3], R.M. ROE[4], M. RUDNICKA[5], T.C. SPARKS[6], and K.D. WING[7]

1. Introduction

In this chapter we will present an evaluation of the role of juvenile hormone (JH) catabolism in the Lepidoptera. This will be accomplished by first describing observations on the regulation of the esterase activity which hydrolyzes JH (JH esterase) during the last larval stadium, then describing the synthesis and kinetics of selective inhibitors of JH esterase, and finally describing the use of these inhibitors as probes for elucidating the biological role of JH esterase in the metamorphosis of selected Lepidoptera. Support for three hypotheses will be advanced. First, that in the Lepidoptera examined, JH ester hydrolysis is a major route of JH catabolism and that in the cabbage looper (Noctuidae, *Trichoplusia ni*) and in the tobacco hornworm (Sphingidae, *Manduca sexta*) this hydrolysis is due largely to a single enzyme at developmentally critical times. Second, this single protein in the early last larval stadium possibly is under the control of neurosecretory factors, while in the prepupal period JH directly results in JH esterase production. Finally, we advance the concept that in contrast to reports in vertebrates, where hormone titer is regulated by constant degradation and variations in biosynthesis, that titer of JH is, in part, controlled by a reduction in rate of biosynthesis and a large increase in the levels of catabolic enzymes.

When viewed as a xenobiotic, many pathways of JH catabolism appear likely. However, with the exception of pharmacological concentrations of JH in some insecticide resistant Diptera, only hydration of the 10,11-epoxide and hydrolysis of the methyl ester of JH have yet been shown to be important in catabolism. This fact does not rule out the possibility that other pathways may be significant in unexplored species or developmental stages [1]. However, in the majority of Lepidoptera examined, ester cleavage predominates over epoxide hydration in the catabolism of JH [2, 3]. For instance in *T. ni*, the maximum rate of epoxide hydration is more than an order of magnitude lower than the JH esterase activity, the tissue distribution of epoxide hydro-

1 Departments of Entomology and Environmental Toxicology, University of California, Davis, CA, USA
2 On sabbatical leave from Department of Plant Protection, Assiut University, Assiut, Egypt
3 Department of Entomology, University of Kentucky, Lexington, KY, USA
4 Department of Entomology, N. Carolina State University, Raleigh, NC, USA
5 Institute of Organic and Physical Chemistry, Technical University of Wroclaw, Wroclaw, Poland
6 Department of Entomology, Louisiana State University, Baton Rouge, LA, USA
7 Research Laboratories, Rohm and Haas Comp., Spring House, PA, USA

Biosynthesis, Metabolism and Mode of Action of Invertebrate Hormones
(ed. by J. Hoffmann and M. Porchet)

lase activity is more limited than that of JH esterase, and maximal epoxide hydrolase activity occurs after the majority of JH apparently has been removed from the system. Thus, epoxide hydrolases may have a role in the further catabolism of JH acid, but we may concentrate on JH esterases when concerned with initial clearance of JH from the larva [4].

2. Factors Influencing JH Titer

Many factors influence hormonal action, titer, and distribution as illustrated by many of the lectures and posters at this meeting. This chapter will emphasize JH esterase activity and is intended to complement the following discussion of carrier proteins. Since Goodman and co-workers earlier established at least one physiological role for the low molecular weight hemolymph binding protein of *M. sexta*, we refer to it as a JH carrier protein to distinguish it from tissue binding and hemolymph binding proteins of other species currently under investigation (see Part III). At least in hemolymph of *T. ni* binding of JH demonstrates a very small dynamic range in the last larval stadium [4]. Thus, we feel that variations in the interaction between the JH carrier protein and JH esterase during the last larval stadium are determined largely by the tremendous fluctuations in JH esterase activity rather than the minor changes observed in JH binding to hemolymph proteins [4].

The rate of catabolism of JH in an individual compartment may be an important factor in determining the action of JH in that compartment. However, the rate of clearance of JH from the entire insect depends upon the specific activity of the JH esterase within the compartment, the mass of the compartment, and the proportion of JH reaching the compartment. Two of the former parameters were evaluated and indicated that the hemolymph and the fat body probably account for the large majority of JH esterase activity [4].

In most of the Lepidoptera examined there are two peaks of JH esterase activity in the hemolymph which we refer to chronologically as the prewandering and the postwandering peaks [5-9]. In *T. ni* the activity of JH esterase in homogenates of fat body correlates well with the prewandering peak of JH esterase activity in the hemolymph. The rise in hemolymph and fat body JH esterase activity appears concomitant in the postwandering stage. However, the hemolymph titers of JH esterase fall off rapidly, while the activity in the fat body remains high. This observation raises the possibility that production and release of JH esterase are under separate control at least in the prepupa [4].

3. Multiplicity of JH Esterase

Before moving into the regulation of JH esterase activity, it is important to consider how many proteins account for this esterase activity. The concept was advanced earlier that in *M. sexta*, two classes of esterases existed. They were separable by gel filtration, of roughly equivalent activity, and could metabolize JH. The "JH specific esterase" could metabolize JH rapidly in the presence of the carrier protein, while "general esterases" could not [10]. Although the concept of general vs specific juvenile hormone esterases may be valid and even directly applicable to some insects, the concept is misleading when applied to the hemolymph of *T. ni* [8, 11]. In this species JH metabolism is due largely to a single enzyme and the contribution to JH metabolism offered by general esterases in the hemolymph of larvae in the last stadium is minor by comparison with the JH esterase whether the carrier molecule is present or not [8, 12-14].

It is difficult to disprove the existence of multiple enzymes, but the evidence to date indicates that one enzyme predominately is responsible for JH metabolism in hemolymph and fat body from the last larval stadium of *T. ni.* This concept was advanced by attempting to find multiple forms of JH esterase using techniques of varying rigor. In no case was firm evidence for multiple forms of JH esterase in these two compartments developed [8, 13]. Similar experiments in other species indicate that the JH esterase activity in the hemolymph is due to one to several acidic proteins [15, 16].

Using analogous techniques we questioned if the JH esterases in the prewandering and postwandering peaks were different and if the JH esterase in fat body cytosol was distinct from the hemolymph enzyme. In no case could differences be demonstrated in *T. ni.* Based on similar experiments it was concluded that the artificially-induced JH esterase activities discussed below were due to an enzyme indistinguishable from that occurring in normal animals [4, 13, 17].

4. Regulation of JH Esterase

The JH esterase activities of *M. sexta* and *T. ni* are nice targets for the study of protein regulation since they appear to be due to a single protein whose activity can be measured readily. Ligation at the neck of either *M. sexta* or *T. ni* prevents the appearance of the prewandering and postwandering JH esterase peaks, but here the similarity of regulation of the two peaks of activity appears to end. Thus, the regulation of the prewandering and postwandering peaks of JH esterase activity will be treated separately.

The obvious factor to test when trying to restore the prewandering JH esterase peak in isolated abdomens is JH. However, the application of JH homologs or juvenoids to isolated abdomens from prewandering last stadium larvae fails to increase the JH esterase activity present in the hemolymph of *T. ni* or *M. sexta*. When applied to normal prewandering larvae JH will prolong and even increase the enzymatic activity of the first JH esterase peak. In allatectomized *M. sexta* the JH esterase activity present during the prewandering peak is reduced, but not eliminated. Similarly, the esterase activity can be somewhat increased by JH application to allatectomized larvae. Thus, JH appears to synergize the appearance or action of other factor(s) yet lacks the ability to induce its own degradation at this stage in isolated abdomens [8, 17].

In contrast to JH, the injection of extracts or implantation of brains and/or subesophageal ganglia leads to a rise in JH esterase activity in isolated abdomens of prewandering *T. ni.* Interestingly, implantation of brains and subesophageal ganglia with connectives intact seems more effective than implantation with the connectives severed.

Aqueous homogenates of brains seem to be most active near the time when JH esterase appears to be highest, yet there is another peak of activity near the time of the molt from the penultimate instar of *T. ni*. By injecting a standard homogenate into isolated abdomens of different stages it was determined that the abdomens were rather insensitive to the action of the head factor in the fourth stadium and wandering stage, while responsive during the time of the prewandering JH esterase peak [18].

These data support the hypothesis that a brain-subesophageal factor leads to the production and release of the prewandering JH esterase peak. However, reimplantation and injection experiments in isolated abdomens have led thus far to only partial restoration of the JH esterase activity present in normal animals.

Commitment to pupation is a crucial step for a larva. The longer it remains in the larval stage, the longer it is suceptible to a variety of biotic and abiotic factors that can destroy it. However, it must not commit itself to pupation without amassing an adequate nutritional reserve. It is not surprising that during the early part of the last larval stadium both the production and degradation of juvenile hormone are under apparent nervous and/or neuroendocrine control. Lack of an adequate diet in either *M. sexta* or *T. ni* leads to an increase in JH titer and a dramatic fall in JH esterase activity [8, 19]. Levinson (unpubl.) demonstrated that the larvae were apparently assessing the quality of their food rather than simply gauging the amount consumed. The most important single quality of the diet appeared to be protein, although the evaluation of the diet by the larva appeared to depend upon a variety of factors.

By the time the postwandering peak of JH esterase activity appears, the larva has changed dramatically, as has the regulation of the JH esterase. The implantation of brains or injection of brain extracts from postwandering larvae has little apparent influence on the levels of JH esterase in the hemolymph [18]. However, application of JH or juvenoids to isolated abdomens or allatectomized larvae of *M. sexta* or *T. ni* or postwandering larvae results in a large, dose-dependent increase in JH esterase activity [4, 8, 17]. This effect can be blocked by the application of JH antagonists thought to act at the receptor site, and can form the basis for a JH antagonist assay [20].

Numerous experiments provide circumstantial evidence that JH stimulates the appearance of the esterase which degrades it. The best evidence to date comes from extirpation-reimplantation experiments in *T. ni*. By removing the corpora allata from progressively later stages of postwandering larvae, one can demonstrate that the corpora allata must be present during a 10 h period on day 3 of the last larval stadium. This period correlates well with an apparent rise in JH levels in the whole body (Baker, Miller, and Schooley, unpubl). In either allatectomized larvae or isolated abdomens prepared before this critical period, reimplantation of corpora allata results in restoration of apparently normal levels of JH esterase [21]. Similar effects can be obtained by applying JH antagonists or biosynthesis inhibitors [20, 22].

There has been a great deal of speculation regarding the role of juvenile hormone in the prepupa. Many of the arguments ultimately return to the apparent prothoracicocotropic effects of JH. In some insects, such as *Bombyx mori*, allatectomy results in apparently normal, precocious pupae [23], in *Hyalophora cecropia* allatectomy leads to precocious adult structures [24], while in *M. sexta* allatectomy delays pupation and results in a low incidence of deformed adult-like structures [25]. In *T. ni* the effects of allatectomy, ligation or anti-JH application are dramatic in that pupation is prevented [21, 22]. Again, the effects of allatectomy can be reversed by the application of JH or juvenoids. Since the JH effects seem to overlap with the release of the prothoracio-tropic hormone, the role of JH does not seem to be explained by stimulation of PTTH production. Whether JH stimulates production or action of ecdysone remains to be determined (Jones, unpubl.) [22].

Not only is the presence of JH in the prepupa critical for further development, it is equally essential that the JH titer is reduced again for normal development. Numerous reports involving the application of JH or juvenoids or the implantation of corpora allata demonstrate that the presence of high levels of JH during the larval pupal transformation is detrimental. In contrast to the prewandering stage, when it is important that the larva adjust its development to its nutritional status and other environmental cues, after wandering it is committed to metamorphosis. The subsequent developmental events must then occur in a precise, temporal sequence. Thus, it would seem advantageous to have a system in which JH stimulates its own degradation in the prepupa. Obviously, such a regulatory mechanism would be detrimental to the insect during stages when it must maintain high JH titers.

The arguments put forward in this chapter are based largely upon data from *M. sexta* and to a greater extent, *T. ni*. How well these hypotheses will explain the development of other lepidopterous larvae is questionable. Some similarities have been observed in *Galleria mellonella* [26, 27], but few studies on regulation have been performed in other species. However, JH esterase activity has been detected in all of the Lepidoptera examined, and the most common pattern is a prewandering and postwandering peak. Both JH esterase and JH binding activities seem due to a limited number of proteins with similar isolectric points, and the K_ms and K_ds of the esterase and binding activity in the various species appear similar. Thus, caution is required for extension of the hypotheses developed here to Lepidoptera in general, but there are certainly obvious similarities among the species studied [3, 9, 15, 16, 28, 29].

5. Inhibition of JH Esterase

The arguments to this point for the involvement of JH esterase in metamorphosis have been based upon correlations. For more direct proof it was necessary to develop chemical probes which would allow the inhibition of JH esterase activity. Numerous common inhibitors of esterases, such as eserine, tri-o-cresyl phosphate, and S,S,S-tri-n-butyl-phosphorotrithioate fail to inhibit JH esterase at high micromolar concentrations [8, 30, 31]. Possibly the most diagnostic of these inhibitors is 0,0-di-isopropyl phosphoro-fluoridate [10]. Although by no means magical, it is a remarkably weak inhibitor of the majority of JH esterase activity during the last stadium, while being moderately active on many of the esterases detected with α-naphthylacetate (general esterase).

Homologs of paraoxon are useful for eliminating JH esterase as well as other esterase activities. However, these compounds are too toxic for in vivo use. Interestingly, some phosphoramidothiolates are good inhibitors of JH esterase. The potency of methamidophos was unexpected since it is only a weak inhibitor of acetylcholinesterase in vitro and some workers hypothesized that it had to be metabolically activated before it could inhibit an esterase. A related compound O-ethyl-S-phenyl phosphoramidothioate (EPPAT) has enough selectively for use in vivo as described below [30]. However, at levels that inhibit the majority of JH esterase, many other esterases are inhibited, and enough acetylcholinesterase is inhibited to yield some mortality in *T. ni*. EPPAT appears more selective in *M. sexta* [8], however, more selective inhibitors are desirable. The use of compounds of several different classes reduces concern over side effects of the inhibitors.

While working on the development of juvenoids active on resistant Diptera, fluorines were placed at key sites of metabolic attack. One of the intermediates in these syntheses was a trifluoromethylketone which proved to be a potent inhibitor of JH esterase, while the corresponding methyl ketone was inactive. This discovery led to the synthesis of a variety of compounds, the most active of which is 1,1,1-trifluorotetradecan-2-one (TFT). This material proved to be a potent competitive inhibitor of JH esterase activity from *T. ni* with a K_i of 3×10^{-9} *M*. Rapidly reversible inhibitors must be very potent in order to obtain good inhibition in vivo, and previous experiments with EPPAT (an irreversible inhibitor) indicated that the majority of the JH esterase activity had to be inhibited in order to observe in vivo effects. Thus, the negligible effects of TFT in vivo were not surprising [30, 32].

We subsequently found that placement of a sulfide bond β to the carbonyl increased the potency of the compounds on a variety of esterases. Possibly with JH esterase this effect is due to the sulfide bond mimicking the 2,3 double bond of JH, but X-ray crystallographic data indicate that at least in the crystal structure the sulfur stabilizes the inhibitory (tetrahedral) form of the compound by strong hydrogen bonding to the hydrate. Possibly, as discussed elsewhere, these compounds are inhibitors due to their mimicking an intermediate or transition state along the reaction coordinate between the substrate and the product. Regardless of the mechanism, these compounds are certainly very potent inhibitors [33-36].

Unlike compounds in the TFT series which yielded pure competitive kinetics and EPPAT which behaved as a phosphorylating agent, trifluoroketones containing a sulfide bond, such as 3-ocytylthio-1,1,1-trifluoropropan-2-one (OTFP), yielded confusing kinetics [31, 33]. It was ultimately determined that the compounds reacted with JH esterase with both a slow on-rate and off rate (thus, giving time dependent inhibition). They also were such potent inhibitors that their reaction with JH esterase approached stoichiometric levels [34]. Thus, the assumptions necessary for a Lineweaver-Burk treatment of their interaction with JH esterase did not hold, but their behavior could be approach using a kinetic treatment discussed by Morrison [37] and others. One advantage of this treatment is that plots of velocity vs total enzyme concentration with several fixed inhibitor concentrations can provide an estimate of total catalytic sites of the enzyme in an incubation medium which allows one to calculate a number of parameters for JH esterase in the hemolymph (Table 1). Although we are far from modeling the dynamics of JH in the hemolymph of *T. ni*, these data strongly support the hypothesis that the JH esterase present in the pre- and postwandering peaks is capable of clearing JH from the hemolymph even when the hormone is present far below the K_m of the enzyme and is in the presence of carrier protein [34]. The efficiency of JH esterase in hydrolyzing JH homologs can be projected from the apparent second order rate constant for the reaction of free enzyme with free substrate (k_{cat}/K_m, Table 1) which is close

Table 1. Kinetic parameters for JH esterase from *T. ni* [34]

Substrate	K_m (*M*)	V_{max} (nmol/min ml^{-1})	k_{cat} (min^{-1})	k_{cat}/K_m (M^{-1} min^{-1})
JH I	6.1×10^{-8}	177	37	6.1×10^{8}
JH III	3.0×10^{-8}	85	19	6.4×10^{8}

to the diffusion controlled encounter of the enzyme and the substrate. Therefore, even if the JH-esterase-JH-binding protein species are present in thermodynamic equilibrium, the enzyme will be capable of shifting the whole equilibrium toward ester hydrolysis.

6. Effects of Inhibitors of JH Esterase Upon Metamorphosis

Nijhout [38] and Fain and Riddiford [39] were quite justified in questioning the role of the JH esterase in *M. sexta* in clearing JH from the hemolymph because the evidence for the involvement of the enzyme was based simply upon the correlation of esterase levels with developmental events. In examining any biological system it is critical to realize that the metabolism of JH by an enzyme system in vitro could simply be another action of an enzyme whose actual biological function may involve the hydrolysis of ester or amide bonds of very different substrates. This is not to say that such JH metabolizing enzymes are unimportant pharmacologically or experimentally, but rather that one should approach the assumption of a significant biological role in JH metabolism cautiously.

The role of JH esterase in the metamorphosis of *T. ni* is not yet conclusively established, but the evidence presented in Table 2 argues strongly for its involvement. The most conclusive independent evidence stems from the observations that inhibition of the JH esterase activity in the pre- and postwandering peaks leads to the biological effects expected from a continued titer of JH. Thus, in *M. sexta* and *T. ni* inhibition of the prewandering JH esterase results in a significant increase in the feeding phase with a subsequent delay in wandering and pupation [8, 30, 35, 36]. Similarly, inhibition of the postwandering peak results in a failure to molt or the retention of juvenile structures [22, 30]. With any inhibitor it may be argued with some validity that inhibition of JH esterase could be secondary to the critical lesion caused by the material. This possibility certainly exists, but a variety of different experiments all indicate that JH esterase is the critical lesion for the site of action of these compounds (Table 2).

7. Conclusion

In this article we have argued that one is justified with *M. sexta* and *T. ni* in considering JH esterase as the major enzyme responsible for JH catabolism during two critical periods in the last larval stadium. It appears that the JH esterase is somewhat unique

Table 2. Evidence for a role for JH esterase in the metamorphosis of *T. ni*

A. The enzyme appears to be a catalyst suited to the hydrolysis of JH
   1. The enoate ester of JH is very stable to base and, thus, enzymatic hydrolysis, yet JH esterase metabolizes it with a high turnover number
   2. Using several criteria the enzyme shows selectivity or specificity for the JHs as substrates over less stable esters, such as α-naphthylacetate
   3. The enzyme shows selectivity for hydrolysis of the natural JHs over closely related compounds, such as the 2Z isomers or ethyl or isopropyl rather than methyl esters

B. The enzyme appears in high titer at two times during the last stadium. In both cases in vivo titer determination as well as biological observations indicate that JH must be reduced to low titers during this time

C. Studies on lepidopterous larvae indicate that JH production by the corpora allata does not cease until the pupal stage. This reduction in biosynthesis alone is not likely to be adequate to reduce the JH titer sufficiently for PTTH release in a short time

D. The molarity and kinetic constants of the JH carrier molecule and JH esterase in the prewandering and postwandering JH esterase peaks are such that it appears the esterase can reduce the JH molarity far below the K_m of the enzyme

E. The most potent "transition state" inhibitors of JH esterase have a remarkable resemblance to JH in terms of their steric properties

F. Inhibition of JH esterase in vivo at critical periods leads to biological effects expected of increased titers of JH
   1. Inhibition of the prewandering JH esterase peak greatly extends the feeding phase of both *M. sexta* and *T. ni*
   2. Inhibition of the postwandering JH esterase peak leads to difficulty in pupating or the appearance of immature structures in the resulting pupa and/or adult
   3. Both of the above effects (1 and 2) have been observed to occur in a dose-dependent fashion following the application of JH esterase inhibitors in two distinct classes (EPPAT and OTFP)
   4. The dose necessary for in vitro and in vivo inhibition of JH esterase correlates nicely with biological effects
   5. At times when partial inhibition of acetylcholinesterase is not lethal, paraoxon (inhibits JH esterase) gives the same effects as EPPAT and OTFP, while DFP (inhibits most general hemolymph esterases) does not
   6. OTFP shows some selectively in that at doses that lead to JH-like effects in vivo, there is only slight inhibition of "general esterases" in the hemolymph, but substantial inhibition of JH esterase
   7. Application of EPPAT to inhibit JH esterase or exposing day 1 last stadium larvae to an inadequate diet results in a significant increase in whole body JH levels as determined by glc-ms

among biological systems in that the production of the same protein appears at two separate times in development to be under two separate mechanisms of control.

The postwandering burst of JH esterase activity which occurs immediately before pupation seems to be critical to remove the last traces of juvenile hormone which occur prior to pupation. The tight coupling of JH production with JH esterase production provides further circumstancial evidence for a biological role for both materials. Extirpation-reimplantation studies establish that in *T. ni* the postwandering burst of JH is absolutely essential for pupation to occur. Thus, *T. ni* may be a nice model system for the elucidation of the exact mechanism by which JH exerts its apparent prothoracicio-tropic or ecdysiotropic role [40].

The events which lead the last larval stadium to undertake metamorphosis have been explored for many years. Clearly a low titer of JH is needed before PTTH and ecdysone can be released. Undoubtedly the insect "knows" it has reached the last larval stadium based on a large number of factors. At this point endocrine changes occur which set into motion a myriad of both simultaneous and sequential events. Yet as scientists we continue to search for single cues for and events in the complex process of metamorphosis.

The experiments summarized above support the hypothesis that a reduction in biosynthesis coupled with a large increase in the rate of catabolism of JH lead to the observed reduction in JH titer. There is evidence for the nervous and humoral control of JH biosynthesis [41] as well as for control of JH esterase [18]. However, at this time the appearance of JH esterase in the hemolymph is the earliest event leading to pupation which can be explained based upon an isolatable, biochemical material. Thus, it seems safe to predict that the biochemistry and regulation of JH esterase will be the subject of intensive research by several laboratories during the next few years.

References

1. Hammock BD, Quistad GB (1981) In: Hutson DH, Roberts TR (eds) Progress in pesticide biochemistry, vol I. Wiley, New York, pp 1–83
2. Slade M, Zibitt CH (1972) In: Menn JJ, Beroza M (eds) Insect juvenile hormones: Chemistry and action. Academic Press, London New York, pp 155–176
3. Hammock BD, Jones D, Jones G, Rudnicka M, Sparks TC, Wing KD (1981) In: Sehnal F, Zabza A, Menn JJ, Cymborowski B (eds) Int Conf Regul Insect Dev Behav. Tech Univ Press, Wroclaw, Wroclaw, Poland, pp 219–235
4. Wing KD, Sparks TC, Lovell VM, Levinson SO, Hammock BD (1981) Insect Biochem 11: 473–485
5. Weirich G, Wren J, Siddall JB (1973) Insect Biochem 3:397–407
6. Vince RK, Gilbert LI (1977) Insect Biochem 7:115–120
7. Sparks TC, Willis WS, Shorey HH, Hammock BD (1979) J Insect Physiol 25:125–132
8. Sparks TC, Hammock BD, Riddiford LM (1983) Insect Biochem 13:529–541
9. Jones D, Jones G, Wing KD, Rudnicka M, Hammock BD (1982) J Comp Physiol 148:1–10
10. Sanburg LL, Kramer KJ, Kézdy FJ, Law JH (1975) J Insect Physiol 21:873–887
11. Hammock BD (in press) In: Kerkut GA, Gilbert LI (eds) Comprehensive insect physiology, biochemistry and pharmacology
12. Coudron TA, Dunn PE, Seballos HL, Wharen RE, Sanburg LL, Law JH (1981) Insect Biochem 11:453–461
13. Sparks TC, Hammock BD (1979) Insect Biochem 9:411–421
14. Yuhas DA, Roe RM, Sparks TC, Hammond AM (1983) Insect Biochem 13:129–136
15. Roe RM, Hammond AM, Sparks TC (1983) Insect Biochem 13:163–170
16. Wing KD, Rudnicka M, Jones G, Jones D, Hammock BD (1984) J Comp Physiol 154:213–223
17. Sparks TC, Hammock BD (1979) J Insect Physiol 25:551–560
18. Jones G, Wing KD, Jones D, Hammock BD (1981) J Insect Physiol 27:85–91
19. Cymborowski B, Bogus M, Beckage NE, Williams CM, Riddiford LM (1982) J Insect Physiol 28:129–135
20. Sparks TC, Wing KD, Hammock BD (1979) Life Sci 25:445–450
21. Jones G, Hammock BD (1983) J Insect Physiol 29:471–475
22. Sparks TC (1984) J Insect Physiol 25:125–132
23. Wigglesworth VB (1970) The physiology of insect metamorphosis. Cambridge Univ Press, New York, p 163
24. Williams CM (1961) Biol Bull Woods Hole 121:572–585

25. Kiguchi K, Riddiford LM (1978) J Insect Physiol 24:673–680
26. Reddy G, Hwang-Hsu K, Kumaran AK (1979) J Insect Physiol 25:65–71
27. McCaleb DC, Kumaran AK (1980) J Insect Physiol 26:171–177
28. Weirich G, Wren J (1976) Physiol Zool 49:341–350
29. Oźyhar A, Wisniewski JR, Sehnal F, Kockman M (1983) Insect Biochem 13:435–441
30. Sparks TC, Hammock BD (1980) Pestic Biochem Physiol 14:290–302
31. Sparks TC, Rose RL (1983) Insect Biochem 13:633–640
32. Hammock BD, Wing KD, McLaughlin J, Lovell VM, Sparks TC (1982) Pestic Biochem Physiol 17:76–88
33. Abdel-Aal YAI, Roe RM, Hammock BD (1984) Pestic Biochem Physiol 21:232–241
34. Abdel-Aal YAI, Hammock BD (1984) Insect Biochem (submitted)
35. Hammock BD, Abdel-Aal YAI, Mullin CA, Hanzlik TN, Roe RM (in press) Pestic Biochem Physiol
36. Prestwich GD, Eng W-S, Roe RM, Hammock BD (1984) Arch Biochem Biophys 228:639–645
37. Morrison JF (1969) Biochim Biophys Acta 185:269–286
38. Nijhout HF (1975) Biol Bull Woods Hole 149:568–579
39. Fain MJ, Riddiford LM (1975) Biol Bull 149:506–521
40. Safranek L, Cymborowski B, Williams CM (1980) Biol Bull 158:248–256
41. de Kort CAD, Granger NA (1981) Annu Rev Entomol 26:1–28

III.8 Hemolymph Transport of the Juvenile Hormones

W.G. GOODMAN[1], J.T. TROST[1], C.T. REITER[1], D.W. BEAN[1,2], and
R.O. CARLSON[1]

1. Functions of Circulatory Hormone Binding Proteins

Despite their evolutionary divergence better than 300 million years ago, insects and vertebrates have evolved many similar mechanisms that promote conservation and maintenance of hormone titers. One prominent feature common to both groups is the presence of plasma or hemolymph binding proteins which transport lipophilic hormones through the circulatory system. In effect, these proteins reduce surface active properties and increase solubility of the bound hormone so that sufficient titers may develop in peripheral tissues. At the target site, the hormone presumably dissociates from the protein, diffuses unaided through the plasma membrane, and binds to receptors within the cell. Thus, the role of the binding proteins in transporting hormones from the synthetic site to distant target cells is essential in the biochemical initiation of physiological events within the organism.

Besides increasing the hormone concentration at the target site, the hormone binding proteins exert a regulatory effect on circulating hormone levels by (1) facilitating the passage of the hydrophobic hormones from the site of synthesis into the circulatory system; (2) protecting the hormone from enzymatic attack; (3) preventing the rapid loss of hormone due to excretion; (4) reducing nonspecific binding; and (5) serving as a reservoir from which the hormone can be rapidly transferred to the target cell [1].

In addition to hormone conservation, the hormone binding proteins serve other functions. The proteins may sequester their respective ligands from areas in which the hormones may have a deleterious effect [2]. Moreover, it has been demonstrated that certain ligands are recognized by target tissues only when complexed to their respective binding proteins [3-5]. The most elegant studies conducted on binding protein-ligand internalization are those dealing with cholesterol transport. Despite its high solubility in membranes, cholesterol must, under physiological conditions, be bound to a plasma low-density lipoprotein to facilitate uptake [6]. Although evidence has yet to be presented, it is possible that certain hormone binding proteins may also be required for recognition at the target site.

1 Department of Entomology, University of Wisconsin, Madison, WI 53706, USA
2 Department of Entomology, Oregon State University, Corvallis, OR 97331, USA

Biosynthesis, Metabolism and Mode of Action of Invertebrate Hormones
(ed. by J. Hoffmann and M. Porchet)

2. Classes and Evolutionary Distribution of JH Transport Proteins

As the structure and physical properties of the juvenile hormones became known, investigators began searching for hemolymph juvenile hormone transport proteins. Evidence for hormone transport by hemolymph proteins was first reported by Whitmore and Gilbert [7] in their studies on *Hyalophora cecropia*. Due to the lipophilic nature of juvenile hormone (JH), it was reasoned that the hormone would be associated with lipoproteins in the hemolymph. Characterization of the proteins by sodium bromide density gradient centrifugation revealed that the major binding protein was a high density lipoprotein with a specific density of 1.15–1.17 g/ml. Following this discovery, several investigators reported that lipoproteins represented the mode of transport for the juvenile hormones in a number of species (Table 1). The interactions are characterized by their low affinity ($K_a \leqslant 10^5\ M^{-1}$) and their ability to bind more than one molecule of hormone per molecule of protein [8, 9]. These proteins have been reported to display little specificity and to bind many types of nonpolar molecules.

Following the discovery of hemolymph transport via lipoproteins, two other forms of JH binding proteins (JHBP) were reported, both displaying relatively high affinity for their respective ligands ($K_a \geqslant 10^7\ M^{-1}$), but differing considerably in their molecular weights. Binding studies using the naturally occurring homolog, JH III, led to the discovery of specific, high affinity, high molecular weight binding proteins in the hemolymph of *Leptinotarsa* [10], *Leucophaea* [11], and *Locusta* [12]. JHBP from all three species display a reduced affinity towards JH I, and *Leptinotarsa* and *Locusta* have now been shown to preferentially bind the naturally occurring enantiomer of JH III. Since JH III is the only homolog found to date in nonlepidopterans [13], hemolymph JH transport via lipoproteins alone may be the pattern in most insect species. In contrast to the high molecular weight, high affinity JHBP, a much smaller protein has been found in over 20 species of Lepidoptera (Table 1). There is a remarkable similarity in molecular weights [9] and isoelectric points [12] which suggests a common ancestral JHBP. It is tempting to speculate that there evolved in Lepidoptera, which contain at least four homologs [14], a specific low molecular weight protein capable of binding more than one type of homolog.

At present, there appear to be two general classes of hemolymph JH binding proteins which can be distinguished on the basis of binding affinity: (1) low affinity, low specificity, high capacity, high molecular weight proteins and (2) high affinity, high specificity binding proteins. The latter category can be divided into two groups based on molecular weight; those greater than 100,000, and those below 30,000. The distribution of these proteins throughout various insect orders is compiled in Table 1. Although certain species utilize lipoproteins as the only mode of JH transport, other species use both high and low affinity proteins. When present in the hemolymph, the high affinity, low capacity, low molecular weight proteins generally contribute more to the overall regulation of JH titers than lipoproteins of low specificity and affinity. On the other hand, the relatively high concentration of lipoprotein binding sites compared with the JH-specific binding sites indicates that the lipoproteins have a larger role in transport than previously thought. Unfortunately, the importance of lipoproteins cannot be determined until quantitative JH distribution studies are performed.

Table 1. Major hemolymph juvenile hormone binding proteins in various insect species. (After [9])

Species	Stage	Molecular weight ($\times 10^{-3}$)	Association constant (K_a) ($M^{-1} \times 10^{-6}$, 4°)	References
Gomphocerus rufus	Adult	220	20	[35]
Locusta migratoria	Adult	220		[41]
Locusta migratoria [a]	Adult		12	[12]
Leucophaea maderae [a]	Adult	>200	45	[11]
Oncopeltus fasciatus	Nymph	100	<0.1	[42]
Leptinotarsa decemlineata	Adult	>200	<0.1	[43]
Leptinotarsa decemlineata [a]	Adult	>200	12	[10]
Tenebrio molitor [b]	Larva			[44]
Tenebrio obscurans	Larva	>100		[20]
Tenebroides mauritanicus	Larva	30		[20]
Drosophila hydei	Larva	44	10	[45]
Hyalophora cecropia	Pupa	200		[7]
Hyalophora gloveri	Pupa	200		[7]
Antheraea polyphemus	Pupa	200		[7]
Hemileuca nevadensis	Larva			[12]
Manduca quinquemaculata	Larva	30	11	[46]
Manduca sexta	Larva,	28	10	[15, 16, 29]
	Pupa,			
	Adult	28	10	[15, 16,29]
Sphinx chersis	Larva	30		[46]
Hyloicus chersis	Larva	30		[46]
Estigmene acraea	Larva	30		[12]
Tricoplusia ni	Larva	19	2	[47, 48]
Spodoptera exigua	Larva	30		[12]
Heliothis virescens	Larva	30		[12]
Agrotis ipsilon	Larva	25		[50]
Orygia vetusta	Larva			[12]
Ephestia elutella	Larva			[12]
Anagasta kuehniella	Larva	30	18	[46]
Caudra cautella	Larva	30	70	[46]
Plodia interpunctella	Larva	25	76	[44, 49]
Galleria mellonella	Larva	25		[27]
Ostrinia nubilalis	Larva	25		[50]
Diatraea crambidoides	Larva	30		[51]
Diatraea grandiosella	Larva	28–30	3	[51, 52]
Paramyelois transitella	Larva	30	4	[46]
Pectinophora gossypiella	Larva	30		[12]
Danaus plexippus	Larva	29		[53]
Colias eurytheme	Larva			[12]
Pieris rapae	Larva			[12]
Junonia coenia	Larva			[12]

[a] Affinity constants determined for JH III
[b] Radiolabeled JH analog used

3. Purification and Characterization

Presently only the JHBP from *Manduca sexta* has been purified to apparent homogeneity [15-17]. The *Manduca* JHBP has been isolated from the hemolymph of both late fourth and mid-fifth instar larvae. Because there are qualitative and quantitative changes in hemolymph proteins from these two developmental stages, different methods of purification were employed. One major problem of any JHBP is the limited amount of hemolymph that can be collected and the low concentration of binding protein in the hemolymph (20-40 μg/ml).

To increase the purification yields of JH binding protein, we have turned to the use of affinity chromatography [18]. Employing JH II as the immobilized ligand, an affinity resin has been developed which can partially purify JHBP from hemolymph. A JHBP fraction partially purified by gel filtration is passed through the affinity column and the column then washed with buffer. The binding fraction eluted with JH I contains approx. 85% JHBP, which represents 75% recovery of the starting material. Although the procedure does not remove all contaminants, chromatofocusing [9] or preparative gel electrophoresis [16] can be used as the last step.

The utilization of this form of chromatography should greatly simplify the isolation of proteins that interact with JH. The success of conventional purification techniques depends upon the composition of the preparation and requires different conditions be tested each time a new protein or protein source is investigated. Affinity chromatography is virtually independent of such considerations. With minor modifications, a single protocol can be used to search for various types of proteins which interact with the ligand, such as intracellular receptors, biosynthetic enzymes, catabolic enzymes, and antisera preparations. Moreover, differences between preparations containing a given protein rarely affect the efficiency of the affinity purification method. The only limitation to JH affinity chromatography is the prohibitive cost of the hormone used to elute the protein from the column. Methods to circumvent this problem are now being studied.

The physical characteristics of the JHBP from the hemolymph of *Manduca sexta* are presented elsewhere [9]. The molecular weight of the protein from *Manduca* is similar to that of other lepidopteran species and it remains to be seen whether this reflects similarities in the primary sequence and binding specificity. Recent sequence studies performed on the amino terminus of *Manduca* JHBP [17] confirmed that the protein is a single polypeptide chain [15, 16], with the initial sequence

1 10
Asp–Gln–Gly–Ala–Leu–Phe–Glu–Pro–Cys–Ser–
20
Thr–Gln–Asp–Ile–Ala–Cys–Leu–Ser–?–Ala– .

According to Peterson et al. [17], the cysteine residues are probably involved in covalent disulfide bonds which would restrict motion at the N-terminus. Further characterization indicates that JHBP is neither a glycoprotein nor a lipoprotein [16].

3.1 Binding Site Specificity

Since a number of compounds exhibit JH-like activity in bioassays, specificity studies have been performed to correlate biological activity with relative binding affinities. These correlations provide insight into the dimensions and conformation of the binding site as well as the forces involved in binding. Using competitive displacement of radiolabeled JH I as an indication of relative binding activity, it was demonstrated that within the homologous series, there are differences in affinities based on the hydrophobicity of the homolog [19,20]. The relative binding affinity decreases with increasing polarity in the order JH 0 > JH I > JH II > JH III over a 40-fold range. These findings are an extension of the polarity rule which governs, in part, the interactions between steroid hormones and their respective hormone binding proteins [21]. The morphogenetically inactive metabolites JH acid and JH diol were unable to displace radiolabeled JH I. Likewise, several analogs of JH which are commercially used as growth regulators displayed no binding activity. Studies with other structurally related compounds indicate that only those with an isoprenoid carbon skeleton similar in length to that of JH are effective in displacement. Although decreasing the polarity of the molecule enhances the activity, some substitutions which reduce polarity further will inhibit binding. Apparently, hydrophilic groups, such as the epoxide and carboxylic ester, must be conserved for significant hormone-protein interactions to occur [22].

From a large body of evidence (summarized in references [8, 9]), the binding site can be envisioned as a hydrophobic groove or pocket on the surface of the protein. The primary interactions between the hormone and the binding site occur along the alkyl side chains and the methyl of the ester group, which together constitute a distinct hydrophobic "face" (Fig. 1). Hydrogen bonding which involves electrophilic groups in the binding domain anchors the ligand at the epoxide and carbomethoxy group, with the distance between these substituents fixed [8, 9, 22].

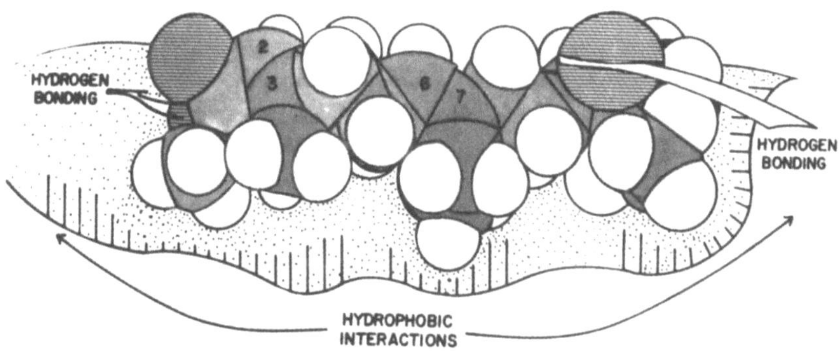

Fig. 1. Space filling model of the naturally occurring JH I isomer and a schematic interpretation of the *Manduca* JHBP binding site. The postulated hydrophobic "face" is directed into the binding site [9]

4. Control of JH Binding Protein Titers

Since JHBP plays an integral role in the overall regulation of JH titers, it is important to understand what factors govern binding protein levels. Studies on *Manduca* showed that the amount of JHBP in the hemolymph changes during the fourth instar [29] for reasons which remain unclear. Three hypotheses have been advanced to account for the fluctuations: (1) JHBP concentrations follow general hemolymph protein concentration; (2) JH induces an increase in JHBP levels; and (3) a decrease in JH titers stimulates an increase in JHBP levels. In order to evaluate these hypotheses, binding site concentrations were measured at numerous points during the life cycle of *Manduca*. These values could then be correlated to overall hemolymph protein concentration and JH titers determined for the same developmental stages.

The levels of JHBP as determined by a modified hydroxylapatite assay are shown in Fig. 2. During the early part of the third and fourth instars, when JH titers are presumably elevated, JHBP titers are relatively low; near the time of PTTH release, when JH titers drop, the levels of JHBP rise. In the early fifth instar there appears to be a similar inverse relationship; the relatively high JH titers at the beginning of the instar decline rapidly, while the JHBP titers climb. At hour 84 of the fifth instar, the binding protein titers begin to climb as does the hemolymph protein concentration. At hour 204, the binding protein titers reach their highest levels during larval life. It is noteworthy that the concentrations of JHBP during the fifth instar, when JH titers are low, are nearly twice that found in the preceding instars, when JH titers are much higher. Initially, it was noted that at pupal commitment (96 h in this population), when JH titers are low for the first burst of ecdysteroid to reprogram development [31], there was a corresponding drop in the titer of binding protein. The apparent decline in JHBP was attributed to the lack of phenyl methyl sulfonyl fluoride inhibition of the JH-specific esterase. If more potent inhibitors of the esterase are used, no decline in the JHBP levels is detected.

The most dramatic increase in binding protein levels are seen in the adult female. At 60 h, when fully matured eggs are present in the ovary, the amount of JHBP is 2–3 times that found in the male. Nonspecific binding by vitellogenin may account for this dimorphism, but as yet no evidence of vitellogenin binding has been reported. The dramatic increase in JH binding capacity of the female probably reflects the need for JH during oogenesis. In both sexes a doubling of the JHBP concentration occurs at adult emergence, but this increase reflects hemolymph volume changes. When JHBP concentrations are traced back to the pupal stage, the levels in female pupae are not significantly different from those of the male.

How then do the fluctuations in JHBP titers during development relate to the proposed mechanisms for control? One hypothesis suggests that changes in binding protein concentration reflect the changes in general hemolymph protein concentration [29]. This possibility cannot be ruled out since in *Manduca* [29] and in other lepidopteran species [30, 32] a close correlation is observed. A second hypothesis suggests that JH induces an increase in JHBP levels and the titers, therefore, are directly proportional. The data do not support this hypothesis. However, as a further test, fourth instar *Manduca* at spiracle apolysis stage were neck-ligated and treated topically with JH I. The animals

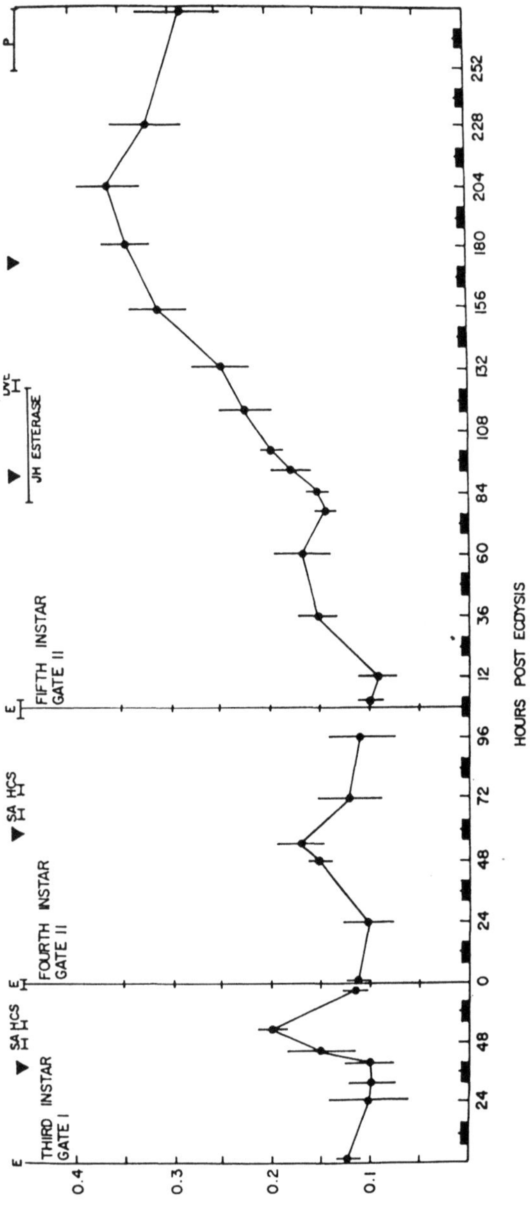

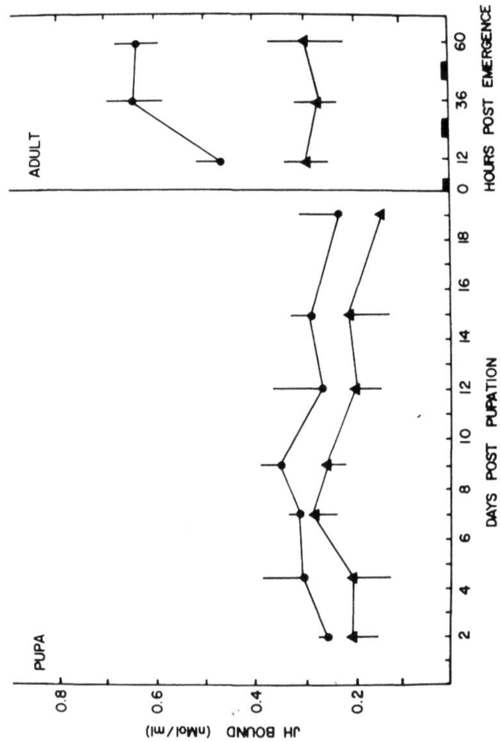

Fig. 2. Relative concentrations of JHBP during larval, pupal, and adult development. JHBP concentrations were determined by a modified hydroxylapatite assay. Each observation represents 6–8 animals. *Top panel: SA* spiracle apolysis: *HCS* head capsule slippage; *E* ecdysis; *DVE* dorsal vessel exposure; *P* pupation. *Solid triangles* midpoints of the head critical period for the third and fourth instars. In the fifth instar the *solid triangle* represents the head critical period for dorsal vessel exposure and pupation. *Bottom panel: circles* female JHBP levels, *triangles* male JHBP levels. *Bars* SD

Fig. 3. Effect of JH I on levels of JHBP. Fourth instar head capsule slippage animals were neck-ligated and treated topically with varying concentrations of JH I. Animals were bled 24 h later and binding protein levels determined using multiple equilibrium dialysis [29]. Relative binding values were determined by radioassaying a small portion of the dialyzed hemolymph

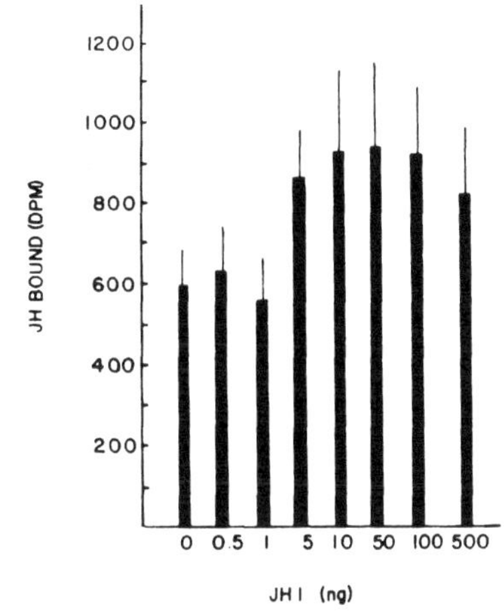

were bled 24 h later and the relative binding protein concentration determined by multiple equilibrium dialysis [29]. Although the levels of binding protein rose concurrent with the amount of hormone applied, the increase was not significantly different from the control (Fig. 3). The evidence does support the possibility that JHBP and JH levels are inversely related during larval life, in *Manduca* and in other Lepidoptera [30, 32]. Such a relationship may reflect an endocrine feedback system to protect the circulating JH remaining at low titer. Unfortunately, much less is known about the titers of JH in pupae and adults, and correlations between JH and JHBP concentrations cannot be made.

One argument against any correlation of hormone to binding protein concentration is the apparent excess of binding sites [29]. For example, in fourth instar *Manduca* the equilibrium concentration of binding sites exceeds that of JH I by a factor of 100 or more at all time points examined. Why then should the insect regulate the concentration of binding sites when such a large excess is always available? It has been established that JH II binds only half as well as JH I [19, 20] and, although precise numbers are lacking, JH II contributes substantially to the juvenile hormone concentration [13]. JH III, which binds to JHBP about a tenth as well as JH I, also makes a small, but significant contribution to the JH titer. Therefore, once the equilibrium concentration of binding sites is calculated for all three homologs in the hemolymph, the apparent excess may in fact vanish.

5. Functions of JHBP

It has become increasingly evident that the hemolymph JHBP serves functions similar to its better known counterparts in the vertebrate circulatory system by providing rapid dispersal and conservation of hormone. The primary mechanism for hormone conservation is the protective role the JHBP performs [23-28]. A number of studies have confirmed that incubating unbound hormone with hemolymph results in a complete and rapid hydrolysis, while incubating bound hormone in the same medium results in only limited hydrolysis. Since these hydrolytic enzymes are present throughout larval life, the protection afforded by the JHBP is crucial to the maintenance of elevated JH titers.

In addition to protecting JH from degradation, the binding proteins prevent adsorption of the highly surface active JH to nonspecific sites, such as basement and plasma membranes. To illustrate this point, radiolabeled JH was incubated in vitro with fat body and epidermis in the presence or absence of JHBP [33]. In cultures containing JHBP the recovery of JH from the medium was significantly enhanced, presumably because JHBP reduced nonspecific adherence of the hormone to tissue and the culture vessel. As a corollary, it has been proposed that JHBP generates a steep hormonal gradient between the corpora allata and the hemolymph. Without this steep gradient, high concentrations of hormone would be localized at the site of synthesis, and any feedback mechanism relying on circulating hormone levels to control synthesis would be severely hampered. In vertebrates it has been established that binding proteins generate and maintain a gradient by transporting the secreted hormone away from the active gland [34].

It has been suggested that the JHBP may scavenge hormone from potentially sensitive tissues; once mobilized in the hemolymph, the hormone could be metabolized or excreted. Although there is a sufficient pool of excess unbound binding sites to perform such a function, the task itself may not be necessary. Implicit in the concept of "scavenging" is the idea that sufficient unbound hormone accumulates in peripheral tissue to elicit a biological response [2]. Simple equilibrium constant calculations which include only the high affinity JHBP indicate that virtually all of the hormone is bound at any given stage [8, 9]. Accumulation of unbound hormone over time could become significant only in the absence of known hemolymph and epidermal esterases. Thus, the possibility that JHBP may act as a scavenger is unlikely.

One of the more intriguing and controversial functions assigned to the circulatory hormone binding proteins is that of involvement in hormone recognitions and/or transfer at the target site. Considerable experimental evidence suggests that lipophilic hormones enter the cell by simple diffusion, yet there are some indications that the circulatory hormone binding proteins may assist. Recognition and uptake of JH at the target cell has been explained as both mediated and unmediated, with evidence to support both hypotheses.

The involvement of JHBP in hormone recognition and/or transfer at the target site is supported by thermodynamic considerations [9] which suggest that virtually all of the hormone is bound throughout larval, pupal, and adult life. As a result, probability would favor recognition of bound rather than unbound hormone at the target cell. An examination of the rates of dissociation (k_d or k_{off}) reveals that the half-life for dissociation of the JH–JHBP complex is 15 min [24]. This dissociation rate is extraordinarily slow when compared to vertebrate hormone binding proteins [1], lending credence to the possibility that a hormone-protein complex may be involved in recognition at the target cell.

Experimental evidence for recognition of a JH–JHBP complex is contradictory and open to some question. In one study, the addition of low concentrations of JHBP to a culture of JH-treated imaginal disks led to a large synergistic effect [24]. In novel immunochemical intervention experiments with *Gomphocercus* [35], antiserum against

JHBP injected into adult females abolished JH-mediated egg maturation. Since the involvement of JHBP in hormone activity is complex, it is difficult to dissect which JHBP functions were disrupted by the antiserum. However, it is possible that removal of JHBP resulted in the loss of recognition at the target cell.

Investigators asserting that the binding protein plays no part in recognition have observed that the JH-specific binding proteins do not bind biologically active JH analogs [36]. Implicit is the assumption that analogs and homologs exert their effects on the target cell in precisely the same manner. Since the naturally occurring hormones and analogs are neither transported nor metabolized in the same way, the mode of action of analogs may not accurately reflect that of the naturally occurring hormones [37, 38]. Using epidermal explants from larval *Manduca*, it was demonstrated that unbound JH I could inhibit commitment to pupation [39]. These data imply that binding is not involved in recognition; however, the concentration of hormone necessary to maintain the uncommitted state is at least an order of magnitude higher than the physiological levels. It could easily be argued that the amounts of hormone used in this study ($>$ 100 n*M*) entered the cell nonspecifically, with or without the protein. This problem was reexamined in the presence of JHBP [40] and it was demonstrated that JHBP had little or no effect on the commitment to pupation. Calculations based on the concentrations of JHBP and JH used in these experiments indicated that the amount of hormone bound does not approach what is observed physiologically and assessment of the role of JHBP remains inconclusive.

Since the first hemolymph binding protein was reported over a decade ago, a number of species have been examined for these important proteins. It is now clear that concepts originally developed to categorize the proteins must be expanded to include the high molecular weight, high affinity JHBP. Moreover, the polarity rule, which defines the binding order of JH homologs in Lepidoptera, must be reexamined in light of the recently discovered proteins. Although only a few species outside Lepidoptera have been studied, the initial survey on binding preference supports the contention that JH III may be the primary homolog of most species. It is hoped that with new techniques, these proteins can be purified and extensively characterized. At present, questions about the control of JHBP titers and recognition at the target cell remain largely unexplored. Answers to these difficult problems will go far in defining parameters of JH distribution and possible modes of action at the target cell.

Acknowledgments. We wish to thank H.L. Goodman for her helpful comments and assistance in preparation of this manuscript. Support for this work was provided in part by the College of Agricultural and Life Sciences, University of Wisconsin and by a grant from the Johnson Wax Fund. We thank Dr. B.D. Hammock of the University of California-Davis for the suggestion and the gift of the trifluoroketone esterase inhibitor.

References

1. Westphal U (1980) In: Wittliff J, Dapunt O (eds) Steroid receptors and hormone dependent neoplasia. Masson, New York, p 1
2. Diamond M, Rust N, Westphal U (1969) Endocrinology 84:1143–1151
3. Heller J (1975) J Biol Chem 250:3613–3619
4. Glass T, Nunez N, Fischer S, Robinson H (1978) Biochem Biophys Acta 542:154–162
5. Youngdahl-Turner P, Mellman I, Allen R, Rosenberg LE (1979) Exp Cell Res 118:127–134

6. Brown MS, Kovanen P, Goldstein JL (1981) Science 212:628–635
7. Whitmore E, Gilbert LI (1972) J Insect Physiol 18:1153–1167
8. Goodman WG (1983) In: Downer RGH, Laufer H (eds) Invertebrate endocrinology, vol. 1. Endocrinology of insects. Liss, New York, p 147
9. Goodman WG, Chang ES (1984) In: Kercut GA, Gilbert LI (eds) Comprehensive insect physiology, biochemistry and pharmacology, vol VII. Pergamon Press, Oxford (in press)
10. DeKort CA, Peter MG, Koopmanschap AB (1983) Insect Biochem 13:481–487
11. Koeppe JK, Kovalick GE, Lapointe MC (1981) In: Pratt GE, Brooks GT (eds) Juvenile hormone biochemistry. Elsevier/North Holland, Amsterdam, p 215
12. Wing KD, Rudnicka M, Jones G, Jones D, Hammock BD (1984) J Comp Physiol (in press)
13. Schooley DA, Baker FC (this vol)
14. Bergot BJ, Baker FC, Cerf DC, Jamieson G, Schooley DA (1981) In: Pratt GE, Brooks GT (eds) Juvenile hormone biochemistry. Elsevier/North Holland, Amsterdam, p 33
15. Kramer KJ, Dunn PE, Peterson RE, Seballos HL, Law JH (1976) J Biol Chem 4979–4985
16. Goodman WG, O'Hern PA, Zaugg RH, Gilbert LI (1978) Mol Cell Endocrinol 11:225–242
17. Peterson RC, Dunn PE, Seballos HL, Barbeau BK, Keim PS, Riley CT, Heinrikson RL, Law JH (1982) Insect Biochem 12:643–650
18. Goodman WG, Goodman HL (1981) In: Pratt GE, Brooks GT (eds) Juvenile hormone biochemistry. Elsevier/North Holland, Amsterdam, p 365
19. Goodman WG, Bollenbacher WE, Zvenko HL, Gilbert LI (1976) In: Gilbert LI (ed) The juvenile hormones. Plenum Press, New York, p 75
20. Kramer KJ, Dunn PE, Peterson RC, Law JH (1976) In: Gilbert LI (ed) The juvenile hormones. Plenum Press, New York, p 327
21. Westphal U (1971) Steroid-protein interactions. Springer, Berlin Heidelberg New York
22. Goodman WG, Schooley DA, Gilbert LI (1978) Proc Natl Acad Sci USA 75:185–189
23. Hammock BD, Nowock J, Goodman W, Stamoudis V, Gilbert LI (1975) Mol Cell Endocrinol 3:167–184
24. Sanburg LL, Kramer KJ, Keźdy FJ, Law JH, Oberlander H (1975) Nature (London) 253: 266–267
25. Sanburg LL, Kramer KJ, Keźdy FJ, Law JH (1975) J Insect Physiol 21:873–888
26. Sams G, Cocchiaro GF, Bell WJ (1978) In Vitro 14:956–960
27. Rudnicka M, Sehnal F, Jarolim V, Kochman M (1979) Insect Biochem 9:569–575
28. Peter MG, Gunawan S, Gellissen G, Emmerich H (1979) Z Naturforsch 34:588–598
29. Goodman W, Gilbert LI (1978) Gen Comp Endocrinol 35:27–34
30. Ozyhar A, Wisniewski JH, Sehnal F, Kochman M (1983) Insect Biochem 13:435–441
31. Riddiford LM (1976) In: Gilbert LI (ed) The juvenile hormones. Plenum Press, New York, p 198
32. Wing KD, Sparks TC, Lovell VM, Levinson SO, Hammock BD (1981) Insect Biochem 11: 473–485
33. Nowock J, Hammock BD, Gilbert LI (1976) In: Gilbert LI (ed) The juvenile hormones. Plenum Press, New York, p 354
34. Ewing L, Chubb C, Robarie B (1976) Nature (London) 264:84–86
35. Hartmann R (1978) Wilhelm Roux' Arch 184:301–324
36. Wyatt GR, Pan M (1978) Annu Rev Biochem 47:779–817
37. Chihara C, Petri W, Fristrom J, King DS (1972) J Insect Physiol 18:1115–1124
38. Slade M, Wilkinson C (1974) Science 181:672–674
39. Riddiford LM (1976) Nature (London) 259:32–42
40. Mitsui T, Riddiford LM, Bellamy G (1979) Insect Biochem 9:637–643
41. Emmerich H, Hartmann R (1973) J Insect Physiol 19:1663–1675
42. Bassi S, Goodman W, Altenhofen C, Gilbert LI (1977) Insect Biochem 7:309–312
43. Kramer S, DeKort CA (1978) Insect Biochem 8:87–92
44. Trautmann KH (1972) Z Naturforsch 27:263–273
45. Klages G, Emmerich H (1979) Insect Biochem 9:23–30
46. Kramer KJ, Childs C (1977) Insect Biochem 7:397–403
47. Sparks TC, Hammock BD (1979) Insect Biochem 9:411–421

48. Hammock BD, Sparks TC, Mumby S (1977) Pestic Biochem Physiol 7:517–530
49. Ferkovich S, Silhacek D, Rutter R (1975) Insect Biochem 5:141–150
50. Goodman WG (unpubl. observations)
51. Turunen S, Chippendale GM (1981) Insect Biochem 11:429–435
52. Turunen S, Chippendale GM (1979) Nature (London) 280:836–838
53. Lessing C, Herman W (1984) Insect Biochem (in press)

III.9 Photoaffinity Labelling of Juvenile Hormone-Binding Proteins in *Leucophaea maderae*

J.K. KOEPPE[1], G.E. KOVALICK[1], and G.D. PRESTWICH[2]

1. Introduction

Juvenile hormone (JH) binding proteins have been identified in tissues from several insect species [1-3]. In the adult female cockroach *Leucophaea maderae*, JH-binding proteins have been found in the fat body [4], ovaries [5], and hemolymph [4, 5]. Although some properties of these binding proteins have been described, further characterization has been hampered by the extremely rapid dissociation of JH from the binding protein under nonequilibrium conditions. In an attempt to circumvent this problem, we have initiated studies on the possible use of a photoactivatable JH III analog (Fig. 1), 10,11-epoxyfarnesyl diazoacetate (EFDA), to form an irreversible covalent attachment at the JH binding site on the hormone binding protein. Our preliminary investigations suggested that EFDA covalently attached to JH-binding proteins in the hemolymph and ovaries [6]. Recent studies have focused on the specificity of [^{3}H]-EFDA binding to both the hemolymph and the ovarian binding proteins.

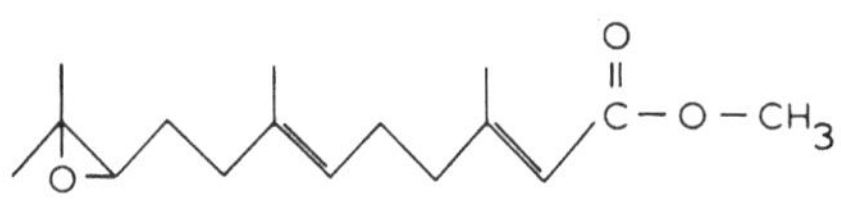

Juvenile hormone III (JH III)

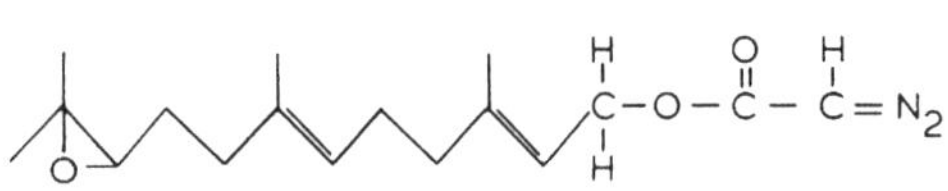

10,11-Epoxy farnesyl diazoacetate (EFDA)

Fig. 1. Structures of JH III and EFDA

1 Department of Biology, University of North Carolina, Chapel Hill, NC 27514, USA
2 Department of Chemistry, SUNY, Stony Brook, NY 11794, USA

Biosynthesis, Metabolism and Mode of Action of Invertebrate Hormones
(ed. by J. Hoffmann and M. Porchet)

2. Methodology

Detailed descriptions of the methodology employed in these studies are published elsewhere [7, 8]. Crude cytosolic extracts of hemolymph or ovarian tissues were prepared by homogenizing tissues in a Tris/KCl/$MgCl_2$ buffer (pH 7.4), followed by centrifugation at 100,000 g. Reaction mixtures containing either JH or EFDA were assayed under equilibrium conditions. JH-binding protein complexes were separated from free JH by polyethylene glycol (PEG) precipitation of the proteins. Photoactivation of EFDA occurred when samples containing EFDA were irradiated with UV light at a wavelength of 254 nm. The amount of EFDA covalently attached to proteins was determined by precipitation of the proteins with trichloroacetic acid (TCA), followed by filtration of the solution to remove unbound [^{3}H]-EFDA. Proteins subjected to electrophoresis were first irradiated in the presence of [^{3}H]-EFDA, precipitated with TCA, then resuspended in SDS sample buffer. SDS-polyacrylamide gel electrophoresis was performed using the buffer system described by Laemmli [9]. The presence of radioactivity in the gel was determined by a modification of the fluorography technique of Bonner and Laskey [10].

3. Results

3.1 EFDA as a JH Analog

To determine the relative affinity of the binding protein for EFDA, competition experiments were performed under equilibrium conditions using increasing concentrations of unlabelled JH III, JH I, JH 0, EFDA, or methoprene to displace [^{3}H]–JH III from the binding protein. Tissue extracts were incubated with 1×10^{-9} *M* [^{3}H]–JH III and various concentrations of unlabelled hormone or analog ranging from 1×10^{-9} *M* to 7×10^{-5} *M*, then assayed by PEG protein precipitation. Our results (Fig. 2) indicated that although EFDA competed less effectively for binding sites than any of the unlabelled JH homologs, EFDA competed much more efficiently than the other JH analog, methoprene. JH III, the naturally occurring homolog in *Leucophaea*, competed most efficiently with [^{3}H]–JH III. When JH I was used as a competitor, 2.5-3 times as much was required to displace the same amount of [^{3}H]–JH III as did JH III; for JH 0 6.5-7 times more was needed, and for EFDA 15-20 times more.

3.2 [^{3}H]-EFDA Attachment to Proteins

Since unlabelled EFDA competed with [^{3}H]–JH III for binding sites, we next determined whether photoactivated [^{3}H]-EFDA would covalently attach to proteins, and if so, whether attachment could be prevented by the presence of excess unlabelled JH III during photolysis. Hemolymph and ovarian extracts were diluted so that the binding protein concentration in each was about $2\text{-}4 \times 10^{-8}$ *M*, then incubated 30 min at 0 °C with 1×10^{-8} *M* [^{3}H]-EFDA in the absence or presence of 1×10^{-5} *M* unlabelled JH III. Samples were irradiated with UV light for 0, 15, 30, or 60 s and then analyzed

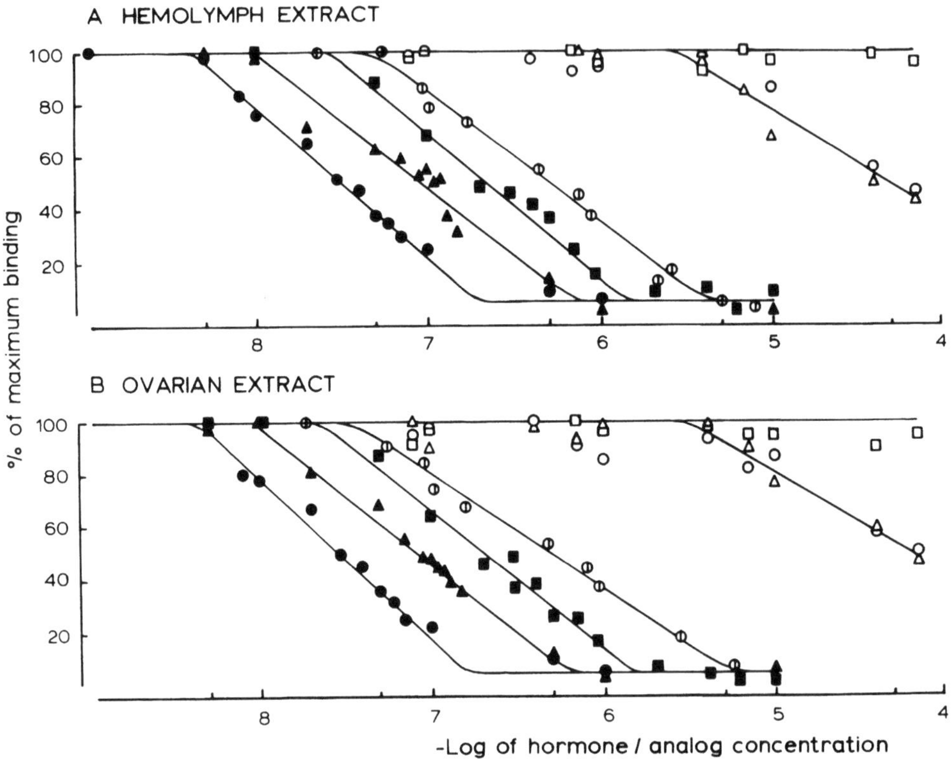

Fig. 2. Displacement of [^{3}H]–JH III from the hemolymph JH-binding protein using increasing concentrations of JH I (▲), JH 0 (■), EFDA (⊕), and methoprene (△). The concentration of [^{3}H]–JH III in each reaction mixture was 1×10^{-9} *M* and the concentration of the JH-binding protein was approx. $1-3 \times 10^{-8}$ *M*. The results are expressed as a percentage of maximum binding found with only [^{3}H]–JH III in the reaction mixture

for binding using the TCA filter assay. When hemolymph extracts containing 1×10^{-8} *M* [^{3}H]-EFDA were irradiated, 4.4% of the labelled analog bound to proteins. The presence of 1×10^{-5} *M* unlabelled JH III reduced covalent attachment of [^{3}H]-EFDA to proteins to only 0.86%. The ability of JH to protect proteins from [^{3}H]-EFDA binding was taken as a measure of [^{3}H]-EFDA binding specificity. Thus, 81% of [^{3}H]-EFDA binding to proteins in the hemolymph was JH specific. Similar data was obtained for ovarian extracts.

3.3 Specificity of [^{3}H]-EFDA Binding

The effectiveness of the various JH homologs in the displacement of radiolabelled ligand from the binding protein can be used to determine if [^{3}H]-EFDA binds to the JH binding protein at the JH binding site. We have already established that the unlabelled homologs JH III and JH I have different affinities for the JH binding protein. If [^{3}H]-EFDA binds at the same site as [^{3}H]–JH III, then the ratio of the concentration of

JH III to JH I needed to displace 50% of the labelled ligand from the binding protein should be identical for both tritiated compounds. In these comparative displacement studies, hemolymph or ovarian extracts were incubated with either [^{3}H]-EFDA or [^{3}H]–JH III in the presence of increasing concentrations of unlabelled JH III or JH I. The amount of [^{3}H]–JH III bound was assayed using PEG precipitation of proteins. [^{3}H]-EFDA was covalently attached to protein with UV irradiation, and the amount bound to proteins was determined by the TCA filter assay. The results (Table 1) indicated that in both hemolymph and ovarian extracts three to four times more JH I than JH III is needed to displace either labelled ligand from the binding site.

Table 1. 50% displacement of labelled ligands with unlabelled JH III or JH I

Labelled ligand	[JH III]	[JH I]	Ratio
[^{3}H]–JH III	1.5×10^{-8} *M*	6.5×10^{-8} *M*	4.3 (Hemolymph)
[^{3}H]–EFDA	2.9×10^{-8} *M*	10.5×10^{-8} *M*	3.6 (Hemolymph)
[^{3}H]–JH III	1.8×10^{-8} *M*	5.8×10^{-8} *M*	3.2 (Ovarian)
[^{3}H]–EFDA	2.4×10^{-8} *M*	10.3×10^{-8} *M*	4.3 (Ovarian)

3.4 Electrophoretic Separation of [^{3}H]-EFDA Labelled Proteins

Using SDS-polyacrylamide gel electrophoresis and fluorescent autoradiography, we have determined to which proteins [^{3}H]-EFDA becomes covalently attached. Hemolymph and ovarian extracts containing 1×10^{-7} *M* [^{3}H]-EFDA were irradiated in the presence of various concentrations of unlabelled JH III. Proteins were precipitated with TCA, filtered, resuspended in SDS sample buffer, and subjected to gel electrophoresis and fluorography. Autoradiograms (Fig. 3) of hemolymph and ovarian proteins separated in this manner confirmed that in each extract increased concentrations of unlabelled JH III markedly inhibited [^{3}H]-EFDA attachment to one protein with an R_f of about 0.25. Comparing this R_f to those of unlabelled markers ranging in molecular weight from 43,000 to 130,000, we determined that the JH-binding protein in each extract had a molecular weight of greater than 130,000. Although several other proteins were also labelled, these proteins appeared to be nonspecifically labelled, since very little inhibition of [^{3}H]-EFDA attachment occurred when the JH III concentration was increased as indicated in Fig. 3.

4. Summary

These experiments establish that [^{3}H]-EFDA binds specifically and with relatively high affinity at the JH-binding site on the JH-binding protein in the hemolymph and in the ovaries. Additionally, we have shown that the JH-binding protein in each tissue has a molecular weight of greater than 130,000. When working with photoactivatable hormone analogs, such as EFDA, studies similar to the ones outlined here must be carried

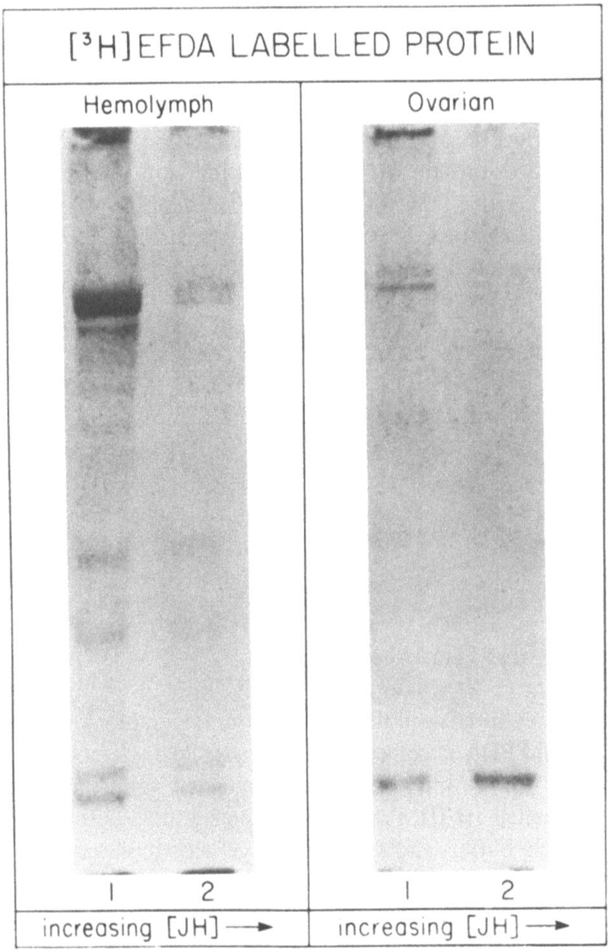

Fig. 3. Fluorescent-autoradiogram of electrophoretically separated hemolymph and ovarian proteins labelled with [^{3}H]-EFDA. Extracts were equilibrated with labelled EFDA and two concentrations of unlabelled JH III and then photolyzed for 40 s. Extracts in *lanes 1* were photolyzed in the presence of 1×10^{-7} *M* labelled EFDA and 1×10^{-8} *M* unlabelled JH III. Extracts in *lanes 2* were photolyzed as for # 1, but the JH III concentration was five times higher (5×10^{-8} *M*). TCA precipitated proteins were subjected to SDS gel electrophoresis as described elsewhere [7, 8]. Each well contained approx. 50–100 μg of protein

out in order to unambiguously establish specificity and affinity of the binding protein for the label and to avoid accumulation of "false positive" binding proteins in the literature. Having shown that [^{3}H]-EFDA covalently attaches at the JH-binding site, we can now use this photoaffinity label in the purification and characterization of JH-binding proteins in the hemolymph and ovaries of *Leucophaea maderae*.

Acknowledgments. We thank J. Carvalho, A. Singh, and R. D'Ambrosio for assistance in the synthesis of the EFDA compounds. We also thank D. Suddreth for his technical assistance with the SDS gels. Partial financial support was provided by the National Science Foundation (Grant PCM-

8021190 to J.K.K. and CHE-7925081 to G.D.P.) and the National Institutes of Health (GM-30899 to G.D.P.). Additional support from the Alfred P. Sloan Foundation and the Camille and Henry Dreyfus Foundation (awards to G.D.P.) is also gratefully acknowledged. We are grateful to S. Jacobs, P. Cuatrecasas, and L. Gilbert for their discussion and helpful suggestions.

References

1. Gilbert LI, Goodman W, Nowock J (1976) In: Durchon M, Joly P (eds) Actualites sur les hormones d'invertebres. CNRS, Paris, pp 413–434
2. Roberts PE, Wyatt GR (1983) Mol Cell Endocrinol 31:53–69
3. Chang ES, Coudron TA, Bruce MJ, Sage BA, O'Connor JD, Law JH (1980) Proc Natl Acad Sci USA 77:4657–4661
4. Engelmann F (1981) Mol Cell Endocrinol 24:103–112
5. Koeppe JK, Kovalick GE, LaPointe MC (1981) In: Pratt GE, Brooks GT (eds) Juvenile hormone biochemistry. Elsevier/North-Holland, Amsterdam, pp 215–231
6. Prestwich GD, Kovalick GE, Koeppe JK (1982) Biochem Biophys Res Commun 107:966–973
7. Kovalick GE, Koeppe JK (1983) Mol Cell Endocrinol 31:271–286
8. Koeppe JK, Kovalick GE, Prestwich GD (1984) J Biol Chem 259:3219–3223
9. Laemmli UK (1971) Nature (London) 227:680–685
10. Bonner WM, Laskey RA (1974) Eur J Biochem 46:83–88

III.10 Regulation of Vitellogenesis in Insects: The Pleiotropic Role of Juvenile Hormones

F. ENGELMANN[1]

1. Introduction

Endocrine controlled egg production in insects has been a major research topic since Wigglesworth [1] discovered the source of a hormone that promotes vitellogenesis in *Rhodnius prolixus*. Emphasis has been directed in recent years towards an understanding of the molecular mode of hormone action, and this has been particularly fruitful with regard to juvenile hormone (JH) directed vitellogenin synthesis [2, 3]. Vitellogenin is the precursor to the major yolk protein of the mature eggs. It is produced in the fat bodies of most species, but in *Drosophila* the follicular epithelial cells of the ovaries also produce an identical protein species [4]. JH-controlled vitellogenin synthesis is a dramatic event, both because it is produced in an all-or-none fashion upon JH exposure and because of the massive quantities synthesized within a short period. Hormonal control is basically via induction of transcriptional events which are amenable to an experimental approach in many species of insects, and in analogy, as well in egg-laying vertebrates, such as amphibians and birds.

While it is generally acknowledged that JH regulates vitellogenin mRNA transcription, it is now also known that several additional aspects associated with the making of a fully grown egg are influenced by this hormone [2, 3]. Among these are the accelerated proliferation of the endoplasmic reticulum in fat bodies, vitellogenin export to the hemolymph, juvenile hormone esterase induction, DNA replication, and vitellogenin uptake by the growing oocytes. The actual role of JH in these processes may be the induction of specific protein synthesis (e.g., vitellogenin), activation of certain enzymes essential in phospholipid biosynthesis, or effects on properties of plasma membranes and basal lamina (most apparent during oocyte growth). JH acts pleiotropically in the control of a multiplicity of tissue activities which synergistically facilitate a rapid egg growth during vitellogenesis. Certain of these effects may be mediated by hormone binders and tissue receptors within target tissues. JH translocation to the nucleus via receptors may be an essential component for the massive vitellogenin synthesis in the fat bodies.

1 Department of Biology, University of California, Los Angeles, CA 90024, USA

Biosynthesis, Metabolism and Mode of Action of Invertebrate Hormones
(ed. by J. Hoffmann and M. Porchet)

2. JH Controlled Vitellogenin Synthesis

For *Leucophaea* it was shown that the corpora allata control vitellogenesis via induction of vitellogenin synthesis [5, 6]. In the majority of species investigated the same appears to be the case [3]. Subsequent to the basic description of endocrine controlled vitellogenin synthesis, translatable mRNA for vitellogenin has been extracted only from vitellogenic females of *Drosophila* [7] and *Locusta* [8] and, thus, providing a convincing correlation between production of JH and induced transcription of specific mRNA. A genetic analysis of vitellogenin production was performed in *Drosophila*, the animal of choice [9]. Accordingly, JH regulates the coordinated transcription of three vitellogenin genes and synthesis of three vitellogenin polypeptides. Similar considerations have been applied to *Locusta* in which two large primary translation products (> 200,000 mol. wt.) are known to be induced coordinately by JH [8, 10]. These large polypeptides are apparently processed proteolytically and several peptides of different sizes result. It will be a formidable task to unravel the details of the molecular mechanisms underlying the induction of vitellogenin in *Locusta* [11] or any other species that has a large primary translation product.

2.1 JH Cytosol Receptors in the Fat Bodies

The effect of several steroid hormones which induce specific cellular responses, such as new protein synthesis, is mediated by cytosolic and nuclear receptor-acceptor molecules [12]. In analogy, one may postulate that translocation of JH to the nucleus in fat bodies occurs by cytosolic receptors as well. Vitellogenin synthesis may follow. Compounds with high affinity for JH (K_d about 10^{-9} *M*) have indeed been identified in cytosol of fat bodies of adult *Leucophaea* [13] and follicular epithelia of vitellogenic ovaries [14]. Both of these tissues are known target tissues for JH. The identified high affinity molecules were saturable (Fig. 1) and specificity was shown [13]. The correlation of the occurrence of the high affinity JH binders in tissues that exhibit a response to the hormone makes it plausible that we are dealing with a physiological phenomenon. These binding compounds may be classified as receptors.

High affinity JH binding compounds have also been extracted from nuclei of fat bodies of vitellogenic females [14]. It remains to be shown whether these molecules are identical to those of the cytosol and, thus, represent the translocated cytosol receptors. Interestingly, they were found in vitellogenic females, whereas nonvitellogenic females had comparatively low concentrations of these JH binding compounds. We see a biological correlation which is highly suggestive for the functional importance of these high affinity binders. Whether these binding compounds are a prerequisite for the initiation of vitellogenin gene transcription remains to be worked out.

2.2 Low Affinity JH Binders in the Fat Bodies

Fat body cytosol of *Leucophaea* contained not only compounds with high affinity for JH, but also a population of binders with affinity one order of magnitude lower (Fig. 1). Follicular epithelia of the ovaries also contained such a population of lower affinity

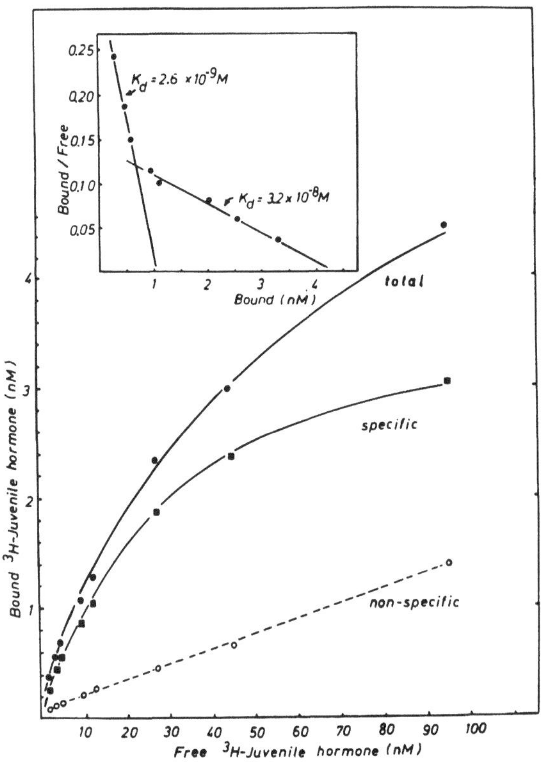

Fig. 1. Saturation analysis of JH binding to compounds in cytosol of fat bodies of vitellogenic females. 100 μl of a cytosol solution (30 ml solution per 100 mg of tissue) were incubated with ^{3}H-JH-III at 5 °C for 90 min before the unbound JH was removed by adsorption to dextran-coated charcoal [13]. Nonspecific binding was determined by incubations in parallel with 100-fold excess of unlabeled hormone. The inset shows the Scatchard analysis of the specific binding

binders for JH [14, 15]. Several species of JH binders with a similar affinity (K_d about 10^{-8} *M*) seem to exist in both of these tissues. Such binding compounds were not only identified in cytosol preparations, but also were extractable with 0.5 *M* KCl from tissue organelles, such as microsomes, protein coated urate crystals, and bacterial symbionts [14]. I have reported earlier [13] that cytosol of a nontarget tissue, midgut epithelia, also contained JH binding compounds of this modest affinity. Unexpectedly, even vitellogenin bound JH with similar affinity as the compounds extracted from tissue organelles. No function has been attributed to JH binders extracted from the tissue organelles and we are left with speculation as to the biological importance. Until such a function can be documented we may assume nonfunctional, but saturable and hormone specific binding. This binding could represent artifactual binding to lipoproteins, analogous to what has been described for the interaction of insulin and talc, which certainly is not a hormone receptor [16]. On the other hand, if such binding indeed occurs in the living tissue – and there is no reason to argue against it – one may hypothesize that this bound JH represents a reservoir of hormone available when the high affinity binders are no longer saturated. From this point of view, the total binding capacity of the several lower affinity binders, and thus, the available JH, is considerable and presumably exceeds that of the high affinity receptors in the same tissues.

2.3 JH Controlled Synthesis of Endoplasmic Reticulum

In conjunction with vitellogenin synthesis in the fat bodies of *Leucophaea*, an extensive network of rER membranes develops. The cisternae contain newly synthesized vitellogenin [17]. The same is seen in other species and has been documented electron-microscopically for *Locusta* [18]. In *Leucophaea* proliferation of the ER is controlled by JH via phospholipid synthesis. Phospholipids are incorporated into the ER in a JH dose-dependent manner (Fig. 2) [19, 20]. It is conjecture whether ER proliferation is a prerequisite for vitellogenin biosynthesis or simply allows an accelerated production of vitellogenin. The latter is easily conceivable, because of the parallel increase in vitellogenin synthesis and ER proliferation (Fig. 2). In addition, it is highly possible that new ER synthesis entails the production of essential specific binding sites for the vitellogenin polysomes. This may be deduced from the observation that ER in nymphal fat bodies is not responsive to JH and the rate of vitellogenin production is only about 1% of that of the adult females for the same hormone dose [20]. This interpretation does, however, not include additional factors important for a normal rate of vitellogenin production. Additional factors must exist, since adult males were not inducible to make vitellogenin at a high rate either, yet their ER responded to applied JH and proliferation was seen as in adult females.

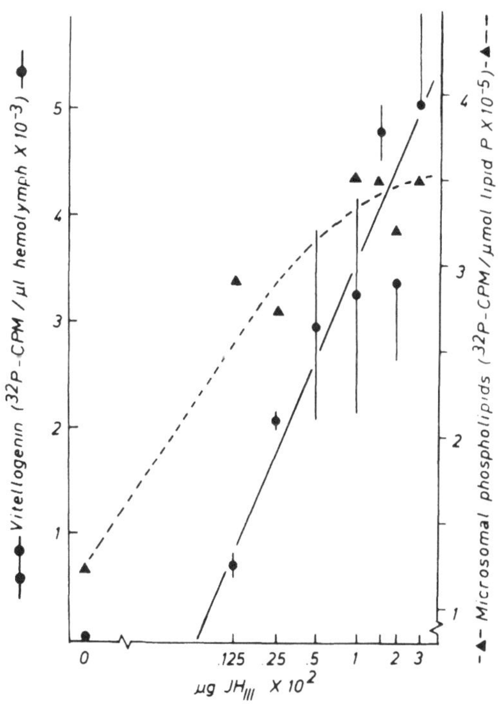

Fig. 2. JH dose-dependent induction of vitellogenin as measured by incorporation of ^{32}P into monospecific antibody precipitable proteins from the hemolymph (n = 3). Incorporation of ^{32}P into phospholipids of the ER membranes was dose-dependent as well. Phospholipids were extracted with chloroform:methanol (2:1, v/v) from microsomes of fat bodies of three animals (about 600 mg of tissue) [20]

3. JH Esterases of the Hemolymph

JH degrading carboxylesterases were first identified in *Schistocerca* and *Rhodnius* [21] and in *Hyalophora* [22]. Later, the same was found in many additional species of other orders [23]. It was often thought that these enzymes participate in the essential regulation of JH titers during various developmental stages. This may well be the case for some species, but may not apply in others, as can be shown in several instances. An analysis of the role of JH esterases can only be made reliably in conjunction with a study on JH binding molecules (carriers) in tissues and hemolymph.

JH specific esterases occur in the hemolymph of adult females of *Leucophaea*. These esterases were isolated and essentially purified in one step by anion exchange column chromatography (QAE, Pharmacia). Nonvitellogenic females, i.e., newly emerged, pregnant, or allatectomized animals, had only one esterase that eluted from a column at about 0.6 *M* NaCl. Vitellogenic females, either during their normal vitellogenesis or after JH treatment of allatectomized females, had a second JH esterase, easily separated from the first by anion exchange chromatography (Fig. 3) [24]. JH appeared to induce this second esterase which was characterized by different K_m and V_{max} values and an apparent lower molecular weight by 10,000. The total JH esterase activity per μl hemolymph fluctuated during a reproductive cycle of this species. The level of fluctuation was, however, modest because one of the esterases was always present in adult females.

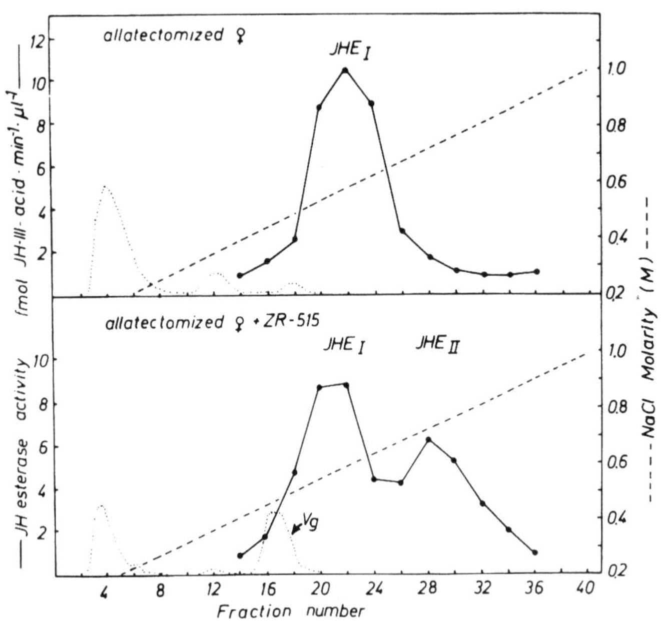

Fig. 3. JH esterases isolated from hemolymph of allatectomized and methoprene (ZR-515) treated females. 300 μl of hemolymph (from three to four animals) were fractionated on a QAE anion exchanger. 10 ml fractions were collected and for establishing the JH esterase profiles 25 μl of these fractions were assayed for esterolytic activity [24]. *JHE I:* $K_m = 0.68 \times 10^{-6}$ *M*, $V_{max} = 4.84$ pmol. $min^{-1}.\mu l^{-1}$ hemolymph. *JHE II:* $K_m = 1.54 \times 10^{-6}$ *M*, $V_{max} = 8.13$ pmol.$min^{-1}.\mu l^{-1}$ hemolymph

It should be pointed out that during periods of high JH concentrations (vitellogenesis), the total available JH esterases was highest, yet JH is obviously not removed from the circulation at a high rate, as evidenced by the continued vitellogenin production. In other words, in adult females of *Leucophaea* JH esterases do not appear to effectively control JH titers of the circulation. No doubt, JH stimulated the production of a new JH esterase in this species, but the functional significance of this observation remains a matter of speculation.

As is documented, a 1,000-fold gain of JH esterase activity was obtained upon purification [24]. This is the result of the removal of several JH binders functioning as JH esterase inhibitors. Evidence for this conclusion will be presented below.

4. The Role of JH Binding Compounds in the Hemolymph

JH binding compounds have been identified in many insect species and invariably have been described as "JH carriers" or "JH protecting" agents [23]. These designations imply the functional role of these hormone binders. A priori there appears to be no need for JH carrier molecules, since JH is soluble at concentration considered necessary for eliciting hormonal responses. The term carrier is, thus, misleading and if the hormone is indeed bound to a circulating protein, it is not of functional importance for distribution of the hormone. On the other hand, JH bound to a protein may cover the site of attack of degrading enzymes and, thus, effectively prevent rapid degradation. However, binding per se does not need to be associated with "protection" of the hormone, i.e., demonstration of binding is just that. The term protection describes the overall event without explaining how this might be achieved.

The availability of relatively pure JH binders and that of the purified JH esterases of the hemolymph of *Leucophaea* [24] allowed an analysis of the interactions of these molecules. Quantitatively two major JH binding compounds were isolated. Both were saturable and hormone specificity was shown in competition assays. The so-called compound I was the predominant binder at all physiological stages of the adult females [14]. Vitellogenin was an additional functional binder in vitellogenic females. Compound I, when mixed with the esterase I or II lowered V_{max} of these enzymes, and from Lineweaver-Burk plots [14] we could read the characteristics of a so-called uncompetitive enzyme inhibitor [25]. This binder, which presumably is analogous to the JH carriers of other species, is indeed an enzyme inhibitor. In other words, hormone protection is achieved by inhibition of the degrading enzymes.

The second major JH binding compound of the hemolymph - vitellogenin - had a somewhat lower binding capacity than compound I, based on μg protein assayed. The affinity for JH III was in the same order of magnitude (K_d about 10^{-8} *M*) as that of compound I (Fig. 4). Vitellogenin inhibited the activity of JH esterase I, and the Lineweaver-Burk plot indicated a mixed enzyme inhibitor (Fig. 5). Certainly, this function of vitellogenin was unexpected; vitellogenin contributes to the ineffectiveness of circulating JH esterases.

As is now recognized, both major JH binding compounds of the hemolymph are JH esterase inhibitors. The in vitro assays also revealed that only a fraction of the com-

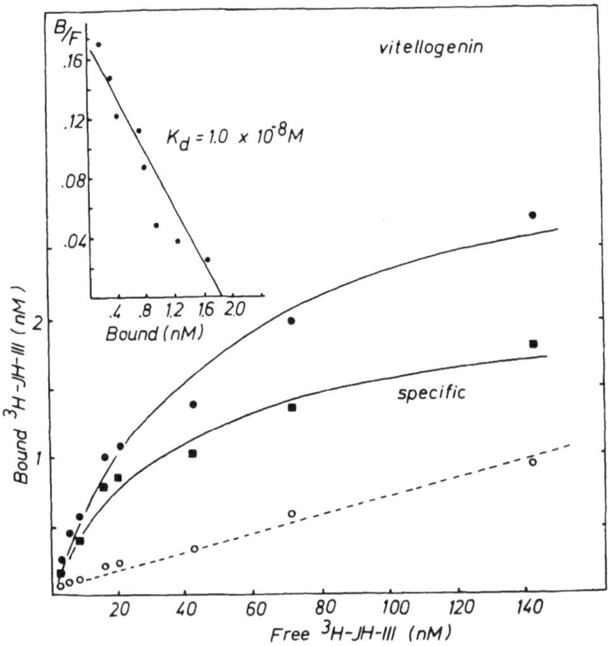

Fig. 4. Analysis of JH–III binding to purified (QAE) vitellogenin. For this 5.4 μg of vitellogenin in 100 μl of buffer were incubated with ^{3}H-JH-III at 5 °C for 90 min. Assay conditions were the same as in Fig. 1. *Inset* shows a Scatchard plot for the specific binding

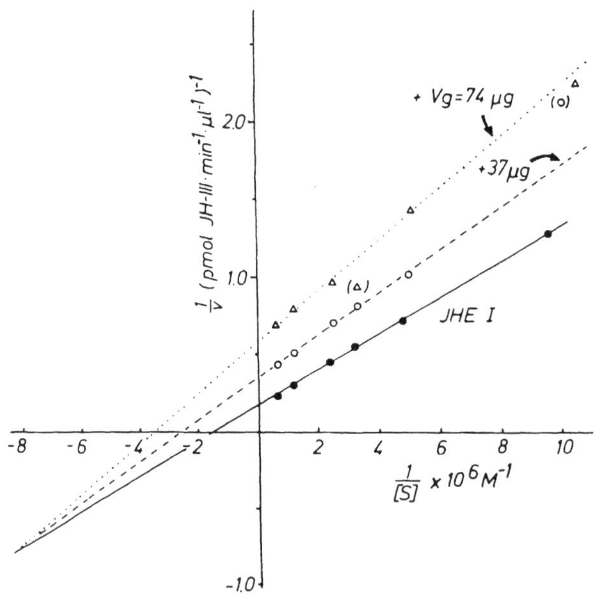

Fig. 5. Lineweaver-Burk plot of JH esterase I showing the effect of added purified vitellogenin at two different concentrations. A representative example taken from four replications. Assay conditions have been published [24]. For this plot it was assumed that binding of JH to vitellogenin lowered the substrate concentrations by removal of the ligand and this was considered in the plotting of the data. The number of the binding sites was determined by Scatchard analysis. With this consideration the curvilinear relation of substrate concentration to product formed (JH-III-acid) was transformed into a straight line. The plot indicates that vitellogenin is a mixed esterase inhibitor

pound I and vitellogenin molecules did bind JH. From batch to batch this fraction varied somewhat. If one assumes that in vivo a higher percentage of molecules bind JH, one may postulate an even greater effectiveness of the inhibition of JH esterases. Whether uncompetitive or mixed inhibition, in both cases V_{max} was lowered proportionally to the amounts of inhibitors within the incubation solutions which represents the circulation (Fig. 5). JH binding to certain protein molecules is, thus, equivalent to JH esterase inhibition. The level of esterase activity in natural hemolymph was about 1/1,000 of that of purified esterases when computed per content of 1 μl of hemolymph; indeed, an effective inhibition of enzyme activity.

5. Synthesis and Outlook

The role of JH in the control of vitellogenesis and particularly of vitellogenin synthesis is multifaceted. JH controls vitellogenin mRNA production, fat body ER proliferation, and vitellogenin uptake by the growing oocytes (Fig. 6). Vitellogenin mRNA transcription and ER proliferation may synergistically facilitate the massive vitellogenin production. JH is presumably bound to macromolecules when circulating in the hemolymph. This binding of JH lowers the JH esterase activity and, thus, prolongs the half-life of JH. JH also induces the production of a new JH esterase which essentially increases the total level of these esterases in the hemolymph, particularly towards the end of vitellogenesis. The subsequent cessation of JH production by the corpora allata, coinciding with ovulation in *Leucophaea*, may allow a more rapid degradation of the remaining

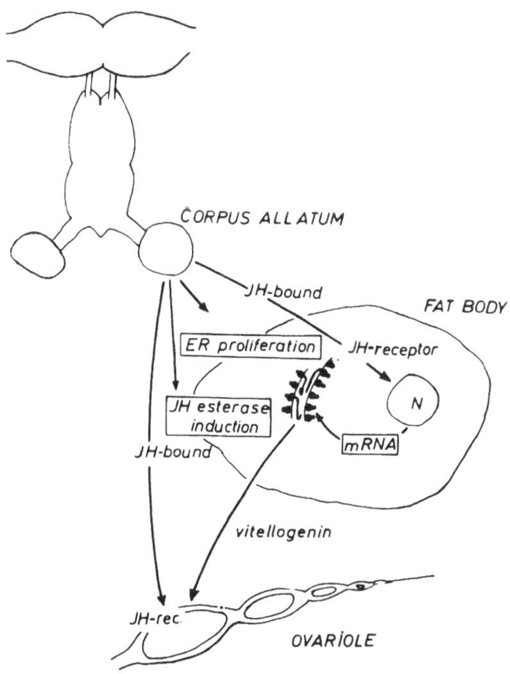

Fig. 6. A composite diagram of the known JH actions in adult females of *Leucophaea* during vitellogenesis

JH in the circulation. At this time JH may also be released from the bound state shortly before ovulation since vitellogenin is no longer produced and, thus, the number of binding molecules in the hemolymph is dramatically reduced.

One of the largely unexplored, but most exciting areas of research on JH regulated vitellogenin synthesis today, is the role of cytosol and nuclear receptor-acceptor molecules in hormone action. Results of research on this aspect may permit us to widen the concept of receptor mediated hormonal activation of specific genes to the sesquiterpenoid JH. The specific question of how transcription of the vitellogenin mRNA is controlled needs to be answered. Tissue specificity, stage, and sex specific responses to JH may be mediated by cytosolic and nuclear receptors indeed. For a complete understanding of JH regulated vitellogenin synthesis advances will have to be made on these aspects. While we have identified the role of hemolymph JH binders as JH esterase inhibitors, these binders may have additional functions. Among these may be the facilitated uptake of the hormone by target tissues. This needs to be explored. An understanding of how JH stimulated ER proliferation is associated with vitellogenin production is equally important. Is it the production of specific binding sites for the vitellogenin polysomes, induction of cytoplasmic signal recognition proteins, or simply the enlargement of ER membranes? We have to explore the JH controlled export of vitellogenin from the fat bodies. And furthermore, our knowledge of the complexity of vitellogenin uptake by the growing oocytes is only in its beginning stages. As stated at the outset, JH is the key agent in all of these events in the majority of insect species known to date. The complexity of the various controls and their understanding will require a technologically integrated effort.

Acknowledgments. The results of the research contained in this communication were made possible through the support by a grant from the US National Institutes of Health (HD-15530).

References

1. Wigglesworth VB (1936) Q J Microsc Sci 79:91–121
2. Engelmann F (1979) Adv Insect Physiol 14:49–108
3. Engelmann F (1983) In: Downer RGH, Laufer H (eds) Endocrinology of insects. Liss, New York, pp 259–270
4. Brennan MD, Weiner AJ, Goralski TJ, Mahowald AP (1982) Dev Biol 89:225–236
5. Engelmann F, Penney D (1966) Gen Comp Endocrinol 7:314–325
6. Engelmann F (1969) Science 165:407–409
7. Bownes M, Hames BD (1978) FEBS Lett 96:327–330
8. Chen TT (1980) Arch Biochem Biophys 201:266–276
9. Bownes M (1982) Q Rev Biol 57:246–274
10. Chinzei Y, White BN, Wyatt GR (1982) Can J Biochem 60:243–251
11. Wyatt GR (this volume)
12. Katzenellenbogen BS (1980) Annu Rev Physiol 42:17–35
13. Engelmann F (1981) Mol Cell Endocrinol 24:103–112
14. Engelmann F (1984) Adv Invertebr Reprod 3:177–187
15. Koeppe JK, Kovalick GE, Lapointe MC (1981) In: Pratt GE, Brooks GT (eds) Juvenile hormone biochemistry. Elsevier/North Holland, Amsterdam New York, pp 215–231
16. Hollenberg MD, Cuatrecasas P (1979) In: O'Brien RD (ed) The receptors. A comprehensive treatise, vol I. Plenum Press, New York London, pp 193–214
17. Engelmann F, Barajas L (1975) Exp Cell Res 92:102–110

18. Chen TT, Couble P, De Lucca FL, Wyatt GR (1976) In: Gilbert LI (ed) The juvenile hormones. Plenum Press, New York London, pp 505–529
19. della-Cioppa G, Engelmann F (1980) Biochem Biophys Res Commun 93:825–832
20. della-Cioppa G, Engelmann F (1984) Roux Arch Dev Biol 193:78–85
21. White AF (1972) Life Sci 11:201–210
22. Whitmore D, Whitmore E, Gilbert LI (1972) Proc Natl Acad Sci USA 69:1592–1595
23. de Kort CAD, Granger NA (1981) Annu Rev Entomol 26:1–28
24. Gunawan S, Engelmann F (1984) Insect Biochem 14:601–607
25. Cornish-Bowden A (1976) Principles in enzyme kinetics. Butterworths, London

III.11 On Titers, Origin, and Functions of Juvenile Hormone III, Methylfarnesoate, and Ecdysteroids in Embryonic Development of the Ovoviviparous Cockroach *Nauphoeta cinerea*

B. LANZREIN[1], H. IMBODEN[1], C. BÜRGIN[1], E. BRÜNING[1], and H. GFELLER[2]

1. Introduction

It has been known for a long time that the endocrine glands of insects become fully differentiated during the later phases of embryonic development. Their being functional and/or essential for normal development, however, is a matter of debate [1, 2].

As far as concerns the presence of juvenile hormone (JH) in embryos, rather high quantities of JH-active material as measured by bioassay have been found in eggs of *Hyalophora cecropia* [3], *Oncopeltus fasciatus* [4], *Macrotermes subhyalinus* [5, 6], and *Locusta migratoria* [7, 8]. Physicochemical analyses have shown the presence of rather low quantities of known and new JHs in some Lepidopteran species [9] and the results obtained were not consistent with the bioassay data in the case of *Oncopeltus fasciatus* [4, 9] and *Locusta migratoria* [7, 8, 10], since only small quantities of JH III were observed. The site of production of the embryonic JH has never been investigated in detail. Nevertheless, the fact that eggs of allatectomized females of *Hyalophora cecropia* do not contain JH-active material when newly laid, but do so when analysed 7 days later [3] and the correlation between the ultrastructural changes in the embryonic corpora allata (CA) concomitant with an increase in JH-active material [4] suggest that the CA become active in the older embryo.

In contrast to the paucity of information available on the presence, function, and origin of JH in the embryo many reports exist on identification, titers, origin, and function of ecdysteroids in insect embryogenesis (see also section entitled Ecdysteroids). Eggs and embryos were always analyzed in toto and since they are a closed system the great variety of ecdysteroids identified includes precursors, hormone(s), and metabolites. Yet it was never investigated whether and which of these ecdysteroids circulate in the embryo hemolymph. With respect to the prothoracic or ventral glands, the major site of ecdysone production in postembryonic immature stages, histological changes [1] as well as ultrastructural variations [11] suggest endocrine activity during later phases of embryonic development, but still these glands do not seem to be indispensable for ecdysteroid production [12].

In no case were ecdysteroids and JHs analyzed simultaneously during embryonic development nor were they sought in the hemolymph of the embryo. We, thus, figured

1 Division of Animal Physiology, Zoological Institute, University of Berne, Erlachstraße 9a, CH-3012 Berne, Switzerland
2 Institute of Organic Chemistry, University of Berne, Freiestraße 3, CH-3012 Berne, Switzerland

Biosynthesis, Metabolism and Mode of Action of Invertebrate Hormones
(ed. by J. Hoffmann and M. Porchet)

in 1976 that these points deserved attention, since the action of the two hormones is closely linked in postembryonic development, and since for a substance to be considered a hormone its presence in the hemolymph is of importance. Our laboratory insect, the ovoviviparous cockroach *Nauphoeta cinerea*, seemed especially suitable for these investigations because we can easily obtain synchronized embryos of known ages by selecting the pregnant females. Here we review the identification of ecdysone, 20-hydroxy-ecdysone, and JH III, their titer fluctuations in the embryo hemolymph, and the search for ecdysteroid conjugates hydrolyzable by *Helix pomatia* enzymes. We report for the first time that the embryonic CA are biosynthetically active and that they synthesize methylfarnesoate and JH III under in vitro conditions at stages after dorsal closure. We also include data on the in vivo synthesis and identification of methylfarnesoate in the embryo and present evidence that this substance is released together with JH III from the CA and that both circulate in stage specific quantities in the embryo hemolymph.

2. Results

In our earlier investigations [13] of embryonic development in *Nauphoeta cinerea*, which lasts 35–40 days, we found, through the *Musca* bioassay, variable and stage specific quantities of ecdysteroid active material and, through the *Galleria* wax text, variable and stage specific quantities of JH active material. The ecdysteroid active material of 32-day-old egg cases was analyzed by gas-liquid chromatography mass spectrometry (GLC–MS) and the results showed the presence of ecdysone, 20-hydroxy-ecdysone, and 26-hydroxy-ecdysone; there was also some indication of the presence of other 20-hydroxy-ecdysteroids. Using high pressure liquid chromatography (HPLC) combined with the *Galleria* wax test we were able tentatively to identify the JH active material as JH III. Figure 1 shows the results of the simultaneous measurement of the titer fluctuations of JH III, ecdysone, and 20-hydroxy-ecdysone in egg content and, after dorsal closure, in the hemolymph of the embryos; these results were obtained by using radioimmunoassay (RIA) for ecdysteroid determinations and the *Galleria* wax test for JH quantification. It would appear that the ecdysone titer is always rather low, whereas the 20-hydroxy-ecdysone titer displays a first slight increase around the time of dorsal closure, followed by two marked peaks at the time when the JH concentration is also high. Since one *Galleria* unit corresponds to approx. 8 pg of JH III, one can calculate that the embryo hemolymph contains up to 1 μg/ml of JH III. The titer fluctuations of ecdysone, 20-hydroxy-ecdysone, and JH III in the hemolymph are similar to those observed in whole egg cases [13], which indicates that these three hormones circulate in the hemolymph after dorsal closure.

In a more detailed analysis of early embryonic development [14] JH and ecdysteroid titers were measured at daily intervals and the ecdysteroid conjugates hydrolyzable by *Helix* enzymes were also investigated. The titers of ecdysone, 20-hydroxy-ecdysone, and of the highly polar products, as measured by RIA, are shown in Fig. 2; there is a distinct peak of 20-hydroxy-ecdysone and a slight increase in ecdysone immediately before dorsal closure, as well as a continuous increase in the quantity of RIA active

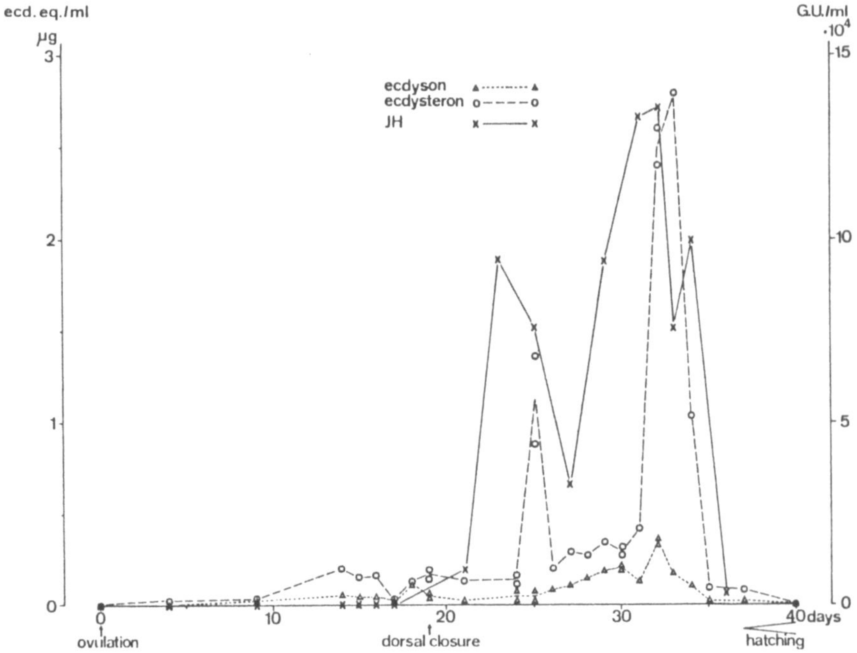

Fig. 1. Ecdysone, 20-hydroxy-ecdysone and JH activity in egg content and hemolymph of embryo during embryogenesis. Ecdysone and 20-hydroxy-ecdysone were determined by RIA after separate elution from TLC plates and the results are expressed in ecdysone or 20-hydroxy-ecdysone equivalents, respectively. JH activity was determined by *Galleria* wax test after TLC purification and the results are expressed in *Galleria* units (GU). 1 GU corresponds to approx. 8 pg JH III (Calbiochem). Each *symbol* represents one determination. (After [13])

highly polar products. Analysis of the latter using *Helix pomatia* enzymes reveals the presence of only some conjugates of 20-hydroxy-ecdysone in day 0 egg cases and of even smaller quantities in day 17 egg cases. Large quantities of RIA active highly polar products, particularly in day 17 egg cases, were not hydrolyzed by the *Helix* enzymes, indicating the presence of polar ecdysteroids which cannot be hydrolyzed by these enzymes. It can be concluded from histological investigations that the peak of free 20-hydroxy-ecdysone before dorsal closure coincides with the synthesis of an embryonic cuticle. During the entire period of early embryonic development, no JH active material, or only traces thereof, could be detected by using the *Galleria* wax test. Using a physicochemical method we then reinvestigated selected stages with respect to the quality and quantity of JH (Table 1). The results thus obtained are fully consistent with those obtained earlier by means of the *Galleria* wax test and confirm the presence of enormous quantities of JH III at stages after dorsal closure. The sudden appearance of large quantities of JH III seemed promising for investigating the origin of the embryonic JH. The CA are differentiated after dorsal closure (Imboden, unpublished) and we tried to determine whether they are capable of synthesizing JH. In a medium composed of larval hemolymph, tissue culture medium M-199, and [methyl-^{14}C]methionine brain corpora cardiaca (CC)–CA complexes were found to synthesize and release labelled

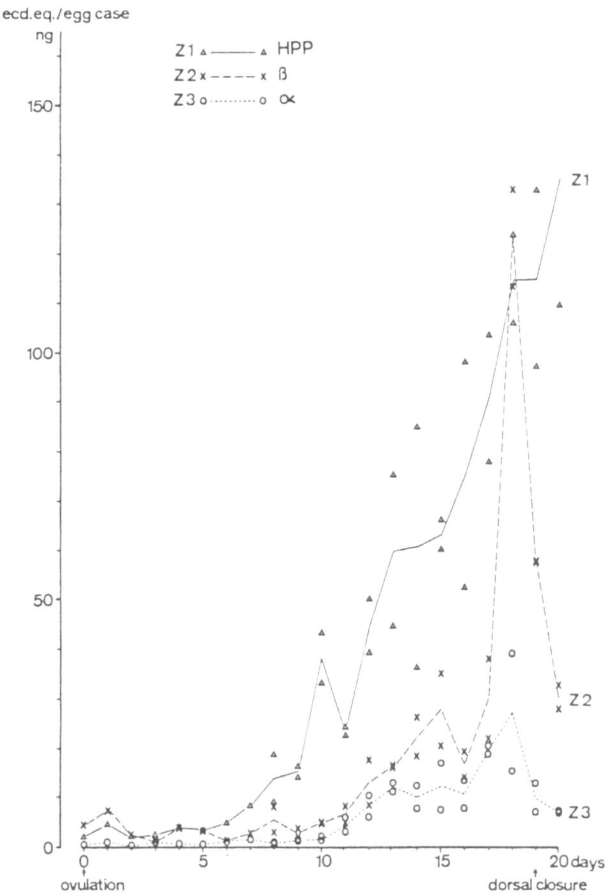

Fig. 2. RIA active material in egg case in TLC *zones 1* (highly polar products), *2* (20-hydroxy-ecdysone), and *3* (ecdysone) from ovulation until dorsal closure. The results are expressed for *zones 1* and *3* in ng ecdysone equivalents/egg case and for *zone 2* in ng 20-hydroxy-ecdysone equivalents/egg case. One egg case weighs between 110 and 220 mg [13]. Each *symbol* represents one determination (4 egg cases were pooled). (After [14])

Table 1. Titers of JH III in ng/g during embryonic development of *Nauphoeta cinerea* determined by the GLC–MS method of Bergot et al. [44]. (After Baker et al. [18])

Developmental stage in days							
16	17	18	21	25	32	34	35
0	2.2	0.1	10.0	838.0	432.0	339.0	1.24
0	3.7	0		745.0	963.0	511.0	
	0.14						
	0						
	0						
	0						

material of similar polarity to JH III and of a lower polarity, as determined by TLC analysis [15]. When brains alone were incubated no radioactivity was observed in these zones, demonstrating that these substances are not produced by the brain. When CC–CA complexes were cultured in M-199, to which farnesoic acid, which is known to stimulate JH synthesis [16, 17], and [methyl-^{14}C]methionine had been added, analysis of the medium together with the CC-CA complexes through reversed phase HPLC revealed the presence of labelled products with retention times corresponding to JH III

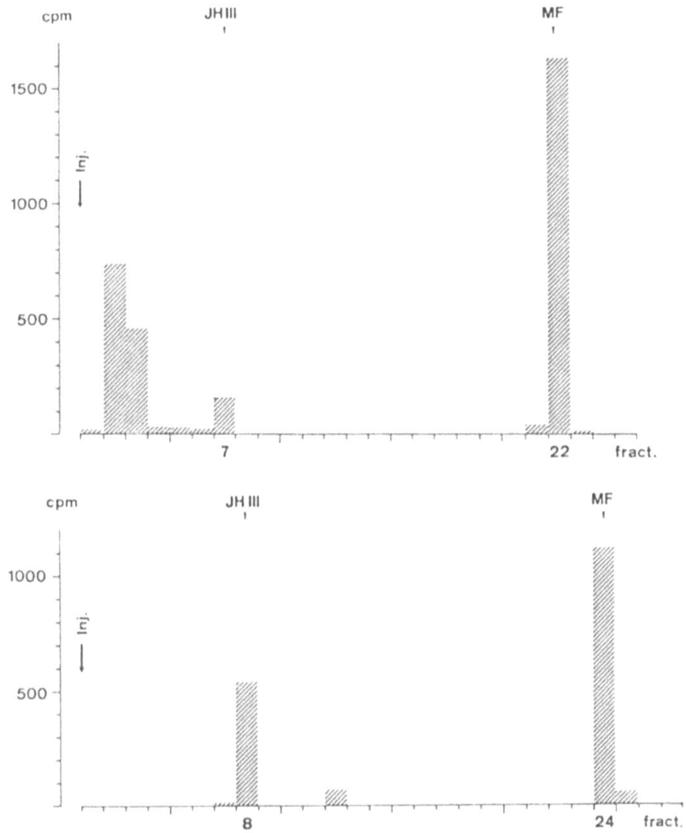

Fig. 3. Radiochromatogram of an in vitro experiment (*top*) culturing CC-CA complexes in the presence of [methyl-^{14}C]methionine and of an in vivo experiment (*bottom*) where [methyl-^{14}C]methionine was injected into egg cases. HPLC conditions: column RP 18 (Merck) and 70% acetonitrile in water as a solvent. *Top:* 30 CC–CA complexes of 29-day-old embryos were incubated for 6 h at 26 °C in 100 μl M-199 (pH 6.6) to which farnesoic (40 *μM*) and [methyl-^{14}C]methionine giving a final specific activity of 36 mCi/mmol was added. Culture medium together with the CC–CA complexes was extracted using ethyl acetate and purified by TLC (ethyl acetate/hexane 1:1). For analysis in HPLC TLC purified material [15] was injected together with reference methylfarnesoate and JH III (retention times given in the figure). Fractions were collected every 2 min and the flow rate was 750 μl/min. *Bottom:* 5 egg cases (28 days old) were injected with 5 μCi of [methyl-^{14}C]-methionine in 10 μl M-199 each and extracted 24 h later using ethyl acetate/ethanol 4:1. After passage through a C_{18} Sep-pak cartridge using acetonitrile as solvent the extracts were purified by TLC and analyzed by HPLC as described above. Fractions were collected every 2 min and the flow rate was 700 μl/min

and methylfarnesoate (Fig. 3, top). The above findings indicate that the embryonic CC–CA complex synthesizes and releases both methylfarnesoate and JH III. In order to test wether the synthesis of important quantities of methylfarnesoate was an artifact of the in vitro system, [methyl-^{14}C]methionine was injected into 28-day-old egg cases, which were then extracted and analyzed using HPLC (Fig. 3, bottom). It is evident that the same radiolabelled components were produced in vivo and in vitro (Fig. 4, top), which means that the production of methylfarnesoate is a true occurrence and not an artifact of the in vitro system. We then investigated whether methylfarnesoate is present in normally developing embryos at stages when the CA were observed to synthesize it. For this purpose 27-day-old egg cases were extracted, purified, and analyzed by HPLC (Fig. 4). The UV absorbance trace shows a peak with the retention time of methylfarnesoate and the material constituting this peak was further analyzed using gas chromatography mass spectrometry/chemical ionization (GCMS/CI). The spectra of reference methylfarnesoate (top) and of fraction no. 6 (bottom) are shown in Fig. 5 and their similarity conclusively demonstrates the presence of methylfarnesoate in 27-day-old egg cases. Through using the same technique fraction 2 was shown to contain JH III together with other substances. Provided that the HPLC peak with the retention time of methylfarnesoate consists only of methylfarnesoate it can be calculated that egg cases contain up to μg quantities/g of methylfarnesoate, an order of magnitude comparable to that of JH III (Table 1).

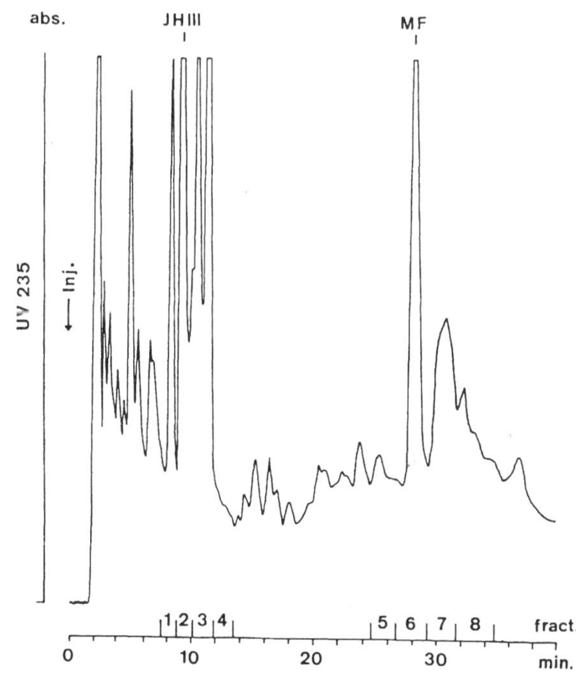

Fig. 4. HPLC trace (UV 235 nm) of an extract from nine 27-day-old egg cases. Column: RP 18 (Merck), solvent: 73% acetonitrile in water, flow rate 850 μl/min. Fractions were collected as indicated and retention times of JH III and methylfarnesoate are marked

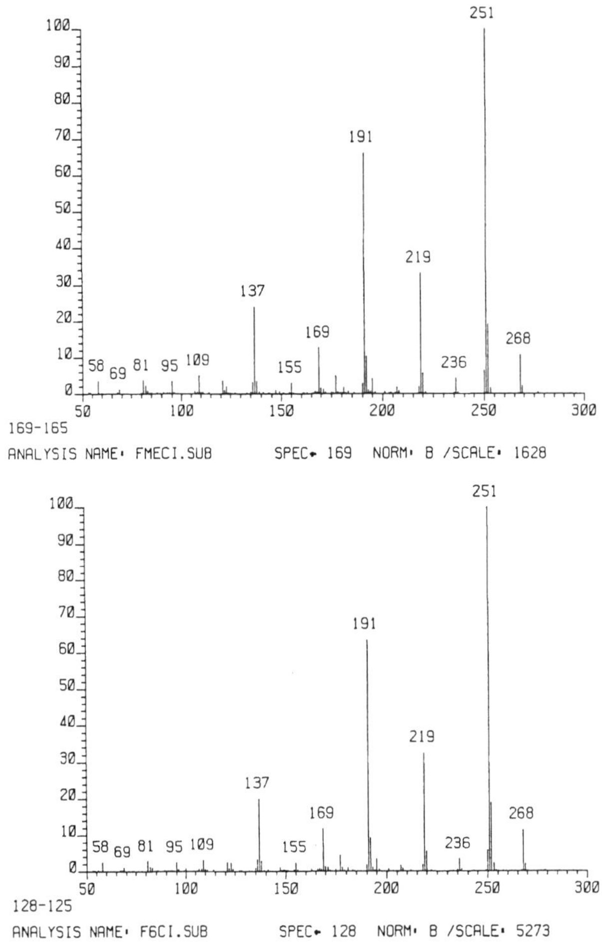

Fig. 5. Mass spectrum of reference methylfarnesoate (*top*) and of HPLC fraction no. 6 from Fig. 4 (*bottom*) using chemical ionization (CI). GCMS was performed on a Varian Mat 44S instrument, equipped with an SS 188 data system, using a 25 m capillary column coated with SE30 with an inner diameter of 0.285 mm. GC conditions: column temperature 168 °C, carrier gas helium (11.5 psi). Interface: Pyrex capillary persilylated, temperature 200 °C. MS conditions: source temperature 200 °C, electron energy 300 eV, emission current 500 μA, reactant gas ammonia (250 μb), mass range 50–300

3. Discussion

3.1 Presence and Origin of JH and Methylfarnesoate in Embryos

Our JH analyses show the presence of enormous quantities of JH III after dorsal closure (Fig. 1, Table 1). This is the same JH as that found in larvae and adult females [18, 19], although the highest concentration in the hemolymph of the embryo is about ten times higher than the highest titer in the hemolymph of the oocyte-maturing female [19].

There are remarkable differences between the insect we have used and other insect species in which physicochemical JH analyses have been performed. The Lepidopteran embryos which have been investigated contain predominantly the higher JH homologs, namely, JH 0, I, and II [9], but at much lower concentrations than those observed for JH III in *Nauphoeta cinerea* (Table 1). In the cricket *Teleogryllus commodus*, where embryogenesis lasts 12-14 days, very low levels (0.05 ng/g) of JH III only are reported for 6-day-old embryos [20] and in *Locusta migratoria* the presence of rather low quantities of JH III (up to approx. 13 ng/g) has been reported recently for stages after dorsal closure only [10]. Using physicochemical techniques JHs could be found only at stages after dorsal closure in all cases investigated so far, whereas the use of biological tests indicated the presence of JH active material already in newly laid eggs in the case of *Hyalophora cecropia* [3], *Oncopeltus fasciatus* [4], *Locusta migratoria* [7, 8], and *Macrotermes subhyalinus* [5, 6]. This difference could be due to the fact that a bioassay, if carried out with crude extracts, could interpret molecules other than the free known JHs as JH active material. Nevertheless, if the extracts are carefully purified before the bioassay is carried out reliable results can be obtained with respect to the free known JHs, as demonstrated by the similarity of the data resulting from *Galleria* wax tests and from physicochemical analyses (Fig. 1, Table 1) in the case of *Nauphoeta cinerea*.

In addition to the large quantities of JH III we also found high amounts of methylfarnesoate in 27-day-old embryos (Figs. 4 and 5). Apparently this substance circulates in the hemolymph and is not found, or only traces of it are found, before dorsal closure and after day 35 [21]. So far we know of no other report mentioning the presence of methylfarnesoate in embryos, but it has very likely never been sought. It would, of course, not be detected by JH analytical techniques based on derivatization of the epoxide moiety, but it exerts some JH activity in biological tests, such as the *Galleria* wax test [21] and possibly the *Tenebrio* test. These facts could explain discrepancies, such as those observed in *Oncopeltus fasciatus* [4, 9], between bioassay results and physicochemical JH analyses.

As far as concerns the origin of the embryonic JH, our findings clearly demonstrate that the embryonic CC–CA complexes at stages after dorsal closure are capable of performing the last steps of synthesis of JH III, since they incorporate [methyl-^{14}C]methionine to form methylfarnesoate and JH III (Fig. 3). It is very likely that the CA are the site of synthesis of these products (a supposition which has been conclusively proved for adult CA of *Schistocerca gregaria* [22]) and not the CC, which remain attached to the CA for technical reasons. This is the first direct demonstration of hormone production by an embryonic endocrine gland. So far only indirect evidence or the results of ultrastructural investigations have suggested that the embryonic CA become active [3, 4, 9, 23]. The quantities of JH synthesized in vitro (Fig. 3) appear low in comparison with the titers found in the embryo hemolymph (Fig. 1). This could be due either to the fact that dissection of the embryonic CC–CA complexes is extremely difficult and/or to deficiencies of the culture medium, since rates of biosynthesis were found to be higher in a hemolymph medium than in M-199 supplemented with farnesoic acid. However, we cannot completely exclude the possibility of JH also being produced in vivo outside the CA. Preliminary analyses of rates of JH synthesis at various stages of embryonic development showed some variations between days 22 and 35 and then a decrease after day 35 [15]. The comparison with the JH titer curve (Fig. 1) reveals some similarity, but the precise relationship between rates of JH synthesis by the CA and the JH titer is not yet clear. Hemolymph from 24- to 36-day-old embryos was found to be almost devoid of JH degrading enzymes [15] and we, thus, do not yet know how the JH titer is regulated in the embryo.

The major product synthesized by the embryonic CA is methylfarnesoate, which is certainly not an artifact of the in vitro system since it is also synthesized in vivo (Fig. 3) and has been identified in the normally developing embryo (Figs. 4 and 5). As to the relative quantities of methylfarnesoate and JH III synthesized, preliminary data indicate that methylfarnesoate makes up 70%-80% between days 22 and 34 and only 20%-25% between days 35 to 39 [15]. This reversal of the ratio of methylfarnesoate to JH III at the end of embryonic development seems to coincide with the breaking up of the chorion and is possibly correlated with changes in consumption and use of oxygen.

3.2 Presence and Origin of Ecdysteroids

As to the embryonic ecdysteroids, we would like to mention only a few comparative aspects, since this topic is certainly extensively discussed in the section Ecdysteroids and since several articles have appeared lately reviewing this subject [24-27]. A striking difference can be observed between locusts and *Nauphoeta cinerea* with respect to both the quality and the quantity of the ecdysteroids in ovaries and embryos. In *Nauphoeta cinerea* we find only small quantities of conjugates hydrolyzable by *Helix pomatia* enzymes in ovaries [28] and egg cases [14] and 20-hydroxy-ecdysone is the predominant free and conjugated ecdysteroid, whereas in *Locusta migratoria* up to 100 μM conjugated 2-deoxyecdysone and 50 μM conjugated ecdysone are observed in mature ovaries [25]. Considering the fact that similar increases in the titer of free ecdysteroids are observed in both species prior to the differentiation of the endocrine glands these differences in the content of conjugates could mean that *Helix* hydrolyzable conjugates in high quantities, such as those found in *Locusta migratoria*, are not prerequisite for an increase in free ecdysteroids, or that *Nauphoeta cinerea* has other conjugates not susceptible to the *Helix* enzymes. The presence of highly polar ecdysteroids refractory to enzymatic hydrolysis has been reported earlier [26, 29]. The increase in RIA active highly polar products that we observed between ovulation and dorsal closure together with the peak of free 20-hydroxy-ecdysone (Fig. 2) and the simultaneous decrease in the quantity of *Helix* hydrolyzable 20-hydroxy-ecdysone conjugates [14] could be interpreted as a consequence of both the formation of free 20-hydroxy-ecdysone from *Helix* hydrolyzable conjugates and the formation of polar ecdysteroid metabolites [30].

The origin of the late peaks in free ecdysteroids is probably due to de novo synthesis [25, 27], but the site of production is not yet clear. In *Carausius* the results of ultrastructural investigations suggest activity of the ventral glands concurrent with the beginning of the third embryonic (= first larval) cuticulogenesis [11]. In *Clitumnus* embryos with microsurgically suppressed ventral glands the ecdysteroid peak at dorsal closure is normal, but the second rise in ecdysteroids is earlier than in normal embryos and the titer never decreases [12], indicating that the ventral glands are not absolutely necessary for the ecdysteroid titer to rise, but seem to be necessary for the timing and the decrease in ecdysteroids [11]. The results of histological investigations of prothoracic glands and oenocytes in *Oncopeltus fasciatus* suggest that the latter are the source of ecdysteroids in late embryonic development in this insect [31] and experiments with embryos of the cockroach *Blabera craniifer* show that tissues of the metathorax or the abdomen produce ecdysteroids [32]. In view of these findings the possibility of ecdysteroid production by the epidermis, as suggested for pupae of *Tenebrio* [33], should also be taken into consideration.

3.3 Function of JH, Methylfarnesoate, and Ecdysteroids

It is known that addition of ecdysteroids to in vitro cultures induces premature cuticle formation in embryos of *Blabera craniifer* [34, 35] and in many cases increases in free ecdysteroids are correlated with cuticle formation [24, 36]. In general, smaller quantities of ecdysteroids are found at the time of dorsal closure and formation of an embryonic cuticle than at the time when the larval cuticle is secreted [12, 31, 36, 37] (Fig. 1). However, comparisons are difficult to make in the cases where only crude extracts were analyzed by RIA or bioassay [12, 31, 37]. The first larval cuticle is always thicker than the embryonic cuticles [14, 34, 38, 39] and in *Nauphoeta cinerea* it is formed in the presence of JH III (Fig. 1). A similar coincidence is indicated in *Locusta migratoria* [10, 36], although the quantities of JH III present are much smaller than in *Nauphoeta cinerea*. Application of JH analogs to embryos of *Schistocerca gregaria* was found to induce a variety of effects including the formation of fully differentiated miniature embryos with more advanced pigmentation and sclerotization of the cuticle [40]. In the same insect application of the JH analog farnesylmethyl ether to preblastokinetic embryos was shown to induce premature larvalization, namely, formation of a cuticle with hairs and setae [41], and application of JH III was seen to affect the deposition of the various cuticle layers [42]. Similarly, in *Clitumnus* embryos application of JH before the ecdysteroid peak at dorsal closure resulted in the formation of a larval-type cuticle [43]. On the other hand, when the JH titer in embryos of *Manduca sexta* was artifically lowered by treatment with fluoromevalonate, an inhibitor of JH biosynthesis, this induced the formation of a pupal-type cuticle in these embryos [9]. These observations and the simultaneous appearance of JH with the 20-hydroxy-ecdysone peak concomitant with the formation of the first larval cuticle [14] (Fig. 1) strongly suggest that JH plays a role in the formation of the larval cuticle by inducing the formation of features of a larval cuticle, such as hairs, setae, and pigmentation. This would mean that JH has a morphogenetic effect in the embryo similar to that in the larval stages. Yet one wonders why there is so much JH III in the embryo of *Nauphoeta cinerea* (up to 1 μg/g, as shown in Table 1, compared to values around 1 ng/g in the first larval instar [18, 19]). Whether this has to do with differences in the sensitivity of the target tissues or whether it implies additional functions for JH in the embryo remains to be established. In *Locusta migratoria*, on the other hand, titers in the embryo [10] and in larval stages [44] are similar. The importance of the presence of JH in late embryonic development is demonstrated by the ovicidal effect of precocene at these stages [45] and by the observation that the development of *Clitumnus* embryos with microsurgically suppressed CA is halted [12]. In this insect additional functions for JH, such as induction of the differentiation of the ovary and of the midgut epithelium, have also been proposed and the fact that the ecdysteroid titer rises prematurely in embryos without CA suggests an additional function of JH in the regulation of ecdysteroid production [12].

As far as concerns the JH active material observed in newly laid eggs and early embryos of *Hyalophora cecropia* [3], *Oncopeltus fasciatus* [4], *Locusta migratoria* [7, 8], and *Macrotermes subhyalinus* [5,6] interpretations are very difficult to make. In *M. subhyalinus* we used the same techniques for JH titer measurement as in *Nauphoeta cinerea* and the JH of newly laid eggs was tentatively identified as JH III using HPLC combined with *Galleria* bioassay [6]. We can, thus, exclude the possibility of having measured only molecules other than JH III. Since JH seems to play a role in caste determination in termites [46] its presence in newly laid eggs of *Macrotermes subhyalinus* might have a function in this process [6] and, thus, be a peculiarity only seen in this or possibly other social insects. In *Oncopeltus fasciatus* application of precocene to eggs resulted in a dorsal closure defect which could be counteracted by JH application [47], suggesting that in this insect JH has a function in the embryo already before dorsal closure. However, the pronounced ovicidal effects observed after application of JH analogs to freshly laid eggs of various species compared to the weak effect when the substances are applied after germ-band formation [48] tend to suggest that JH is not normally present in early embryonic stages.

With respect to the role of methylfarnesoate we do not know whether it fulfills hormonal functions or whether it is an accidental by-product of the JH synthesis. This could be due to a lack of epoxydases or also to a limited oxygen supply since epoxydation of methylfarnesoate was shown to require molecular oxygen in a cockroach CA system [49]. Methylfarnesoate shows JH activity, such as induction of vitellogenin or oocyte growth in decapitated females, or prevention of metamorphosis when applied to last instar larvae, in several assay systems of *Nauphoeta cinerea* (Fehr and Lanzrein, unpubl.). We do not know, however, whether this activity is due to conversion of methylfarnesoate into JH III by nonspecific epoxydases. In this context it is interesting to note that farnesyl methyl ether [41, 50] and other farnesol derivatives [50] were found to be morphogenetically very active when applied to embryos of *Schistocerca gregaria*.

In conclusion, we can say that JH and possibly methylfarnesoate seem to act in conjunction with the ecdysteroids in late embryonic development in steering larval cuticle formation. However, from the data shown and discussed here it is evident that much more information, particularly on JHs and methylfarnesoate, is necessary to establish fully the roles of JH, methylfarnesoate, and ecdysteroids in insect embryogenesis.

Acknowledgments. We should like to express our thanks to Mrs. A. Tschan for technical assistance and to the Swiss National Science Foundation for financial support (grants 3.714–0.80 and 3.291–0.82).

References

1. Sbrenna-Micciarelli A, Sbrenna G (1972) J Insect Physiol 18:1027–1037
2. Doane WW (1973) In: Counce SJ, Waddington CH (eds) Developmental systems: Insects. Academic Press, London New York, pp 291–497
3. Gilbert LI, Schneiderman HA (1961) Gen Comp Endocrinol 1:453–472
4. Dorn A (1975) Verh Dtsch Zool Ges 1974:85–89
5. Lanzrein B, Gentinetta V, Lüscher M (1977) Proc VIII Congr IUSSI, Wageningen, pp 265–268

6. Lanzrein B, Gentinetta V, Fehr R (in press) Current themes in torpical science, vol III
7. Roussel JP, Schneider M (1979) CR Acad Sci Ser D 289:29–31
8. Roussel JP, Aubry R (1981) Arch Zool exp Gén 122:91–98
9. Bergot JB, Baker FC, Cerf DC, Jamieson G, Schooley DA (1981) In: Pratt GE, Brooks GT (eds) Juvenile hormone biochemistry. Developments in endocrinology, vol 15. Elsevier/North-Holland Biomed Press, Amsterdam New York, pp 33–48
10. Zander JM, Wolff R, Temim G, Kappler Ch, Hoffmann JA (1983) Poster Commun 2nd Int CNRS Symp Biosynth Metab Mode Action Inverteb Horm, Strasbourg, August 1983
11. Haget A, Ressouches A, Rogueda J (1982) Int J Insect Morphol Embryol 11:99–108
12. Cavallin M, Fournier B (1981) J Insect Physiol 27:527–534
13. Imboden H, Lanzrein B, Delbecque JP, Lüscher M (1978) Gen Comp Endocrinol 36:628–635
14. Imboden H, Lanzrein B (1982) J Insect Physiol 28:37–46
15. Bürgin C, Lanzrein B (1983) Poster Commun 2nd Int CNRS Symp Biosynth Metab Mode Action Invertebr Horm, Strasbourg, August 1983
16. Tobe SS, Pratt GE (1974) Biochem J 144:107–113
17. Feyereisen R, Friedel T, Tobe SS (1981) Insect Biochem 11:401–410
18. Baker F, Jamieson G, Lanzrein B, Miller CA, Tsai LW, Schooley DA (in preparation)
19. Lanzrein B, Gentinetta V, Baker F (in preparation)
20. Loher W, Ruzo L, Baker FC, Miller CA, Schooley DA (1983) J Insect Physiol 29:585–589
21. Brüning E, Lanzrein B (in preparation)
22. Tobe SS, Saleuddin AS (1977) Cell Tissue Res 183:25–32
23. Haget A, Ressouches A, Rogueda J (1981) Int J Insect Morphol Embryol 10:65–81
24. Hoffmann JA, Lagueux M, Hetru Ch, Charlet M, Goltzené F (1980) In: Hoffmann JA (ed) Progress in ecdysone research. Elsevier/North Holland Biomed Press, Amsterdam New York, pp 431–465
25. Lagueux M, Sall C, Hoffmann JA (1981) Am Zool 21:715–726
26. Rees HH, Greenwood DR, Dinan LN, Isaac RE, Goodwin TW (1981) In: Sehnal F, Zabza A, Renn JJ, Cymborowski B (eds) Int Conf Regul Insect Dev Behav, Wroclaw Tech Univ Press, Wroclaw, Poland, pp 71–92
27. Sall C, Tsoupras G, Kappler C, Lagueux M, Zachary D, Luu B, Hoffmann JA (1983) J Insect Physiol 29:491–507
28. Zhu XX, Gfeller H, Lanzrein B (1983) J Insect Physiol 29:225–235
29. Dinan LN, Rees HH (1981) Insect Biochem 11:255–265
30. Isaac RE, Milner NP, Rees HH (1983) Biochem J 213:261–265
31. Dorn A, Romer F (1976) Cell Tissue Res 171:331–350
32. Bullière D, Bullière F (1974) CR Acad Sci Ser D 278:377–380
33. Delbecque JP, Quennedey A, Besson MT, Delachambre J (1983) Poster Commun 2nd Int CNRS Symp Biosynth Metab Mode Action Invertebr Hormones, Strasbourg, August 1983
34. Bullière F (1973) J Insect Physiol 19:1465–1479
35. Bullière F, Bullière D (1975) CR Acad Sci Ser D 281:925–928
36. Lagueux M, Hetru C, Goltzené F, Kappler C, Hoffmann JA (1979) J Insect Physiol 25:709–723
37. Bullière D, Bullière F, De Reggi M (1979) W Roux' Arch 186:103–114
38. Dorn A (1972) Z Morphol Tiere 71:52–104
39. Rinterknecht E, Matz G (1977) CR Acad Sci Ser D 284:2395–2398
40. Novak VJA (1972) Endocrinol Exp 6:251–269
41. Sbrenna-Micciarelli A (1977) Acta Embryol Exp 155:295–303
42. Injeyan HS, Tobe SS, Rapport E (1979) Can J Zool 57:838–845
43. Kadiri Z, Louvet JP (1983) Poster Commun 2nd Int CNRS Symp Biosynth Metab Mode Action Invertebr Horm, Strasbourg, August 1983
44. Bergot BJ, Schooley DA, De Kort CAD (1981) Experientia 37:909–910
45. Bowers WJ, Outa T, Cleere JS, Marsella P (1976) Science 193:542–547
46. Lüscher M (1976) Phase and caste determination in insects, endocrine aspects. Pergamon Press, New York Oxford, p 91
47. Dorn A (1982) Gen Comp Endocrinol 46:42–52
48. Riddiford LM (1972) In: Menn JJ, Beroza M (eds) Insect juvenile hormones. Academic Press, London New York, pp 95–111
49. Hammock BD (1975) Life Sci 17:323–328
50. Novak VJA (1969) J Embryol Exp Morphol 21:1–21

III.12 Defining the Anti-Metamorphic Action of Juvenile Hormone

J.H. WILLIS and D.L. COX[1]

1. Introduction

The morphological changes which accompany insect metamorphosis are frequently dramatic. Yet metamorphosis is controlled by the interaction of but two hormones. An ecdysteroid initiates all molts and promotes metamorphosis unless a juvenoid is also present. In the presence of both juvenoids and ecdysteroids insects molt without changing metamorphic stage.

What is the basis for this temporal polymorphism? Is it as Wigglesworth suggested [1] and most investigators since have accepted [2-5], that the genome is divided into distinct larval, pupal, and adult gene sets? If so, the evolution of metamorphosis must have been accompanied by the establishment of such stage specific gene sets. This could have occurred by acquisition of new genes (possibly via gene duplication and subsequent alteration) or by delegation of genes into metamorphic sets by providing them with new stage specific controlling elements. Certainly recent recognition of embryonic forms of enzymes and structural proteins [6], indeed the entire fetal antigen story [7], lends support to the concept that different genes are used in different developmental stages.

The goal of our laboratory has been to describe and understand the biochemical differences among the three metamorphic stages of the giant silkmoth, *Hyalophora cecropia*. The original intent was to identify molecules which would serve as indicators of each metamorphic stage so we could pursue the basis of juvenoid action. Surprisingly, our data have revealed that the generally accepted paradigm of hormonally regulated stage specific gene sets has no foundation. It is these data and their implications for the hormonal control of metamorphosis which will be the basis of this communication.

The molecules we have chosen to examine are cuticular proteins. As we showed several years ago, these are made by the epidermis [8]. The epidermis as demonstrated in tissue culture is a direct target for both juvenoids and ecdysteroids [9-11]. In Cecropia, the cuticle of each stage has distinct gross and microscopic anatomy, and a unique color and texture. Thus, cuticular proteins are obvious candidates for organization into stage specific gene sets. Furthermore, in Lepidoptera continuous populations of cells in the abdomen make, in turn, larval, pupal, and adult cuticles. This is unlike the situation in *Drosophila* where, with the exception of histoblasts, cell populations change

1 Department of Entomology, 320 Morrill Hall, University of Illinois, 505 S. Goodwin Avenue, Urbana, IL 61801, USA

Biosynthesis, Metabolism and Mode of Action of Invertebrate Hormones
(ed. by J. Hoffmann and M. Porchet)

between two of the three metamorphic stages. In Lepidoptera, then, we are asking whether a fixed population of cells makes its successive cuticles from a succession of stage specific gene sets.

Our work mirrors the history of developments in electrophoresis. We began with simple tube gels for DISC electrophoresis, moved to SDS gels, then into slabs, and isoelectric focusing and now 2D gels and immunoelectrophoresis. Along the way we purified a dozen bands sufficiently to determine their amino acid compositions and to be assured that different bands are, for the most part, distinct proteins [8]. Multiplicity of cuticular protein bands on gels reflects a diversity of primary sequence [12].

The presence on a gel of bands from cuticles of two different metamorphic stages with identical electrophoretic mobilities can mean one of two things:

1. The same gene is used in both stages.
2. Different genes are used, but have "indistinguishable" products.

If the first possibility is assumed, one must explain how the same gene comes to be activated in different hormonal environments. With the second situation, this is easier to explain, for one could postulate that gene duplication had occurred, with subsequent modification of controlling regions. (Of course, one would never engage in such speculation without first demonstrating more similarity than coincidence of mobility.)

If different proteins are found in cuticles taken from a single anatomical region at different metamorphic stages, then two interpretations are possible:

1. The same genes are active in both metamorphic stages, but different pre- and/or post-transcriptional modifications have occurred.
2. Different genes are expressed, reflecting stage specificity.

If modifications can be ruled out, then one has evidence for stage specific gene sets. Suppose, however, that examination of cuticular proteins from several anatomical regions revealed that proteins previously identified as belonging to a particular metamorphic stage were expressed in another stage. Perhaps then the proper interpretation would be that the cuticular properties (functions) of an anatomical region are more of a determinant of gene expression than metamorphic stage.

We will summarize recent work from our laboratory which shows that distinct sets of genes are not set aside for exclusive use in a single metamorphic stage. Rather, functional requirements appear to dictate which proteins will be secreted into the cuticle.

2. Methods and Guide to Interpretation

One needs variants as well as breeding tests for genetic analyses. Fortunately, we can meet these requirements. There are four species in the *Hyalophora* genus of giant North American silkworms, and we have used two of them, the standard *Hyalophora cecropia* and its more western and slightly smaller relative, *H. gloveri*. They will mate and form viable hybrids. Most important, species-specific variants in a limited number of cuticular proteins are seen when guanidine extracts are analyzed on isoelectric focusing slab gels.

Our analysis in this paper will emphasize a homologous set of proteins in the very acidic region of the gel, one from Cecropia and the other from Gloveri. Each species has three spots from this region on two dimensional gels; those from Gloveri are all more acidic. 2 D gels reveal identical molecular weights for homologous spots, the most prominent about 20,000. Strain differences also occur which complicate the picture somewhat, and which suggested at first that we might be dealing with a multigene cluster. However, the mode of inheritance you will see and the work reported by the Cherbases [13] suggests that in each species all three spots may come from a single genetic locus.

In addition to the information from the gel on isoelectric point and molecular weight, there is another character that can be followed; namely, the color of the band when viewed with fiber optic light applied parallel to the gel face. Under these conditions, some bands appear red after staining with Coomassie, while neighboring bands remain blue. The effect is very reproducible and not dependent on concentration; if the band is made visible with the stain, the red color is there.

3. Results and Discussion

3.1 Genetic Analyses

In the upper panel of Fig. 1, you can see that the larval cuticle of Gloveri has a distinct acidic band (G); it gives the red color under fiber optic light, and is absent in pupal sclerites, but present in the intersegmental membranes of pupae and adults. There is no information available at present on this band in the sclerite regions of adult cuticle. In Cecropia, different stocks vary in the intensity of the band in this region. Sometimes it is totally absent, in others it is faint, and in a western stock which came from Colorado (depicted in the center panel), it is relatively intense. All Cecropia, irrespective of origin, have a heavy band (C) in a slightly more basic (about 0.1 pH unit) position. Other anatomical regions in Cecropia have been analyzed, and this lower red band is present in larval cuticle, and in pupal and adult intersegmental membranes. It is also present in the lateral scale-bearing region of the adult sclerite.

The band variability seen in some Cecropia should not detract from the major finding. The same species-specific electrophoretic bands are found in all three metamorphic stages. Extracts of different regions of Gloveri have only the "Gloveri" band. Could

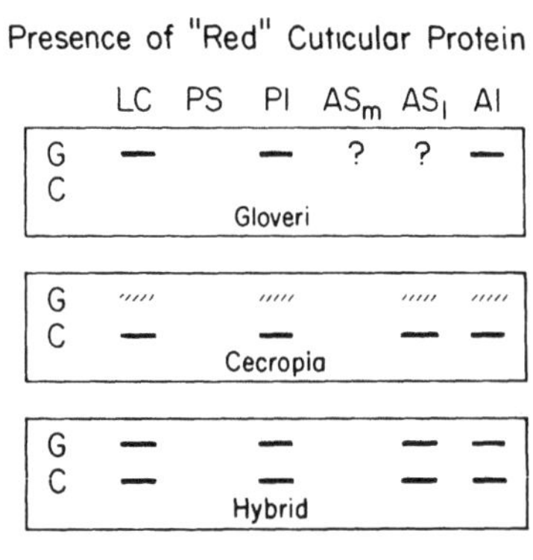

Fig. 1. Cuticular proteins discussed in the text from *H. gloveri, H. cecropia*, and their hybrid. Extracts were made from cuticles taken from the anatomical regions listed across the top: *LC* larval cuticle; *PS* pupal sclerite; *PI* pupal intersegmental membrane (polyglonal field zone); *ASm* medialdorsal region of adult sclerite; *ASl* lateral region of adult sclerites; *AI* adult intersegmental membrane

these data be consistent with the supposition that different genes in different stages make indistinguishable products? If so, one would have to postulate that the assignment of these genes to stage-specific batteries occurred after the modification which led to the acquisition of a different mobility form by one of the species. Since Gloveri and Cecropia are indistinguishable in the anatomical and stage distribution of these proteins one would have to postulate that some highly unlikely convergence took place when these different genes became associated with identical sets of controls. The simplest explanation consistent with these data, then, is that the same allele is used in different anatomical regions and different metamorphic stages. This in itself would not be heretical if the gene were constitutive, for there may be cuticular proteins common to all cuticles. Our data, however, show that the "red" gene is not constitutive, for certain anatomical regions (e.g., lateral sclerites), made by fixed populations of cells, fail to express the protein in all stages.

Even more definitive evidence that the same gene is working in different stages comes from analysis of hybrids (Fig. 1, lower panel). In hybrids, both proteins are present in the cuticles from all the appropriate regions. Heavy bands appear in the extracts of larval cuticle, pupal and adult intersegmental membranes, and in the adult lateral sclerite. The presence of both proteins in the hybrids also makes very unlikely any possibility that the differences in mobility between species are caused by modifying enzymes [14]. The banding pattern we found in hybrids can best be interpreted as showing that the same gene is used in all three developmental stages.

Table 1 describes these results in terms of gene activity, using gene activity in its broadest sense, i.e., the presence of the final product of transcription and translation of the gene. Is the "red" gene a larval gene? From the point of view of the medial sclerite it is. For the lateral sclerite, it is larval and adult; for the intersegmental membrane it is a constitutive protein. From the point of view of the perceptive insect physiologist, it certainly is not a "larval" protein.

What is the role of juvenile hormone in all of this? Pupal abdomens of Cecropia are extremely sensitive to JH; it is easy to obtain second pupae whose intersegmental membranes and sclerites have good pupal morphology. We have previously shown that second pupae produced with JH have the cuticular proteins which are appropriate for each specific pupal anatomical area [15]. The red band is reexpressed in the second pupal cuticle of the intersegmental membrane. It is not turned on again in the sclerite (Table 1). The influence of JH is to cause the protein to appear in the intersegmental membrane and to prevent its reappearance in the lateral sclerite. This evidence further negates the hypothesis that JH is selecting different genes for activation and repression.

One of the gels which laid the basis for the diagrams and conclusions is shown in Fig. 2. In this figure you can see the banding patterns from extracts of larval and adult

Table 1. Activity of the "red" gene

	Larval	Pupal	Adult	Second pupa
M SCL	On	Off	Off	Off
L SCL	On	Off	On	Off
ISM	On	On	On	On

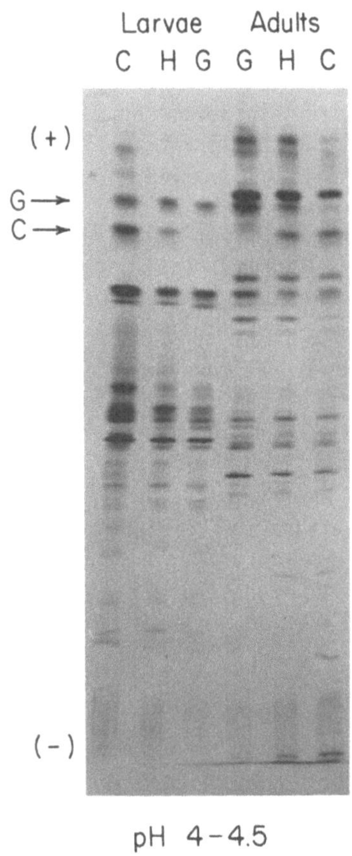

Fig. 2. Isoelectric focusing slab gel containing extracts of abdominal cuticular proteins. The *lanes* come from *C* Cecropia; *G* Gloveri; *H* their hybrid. *G* and *C* on the *left side* designate the positions of the "red" proteins discussed in the text and shown in Fig. 1

cuticles. These have been run on an extremely narrow range gel, pH 4-4.5, which produces good separation of the red bands. This figure shows extracts from larvae and adults of Gloveri and Cecropia and their hybrid. The bands which correspond in mobility to the major red band of each species are indicated with a labelled arrow. In these western Cecropia larvae the upper, Gloveri red band is quite prominent. Densitometry scans have revealed a ratio of the upper (G) to lower (C) band of 0.5. Gloveri lacks the Cecropia band, but it is expressed in the hybrid. In a series of hybrids the densitometry ratio was 1-1.6.

So this is the genetic evidence that a particular gene cannot be assigned to a larval, pupal, or adult gene battery - rather the translational product of that gene may be called into play in different tissues at different metamorphic stages.

3.2 Comparisons of Rigid and Flexible Cuticles

Further evidence against stage specific gene sets comes from work with a wider range of anatomical regions. While not rooted in genetic variants, it involves a larger number of proteins, and it will demonstrate that we have not selected an isolated protein or two on which to base our conclusions.

As you are well aware caterpillars are not mere worms, but as their French name, *la chenille*, suggests, fuzzy bodies. Rather than being fuzzy, Cecropia is ornamented with a magnificient array of hard tubercles which are brightly colored. These are common in many saturniids.

After noting subtle variations in cuticular protein banding patterns, experiments were undertaken to follow a single individual from larva to adult in order to see if the same gene was used in two stages. Larval tubercles were extracted and their cuticular proteins separated on gels. If we detected variants of potential interest, we had planned to raise the caterpillar through to adulthood in order to examine its adult cuticular proteins. You can imagine our surprise when we saw that the extracts of tubercles had none of the characteristic larval acidic proteins. Rather, as we subsequently learned, tubercles are made of what we had described previously as pupal proteins (Fig. 3). Further demonstration of similarity required other approaches.

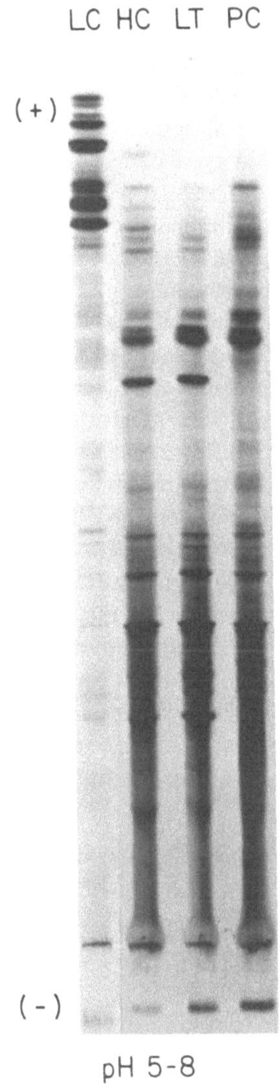

Fig. 3. Isoelectric focusing slab gel containing extracts of cuticular proteins from diverse regions: *LC* larval abdominal cuticle; *HC* larval head capsule; *LT* larval tubercles; *PC* pupal sclerites

Since tubercle cuticle is harder than cuticle from other regions, there was concern that there might be solubility problems. There is as yet no way to separate and examine the cuticular proteins which are insoluble in strong denaturing agents, such as guanidine. However, they can be made part of a gross analysis of the cuticle by hydrolyzing the entire cuticle and studying its amino acid composition. The results, of course, do no more than sum the amino acid compositions of various proteins, so that if two cuticles have identical proteins, but these are present in different relative concentrations, the total amino acid composition would be quite different. However, on the basis of our past analyses there was reason to believe that useful differences might be established. The results are gratifying.

Table 2 shows that even though alanine is a major amino acid in pupal cuticle, it is minor in larval. Alanine is the major amino acid in larval tubercles. This then is the first indication that the bulk of the protein is similar in larval tubercles and pupal cuticle. There is also a more sensitive indicator using this kind of information. A statistical test for comparing amino acid compositions of individual proteins was developed by Marchalonis and Weltman [16]. We have employed it to compare hydrolysates of whole cuticle. The larger the SΔQ generated by this test, the more dissimilar the compositions. Larval cuticle differs from pupal cuticle by 257 SΔQ units, whereas larval tubercles and pupal cuticle differ by only 17 units.

These proteins have also been examined on 2D gels. In Fig. 4 seven spots with identical mobilities may be seen, one set from extracts of larval tubercles, the other from pupal sclerites. None of these spots is present in 2D gels of cuticular proteins from larval sclerites.

Thus, alanine content, gross amino acid composition, protein banding patterns, and 2D gel analyses are all in agreement: larval tubercle proteins resemble pupal cuticle proteins. Also identical mobility of individual proteins on IF and 2D gels is consistent with the interpretation that the same genes are used for larval and pupal cuticles, but in different anatomical regions.

A simple interpretation of the above is that when insects began to build hard projections on their flexible larval cuticles during the course of evolution, new structural genes were not invented, rather pupal cuticular proteins were impressed into use as the building blocks of the new structures. Certainly new morphogenetic instructions had to evolve, so that the appropriate changes in division patterns and cellular morphogenesis would develop to give rise to new structures, but the structural proteins were conserved.

Here in the larva we have seen that under larval hormonal conditions, a group of what we thought were pupal proteins is expressed. You might care to believe that these cells had just lost their sensitivity to JH, and that is why they make pupal proteins.

Table 2. Comparisons of amino acid compositions

% ALA		SΔQ	
LC	8	LC vs PC	257
LT	21	LT vs PC	17
PC	22		

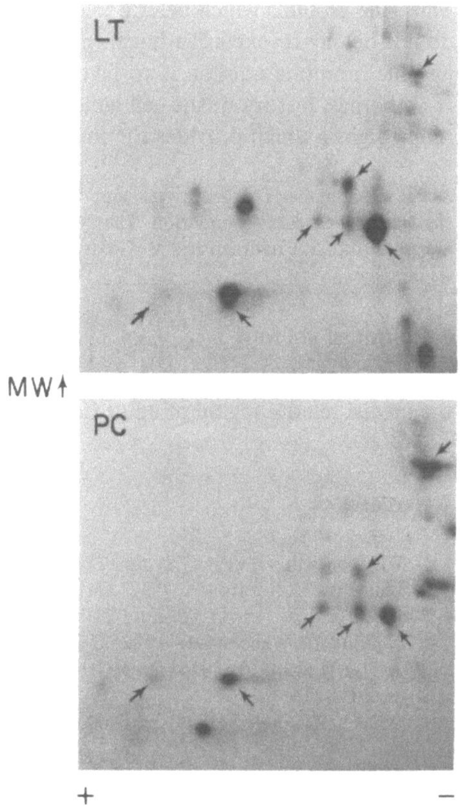

Fig. 4. Regions from two 2D gels, showing seven spots (*arrows*) with identical mobility in extracts from larval tubercles and pupal sclerites. The first dimension was pH 3.5–10; the second was 12% acrylamide

But we already know that tubercle cells are among the most sensitive of larval tissues to JH [17, 18].

Although we would like to be able to have a clear argument that pupal "genes" have been impressed into service for the larva, we must confess that after some thought we realized that these proteins might also be used in a conventional larval structure – the rigid head capsule. As you can see in Fig. 3, head capsule proteins are almost identical in electrophoretic mobility to those found in the tubercles. So it is inappropriate to refer to these as pupal proteins, proteins for hard structures would be a more appropriate desgination.

4. Summary and Interpretation

1. In comparing two different anatomical regions, sets of cuticular proteins which are pupal for one region may be larval for another.
2. Genetic analysis of the red band of hybrids indicates that when a protein appears in more than one stage, it is there because the same genetic region is being transcribed.

Thus, the proteins which belong to each region and stage, and by inference their genes, do not belong to exclusive larval, pupal, or adult batteries. In all cases the hallmark of juvenile hormone's action is to favor reexpression of a particular pattern of syntheses. It is the past history of the cell which has been set in response to prior spatial and hormonal signals which provides the foundation against which JH works.

Acknowledgments. This work was supported by grants from the Natioanl Institutes of Health and the National Science Foundation. The assistance of L. Koryta and G. McNamara is appreciated. We are grateful to S. Friedman and M. Goldsmith whose comments on the manuscript helped to clarify its message.

Note Added in Proof

We have recently found that band C is present in low concentration in extracts from the medial sclerite region of adult Cecropia cuticle.

References

1. Wigglesworth VB (1961) Symp R Entomol Soc London 1:103–114
2. Williams CM, Kafatos FC (1971) Mitt Schweiz Entomol Ges 44:151–162
3. Riddiford LM (1981) Am Zool 21:751–762
4. Nijhout HF, Wheeler DE (1982) Q Rev Biol 57:109–133
5. Laufer H (1983) In: Downer RGH, Laufer H (eds) Endocrinology of insects. Liss, New York, pp 203–216
6. Whalen RG, Sell SM, Butler-Browne GS, Schwartz K, Bouvert P, Pinset-Härström I (1981) Nature (London) 292:805–809
7. Uriel J (1979) Adv Cancer Res 29:127–174
8. Willis JH, Regier JC, Debrunner BA (1981) In: Current topics in insect endocrinology and nutrition. Plenum Press, New York, pp 27–46
9. Willis JH, Hollowell MP (1976) In: Gilbert LI (ed) The juvenile hormones. Plenum Press, New York, pp 270–287
10. Riddiford LM (1976) In: Gilbert LI (ed) The juvenile hormones. Plenum Press, New York, pp 198–219
11. Oberlander H, Lynn DE (1982) Adv Cell Cult 2:237–265
12. Snyder M, Hunkapiller M, Yuen D, Silvert D, Fristrom J, Davidson N (1982) Cell 29:1027–1040
13. Cherbas P, Savakis C, Cherbas L, Koehler MMD (1983) Symp Soc Dev Biol 41:95–111
14. Finnerty V, Johnson G (1979) Genetics 91:695–722
15. Willis JH (1981) Am Zool 21:763–773
16. Marchalonis JJ, Weltman JE (1971) Comp Biochem Physiol 38B:609–625
17. Willis JH (1969) J Embryol Exp Morphol 22:27–44
18. Riddiford LM, Williams CM (1967) Proc Natl Acad Sci USA 57:595–601

III.13 Vitellogenin Synthesis in Locust Fat Body: Juvenile Hormone-Stimulated Gene Expression

G.R.WYATT[1], T.S. DHADIALLA[1,2], and P.E. ROBERTS[3]

1. A System for Studying Juvenile Hormone Action

In contrast to the extensive research on the cellular, subcellular, and molecular actions of the ecdysteroids, rather little has been done on the mode of action of the juvenile hormones (JHs). This is surprising and regrettable, in view of the unique chemical nature of the JHs, their powerful actions in regulating insect development, and the use of synthetic juvenoids in practical insect control. The most dramatic role of JH, its control over metamorphosis, involves modulation of gene expression in tissues under concurrent or subsequent stimulation by ecdysterone. This is the subject of two contributions to this symposium [1, 2] and the picture that is coming to light is an intriguing but complex one, with patterns of response depending on the precise concentrations and timing of exposure to the two hormones, as well as the localization of the responding cells, even within a tissue such as the epidermis.

The second major role of JH, in regulating specific functions required for reproduction in certain organs of adult insects, is less distinctive but may be more readily accessible to study at the molecular level. These functions include the synthesis of vitellogenins (Vgs) in the fat body of many kinds of insects [3], though dependence on JH cannot be assumed without question, for Vg synthesis is independent of JH in some Lepidoptera [4, 5], while in certain Diptera, including *Aedes* [6] and *Drosophila* [7], ecdysterone may play the major role. Where JH does regulate Vg production in the fat body, associated with this is proliferation of ribosomes and endoplasmic reticulum – cell components needed for enhanced protein synthesis, processing, and export [8], and the relations between activation of the genes that encode the major export protein and build-up of the supporting cellular machinery will be important to determine. A comparable JH-dependent function, now under study at the molecular level [9], is oothecin synthesis in the female cockroach collaterial gland.

For an attack on the problem of JH action, we chose to investigate the regulation of Vg synthesis in the fat body of *Locusta migratoria* [10, 11]. This species was selected because of its large size, easy rearing, and the wealth of available information on its physiology and endocrinology. To begin with, we characterized the Vg of *Locusta* and

1 Department of Biology, Queen's University, Kingston, Ontario K7L 3N6, Canada
2 Present address: International Centre of Insect Physiology and Ecology, Nairobi, Kenya
3 Department of Zoology and Entomology, Colorado State University, Fort Collins, CO 80523, USA

Biosynthesis, Metabolism and Mode of Action of Invertebrate Hormones
(ed. by J. Hoffmann and M. Porchet)

quantitative aspects of the control of its production by the JH analog, methoprene [12-14]. Methoprene has been used because of its stability, which (in contrast to the natural JHs) allows full induction by a single applied dose. We then characterized the Vg messenger RNA ($mRNA_{Vg}$) and took advantage of its exceptionally large size and abundance in vitellogenic fat body to measure the changes in its tissue level by gel electrophoresis [15]. The results indicated that JH causes accumulation of $mRNA_{Vg}$, in part at least by stimulation of transcription. Recent work has, therefore, been directed toward study of gene activity.

During the decade that this project has been in progress, the study of gene expression has been revolutionized by the introduction of recombinant DNA techniques. The locust, with its large genome size and lack of known mutants, inbred lines, or genetic characterization, is not ideal for molecular genetic study. The isolation of cloned Vg genes from *Locusta* has been much more laborious than the comparable achievement with *Drosophila melanogaster* [16, 17]. We have been encouraged to persist, however, by the belief that it is important to gain knowledge of the molecular genetics of insects in addition to *Drosophila*, and by the virtual absence of such studies on any hemimetabolous species. At present, two Vg genes have been cloned from the *L. migratoria* genome [18-20]. From restriction mapping and cross-hybridization experiments, no homology between the two is apparent, although both encode mRNAs of the same size (6,300 nucleotides). Cloned probes have been used to show, by assay of gene dosage and restriction fragment length polymorphism, that both Vg genes, like the Vg genes of *Drosophila*, are located on the X chromosome [21].

We now summarize some recent progress in the study of the expression of Vg genes in *Locusta* under stimulation by the JH analog: stage- and sex-related programming of the genes for activity; the kinetics of induction; and initial characterization of a fat body JH-binding fraction that may include the JH receptor.

2. Sites of Vg Synthesis in Locusta; Stimulation of Fat Body In Vivo and In Vitro

The chief site of Vg synthesis in *Locusta* is the bulky perivisceral abdominal fat body, which, when fully induced, produces 60%-65% of its secreted protein, or more than 15% of its total protein synthesis, as Vg [13, 22]. We also assayed the paragonadal fat body, which is sandwiched between the ovaries and the gut, and found evidence for regional specialization: paragonadal fat body yielded 81% of export protein as Vg (compared with 67% from perivisceral tissue of the same animals) and was responsible for about 20% of the total Vg production [23]. Because of the small size and relatively difficult dissection of paragonadal fat body, however, we did not continue to work with it. *Locusta* ovaries, carefully dissected free of fat body and assayed in vitro, produced no measurable Vg, in contrast to the situation in *Drosophila*, where the ovaries produce significant endogenous Vg [7].

The biosynthetic activity of fat body has been routinely assayed by incorporation of ^{3}H-leucine into immunoprecipitable Vg during 3 h culture in vitro, after hormonal stimulation in vivo [13]. Vg synthesis can also be induced in fat bodies from surgically or chemically allatectomized locusts by culturing in the presence of methoprene. Despite some encouraging data [24], however, we have experienced difficulty in the routine culture of fat body for sufficient lengths of time to permit analysis of induction in vitro. Because of the great advantages of a fully in vitro system, experiments with this objective are continuing.

3. Competence of Fat Body for Vg Synthesis in Female and Male Locusts

In *Locusta*, Vg is normally made only in the adult female, although the genes and the inducer, JH, are present also in males and in larval stages. To lay a foundation for examining molecular mechanisms, it was important to know more about this restriction of expression by sex and stage [22, 25].

Adult females become competent to make Vg, after treatment with methoprene, on day 4 or 5 after eclosion. Adult males, or incubated adult male fat body, produce no immunochemically identified Vg even after high doses of methoprene, and no $mRNA_{Vg}$ can be detected by hybridization with a cloned DNA probe [23]. Adults of both sexes, however, do respond to the JH analog by fat body DNA replication [26] and a low level (about twofold) increase in general protein synthesis [22, 27]. These results show that the adult male fat body is a target tissue capable of responding to JH stimulation, and that the restriction of Vg synthesis to females must be due to programming at the gene level.

In larval *Locusta*, this sex differentiation of the fat body cells is not yet complete. Although Vg is not found in normal larvae, synthesis can be induced by applying high doses of methoprene to larvae of both sexes. In mid-fifth stage females, full induction of Vg synthesis (64% of total export protein) was obtained, but with an ED_{50} for methoprene 5 × higher than in adults, while in fifth stage males the maximum production was only 20% of export protein (Fig. 1). In the final larval stage, therefore, sexual differentiation of the response to JH was partially, but not completely, established. Toward the end of the fifth instar, Vg synthesis could not be induced in either sex.

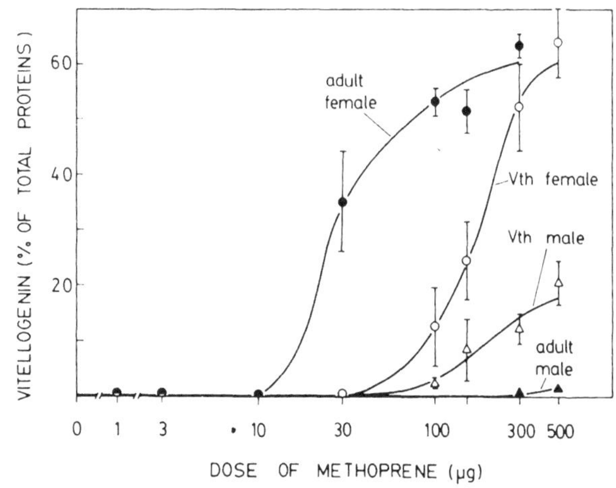

Fig. 1. Dose-response curves for induction of Vg synthesis by methoprene in fat bodies of adult and fifth instar locusts of both sexes. Methoprene was injected in mineral oil and after 72 h fat bodies were excised and assayed for Vg and total protein output during 3 h in culture with ^{3}H-leucine. Means and SEM from four fat bodies are shown. In the adult male, the identity of the immunoprecipitated material was not confirmed and (in later experiments) no $mRNA_{Vg}$ could be detected by hybridization with cloned Vg gene DNA [22]

Then, after the larval-adult transition, competence for Vg synthesis in response to methoprene increased in the female fat body, while in the male it remained at zero.

To test whether the differentiation in the adult depended on humoral differences between the sexes during the final molt, or on differences intrinsic to the fat body, we transplanted female fat body into fifth stage males and allowed it to pass through metamorphosis in the male milieu. When assayed after the molt it behaved like normal adult female tissue, showing that the sex difference was determined in the fat body cells at least by the fifth instar. In a further experiment, the competence of fat body taken from precocious adults, produced after only four molts by treating third stage larvae with precocene [28, 29], showed normal sexual differentiation of function between females and males. This shows that a molt in the absence of JH, and not a set number of molts, is the signal needed to evoke full sexual differentiation. These conclusions with respect to the development of sex-differentiated programming of the Vg genes in *Locusta* fat body are summarized in Fig. 2.

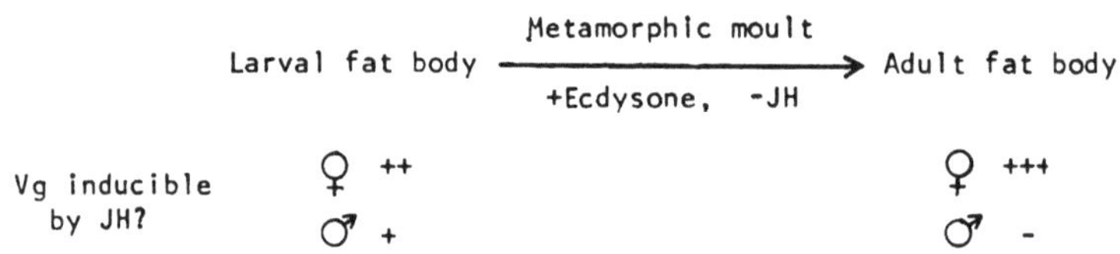

Fig. 2. The differentiation of expressibility, in response to stimulation by JH, of Vg genes in fat body during the metamorphic molt in female and male locusts

A similar pattern of stage- and sex-related competence for Vg gene expression appears to exist also in *Blattella germanica* [30, 31], and perhaps in many insects, but not in all. In *Rhodnius prolixus* Vg is found in normal adult males [32], and in *Diploptera punctata* Vg synthesis can be induced in adult males with sufficient doses of JH or an analog [33]. In *Drosophila* [34] and in *Sarcophaga* [35], the controlling hormone, ecdysterone, stimulates Vg production in males as well as in females, and the normal limitation of expression probably depends on differences in circulating levels of the hormone. This would resemble the picture that is general in the oviparous vertebrates, where Vg synthesis can be induced by estrogen equally in female and male liver, but only the female normally has sufficient titer of the hormone to activate transcription in vivo [36]. Thus, the system used by *Locusta migratoria* is not the only one that can serve the biological need for sex-limited regulation.

4. $mRNA_{Vg}$ in Fat Body Cells After Stimulation by Methoprene

For analysis of gene regulation, it is essential to assay specific mRNA. Because of the distinctive size and exceptional abundance of *Locusta* $mRNA_{Vg}$, we could estimate its amount in induced fat body by photometric scanning after electrophoretic separation [15]. This technique, however, which can detect about 5 ng of $mRNA_{Vg}$, is not sufficiently sensitive to assay its first appearance and rate of early accumulation after hormonal stimulation of the cells. With the availability of cloned Vg gene sequences as probes, the much more sensitive method of RNA:DNA hybridization could be used [23, 37].

The probe used was LmVg144p4.6, a plasmid subclone of *Locusta* genomic DNA that includes the 3′ end and about 3 kb of coding sequence of Vg gene A, and is free of repetitive sequences [19]. Probing of Northern blots of vitellogenic fat body RNA with this DNA fragment showed hybridization only to the $mRNA_{Vg}$ band. For the quantitative assays, we have used the rather simple technique of dot hybridization [38]. RNA samples prepared from individual locust fat bodies are bound as dots on a sheet of nitrocellulose filter membrane, along with $mRNA_{Vg}$ standards, and the sheet is then incubated under hybridizing conditions with the ^{32}P-labelled DNA probe, washed, and autoradiographed on X-ray film to reveal the positions where radioactivity is fixed. The dots are cut from the filter sheet and counted by liquid scintillation. A standard curve, plotted on a log-log scale, was linear from 10 pg up to 8 ng of $mRNA_{Vg}$, and the sensitivity was, therefore, about 500 × greater than that of the photometric method mentioned above. This permits the detection of $mRNA_{Vg}$ down to a level of about two molecules per fat body cell.

In adult female locusts that were allatectomized by treatment with precocene on the first day after eclosion and assayed 2 weeks later, no $mRNA_{Vg}$ could be detected in the fat body RNA. An experiment to follow the decay of $mRNA_{Vg}$ after allatectomy of vitellogenic females showed a half-life of about 24 h. This evidence that $mRNA_{Vg}$ is relatively stable rules out rapid turnover as an explanation for its absence in the absence of JH. We, therefore, conclude that without JH the Vg genes are silent.

After administration of methoprene to precocene-treated locusts (primary stimulation), $mRNA_{Vg}$ was first detected at 24 h, accumulated slowly during the next 12 h and then much more rapidly between 36 and 48 h (Fig. 3). It had previously been observed that the response to a secondary stimulation with methoprene, after allowing 5 weeks for complete decay of the effects of the primary treatment [15], is more rapid. To obtain animals for secondary stimulation more quickly, we allatectomized mature females during their first natural vitellogenic cycle and found that after 2 weeks Vg was no longer detectable in the hemolymph. In the secondary response obtained by treating these locusts with methoprene, an increase in $mRNA_{Vg}$ could be detected by 12 h and accumulation proceeded with much less lag than in the primary response (Fig. 3).

The activity of the cells in Vg synthesis was measured by incubating these fat bodies with ^3H-leucine before they were used for preparation of RNA. In both the primary and secondary responses to methoprene, Vg synthesis was first detected at the same times when $mRNA_{Vg}$ was first found, and then increased at rates roughly in parallel with the increasing mRNA levels (Fig. 3). This indicates that, initially at least, specific mRNA is the limiting factor in Vg synthesis. A shift in this relationship at later times suggests that translational factors may become limiting.

In female fifth instar larval *Locusta*, after treatment with 300 μg of methoprene, $mRNA_{Vg}$ first appeared after 24 h, accumulated with a pronounced lag up to 36 h and then much more rapidly, just as in primary stimulation in the adult [23]. During normal larval development JH is, of course, present in the hemolymph, but not at sufficient titers to activate the Vg genes in their larval state of programming. This result suggests, therefore, that activity of these genes, not mere exposure to JH, is required for conversion to the primed state which exhibits the accelerated secondary response.

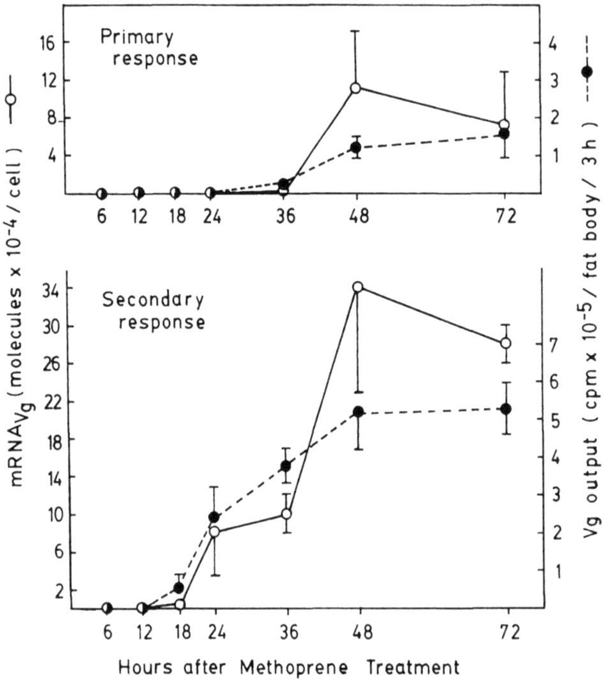

Fig. 3. Levels of $mRNA_{Vg}$ and rates of Vg synthesis in adult female locust fat bodies after primary and secondary stimulation by methoprene in vivo. Locusts, pretreated as described in the text, were injected with 150 μg of methoprene in mineral oil and, after the times indicated, were assayed in vitro for incorporation of ^{3}H-leucine into Vg and then used for extraction of RNA. $mRNA_{Vg}$ was assayed by dot hybridization. *Points* are means and SEM from three fat bodies

5. Binding of JH to a Possible Receptor from Fat Body

The responses of locust fat body to JH generally resemble tissue responses to steroid hormones, including the estrogenic induction of Vg synthesis in the livers of oviparous vertebrates [36]. The analogy suggests that JH, like a steroid, may be bound to a cytoplasmic receptor and transported in its company into the nucleus [39]. JH-binding proteins have been reported from several JH-responsive cell types: *Drosophila* epidermis [40] and culture cells [41], and *Leucophaea* fat body [42] and ovary [43]. We have, therefore, sought to identify the JH receptor in locust fat body. The following summarizes our chief initial findings [44].

Fat body cytosol or cytosol fractions were incubated at 0° for 15 min with ^{3}H–JH–I, followed by separation of bound from unbound hormone by adsorption of proteins to hydroxyapatite [45]. Separation with dextran-treated charcoal, which is widely used in steroid receptor studies, gives low and variable values for binding of JH because of dissociation of the JH-protein complex. Specific binding is taken as that portion of total binding that is competed by 100-fold excess of unlabelled JH–I. Although locust fat body contains JH-degrading enzyme activity, the addition of esterase inhibitors was

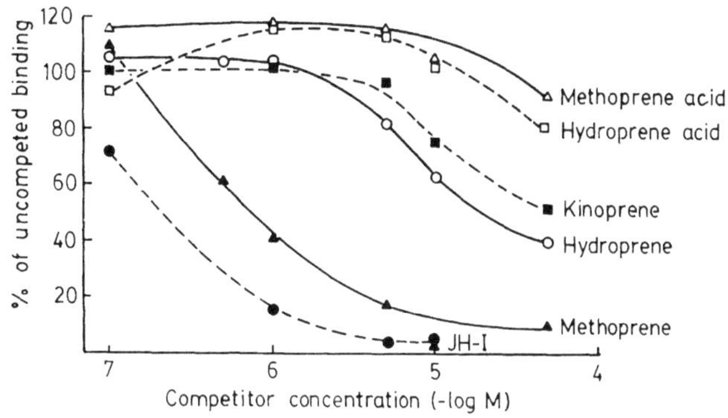

Fig. 4. Binding of ^{3}H–JH–I by fat body cytosol fraction BP–I: competition by JH and several synthetic analogs [45]

not found to improve results, and the radioactivity extracted after binding to the putative receptor fraction was identified by thin-layer chromatography as unaltered JH–I.

After binding was found in crude cytosol, we used DEAE-cellulose to separate three JH-binding fractions. One of these (BP–I) had characteristics expected of a hormone receptor: stable binding during several hours, high affinity ($K_d = 1.7 \times 10^{-8}$ *M*), and relative competition by several JH analogs approximately in the order of their hormonal effectiveness (Fig. 4). That methoprene has greater hormonal activity in vivo than JH–I, although it competes less effectively with the proposed receptor, can be explained by its greater biological stability. Hydroprene, which is about 5 × less effective than methoprene in inducing Vg synthesis [22], and kinoprene are also weaker competitors, and the hormonally inactive acids scarcely compete. JH–III, the natural form of the hormone in *Locusta*, binds and competes equally as JH–I [44].

The component BP–I, which we believe may be the cytosol JH receptor, sediments in sucrose gradients at 13S, which indicates a greater molecular weight than has been reported for various steroid receptors. It is distinct from the JH-binding components found in *Locusta* hemolymph [46, 47]. Locust vitellin, in contrast to *Leucophaea* Vg and vitellin [8], does not bind JH–I appreciably.

To confirm the role of BP–I, further characterization and demonstration of transport into the nucleus will be important. Changes in receptor levels and distribution between cytoplasm and nucleus may affect the rate and magnitude of the response to hormonal stimulation.

6. JH Action and Steroid Hormone Action

As open chain sesquiterpenoids, the insect JHs are produced by a branch of the biosynthetic pathway that, in plants and vertebrate animals, leads to the steroids, and the information now at hand on the mode of action of JH is consistent with current concepts of steroid hormone action. This model includes transport into the nucleus in complex with a receptor protein, binding to the chromatin, and modulation of the activity of selected genes [39]. Even for the steroids, however, much remains unsettled. Some genes appear to be direct targets of hormones and some indirect, responding after a time lag possibly via protein products of early responding genes [48]. For both early and late responding genes there is evidence for locus-specific hormone binding [49]. Recently, selective binding of steroid-receptor complexes to specific DNA sequences associated with responsive genes has been demonstrated, but interpretation is complicated by binding to multiple sites [50, 51] and it is not clear that any of these has sufficiently high affinity to account for the efficient on-off control that is observed in vivo. The mechanism may involve specific chromosomal proteins, the nature and roles of which remain to be identified.

The Vg genes in locust fat body qualify as hormone-responsive genes. From a ground state of apparently total inactivity they are turned on in the presence of JH or an active analog so that $mRNA_{Vg}$ accumulates and becomes the most abundant cellular message. While the fat body cells exhibit other quantitative responses to JH - stimulation of DNA replication [26], synthesis of ribosomal RNA [15], and moderate enhancement of general protein synthesis [27] - complete dependence on JH appears to characterize only a very small gene set. Northern blots of mature female fat body RNA, probed with cDNA that has been absorbed with male fat body RNA and, thus, made specific to the JH-induced female, reveal only $mRNA_{Vg}$ and one other mRNA band of much lower molecular weight [20].

To respond to a hormone, genes must be in an expressible state which is established during ontogeny. This is exemplified by the programming of the Vg genes for sex-limited expression in locust fat body. Sexual differentiation of function is determined and partially expressed in the last larval instar, but fully brought out only after the metamorphic molt, when the Vg genes attain full responsiveness to JH in the female and are totally repressed in the male. These differences in the expressibility of specific genes, brought about without cell division (but with DNA replication) in a tissue that otherwise differs little in function, give an excellent opportunity to study the basis of developmental gene programming. This may involve changes in DNA methylation [52] or in chromatin structure manifest as differences in sensitivity to nucleases [53] possibly involving specific non-histone proteins [54]. Now that an adequate set of Vg gene probes have become available, studies on these possibilities are being commenced.

Once the Vg genes are in the expressible state, their activity is modulated by JH or an analog. Primary stimulation involves a lag of about 24 h before hybridizing transcripts can be detected, and a further lag before the full rate of transcription is attained. The genes then remain in a primed state and respond to secondary stimulation with much reduced lag, although it still takes about 12 h before new transcripts appear. A lag phase between the time when hormone-receptor complexes enter the nucleus and the start of induced transcription is characteristic of steroid hormone action [55], but the lag in the response to methoprene in *Locusta* is exceptionally long. In *Xenopus*, transcription of Vg genes begins only 4 h after primary exposure or 1 h after secondary exposure of liver cells to estradiol [36]. In the locust, part of the lag may represent time required for diffusion of hormone analog from the injected oil vehicle, but this component would be constant for primary and secondary induction, and the difference between these must reflect significant processes in the responding cells. Knowledge of events during the lag phase seems to be crucial to understanding of the mechanism of gene activation.

What applicability may these results and conjectures have to the problem of JH action in larval insects, where this hormone prevents the transition from larva to adult during ecdysteroid-induced molting? Recent evidence from *Manduca sexta* indicates that the antimetamorphic role of JH involves maintenance of activity of certain genes and repression of activity of others [1]. Again, steroid action is similar, for steroid hormones (e.g., glucocorticoids [56]), can both activate and repress genes, and multiple steroids can interact in the regulation of genes (e.g., egg white proteins [57]). We suggest, therefore, that although more than one molecular mechanism is probably used, and none is yet fully known, the actions of JH in larval and adult insects belong to one family together with the diverse actions of steroid hormones, the nature of which will become understood through continued study of a variety of experimental systems.

References

1. Riddiford LM (1984) In: Hoffmann JA, Porchet M (eds) Biosynthesis, metabolism and mode of action of invertebrate hormones. Springer, Berlin Heidelberg New York, pp 265–272
2. Willis JH, Cox DL (1984) In: Hoffmann JA, Porchet M (eds) Biosynthesis, metabolism and mode of action of invertebrate hormones. Springer, Berlin Heidelberg New York, pp 466–474
3. Engelmann F (1979) Adv Insect Physiol 14:49–108
4. Pan ML (1977) Biol Bull 153:336–345
5. Nijhout MM, Riddiford LM (1974) Biol Bull 146:377–392
6. Hagedorn HH (1984) In: Gilbert LI, Kerkut GA (eds) Comprehensive insect physiology, biochemistry and pharmacology, vol VIII, chap. 7. Pergamon Press, Oxford New York
7. Bownes M (1982) Q Rev Biol 57:247–274
8. Engelmann F (1984) In: Hoffmann JA, Porchet M (eds) Biosynthesis, metabolism and mode of action of invertebrate hormones. Springer, Berlin Heidelberg New York, pp 444–453
9. Pau RN (1981) In: Pratt GE, Brooks GT (eds) Juvenile hormone biochemistry. Elsevier/North Holland, Amsterdam, pp 285–293
10. Chen TT, Couble P, DeLucca FL, Wyatt GR (1976) In: Gilbert LI (ed) The juvenile hormones. Plenum Press, New York, pp 505–529
11. Chen TT, Wyatt GR (1981) Sci Pap Inst Org Phys Chem Wroclaw Tech Univ 22:535–566
12. Chen TT, Strahlendorf PW, Wyatt GR (1978) J Biol Chem 253:5325–5331
13. Chen TT, Couble P, Abu-Hakima R, Wyatt GR (1979) Dev Biol 69:59–72
14. Chinzei Y, Chino H, Wyatt GR (1981) Insect Biochem 11:1–7
15. Chinzei Y, White BN, Wyatt GR (1982) Can J Biochem 60:243–251
16. Hovemann B, Galler R (1982) Nucleic Acids Res 10:2261–2275
17. Hung MC, Wensink PC (1983) J Mol Biol 164:481–492
18. Wyatt GR, Locke J, Bradfield JY, White BN, Deeley RG (1981) In: Pratt GE, Brooks GT (eds) Juvenile hormone biochemistry. Elsevier/North Holland, Amsterdam, pp 299–307
19. Wyatt GR, Locke J, Bradfield JY (1984) In: Engels W (ed) Invertebr Reprod Proc III Int Symp. Elsevier, Amsterdam, pp 73–80
20. Locke J, White BN, Wyatt GR (in preparation)
21. Bradfield JY, Wyatt GR (1983) Chromosoma 88:190–193
22. Dhadialla TS, Wyatt GR (1983) Dev Biol 96:436–444
23. Dhadialla TS (1983) Juvenile hormone regulation of vitellogenin genes in the African migratory locust. PhD thesis, Queen's Univ, Kingston, Ontario
24. Abu-Hakima R (1981) Experientia 37:1309–1311
25. Dhadialla TS, Wyatt GR (1981) In: Pratt GE, Brooks GT (eds) Juvenile hormone biochemistry. Elsevier/North Holland, Amsterdam, pp 257–262
26. Nair KK, Chen TT, Wyatt GR (1979) Dev Biol 81:356–360
27. Gellissen G, Wyatt GR (1981) Can J Biochem 59:648–654
28. Nemec V, Chen TT, Wyatt GR (1978) Acta Ent Bohemoslov 75:285–296

29. Pener MP, Orshan L, DeWilde J (1978) Nature (London) 272:350–353
30. Kunkel JG (1981) In: Bhaskaran G, Friedman S, Rodriguez JG (ed) Current topics in insect endocrinology and nutrition. Plenum Press, New York, pp 107–129
31. Kunkel JG (pers. comm.)
32. Chalaye D (1979) Can J Zool 57:329–336
33. Mundall EC, Tobe SS, Stay B (1979) Nature (London) 282:97–98
34. Shirk PD, Minoo P, Postlethwait JH (1983) Proc Natl Acad Sci USA 80:186–190
35. Huybrechts R, De Loof A (1981) Int J Invert Reprod 3:157–168
36. Shapiro D (1982) CRC Crit Rev Biochem 12:187–203
37. James TG, Bond UM, Maack CA, Applebaum SW, Tata JR (1982) Biochem J 205:521–528
38. Thomas PS (1980) Proc Natl Acad Sci USA 77:5201–5205
39. Yamamoto KP, Alberts BM (1976) Annu Rev Biochem 45:721–746
40. Klages G, Emmerich H, Peter MG (1980) Nature (London) 286:282–285
41. Chan ES, Coudron TA, Bruce MJ, Sage BA, O'Connor JD, Law JH (1980) Proc Natl Acad Sci USA 77:4657–4661
42. Engelmann F (1981) Mol Cell Endocrinol 24:103–112
43. Koeppe JK, Kovalick GE, Lapointe MC (1981) In: Pratt GE, Brooks GT (eds) Juvenile hormone biochemistry. Elsevier/North Holland, Amsterdam, pp 215–231
44. Roberts PE, Wyatt GR (1983) Mol Cell Endocrinol 31:53–69
45. Goodman W, Schooley DA, Gilbert LI (1978) Proc Natl Acad Sci USA 75:185–189
46. Peter MG, Gunawan S, Gellissen G, Emmerich H (1979) Z Naturforsch 34c:588–598
47. Roberts PE (1984) Arch Insect Biochem Physiol (in press)
48. Ashburner M, Chihara C, Meltzer P, Richards G (1974) Cold Spring Harbor Symp Quant Biol 38:655–662
49. Dworniczak B, Seidel R, Pongs O (1983) EMBO J 2:1323–1330
50. Compton JG, Schrader WT, O'Malley BW (1983) Proc Natl Acad Sci USA 80:16–20
51. Payvar F, Firestone GL, Ross SR, Chandler CL, Wrange O, Carlstedt-Duke J, Gustafsson JA, Yamamoto KR (1983) J Cell Biochem 19:241–247
52. Colgan V, Elbrecht A, Goldman P, Lazier CB, Deeley R (1982) J Biol Chem 257:14453–14460
53. Burch JBE, Weintraub H (1983) Cell 33:65–76
54. Weisbrod S (1982) Nature (London) 297:289–295
56. Shall S, Cutroneo KR (1983) J Biol Chem 258:3364–3369
57. Palmiter RD, Mulvihill ER, Shepherd JH, McKnight GS (1981) J Biol Chem 256:7910–7916

Part IV Hormonal Mechanisms in Echinoderms

Introduction

J.A. HOFFMANN

In the last section of this volume, hormonal mechanisms in the deuterostomian echinoderms are considered. It has been established over the last 15 years that neurosecretory cells from the radial nerves release a peptide hormone, GSH (gonad-stimulating hormone), which induces the follicle cells to produce the meiosis-inducing substance 1-methyl-adenine. This molecule, which is referred to as a hormone in this context, binds to receptors of the oocyte membrane and, via a cascade of events, induces the formation of a cytoplasmic factor, MPF (maturation-promoting factor) which promotes, in a relatively short time, the breakdown of the germinal vesicle and the accomplishment of maturation with extrusion of the two polar bodies. The contributions by P. Guerrier's and M. Dorée's groups center on the events leading to the production of MPF, which appears at present to be a non-species-specific and universal factor capable of controlling both meiotic and mitotic divisions.

Oocyte maturation is normally followed in sea-urchins by spawning, a process which was believed to be directly controlled by 1-methyl-adenine. The aim of the contribution of H. Kanatani was to show that, in fact, the action of 1-methyl-adenine in inducing spawning is indirect and proceeds via MPF, in a manner similar to the induction of germinal vesicle breakdown. The data which H. Kanatani intended to present at the Symposium are included in this volume in abstract form. They have now been published in detail in the January 1984 issue of Developmental Biology, Vol 101. H. Kanatani's illness prevented him from attending the Symposium and he died on the 18th of February 1984. This fourth part of the Symposium volume is dedicated by the authors and by the editors to the memory of this most distinguished scientist who has left his profound mark in this field.

Biosynthesis, Metabolism and Mode of Action of Invertebrate Hormones
(ed. by J. Hoffmann and M. Porchet)

IV.1 Early Steps in Transduction of the Hormonal Signal Controlling Starfish Oocyte Maturation

P. GUERRIER, L. MEIJER, M. MOREAU, and J. A. RENDER[1]

1. Introduction

Starfish oocytes remain arrested in the first prophase stage of meiosis in the ovary (Fig. 1). The resumption of meiotic divisions depends on at least three substances. First, a neurosecretory product known as gonad stimulating substance (GSS) stimulates the follicle cells to produce and release a relay hormone, the meiosis inducing substance (MIS). Then, this hormone, which has been identified as l-methyladenine (l-MeAde) induces the formation of a cytoplasmic factor, the maturation promoting factor (MPF) that promotes breakdown of the germinal vesicle (GVBD) and achievement of maturation with extrusion of the two polar bodies. Since we showed that a simple mechanical breakdown of the germinal vesicle without any hormone treatment leads to full cytoplasmic and nuclear maturation, including fertilizability and the capacity of undergo further development, these terminal processes seem to be preloaded in the oocyte structure. The main problems to be resolved, thus, concern the mechanisms by which the hormone signal is recognized and transduced up to the time when GVBD finally occurs. Such a sequence of events has been shown to be independent on macromolecular synthesis. It thus results that the starfish oocyte constitutes an excellent model of cell activation controlled at the posttranslational level. This implies that the process of meiosis reinitiation depends upon the modification of preexisting elements, such as specific enzymes which may be regulated via internal ionic changes.

It has been previously established that l-MeAde is recognized by specific structures localized on the external side of the plasma membrane and that it triggers a quite early and transient surge in intracellular free calcium, followed by an increase in endogenous protein phosphorylation.

In this article, we review all these data and present furthermore some new findings which undoubtedly, will contribute to the final elucidation of the mode of action of l-MeAde. First, we report that a mimetic of the hormone can be attached to sepharose beads without loosing its biological activity, a feature which may help in isolating the hormone receptor. Second, we show that calcium may exert its effect through activation of a calmodulin- dependent and trypsin activable enzyme, since externally applied

1 Station Biologique, 29211 Roscoff, France

Biosynthesis, Metabolism and Mode of Action of Invertebrate Hormones
(ed. by J. Hoffmann and M. Porchet)

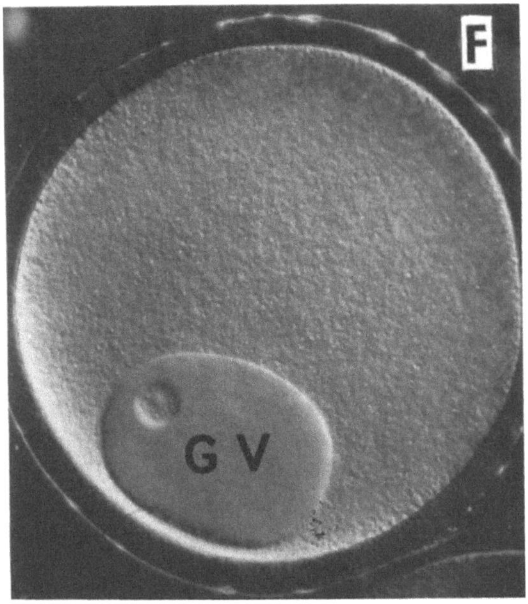

Fig. 1. A prophase-blocked oocyte of *Marthasterias glacialis* showing the germinal vesicle (*GV*) and the surrounding follicle cells layer (*F*)

anticalmodulin drugs block maturation unless the oocytes are previously treated with trypsin. Third, we present decisive indirect arguments suggesting that this enzyme may be a membranal phospholipase and that lipoxygenase metabolites of arachidonic acid are involved in the processes which lead to MPF formation and GVBD.

2. The Starfish Hormonal Control System

The first results concerning the endocrinology of the Asteroidea appeared in 1959, when Chaet and Mc Connaughy [1] described that the injection of an aqueous extract of radial nerves induced shedding of the oocytes. The responsible substance was described as GSS or "gamete shedding substance" and was further purified by Kanatani et al. [2]. It appeared to be a small peptide with a molecular weight of about 2,100. Since GSS was also observed to promote oocyte maturation in ligated fragments of the ovary, it was first considered as the actual hormonal factor which triggered maturation of the prophase blocked oocytes [3]. In 1967, however, it was recognized that GSS only acted as a "gonad stimulating substance", producing a lower molecular weight factor [4, 5]. This maturation inducing substance or MIS has been definitively purified and identified as l-Methyladenine (Fig. 2) by Kanatani et al. [6] and has been shown to directly result from the action of GSS upon the follicle cells [7]. The cytological studies of P.C. Schroeder et al. [8] and T.E. Schroeder [9] suggest that l-MeAde may transit intracellularly along the follicle cell processes and through a junctional complex up the oocyte surface. There are, thus, great convergences between the hormonal control systems of the starfish and the amphibian where pituitary gonadotropins induce the follicle cells to produce and release progesterone, the hormone which acts at the oocyte level.

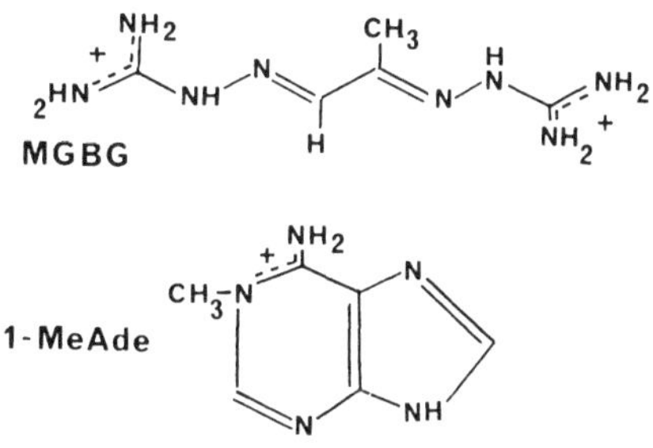

Fig 2. Compared structures of two agonists: MGBG and 1-MeAde

3. The Maturation Promoting Factor

The action of 1-MeAde (threshold of about 0.1 μM) is particularly rapid since breakdown of the germinal vesicle (GBVD) occurs within 20 min following hormone addition. A minimum contact time with the hormone is required for this event to occur. Such a hormone dependent period (HDP) is independent on the hormone concentration, but highly dependent upon the temperature. It accounts for about 5 min at 24° C [10, 11]. Repeated treatments with the hormone for periods shorter than the HDP are cumulative up to the state of irreversibility [12, 13]. This peculiarity as well as the fact that the hormone is only efficient when applied externally [14, 15] speak for the existence of a third mediator acting intracellularly and which could be homologous to the so-called maturation promoting factor or MPF first described by Masui and Markert [16] in the amphibian maturing oocyte. And, indeed, this factor has been also demonstrated to occur in the cytoplasm of maturing starfish oocytes [17]. It shares the same absence of species specificity [18] and the same autocatalytic amplification properties as the amphibian MPF since it does not lose its efficacy through multiple successive transfers into recipient oocytes. After hormone stimulation, it is note worthy that GVBD always occurs normally, even in the presence of specific inhibitors that completely suppress RNA or protein synthesis [12]. This indicates that the production of MPF is controlled at a posttranslational level and may only result from the modification of endogenous preexisting elements. In any case, it has been shown that meiosis reinitiation is always accompanied by an increased phosphorylation of endogenous proteins, both in the starfish [19, 20] and in the amphibian oocyte [21]. Moreover, changes in protein phosphorylation and MPF activity have been shown to be highly related [22].

Until recently, the exact role of MPF in meiosis reinitiation was not fully understood since one could not definitively discriminate between the effects of MPF (resulting from 1-MeAde stimulation) or the nucleoplasm (liberated following MPF action) upon further events in cytoplasmic maturation [49]. By inducing a purely mechanical disruption of the germinal vesicle without any hormonal treatment [23], we have now shown that all the characteristics of cytoplasmic maturity, e.g., sperm aster differentiation, maturation spindle formation, polar body extrusion, and male and female pronuclei development are independant on MPF and only rely upon mixing of the nucleoplasm and the cytoplasm. It results that the only specific and evident role for the MPF remains the induction of GBVD. This justifies that we focus all our attention on the earliest steps in the transduction of the hormonal signal.

4. The Signal Recognition Step

From complementary experiments performed either in Japan [14] or in Roscoff [15, 24], it appeared that the hormone could not act by the way of one of its metabolites, but rather by stimulating some recognition sites localized on the plasma membrane. These structures exhibit a rather strict specificity since they can only respond to structural analogues substituted in position 1 and 6, but not in 1–7 and 1–9 [11, 25]. The localization of these sites at the level of the plasma membrane has been confirmed by the fact that they resist a proteolytic treatment sufficient to digest the vitelline membrane, whereas they are extracted by a moderate treatment with Triton X 100 [26–28]. We also found recently that methylglyoxal-bis (guanylhydrazone) or MGBG (Fig. 2), covalently linked to CH-Sepharose 4B beads, could induce maturation in oocytes whose jelly layer and vitelline membrane have been removed by a mild pronase treatment, allowing a close contact with the plasma membrane [29].

Besides such compounds, which must be considered as active structural analogues of the hormone, only three types of agents have been found to promote starfish oocyte maturation, which may either induce or participate in the normal sequence of events triggered by the hormone. These are some disulfide reducing agents, such as dithiothreitol or 2,3 dimercaptopropanol [30, 31], the calcium ions which may act from the external medium or after moderate iontophoresis [32] and finally the enzyme phospholipase A_2 and its products, namely, the arachidonic acid and some of its derivatives of the hydroxyeicosatetraenoic acid series [33].

5. Importance of the Calcium Release Step

Our previous studies already stressed a pivotal role of this ion in the process of meiosis reinitiation. First, we have shown that the hormone, its active structural analogues, an efficient increased external calcium concentration, and dithiothreitol triggered a transient surge of intracellular calcium on the internal side of the plasma membrane [34, 35]. This response which begins within 1 s after external hormone addition and lasts for only 30 s, appears moreover indispensable for obtaining the biological response. We found, indeed, that efficient inhibitors of the maturation actually acted by decreasing the extent of the calcium surge below its threshold value [34, 36]. In contrast, the intracellular injection of dbc-AMP increased the calcium surge and facilitated maturation [37, 38]. Moreover, in experiments where EGTA was pressure injected into a single aequorin injected oocyte and the calcium response simultaneously monitored, we found that only those oocytes which did not present a normal threshold calcium response were unable to mature [24, 34, 36, 37]. These experiments have been, however, recently questioned by Picard and Doree [39] who could not reproduce our results. In our opinion, this discrepancy is by no means significant since (1) these authors used a different microinjection device; (2) they did not work in low calcium sea water (10 m*M* instead of 4 m*M*), injected about 3 times less EGTA than we did; and (3) they could not follow our rather fast kinetics where EGTA injection occured within milliseconds following l-MeAde addition. In fact, the major criticism raised by our contradictors concerns the fact that their EGTA injections where only efficient in blocking the cortical reaction induced by sperm, ionophore A23187, or calcium buffers, whereas no effect was observed upon meiosis reinitiation. It looks, however, quite unlikely that the same sites may be involved or similarly accessible for the same time period when dealing with such different processes. The same authors [40] further described without any trouble that the hormone l-MeAde could also trigger a normal cortical reaction, but that only at a later stage, after the completion of maturation. It is, thus, quite possible that the efficacy of the EGTA clamp may be maintained for a longer period in the peripheral cortex involved in the cortical reaction than in the membrane itself where the significant events for nuclear maturation are supposed to take place, possibly through activation of some relevant enzyme.

Even when the calcium response has been shown to be indispensable for maturation, it remains, however, that it is not sufficient in itself since the hormone, the calcium, or the dithiothreitol-dependent periods exceed by far the duration of the calcium releasing period. Similarly, nearly all of the known maturation inhibitors proved to be efficient all along the hormone dependent period. It, thus, results that calcium, as a potential second messenger, must induce some irreversible event which also requires that the membrane must be held in its responsible state (enzyme activation or substrate availability).

6. Possible Involvement of Calmodulin

Calmodulin is known as a small (16,723 mol. wt.) acidic and hydrophobic protein present in all eukaryotes so far investigated, which is able to bind four calcium ions and to activate a great variety of enzymes, including phospholipases [41]. Calmodulin has been identified in the starfish oocyte and it has been also shown that various calmodulin inhibitors, applied externally, inhibit l-MeAde dependent meiosis reinitiation [42, 44]. Despite the fact that maturation cannot be induced in the starfish oocyte after injection of the calmodulin-calcium complex [44, 45], another important observation may speak in favor of its involvement. Thus, we found recently [46] that oocytes, exposed for 10 min to trypsin, lose their sensitivity to calmodulin antagonists. Since it has been already shown that most calmodulin-dependent enzymes can be activated by trypsin and that they then become fully activated and insensitive to calmodulin, calcium, and calmodulin inhibitors [47], this strongly suggest that some calmodulin-dependent step might be involved in the process of meiosis reinitiation.

This step is likely to occur at the plasma membrane level. Here again, as for the calcium surge step, it is clear that such a trypsin effect appears insufficient by itself to engage full maturation and that some other modifications are also needed.

7. A Role for a Membranal Calcium or/and Calmodulin-Dependent Phospholipase During Transduction of the Hormonal Signal

This hypothesis is based on the indirect finding that arachidonic acid, which does not occur free in the cell, but rather covalently linked in the middle position of membrane phospholipids, is able to trigger full maturation of starfish oocytes at rather low concentrations of the order of 1 μM [33]. This effect appears to be facilitated in the presence of external calcium (Fig. 3). Moreover, the kinetics of the maturations induced by arachidonic acid or 1-MeAde look quite identical, suggesting that arachidonic acid acts at a very proximal step in the sequence of events leading to GVBD. Arachidonic acid release may occur through activation of phospholipase C plus diglyceride lipase or the activation of phospholipase A_2.

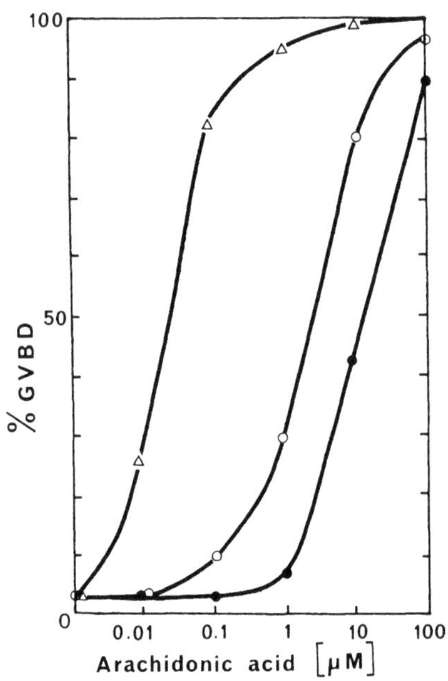

Fig. 3. Effect of external calcium upon the efficiency of meiosis reinitiation as triggered by arachidonic acid (*Marthasterias glacialis*). △10 mM Ca^{2+} seawater; ○ Ca^{2+} free seawater, ● Ca^{2+} free + 5 mM EGTA. Normal development was obtained after fertilization

In fact, we found that phospholipase A_2 from bee venom or Naja venom (0.1 to 1 U/ml) also triggered meiosis reinitiation, provided that calcium was present in the incubation medium. At an efficient enzyme concentration, the required contact times for phospholipase A_2 and 1-MeAde-induced maturations proved moreover to be identical, suggesting that a certain threshold of released arachidonic acid was required to produce the expected biological effect. In contrast, neither phospholipase C alone nor phospholipase D could induce maturation, although the latter enzyme slightly facilitated maturation in the presence of 1-MeAde.

We found, moreover, that the lysogenic polypeptide and phospholipase A_2 activator, melittin, also triggered GVBD at a concentration of about 5μM. This compound has, however, quite deleterious effects since maturation cannot proceed further in its presence. Finally, more direct evidence for the possible involvement of an endogenous phospholipase A_2 in meiosis reinitiation is given by the finding that two phospholipase A_2 inhibitors, quinacrine and bromophenacyl bromide, inhibit maturation at I_{50} concentrations of the order of 10 and 20 μM. In addition, it appeared that the sensitive periods for these two inhibitors exactly correspond to the hormone-dependent period.

Such results as well as the fact that light facilitated 1-MeAde dependent maturation lead us, moreover, to suspect that the postulated 1-MeAde dependent released arachidonic acid might be further oxidized. The two main pathways for such an oxidation were, thus, explored. From this study, it resulted that the cyclooxygenase pathway could not be taken into consideration since neither its final products (prostaglandins and thromboxanes) nor its inhibitors had any effect upon maturation. In contrast, the lipoxygenase pathway appears likely to be involved since at least three of its inhibitors were found to preclude the biological response and since several hydroxyeicosatetraenoic acids could also trigger meiosis reinitiation at concentrations rather close to that found for the arachidonic acid.

8. Conclusion

The maturation of the starfish oocyte depends upon the sequential intervention of at least four factors: two hormones, the GSS and the 1-MeAde, an early second messenger, the calcium ion, and an intracellular effector, the MPF. Our efforts have been specifically directed to elucidate those mechanisms which occur between the intracellular calcium surge and the appearance of the MPF. In our opinion, these mechanisms are likely to involve the activation of a membrane calcium and calmodulin-dependent phospholipase which would release arachidonic acid from the membrane phospholipids. This acid would be further oxidized through the lipoxygenase pathway up to the already effective hydroxyeicosatetraenoic acids. The possibility must thus be considered that such final products may act by activating some specific lipase, associated with control protein kinases, which would finally result in disruption of the nuclear envelope.

Much work remains to be performed on such a model system which has now acquired a quite general significance since it has been discovered that MPF actually represents a nonspecific and universal factor responsible for controlling both meiotic and mitotic divisions (see [48] for a review). This constitutes at least a good justification for those patient studies which have been directed to elucidate the intimate mechanisms which control the reproduction of some of our lower relatives, the invertebrates.

Acknowledgments. This work is dedicated to Prof. J. Brachet, M. Durchon, H. Kanatani, and J. Pasteels who so brilliantly opened us the way and to Prof. J. Bergerard who devoted himself to the Roscoff Marine Station of Biology for the past 10 years. It would not have been feasible without the constant help of the DGRST (ACC 7570783, 7770259, 7970777, 8070341), the INSERM (PRC 135019), and the CNRS (ATP 982112). We would like to thank Mrs C. Guerrier for her excellent technical assistance, Mrs. Guyard for typing the text, the fishermen of the Biological Station in Roscoff for their enthusiastic readiness to help, and our closest steady rivals in the field for not having discouraged our research competitivity.

References

1. Chaet AB, Mc Connaughy RA (1959) Physiologic activity of nerve extracts. Biol Bull 117:407
2. Kanatani H, Ikegami S, Shirai H, Oide H, Tamura S (1971) Purification of gonadstimulating substance obtained from radial nerve of the starfish *Asterias amurensis*. Dev Growth Differ 13:151-164
3. Kanatani H (1964) Spawning of starfish: Action of gamete-shedding substance obtained from radial nerves. Science 146:1177-1179
4. Kanatani H, Shirai H (1967) In vitro production of meiosis inducing substance by nerve extract in ovary of starfish. Nature (London) 216:284-286
5. Schuetz AW, Biggers JD (1967) Regulation of germinal vesicle break-down in starfish oocytes. Exp Cell Res 46:624-628
6. Kanatani H, Shirai H, Nakanishi K, Kurokawa T (1969) Isolation and identification of meiosis inducing substance in starfish *Asterias amurensis*. Nature (London) 221:273-274
7. Hirai S, Kanatani H (1971) Site of production of meiosis-inducing substance in ovary of starfish. Exp Cell Res 67:224-227
8. Schroeder PC, Larsen JH, Waldo AE (1979) Oocyte follicle cell relationships in a starfish. Cell Tissue Res 203:249-256
9. Schroeder TE (1981) Microfilaments-mediated surface change in starfish oocytes in response to l-methyladenine: implications for identifying the pathway and receptor sites for maturation-inducing hormones. J Cell Biol 90:362-371
10. Guerrier P, Dorée M, Moreau M, Cheval J, Freyssinet G (1976) Contrôle hormonal de la méiose de ovocytes d'étoiles de mer. I - Analyse des événements précoces induits par la l-méthyladénine. Actualités sur les hormones d'invertébrés. Colloq Int CNRS, Lille, CNRS, pp 191-198
11. Dorée M, Guerrier P, Leonard NJ (1976 a) Hormonal control of meiosis. The specificity of l-methyladenine receptors in starfish oocytes. Proc Natl Acad Sci USA 73:1669-1673
12. Guerrier P, Doree M (1975) Hormonal control of reinitiation of meiosis in starfish: the requirement of l-methyladenine during nuclear maturation. Dev Diol 47:341-348
13. Nemoto SE (1982) Nature of the l-methyladenine-requiring phase in maturation of starfish oocytes. Dev Growth Differ 24:429-442
14. Kanatani H, Hiramoto Y (1970) Site of action of l-methyladenine in inducing oocyte maturation in starfish. Exp Cell Res 61:280-284
15. Dorée M, Guerrier P (1975) Site of action of l-methyladenine in inducing oocyte maturation in starfish: kinetic evidence for receptors localized on cell membrane. Exp Cell Res 91:296 to 300
16. Masui Y, Markert CI (1971) Cytoplasmic control of nucelar behavior during meiotic maturation of frog oocytes. J Exp Zool 177:129-146
17. Kishimoto T, Kanatani H (1976) Cytoplasmic factor responsible for germinal vesicle breakdown and meiotic maturation in starfish oocyte. Nature (London) 260:321-322
18. Kishimoto T, Kuriyama R, Kondo H, Kanatani H (1982) Generality of the action of various maturation promoting factors. Exp Cell Res 137:121-126
19. Guerrier P, Moreau M, Dorée M (1977) Hormonal control of meiosis in starfish: stimulation of protein phosphorylation induced by l-methyladenine. Mol Cell Endocrinol 7:137-150

20. Mazzei G, Guerrier P (1982) Changes in the pattern of protein phosphorylation during meiosis reinitiation in starfish oocytes. Dev Biol 91:246-256
21. Wu M, Gerhardt JC (1980) Partial purification and characterization of the maturation promoting factor from eggs of *Xenopus laevis*. Dev Biol 79:465-477
22. Schatt P, Moreau M, Guerrier P (1983) Cyclic changes in protein phosphorylation and MPF activity during early cleavage in the sea urchin egg. CR Acad Sci 296:551-554
23. Guerrier P, Meijer L, Moreau M, Longo FJ (1983) Hormone independent GVBD induces cytoplasmic maturity in the starfish oocyte. J Exp Zool 226:303-309
24. Moreau M, Guerrier P, Dorée M (1978 a) Hormonal control of meiosis reinitiation in starfish oocytes. New evidence for the absence of efficient intra-cellular receptors for 1-methyladenine recognition. Exp Cell Res 115:245-249
25. Kanatani H, Shirai H (1971) Chemical structural requirements for induction of oocyte maturation and spawning in starfishes. Dev Growth Differ 13:53-64
26. Dorée M, Guerrier P, Moreau M (1976 b) Contrôle hormonal de la méiose des ovocytes d'Etoile de mer. II - Etude de quelques propriétés fondamentales du récepteur de la l-méthyladénine. Actualités sur les hormones d'invertébrés. Colloq Int CNRS, Lille, CNRS, pp 199 to 205.
27. Morisawa M, Kanatani H (1978) Oocyte-surface factor responsible for 1-methyladenine induced oocyte maturation in starfish. Gamete Res 1:157-164
28. Ikadai H, Kanatani H (1982) Regeneration of oocyte-surface factor responsible for l-methyladenine-induced oocyte maturation in starfish. Biomed Res 3:378-383
29. Meijer L, Guerrier P (1983) Immobilized methylglyoxal bis-(guanylhydrazone) induces starfish oocytes maturation. Dev Diol 100, 308-317
30. Kishimoto T, Kanatani H (1973) Induction of starfish oocyte maturation by disulfide reducing agents. Exp Cell Res 82:296–302
31. Kishimoto T, Cayer MI, Kanatani H (1976) Starfish oocyte maturation and reduction of disulfide-bond on oocyte surface. Exp Cell Res 101:104-110
32. Guerrier P, Moreau M, Dorée M (1978) Control of meiosis reinitiation in starfish: calcium ions as the primary effective trigger. Ann Biol Anim Biochem Biophys 18:441-452
33. Meijer L, Guerrier P (1984) Arachidonic acid in starfish oocytes. I. Arachidonic acid, 12 and 15- hydroxyeicosatetraenoic acids and phospholopase A_2 induce oocyte maturation. Dev Biol (in press)
34. Moreau M, Guerrier P, Dorée M, Ashley CC (1978 b) l-methyladenine-induced release of intracellular calcium triggers meiosis in starfish oocytes. Nature (London) 272:251-253
35. Dorée M, Moreau M, Guerrier P (1978) Hormonal control of meiosis. In vitro induced release of calcium ions from the plasma membrane in starfish oocytes. Exp Cell Res 115:251-260
36. Moreau M, Guerrier P, Dorée M (1978 c) Membrane recognition and transduction during the hormone induced meiosis reinitiation in starfish oocytes. Biol Cell 32:69-74
37. Moreau M, Guerrier P (1979) Free calcium changes associated with hormone action in oocytes. In: Ashley CC, Campbell AK (eds) The measurement of free calcium in cells. Elsevier, North-Holland, pp 219-226
38. Moreau M, Guerrier P (1980) In vitro interaction between membrane, hormone and cyclic nucleotides as revealed with aequorin. Dev Biol 79:488-492
39. Picard A, Dorée M (1983 b) Is calcium the second messenger of 1-methyladenine in meiosis reinitiation of starfish oocytes? Exp Cell Res 145:325-337
40. Picard A, Dorée M (1983 a) Hormone-induced parthenogenetic activation of mature starfish oocytes. Exp Cell Res 145:315-323
41. Craven PA, De Rubertis FR (1983) Ca^{2+} calmodulin-dependent release of arachidonic acid for renal medullary protaglandin synthesis. Evidence for involvement of phospholipases A_2 and C. J Biol Chem 258:4814–4823
42. Meijer L, Guerrier P (1981) Calmodulin in starfish oocytes. I. Calmodulin antagonists inhibit meiosis reinitiation. Dev Biol 88:318-324
43. Dorée M, Kishimoto T, Le Peuch CJ, Demaille JG, Kanatani H (1981) Calciummediated transduction of the hormonal message in meiosis reinitiation of starfish oocytes. Exp Cell Res 135: 237-249

44. Dorée M, Picard A, Cavadore JC, Le Peuch C, Demaille JG (1982) Calmodulin antagonists and hormonal control of meiosis in starfish oocytes. Exp Cell Res 139:135-144
45. Mazzei G, Meijer L, Moreau M, Guerrier P (1981) Role of calcium and cyclic nucleotides during meiosis reinitiation in starfish oocytes. Cell Differ 10:139-145
46. Meijer L, Guerrier P (1984 a) Calmodulin in starfish oocytes. II. Trypsin treatment suppresses the trifluoperazine-sensitive step. Dev Biol 101, 257-262
47. Meijer L, Guerrier P (1982) Activation of calmodulin-dependent NAD-kinase by trypsin. Biochim Biophys Acta 702:143-146
48. Meijer L, Guerrier P (1984 b) Maturation and fertilization in starfish ocoytes. Int Rev Cytol 86:129–196
49. Hirai S, Nagahama Y, Kishimoto T, Kanatani H (1981) Cytoplasmic maturity revealed by the structural changes in incorporated spermatozoon during the course of starfish oocyte maturation. Dev Growth Differ 23:465-478

IV.2 Production of the Maturation-Promoting Factor (MPF) Following Hormonal Stimulation of Starfish Oocytes

M. DORÉE[1], G. PEAUCELLIER[2], A. PICARD[1], and F. LE BOUFFANT[1]

1. Introduction

As long as they remain in the ovary, fully-grown starfish oocytes are arrested in prophase of the first meiotic division. Spawning and release from prophase block are triggered by a hormone originating from the follicle cells [1], which has been identified as 1-methyladenine (l-MeAde), [2]. In starfish the process of meiotic maturation triggered by l-MeAde is not arrested at metaphase II as in vertebrates, but proceeds directly to the formation of the second polar body and completion of meiosis. It has been shown that l-MeAde acts at the cell surface [3,4] to produce a cytoplasmic maturation-promoting factor (MPF), which induces breakdown of the germinal vesicle (GVBD) and the subsequent events of meiotic maturation when transferred into full-grown immature oocytes [5]. MPF is reported not to be species specific and, in fact, not even phylum specific: for example, MPF extracted from mature oocytes of the toad, *Bufo bufo,* induces maturation following injection in starfish oocytes [6]. Moreover, extracts of synchronized mammalian cells prepared at the late G_2 and during the mitosis phase induce GVBD and chromosome condensation when injected in immature oocytes of starfish or amphibians [6, 7]. These data suggest that the same factor (MPF) might be involved in the control of the G_2 to M transition in mitosis and in the equivalent stage of meiosis reinitiation throughout the animal kingdom. Cytoplasmic MPF has autocatalytic and amplifying activity since its biological activity is sustained following serial transfer through numerous non hormone-treated oocytes of both starfish [5] or amphibians [8, 9].

In the present work, MPF activity and the pattern of protein phosphorylation were monitored throughout the time course of meiotic maturation following hormonal stimulation of prophase-arrested oocytes. High and low levels of MPF activities were found to be respectively associated with high and low levels of protein phosphorylation. On the other hand, new experimental results strongly support the view that nuclear activities play a key role in both the initial production and amplification of MPF in starfish oocytes. This contrasts with the case of amphibians, where the nucleus is believed not to be involved in MPF production [8, 9]

1 INSERM U. 249 and CNRS CRBM, B.P. 5015, 34033 Montpellier Cedex, France
2 Station Biologique, 29211 Roscoff, France

Biosynthesis, Metabolism and Mode of Action of Invertebrate Hormones
(ed. by J. Hoffmann and M. Porchet)

2. Changes of MPF Activity During Meiotic Maturation

When cytoplasm taken from oocytes before hormone application is injected into immature oocytes, GVBD never occur in recipient oocytes. In contrast when enough cytoplasm is taken at the time of GVBD, recipient oocytes are readily released from prophase block. Threshold volume of cytoplasm required to bring about GVBD in recipient oocyte (120 pl) does not significantly differ whether cytoplasm is taken from maturing oocytes either 18 or 30 min after l-MeAde addition. We investigated the effect of transferring a slightly higher amount of cytoplasm (160 pl) taken from maturing oocytes from 20 to 150 min after hormone application. Figure 1 shows the result of such an experiment performed with a single batch of oocytes receiving 2 x $10^{-6} M$ 1-MeAde at zero time. Emission of the first polar body was first observed in the oocyte population at time 67 min. When cytoplasm was taken from maturing oocytes from 20 to 81-85 min posthormone addition (p.h.a.), all oocytes underwent GVBD, with exception of those receiving cytoplasm taken from 66 to 71 min p.h.a. Oocytes injected with cytoplasm taken from maturing oocytes later then 85 min p.h.a. did not undergo GVBD.

We conclude that MPF keeps a high activity in the cytoplasm of maturing starfish oocytes from 18 to 85 min after hormone addition, with the exception of a very short period (about 5 min using oocytes from the same ovary) which closely corresponds to the cleavage step, when the first polar body is emitted. At the second meiotic cleavage, MPF activity again decreases considerably, and remains very low during formation of the female pronucleus and in fully-matured oocytes.

In amphibians as in starfishes MPF has been shown to cycle as a function of the mitotic cycle in the early embryo [6, 10]. MPF reaches a peak in activity at about the time of nuclear membrane breakdown, chromosome condensation, and formation of the spindle, i.e., at the G_2-M transition of the mitotic cell cycle, and it drops or

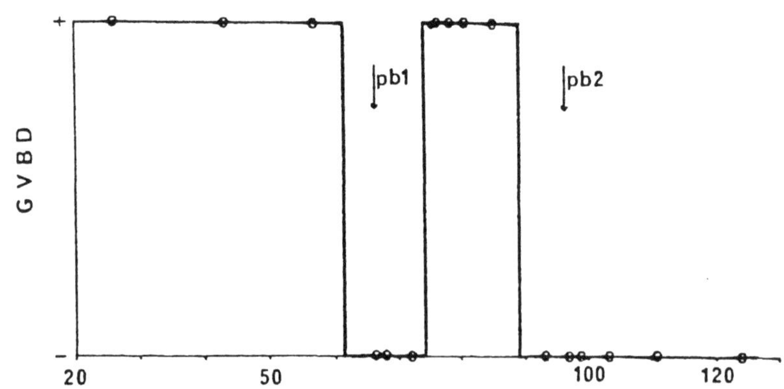

Fig. 1. Changes of MPF activity in oocytes during the course of 1-MeAde-induced meiotic maturation. A constant volume of cytoplasm (160 pl) taken from maturing oocytes at the indicated times after 1-MeAde addition was transferred into prophase-blocked oocytes. Recipient oocytes with GVBD (+) or without GVBD (−) were scored 1 h after cytoplasm transfer (results did not change later). *Arrows* point to the times when emission of the first (p.b.1) and the second polar body (p.b.2) were observed in this homogenous population of donor oocytes

even disappears at the time of cytokinesis. In contrast MPF activity has been reported in amphibians to remain at a constant and high level from GVBD to metaphase II arrest [8, 9]. Absence of cyclic behavior of oocyte MPF was considered to depend on the fact that after the first meiotic division a nuclear membrane does not reform. Rather the chromosomes remain condensed and realign rapidly on the second meiotic spindle. This led to consider that the more general role of MPF was in controlling nuclear membrane breakdown [10]. Nevertheless the present results show that at least in starfish, MPF activity does not only cycle as a function of the mitotic cell cycle, but also in the same way during both meiotic divisions, although nuclear membrane also does not reform and chromosomes remain condensed after GVDB, until formation of the female pronucleus, which occurs once meiosis has been completed.

3. Does Germinal Vesicle Contain Meiosis-Inducing Activity Before Hormonal Treatment?

When the whole germinal vesicle was sucked up into a micropipette from an immature fully-grown oocyte of the starfish *Marthasterias glacialis* captured at the beginning of its breeding season at Roscoff (from early May to mid-June) and injected into intact immature oocytes taken from the same female, the germinal vesicle of the recipient oocytes never broke. In contrast GVBD and the subsequent events of meiotic maturation frequently occurred in the recipient oocytes when the same experiment was performed later on during the breeding season. At this period of the year, it was possible to discriminate between "competent" and "incompetent" oocytes. Competent but not incompetent underwent GVBD following injection of germinal vesicle material. Minimal volumes of germinal vesicle material taken from nonhormone-treated oocytes to be injected into competent oocytes for meiosis to be reinitated were always smaller than minimal volumes of cytoplasm taken from maturing oocytes. Even without injecting additional germinal vesicle material, MPF was produced in cytoplasm of competent oocytes when the germinal vesicle was sucked up with a micropipette and immediately reinjected into the oocyte, which resulted in the mixing of nucleoplasm with cytoplasm.

We were unable to detect any morphological difference between competent and incompetent oocytes. Germinal vesicle material taken from incompetent oocytes was just as efficient as germinal vesicle material taken from competent when injected in cytoplasm of competent oocytes. Thus, we are inclined to believe that "competence" is rather related to some special property of the cytoplasm (or eventually: of the nuclear envelope) of the recipient oocytes than to some particularity of their germinal vesicle material. The meiosis-inducing factor present in the germinal vesicle before hormone action is neither associated with the nucleolus nor the nuclear envelope. On the other hand, we do not believe that it might be associated with chromosomes, since GVBD did not occur following the injection of four female pronuclei, although they would be expected to introduce in recipient competent occytes an amount of chromosomic material comparable to that of a single germinal vesicle (it remains, however, possible that chromosome-associated proteins in the germinal vesicle differ from those contained in female pronuclei). We rather favor the hypothesis that the meiosis-inducing factor originates from germinal vesicle nucleoplasm. Nuclei isolated from the oocytes in aqueous solution [11] failed to induce GVBD when injected into competent oocytes. Thus, the active factor might be a diffusible component of the nucleoplasm, readily lost during the isolation procedure.

In the present state of our investigations, the nature of the meiosis-inducing factor present in the germinal vesicle before l-MeAde application remains unkown. That the nuclear meiosis-inducing factor present *prior to hormonal stimulation* in the germinal vesicle is *not* MPF is supported by the following observations. First, this nuclear factor induced GVBD only when transferred into competent oocytes. In contrast cytoplasm taken at the time of GVBD from hormone-treated oocytes (thus, MPF-containing cytoplasm) *always* induced GVBD in recipient oocytes of both categories. Moreover, minimal volumes of cytoplasm to be transferred for meiosis to be reinitiated were not found to differ whether recipient oocytes belong to the competent of incompetent category of oocytes. Second, the mean time required for GVBD to occur following injection of germinal vesicle material (19.5 min $\pm$ 3.7) by far exceeded the mean time required for GVBD to occur following injection of MPF containing cytoplasm (13.2 min $\pm$ 3.8). Therefore, it seems likely that the germinal vesicle of nonhormone-treated oocytes contains a factor different to MPF and inefficient by itself to induce GVBD, but which triggers MPF production when injected in the cytoplasm of competent, but not of incompetent oocytes.

Kishimoto et al, reported that a small but significant proportion (23%) of enucleated oocytes of the starfish *Asterina pectinifera* produced MPF following l-MeAde addition [12] We never observed MPF production in hormone-treated incompetent oocytes of the starfishes *Asterias rubens* and *Marthasterias glacialis* following enucleation, even when enucleation was performed after hormone addition but before the time of appearance of MPF in intact control oocytes. Therefore, we suspected that production of MPF in enucleated oocytes upon hormonal stimulation was perhaps restricted to competent oocytes. Oocytes were prepared from seven females of the starfish *Marthasterias glacialis*. In each batch of oocytes the frequency of MPF production after enucleation was plotted against the frequency of GVBD in nonhormone-treated intact oocytes injected with germinal vesicle material. As shown on Fig. 2, ability of oocytes to produce MPF after enucleation and upon hormonal stimulation seems actually to be correlated with their ability to undergo GVBD following injection of germinal vesicle material (slope of the linear regression: 0.93; correlation coefficient: 0.71).

Using part of a batch of competent oocytes of the starfish *Marthasterias glacialis*, we determined in another set of experiments the threshold concentration of l-MeAde required for inducing 100% GVBD in intact oocytes. Then enucleated oocytes were prepared from the remaining oocytes and transferred either in seawater containing the threshold (2×10^{-7} *M*) or a much higher concentration of the hormone (2×10^{-5} *M*). Cytoplasm (220 pl) was taken from the enucleated oocytes 20 to 40 min after hormone addition and transferred into intact immature oocytes. GVBD never occurred in recipient oocytes when enucleated oocytes were treated with l-MeAde at the lower concentration. In contrast a significant proportion of recipient oocytes (4/15) underwent GVBD and the subsequent events of meiosis reinitiation when the enucleated oocytes were treated with l-MeAde at the higher concentration. Although competent

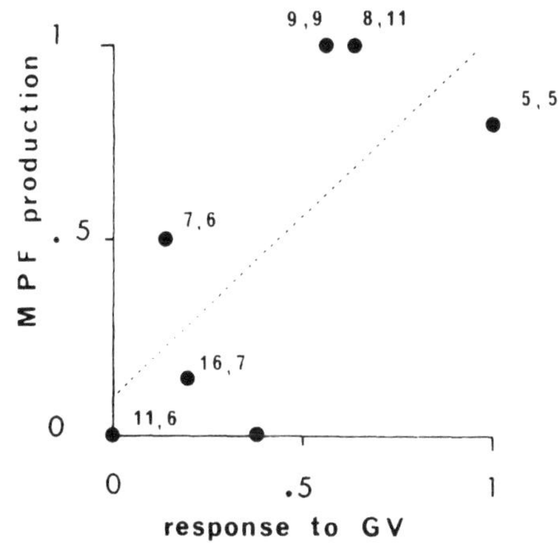

Fig. 2. Relationship between the frequency of GVBD induction in nonhormone-treated following the injection of germinal vesicle material *(abscissa)* and the frequency of MPF production following hormone application to enucleated oocytes from the same batch *(ordinate)*. For each experiment, the first number points to the number of oocytes injected with germinal vesicle material, the second to the number of enucleated oocytes checked for MPF production (260 pl of cytoplasm were transferred)

oocytes were shown to be capable of producing MPF after enucleation, the dose of hormone required to trigger MPF production was higher than for intact oocytes. In addition, the amount of cytoplasm to be taken for inducing GVBD in recipient oocytes was significantly higher from enucleated than from intact oocytes.

In conclusion, while enucleated incompetent oocytes never produce MPF upon 1-MeAde stimulation or MPF microinjection, enucleated competent oocytes produce a cytoplasmic factor which induces GVBD in both competent and incompetent oocytes. Thus, competent oocytes must be considered to be capable of producing MPF even after enucleation. From this point of view, competent starfish oocytes behave like amphibian oocytes, which can also produce MPF following enucleation. We believe, however, that even in competent starfish oocytes, involvement of the germinal vesicle in producing MPF cannot be ruled out, since the amount of MPF produced is lower in enucleated than in nucleated oocytes. Moreover, a higher dose of hormone is required to produce MPF following enucleation.

4. Time Course of the Appearance of MPF Activity in the Germinal Vesicle and in the Cytoplasm of Starfish Oocytes Following 1-MeAde Addition or MPF Microinjection

For the following experiments, only incompetent oocytes served as recipient oocytes, since competent oocytes would mature following injection of germinal vesicle material, even in the absence of hormonal treatment.

When the germinal vesicle material from nonhormone-treated oocytes was injected into such control immature oocytes, the germinal vesicle of the recipient oocytes never broke. In contrast, when donor oocytes were first treated with 10^{-6} *M* 1-MeAde, the germinal vesicle was found to contain meiosis-inducing activity, as demonstrated by its ability to release recipient oocytes from prophase block. In fact meiosis-inducing activity appeared in the germinal vesicle before presence of MPF could be demonstrated in cytoplasm. Meiosis-inducing activity was already detected in the germinal vesicle,

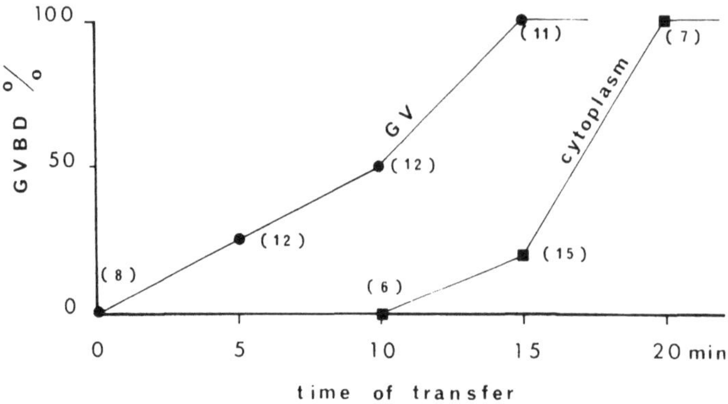

Fig. 3. Time course of the appearance of meiosis-inducing activity in the germinal vesicle and in the cytoplasm of *A. rubens* oocytes following 1-MeAde addition. *In parentheses:* the total number of recipient oocytes

but not in the cytoplasm of the same donor oocyte 5 min after hormone addition in both *Marthasterias glacialis* and *Asterias rubens.* Then the nuclear meiosis-inducing activity increased with time and was maximal after 15 min, thus, a few min before GVBD. Only a small MPF activity was detected in cytoplasm 15 min after l-MeAde addition, when nuclear meiosis-inducing activity was already maximal. Hundred percent GVBD occurred in cytoplasm-injected oocytes only when cytoplasm was taken from donor oocytes at the time of GVBD or later (Fig. 3).

In another set of experiments ten oocytes of the starfish *Marthasterias glacialis* were removed from prophase block by the injection of cytoplasm taken from maturing oocytes at the time of GVBD. 7 to 15 min later, the germinal vesicle (150 pl) was sucked up from each recipient oocyte and injected into a second nonhormone-treated recipient oocyte: 27 out of 30 oocytes injected with germinal vesicle material underwent GVBD 20 min later. In contrast only 9 out of 30 recipient oocytes injected with cytoplasm underwent GVBD, even when cytoplasm was transferred as late as 30 min after the removal of the germinal vesicle. For this experiment, no significant difference was observed whether competent or incompetent oocytes were used as both donor or recipient oocytes.

5. MPF Activity Oscillates Simultaneously with Extent of Protein Phosphorylation During Meiotic Maturation

The rate of ^{32}P-phosphate incorporation into proteins has been shown to increase considerably, before GVDB, following l-MeAde addition to full-grown prophase-blocked oocytes of various starfish species [13–16]. Studies about the changes in protein phosphorylation associated with the release of prophase block, following stimulation by hormones in starfishes [13–16] and amphibians [17, 18] by fertilization in echiurians [19] by proteases, fertilization or ionophore A 23187 in annelids

[20] were restricted to meiosis reinitiation, roughly from stimulus application to metaphase I. With the exception of proteases in annelids, all the other meiosis triggering signals allow, however, meiotic maturation to continue after metaphase I, and possible changes in the pattern of protein phosphorylation associated with progression of meiosis beyond this step had not been studied previously. In a recent work changes in protein phosphorylation were investigated throughout the process of meiotic maturation [21]. We confirmed the earlier results and showed in addition that the rate of ^{32}P-phosphate incorporation into proteins remains higher in maturing than in prophase-blocked oocytes for at least 55 min, although it somewhat decreased around GVBD, about 20 min after hormone addition. Emission of the first polar body (71–77 min after hormone addition) was associated with a dramatic and transient decrease in the amount of ^{32}P-phosphate incorporation into proteins. Thereafter incorporation of ^{32}P-phosphate increased again to reach nearly the maximal level observed before emission of the first polar body. Emission of the second polar body (103–113 min after hormone addition) was again associated with a net dephosphorylation of proteins, but the amount of incorporated radioactivity did not increase again later. It rather continued to decrease very slowly during formation of the female pronucleus. The main result of this work was that changes on the extent of protein phosphorylation occurred *simultaneously* with changes of MPF activity during meiotic maturation. In agreement with these results, we found that the absolute amount of alkali-labile phosphate covalently bound to proteins decreased significantly, with a confidence level higher than 95% (student's test = 2.82 with 4 degrees of freedom) in the time interval from 80 to 120 min after l-MeAde addition.

Since incorporation of phosphate into proteins has been shown to be already stimulated less than 3 min after hormone addition [16], well before MPF activity becomes sufficient for successful transfer of cytoplasm into immature oocytes (about 13 min when GVBD occurs 20 min after hormone addition) [5], an attractive speculation would be that MPF activation results from protein phosphorylation and MPF inactivation from protein dephosphorylation. This would agree with the previous demonstration that both initiation and amplification of MPF are under post-translational control in starfish oocytes [22]. Preliminary attempts to purify MPF from both amphibian and starfish oocytes also provided some arguments in favor of a role for protein phosphorylation in MPF activation or activity: first, the successful extraction and stabilization of MPF requires the presence of phosphorylated small molecules which may inhibit oocyte phosphoprotein phosphatases; second, ATP enhances the activity of the extracted MPF [23, 24]. Possible mechanisms for the enhanced overall phosphorylation of proteins associated with MPF activation includes stimulation of the intrinsic activity of protein kinases or inhibition of protein phosphatases, increased concentration of free phosphoryl group donors, as well as unmasking of protein substrates. Endogenous protein kinase activity has been reported to be about 20% higher in homogenates prepared from hormone-stimulated than in those from control oocytes [25], but this slight increase contrasts with the considerable increase (about 25 times) of the rate of ^{32}P-phosphate incorporation into proteins observed in vivo following l-MeAde addition.

Most phosphoproteins of the starfish *Marthasterias glacialis* were resolved using a pHi gradient in the range 3.8–6.2. More than 100 labelled proteins were found to be

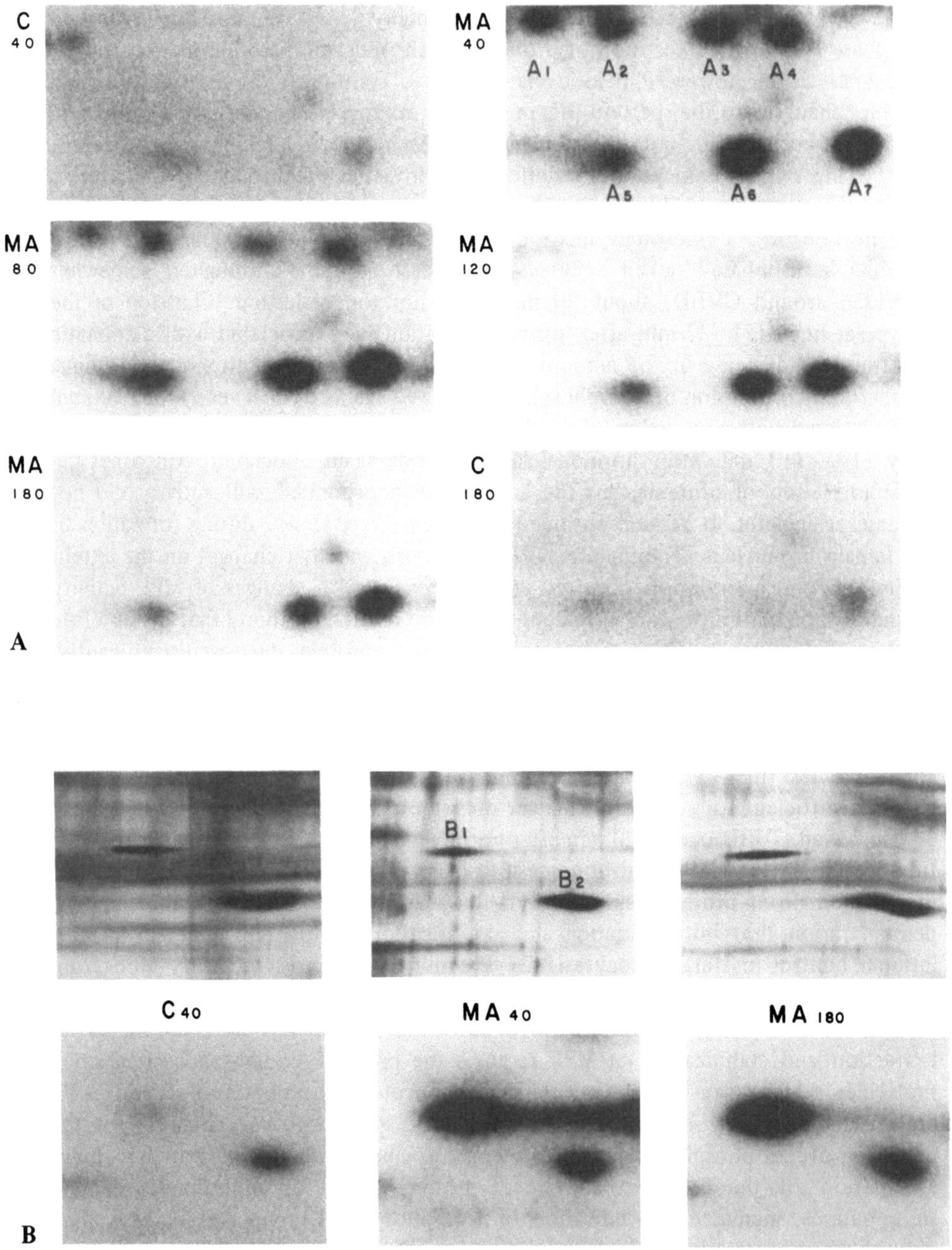

Fig. 4 A, B. Detail from bidimensional slab gel electrophoretograms of homogenates from ^{32}P-preloaded oocytes at various times of meiotic maturation and in control oocytes. **A** (autoradiogram) *Left to right:* 5.2 < pHi < 5.7. *Top to bottom:* 35,000 > Mr > 25,000. **B** 4.2 < pHi < 4.4 and 50,000 > Mr > 35,000. *Upper:* silver stain; *lower:* corresponding autoradiogram

resolved following hormonal stimulation in a bidimensional system involving isoelectric focusing in the first dimension and a 5%-20% polyacrylamide SDS gel electrophoresis in the second dimension. No difference in Coomassie blue staining was observed between bidimensional electrophoretograms obtained from hormone-treated or control oocytes. In a majority of phosphoproteins, labelling was increased with respect to control unstimulated oocytes in homogenates fixed 40 and 80 min after hormone application, then it decreased toward the level of unstimulated oocytes, as shown on Fig. 4A for phosphoproteins A 1, A 5, A 6, and A 7. In contrast, spots A 2, A 3, and A 4 not visible at all on autoradiograms from prophase-blocked oocytes, appeared in samples fixed between GVBD and first polar body emission and disappeared after emission of the second polar body (Fig. 4A). Finally some spots, as B_1 (Fig. 4B), were found to increase at the beginning of maturation, but did not decrease later, whereas others like B_2 (Fig. 4B) showed little variation upon both initiation and completion of maturation.

It was often difficult to correlate spots on autoradiograms and those on stained gels since high radioactivities were often associated with faint spots, which were frequently overlapped by more abundant, but unlabelled proteins. It was observed, however, in same cases that there was no variation (B_1, Fig. 4B) or little variaton (C_1, Fig. 5B) in the abundancy of some phosphoproteins during the course of meiotic maturation, although their ^{32}P-phosphate content increased considerably. In contrast, intensity of staining closely corresponded to incorporated radioactivity in new kinds of phosphoprotein like C 2 (Fig. 5). Due to the high number of nonlabelled spots detected, it was in most cases not possible to associate the appearance of new phosphoproteins with the disappearance of nonlabelled spots of higher isoelectric point. However, some unlabelled proteins like C_3 (Fig. 5) were effectively found to disappear at the beginning of maturation and to reappear later on.

6. Conclusion

In the present state of our knowledge, the cascade of events leading from hormonal stimulation of MPF production can be viewed as including the following steps in starfish oocytes.

1-MeAde, produced by the follicle cells under the influence of a hormonal peptide of neutral origin, first interacts with stereospecific receptors [26] localized on the oocyte plasma membrane [4] and accessible to the hormone only from the outside of the oocyte [3]. Within a few seconds, 1-MeAde induces the release of Ca^{2+} into the cortical region of the oocytes [27, 28]. Although any treatment which suppresses the release of Ca^{2+} or decreases it under a critical value either in living cells or from plasma membranes – rich fractions also abolish meiosis reinitiation, the previous intracellular microinjection of the calcium chelator EGTA does not inhibit meiosis reinitiation [29]. It seems, therefore, that elimination of Ca^{2+} from some component of the oocyte cortex, perhaps the plasma membrane, is an essential event for meiosis reinitiation, although the released Ca^{2+} does not act as a second messenger of the

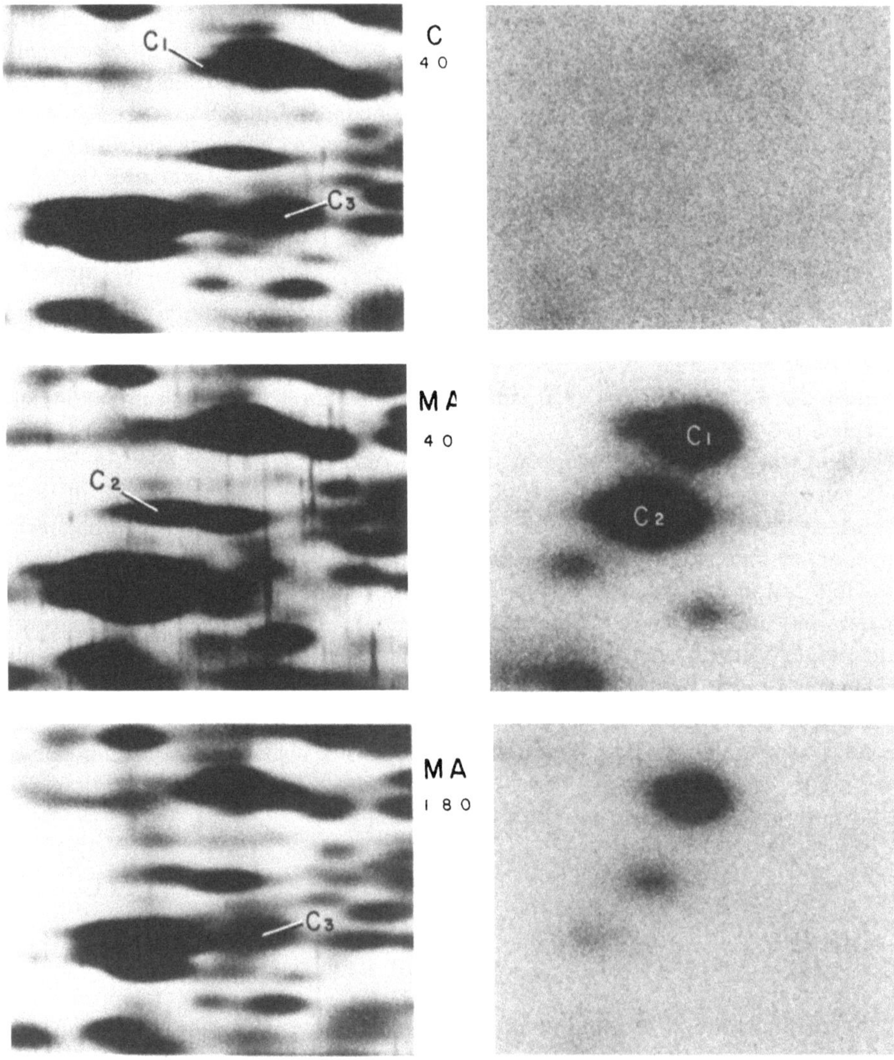

Fig. 5. Details from bidimensional electrophoretograms 4.8 < pHi < 5.2 and 30,000 > Mr > 20,000 at various times of meiotic maturation (40 min and 180 min after hormone addition) and in control prophase-blocked oocytes. *Left:* silver stains, *right:* the corresponding autoradiograms

hormone in this biological response (at least in starfish). Less than 5 min after 1-MeAde addition, the absolute rate of protein phosphorylation increases [21]. Dose responses are identical for enhancement of protein phosphorylation and meiosis reinitiation [13] suggesting that protein phosphorylation plays a key role in this process. In contrast protein synthesis is not required for oocytes to be removed from prophase block and to undergo GVBD [22]. Although it is delayed and reduced with respect to intact oocytes, stimulation of protein phosphorylation is not suppressed in enucleated fragments following hormonal stimulation [15].

Less than 5 min after hormone addition, a factor appears in the germinal vesicle, which triggers without exception the production of cytoplasmic MPF, GVBD, and the subsequent events of meiotic maturation when transferred in the cytoplasm of any full-grown oocyte. We propose to call this factor "nuclear MPF", until its identity to cytoplasmic MPF can be demonstrated. Before hormone action, the germinal vesicle also contains a factor capable of inducing meiosis initiation in recipient oocytes, but in contrast with nuclear MPF, this factor acts exclusively when transferred in the cytoplasm of a special category of oocytes (the "competent" oocytes). It is possible, but not proved, that this factor is a precursor of nuclear MPF. Shortly before GVBD (thus, about 10 min after nuclear MPF) MPF becomes detectable in cytoplasm. It is not known whether cytoplasmic MPF results from the translocation of nuclear MPF across the nuclear envelope, but production of cytoplasmic MPF upon hormonal stimulation does not occur (incompetent oocytes) or is strongly reduced (competent oocytes) following enucleation, and it is restored to a high level following injection of germinal vesicle material. On the other hand, cytoplasmic MPF initially produced in maturing oocytes following hormonal treatment triggers production of nuclear MPF in nonhormone-treated recipient oocytes. Activity of MPF in homogenates prepared from maturing oocytes (thus, containing a mixture of nuclear and cytoplasmic MPF) seems to depend on the activity of protein kinases. Indeed preliminary attempts to purify MPF from starfish homogenates indicated that successful extraction and stabilization of MPF requires the presence of phoshorylated small molecules which may act to inhibit phosphoprotein phosphatases, and the presence of ATP or its nonhydrolysable analog gamma-thio-ATP (Picard and Dorée, unpub. results). Similar features have been reported previously for amphibians MPF [24]. MPF keeps a high activity in maturing oocytes, excepted at the time of first and second meiotic cleavages, where it drops dramatically. MPF activity comes back to a very high level after the first, but not the second cleavage (unless oocytes are fertilized).

Thus, not only in vitro but also in ovo MPF activity remains closely correlated with the overall state of protein phosphorylation. Taken together these results strongly suggest that MPF is a protein able to induce the breakdown of the nuclear envelope when transferred in its phosphorylated form in G2 -arrested oocytes.

References

1. Hirai S, Chida K, Kanatani H (1973) Dev Growth Differ 15:21–31
2. Kanatani H, Shirai H, Nakanishi K, Kurokawa T (1969) Nature (London) 211:273–277
3. Kanatani H, Hiramoto Y (1970) Exp Cell Res 61:280–284
4. Dorée M, Guerrier P (1975) Exp Cell Res 91:296–300
5. Kishimoto T, Kanatani H (1976) Nature (London) 260:321–322
6. Kishimoto T, Kuriyama R, Kondo H, Kanatani H (1982) 137:121–126
7. Sunkara PS, Wright DA, Rao PN (1979) Proc Natl Acad Sci USA 76:2799–2802
8. Masui Y, Markert CL (1971) J Exp Zool 177:129–146
9. Reynhout JK, Smith LD (1974) Dev Biol 38:394–400
10. Wasserman WJ, Smith LD (1978) J Cell Biol 78:R15–R22
11. Dorée M (1981) CR Acad Sci 293:519–522
12. Kishimoto T, Hirai S, Kanatani H (1981) Dev Biol 81:177–181

13. Guerrier P, Moreau M, Dorée M (1977) Mol Cell Endocrinol 7:137–150
14. Mazzei G, Guerrier P (1982) Dev Biol 91:246–256
15. Dorée M (1981) J Exp Zool 217:147–150
16. Dorée M, Kishimoto T, Le Peuch CJ, Demaille JG, Kanatani H (1981) Exp Cell Res 135: 237–249
17. Maller JL, Wu M, Gerhart JC (1977) Dev Biol 58:295–312
18. Bellé R, Boyer J, Ozon R (1979) Gamete Res 2:137–145
19. Meijer L, Paul M, Epel D (1982) Dev Biol 94:62–70
20. Peaucellier G, Dorée M, Demaille JG (1982) Gamete Res 5:115–123
21. Dorée M, Peaucellier G, Picard A (1983) Dev Biol 99:489–501
22. Dorée M (1982) Exp Cell Res 139:127–134
23. Drury K (1978) Differentiation 10:181–186
24. Wu M, Gerhart JC (1980) Dev Biol 79:465–477
25. Guerrier P, Dorée M, Freyssinet G (1975) CR Acad Sc: Par D 281:1475–1478
26. Dorée M, Guerrier P, Leonard NJ (1976) Proc Natl Acad Sci USA 73:1669–1673
27. Moreau M, Guerrier P, Dorée M, Ashley CC (1978) Nature (London) 272:251–253
28. Dorée M, Guerrier P, Moreau M (1978) Exp Cell Res 115:251–260
29. Picard A, Dorée M (1983) Exp Cell Res 145:325–337

IV.3 Hormonal Mechanism of Starfish Spawning

H. KANATANI[1]

Both spawning and oocyte maturation are induced by l-MeAde injected into the coelomic cavity of ripe female starfish [1]. In inducing oocyte maturation the action of l-MeAde is indirect; it acts on the oocyte surface to produce a cytoplasmic MPF, direct trigger of germinal vesicle breakdown (GVBD). In inducing spawning, the action of l-MeAde needs further clarification. Follicular envelope breakdown (FEBD), which is prerequisite for spawning has been considered to result from the direct action of l-MeAde, since FEBD takes place when isolated *Asterina* oocytes with their follicular envelopes are placed in seawater containing l-MeAde [2, 3]. However, the present study indicates that the action of l-MeAde in inducing spawning, as in inducing GVBD, is also indirect, since injection of MPF into isolated oocytes with their follicles induces FEBD in the absence of l-MeAde. MPF acts from the inside of the oocytes on the oocyte plasma membrane to disrupt desmosomes between follicle cell processes and the oocyte surface [4].

Further, the fact that cytochalasin D prevents the movement of follicle cells along the oocyte surface and the clumping which occurs after dissociation of the connection between follicle cells and oocyte surface suggests that an active contraction of the follicle cells mediated by microfilaments is involved in FEBD [4, 5].

Once each oocyte becomes freely movable within the ovary following FEBD, spawning can be induced by the contraction of the ovarian wall. Neither GSS nor l-MeAde alone is able to induce contraction of the ovarian wall [6, 7]. The jelly coat surrounding the oocyte surface contains a substance inducing contraction of the isolated ovarian wall [7]. Since the jelly coat around the oocyte is surrounded by the follicular envelope, the jelly comes into contact with the ovarian wall after FEBD, generating tension in the ovarian wall leading to shedding of oocytes. Although follicle cells can be separated from the oocytes in calcium-free seawater, in this condition wall never contracts even after contact with the jelly [6–9]. Therefore, calcium is necessary for the actual contraction of the ovarian wall. In view of these considerations, starfish spawning is considered to be brought about in the following manner: (1) GSS released from the nervous tissue acts on the follicle cells to produce l-MeAde. (2) l-MeAde released from the follicle cells acts on the surface of the oocytes to produce MPF in the oocyte cytoplasm. (3) MPF acts from the inside of the oocytes to disrupt the desmosomes between follicle cell processes and oocyte plasma membrane, resulting in FEBD. (4) The follicle cells then clump as a result of microfilament-

1 National Institute for Basic Biology, Okazaki, Japan

Biosynthesis, Metabolism and Mode of Action of Invertebrate Hormones
(ed. by J. Hoffmann and M. Porchet)

mediated contraction. (5) The jelly coat containing a contraction-inducing substance now comes into direct contact with the ovarian wall to induce contraction of the wall in the presence of calcium ions, leading to extrusion of the now freely movable oocytes from the ovary through the short gonoducts open to the gonopores.

References

1. Kanatani H (1969) Induction of spawning and oocyte maturation by 1-methyladenine in starfish. Exp Cell Res 57:333–337
2. Kanatani H, Shirai H (1970) Mechanism of starfish spawning. III. Properties and action of meiosis-inducing substance produced in gonad under influence of gonad-stimulating substance. Dev Growth Differ 12:119–140
3. Kanatani H (1973) Maturation-inducing substance in starfishes. Int Rev Cytol 35:253–298
4. Kishimoto T, Usui N, Kanatani H (1984). Breakdown of starfish ovarian follicle induced by maturation-promoting factor. Develop Biol 101:28–34
5. Schroeder PC (1971) Active contraction of starfish oocyte follicle cells after treatment with 1-methyladenine. Naturwissenschaften 58:270–271
6. Kanatani H (1967) Neural substances responsible for maturation of oocytes and shedding of gametes in starfish. Gunma Symp Endocrinol 4:65–78
7. Shirai H, Yoshimoto Y, Kanatani H (1981) Mechanism at the time of spawning. IV. Tension generation in the ovarian wall by 1-methyladenine at the time of spawning. Biol Bull 161: 172–179
8. Schuetz AW, Biggers JD (1968) Effect of calcium on the structure and functional response of the starfish ovary to radial nerve factor. J Exp Zool 168:1–10
9. Kanatani H, Shirai H (1969) Mechanism of starfish spawning. II. Some aspects of action of neural substance obtained from radial nerves. Biol Bull 137:297–311

Genera Index

Acanthonyx 258
Achatina 26
Acheta 172, 361
Aedes 168, 378–398, 473
Agrotis 428
Anagasta 428
Anas 31
Ancylostoma 326
Angyostrongylus 326
Anodonta 26
Antheraea 76, 142, 428
Aphelenchus 324, 339, 340
Aplysia 19, 21, 24–26, 36ff., 51
Ariolimax 27
Arion 27
Armadillidium 155–160
Artemia 256–262
Ascaridia 341
Ascaris 324, 326, 339–341
Astacus 217, 256–262
Asterias 500, 502
Asterina 500, 509
Attacus 374–377

Blabera 463
Blaberus 172
Blatella 478
Bombyx 64, 98, 169, 172, 173, 184, 196, 202, 247, 361
Brugia 326, 327, 343
Buccinum 26
Bufo 497
Bulinus 31

Caenorhabditis 326, 339, 340, 341, 343
Callinectes 155, 202, 203, 208, 257, 258
Calliphora 98, 142, 1698, 169, 172, 173, 182, 200–217, 247, 341, 378
Cancer 217, 227
Carausius 123, 200, 208, 462
Carcinus 145, 154, 155, 156, 200, 202, 217, 256–261
Caudra 428
Cepaea 200
Cephalobus 324, 343
Chironomus 239
Chlamys 26
Clitumnus 173, 463
Colias 428
Crepidula 28, 29

Danaus 428
Diabrotica 378
Diatraea 428
Dipetalonema 326, 328, 329
Diploptera 172, 360, 362, 379, 381, 396–405
Dirofilaria 339–341
Drosophila 172, 217, 247, 260, 341, 378, 428, 479, 476, 478, 480

Echinococcus 344
Ephestia 247, 428
Eriocheir 145–154
Eristalis 98
Estigmene 428
Euhadra 31

Fasciola 324

Galleria 172, 173, 247, 420, 428, 455ff.
Gecarcinus 201
Gomphocerus 428, 434
Gromphadorhina 216

Haemonchus 324, 340, 343
Haliothis 341, 428
Helix 22, 23, 171, 181–184, 339, 455ff.
Helleria 155
Hemileuca 428
Heterodera 343

Homarus 155
Hyalophora 76, 369, 374–376, 391, 419, 428, 448, 454, 461, 464, 466ff.
Hydra 22
Hyloicus 428
Hymenolepis 326, 343, 344

Jasus 202
Junonia 428

Leptinotarsa 359–362, 378, 397–398, 407–414, 427, 428
Leucophaea 173, 217, 404, 427, 428, 438ff., 445ff., 480
Limax 29, 30
Litomosoides 324, 326, 327
Locusta 92, 96, 102, 118, 125, 126, 135, 168–180, 181–195, 200–217, 227, 233, 236, 240, 261, 366–371, 380–385, 387, 388, 390, 391, 393, 397, 398, 428, 445, 454, 461ff., 475ff.
Lymnaea 19–24, 26, 29–33, 44ff., 45, 47, 48, 49, 51ff., 53, 57ff., 58

Macrobrachium 155
Macrocallista 22, 23
Macrotermes 172, 381, 454, 461, 463, 464
Manduca 79, 91, 98, 136, 144, 200–217, 227–233, 374–376, 416ff., 428ff., 483
Marthasterias 487, 499, 500, 502, 503
Mastomys 325–329
Megoura 381
Melanoplus 169, 397
Mesocricetus 332, 333
Moniezia 343
Musca 211, 216, 378, 455
Mya 26
Mytilus 23, 26, 27

Nauphoeta 169, 172, 173, 375, 379, 454ff., 462, 464
Nematospiroides 324, 342, 343

Octopus 23, 24, 27
Ocypoda 155
Oncopeltus 203, 247, 380, 428, 454, 461, 464
Opisthorchis 326
Orchestia 155–156, 258, 259
Orconectes 145, 154, 256, 260–262
Ornithodoros 203, 217
Orygia 428
Ostrinia 98, 428

Pachygrapsus 217
Panagrellus 324, 340, 343
Pandalus 145–154
Papilio 77
Paramyelois 428
Parapenaeopsis 155
Patas 332, 334
Pectinophora 428
Periplaneta 98, 123, 380–385, 386, 387, 388, 390, 393, 394, 397
Phocanema 340, 343
Pieris 191, 200–217, 370–371, 428
Plodia 428
Polistes 361
Porcellio 145–154, 155
Porrocaecum 340, 343
Prodenia 211, 217

Rhodnius 172, 386, 444, 448, 470

Samia 64, 66, 375
Sarcophaga 172, 206, 211, 378, 478
Schistocerca 172, 173, 181–195, 200–217, 227–233, 236–240, 247, 384–385, 388–397, 448, 461, 464
Schistosoma 31, 324, 327, 331ff., 333, 334
Sphinx 428
Spodoptera 191, 200, 211, 213, 428
Strophocheilus 27

Taenia 326
Talitrus 156
Teleogryllus 380, 398, 461
Tenebrio 138, 200, 208, 239, 242, 245–254, 378–381, 393, 397, 428, 463
Tenebriodes 428
Thermobia 172, 380
Toxocara 341
Trichinella 326, 340, 343
Trichobilharzia 31, 32
Trichoplusia 378, 416ff., 428, 483

Unio 26

Xenopus 482
Xyleborus 172

Subject Index

Actin genes
ecdysteroid regulation of 299
in Kc cells 298ff.
Adenosine monophosphate in ecdysteroid conjugates 173, 186, 205
Adenylate cyclase, role of, in biosynthesis and release of caudo-dorsal cell hormone in Molluscs 55
Adipokinetic homrone AKH 122ff.
Adipokinetic peptides
mode of action 131
release during flight 128–129
A^+ Lipoprotein
formation of 132
substrate for flight muscles 130
Allatohibins 78ff., 85, 90
Allatotropic factors (in *Locusta*)
bioassay 93
isolation procedure 93, 94
Allatotropic factors (in *Manduca*) 85ff.
Androgenic gland 155ff.
Androgens (see also steroid hormones)
in Molluscs 32, 33
Arachidonic acid, effect of, on oocyte maturation in starfish 449, 488, 492
Arginine-vasopressin (AVP)
material immunologically related to 100
release into hemolymph 103
Arylsulphatase in hydrolysis of conjugates 181–182, 185
Atrial gland 24, 25, 36, 41
A-yellow (resting lipoprotein) 131–132
Azadirachtin 325, 326

Bag cell system (Bag cells) 19, 21, 24, 25, 38–41, 51
Brain
hormone 63
as a regulator of JH esterase activity 418, 419
Bursicon 137

Calcium
effect on meiosis reinitiation in starfish oocyte 488, 490ff., 492, 505
effect on ovarian wall contraction in starfish 509
and exocytotic activity in the caudo-dorsal cells of Molluscs 52ff.
Calmodulin 488, 491ff.
cAMP
dependent protein kinase in fat body 131
in egg-laying hormone release 39–41
intracellular concentration in prothoracic glands 83–84
involved in the process of caudo dorsal cell hormone release 54
a possible mediator of insect neurohormones 102
and steroidogenesis 84
Catecholamines 98
Caudo-dorsal cell hormone (CDCH)
(= ovulation hormone) 26, 29, 30, 32, 44ff., 51ff., 57ff.
Caudo-dorsal cells (CDC) 19, 21, 26, 29, 30, 44ff., 51, 53, 55, 57ff., 59
Cell lines (see cultured insect cells)
Cellular determination, maturation or committment
in ecd 1 mutant of *Drosophila* 277
in imaginal discs of *Drosophila* 307
in Kc cells of *Drosophila* 295
in *Manduca* epidermis 267
Cestodes
ecdysteroids during development of, 342ff.
ecdysteroids in 323ff., 342ff.
neurosecretory cells 344
Cholesterol 183
in ecdysteroid synthesis 175, 182, 192
uptake 426
Copulation, control of behaviour of, in Molluscs 59

Corpora allata
action on JH esterase activity 419
action in vitellogenesis 385, 386
activation in *Leptinotarsa* 412
activity in vitro 384–387
chemical inhibition of 387–390
differentiation of 456
direct measurements of activity 385
incorporation of methionine in 384
innervation in *Leptinotarsa* 413
neuroendocrine regulation of 85–90
regulation of activity 95–96
in vitro assay in *Manduca* 86
Cultured insects cells
division 294, 311
elongation 294
enzyme induction in 294, 317
hormone responses 311
karyotype 293, 307
Kc cells of *Drosophila* 293ff., 305ff.
morphology 294
oxygen consumption in 296
Cuticle
cuticular lamellae 268
in *Drosophila* imaginal discs 306
melanin granules 267
Cuticle proteins in *Hyalophora* 466ff.
Cuticulogenesis
embryonic 169
in vitro 236–239
Cytochrome P-450
in ecdysone-20-monooxygenase 210, 227–231
inhibitors on JH biosynthesis 393, 394

3-Dehydroecdysone reductase 213–215
3-Dehydroecdysteroids 201
Diapause
breaking of 77
environmental control of 407
Diet as a factor of JH titer and JH esterase activity 419
Diuresis in locusts 101, 102
Diuretic hormone (DH) 119ff.
Dorsal body hormone (DBH) 29, 32

Ecdysis, circadian gating of 137
Ecdysone
in embryonic development of cockroach 456, 457
in embryos 192
epimerase 213–215
induction of cellular activation 236
metabolism 216–218
Ecdysone-20-monooxygenase 210, 212, 228–233
activity related to development 232
CO difference spectra 231
cytochrome P-450 content 231
inhibitors 231
oxidase 212–213
properties 230
subcellular localization 210–211, 227–228
Ecdysonoic acids 191–192, 200–201, 204, 217–218
Ecdysteroids
acetates of 189–191, 205, 207
binding proteins 260–263
to polytene chromosomes 287
to vitellin 172–185
biological activity of 218–221
biosynthesis of 176, 182–184, 341
cell permeability to 221
conjugates of 173, 206–207, 455, 462
in eggs 173–178, 184–186
function 173–178, 186–192, 221
hydrolysis in embryo 186–189
conjugating enzymes of 215–216
in diagnosis and treatment of infections with helminths 31, 325–328
effects on enzyme activities
acetylcholinesterase 317
catalase 296
dopa-decarboxylase 298, 315, 318
β-galactosidase 318
superoxide dismutase 296
effects on the epidermal cell cycle 250
in eggs 175, 176
3-epi- 201
epidermal cell cycle, inhibition of 251
functions of, in embryonic development 173–178, 186–192, 221, 454, 463
HPLC separation of 209, 214, 256–257
26-hydroxy 190–192, 200–201
inactivation 189–192
metabolic patterns 208, 210
metabolism 176, 216–218
morphogenetic role on the integument 236–239
mutants in *Drosophila* 277, 286
in nemathelminths and platyhelminths 323ff.

in oogenesis and early development of crustaceans 256–259
origin of, in embryonic development 173–178, 186–189, 463
ovarian 169–172, 176, 181–182
phosphates 173, 184–186, 188–190, 193, 205, 207
receptors 260–263, 288, 312ff.
release, in vitro,
by follicle cells 169–171
by integument 239, 242
and schistosomiasis 331ff.
sequential gene expression
in epidermis 265, 267, 268
in fat body 273
in imaginal discs 315ff.
in Kc cells 293, 299ff., 319ff.
in salivary glands 273, 285ff.
tissue specificity 276, 290, 381
in serum of infected mammals 332, 333, 334, 336
technics used in identification and purification of 198–199
titers of
in embryonic development 174, 175, 456, 457
in haemolymph 171, 172, 174, 235, 247
intermoult levels 270
in *Manduca* 266
at pupation in *Drosophila* 308
triggered degeneration of motoneurons 141
uptake 259–260
in urine of infected humans 333–336
vertebrate antibody against 327–329, 336
zoo 202, 205
Eclosion hormone 137ff.
EFDA, 10,11-Epoxyfarnesyl diazoacetate, JH_{III} analog 438ff.
Egg-laying hormone (ELH)
antibodies against 36
from *Aplysia* bag cells 72
in gastropods 24–26, 36ff., 57ff.
gene for ELH 38
releasing 36, 38–41
structure 24, 25, 36, 37
synthetic ELH 37
Egg-laying in Molluscs
behaviour 51
and caudo-dorsal cell hormone (CDCH) 45
control of 19, 24–26, 36ff., 44ff.
control of behaviour 57ff.
photoperiodic effects on 29–30
Egg-releasing hormone (ERH) 26, 37, 42
Electrical activity, of caudo-dorsal cells in Molluscs 52
Electro-physiological studies, in Molluscs, on caudo-dorsal cell system 48, 49, 57ff., 58, 60, 61
Embryo
cuticulogenesis 169
ecdysteroid functions in 186–189
ecdysteroids in 174–175
enzymic hydrolysis of ecdysteroid conjugates in 181, 182, 185–189
phosphatase in 187, 188
prothoracic glands of 178
β-endorphin in *Calliphora* 112, 113
α-endorphin-like material 98
α-endorphin-like peptides, in *Calliphora* 114
Epidermal cell cycle 245–248
Epidermal cell cycle inhibition 252–253
Epidermis
cuticle deposition 268, 310, 317
cuticle proteins 266
cuticle synthesis 265
developmental changes in 165
Epoxide hydratation in JH catabolism 416ff.
Epoxide hydrolase activity in JH catabolism 417
Exocytosis
in caudo-dorsal cells (CDC) 52–54
role of calcium ions 52–54
of secretory material, control of, in Molluscs 52

Farnesenic acid 94
Farnesoic acid 458
Farnesoic acid, in JH biosynthesis 394
Fat-body
ecdysteroid responses in 273ff.
larval serum proteins of 277
mRNA synthesis in (see vitellogenin)
P1 and P6 genes 277
phospholipid synthesis in 447
regional specialization of 476
sexual differentiation 477, 478, 482
site of vitellogenin synthesis
in *Leucophaea* 445
in locust 475ff.
synthesis of endoplasmic reticulum in (see JH)
vitellogenin synthesis in (see vitellogenin)
Follicle cell epithelium 169, 170

Follicle cells and oocyte maturation in starfish 487
Follicular envelope breakdown 509
FRMF amide 22–24

Gastrin and insulin-like materials 98
Gastrin/CCK-like peptides 107
Gastrin/CCK-like peptides of *Calliphora*: biochemical studies 111
Gastrointestinal peptides in *Calliphora* 107
Gene regulation by ecdysteroids (see ecdysteroids and sequential gene expression)
Gene size in *Drosophila* 276, 280
Gene transcription in *Drosophila*
  relation to chromosome puffs 288, 291
  unit of 281
Gene transfer in *Drosophila* 273, 279
Germinal vesicle breakdown 487ff., 487ff., 509
  induction of 501
Glucose in Mollusc hemolymph 26, 27
Glucosyl-transferase 216
Gonad stimulating substance (gamete shedding substance) 471ff., 509
Gonadotropic hormones in Molluscs 29, 30

Head activator, *hydra* 72
20-Hydroxyecdysone
  in embryonic development of cockroach 456, 457, 462
  production of secretion and deposition of cuticle 236–239

Imaginal discs
  in *Drosophila* 305ff.
  evagination 310, 314, 317
Indolamines 98
Insecticyanin 266
Insulin-like factors in Molluscs 26, 27, 98
Integument
  culture with or without hormones 237
  in vitro release of ecdysteroids 239–242
Internal standard in JH determination 368, 370, 375
In vitro systems for studies of hormone action
  Drosophila
    fat body 273
    imaginal disc 305, 309
    Kc cells 293, 305
    salivary glands 273, 285
  Manduca epidermis 265ff.

Juvenile hormones (JH_S)
JH action
  on liberation of ecdysone 419, 420
  on specific genes 466ff.
  on steroids 480, 481
  in the synthesis of endoplasmic reticulum in the fat body 447
  on vitellogenin gene expression 479, 482
JH analogs, mode of action of 435
JH binding proteins
  binding site specificity of 430
  characterization of 427, 429
  comparison of affinity for the different JH_S 410, 411
  different affinity for JH_I and JH_{III} 382
  distribution of 424, 428
  function of 426, 433
  labelling of 438ff.
  molecular weight of 441
  physical characteristics of 429
  purification of 429
  regulatory effect of 426
  titers, control of 431, 432
JH biological potency of the different JH_S 381, 382
JH biosynthesis
  action of enzymes in 390, 395
  neural inhibition of 96
  during oocyte maturation 397ff.
  radiochemical assay 94
  regulation of 95
  stimulation of 94
JH biosynthetic pathway 393
JH carrier in the haemolymph 449, 450
JH catabolism in Lepidoptera 416ff.
JH degradation
  general 434
  in *Leptinotarsa* 408ff.
JH determination
  by mass spectrometry 363ff., 375
  by radioimmunoassay 355ff.
  sensivity of 369
JH diol conversion 357, 358
JH enantiomers, differential binding of 410
JH esterase inhibitors 405, 420, 421, 463

JH esterases
  activity in the haemolymph of cockroach 448, 449
  activity during larval stages in Lepidoptera 417
  activity in *Leptinotarsa* 408, 411
  effects on metamorphosis in Lepidoptera 423
  kinetic parameters in Lepidoptera 421, 422
  localisation of activity 417
  regulation of activity 418
  regulation by suboesophageal ganglion 418, 419
  specific vs general 418
  specificity 405
JH extra haemolymph pool 404
JH function in embryonic development 454, 463, 464
JH half life
  general 400, 402
  in *Leptinotarsa* 411
JH interaction with ecdysteroids in development 265
JH-JH binding protein complex
  dissociation rate of 434
  recognition at the target cell 434
JH origin during embryonic development (see copora allata)
JH persistent effects 402, 404
JH presence in various insects 373ff., 378ff., 407ff.
JH production by embryonic corpora allata 454, 461
JH prothoracicotropic action 419, 420
JH quantification by mass spectrometry 367ff.
JH receptors in the fat body 445, 446, 451
JH synthesis
  by embryonic corpora allata 458, 459
  in vitro assay 86
JH titers
  comparison of physiological titers 381
  control of
    by corpora allata in *Leptinotarsa* 411
    by JH carrier protein and JH esterase 417
  in embryonic development 456, 457, 461
  factors regulating 404, 405
  during the first gonotrophic cycle 399, 400
  in the following insects
    Aedes 378
    Attacus 377
    Diploptera 379, 398ff.
    Hyalophora 376, 377
    Manduca 266, 376
    Megoura 381
    Nauphoeta 397
    Teleogryllus 380
    Thermobia 380
  in the haemolymph of egg maturating female 460
  knowledge of 373
Juvenoids, effects on the epidermal cell cycle 249

Kc cells (see cultured insect cells)

Larval serum proteins, molecular studies in fat bodies 277

Maturation promoting factor (MPF) 487ff.
Median neurosecretory cells of the pars intercerebralis 97
Meiosis inducing substance (MIS) (see 1-Methyladenine)
Melanophore dispersing hormone 145, 146–148
Metabolic studies: techniques of 198–199
Met-enkephalin-immunoreactive materials 114
Methoprene (JH analog)
  control of vitellogenin production by 476, 479, 480
  effects on cuticular protein synthesis 268
  induction of vitellogenin mRNA by 479, 480
Methyl farnesoate
  in JH biosynthesis 393, 394
  mass spectrum from egg extracts 460
  synthesis by embryonic corpora allata 454, 462
1-Methyladenine 487ff., 497ff., 501–507
Microsomal enzymes in JH biosynthesis 391
Microsomes, ecdysone 20-monooxygenase in 227–228, 210–211
Mitochondria, ecdysone 20-monooxygenase in 210–211, 227–228
Molt in Nematodes 324 (see also ecdysteroids)

Molting in Insects
  molting fluid resorption 142
  process 136
Morphogenetic factors in *Crepidula* 28, 29

Nematodes
  biological effect of exogenous ecdysteroids 342, 343
  biosynthesis of ecdysteroids 341
  control of molt 324, 338ff.
  development of entomophilic species 342
  ecdysteroids during development 341, 342
  ecdysteroids in 323ff., 338ff.
Neuropeptides
  related to arginin-vasopressin 99
  structure and function of, evolutionary conservation 107 (see also neurophysins)
  of vertebrate type 106
Neurophysin-like material 98, 100, 101
Neurosecretion 97, 102
Neurosecretory cells
  in Cestodes 344
  in Insects 97ff.
    multineuroendocrine regulation 102
    of the suboesophagal ganglion 99
  in Molluscs 19, 21, 30 (see also bag cells, caudo-dorsal cells)
Neurosecretory proteins 98

Octopamine: effect on lipid mobilisation 129
Oocyte maturation 487ff.
  role of JH in 397
  titer of JH_{III} during 398–400
Opioid peptides, in Molluscs 23, 24
Opioid-like peptides, in insects 112
Ovarian wall, contraction during spawning in starfish 509
Ovary
  biosynthesis of ecdysteroids in 176
  ecdysteroids in vitellogenic ovaries 168, 169–172, 181–182
  transformation of ovary to testis in crustaceans 157
Oviposition in Molluscs 52
Ovulation hormone in Molluscs 51, 57 (see also caudo-dorsal cell hormone CDCH)
Ovulation in Molluscs 44ff.

P elements (see gene transfer in *Drosophila*)
Pancreatic polypeptide 98
Pars intercerebralis (PI)
  crude antibody solution 96
  electrostimulation of 96
Peptide A from *Aplysia* atrial glands 72
Peptide B from *Aplysia* atrial glands 72
Peptidergic neurons in Molluscs 20–22
Phosphatase
  activity in eggs 187–188
  in mitochondrial-lysosomal fractions 188
Phospholipases
  action during meiosis reinitiation 488, 490, 492
  inhibition of oocyte maturation by 493
Photo-affinity labelling of ecdysteroids 287
  of JH_S 438ff.
Photoperiod 79
  control of diapause by 407
Photoperiodic effects, in gastropod reproduction 29, 30
Polypodine B, effects on Kc cells 294
Polytene chromosomes
  ecdysteroid binding to 287
  microdissection for cloning 291
  puffs 273, 277, 285, 316
Ponasterone A
  in binding studies 313
  imaginal disc responses 315
Precocenes
  mode of action 388
  stereospecificity of 388, 389
  synthesis of vitellogenin mRNA in female locust treated with 479
Progestagens (see also steroid hormones)
  in Molluscs 32, 33
Prothoracic glands (see also PTTH):
  ecdysone synthesis by 79ff.
Prothoracicotropins = PTTH 72ff., 80ff.
Pupariation in *Drosophila* 308

Receptors
  ecdysteroid 260–263, 288, 312
  juvenile hormone 445ff.
Red pigment concentrating hormone (RPCH) 72, 127, 128, 145–147

S-adenosyl methionine 390
Salivary glands
  in *Drosophila* 273ff., 285ff.
  "glue" genes 274, 275

Schistosomiasis (bilharsiosis) 331ff.
Sex reversal, control of, in Molluscs 28, 29
Sinus gland, analytical separation of neuropeptides 146–152
Somatostatin-like neuropeptides 98
steroids
  C19 and C20 in arthropods 206
  hormones in gastropods 30–33
Substance P-like peptide 98
Sulfotransferase 216

Transformation vectors 280
Transposable elements in *Drosophila*
  ecdysteroid regulation of 299
  and gene transfer 279ff.

Vasoactive intestinal polypeptide-like materials 98
Vasopressin-like substances in Molluscs 27
Vasotocin-like substances in Molluscs 27
Vitellin, ecdysteroids bound to 172, 185
Vitellogenesis
  in Insects 444ff.
  in Molluscs 168
Vitellogenin
  genes of 476
  JH binding to 450
  JH control of 444, 445, 447
  mRNA 445, 451, 478, 479, 480
  stimulating hormone 168
  synthesis of 172, 475ff.

MIX
Papier aus verantwortungsvollen Quellen
Paper from responsible sources
FSC® C105338

If you have any concerns about our products,
you can contact us on
ProductSafety@springernature.com

In case Publisher is established outside the EU,
the EU authorized representative is:
Springer Nature Customer Service Center GmbH
Europaplatz 3, 69115 Heidelberg, Germany

Printed by Libri Plureos GmbH
in Hamburg, Germany